Accession no.
36265028

D1326326

Practical Extrapolation Methods

An important problem that arises in many scientific and engineering applications is that of approximating limits of infinite sequences $\{A_m\}$. In most cases, these sequences converge very slowly. Thus, to approximate their limits with reasonable accuracy, one must compute a large number of the terms of $\{A_m\}$, and this is generally costly. These limits can be approximated economically and with high accuracy by applying suitable *extrapolation* (or *convergence acceleration*) methods to a small number of terms of $\{A_m\}$.

This book is concerned with the coherent treatment, including derivation, analysis, and applications, of the most useful scalar extrapolation methods. The methods it discusses are geared toward problems that commonly arise in scientific and engineering disciplines. It differs from existing books on the subject in that it concentrates on the most powerful nonlinear methods, presents in-depth treatments of them, and shows which methods are most effective for different classes of practical nontrivial problems; it also shows how to fine-tune these methods to obtain best numerical results.

This book is intended to serve as a state-of-the-art reference on the theory and practice of extrapolation methods. It should be of interest to mathematicians interested in the theory of the relevant methods and serve applied scientists and engineers as a practical guide for applying speed-up methods in the solution of difficult computational problems.

Avram Sidi is Professor of Numerical Analysis in the Computer Science Department at the Technion–Israel Institute of Technology and holds the Technion Administration Chair in Computer Science. He has published extensively in various areas of numerical analysis and approximation theory and in journals such as *Mathematics of Computation*, *SIAM Review*, *SIAM Journal on Numerical Analysis*, *Journal of Approximation Theory*, *Journal of Computational and Applied Mathematics*, *Numerische Mathematik*, and *Journal of Scientific Computing*. Professor Sidi's work has involved the development of novel methods, their detailed mathematical analysis, design of efficient algorithms for their implementation, and their application to difficult practical problems. His methods and algorithms are successfully used in various scientific and engineering disciplines.

CAMBRIDGE MONOGRAPHS ON APPLIED AND COMPUTATIONAL MATHEMATICS

Series Editors

P. G. CIARLET, A. ISERLES, R. V. KOHN, M. H. WRIGHT

10   Practical Extrapolation Methods

The *Cambridge Monographs on Applied and Computational Mathematics* reflect the crucial role of mathematical and computational techniques in contemporary science. The series presents expositions on all aspects of applicable and numerical mathematics, with an emphasis on new developments in this fast-moving area of research.

State-of-the-art methods and algorithms as well as modern mathematical descriptions of physical and mechanical ideas are presented in a manner suited to graduate research students and professionals alike. Sound pedagogical presentation is a prerequisite. It is intended that books in the series will serve to inform a new generation of researchers.

Practical Extrapolation Methods

Theory and Applications

AVRAM SIDI

Technion–Israel Institute of Technology

PUBLISHED BY THE PRESS SYNDICATE OF THE UNIVERSITY OF CAMBRIDGE
The Pitt Building, Trumpington Street, Cambridge, United Kingdom

CAMBRIDGE UNIVERSITY PRESS
The Edinburgh Building, Cambridge CB2 2RU, UK
40 West 20th Street, New York, NY 10011-4211, USA
477 Williamstown Road, Port Melbourne, VIC 3207, Australia
Ruiz de Alarcón 13, 28014 Madrid, Spain
Dock House, The Waterfront, Cape Town 8001, South Africa

http://www.cambridge.org

First published 2003

Printed in the United Kingdom at the University Press, Cambridge

Typeface Times Roman 10/13 pt.   *System* LaTeX 2_ε [TB]

A catalog record for this book is available from the British Library.

Library of Congress Cataloging in Publication Data
Sidi, Avram.
Practical extrapolation methods : theory and applications / Avram Sidi.
p.   cm. – (Cambridge monographs on applied and computational mathematics)
Includes bibliographical references and index.
ISBN 0-521-66159-5 (hb)
1. Extrapolation.   I. Title.   II. Series.
QA281 .S555   2002
511′.42 – dc21   2002024669

ISBN 0 521 66159 5

Contents

Preface

An important problem that arises in many scientific and engineering applications is that of finding or approximating limits of infinite sequences $\{A_m\}$. The elements A_m of such sequences can show up in the form of partial sums of infinite series, approximations from fixed-point iterations of linear and nonlinear systems of equations, numerical quadrature approximations to finite- or infinite-range integrals, whether simple or multiple, etc. In most applications, these sequences converge very slowly, and this makes their direct use to approximate limits an expensive proposition. There are important applications in which they may even diverge. In such cases, the direct use of the A_m to approximate their so-called "antilimits" would be impossible. (Antilimits can be interpreted in appropriate ways depending on the nature of $\{A_m\}$. In some cases they correspond to analytic continuation in some parameter, for example.)

An effective remedy for these problems is via application of *extrapolation methods* (or *convergence acceleration methods*) to the given sequences. (In the context of infinite sequences, extrapolation methods are also referred to as *sequence transformations*.) Loosely speaking, an extrapolation method takes a finite and hopefully small number of the A_m and processes them in some way. A good method is generally nonlinear in the A_m and takes into account, either explicitly or implicitly, their asymptotic behavior as $m \to \infty$ in a clever fashion.

The importance of extrapolation methods as effective computational tools has long been recognized. Indeed, the Richardson extrapolation and the Aitken Δ^2-process, two popular representatives, are discussed in some detail in almost all modern textbooks on numerical analysis, and Padé approximants have become an integral part of approximation theory. During the last thirty years a few books were written on the subject and various comparative studies were done in relation to some important subclasses of convergence acceleration problems, pointing to the most effective methods. Finally, since the 1970s, international conferences partly dedicated to extrapolation methods have been held on a regular basis.

The main purpose of this book is to present a unified account of the existing literature on nonlinear extrapolation methods for scalar sequences that is as comprehensive and up-to-date as possible. In this account, I include much of the literature that deals with methods of practical importance whose effectiveness has been amply verified in various surveys and comparative studies. Inevitably, the contents reflect my personal interests and taste. Therefore, I apologize to those colleagues whose work has not been covered.

I have left out completely the important subject of extrapolation methods for vector sequences, even though I have been actively involved in this subject for the last twenty years. I regret this, especially in view of the fact that vector extrapolation methods have had numerous successful applications in the solution of large-scale nonlinear as well as linear problems. I believe only a fully dedicated book would do justice to them.

The Introduction gives an overview of convergence acceleration within the framework of infinite sequences. It includes a discussion of the concept of *antilimit* through examples. Following that, it discusses, both in general terms and by example, the development of methods, their analysis, and accompanying algorithms. It also gives a detailed discussion of stability in extrapolation. A proper understanding of this subject is very helpful in devising effective strategies for extrapolation methods in situations of inherent instability. The reader is advised to study this part of the Introduction carefully.

Following the Introduction, the book is divided into three main parts:

(i) Part I deals with the Richardson extrapolation and its generalizations. It also prepares some of the background material and techniques relevant to Part II. Chapters 1 and 2 give a complete treatment of the *Richardson extrapolation process* (REP) that is only partly described in previous works. Following that, Chapter 3 gives a first generalization of REP with some amount of theory. The rest of Part I is devoted to the *generalized Richardson extrapolation process* (GREP) of Sidi, the Levin–Sidi D-transformation for infinite-range integrals, the Levin–Sidi d-transformation for infinite series, Sidi's variants of the D-transformation for oscillatory infinite-range integrals, the Sidi–Levin rational d-approximants, and efficient summation of power series and (generalized) Fourier series by the d-transformation. (Two important topics covered in connection with these transformations are the class of functions denoted $\mathbf{B}^{(m)}$ and the class of sequences denoted $\mathbf{b}^{(m)}$. Both of these classes are more comprehensive than those considered in other works.) Efficient implementation of GREP is the subject of a chapter that includes the W-algorithm of Sidi and the $\mathrm{W}^{(m)}$-algorithm of Ford and Sidi. Also, there are two chapters that provide a detailed convergence and stability analysis of GREP$^{(1)}$, a prototype of GREP, whose results point to effective strategies for applying these methods in different situations. The strategies denoted *arithmetic progression sampling* (APS) and *geometric progression sampling* (GPS) are especially useful in this connection.

(ii) Part II is devoted to development and analysis of a number of effective sequence transformations. I begin with three classical methods that are important historically and that have been quite successful in many problems: the Euler transformation (the only linear method included in this book), the Aitken Δ^2-process, and the Lubkin transformation. I next give an extended treatment of the Shanks transformation along with Padé approximants and their generalizations, continued fractions, and the qd-algorithm. Finally, I treat other important methods such as the G-transformation, the Wynn ρ-algorithm and its modifications, the Brezinski θ-algorithm, the Levin $\mathcal{L}$- and Sidi $\mathcal{S}$-transformations, and the methods of Overholt and of Wimp. I also give the confluent forms of some of these methods. In this treatment, I include known results on the application of these transformations to so-called linear and logarithmic sequences and quite a few new results, including some pertaining to stability. I use the latter to draw conclusions on how to apply sequence transformations more effectively in different situations. In this respect APS of Part I turns out to be an effective strategy when applying the methods of Part II to linear sequences.

(iii) Part III comprises a single chapter that provides a number of applications to problems in numerical analysis that are not considered in Parts I and II.

Following these, I have also included as Part IV a sequence of appendices that provide a lot of useful information recalled throughout the book. These appendices cover briefly several important subjects, such as asymptotic expansions, Euler–Maclaurin expansions, the Riemann Zeta function, and polynomial approximation theory, to name a few. In particular, Appendix D puts together, for the first time, quite a few Euler–Maclaurin expansions of importance in applications. Appendices G and H include a summary of the extrapolation methods covered in the book and important tips on when and how to apply them. Appendix I contains a FORTRAN 77 code that implements the d-transformation and that is applied to a number of nontrivial examples. The user can run this code as is.

The reader may be wondering why I separated the topic of REP and its generalizations, especially GREP, from that of sequence transformations and treated it first. There are two major reasons for this decision. First, the former is more general, covers more cases of interest, and allows for more flexibility in a natural way. For most practical purposes, the methods of Part I have a larger scope and achieve higher accuracy than the sequence transformations studied in Part II. Next, some of the conclusions from the theory of convergence and stability of extrapolation methods of Part I turn out to be valid for the sequence transformations of Part II and help to improve their performance substantially.

Now, the treatment of REP and its generalizations concerns the problem of finding the limit (or antilimit) as $y \to 0+$ of a function $A(y)$ known for $y \in (0, b]$, $b > 0$. Here y may be a continuous or discrete variable. Such functions arise in many applications in a natural way. The analysis provided for them suggests to which problems they should be applied, how they should be applied for best possible outcome in finite-precision arithmetic, and what type of convergence and accuracy one should expect. All this also has an impact on how sequence transformations should be applied to infinite sequences $\{A_m\}$ that are not directly related to a function $A(y)$. In such a case, we can always draw the analogy $A(y) \leftrightarrow A_m$ for some function $A(y)$ with $y \leftrightarrow m^{-1}$, and view the relevant sequence transformations as extrapolation methods. When treated this way, it becomes easier to see that APS improves the performance of these transformations on linear sequences.

The present book differs from already existing works in various ways:

1. The classes of (scalar) sequences treated in it include many of those that arise in scientific and engineering applications. They are more comprehensive than those considered in previous books and include the latter as well.
2. Divergent sequences (such as those that arise from divergent infinite series or integrals, for example) are treated on an equal footing with convergent ones. Suitable interpretations of their antilimits are provided. It is shown rigorously, at least in some cases, that extrapolation methods can be applied to such sequences, *with no changes*, to produce excellent approximations to their antilimits.

3. A thorough asymptotic analysis of "tails" of infinite series and infinite-range integrals is given. Based on this analysis, detailed convergence studies for the different methods are provided. Considerable effort is made to obtain the best possible results under the smallest number of realistic conditions. These results either come in the form of bona fide asymptotic expansions of the errors or they provide tight upper bounds on errors. They are accompanied by complete proofs in most cases. The proofs chosen for presentation are generally those that can be applied in more than one situation and hence deserve special attention. When proofs are not provided, the reader is referred to the relevant papers and books.

4. The stability issue is formalized and given full attention for the first time. Conclusions are drawn from stability analyses to devise effective strategies that enable one to obtain the best possible accuracy in finite-precision arithmetic, also in situations where instabilities are built in. Immediate applications of this are to the summation of so-called "very oscillatory" infinite-range integrals, logarithmically convergent (or divergent) series, power series, Fourier series, etc., where most methods have not done well in the past.

I hope the book will serve as a reference for researchers in the area of extrapolation methods and for scientists and engineers in different computational disciplines and as a textbook for students interested in the subject. I have kept the mathematical background needed to cope with the material to a minimum. Most of what the reader needs and is not covered by the standard academic curricula is summarized in the appendices. I would like to emphasize here that the subject of asymptotics is of utmost importance to any treatment of extrapolation methods. It would be impossible to appreciate the beauty of the subject of extrapolation – the development of the methods and their mathematical analyses – without it. It is certainly impossible to produce any honest theory of extrapolation methods without it. I urge the reader to acquire a good understanding of asymptotics before everything else. The essentials of this subject are briefly discussed in Appendix A.

I wish to express my gratitude to my friends and colleagues David Levin, William F. Ford, Doron S. Lubinsky, Moshe Israeli, and Marius Ungarish for the interest they took in this book and for their criticisms and comments in different stages of the writing. I am also indebted to Yvonne Sagie and Hadas Heier for their expert typing of part of the book.

Lastly, I owe a debt of gratitude to my wife Carmella for her constant encouragement and support during the last six years while this book was being written. Without her understanding of, and endless patience for, my long absences from home this book would not have seen daylight. I dedicate this book to her with love.

Avram Sidi
Technion, Haifa
December 2001

Introduction

0.1 Why Extrapolation–Convergence Acceleration?

In many problems of scientific computing, one is faced with the task of finding or approximating limits of infinite sequences. Such sequences may arise in different disciplines and contexts and in various ways. In most cases of practical interest, the sequences in question converge to their limits very slowly. This may cause their direct use for approximating their limits to become computationally expensive or impossible.

There are other cases in which these sequences may even diverge. In such a case, we are left with the question of whether the divergent sequence represents anything, and if so, what it represents. Although in some cases the elements of a divergent sequence can be used as approximations to the quantity it represents subject to certain conditions, in most other cases it is meaningless to make direct use of the sequence elements for this purpose.

Let us consider two very common examples:

(i) Summation of infinite series: This is a problem that arises in many scientific disciplines, such as applied mathematics, theoretical physics, and theoretical chemistry. In this problem, the sequences in question are those of partial sums. In some cases, the terms a_k of a series $\sum_{k=0}^{\infty} a_k$ may be known analytically. In other cases, these terms may be generated numerically, but the process of generating more and more terms may become very costly. In both situations, if the series converges very slowly, the task of obtaining good approximations to its sum only from its partial sums $A_n = \sum_{k=0}^{n} a_k$, $n = 0, 1, \ldots$, may thus become very expensive as it necessitates a very large number of the terms a_k. In yet other cases, only a finite number of the terms a_k, say $a_0, a_1, \ldots, a_N$, may be known. In such a situation, the accuracy of the best available approximation to the sum of $\sum_{k=0}^{\infty} a_k$ is normally that of the partial sum A_N and thus cannot be improved further. If the series diverges, then its partial sums have only limited direct use. Divergent series arise naturally in different fields, perturbation analysis in theoretical physics being one of them. Divergent power series arise in the solution of homogeneous ordinary differential equations around irregular singular points.

(ii) Iterative solution of linear and nonlinear systems of equations: This problem occurs very commonly in applied mathematics and different branches of engineering. When continuum problems are solved by methods such as finite differences and finite elements, large and sparse systems of linear and/or nonlinear equations are obtained. A

1

very attractive way of solving these systems is by iterative methods. The sequences in question for this case are those of the iteration vectors that have a large dimension in general. In most cases, these sequences converge very slowly. If the cost of computing one iteration vector is very high, then obtaining a good approximation to the solution of a given system of equations may also become very high.

The problems of slow convergence or even divergence of sequences can be overcome under suitable conditions by applying *extrapolation methods* (equivalently, *convergence acceleration methods* or *sequence transformations*) to the given sequences. When appropriate, an extrapolation method produces from a given sequence $\{A_n\}$ a new sequence $\{\hat{A}_n\}$ that converges to the former's limit more quickly when this limit exists. In case the limit of $\{A_n\}$ does not exist, the new sequence $\{\hat{A}_n\}$ produced by the extrapolation method either diverges more slowly than $\{A_n\}$ or converges to some quantity called the *antilimit* of $\{A_n\}$ that has a useful meaning and interpretation in most applications. We note at this point that the precise meaning of the antilimit may vary depending on the type of the divergent sequence, and that several possibilities exist. In the next section, we shall demonstrate through examples how antilimits may arise and what exactly they may be.

Concerning divergent sequences, there are three important messages that we would like to get across in this book: (i) Divergent sequences can be interpreted appropriately in many cases of interest, and useful antilimits for them can be defined. (ii) Extrapolation methods can be used to produce good approximations to the relevant antilimits in an efficient manner. (iii) Divergent sequences can be treated on an equal footing with convergent ones, both computationally and theoretically, and this is what we do throughout this book. (However, everywhere-divergent infinite power series, that is, those with zero radius of convergence, are not included in the theoretical treatment generally.)

It must be emphasized that each $\hat{A}_n$ is determined from only a *finite* number of the A_m. This is a basic requirement that extrapolation methods must satisfy. Obviously, an extrapolation method that requires knowledge of all the A_m for determining a given $\hat{A}_n$ is of no practical value.

We now pause to illustrate the somewhat abstract discussion presented above with the Aitken Δ^2-process that is one of the classic examples of extrapolation methods. This method was first described in Aitken [2], and it can be found in almost every book on numerical analysis. See, for example, Henrici [130], Ralston and Rabinowitz [235], Stoer and Bulirsch [326], and Atkinson [13].

Example 0.1.1 Let the sequence $\{A_n\}$ be such that

$$A_n = A + a\lambda^n + r_n \text{ with } r_n = b\mu^n + o(\min\{1, |\mu|^n\}) \text{ as } n \to \infty, \quad (0.1.1)$$

where A, a, b, λ, and μ are in general complex scalars, and

$$a, b \neq 0, \quad \lambda, \mu \neq 0, 1, \quad \text{and } |\lambda| > |\mu|. \quad (0.1.2)$$

As a result, $r_n \sim b\mu^n = o(\lambda^n)$ as $n \to \infty$. If $|\lambda| < 1$, then $\lim_{n\to\infty} A_n = A$. If $|\lambda| \geq 1$, then $\lim_{n\to\infty} A_n$ does not exist, A being the antilimit of $\{A_n\}$ in this case. Consider now the Aitken Δ^2-process, which is an extrapolation method that, when applied to $\{A_n\}$, produces a sequence $\{\hat{A}_n\}$ with

$$\hat{A}_n = \frac{A_n A_{n+2} - A_{n+1}^2}{A_n - 2A_{n+1} + A_{n+2}} = \frac{\begin{vmatrix} A_n & A_{n+1} \\ \Delta A_n & \Delta A_{n+1} \end{vmatrix}}{\begin{vmatrix} 1 & 1 \\ \Delta A_n & \Delta A_{n+1} \end{vmatrix}}, \tag{0.1.3}$$

where $\Delta A_m = A_{m+1} - A_m$, $m \geq 0$. To see how $\hat{A}_n$ behaves for $n \to \infty$, we substitute (0.1.1) in (0.1.3). Taking into account the fact that $r_{n+1} \sim \mu r_n$ as $n \to \infty$, after some simple algebra it can be shown that

$$|\hat{A}_n - A| \leq \alpha |r_n| = O(\mu^n) = o(\lambda^n) \text{ as } n \to \infty, \tag{0.1.4}$$

for some positive constant α that is independent of n. Obviously, when $\lim_{n \to \infty} r_n = 0$, the sequence $\{\hat{A}_n\}$ converges to A whether $\{A_n\}$ converges or not. [If $r_n = 0$ for $n \geq N$, then $\hat{A}_n = A$ for $n \geq N$ as well, as implied by (0.1.4). In fact, the formula for $\hat{A}_n$ in (0.1.3) is obtained by requiring that $\hat{A}_n = A$ when $r_n = 0$ for all large n, and it is the solution for A of the equations $A_m = A + a\lambda^m$, $m = n, n+1, n+2$.] Also, in case $\{A_n\}$ converges, $\{\hat{A}_n\}$ converges more quickly and to $\lim_{n \to \infty} A_n = A$, because $A_n - A \sim a\lambda^n$ as $n \to \infty$ from (0.1.1) and (0.1.2). Thus, the rate of convergence of $\{A_n\}$ is enhanced by the factor

$$\frac{|\hat{A}_n - A|}{|A_n - A|} = O(|\mu/\lambda|^n) = o(1) \text{ as } n \to \infty. \tag{0.1.5}$$

A more detailed analysis of $\hat{A}_n - A$ yields the result

$$\hat{A}_n - A \sim b\frac{(\lambda - \mu)^2}{(\lambda - 1)^2}\mu^n \text{ as } n \to \infty, \tag{0.1.6}$$

that is more refined than (0.1.4) and asymptotically best possible as well. It is clear from (0.1.6) that, when the sequence $\{r_n\}$ does not converge to 0, which happens when $|\mu| \geq 1$, both $\{A_n\}$ and $\{\hat{A}_n\}$ diverge, but $\{\hat{A}_n\}$ diverges more slowly than $\{A_n\}$.

In view of this example and the discussion that preceded it, we now introduce the concepts of *convergence acceleration* and *acceleration factor*.

Definition 0.1.2 Let $\{A_n\}$ be a sequence of in general complex scalars, and let $\{\hat{A}_n\}$ be the sequence generated by applying the extrapolation method ExtM to $\{A_n\}$, $\hat{A}_n$ being determined from A_m, $0 \leq m \leq L_n$, for some integer L_n, $n = 0, 1, \ldots$. Assume that $\lim_{n \to \infty} \hat{A}_n = A$ for some A and that, if $\lim_{n \to \infty} A_n$ exists, it is equal to this A. We shall say that $\{\hat{A}_n\}$ *converges more quickly* than $\{A_n\}$ if

$$\lim_{n \to \infty} \frac{|\hat{A}_n - A|}{|A_{L_n} - A|} = 0, \tag{0.1.7}$$

whether $\lim_{n \to \infty} A_n$ exists or not. When (0.1.7) holds we shall also say that the extrapolation method ExtM *accelerates the convergence* of $\{A_n\}$. The ratio $R_n = |\hat{A}_n - A|/|A_{L_n} - A|$ is called the *acceleration factor* of $\hat{A}_n$.

The ratios R_n measure the extent of the acceleration induced by the extrapolation method ExtM on $\{A_n\}$. Indeed, from $|\hat{A}_n - A| = R_n |A_{L_n} - A|$, it is obvious that R_n is the factor by which the acceleration process reduces $|A_{L_n} - A|$ in generating $\hat{A}_n$. Obviously, a good extrapolation method is one whose acceleration factors tend to zero quickly as $n \to \infty$.

In case $\{A_n\}$ is a sequence of vectors in some general vector space, the preceding definition is still valid, provided we replace $|A_{L_n} - A|$ and $|\hat{A}_n - A|$ everywhere with $\|A_{L_n} - A\|$ and $\|\hat{A}_n - A\|$, respectively, where $\|\cdot\|$ is the norm in the vector space under consideration.

0.2 Antilimits Versus Limits

Before going on, we would like to dwell on the concept of antilimit that we mentioned briefly above. This concept can best be explained by examples to which we now turn. These examples do not exhaust all the possibilities for antilimits by any means. We shall encounter more later in this book.

Example 0.2.1 Let A_n, $n = 0, 1, 2, \dots$, be the partial sums of the power series $\sum_{k=0}^{\infty} a_k z^k$, that is, $A_n = \sum_{k=0}^{n} a_k z^k$, $n = 0, 1, \dots$. If the radius of convergence ρ of this series is finite and positive, then $\lim_{n \to \infty} A_n$ exists for $|z| < \rho$ and is a function $f(z)$ that is analytic for $|z| < \rho$. Of course, $\sum_{k=0}^{\infty} a_k z^k$ diverges for $|z| > \rho$. If $f(z)$ can be continued analytically to $|z| = \rho$ and $|z| > \rho$, then the analytic continuation of $f(z)$ is the antilimit of $\{A_n\}$ for $|z| \geq \rho$.

As an illustration, let us pick $a_0 = 0$ and $a_k = -1/k$, $k = 1, 2, \dots$, so that $\rho = 1$ and $\lim_{n \to \infty} A_n = \log(1 - z)$ for $|z| \leq 1$, $z \neq 1$. The principal branch of $\log(1 - z)$ that is analytic for all complex $z \notin [1, +\infty)$ serves as the antilimit of $\{A_n\}$ in case $|z| > 1$ but $z \notin [1, +\infty)$.

Example 0.2.2 Let A_n, $n = 0, 1, 2, \dots$, be the partial sums of the Fourier series $\sum_{k=-\infty}^{\infty} a_k e^{ikx}$; that is, $A_n = \sum_{k=-n}^{n} a_k e^{ikx}$, $n = 0, 1, 2, \dots$, and assume that $C_1 |k|^\alpha \leq |a_k| \leq C_2 |k|^\alpha$ for all large $|k|$ and some positive constants C_1 and C_2 and for some $\alpha \geq 0$, so that $\lim_{n \to \infty} A_n$ does not exist. This Fourier series represents a 2π-periodic *generalized function*; see Lighthill [167]. If, for x in some interval I of $[0, 2\pi]$, this generalized function coincides with an ordinary function $f(x)$, then $f(x)$ is the antilimit of $\{A_n\}$ for $x \in I$. (Recall that $\lim_{n \to \infty} A_n$, in general, exists when $\alpha < 0$ and a_n is monotonic in n. It exists unconditionally when $\alpha < -1$.)

As an illustration, let us pick $a_0 = 0$ and $a_k = 1$, $k = \pm 1, \pm 2, \dots$. Then the series $\sum_{k=-\infty}^{\infty} a_k e^{ikx}$ represents the generalized function $-1 + 2\pi \sum_{m=-\infty}^{\infty} \delta(x - 2m\pi)$, where $\delta(z)$ is the Dirac delta function. This generalized function coincides with the ordinary function $f(x) = -1$ in the interval $(0, 2\pi)$, and $f(x)$ serves as the antilimit of $\{A_n\}$ for $n \to \infty$ when $x \in (0, 2\pi)$.

Example 0.2.3 Let $0 < x_0 < x_1 < x_2 < \cdots$, $\lim_{n \to \infty} x_n = \infty$, $s \neq 0$ and real, and let A_n be defined as $A_n = \int_0^{x_n} g(t) e^{ist} \, dt$, $n = 0, 1, 2, \dots$, where $C_1 t^\alpha \leq |g(t)| \leq C_2 t^\alpha$ for all large t and some positive constants C_1 and C_2 and for some $\alpha \geq 0$, so that

$\lim_{n\to\infty} A_n$ does not exist. In many such cases, the antilimit of $\{A_n\}$ is the *Abel sum* of the divergent integral $\int_0^\infty g(t)e^{ist}\,dt$ (see, e.g., Hardy [123]) that is defined by $\lim_{\epsilon\to 0+}\int_0^\infty e^{-\epsilon t}g(t)e^{ist}\,dt$. [Recall that $\int_0^\infty g(t)e^{ist}\,dt$ exists and $\lim_{n\to\infty} A_n = \int_0^\infty g(t)e^{ist}\,dt$, in general, when $\alpha < 0$ and $g(t)$ is monotonic in t for large t. This is true unconditionally when $\alpha < -1$.]

As an illustration, let us pick $g(t) = t^{1/2}$. Then the Abel sum of the divergent integral $\int_0^\infty t^{1/2}e^{ist}\,dt$ is $e^{i3\pi/4}\sqrt{\pi}/(2s^{3/2})$, and it serves as the antilimit of $\{A_n\}$.

Example 0.2.4 Let $\{h_n\}$ be a sequence in $(0, 1)$ satisfying $h_0 > h_1 > h_2 > \cdots$, and $\lim_{n\to\infty} h_n = 0$, and define $A_n = \int_{h_n}^1 x^\alpha g(x)\,dx$, $n = 0, 1, 2, \ldots$, where $g(x)$ is continuously differentiable on $[0, 1]$ a sufficient number of times with $g(0) \neq 0$ and α is in general complex and $\Re\alpha \leq -1$ but $\alpha \neq -1, -2, \ldots$. Under these conditions $\lim_{n\to\infty} A_n$ does not exist. The antilimit of $\{A_n\}$ in this case is the *Hadamard finite part* of the divergent integral $\int_0^1 x^\alpha g(x)\,dx$ (see Davis and Rabinowitz [63]) that is given by the expression

$$\sum_{i=0}^{m-1} \frac{1}{\alpha + i + 1}\frac{g^{(i)}(0)}{i!} + \int_0^1 x^\alpha \left[g(x) - \sum_{i=0}^{m-1} \frac{g^{(i)}(0)}{i!}x^i \right] dx$$

with $m > -\Re\alpha - 1$ so that the integral in this expression exists as an ordinary integral. [Recall that $\int_0^1 x^\alpha g(x)\,dx$ exists and $\lim_{n\to\infty} A_n = \int_0^1 x^\alpha g(x)\,dx$ for $\Re\alpha > -1$.]

As an illustration, let us pick $g(x) = (1 + x)^{-1}$ and $\alpha = -3/2$. Then the Hadamard finite part of $\int_0^1 x^{-3/2}(1 + x)^{-1}\,dx$ is $-2 - \pi/2$, and it serves as the antilimit of $\{A_n\}$. Note that $\lim_{n\to\infty} A_n = +\infty$ but the associated antilimit is negative.

Example 0.2.5 Let s be the solution to the nonsingular linear system of equations $(I - T)x = c$, and let $\{x_n\}$ be defined by the iterative scheme $x_{n+1} = Tx_n + c$, $n = 0, 1, 2, \ldots$, with x_0 given. Let $\rho(T)$ denote the spectral radius of T. If $\rho(T) > 1$, then $\{x_n\}$ diverges in general. The antilimit of $\{x_n\}$ in this case is the solution s itself. [Recall that $\lim_{n\to\infty} x_n$ exists and is equal to s when $\rho(T) < 1$.]

As should become clear from these examples, the antilimit may have different meanings depending on the nature of the sequence $\{A_n\}$. Thus, it does not seem to be possible to define antilimits in a unique way, and we do not attempt to do this. It appears, though, that studying the asymptotic behavior of A_n for $n \to \infty$ is very helpful in determining the meaning of the relevant antilimit. We hope that what the antilimit of a given divergent sequence is will become more apparent as we proceed to the study of extrapolation methods.

0.3 General Algebraic Properties of Extrapolation Methods

We saw in Section 0.1 that an extrapolation method operates on a given sequence $\{A_n\}$ to produce a new sequence $\{\hat{A}_n\}$. That is, it acts as a mapping from $\{A_n\}$ to $\{\hat{A}_n\}$. In all cases of interest, this mapping has the general form

$$\hat{A}_n = \Phi_n(A_0, A_1, \ldots, A_{L_n}), \tag{0.3.1}$$

where L_n is some *finite* positive integer. (As mentioned earlier, methods for which $L_n = \infty$ are of no use, because they require knowledge of all the A_m to obtain $\hat{A}_n$ with finite n.) In addition, for most extrapolation methods there holds

$$\hat{A}_n = \sum_{i=0}^{K_n} \theta_{ni} A_i, \qquad (0.3.2)$$

where K_n are some nonnegative integers and the θ_{ni} are some scalars that satisfy

$$\sum_{i=0}^{K_n} \theta_{ni} = 1 \qquad (0.3.3)$$

for each n. (This is the case for all of the extrapolation methods we consider in this work.) A consequence of (0.3.2) and (0.3.3) is that such extrapolation methods act as *summability methods* for the sequence $\{A_n\}$.

When the θ_{ni} are independent of the A_m, the approximation $\hat{A}_n$ is linear in the A_m, thus the extrapolation method that generates $\{\hat{A}_n\}$ becomes a *linear summability method*. That is to say, this extrapolation method can be applied to every sequence $\{A_n\}$ with the same θ_{ni}. Both numerical experience and the different known convergence analyses suggest that linear methods are of limited scope and not as effective as nonlinear methods.

As the subject of linear summability methods is very well-developed and is treated in different books, we are not going to dwell on it in this book; see, for example, the books by Knopp [152], Hardy [123], and Powell and Shah [231]. We only give the definition of linear summability methods at the end of this section and recall the Silverman–Toeplitz theorem, which is one of the fundamental results on linear summability methods. Later in this work, we also discuss the Euler transformation that has been used in different practical situations and that is probably the most successful linear summability method.

When the θ_{ni} depend on the A_m, the approximation $\hat{A}_n$ is *nonlinear* in the A_m. This implies that if $C_m = \alpha A_m + \beta B_m$, $m = 0, 1, 2, \ldots$, for some constants α and β, and $\{\hat{A}_n\}$, $\{\hat{B}_n\}$, and $\{\hat{C}_n\}$ are obtained by applying a given nonlinear extrapolation method to $\{A_n\}$, $\{B_n\}$, and $\{C_n\}$, respectively, then $\hat{C}_n \neq \alpha \hat{A}_n + \beta \hat{B}_n$, $n = 0, 1, 2, \ldots$, in general. (Equality prevails for all n when the extrapolation method is linear.) Despite this fact, most nonlinear extrapolation methods enjoy a "sort of linearity" property that can be described as follows: Let $\alpha \neq 0$ and β be arbitrary constants and consider $C_m = \alpha A_m + \beta$, $m = 0, 1, 2, \ldots$. Then

$$\hat{C}_n = \alpha \hat{A}_n + \beta, \quad n = 0, 1, 2, \ldots . \qquad (0.3.4)$$

In other words, $\{C_n\} = \alpha\{A_n\} + \beta$ implies $\{\hat{C}_n\} = \alpha\{\hat{A}_n\} + \beta$. This is called the *quasi-linearity* property and is a useful property that we want every extrapolation method to have. (All extrapolation methods treated in this book are quasi-linear.) A sufficient condition for this to hold is given in Proposition 0.3.1.

Proposition 0.3.1 *Let a nonlinear extrapolation method be such that the sequence $\{\hat{A}_n\}$ that it produces from $\{A_n\}$ satisfies (0.3.2) with (0.3.3). Then the sequence $\{\hat{C}_n\}$ that it produces from $\{C_n = \alpha A_n + \beta\}$ for arbitrary constants $\alpha \neq 0$ and β satisfies the* *quasi-linearity property in (0.3.4) if the θ_{ni} in (0.3.2) depend on the A_m through the $\Delta A_m = A_{m+1} - A_m$ only and are homogeneous in the ΔA_m of degree 0.*

Remark. We recall that a function $f(x_1, \ldots, x_p)$ is *homogeneous of degree r* if, for every $\lambda \neq 0$, $f(\lambda x_1, \ldots, \lambda x_p) = \lambda^r f(x_1, \ldots, x_p)$.

Proof. We begin by rewriting (0.3.2) in the form $\hat{A}_n = \sum_{i=0}^{K_n} \theta_{ni}(\{A_m\})A_i$. Similarly, we have $\hat{C}_n = \sum_{i=0}^{K_n} \theta_{ni}(\{C_m\})C_i$. From (0.3.1) and the conditions imposed on the θ_{ni}, there exist functions $D_{ni}(\{u_m\})$ for which

$$\theta_{ni}(\{A_m\}) = D_{ni}(\{\Delta A_m\}) \text{ and } \theta_{ni}(\{C_m\}) = D_{ni}(\{\Delta C_m\}), \quad (0.3.5)$$

where the functions D_{ni} satisfy for all $\lambda \neq 0$

$$D_{ni}(\{\lambda u_m\}) = D_{ni}(\{u_m\}). \quad (0.3.6)$$

This and the fact that $\{\Delta C_m\} = \{\alpha \Delta A_m\}$ imply that

$$\theta_{ni}(\{C_m\}) = D_{ni}(\{\Delta C_m\}) = D_{ni}(\{\Delta A_m\}) = \theta_{ni}(\{A_m\}). \quad (0.3.7)$$

From (0.3.2) and (0.3.7) we have, therefore,

$$\hat{C}_n = \sum_{i=0}^{K_n} \theta_{ni}(\{A_m\})(\alpha A_i + \beta) = \alpha \hat{A}_n + \beta \sum_{i=0}^{K_n} \theta_{ni}(\{A_m\}). \quad (0.3.8)$$

The result now follows by invoking (0.3.3). ∎

Example 0.3.2 Consider the Aitken Δ^2-process that was given by (0.1.3) in Example 0.1.1. We can reexpress $\hat{A}_n$ in the form

$$\hat{A}_n = \theta_{n,n} A_n + \theta_{n,n+1} A_{n+1}, \quad (0.3.9)$$

with

$$\theta_{n,n} = \frac{\Delta A_{n+1}}{\Delta A_{n+1} - \Delta A_n}, \quad \theta_{n,n+1} = \frac{-\Delta A_n}{\Delta A_{n+1} - \Delta A_n}. \quad (0.3.10)$$

Thus, $\theta_{ni} = 0$ for $0 \leq i \leq n - 1$. It is easy to see that the θ_{ni} satisfy the conditions of Proposition 0.3.1 so that the Δ^2-process has the quasi-linearity property described in (0.3.4). Note also that for this method $L_n = n + 2$ in (0.3.1) and $K_n = n + 1$ in (0.3.2).

0.3.1 Linear Summability Methods and the Silverman–Toeplitz Theorem

We now go back briefly to linear summability methods. Consider the infinite matrix

$$M = \begin{bmatrix} \mu_{00} & \mu_{01} & \mu_{02} & \cdots \\ \mu_{10} & \mu_{11} & \mu_{12} & \cdots \\ \mu_{20} & \mu_{21} & \mu_{22} & \cdots \\ \vdots & \vdots & \vdots & \end{bmatrix}, \quad (0.3.11)$$

where μ_{ni} are some fixed scalars. The linear summability method associated with M is the linear mapping that transforms an arbitrary sequence $\{A_n\}$ to another sequence $\{A'_n\}$ through

$$A'_n = \sum_{i=0}^{\infty} \mu_{ni} A_i, \quad n = 0, 1, 2, \ldots . \qquad (0.3.12)$$

This method is *regular* if $\lim_{n\to\infty} A_n = A$ implies $\lim_{n\to\infty} A'_n = A$. The Silverman–Toeplitz theorem that we state next gives necessary and sufficient conditions for a linear summability method to be regular. For proofs of this fundamental result see, for example, the books by Hardy [123] and Powell and Shah [231].

Theorem 0.3.3 *(Silverman–Toeplitz theorem). The summability method associated with the matrix M in (0.3.11) is regular if and only if the following three conditions are fulfilled simultaneously:*

  (i) $\lim_{n\to\infty} \sum_{i=0}^{\infty} \mu_{ni} = 1$.
  (ii) $\lim_{n\to\infty} \mu_{ni} = 0$, $i = 0, 1, 2, \ldots$.
  (iii) $\sup_n \sum_{i=0}^{\infty} |\mu_{ni}| < \infty$.

Going back to the beginning of this section, we see that (0.3.3) is analogous to condition (i) of Theorem 0.3.3. The issue of numerical stability discussed in Section 0.5 is very closely related to condition (iii), as will become clear shortly.

The linear summability methods that have been of practical use are those whose associated matrices M are lower triangular, that is, those for which $A'_n = \sum_{i=0}^{n} \mu_{ni} A_i$. Excellent treatments of these methods from the point of view of convergence acceleration, including an extensive bibliography, have been presented by Wimp [363], [364], [365], [366, Chapters 2–4].

0.4 Remarks on Algorithms for Extrapolation Methods

A relatively important issue in the subject of extrapolation methods is the development of efficient algorithms (computational procedures) for implementing existing extrapolation methods. An efficient algorithm is one that involves a small number of arithmetic operations and little storage when storage becomes a problem.

Some extrapolation methods already have known closed-form expressions for the sequences $\{\hat{A}_n\}$ they generate. This is the case, for example, for the Aitken Δ^2-process. One possible algorithm for such methods may be the direct computation of the closed-form expressions. This is also the most obvious, but not necessarily the most economical, approach in all cases.

Many extrapolation methods are defined through systems of linear or nonlinear equations, that is, they are defined implicitly by systems of the form

$$\Psi_{n,i}(\hat{A}_n, \alpha_1, \alpha_2, \ldots, \alpha_{q_n}; \{A_m\}) = 0, \quad i = 0, 1, \ldots, q_n, \qquad (0.4.1)$$

in which $\hat{A}_n$ is the main quantity we are after, and $\alpha_1, \alpha_2, \ldots, \alpha_{q_n}$ are additional auxiliary unknowns. As we will see in the next chapters, the better sequences $\{\hat{A}_n\}$ are generated by those extrapolation methods with large q_n, in general. This means that we actually want to solve large systems of equations, which may be a computationally expensive proposition. In such cases, the development of good algorithms becomes especially important. The next example helps make this point clear.

Example 0.4.1 The Shanks [264] transformation of order k is an extrapolation method, which, when applied to a sequence $\{A_n\}$, produces the sequence $\{\hat{A}_n = e_k(A_n)\}$, where $e_k(A_n)$ satisfies the nonlinear system of equations

$$A_r = e_k(A_n) + \sum_{i=1}^{k} \alpha_i \lambda_i^r, \quad n \leq r \leq n + 2k, \tag{0.4.2}$$

where α_i and λ_i are additional (auxiliary) $2k$ unknowns. Provided this system has a solution with $\alpha_i \neq 0$ and $\lambda_i \neq 0, 1$ and $\lambda_i \neq \lambda_j$ if $i \neq j$, then $e_k(A_n)$ can be shown to satisfy the linear system

$$A_r = e_k(A_n) + \sum_{i=1}^{k} \beta_i \Delta A_{r+i-1}, \quad n \leq r \leq n + k, \tag{0.4.3}$$

where β_i are additional (auxiliary) k unknowns. Here $\Delta A_m = A_{m+1} - A_m$, $m = 0, 1, \dots$, as before. [In any case, we can start with (0.4.3) as the definition of $e_k(A_n)$.] Now, this linear system can be solved using Cramer's rule, giving us $e_k(A_n)$ as the ratio of two $(k+1) \times (k+1)$ determinants in the form

$$e_k(A_n) = \frac{\begin{vmatrix} A_n & A_{n+1} & \cdots & A_{n+k} \\ \Delta A_n & \Delta A_{n+1} & \cdots & \Delta A_{n+k} \\ \vdots & \vdots & & \vdots \\ \Delta A_{n+k-1} & \Delta A_{n+k} & \cdots & \Delta A_{n+2k-1} \end{vmatrix}}{\begin{vmatrix} 1 & 1 & \cdots & 1 \\ \Delta A_n & \Delta A_{n+1} & \cdots & \Delta A_{n+k} \\ \vdots & \vdots & & \vdots \\ \Delta A_{n+k-1} & \Delta A_{n+k} & \cdots & \Delta A_{n+2k-1} \end{vmatrix}}. \tag{0.4.4}$$

We can use this determinantal representation to compute $e_k(A_n)$, but this would be very expensive for large k and thus would constitute a bad algorithm. A better algorithm is one that solves the linear system in (0.4.3) by Gaussian elimination. But this algorithm too becomes costly for large k. The ε-algorithm of Wynn [368], on the other hand, is very efficient as it produces all of the $e_k(A_n)$, $0 \leq n + 2k \leq N$, that are defined by $A_0, A_1, \dots, A_N$ in only $O(N^2)$ operations. It reads

$$\varepsilon_{-1}^{(n)} = 0, \quad \varepsilon_0^{(n)} = A_n, \quad n = 0, 1, \dots,$$

$$\varepsilon_{k+1}^{(n)} = \varepsilon_{k-1}^{(n+1)} + \frac{1}{\varepsilon_k^{(n+1)} - \varepsilon_k^{(n)}}, \quad n, k = 0, 1, \dots, \tag{0.4.5}$$

and we have

$$e_k(A_n) = \varepsilon_{2k}^{(n)}, \quad n, k = 0, 1, \dots. \tag{0.4.6}$$

Incidentally, $\hat{A}_n$ in (0.1.3) produced by the Aitken Δ^2-process is nothing but $e_1(A_n)$. [Note that the Shanks transformations are quasi-linear extrapolation methods. This can be seen either from the equations in (0.4.2), or from those in (0.4.3), or from the determinantal representation of $e_k(A_n)$ in (0.4.4), or even from the ε-algorithm itself.]

Finally, there are extrapolation methods in the literature that are defined exclusively by recursive algorithms from the start. The θ-algorithm of Brezinski [32] is such an extrapolation method, and it is defined by recursion relations very similar to those of the ε-algorithm.

0.5 Remarks on Convergence and Stability of Extrapolation Methods

The analysis of convergence and stability is the most important subject in the theory of extrapolation methods. It is also the richest in terms of the variety of results that exist and still can be obtained for different extrapolation methods and sequences. Thus, it is impossible to make any specific remarks about convergence and stability at this stage. We can, however, make several remarks on the approach to these topics that we take in this book. We start with the topic of convergence analysis.

0.5.1 Remarks on Study of Convergence

The first stage in the convergence analysis of extrapolation methods is formulation of conditions that we impose on the $\{A_n\}$. In this book, we deal with sequences that arise in common applications. Therefore, we emphasize mainly conditions that are relevant to these applications. Also, we keep the number of the conditions imposed on the $\{A_n\}$ to a minimum as this leads to mathematically more valuable and elegant results. The next stage is analysis of the errors $\hat{A}_n - A$ under these conditions. This analysis may lead to different types of results depending on the complexity of the situation. In some cases, we are able to give a full asymptotic expansion of $\hat{A}_n - A$ for $n \to \infty$; in other cases, we obtain only the most dominant term of this expansion. In yet other cases, we obtain a realistic upper bound on $|\hat{A}_n - A|$ from which powerful convergence results can be obtained. An important feature of our approach is that we are not content only with showing that the sequence $\{\hat{A}_n\}$ converges more quickly than $\{A_n\}$, that is, that convergence acceleration takes place in accordance with Definition 0.1.2, but instead we aim at obtaining the precise asymptotic behavior of the corresponding acceleration factor or a good upper bound for it.

0.5.2 Remarks on Study of Stability

We now turn to the topic of stability in extrapolation. Unlike convergence, this topic may not be common knowledge, so we start with some rather general remarks on what we mean by stability and how we analyze it. Our discussion here is based on those of Sidi [272], [300], [305], and is recalled in relevant places throughout the book.

When we compute the sequence $\{\hat{A}_n\}$ in finite-precision arithmetic, we obtain a sequence $\{\tilde{A}_n\}$ that is different from $\{\hat{A}_n\}$, the exact transformed sequence. This, of course, is caused mainly by errors (roundoff errors and errors of other kinds as well) in the A_n. Naturally, we would like to know by how much $\tilde{A}_n$ differs from $\hat{A}_n$, that is, we want to be able to estimate $|\tilde{A}_n - \hat{A}_n|$. This is important also since knowledge of $|\tilde{A}_n - \hat{A}_n|$ assists in assessing the cumulative error $|\tilde{A}_n - A|$ in $\tilde{A}_n$. To see this, we start with

$$\left| |\tilde{A}_n - \hat{A}_n| - |\hat{A}_n - A| \right| \leq |\tilde{A}_n - A| \leq |\tilde{A}_n - \hat{A}_n| + |\hat{A}_n - A|. \qquad (0.5.1)$$

Next, let us assume that $\lim_{n \to \infty} |\hat{A}_n - A| = 0$. Then (0.5.1) implies that $|\tilde{A}_n - A| \approx |\tilde{A}_n - \hat{A}_n|$ for all sufficiently large n, because $|\tilde{A}_n - \hat{A}_n|$ remains nonzero.

We have observed numerically that, for many extrapolation methods that satisfy (0.3.2) with (0.3.3), $|\tilde{A}_n - \hat{A}_n|$ can be estimated by the product $\Gamma_n \epsilon^{(n)}$, where

$$\Gamma_n = \sum_{i=0}^{K_n} |\theta_{ni}| \geq 1 \quad \text{and} \quad \epsilon^{(n)} = \max\{|\epsilon_i| : \theta_{ni} \neq 0\}, \qquad (0.5.2)$$

and, for each i, ϵ_i is the error in A_i. The idea behind this is that the θ_{ni} and hence Γ_n do not change appreciably with small errors in the A_i. Thus, if $A_i + \epsilon_i$ are the computed A_i, then $\tilde{A}_n$, the computed $\hat{A}_n$, is very nearly given by $\sum_{i=0}^{K_n} \theta_{ni}(A_i + \epsilon_i) = \hat{A}_n + \sum_{i=0}^{K_n} \theta_{ni}\epsilon_i$. As a result,

$$|\tilde{A}_n - \hat{A}_n| \approx \left| \sum_{i=0}^{K_n} \theta_{ni}\epsilon_i \right| \leq \Gamma_n \epsilon^{(n)}. \qquad (0.5.3)$$

The meaning of this is that the quantity Γ_n [that always satisfies $\Gamma_n \geq 1$ by (0.3.3)] controls the propagation of errors in $\{A_n\}$ into $\{\hat{A}_n\}$, in the sense that the absolute computational error $|\tilde{A}_n - \hat{A}_n|$ is practically the maximum of the absolute errors in the A_i, $0 \leq i \leq K_n$, magnified by the factor Γ_n. Thus, combining (0.5.1) and (0.5.3), we obtain

$$|\tilde{A}_n - A| \lesssim \Gamma_n \epsilon^{(n)} + |\hat{A}_n - A| \qquad (0.5.4)$$

for the absolute errors, and

$$\frac{|\tilde{A}_n - A|}{|A|} \lesssim \Gamma_n \frac{\epsilon^{(n)}}{|A|} + \frac{|\hat{A}_n - A|}{|A|}, \quad \text{provided } A \neq 0, \qquad (0.5.5)$$

for the relative errors.

The implication of (0.5.4) is that, practically speaking, the cumulative error $|\tilde{A}_n - A|$ is at least of the order of the corresponding theoretical error $|\hat{A}_n - A|$ but it may be as large as $\Gamma_n \epsilon^{(n)}$ if this quantity dominates. [Note that because $\lim_{n \to \infty} |\hat{A}_n - A| = 0$, $\Gamma_n \epsilon^{(n)}$ will dominate $|\tilde{A}_n - A|$ for sufficiently large n.] The approximate inequality in (0.5.5) reveals even more in case $\{A_n\}$ is convergent and hence $\hat{A}_n$, A, and the A_i, $0 \leq i \leq K_n$, are all of the same order of magnitude and the latter are known correctly to r significant decimal digits and Γ_n is of order 10^s, $s < r$. Then $\epsilon^{(n)}/|A|$ will be of order 10^{-r}, and, therefore, $|\tilde{A}_n - A|/|A|$ will be of order 10^{s-r} for sufficiently large n. This implies that $\tilde{A}_n$ will have approximately $r - s$ correct significant decimal digits when n is sufficiently large. When $s \geq r$, however, $\tilde{A}_n$ may be totally inaccurate in the sense that it may be completely different from $\hat{A}_n$. In other words, in such cases Γ_n is also a measure of the loss of relative accuracy in the computed $\hat{A}_n$.

One conclusion that can be drawn from this discussion is that it is possible to achieve sufficient accuracy in $\tilde{A}_n$ by increasing r, that is, by computing the A_n with high accuracy. This can be accomplished on a computer by doubling the precision of the floating-point arithmetic used for computing the A_n.

When applying an extrapolation method to a convergent sequence $\{A_n\}$ numerically, we would like to be able to compute the sequence $\{\tilde{A}_n\}$ without $|\tilde{A}_n - \hat{A}_n|$ becoming unbounded for increasing n. In view of this and the discussion of the previous paragraphs, we now give a formal definition of stability.

Definition 0.5.1 If an extrapolation method that generates from $\{A_n\}$ the sequence $\{\hat{A}_n\}$ satisfies (0.3.2) with (0.3.3), then we say that it is *stable* provided $\sup_n \Gamma_n < \infty$, where $\Gamma_n = \sum_{i=0}^{K_n} |\theta_{ni}|$. Otherwise, it is *unstable*.

The mathematical treatment of stability then evolves around the analysis of $\{\Gamma_n\}$. Note also the analogy between the definition of a stable extrapolation method and condition (iii) in the Silverman–Toeplitz theorem (Theorem 0.3.3).

It is clear from our discussion above that we need Γ_n to be of reasonable size relative to the errors in the A_i for $\tilde{A}_n$ to be an acceptable representation of $\hat{A}_n$. Obviously, the ideal case is one in which $\Gamma_n = 1$, which occurs when all the θ_{ni} are nonnegative. This case does arise, for example, in the application of some extrapolation methods to oscillatory sequences. In most other situations, however, direct application of extrapolation methods without taking into account the asymptotic nature of $\{A_n\}$ results in large Γ_n and even unbounded $\{\Gamma_n\}$. It then follows that, even though $\{\hat{A}_n\}$ may be converging, $\tilde{A}_n$ may be entirely different from $\hat{A}_n$ for all large n. This, of course, is a serious drawback that considerably reduces the effectiveness of extrapolation methods that are being used. This problem is inherent in some methods, and it can be remedied in others by proper tuning. We show later in this book how to tune extrapolation methods to reduce the size of Γ_n and even to stabilize the methods completely.

Numerical experience and some theoretical results suggest that *reducing Γ_n not only stabilizes the extrapolation process but also improves the theoretical quality of the sequence $\{\hat{A}_n\}$.*

Example 0.5.2 Let us see how the preceding discussion applies to the Δ^2-process on the sequences $\{A_n\}$ discussed in Example 0.1.1. First, from Example 0.3.2 it is clear that

$$\Gamma_n = \frac{1 + |g_n|}{|1 - g_n|}, \quad g_n = \frac{\Delta A_{n+1}}{\Delta A_n}. \tag{0.5.6}$$

Next, from (0.1.1) and (0.1.2) we have that $\lim_{n\to\infty} g_n = \lambda \neq 1$. Therefore,

$$\lim_{n\to\infty} \Gamma_n = \frac{1 + |\lambda|}{|1 - \lambda|} < \infty. \tag{0.5.7}$$

This shows that the Δ^2-process on such sequences is stable. Note that, for all large n, $|\hat{A}_n - A|$ and Γ_n are proportional to $|1 - \lambda|^{-2}$ and $|1 - \lambda|^{-1}$, respectively, and hence are large when λ is too close to 1 in the complex plane. It is not difficult to see that they can be reduced simultaneously in such a situation if the Δ^2-process is applied to a subsequence $\{A_{\kappa n}\}$, where $\kappa \in \{2, 3, \dots\}$, since, for even small κ, λ^κ is farther away from 1 than λ is.

We continue our discussion of $|\tilde{A}_n - A|$ assuming now that the A_i have been computed with relative errors not exceeding η. In other words, $\epsilon_i = \eta_i A_i$ and $|\eta_i| \leq \eta$ for all i. (This is the case when the A_i have been computed to maximum accuracy that is possible in finite-precision arithmetic with rounding unit $\mathbf{u}$; we have $\eta = \mathbf{u}$ in this situation.) Then (0.5.4) becomes

$$|\tilde{A}_n - A| \lesssim \eta \, \Gamma_n I_n(\{A_s\}) + |\hat{A}_n - A|, \quad I_n(\{A_s\}) \equiv \max\{|A_i| : \theta_{ni} \neq 0\}. \quad (0.5.8)$$

Obviously, when $\{A_n\}$ converges, or diverges but is bounded, the term $I_n(\{A_s\})$ remains bounded as $n \to \infty$. In this case, it follows from (0.5.8) that, provided Γ_n remains bounded, $|\tilde{A}_n - A|$ remains bounded as well. It should be noted, however, that when $\{A_n\}$ diverges and is unbounded, $I_n(\{A_s\})$ is unbounded as $n \to \infty$, which causes the right-hand side of (0.5.8) to become unbounded as $n \to \infty$, even when Γ_n is bounded. In such cases, $|\tilde{A}_n - A|$ becomes unbounded, as we have observed in all our numerical experiments. The hope in such cases is that the convergence rate of the exact transformed sequence $\{\hat{A}_n\}$ is much greater than the divergence rate of $\{A_n\}$ so that sufficient accuracy is achieved by $\hat{A}_n$ before $I_n(\{A_s\})$ has grown too much.

We also note that, in case $\{A_n\}$ is divergent and the A_i have been computed with relative errors not exceeding η, numerical stability can be assessed more accurately by replacing (0.5.3) and (0.5.4) by

$$|\tilde{A}_n - \hat{A}_n| \lesssim \eta \sum_{i=0}^{K_n} |\theta_{ni}| \, |A_i| \qquad (0.5.9)$$

and

$$|\tilde{A}_n - A| \lesssim \eta \sum_{i=0}^{K_n} |\theta_{ni}| \, |A_i| + |\hat{A}_n - A|, \qquad (0.5.10)$$

respectively. Again, when the A_i have been computed to maximum accuracy that is possible in finite-precision arithmetic with rounding unit $\mathbf{u}$, we have $\eta = \mathbf{u}$ in (0.5.9) and (0.5.10). [Of course, (0.5.9) and (0.5.10) are valid when $\{A_n\}$ converges too.]

0.5.3 Further Remarks

Finally, in connection with the studies of convergence and stability, we have found it very useful to relate the given infinite sequences $\{A_m\}$ to some suitable functions $A(y)$, where y may be a discrete or continuous variable. These relations take the form $A_m = A(y_m)$, $m = 0, 1, \dots$, for some positive sequences $\{y_m\}$ that tend to 0, such that $\lim_{m \to \infty} A_m = \lim_{y \to 0+} A(y)$ when $\lim_{m \to \infty} A_m$ exists. In some cases, a sequence $\{A_m\}$ is derived directly from a known function $A(y)$ exactly as described. The sequences

14 *Introduction*

discussed in Examples 0.2.3 and 0.2.4 are of this type. In certain other cases, we can show the existence of a suitable function $A(y)$ that is associated with a given sequence $\{A_m\}$ even though $\{A_m\}$ is not provided by a relation of the form $A_m = A(y_m)$, $m = 0, 1, \ldots$, a priori.

This kind of an approach is obviously of greater generality than that dealing with infinite sequences alone. First, for a given sequence $\{A_m\}$, the related function $A(y)$ may have certain asymptotic properties for $y \to 0+$ that can be very helpful in deciding what kind of an extrapolation method to use for accelerating the convergence of $\{A_m\}$. Next, in case $A(y)$ is known a priori, we can choose $\{y_m\}$ such that (i) the convergence of the derived sequence $\{A_m = A(y_m)\}$ will be easier to accelerate by some extrapolation method, and (ii) this extrapolation method will also enjoy good stability properties. Finally, the *function* $A(y)$, in contrast to the *sequence* $\{A_m\}$, may possess certain analytic properties in addition to its asymptotic properties for $y \to 0+$. The analytic properties may pertain, for example, to smoothness and differentiability in some interval $(0, b]$ that contains $\{y_m\}$. By taking these properties into account, we are able to enlarge considerably the scope of the theoretical convergence and stability studies of extrapolation methods. We are also able to obtain powerful and realistic results on the behavior of the sequences $\{\hat{A}_n\}$.

We shall use this approach to extrapolation methods in many places throughout this book, starting as early as Chapter 1.

Historically, those convergence acceleration methods associated with functions $A(y)$ and derived from them have been called *extrapolation methods*, whereas those that apply to infinite sequences and that are derived directly from them have been called *sequence transformations*. In this book, we also make this distinction, at least as far as the order of presentation is concerned. Thus, we devote Part I of the book to the Richardson extrapolation process and its various generalizations and Part II to sequence transformations.

0.6 Remark on Iterated Forms of Extrapolation Methods

One of the ways to apply extrapolation methods is by simply *iterating* them. Let us again consider an arbitrary extrapolation method ExtM. The iteration of ExtM is performed as follows: We first apply ExtM to $\{C_0^{(n)} = A_n\}$ to obtain the sequence $\{C_1^{(n)}\}$. We next apply it to $\{C_s^{(n)}\}_{n=0}^\infty$ to obtain $\{C_{s+1}^{(n)}\}_{n=0}^\infty$, $s = 1, 2, \ldots$. Let us organize the $C_s^{(n)}$ in a two-dimensional array as in Table 0.6.1. In general, columns of this table converge, each column converging at least as quickly as the one preceding it. Diagonals converge as well, and they converge much quicker than the columns.

It is thus obvious that every extrapolation method can be iterated. For example, we can iterate e_k, the Shanks transformation of order k discussed in Example 0.4.1, exactly as explained here. In this book, we consider in detail the iteration of only two classic methods: The Δ^2-process, which we have discussed briefly in Example 0.1.1, and the Lubkin transformation. Both of these methods and their iterations are considered in detail in Chapter 15. We do not consider the iterated forms of the other methods discussed in this book, mainly because, generally speaking, their performance is not better than that

Table 0.6.1: *Iteration of an extrapolation method ExtM.*
Each column is obtained by applying ExtM to the
preceding column

$C_0^{(0)}$				
$C_0^{(1)}$	$C_1^{(0)}$			
$C_0^{(2)}$	$C_1^{(1)}$	$C_2^{(0)}$		
$C_0^{(3)}$	$C_1^{(2)}$	$C_2^{(1)}$	$C_3^{(0)}$	
$\vdots$	$\vdots$	$\vdots$	$\vdots$	$\ddots$

provided by the straightforward application of the corresponding methods; in some cases they even behave in undesired peculiar ways. In addition, their analysis can be done very easily once the techniques of Chapter 15 (on the convergence and stability of their column sequences) are understood.

0.7 Relevant Issues in Extrapolation

We close this chapter by giving a list of issues we believe are relevant to both the theory and the practice of extrapolation methods. Inevitably, some of these issues are more important than others and deserve more attention.

The first issue we need to discuss is that of development and design of extrapolation methods. Being the start of everything, this is a very important phase. It is best to embark on the project of development with certain classes of sequences $\{A_n\}$ in mind. Once we have developed a method that works well on those sequences for which it was designed, we can always apply it to other sequences as well. In some cases, this may even result in success. The Aitken Δ^2-process may serve as a good example to illustrate this point in a simple way.

Example 0.7.1 Let A, λ, a, and b be in general complex scalars and let the sequence $\{A_n\}$ be such that

$$A_n = A + \lambda^n \left[an^s + bn^{s-1} + O(n^{s-2}) \right] \text{ as } n \to \infty; \ \lambda \neq 0, 1, \ a \neq 0, \text{ and } s \neq 0. \tag{0.7.1}$$

Even though the Δ^2-process was designed to accelerate the convergence of sequences $\{A_n\}$ whose members satisfy (0.1.1), it turns out that it accelerates the convergence of sequences $\{A_n\}$ that satisfy (0.7.1) as well. Substituting (0.7.1) in (0.1.3), it can be shown after tedious algebraic manipulations that

$$\hat{A}_n - A \sim K\lambda^n n^{s-2} \text{ as } n \to \infty; \ K = -as\left(\frac{\lambda}{\lambda - 1}\right)^2, \tag{0.7.2}$$

as opposed to $A_n - A \sim a\lambda^n n^s$ as $n \to \infty$. Thus, for $|\lambda| < 1$ both $\{A_n\}$ and $\{\hat{A}_n\}$ converge, and $\{\hat{A}_n\}$ converges more quickly.

Naturally, if we have to make a choice about which extrapolation methods to employ regularly as users, we would probably prefer those methods that are effective accelerators for more than one class of sequences. Finally, we would like to remark that, as the design of extrapolation methods is based on the asymptotic properties of A_n for $n \to \infty$ in many important cases, there is great value to analyzing A_n asymptotically for $n \to \infty$. In addition, this analysis also produces the conditions on $\{A_n\}$ that we need for the convergence study, as discussed in the second paragraph of Section 0.5. In many cases of interest, this study may turn out to be very nontrivial and challenging.

The next issue is the design of good algorithms for implementing extrapolation methods. Recall that there may be more than one algorithm for implementing a given method. As mentioned before, a good algorithm is one that requires a small number of operations and little storage. In addition, it should be as stable as possible numerically. Needless to say, from the point of view of a user, we should always prefer the most efficient algorithm for a given method. Note that the development of algorithms, because it relies on algebraic manipulations only, is not in general as important and illuminating as either the development of extrapolation methods or the analysis of these methods, an issue we discuss below, or the asymptotic study of $\{A_n\}$, which precedes all this. After all, the quality of the sequence $\{\hat{A}_n\}$ is determined exclusively by the extrapolation method that generates it and not by whichever algorithm is used for implementing this extrapolation method. Algorithms are only numerical means by which we obtain the $\{\hat{A}_n\}$ already uniquely determined by the extrapolation method. For these reasons, we reduce our treatment of algorithms to a minimum.

Some of the literature deals with so-called kernels of extrapolation methods. The *kernel* of an extrapolation method is the class of sequences $\{A_n\}$ for which $\hat{A}_n = A$ for all n, where A is the limit or antilimit of $\{A_n\}$. Unfortunately, there is very little one can learn about the convergence behavior and stability of the method in question by looking at its kernel. For this reason, we have left out the treatment of kernels almost completely. We have considered in passing only those kernels that are obtained in a trivial way.

Given an extrapolation method and a class of sequences to which it is applied, as we mentioned in the preceding section, the most important and challenging issues are those of convergence and stability. Proposing realistic and useful analyses for convergence and stability has always been a very difficult task because most of the practical extrapolation methods are highly nonlinear. As a result, very few papers have dealt with convergence and stability, where we believe more efforts should be spent. A variety of mathematical tools are needed for this task, asymptotic analysis being the most crucial of them.

As for the user, we believe that he should be at least familiar with the existing convergence and stability results, as these may be very helpful in deciding which extrapolation method should be applied, to which problem, how it should be applied, and how it should be tuned for good stability properties. In addition, having at least some idea about the asymptotic behavior of the elements of the sequence whose convergence is being accelerated is of great assistance in efficient application of extrapolation methods.

We would like to make one last remark about extrapolation *methods* as opposed to *algorithms* by which methods are implemented. In recent years, there has been an unfortunate confusion of terminology, in that many papers use the concepts of method and algorithm interchangeably. The result of this confusion has been that effectiveness due to extrapolation methods has been incorrectly assigned to extrapolation algorithms. As noted earlier, the sequences $\{\hat{A}_n\}$ are uniquely defined and their properties are determined only by extrapolation methods and not by algorithms that implement the methods. In this book, we are very careful to avoid this confusion by distinguishing between methods and algorithms.

Part I

The Richardson Extrapolation Process
and Its Generalizations

1

The Richardson Extrapolation Process

1.1 Introduction and Background

In many problems of practical interest, a given infinite sequence $\{A_n\}$ can be related to a function $A(y)$ that is known, and hence is computable, for $0 < y \leq b$ with some $b > 0$, the variable y being continuous or discrete. This relation takes the form $A_n = A(y_n)$, $n = 0, 1, \ldots$, for some monotonically decreasing sequence $\{y_n\} \subset (0, b]$ that satisfies $\lim_{n\to\infty} y_n = 0$. Thus, in case $\lim_{y\to 0+} A(y) = A$, $\lim_{n\to\infty} A_n = A$ as well. Consequently, computing $\lim_{n\to\infty} A_n$ amounts to computing $\lim_{y\to 0+} A(y)$ in such a case, and this is precisely what we want to do.

Again, in many cases of interest, the function $A(y)$ may have a well-defined expansion for $y \to 0+$ whose *form* is known. For example – and this is the case we treat in this chapter – $A(y)$ may satisfy for some positive integer s

$$A(y) = A + \sum_{k=1}^{s} \alpha_k y^{\sigma_k} + O(y^{\sigma_{s+1}}) \text{ as } y \to 0+, \qquad (1.1.1)$$

where $\sigma_k \neq 0$, $k = 1, 2, \ldots, s+1$, and $\Re\sigma_1 < \Re\sigma_2 < \cdots < \Re\sigma_{s+1}$, and where α_k are constants *independent of* y. Obviously, $\Re\sigma_1 > 0$ guarantees that $\lim_{y\to 0+} A(y) = A$. When $\lim_{y\to 0+} A(y)$ does not exist, A is the *antilimit* of $A(y)$ for $y \to 0+$, and in this case $\Re\sigma_i \leq 0$ at least for $i = 1$. If (1.1.1) is valid for all $s = 1, 2, 3, \ldots$, and $\Re\sigma_1 < \Re\sigma_2 < \cdots$, with $\lim_{k\to\infty} \Re\sigma_k = +\infty$, then $A(y)$ has the true asymptotic expansion

$$A(y) \sim A + \sum_{k=1}^{\infty} \alpha_k y^{\sigma_k} \text{ as } y \to 0+, \qquad (1.1.2)$$

whether the infinite series $\sum_{k=1}^{\infty} \alpha_k y^{\sigma_k}$ converges or not. (In most cases of interest, this series diverges strongly.) The σ_k are assumed to be known, but the coefficients α_k need not be known; generally, the α_k are not of interest to us. We are interested in finding A whether it is the limit or the antilimit of $A(y)$ for $y \to 0+$.

Suppose now that $\Re\sigma_1 > 0$ so that $\lim_{y\to 0+} A(y) = A$. Then A can be approximated by $A(y)$ with sufficiently small values of y, the error in this approximation being $A(y) - A = O(y^{\sigma_1})$ as $y \to 0+$ by (1.1.1). If $\Re\sigma_1$ is sufficiently large, $A(y)$ can approximate A well even for values of y that are not too small. If this is not the case, however, then we may have to compute $A(y)$ for very small values of y to obtain reasonably good approximations to A. Unfortunately, this straightforward idea of reducing y to very small values is not always applicable. In most cases of interest, computing $A(y)$ for very small values of y either is very costly or suffers from loss of significance in finite-precision arithmetic. The deeper idea of the Richardson extrapolation, on the other hand, is to somehow eliminate the y^{σ_1} term from the expansion in (1.1.1) and to obtain a new approximation $A_1(y)$ to A whose error is $A_1(y) - A = O(y^{\sigma_2})$ as $y \to 0+$. Obviously, $A_1(y)$ will be a better approximation to A than $A(y)$ for small y since $\Re\sigma_2 > \Re\sigma_1$. In addition, if $\Re\sigma_2$ is sufficiently large, then we expect $A_1(y)$ to approximate A well also for values of y that are not too small, independently of the size of $\Re\sigma_1$. At this point, we mention only that the Richardson extrapolation is achieved by taking an appropriate "weighted average" of $A(y)$ and $A(\omega y)$ for some $\omega \in (0, 1)$. We give the precise details of this procedure in the next section.

From (1.1.1), it is clear that $A(y) - A = O(y^{\sigma_1})$ as $y \to 0+$, whether $\Re\sigma_1 > 0$ or not. Thus, the function $A_1(y)$ that results from the Richardson extrapolation can be a useful approximation to A for small values of y also when $\Re\sigma_1 \leq 0$, provided $\Re\sigma_2 > 0$. That is to say, $\lim_{y\to 0+} A_1(y) = A$ provided $\Re\sigma_2 > 0$ whether $\lim_{y\to 0+} A(y)$ exists or not. This is an additional fundamental and useful feature of the Richardson extrapolation.

In the following examples, we show how functions $A(y)$ exactly of the form we have described here come about naturally. In these examples, we treat the classic problems of computing π by the method of Archimedes, numerical differentiation by differences, numerical integration by the trapezoidal rule, summation of an infinite series that is used in defining the Riemann Zeta function, and the Hadamard finite parts of divergent integrals.

Example 1.1.1 The Method of Archimedes for Computing π The method of Archimedes for computing π consists of approximating the area of the unit disk (that is nothing but π) by the area of an inscribed or circumscribing regular polygon. If this polygon is inscribed in the unit disk and has n sides, then its area is simply $S_n = (n/2)\sin(2\pi/n)$. Obviously, S_n has the (convergent) series expansion

$$S_n = \pi + \frac{1}{2}\sum_{i=1}^{\infty} \frac{(-1)^i (2\pi)^{2i+1}}{(2i+1)!} n^{-2i}, \tag{1.1.3}$$

and the sequence $\{S_n\}$ is monotonically increasing and has π as its limit.

If the polygon circumscribes the unit disk and has n sides, then its area is $S_n = n\tan(\pi/n)$, and S_n has the (convergent) series expansion

$$S_n = \pi + \sum_{i=1}^{\infty} \frac{(-1)^i 4^{i+1}(4^{i+1} - 1)\pi^{2i+1} B_{2i+2}}{(2i+2)!} n^{-2i}, \tag{1.1.4}$$

where B_k are the Bernoulli numbers (see Appendix D), and the sequence $\{S_n\}$ this time is monotonically decreasing and has π as its limit.

As the expansions given in (1.1.3) and (1.1.4) are also asymptotic as $n \to \infty$, S_n in both cases is analogous to the function $A(y)$. This analogy is as follows: $S_n \leftrightarrow A(y)$, $n^{-1} \leftrightarrow y$, $\sigma_k = 2k$, $k = 1, 2, \ldots$, and $\pi \leftrightarrow A$. The variable y is discrete and assumes the values $1/3, 1/4, \ldots$.

Finally, the subsequences $\{S_{2^m}\}$ and $\{S_{3 \cdot 2^m}\}$ can be computed recursively without having to know π, their computation involving only square roots. (See Example 2.2.2 in Chapter 2.)

Example 1.1.2 Numerical Differentiation by Differences Let $f(x)$ be continuously differentiable at $x = x_0$, and assume that $f'(x_0)$, the first derivative of $f(x)$ at x_0, is needed. Assume further that the only thing available to us is $f(x)$, or a procedure that computes $f(x)$, for all values of x in a neighborhood of x_0.

If $f(x)$ is known in the neighborhood $[x_0 - a, x_0 + a]$ for some $a > 0$, then $f'(x_0)$ can be approximated by the centered difference $\delta_0(h)$ that is given by

$$\delta_0(h) = \frac{f(x_0 + h) - f(x_0 - h)}{2h}, \ 0 < h \leq a. \tag{1.1.5}$$

Note that h here is a continuous variable. Obviously, $\lim_{h \to 0} \delta_0(h) = f'(x_0)$. The accuracy of $\delta_0(h)$ is quite low, however. When $f \in C^3[x_0 - a, x_0 + a]$, there exists $\xi(h) \in [x_0 - h, x_0 + h]$, for which the error in $\delta_0(h)$ satisfies

$$\delta_0(h) - f'(x_0) = \frac{f'''(\xi(h))}{3!} h^2 = O(h^2) \text{ as } h \to 0. \tag{1.1.6}$$

When the function $f(x)$ is continuously differentiable a number of times, the error $\delta_0(h) - f'(x_0)$ can be expanded in powers of h^2. For $f \in C^{2s+3}[x_0 - a, x_0 + a]$, there exists $\xi(h) \in [x_0 - h, x_0 + h]$, for which we have

$$\delta_0(h) = f'(x_0) + \sum_{k=1}^{s} \frac{f^{(2k+1)}(x_0)}{(2k+1)!} h^{2k} + R_s(h), \tag{1.1.7}$$

where

$$R_s(h) = \frac{f^{(2s+3)}(\xi(h))}{(2s+3)!} h^{2s+2} = O(h^{2s+2}) \text{ as } h \to 0. \tag{1.1.8}$$

The proof of (1.1.7) and (1.1.8) can be achieved by expanding $f(x_0 \pm h)$ in a Taylor series about x_0 with remainder.

The difference $\delta_0(h)$ is thus seen to be analogous to the function $A(y)$. This analogy is as follows: $\delta_0(h) \leftrightarrow A(y)$, $h \leftrightarrow y$, $\sigma_k = 2k$, $k = 1, 2, \ldots$, and $f'(x_0) \leftrightarrow A$.

When $f \in C^\infty[x_0 - a, x_0 + a]$, the expansion in (1.1.7) holds for all $s = 0, 1, \ldots$. As a result, we can replace it by the genuine asymptotic expansion

$$\delta_0(h) \sim f'(x_0) + \sum_{k=1}^{\infty} \frac{f^{(2k+1)}(x_0)}{(2k+1)!} h^{2k} \text{ as } h \to 0, \tag{1.1.9}$$

whether the infinite series on the right-hand side of (1.1.9) converges or not.

As is known, in finite-precision arithmetic, the computation of $\delta_0(h)$ for very small values of h is dominated by roundoff. The reason for this is that as $h \to 0$ both $f(x_0 + h)$ and $f(x_0 - h)$ tend to $f(x_0)$, which causes the difference $f(x_0 + h) - f(x_0 - h)$ to have fewer and fewer correct significant digits. Thus, it is meaningless to carry out the computation of $\delta_0(h)$ beyond a certain threshold value of h.

Example 1.1.3 Numerical Quadrature by Trapezoidal Rule Let $f(x)$ be defined on $[0, 1]$, and assume that $I[f] = \int_0^1 f(x) \, dx$ is to be computed by numerical quadrature. One of the simplest numerical quadrature formulas is the trapezoidal rule. Let $T(h)$ be the trapezoidal rule approximation to $I[f]$, with $h = 1/n$, n being a positive integer. Then, $T(h)$ is given by

$$T(h) = h \left[\frac{1}{2} f(0) + \sum_{j=1}^{n-1} f(jh) + \frac{1}{2} f(1) \right]. \qquad (1.1.10)$$

Note that h for this problem is a discrete variable that takes on the values $1, 1/2, 1/3, \ldots$. It is well known that $T(h)$ tends to $I[f]$ as $h \to 0$ (or $n \to \infty$), whenever $f(x)$ is Riemann integrable on $[0, 1]$. When $f \in C^2[0, 1]$, there exists $\xi(h) \in [0, 1]$, for which the error in $T(h)$ satisfies

$$T(h) - I[f] = \frac{f''(\xi(h))}{12} h^2 = O(h^2) \text{ as } h \to 0. \qquad (1.1.11)$$

When the integrand $f(x)$ is continuously differentiable a number of times, the error $T(h) - I[f]$ can be expanded in powers of h^2. For $f \in C^{2s+2}[0, 1]$, there exists $\xi(h) \in [0, 1]$, for which

$$T(h) = I[f] + \sum_{k=1}^{s} \frac{B_{2k}}{(2k)!} \left[f^{(2k-1)}(1) - f^{(2k-1)}(0) \right] h^{2k} + R_s(h), \qquad (1.1.12)$$

where

$$R_s(h) = \frac{B_{2s+2}}{(2s+2)!} f^{(2s+2)}(\xi(h)) h^{2s+2} = O(h^{2s+2}) \text{ as } h \to 0. \qquad (1.1.13)$$

Here B_p are the Bernoulli numbers as before. The expansion in (1.1.12) with (1.1.13) is known as the Euler–Maclaurin expansion (see Appendix D) and its proof can be found in many books on numerical analysis.

The approximation $T(h)$ is analogous to the function $A(y)$ in the following sense: $T(h) \leftrightarrow A(y)$, $h \leftrightarrow y$, $\sigma_k = 2k$, $k = 1, 2, \ldots$, and $I[f] \leftrightarrow A$.

Again, for $f \in C^{2s+2}[0, 1]$, an expansion that is identical in form to (1.1.12) with (1.1.13) exists for the midpoint rule approximation $M(h)$, where

$$M(h) = h \sum_{j=1}^{n} f(jh - \tfrac{1}{2}h). \qquad (1.1.14)$$

This expansion is

$$M(h) = I[f] + \sum_{k=1}^{s} \frac{B_{2k}(\frac{1}{2})}{(2k)!} \left[f^{(2k-1)}(1) - f^{(2k-1)}(0) \right] h^{2k} + R_s(h), \qquad (1.1.15)$$

where, again for some $\xi(h) \in [0, 1]$,

$$R_s(h) = \frac{B_{2s+2}(\frac{1}{2})}{(2s+2)!} f^{(2s+2)}(\xi(h)) h^{2s+2} = O(h^{2s+2}) \text{ as } h \to 0. \qquad (1.1.16)$$

Here $B_p(x)$ is the Bernoulli polynomial of degree p and $B_{2k}(\frac{1}{2}) = -(1 - 2^{1-2k}) B_{2k}$, $k = 1, 2, \ldots$.

When $f \in C^\infty[0, 1]$, both expansions in (1.1.12) and (1.1.15) hold for all $s = 0, 1, \ldots$. As a result, we can replace both by genuine asymptotic expansions of the form

$$Q(h) \sim I[f] + \sum_{k=1}^{\infty} c_k h^{2k} \quad \text{as } h \to 0, \tag{1.1.17}$$

where $Q(h)$ stands for $T(h)$ or $M(h)$, and c_k is the coefficient of h^{2k} in (1.1.12) or (1.1.15). Generally, when $f(x)$ is not analytic in $[0, 1]$, or even when it is analytic there but is not entire, the infinite series $\sum_{k=1}^{\infty} c_k h^{2k}$ in (1.1.17) diverges very strongly.

Finally, by $h = 1/n$, the computation of $Q(h)$ for very small values of h involves a large number of integrand evaluations and hence is very costly.

Example 1.1.4 Summation of the Riemann Zeta Function Series Let $A_n = \sum_{m=1}^{n} m^{-z}$, $n = 1, 2, \ldots$. When $\Re z > 1$, $\lim_{n \to \infty} A_n = \zeta(z)$, where $\zeta(z)$ is the Riemann Zeta function. For $\Re z \leq 1$, on the other hand, $\lim_{n \to \infty} A_n$ does not exist. Actually, the infinite series $\sum_{m=1}^{\infty} m^{-z}$ is taken as the definition of $\zeta(z)$ for $\Re z > 1$. With this definition, $\zeta(z)$ is an analytic function of z for $\Re z > 1$. Furthermore, it can be continued analytically to the whole z-plane with the exception of the point $z = 1$, where it has a simple pole with residue 1.

For all $z \neq 1$, i.e., whether $\lim_{n \to \infty} A_n$ exists or not, we have the well-known asymptotic expansion (see Appendix E)

$$A_n \sim \zeta(z) + \frac{1}{1-z} \sum_{i=0}^{\infty} (-1)^i \binom{1-z}{i} B_i n^{-z-i+1} \quad \text{as } n \to \infty, \tag{1.1.18}$$

where B_i are the Bernoulli numbers as before and $\binom{a}{i}$ are the binomial coefficients. We also recall that $B_3 = B_5 = B_7 = \cdots = 0$, and that the rest of the B_i are nonzero.

The partial sum A_n is thus analogous to the function $A(y)$ in the following sense: $A_n \leftrightarrow A(y)$, $n^{-1} \leftrightarrow y$, $\sigma_1 = z - 1$, $\sigma_2 = z$, $\sigma_k = z + 2k - 5$, $k = 3, 4, \ldots$, and $\zeta(z) \leftrightarrow A$ provided $z \neq -m + 1$, $m = 0, 1, 2, \ldots$. Thus, $\zeta(z)$ is the limit of $\{A_n\}$ when $\Re z > 1$, and its antilimit otherwise, provided $z \neq -m + 1$, $m = 0, 1, 2, \ldots$. Obviously, the variable y is now discrete and takes on the values $1, 1/2, 1/3, \ldots$.

Note also that the infinite series on the right-hand side of (1.1.18) is strongly divergent.

Example 1.1.5 Numerical Integration of Periodic Singular Functions Let us now consider the integral $I[f] = \int_0^1 f(x)\,dx$, where $f(x)$ is a 1-periodic function that is infinitely differentiable on $(-\infty, \infty)$ except at the points $t + k$, $k = 0, \pm 1$, $\pm 2, \ldots$, where it has logarithmic singularities, and can be written in the form $f(x) = g(x) \log|x - t| + \tilde{g}(x)$ when $x, t \in [0, 1]$. For example, with $u \in C^\infty(-\infty, \infty)$ and periodic with period 1, and with c some constant, $f(x) = u(x) \log(c|\sin \pi(x - t)|)$ is such a function. For this $f(x)$, we have $g(t) = u(t)$ and $\tilde{g}(t) = u(t) \log(\pi c)$. Sidi and Israeli [310] derived the "corrected" trapezoidal rule approximation

$$T(h; t) = h \sum_{i=1}^{n-1} f(t + ih) + \tilde{g}(t)h + g(t)h \log\left(\frac{h}{2\pi}\right), \quad h = 1/n, \tag{1.1.19}$$

for $I[f]$, and showed that $T(h;t)$ has the asymptotic expansion

$$T(h;t) \sim I[f] - 2\sum_{k=1}^{\infty} \frac{\zeta'(-2k)}{(2k)!} g^{(2k)}(t) h^{2k+1} \quad \text{as } h \to 0. \qquad (1.1.20)$$

Here $\zeta'(z) = \frac{d}{dz}\zeta(z)$. (See Appendix D.)

The approximation $T(h;t)$ is analogous to the function $A(y)$ in the following sense: $T(h;t) \leftrightarrow A(y)$, $h \leftrightarrow y$, $\sigma_k = 2k+1$, $k = 1, 2, \ldots$, and $I[f] \leftrightarrow A$. In addition, y takes on the discrete values $1, 1/2, 1/3, \ldots$.

Example 1.1.6 Hadamard Finite Parts of Divergent Integrals Consider the integral $\int_0^1 x^\rho g(x)\,dx$, where $g \in C^\infty[0,1]$ and ρ is generally complex such that $\rho \neq -1, -2, \ldots$. When $\Re\rho > -1$, the integral exists in the ordinary sense. In case $g(0) \neq 0$ and $\Re\rho \leq -1$, the integral does not exist in the ordinary sense since $x^\rho g(x)$ is not integrable at $x = 0$, but its Hadamard finite part exists, as we mentioned in Example 0.2.4. Let us define $Q(h) = \int_h^1 x^\rho g(x)\,dx$. Obviously, $Q(h)$ is well-defined and computable for $h \in (0,1]$. Let m be any nonnegative integer. Then, there holds

$$Q(h) = \int_h^1 x^\rho \left[g(x) - \sum_{i=0}^{m-1} \frac{g^{(i)}(0)}{i!} x^i \right] dx + \sum_{i=0}^{m-1} \frac{g^{(i)}(0)}{i!} \frac{1 - h^{\rho+i+1}}{\rho+i+1}. \qquad (1.1.21)$$

Now let $m > -\Re\rho - 1$. Expressing the integral term in (1.1.21) in the form $\int_h^1 = \int_0^1 - \int_0^h$, using the fact that

$$g(x) - \sum_{i=0}^{m-1} \frac{g^{(i)}(0)}{i!} x^i = \frac{g^{(m)}(\xi(x))}{m!} x^m, \quad \text{for some } \xi(x) \in (0,x),$$

and defining

$$I(\rho) = \int_0^1 x^\rho \left[g(x) - \sum_{i=0}^{m-1} \frac{g^{(i)}(0)}{i!} x^i \right] dx + \sum_{i=0}^{m-1} \frac{1}{\rho+i+1} \frac{g^{(i)}(0)}{i!}, \qquad (1.1.22)$$

and $\|g^{(m)}\| = \max_{0 \leq x \leq 1} |g^{(m)}(x)|$, we obtain from (1.1.21)

$$Q(h) = I(\rho) - \sum_{i=0}^{m-1} \frac{g^{(i)}(0)}{i!} \frac{h^{\rho+i+1}}{\rho+i+1} + R_m(h); \quad |R_m(h)| \leq \frac{\|g^{(m)}\|}{m!} \frac{h^{\Re\rho+m+1}}{\Re\rho+m+1}, \qquad (1.1.23)$$

[Note that, with $m > -\Re\rho - 1$, the integral term in (1.1.22) exists in the ordinary sense and $I(\rho)$ is independent of m.] Since m is also arbitrary in (1.1.23), we conclude that $Q(h)$ has the asymptotic expansion

$$Q(h) \sim I(\rho) - \sum_{i=0}^{\infty} \frac{g^{(i)}(0)}{i!} \frac{h^{\rho+i+1}}{\rho+i+1} \quad \text{as } h \to 0. \qquad (1.1.24)$$

Thus, $Q(h)$ is analogous to the function $A(y)$ in the following sense: $Q(h) \leftrightarrow A(y)$, $h \leftrightarrow y$, $\sigma_k = \rho + k$, $k = 1, 2, \ldots$, and $I(\rho) \leftrightarrow A$. Of course, y is a continuous variable in this case. When the integral exists in the ordinary sense, $I(\rho) = \lim_{h\to 0} Q(h)$; otherwise, $I(\rho)$ is the Hadamard finite part of $\int_0^1 x^\rho g(x)\,dx$ and serves as the antilimit of $Q(h)$

as $h \to 0$. Finally, $I(\rho) = \int_0^1 x^\rho g(x) \, dx$ is analytic in ρ for $\Re\rho > -1$ and, by (1.1.22), can be continued analytically to a meromorphic function with simple poles possibly at $\rho = -1, -2, \ldots$. Thus, the Hadamard finite part is nothing but the analytic continuation of the function $I(\rho)$ that is defined via the convergent integral $\int_0^1 x^\rho g(x) \, dx$, $\Re\rho > -1$, to values of ρ for which $\Re\rho \leq -1$, $\rho \neq -1, -2, \ldots$.

Before going on, we mention that many of the developments of this chapter are due to Bulirsch and Stoer [43], [45], [46]. The treatment in these papers assumes that the σ_k are real and positive. The case of generally complex σ_k was considered recently in Sidi [298], where the function $A(y)$ is allowed to have a more general asymptotic behavior than in (1.1.2). See also Sidi [301].

1.2 The Idea of Richardson Extrapolation

We now go back to the function $A(y)$ discussed in the second paragraph of the preceding section. We do not assume that $\lim_{y \to 0+} A(y)$ necessarily exists. We recall that, when it exists, this limit is equal to A in (1.1.1) ; otherwise, A there is the antilimit of $A(y)$ as $y \to 0+$. Also, the nonexistence of $\lim_{y \to 0+} A(y)$ immediately implies that $\Re\sigma_i \leq 0$ at least for $i = 1$.

As mentioned in the third paragraph of the preceding section, $A(y) - A = O(y^{\sigma_1})$ as $y \to 0+$, and we would like to eliminate the y^{σ_1} term from (1.1.1) and thus obtain a new approximation to A that is better than $A(y)$ for $y \to 0+$. Let us pick a constant $\omega \in (0, 1)$, and set $y' = \omega y$. Then, from (1.1.1) we have

$$A(y') = A + \sum_{k=1}^{s} \alpha_k \omega^{\sigma_k} y^{\sigma_k} + O(y^{\sigma_{s+1}}) \text{ as } y \to 0+ . \tag{1.2.1}$$

Multiplying (1.1.1) by ω^{σ_1} and subtracting from (1.2.1), we obtain

$$A(y') - \omega^{\sigma_1} A(y) = (1 - \omega^{\sigma_1})A + \sum_{k=2}^{s} (\omega^{\sigma_k} - \omega^{\sigma_1}) \alpha_k y^{\sigma_k} + O(y^{\sigma_{s+1}}) \text{ as } y \to 0+ . \tag{1.2.2}$$

Obviously, the term y^{σ_1} is missing from the summation in (1.2.2). Dividing both sides of (1.2.2) by $(1 - \omega^{\sigma_1})$, and identifying

$$A(y, y') = \frac{A(y') - \omega^{\sigma_1} A(y)}{1 - \omega^{\sigma_1}} \tag{1.2.3}$$

as the new approximation to A, we have

$$A(y, y') = A + \sum_{k=2}^{s} \frac{\omega^{\sigma_k} - \omega^{\sigma_1}}{1 - \omega^{\sigma_1}} \alpha_k y^{\sigma_k} + O(y^{\sigma_{s+1}}) \text{ as } y \to 0+, \tag{1.2.4}$$

so that $A(y, y') - A = O(y^{\sigma_2})$ as $y \to 0+$, as was required. It is important to note that (1.2.4) is exactly of the form (1.1.1) with $A(y)$ and the α_k replaced by $A(y, y')$ and the $\frac{\omega^{\sigma_k} - \omega^{\sigma_1}}{1 - \omega^{\sigma_1}} \alpha_k$, respectively.

We can now continue along the same lines and eliminate the y^{σ_2} term from (1.2.4). This can be achieved by combining $A(y, y')$ and $A(y', y'')$ with $y'' = \omega y' = \omega^2 y$. The resulting new approximation is

$$A(y, y', y'') = \frac{A(y', y'') - \omega^{\sigma_2} A(y, y')}{1 - \omega^{\sigma_2}}, \qquad (1.2.5)$$

and we have

$$A(y, y', y'') = A + \sum_{k=3}^{s} \frac{\omega^{\sigma_k} - \omega^{\sigma_1}}{1 - \omega^{\sigma_1}} \frac{\omega^{\sigma_k} - \omega^{\sigma_2}}{1 - \omega^{\sigma_2}} \alpha_k y^{\sigma_k} + O(y^{\sigma_{s+1}}) \text{ as } y \to 0+, \quad (1.2.6)$$

so that $A(y, y', y'') - A = O(y^{\sigma_3})$ as $y \to 0+$.

This process, which is called the *Richardson extrapolation process*, can be repeated to eliminate the terms $y^{\sigma_3}, y^{\sigma_4}, \ldots$, from the summation in (1.2.6). We do this in the next section by formalizing the preceding procedure, where we give a very efficient recursive algorithm for the Richardson extrapolation process as well.

Before we end this section, we would like to mention that the preceding procedure was first described [for the case $\sigma_k = 2k$, $k = 1, 2, \ldots$, in (1.1.1)] by Richardson [236], who called it *deferred approach to the limit*. Richardson applied this approach to improve the finite difference solutions of some partial differential equations, such as Laplace's equation in a square. Later, Richardson [237] used it to improve the numerical solutions of an integral equation. Finally, Richardson [238] used the idea of extrapolation to accelerate the convergence of sequences, to compute Fourier coefficients, and to solve differential eigenvalue problems. For all these details and many more of the early references on this subject, we refer the reader to the excellent survey by Joyce [145]. In the remainder of this work, we refer to the extrapolation procedure concerning $A(y) \sim A + \sum_{k=1}^{\infty} \alpha_k y^{kr}$ as $y \to 0+$, where $r > 0$, as the *polynomial Richardson extrapolation process*, and we discuss it in some detail in the next chapter.

1.3 A Recursive Algorithm for the Richardson Extrapolation Process

Let us pick a constant $\omega \in (0, 1)$ and $y_0 \in (0, b]$ and let $y_m = y_0 \omega^m$, $m = 1, 2, \ldots$. Obviously, $\{y_m\}$ is a decreasing sequence in $(0, b]$ and $\lim_{m \to \infty} y_m = 0$.

Algorithm 1.3.1

1. Set $A_0^{(j)} = A(y_j)$, $j = 0, 1, 2, \ldots$.
2. Set $c_n = \omega^{\sigma_n}$ and compute $A_n^{(j)}$ by the recursion

$$A_n^{(j)} = \frac{A_{n-1}^{(j+1)} - c_n A_{n-1}^{(j)}}{1 - c_n}, \quad j = 0, 1, \ldots, \ n = 1, 2, \ldots.$$

The $A_n^{(j)}$ are approximations to A produced by the Richardson extrapolation process. We have the following result concerning the $A_n^{(j)}$.

Table 1.3.1: *The Romberg table*

$$
\begin{array}{lllll}
A_0^{(0)} & & & & \\
A_0^{(1)} & \searrow & A_1^{(0)} & & \\
& \searrow & & \searrow & \\
A_0^{(2)} & \to & A_1^{(1)} & \to & A_2^{(0)} \\
& \searrow & & \searrow & & \searrow \\
A_0^{(3)} & \to & A_1^{(2)} & \to & A_2^{(1)} & \to & A_3^{(0)} \\
\vdots & & \vdots & & \vdots & & \vdots & & \ddots
\end{array}
$$

Theorem 1.3.2 *In the notation of the previous section, the $A_n^{(j)}$ satisfy*

$$A_n^{(j)} = A(y_j, y_{j+1}, \dots, y_{j+n}). \tag{1.3.1}$$

The proof of (1.3.1) can be done by induction and is left to the reader.

The $A_n^{(j)}$ can be arranged in a two-dimensional array, called the *Romberg table*, as in Table 1.3.1. The arrows in the table show the flow of computation.

Given the values $A(y_m)$, $m = 0, 1, \dots, N$, and $c_m = \omega^{\sigma_m}$, $m = 1, 2, \dots, N$, this algorithm produces the $\frac{1}{2}N(N+1)$ approximations $A_n^{(j)}$, $1 \le j + n \le N$, $n \ge 1$. The computation of these $A_n^{(j)}$ can be achieved in $\frac{1}{2}N(N+1)$ multiplications, $\frac{1}{2}N(N+1)$ divisions, and $\frac{1}{2}N(N+3)$ additions. This computation also requires $\frac{1}{2}N^2 + O(N)$ storage locations. When only the diagonal approximations $A_n^{(0)}$, $n = 1, 2, \dots, N$, are required, the algorithm can be implemented with $N + O(1)$ storage locations. This can be achieved by computing Table 1.3.1 columnwise and letting $\{A_n^{(j)}\}_{j=0}^{N-n}$ overwrite $\{A_{n-1}^{(j)}\}_{j=0}^{N-n+1}$. It can also be achieved by computing the table row-wise and letting the row $\{A_n^{(l-n)}\}_{n=0}^{l}$ overwrite the row $\{A_n^{(l-1-n)}\}_{n=0}^{l-1}$. The latter approach enables us to introduce $A(y_m)$, $m = 0, 1, \dots$, one by one. As we shall see in Section 1.5, the diagonal sequences $\{A_n^{(j)}\}_{n=0}^{\infty}$ have excellent convergence properties.

1.4 Algebraic Properties of the Richardson Extrapolation Process

1.4.1 A Related Set of Polynomials

As part of the input to Algorithm 1.3.1 is the sequence $\{A_0^{(m)} = A(y_m)\}$, the $A_n^{(j)}$ are, of course, functions of the $A(y_m)$. The relationship between $\{A(y_m)\}$ and the $A_n^{(j)}$, however, is hidden in the algorithm. We now investigate the precise nature of this relationship.

We start with the following simple lemma.

Lemma 1.4.1 *Given the sequence $\{B_0^{(j)}\}$, define the quantities $\{B_n^{(j)}\}$ by the recursion*

$$B_n^{(j)} = \lambda_n^{(j)} B_{n-1}^{(j+1)} + \mu_n^{(j)} B_{n-1}^{(j)}, \quad j = 0, 1, \dots, \ n = 1, 2, \dots, \tag{1.4.1}$$

where the scalars $\lambda_n^{(j)}$ and $\mu_n^{(j)}$ satisfy

$$\lambda_n^{(j)} + \mu_n^{(j)} = 1, \quad j = 0, 1, \dots, \ n = 1, 2, \dots. \tag{1.4.2}$$

Then there exist scalars $\gamma_{ni}^{(j)}$ that depend on the $\lambda_k^{(m)}$ and $\mu_k^{(m)}$, such that

$$B_n^{(j)} = \sum_{i=0}^{n} \gamma_{ni}^{(j)} B_0^{(j+i)} \quad and \quad \sum_{i=0}^{n} \gamma_{ni}^{(j)} = 1. \tag{1.4.3}$$

Proof. The proof of (1.4.3) is by induction on n.   ■

Lemma 1.4.1 and Algorithm 1.3.1 together imply that $A_n^{(j)}$ is of the form $A_n^{(j)} = \sum_{i=0}^{n} \gamma_{ni}^{(j)} A_0^{(j+i)}$ for some $\gamma_{ni}^{(j)}$ that satisfy $\sum_{i=0}^{n} \gamma_{ni}^{(j)} = 1$. Obviously, this does not reveal anything fundamental about the relationship between $A_n^{(j)}$ and $\{A(y_m)\}$, aside from the assertion that $A_n^{(j)}$ is a "weighted average" of some sort of $A(y_l)$, $j \leq l \leq j + n$. The following theorem, on the other hand, gives a complete description of this relationship.

Theorem 1.4.2 *Let $c_i = \omega^{\sigma_i}$, $i = 1, 2, \ldots$, and define the polynomials $U_n(z)$ by*

$$U_n(z) = \prod_{i=1}^{n} \frac{z - c_i}{1 - c_i} \equiv \sum_{i=0}^{n} \rho_{ni} z^i. \tag{1.4.4}$$

Then $A_n^{(j)}$ is related to the $A(y_m)$ through

$$A_n^{(j)} = \sum_{i=0}^{n} \rho_{ni} A(y_{j+i}). \tag{1.4.5}$$

Obviously,

$$\sum_{i=0}^{n} \rho_{ni} = 1. \tag{1.4.6}$$

Proof. From (1.4.4), we have

$$U_n(z) = \frac{z - c_n}{1 - c_n} U_{n-1}(z), \tag{1.4.7}$$

from which we also have, with $\rho_{ki} = 0$ for $i < 0$ or $i > k$ for all k,

$$\rho_{ni} = \frac{\rho_{n-1,i-1} - c_n \rho_{n-1,i}}{1 - c_n}, \quad 0 \leq i \leq n. \tag{1.4.8}$$

Now we can use (1.4.8) to show that $A_n^{(j)}$, as given in (1.4.5), satisfies the recursion relation in Algorithm 1.3.1. This completes the proof.   ■

From this theorem we see that, for the Richardson extrapolation process we are discussing now, the $\gamma_{ni}^{(j)}$ alluded to above are simply ρ_{ni}; therefore, they are *independent of j* as well.

The following result concerning the ρ_{ni} will be of use in the convergence and stability analyses that we provide in the next sections of this chapter.

Theorem 1.4.3 *The coefficients ρ_{ni} of the polynomial $U_n(z)$ defined in Theorem 1.4.2 are such that*

$$\sum_{i=0}^{n} |\rho_{ni}|\, |z|^i \leq \prod_{i=1}^{n} \frac{|z| + |c_i|}{|1 - c_i|}. \tag{1.4.9}$$

In particular,

$$\sum_{i=0}^{n} |\rho_{ni}| \leq \prod_{i=1}^{n} \frac{1 + |c_i|}{|1 - c_i|}. \tag{1.4.10}$$

If c_i, $1 \leq i \leq n$, all have the same phase, which occurs when σ_i, $1 \leq i \leq n$, all have the same imaginary part, then equality holds in both (1.4.9) and (1.4.10). This takes place, in particular, when c_i, $1 \leq i \leq n$, are all real positive or all real negative. Furthermore, we have $\sum_{i=0}^{n} |\rho_{ni}| = 1$ for the case in which c_i, $1 \leq i \leq n$, are all real negative.

Theorem 1.4.3 is stated in Sidi [298] and its proof can be achieved by using the following general result that is also given there.

Lemma 1.4.4 *Let $Q(z) = \sum_{i=0}^{n} a_i z^i$, $a_n = 1$. Denote the zeros of $Q(z)$ by $z_1, z_2, \ldots, z_n$. Then*

$$\sum_{i=0}^{n} |a_i|\, |z|^i \leq \prod_{i=1}^{n} (|z| + |z_i|), \tag{1.4.11}$$

whether the a_n and/or z_i are real or complex. Equality holds in (1.4.11) when $z_1, z_2, \ldots, z_n$ all have the same phase. It holds, in particular, when $z_1, z_2, \ldots, z_n$ are all real positive or all real negative.

Proof. Let $\tilde{Q}(z) = \prod_{i=1}^{n}(z + |z_i|) = \sum_{i=0}^{n} \tilde{a}_i z^i$, $\tilde{a}_n = 1$. From $(-1)^i a_{n-i} = \sum_{k_1 < k_2 < \cdots < k_i} \prod_{s=1}^{i} z_{k_s}$, $i = 1, 2, \ldots, n$, we have

$$|a_{n-i}| \leq \sum_{k_1 < k_2 < \cdots < k_i} \prod_{s=1}^{i} |z_{k_s}| = \tilde{a}_{n-i}, \; i = 1, \ldots, n. \tag{1.4.12}$$

Thus,

$$\sum_{i=0}^{n} |a_i|\, |z|^i \leq \sum_{i=0}^{n} \tilde{a}_i |z|^i = \tilde{Q}(|z|), \tag{1.4.13}$$

from which (1.4.11) follows. When the z_i all have the same phase, equality holds in (1.4.12) and hence in (1.4.13). This completes the proof. ∎

1.4.2 An Equivalent Alternative Definition of Richardson Extrapolation

Finally, we give a different but equivalent formulation of the Richardson extrapolation process in which the $A_n^{(j)}$ are defined by linear systems of equations. In Section 1.2, we showed how the approximation $A(y, y')$ that is given by (1.2.3) and satisfies (1.2.4) can be obtained by eliminating the y^{σ_1} term from (1.1.1). The procedure used to this end is equivalent to solving the linear system

$$A(y) = A(y, y') + \bar{\alpha}_1 y^{\sigma_1}$$

$$A(y') = A(y, y') + \bar{\alpha}_1 y'^{\sigma_1}.$$

Generalizing this, we have the following interesting result for the $A_n^{(j)}$.

Theorem 1.4.5 *For each j and n, $A_n^{(j)}$ with the additional parameters $\bar{\alpha}_1, \ldots, \bar{\alpha}_n$ satisfy the linear system*

$$A(y_l) = A_n^{(j)} + \sum_{k=1}^{n} \bar{\alpha}_k y_l^{\sigma_k}, \quad j \leq l \leq j+n. \tag{1.4.14}$$

Proof. Letting $l = j + i$ and multiplying both sides of (1.4.14) by ρ_{ni} and summing from $i = 0$ to $i = n$ and invoking (1.4.6), we obtain

$$\sum_{i=0}^{n} \rho_{ni} A(y_{j+i}) = A_n^{(j)} + \sum_{k=1}^{n} \bar{\alpha}_k \sum_{i=0}^{n} \rho_{ni} y_{j+i}^{\sigma_k}. \tag{1.4.15}$$

By $y_{j+i} = y_j \omega^i$ and $c_k = \omega^{\sigma_k}$, we have

$$\sum_{i=0}^{n} \rho_{ni} y_{j+i}^{\sigma_k} = y_j^{\sigma_k} \sum_{i=0}^{n} \rho_{ni} c_k^i = y_j^{\sigma_k} U_n(c_k). \tag{1.4.16}$$

The result now follows from this and from the fact that

$$U_n(c_k) = 0, \quad k = 1, \ldots, n. \tag{1.4.17}$$

∎

Note that the new formulation of (1.4.14) is expressed only in terms of the y_m and without any reference to ω. So it can be used to define an extrapolation procedure not only for $y_m = y_0 \omega^m$, $m = 0, 1, \ldots$, but also for *any* sequence $\{y_m\} \subseteq (0, b]$. This makes the Richardson extrapolation more practical and useful in applications, including numerical integration. We come back to this point in Chapter 2.

Comparing the equations in (1.4.14) that define $A_n^{(j)}$ with the asymptotic expansion of $A(y)$ for $y \rightarrow 0+$ that is given in (1.1.2), we realize that the former are obtained from the latter by truncating the asymptotic expansion at the term $\alpha_n y^{\sigma_n}$, replacing $\sim$ by $=$, A by $A_n^{(j)}$, and α_k by $\bar{\alpha}_k$, $k = 1, \ldots, n$, and finally collocating at $y = y_l$, $l = j, j + 1, \ldots, j + n$. This forms the basis for the different generalizations of the Richardson extrapolation process in Chapters 3 and 4 of this work.

Finally, the parameters $\bar{\alpha}_1, \bar{\alpha}_2, \ldots, \bar{\alpha}_n$ in (1.4.14) turn out to be approximations to $\alpha_1, \alpha_2, \ldots, \alpha_n$ in (1.1.1) and (1.1.2). In fact, that $\bar{\alpha}_k$ tends to α_k, $k = 1, \ldots, n$, as $j \rightarrow \infty$ with n fixed can be proved rigorously. In Chapter 3, we prove a theorem on the convergence of the $\bar{\alpha}_k$ to the respective α_k within the framework of a generalized Richardson extrapolation process, and this theorem covers the present case. Despite this positive result, the use of $\bar{\alpha}_k$ as an approximation to α_k, $k = 1, \ldots, n$, is not recommended in finite-precision arithmetic. When computed in finite-precision arithmetic, the $\bar{\alpha}_k$ turn out to be of very poor quality. This appears to be the case in all generalizations of the Richardson extrapolation process as well. Therefore, if the α_k are required, an altogether different approach needs to be adopted.

1.5 Convergence Analysis of the Richardson Extrapolation Process

Because the Richardson extrapolation process, as described in the preceding section, produces a two-dimensional array of approximations to A, we may have an infinite number of sequences with elements in this array that we may analyze. In particular, each column or each diagonal in Table 1.3.1 is a bona fide sequence. Actually, columns and diagonals are the most widely studied sequences in Richardson extrapolation and its various generalizations. In addition, diagonal sequences appear to have the best convergence properties. In this section, we give a thorough analysis of convergence of columns and diagonals.

1.5.1 Convergence of Columns

By Theorem 1.3.2 and by (1.1.1), (1.2.4), and (1.2.6), we already know that $A_n^{(j)} - A = O(y_j^{\sigma_{n+1}})$ as $j \to \infty$, for $n = 0, 1, 2$. The following theorem gives an asymptotically optimal result on the behavior of the sequence $\{A_n^{(j)}\}_{j=0}^{\infty}$ for arbitrary fixed n.

Theorem 1.5.1 *Let the function $A(y)$ be as described in the second paragraph of Section 1.1.*

(i) *In case the integer s in (1.1.1) is finite and largest possible, $A_n^{(j)} - A$ has the complete expansion*

$$A_n^{(j)} - A = \sum_{k=n+1}^{s} U_n(c_k)\alpha_k y_j^{\sigma_k} + O(y_j^{\sigma_{s+1}}) \ as \ j \to \infty,$$

$$= O(\omega^{(\Re\sigma_{n+1})j}) \ as \ j \to \infty, \tag{1.5.1}$$

for $1 \leq n \leq s$, where $U_n(z)$ is as defined in Theorem 1.4.2 For $n \geq s + 1$, $A_n^{(j)} - A$ satisfies

$$A_n^{(j)} - A = O(y_j^{\sigma_{s+1}}) = O(\omega^{(\Re\sigma_{s+1})j}) \ as \ j \to \infty. \tag{1.5.2}$$

(ii) *In case (1.1.2) holds, that is, (1.1.1) holds for all $s = 1, 2, 3, \ldots$, $A_n^{(j)} - A$ has the complete asymptotic expansion*

$$A_n^{(j)} - A \sim \sum_{k=n+1}^{\infty} U_n(c_k)\alpha_k y_j^{\sigma_k} \ as \ j \to \infty,$$

$$= O(\omega^{(\Re\sigma_{n+1})j}) \ as \ j \to \infty. \tag{1.5.3}$$

All these results are valid whether $\lim_{y\to 0+} A(y)$ and $\lim_{y\to\infty} A_n^{(j)}$ exist or not.

Proof. The result in (1.5.1) can be seen to hold true by induction starting with (1.1.1), (1.2.4), and (1.2.6). However, we use a different technique to prove (1.5.1). Invoking Theorem 1.4.2 and (1.1.1), we have

$$A_n^{(j)} = \sum_{i=0}^n \rho_{ni} \left[A + \sum_{k=1}^s \alpha_k y_{j+i}^{\sigma_k} + O(y_{j+i}^{\sigma_{s+1}}) \right] \text{ as } j \to \infty,$$

$$= A + \sum_{k=1}^s \alpha_k \sum_{i=0}^n \rho_{ni} y_{j+i}^{\sigma_k} + O(y_j^{\sigma_{s+1}}) \text{ as } j \to \infty. \tag{1.5.4}$$

The result follows by invoking (1.4.16) and (1.4.17) in (1.5.4). The proof of the rest of the theorem is easy and is left to the reader.   ∎

Corollary 1.5.2 *If $\alpha_{n+\mu}$ is the first nonzero α_{n+i} with $i \geq 1$ in (1.5.1) or (1.5.3), then we have the asymptotic equality*

$$A_n^{(j)} - A \sim U_n(c_{n+\mu})\alpha_{n+\mu} y_j^{\sigma_{n+\mu}} \text{ as } j \to \infty. \tag{1.5.5}$$

The meaning of Theorem 1.5.1 and its corollary is that every column is at least as good as the one preceding it. In particular, if column n converges, then column $n + 1$ converges at least as quickly as column n. If column n diverges, then either column $n + 1$ converges or it diverges at worst as quickly as column n. In any case, $\lim_{j \to \infty} A_n^{(j)} = A$ if $\Re \sigma_{n+1} > 0$. Finally, if $\alpha_k \neq 0$ for each $k = 1, 2, 3, \ldots$, and $\lim_{y \to 0+} A(y) = A$, then each column converges more quickly than all the preceding columns.

1.5.2 Convergence of Diagonals

The convergence theory for the diagonals of Table 1.3.1 turns out to be much more involved than that for the columns. The results pertaining to diagonals show, however, that diagonals enjoy much better convergence than columns when $A(y)$ satisfies (1.1.2) or, equivalently, when $A(y)$ satisfies (1.1.1) for all s.

We start by deriving an upper bound on $|A_n^{(j)} - A|$ that is valid for arbitrary j and n.

Theorem 1.5.3 *Let the function $A(y)$ be as described in the second paragraph of Section 1.1. Let*

$$\hat{\alpha}_{s+1} = \max_{0 \leq y \leq y_0} \left| \left[A(y) - A - \sum_{k=1}^s \alpha_k y^{\sigma_k} \right] / y^{\sigma_{s+1}} \right|. \tag{1.5.6}$$

Then, for each j and each $n \geq s$, we have

$$\left| A_n^{(j)} - A \right| \leq \hat{\alpha}_{s+1} |y_j^{\sigma_{s+1}}| \left(\prod_{i=1}^n \frac{|c_{s+1}| + |c_i|}{|1 - c_i|} \right), \tag{1.5.7}$$

with $c_i = \omega^{\sigma_i}$, $i = 1, 2, \ldots$.

Proof. From (1.5.6), we see that if we define

$$R_s(y) = A(y) - A - \sum_{k=1}^{s} \alpha_k y^{\sigma_k}, \tag{1.5.8}$$

then $|R_s(y)| \leq \hat{\alpha}_{s+1} |y^{\sigma_{s+1}}|$ for all $y \in (0, y_0]$. Now, substituting (1.1.1) in (1.4.5), and using (1.4.6), and proceeding exactly as in the proof of Theorem 1.5.1, we have on account of $s \leq n$

$$A_n^{(j)} = A + \sum_{i=0}^{n} \rho_{ni} R_s(y_{j+i}). \tag{1.5.9}$$

Therefore,

$$|A_n^{(j)} - A| \leq \sum_{i=0}^{n} |\rho_{ni}| \, |R_s(y_{j+i})| \leq \hat{\alpha}_{s+1} \sum_{i=0}^{n} |\rho_{ni}| \, |y_{j+i}^{\sigma_{s+1}}|, \tag{1.5.10}$$

which, by $y_{j+i} = y_j \omega^i$ and $c_{s+1} = \omega^{\sigma_{s+1}}$, becomes

$$|A_n^{(j)} - A| \leq \hat{\alpha}_{s+1} |y_j^{\sigma_{s+1}}| \sum_{i=0}^{n} |\rho_{ni}| \, |c_{s+1}|^i. \tag{1.5.11}$$

Invoking now Theorem 1.4.3 in (1.5.11), we obtain (1.5.7). ∎

Interestingly, the upper bound in (1.5.7) can be computed numerically since the y_m and the c_k are available, provided that $\hat{\alpha}_{s+1}$ can be obtained. If a bound for $\hat{\alpha}_{s+1}$ is available, then this bound can be used in (1.5.7) instead of the exact value.

The upper bound of Theorem 1.5.3 can be turned into a powerful convergence theorem for diagonals, as we show next.

Theorem 1.5.4 *In Theorem 1.5.3, assume that*

$$\Re\sigma_{i+1} - \Re\sigma_i \geq d > 0 \quad \text{for all } i, \text{ with } d \text{ fixed.} \tag{1.5.12}$$

(i) *If the integer s in (1.1.1) is finite and largest possible, then, whether $\lim_{y \to 0+} A(y)$ exist or not,*

$$A_n^{(j)} - A = O(\omega^{(\Re\sigma_{s+1})n}) \quad \text{as } n \to \infty. \tag{1.5.13}$$

(ii) *In case (1.1.2) holds, that is, (1.1.1) holds for all $s = 0, 1, 2, \ldots$, for each fixed j, the sequence $\{A_n^{(j)}\}_{n=0}^{\infty}$ converges to A whether $\lim_{y \to 0+} A(y)$ exists or not. We have at worst*

$$A_n^{(j)} - A = O(\omega^{\mu n}) \quad \text{as } n \to \infty, \text{ for every } \mu > 0. \tag{1.5.14}$$

(iii) *Again in case (1.1.2) holds, if also $\hat{\alpha}_k y_0^{\Re\sigma_k} = O(e^{\beta k^{\eta}})$ as $k \to \infty$ for some $\eta < 2$ and $\beta \geq 0$, then the result in (1.5.14) can be improved as follows: For any $\epsilon > 0$ such that $\omega + \epsilon < 1$, there exists a positive integer n_0 that depends on ϵ, such that*

$$|A_n^{(j)} - A| \leq (\omega + \epsilon)^{dn^2/2} \quad \text{for all } n \geq n_0. \tag{1.5.15}$$

Proof. For the proof of part (i), we start by rewriting (1.5.7) in the form

$$|A_n^{(j)} - A| \leq \hat{\alpha}_{s+1} |y_j^{\sigma_{s+1}}| |c_{s+1}|^n \left(\prod_{i=1}^{n} \frac{1 + |c_i/c_{s+1}|}{|1 - c_i|} \right), \quad n \geq s. \quad (1.5.16)$$

By (1.5.12), we have $|c_{i+1}/c_i| \leq \omega^d < 1$, $i = 1, 2, \dots$. This implies that the infinite series $\sum_{i=1}^{\infty} c_i$ and hence the infinite products $\prod_{i=1}^{\infty} |1 \pm c_i|$ converge absolutely, which guarantees that $\inf_k \prod_{i=1}^{k} |1 - c_i|$ is bounded away from zero. Again, by (1.5.12), we have $|c_i/c_{s+1}| \leq \omega^{(i-s-1)d}$ for all $i \geq s + 1$ so that

$$\prod_{i=1}^{n} (1 + |c_i/c_{s+1}|) \leq \prod_{i=1}^{s+1} (1 + |c_i/c_{s+1}|) \prod_{i=1}^{n-s-1} (1 + \omega^{id})$$

$$< \prod_{i=1}^{s+1} (1 + |c_i/c_{s+1}|) \prod_{i=1}^{\infty} (1 + \omega^{id}) \equiv K_s < \infty. \quad (1.5.17)$$

Consequently, (1.5.16) gives

$$A_n^{(j)} - A = O(|c_{s+1}|^n) \quad \text{as } n \to \infty, \quad (1.5.18)$$

which is the same as (1.5.13). This proves part (i).

To prove part (ii), we observe that when s takes on arbitrary values in (1.1.1), (1.5.13) is still valid because $n \to \infty$ there. The result in (1.5.14) now follows from the fact that $\Re\sigma_{s+1} \to +\infty$ as $s \to \infty$.

For the proof of part (iii), we start by rewriting (1.5.7) with $s = n$ in the form

$$|A_n^{(j)} - A| \leq \hat{\alpha}_{n+1} |y_j^{\sigma_{n+1}}| \left(\prod_{i=1}^{n} |c_i| \right) \left(\prod_{i=1}^{n} \frac{1 + |c_{n+1}/c_i|}{|1 - c_i|} \right). \quad (1.5.19)$$

Again, by (1.5.12), we have $|c_{n+1}/c_i| \leq \omega^{(n-i+1)d}$, so that

$$\prod_{i=1}^{n} (1 + |c_{n+1}/c_i|) \leq \prod_{i=1}^{n} (1 + \omega^{id}) < \prod_{i=1}^{\infty} (1 + \omega^{id}) \equiv K' < \infty. \quad (1.5.20)$$

Thus, the product inside the second pair of parentheses in (1.5.19) is bounded in n. Also, from the fact that $\Re\sigma_i \geq \Re\sigma_1 + (i - 1)d$, there follows

$$\prod_{i=1}^{n} |c_i| = \omega^{\sum_{i=1}^{n} \Re\sigma_i} \leq \omega^{n\Re\sigma_1 + dn(n-1)/2}. \quad (1.5.21)$$

Invoking now the condition on the growth rate of the $\hat{\alpha}_k$, the result in (1.5.15) follows. ∎

Parts (i) and (iii) of this theorem are essentially due to Bulirsch and Stoer [43], while part (ii) is from Sidi [298] and [301].

The proof of Theorem 1.5.4 and the inequality in (1.5.19) suggest that $A_n^{(j)} - A$ is $O(\prod_{i=1}^{n} |c_i|)$ as $n \to \infty$ for all practical purposes. More realistic information on the convergence of $A_n^{(j)}$ as $n \to \infty$ can be obtained by analyzing the product $\prod_{i=1}^{n} |c_i|$ carefully.

Let us return to the case in which (1.1.2) is satisfied. Clearly, part (ii) of Theorem 1.5.4 says that all diagonal sequences converge to A *superlinearly* in the sense that, for fixed j, $A_n^{(j)} - A$ tends to 0 as $n \to \infty$ like $e^{-\lambda n}$ for *every* $\lambda > 0$. Part (iii) of Theorem 1.5.4 says that, under suitable growth conditions on the $\hat{\alpha}_k$, $A_n^{(j)} - A$ tends to 0 as $n \to \infty$ like $e^{-\kappa n^2}$ for *some* $\kappa > 0$. These should be compared with Theorem 1.5.1 that says that column sequences, when they converge, do so only linearly, in the sense that, for fixed n, $A_n^{(j)} - A$ tends to 0 as $j \to \infty$ precisely like $\omega^{(\Re \sigma_{n+\mu})j}$ for some integer $\mu \geq 1$. Thus, the diagonals have much better convergence than the columns.

1.6 Stability Analysis of the Richardson Extrapolation Process

From the discussion on stability in Section 0.5, it is clear that the propagation of errors in the $A(y_m)$ into $A_n^{(j)}$ is controlled by the quantity $\Gamma_n^{(j)}$, where

$$\Gamma_n^{(j)} = \sum_{i=0}^{n} |\gamma_{ni}^{(j)}| = \sum_{i=0}^{n} |\rho_{ni}| \leq \prod_{i=1}^{n} \frac{1 + |c_i|}{|1 - c_i|}, \tag{1.6.1}$$

that turns out to be independent of j in the present case. In view of Definition 0.5.1, we have the following positive result.

Theorem 1.6.1

(i) *The process that generates $\{A_n^{(j)}\}_{j=0}^{\infty}$ is stable in the sense that*

$$\sup_j \Gamma_n^{(j)} = \sum_{i=0}^{n} |\rho_{ni}| < \infty. \tag{1.6.2}$$

(ii) *Under the condition (1.5.12) and with $c_i = \omega^{\sigma_i}$, $i = 1, 2, \ldots$, we have*

$$\limsup_{n \to \infty} \sum_{i=0}^{n} |\rho_{ni}| \leq \prod_{i=1}^{\infty} \frac{1 + |c_i|}{|1 - c_i|} < \infty. \tag{1.6.3}$$

Consequently, the process that generates $\{A_n^{(j)}\}_{n=0}^{\infty}$ is also stable in the sense that

$$\sup_n \Gamma_n^{(j)} < \infty. \tag{1.6.4}$$

Proof. The validity of (1.6.2) is obvious. That (1.6.3) is valid follows from (1.4.10) in Theorem 1.4.3 and the absolute convergence of the infinite products $\prod_{i=1}^{\infty} |1 \pm c_i|$ that was demonstrated in the proof of Theorem 1.5.4. The validity of (1.6.4) is a direct consequence of (1.6.3). ∎

Remark. In case σ_i all have the same imaginary part, (1.6.3) is replaced by

$$\lim_{n \to \infty} \sum_{i=0}^{n} |\rho_{ni}| = \prod_{i=1}^{\infty} \frac{1 + |c_i|}{|1 - c_i|}. \tag{1.6.5}$$

Table 1.7.1: *Richardson extrapolation on the Zeta function series with $z = 1 + 10i$.*
Here $E_n^{(j)} = |A_n^{(j)} - A|/|A|$

j	$E_0^{(j)}$	$E_1^{(j)}$	$E_2^{(j)}$	$E_3^{(j)}$	$E_4^{(j)}$	$E_5^{(j)}$	$E_6^{(j)}$
0	$2.91D - 01$						
1	$1.38D - 01$	$6.46D - 01$					
2	$8.84D - 02$	$2.16D - 01$	$6.94D - 01$				
3	$7.60D - 02$	$7.85D - 02$	$1.22D - 01$	$2.73D - 01$			
4	$7.28D - 02$	$3.57D - 02$	$2.01D - 02$	$2.05D - 02$	$2.84D - 02$		
5	$7.20D - 02$	$1.75D - 02$	$4.33D - 03$	$1.06D - 03$	$8.03D - 04$	$8.93D - 04$	
6	$7.18D - 02$	$8.75D - 03$	$1.04D - 03$	$5.97D - 05$	$1.28D - 05$	$8.79D - 06$	$8.82D - 06$
7	$7.17D - 02$	$4.38D - 03$	$2.57D - 04$	$3.62D - 06$	$1.89D - 07$	$4.25D - 08$	$2.87D - 08$
8	$7.17D - 02$	$2.19D - 03$	$6.42D - 05$	$2.24D - 07$	$2.89D - 09$	$1.67D - 10$	$4.18D - 11$
9	$7.17D - 02$	$1.10D - 03$	$1.60D - 05$	$1.40D - 08$	$4.50D - 11$	$6.47D - 13$	$4.37D - 14$
10	$7.17D - 02$	$5.49D - 04$	$4.01D - 06$	$8.74D - 10$	$7.01D - 13$	$2.52D - 15$	$4.31D - 17$
11	$7.17D - 02$	$2.75D - 04$	$1.00D - 06$	$5.46D - 11$	$1.10D - 14$	$9.84D - 18$	$4.21D - 20$
12	$7.17D - 02$	$1.37D - 04$	$2.50D - 07$	$3.41D - 12$	$1.71D - 16$	$3.84D - 20$	$4.11D - 23$

This follows from the fact that now

$$\sum_{i=0}^{n} |\rho_{ni}| = \prod_{i=1}^{n} \frac{1 + |c_i|}{|1 - c_i|}, \tag{1.6.6}$$

as stated in Theorem 1.4.3. Also, $\sum_{i=0}^{n} |\rho_{ni}|$ is an increasing function of ω for $\sigma_i > 0$, $i = 1, 2, \ldots$. We leave the verification of this fact to the reader.

As can be seen from (1.6.1), the upper bound on $\Gamma_n^{(j)}$ is inversely proportional to the product $\prod_{i=1}^{n} |1 - c_i|$. Therefore, the processes that generate the row and column sequences will be increasingly stable from the numerical viewpoint when the c_k are as far away from unity as possible in the complex plane. The existence of even a few of the c_k that are too close to unity may cause $\Gamma_n^{(j)}$ to be very large and the extrapolation processes to be prone to roundoff even though they are stable mathematically. Note that we can force the c_k to stay away from unity by simply picking ω small enough. Let us also observe that the upper bound on $|A_n^{(j)} - A|$ given in Theorem 1.5.3 is inversely proportional to $\prod_{i=1}^{n} |1 - c_i|$ as well. It is thus very interesting that, by forcing $\Gamma_n^{(j)}$ to be small, we are able to improve not only the numerical stability of the approximations $A_n^{(j)}$, but their mathematical quality too.

1.7 A Numerical Example: Richardson Extrapolation on the Zeta Function Series

Let us consider the Riemann Zeta function series considered in Example 1.1.4. We apply the Richardson extrapolation process to $A(y)$, with y, $A(y)$, and the σ_k exactly as in Example 1.1.4, and with $y_0 = 1$, $y_l = y_0 \omega^l$, $l = 1, 2, \ldots$, and $\omega = 1/2$. Thus, $A(y_l) = A_{2^l}$, $l = 0, 1, \ldots$.

Table 1.7.1 contains the relative errors $|A_n^{(j)} - A|/|A|$, $0 \leq n \leq 6$, for $\zeta(z)$ with $z = 1 + 10i$, and allows us to verify the result of Theorem 1.5.1 concerning column sequences numerically. For example, it is possible to verify that $E_n^{(j+1)}/E_n^{(j)}$ tends to $|c_{n+1}| = \omega^{\Re \sigma_{n+1}}$

Table 1.7.2: *Richardson extrapolation on the Zeta function series with*
$z = 2$ *(convergent),* $z = 1 + 10\mathrm{i}$ *(divergent but bounded), and* $z = 0.5$
(divergent and unbounded). Here $E_n^{(j)}(z) = |A_n^{(j)} - A|/|A|$

n	$E_n^{(0)}(2)$	$E_n^{(0)}(1 + 10\mathrm{i})$	$E_n^{(0)}(0.5)$
0	$3.92D - 01$	$2.91D - 01$	$1.68D + 00$
1	$8.81D - 02$	$6.46D - 01$	$5.16D - 01$
2	$9.30D - 03$	$6.94D - 01$	$7.92D - 02$
3	$3.13D - 04$	$2.73D - 01$	$3.41D - 03$
4	$5.33D - 06$	$2.84D - 02$	$9.51D - 05$
5	$3.91D - 08$	$8.93D - 04$	$1.31D - 06$
6	$1.12D - 10$	$8.82D - 06$	$7.60D - 09$
7	$1.17D - 13$	$2.74D - 08$	$1.70D - 11$
8	$4.21D - 17$	$2.67D - 11$	$1.36D - 14$
9	$5.02D - 21$	$8.03D - 15$	$3.73D - 18$
10	$1.93D - 25$	$7.38D - 19$	$3.36D - 22$
11	$2.35D - 30$	$2.05D - 23$	$9.68D - 27$
12	$9.95D - 33$	$1.70D - 28$	$8.13D - 30$

as $j \to \infty$. For $z = 1 + 10\mathrm{i}$ we have $|c_1| = 1$, $|c_2| = 1/2$, $|c_3| = 1/2^2$, $|c_4| = 1/2^4$, $|c_5| = 1/2^6$, $|c_6| = 1/2^8$, $|c_7| = 1/2^{10}$. In particular, the sequence of the partial sums $\{A_n\}_{n=0}^{\infty}$ that forms the first column in the Romberg table diverges but is bounded.

Table 1.7.2 contains the relative errors in the diagonal sequence $\{A_n^{(0)}\}_{n=0}^{\infty}$ for $\zeta(z)$ with $z = 2$ (convergent series), $z = 1 + 10\mathrm{i}$ (divergent but bounded series), and $z = 0.5$ (divergent and unbounded series). The rate of convergence of this sequence in every case is remarkable.

Note that the results of Tables 1.7.1 and 1.7.2 have all been obtained in quadruple-precision arithmetic.

1.8 The Richardson Extrapolation as a Summability Method

In view of the fact that $A_n^{(j)}$ is of the form described in Theorem 1.4.2 and in view of the brief description of linear summability methods given in Section 0.3, we realize that the Richardson extrapolation process is a summability method for both its row and column sequences. Our purpose now is to establish the regularity of the related summability methods as these are applied to arbitrary sequences $\{B_m\}$ and not only to $\{A(y_m)\}$.

1.8.1 Regularity of Column Sequences

Let us consider the polynomials $U_n(z)$ defined in (1.4.4). From (1.4.5), the column sequence $\{A_n^{(j)}\}_{j=0}^{\infty}$ with fixed n is that generated by the linear summability method whose associated matrix $M = [\mu_{jk}]_{j,k=0}^{\infty}$ [cf. (0.3.11)], has elements given by

$\mu_{jk} = 0$ for $0 \le k \le j - 1$ and $k \ge j + n + 1$, and

$$\mu_{j,j+i} = \rho_{ni} \text{ for } 0 \le i \le n, \quad j = 0, 1, 2, \dots . \quad (1.8.1)$$

Thus, the matrix M is the following upper triangular band matrix with band width $n + 1$:

$$M = \begin{bmatrix} \rho_{n0} & \rho_{n1} & \cdot & \cdot & \cdots & \rho_{nn} & 0 & 0 & 0 & \cdots \\ 0 & \rho_{n0} & \rho_{n1} & \cdot & \cdots & \cdot & \rho_{nn} & 0 & 0 & \cdots \\ 0 & 0 & \rho_{n0} & \rho_{n1} & \cdots & \cdot & & \rho_{nn} & 0 & \cdots \\ \hline \end{bmatrix}.$$

Let us now imagine that this summability method is being applied to an arbitrary sequence $\{B_m\}$ to produce the sequence $\{B'_m\}$ with $B'_j = \sum_{k=0}^{\infty} \mu_{jk} B_k = \sum_{i=0}^{n} \rho_{ni} B_{j+i}$, $j = 0, 1, \ldots$. From (1.4.6), (1.6.2), and (1.8.1), we see that all the conditions of Theorem 0.3.3 (Silverman–Toeplitz theorem) are satisfied. Thus, we have the following result.

Theorem 1.8.1 *The summability method whose matrix M is as in (1.8.1) and generates also the column sequence $\{A_n^{(j)}\}_{j=0}^{\infty}$ is regular. Thus, for every convergent sequence $\{B_m\}$, the sequence $\{B'_m\}$ generated from it through $\{B'_m\} = \sum_{k=0}^{\infty} \mu_{mk} B_k$, $m = 0, 1, \ldots$, converges as well and $\lim_{m \to \infty} B'_m = \lim_{m \to \infty} B_m$.*

1.8.2 Regularity of Diagonal Sequences

Let us consider again the polynomials $U_n(z)$ defined in (1.4.4). From (1.4.5), the diagonal sequence $\{A_n^{(j)}\}_{n=0}^{\infty}$ with fixed j is that generated by the linear summability method whose associated matrix $M = [\mu_{nk}]_{n,k=0}^{\infty}$ has elements given by

$$\mu_{nk} = 0 \text{ for } 0 \le k \le j - 1 \text{ and } k \ge j + n + 1, \text{ and}$$

$$\mu_{n,j+i} = \rho_{ni} \text{ for } 0 \le i \le n, \quad n = 0, 1, 2, \ldots . \quad (1.8.2)$$

Thus, the matrix M is the following shifted lower triangular matrix with zeros in its first j columns:

$$M = \begin{bmatrix} 0 \cdots 0 & \rho_{00} & 0 & 0 & 0 & \cdots \\ 0 \cdots 0 & \rho_{10} & \rho_{11} & 0 & 0 & \cdots \\ 0 \cdots 0 & \rho_{20} & \rho_{21} & \rho_{22} & 0 & \cdots \\ \hline \end{bmatrix}.$$

Let us imagine that this summability method is being applied to an arbitrary sequence $\{B_m\}$ to produce the sequence $\{B'_m\}$ with $B'_n = \sum_{k=0}^{\infty} \mu_{nk} B_k = \sum_{i=0}^{n} \rho_{ni} B_{j+i}$, $n = 0, 1, \ldots$. We then have the following result.

Theorem 1.8.2 *The summability method whose matrix M is as in (1.8.2) and generates also the diagonal sequence $\{A_n^{(j)}\}_{n=0}^{\infty}$ is regular provided (1.5.12) is satisfied. Thus, for every convergent sequence $\{B_m\}$, the sequence $\{B'_m\}$ generated from it through $B'_m = \sum_{k=0}^{\infty} \mu_{mk} B_k$, $m = 0, 1, \ldots$, converges as well and $\lim_{m \to \infty} B'_m = \lim_{m \to \infty} B_m$.*

Proof. Obviously, conditions (i) and (iii) of Theorem 0.3.3 (Silverman–Toeplitz theorem) are satisfied by (1.4.6) and (1.6.4), respectively. To establish that condition (ii) is satisfied, it is sufficient to show that

$$\lim_{n \to \infty} \rho_{ni} = 0 \text{ for finite } i = 0, 1, \dots . \tag{1.8.3}$$

We do this by induction on i. From (1.4.4), we have that $\rho_{n0} = (-1)^n \left(\prod_{i=1}^n c_i \right) / \prod_{i=1}^n (1 - c_i)$. Under (1.5.12), $\prod_{i=1}^n (1 - c_i)$ has a nonzero limit for $n \to \infty$ and $\lim_{n \to \infty} \prod_{i=1}^n c_i = 0$ as shown previously. Thus, (1.8.3) holds for $i = 0$. Let us assume that (1.8.3) holds for $i - 1$. Now, from (1.6.4), $|\rho_{ni}|$ is bounded in n for each i. Also $\lim_{n \to \infty} c_n = 0$ from (1.5.12). Invoking in (1.4.8) these facts and the induction hypothesis, (1.8.3) follows. This completes the proof. ∎

1.9 The Richardson Extrapolation Process for Infinite Sequences

Let the infinite sequence $\{A_m\}$ be such that

$$A_m = A + \sum_{k=1}^s \alpha_k c_k^m + O(c_{s+1}^m) \text{ as } m \to \infty, \tag{1.9.1}$$

where $c_k \neq 0$, $k = 1, 2, \dots$, and $|c_1| > |c_2| > \cdots > |c_{s+1}|$. If (1.9.1) holds for all $s = 1, 2, \dots$, with $\lim_{k \to \infty} c_k = 0$, then we have the true asymptotic expansion

$$A_m \sim A + \sum_{k=0}^\infty \alpha_k c_k^m \text{ as } m \to \infty. \tag{1.9.2}$$

We assume that the c_k are known. We do not assume any knowledge of the α_k, however. It should be clear by now that the Richardson extrapolation process of this chapter can be applied to obtain approximations to A, the limit or antilimit of $\{A_m\}$. It is not difficult to see that *all* of the results of Sections 1.3–1.8 pertaining to the $A_n^{(j)}$ apply to $\{A_m\}$ with no changes, provided the following substitutions are made everywhere: $A(y_m) = A_m$, $\omega^{\sigma_k} = c_k$, $y_0 = 1$, and $y_m^{\sigma_k} = c_k^m$. In addition, $\hat{\alpha}_{s+1}$ should now be defined by

$$\hat{\alpha}_{s+1} = \max_m \left| \left[A_m - A - \sum_{k=1}^s \alpha_k c_k^m \right] / c_{s+1}^m \right|, \quad s = 1, 2, \dots .$$

We leave the verification of these claims to the reader.

2

Additional Topics in Richardson Extrapolation

2.1 Richardson Extrapolation with Near Geometric and Harmonic $\{y_l\}$

In Theorem 1.4.5, we showed that the Richardson extrapolation process can be defined via the linear systems of equations in (1.4.14) and that this definition allows us to use arbitrary $\{y_l\}$. Of course, with arbitrary $\{y_l\}$, the $A_n^{(j)}$ will have different convergence and stability properties than those with $y_l = y_0 \omega^l$ (geometric $\{y_l\}$) that we discussed at length in Chapter 1. In this section, we state without proof the convergence and stability properties of the column sequences $\{A_n^{(j)}\}_{j=0}^\infty$ for two essentially different types of $\{y_l\}$. In both cases, the stated results are best possible asymptotically.

The results in the following theorem concerning the column sequences in the extrapolation table with near geometric $\{y_l\}$ follow from those given in Sidi [290], which are the subject of Chapter 3.

Theorem 2.1.1 *Let $A(y)$ be exactly as in Section 1.1, and choose $\{y_l\}$ such that $\lim_{l\to\infty}(y_{l+1}/y_l) = \omega \in (0, 1)$. Set $c_k = \omega^{\sigma_k}$ for all k. If $\alpha_{n+\mu}$ is the first nonzero α_{n+i} with $i \geq 1$, then*

$$A_n^{(j)} - A \sim \left(\prod_{i=1}^n \frac{c_{n+\mu} - c_i}{1 - c_i} \right) \alpha_{n+\mu} y_j^{\sigma_{n+\mu}} \quad as \ j \to \infty,$$

and $\lim_{j\to\infty} \Gamma_n^{(j)}$ exists and

$$\lim_{j\to\infty} \Gamma_n^{(j)} = \sum_{i=0}^n |\rho_{ni}| \leq \prod_{i=1}^n \frac{1 + |c_i|}{|1 - c_i|}; \quad \sum_{i=0}^n \rho_{ni} z^i \equiv \prod_{i=1}^n \frac{z - c_i}{1 - c_i},$$

with equality when all the σ_k have the same imaginary part.

The results of the next theorem that concern harmonic $\{y_l\}$ have been given recently in Sidi [305]. Their proof is achieved by using Lemma 16.4.1 and the technique developed following it in Section 16.4.

Theorem 2.1.2 *Let $A(y)$ be exactly as in Section 1.1, and choose*

$$y_l = \frac{c}{(l + \eta)^q}, \quad l = 0, 1, \ldots, \quad for \ some \ c, \eta, q > 0.$$

If $\alpha_{n+\mu}$ is the first nonzero α_{n+i} with $i \geq 1$, then

$$A_n^{(j)} - A \sim \left[\alpha_{n+\mu} \left(\prod_{i=1}^{n} \frac{\sigma_i - \sigma_{n+\mu}}{\sigma_i} \right) c^{\sigma_{n+\mu}} \right] j^{-q\sigma_{n+\mu}} \ \text{as} \ j \to \infty,$$

and

$$\Gamma_n^{(j)} \sim \left(\prod_{i=1}^{n} |\sigma_i| \right)^{-1} \left(\frac{2j}{q} \right)^n \ \text{as} \ j \to \infty.$$

Thus, provided $\Re\sigma_{n+1} > 0$, there holds $\lim_{j\to\infty} A_n^{(j)} = A$ in both theorems, whether $\lim_{y\to 0+} A(y)$ exists or not. Also, each column is at least as good as the one preceding it. Finally, the column sequences are all stable in Theorem 2.1.1. They are unstable in Theorem 2.1.2 as $\lim_{j\to\infty} \Gamma_n^{(j)} = \infty$. For proofs and more details on these results, see Sidi [290], [305].

Note that the Richardson extrapolation process with y_l as in Theorem 2.1.2 has been used successfully in multidimensional integration of singular integrands. When used with high-precision floating-point arithmetic, this strategy turns out to be very effective despite its being unstable. For these applications, see Davis and Rabinowitz [63] and Sidi [287].

2.2 Polynomial Richardson Extrapolation

In this section, we give a separate treatment of *polynomial Richardson extrapolation* that we mentioned in passing in the preceding chapter. This method deserves independent treatment as it has numerous applications and a special theory.

The problem we want to solve is that of determining $\lim_{y\to 0} A(y) = A$, where $A(y)$, for some positive integer s, satisfies

$$A(y) \sim A + \sum_{k=1}^{s} \alpha_k y^{rk} + O(y^{r(s+1)}) \ \text{as} \ y \to 0. \tag{2.2.1}$$

Here α_k are constants independent of y, and $r > 0$ is a known constant. The α_k are not necessarily known and are not of interest. We assume that $A(y)$ is defined (i) either for $y \geq 0$ only, in which case r may be arbitrary and $y \to 0$ in (2.2.1) means $y \to 0+$, (ii) or for both $y \geq 0$ and $y \leq 0$, in which case r may be only a positive integer and $y \to 0$ from both sides in (2.2.1).

In case (2.2.1) holds for every s, $A(y)$ will have the genuine asymptotic expansion

$$A(y) \sim A + \sum_{k=1}^{\infty} \alpha_k y^{rk} \ \text{as} \ y \to 0. \tag{2.2.2}$$

We now use the alternative definition of the Richardson extrapolation process that was given in Section 1.4. For this, we choose $\{y_l\}$ such that y_l are distinct and satisfy

$$y_0 > y_1 > \cdots > 0; \ \lim_{l\to\infty} y_l = 0, \ \text{if} \ A(y) \ \text{defined for} \ y \geq 0 \ \text{only},$$

$$|y_0| \geq |y_1| \geq \cdots ; \ \lim_{l\to\infty} y_l = 0, \ \text{if} \ A(y) \ \text{defined for} \ y \geq 0 \ \text{and} \ y \leq 0. \tag{2.2.3}$$

Following that, we define the approximation $A_n^{(j)}$ to A via the linear equations

$$A(y_l) = A_n^{(j)} + \sum_{k=1}^{n} \bar{\alpha}_k y_l^{rk}, \quad j \leq l \leq j + n. \tag{2.2.4}$$

This system has a very elegant solution that goes through polynomial interpolation. For convenience, we set $t = y^r$, $a(t) = A(y)$, and $t_l = y_l^r$ everywhere. Then the equations in (2.2.4) assume the form

$$a(t_l) = A_n^{(j)} + \sum_{k=1}^{n} \bar{\alpha}_k t_l^k, \quad j \leq l \leq j + n. \tag{2.2.5}$$

It is easy to see that $A_n^{(j)} = p_{n,j}(0)$, where $p_{n,j}(t)$ is the polynomial in t of degree at most n that interpolates $a(t)$ at the points t_l, $j \leq l \leq j + n$.

Now the polynomials $p_{n,j}(t)$ can be computed recursively by the Neville–Aitken interpolation algorithm (see, for example, Stoer and Bulirsch [326]) as follows:

$$p_{n,j}(t) = \frac{(t - t_{j+n}) p_{n-1,j}(t) - (t - t_j) p_{n-1,j+1}(t)}{t_j - t_{j+n}}, \quad p_{0,j}(t) = a(t_j). \tag{2.2.6}$$

Letting $t = 0$ in this formula, we obtain the following elegant algorithm, one of the most useful algorithms in extrapolation, due to Bulirsch and Stoer [43]:

Algorithm 2.2.1

1. Set $A_0^{(j)} = a(t_j)$, $j = 0, 1, \dots$.
2. Compute $A_n^{(j)}$ by the recursion

$$A_n^{(j)} = \frac{t_j A_{n-1}^{(j+1)} - t_{j+n} A_{n-1}^{(j)}}{t_j - t_{j+n}}, \quad j = 0, 1, \dots, \quad n = 1, 2, \dots .$$

From the theory of polynomial interpolation, we have the error formula

$$a(t) - p_{n,j}(t) = a[t, t_j, t_{j+1}, \dots, t_{j+n}] \prod_{i=0}^{n} (t - t_{j+i}),$$

where $f[x_0, x_1, \dots, x_s]$ denotes the divided difference of order s of $f(x)$ over the set of points $\{x_0, x_1, \dots, x_s\}$. Letting $t = 0$ in this formula, we obtain the following error formula for $A_n^{(j)}$:

$$A_n^{(j)} - A = (-1)^n a[0, t_j, t_{j+1}, \dots, t_{j+n}] \prod_{i=0}^{n} t_{j+i}. \tag{2.2.7}$$

We also know that in case $f(x)$ is real and $f \in C^s(I)$, where I is some interval containing $\{x_0, x_1, \dots, x_s\}$, then $f[x_0, x_1, \dots, x_s] = f^{(s)}(\xi)/s!$ for some $\xi \in I$. Thus, when $a(t)$ is real and in $C^{n+1}(I)$, where I is an interval that contains all the points t_l, $l = j$, $j + 1, \dots, j + n$, and $t = 0$, (2.2.7) can be expressed as

$$A_n^{(j)} - A = (-1)^n \frac{a^{(n+1)}(\hat{t}_{j,n})}{(n+1)!} \prod_{i=0}^{n} t_{j+i}, \quad \text{for some } \hat{t}_{j,n} \in I. \tag{2.2.8}$$

By approximating $a[0, t_j, t_{j+1}, \ldots, t_{j+n}]$ by $a[t_j, t_{j+1}, \ldots, t_{j+n+1}]$, which is acceptable, we obtain the practical error estimate

$$|A_n^{(j)} - A| \approx \left| a[t_j, t_{j+1}, \ldots, t_{j+n+1}] \right| \left(\prod_{i=0}^{n} |t_{j+i}| \right).$$

Note that if $a(t)$ is real and $a^{(n+1)}(t)$ is known to be of one sign on I, and $t_l > 0$ for all l, then the right-hand side of (2.2.8) gives important information about $A_n^{(j)}$. Therefore, let us assume that, for each n, $a^{(n)}(t)$ does not change sign on I, and that $t_l > 0$ for all l. It then follows that (i) if $a^{(n)}(t)$ has the same sign for all n, then $A_n^{(j)} - A$ alternates in sign as a function of n, which means that A is between $A_n^{(j)}$ and $A_{n+1}^{(j)}$ for all n, whereas (ii) if $(-1)^n a^{(n)}(t)$ has the same sign for all n, then $A_n^{(j)} - A$ is of one sign for all n, which implies that $A_n^{(j)}$ are all on the same side of A.

Without loss of generality, in the remainder of this chapter we assume that $A(y) [a(t)]$ is real.

Obviously, the error formula in (2.2.8) can be used to make statements on the convergence rates of $A_n^{(j)}$ both as $j \to \infty$ and as $n \to \infty$. For example, it is easy to see that, for *arbitrary* t_l,

$$A_n^{(j)} - A = O(t_j t_{j+1} \cdots t_{j+n}) \text{ as } j \to \infty. \tag{2.2.9}$$

Of course, when $y_l = y_0 \omega^l$ for all l, the theory of Chapter 1 applies with $\sigma_k = rk$ for all k. Similarly, Theorems 2.1.1 and 2.1.2 apply when $\lim_{l\to\infty}(y_{l+1}/y_l) = \omega$ and $y_l = c/(l+\eta)^q$, respectively, again with $\sigma_k = rk$ for all k. In all three cases, it is not necessary to assume that $a(t)$ is differentiable. For other choices of $\{t_l\}$, the analysis of the $A_n^{(j)}$ turns out to be much more involved. This analysis is the subject of Chapter 8.

We end this section by presenting a recursive method for computing the $\Gamma_n^{(j)}$ for the case in which $t_l > 0$ for all l. As before,

$$A_n^{(j)} = \sum_{i=0}^{n} \gamma_{ni}^{(j)} a(t_{j+i}) \text{ and } \Gamma_n^{(j)} = \sum_{i=0}^{n} |\gamma_{ni}^{(j)}|,$$

and from the recursion relation among the $A_n^{(j)}$ it is clear that

$$\gamma_{ni}^{(j)} = \frac{t_j \gamma_{n-1,i-1}^{(j+1)} - t_{j+n} \gamma_{n-1,i}^{(j)}}{t_j - t_{j+n}}, \quad i = 0, 1, \ldots, n,$$

with $\gamma_{0,0}^{(j)} = 1$ and $\gamma_{ni}^{(j)} = 0$ for $i < 0$ and $i > n$. From this, we can see that $(-1)^{n+i} \gamma_{ni}^{(j)} > 0$ for all n and i when $t_0 > t_1 > \cdots > 0$. Thus,

$$|\gamma_{ni}^{(j)}| = \frac{t_j |\gamma_{n-1,i-1}^{(j+1)}| + t_{j+n} |\gamma_{n-1,i}^{(j)}|}{t_j - t_{j+n}}, \quad i = 0, 1, \ldots, n.$$

Summing both sides over i, we finally obtain, for $t_0 > t_1 > \cdots > 0$,

$$\Gamma_n^{(j)} = \frac{t_j \Gamma_{n-1}^{(j+1)} + t_{j+n} \Gamma_{n-1}^{(j)}}{t_j - t_{j+n}}, \quad j \geq 0, \ n \geq 1; \quad \Gamma_0^{(j)} = 1, \ j \geq 0. \tag{2.2.10}$$

Obviously, this recursion for the $\Gamma_n^{(j)}$ can be incorporated in Algorithm 2.2.1 in a straightforward manner.

Table 2.2.1: *Polynomial Richardson extrapolation on the Archimedes method for approximating π by inscribed regular polygons. Here $E_{n,i}^{(j)} = (\pi - A_n^{(j)})/\pi$*

j	$E_{0,i}^{(j)}$	$E_{1,i}^{(j)}$	$E_{2,i}^{(j)}$	$E_{3,i}^{(j)}$	$E_{4,i}^{(j)}$	$E_{5,i}^{(j)}$	$E_{6,i}^{(j)}$
0	$3.63D-01$						
1	$9.97D-02$	$1.18D-02$					
2	$2.55D-02$	$7.78D-04$	$4.45D-05$				
3	$6.41D-03$	$4.93D-05$	$7.20D-07$	$2.42D-08$			
4	$1.61D-03$	$3.09D-06$	$1.13D-08$	$9.67D-11$	$2.14D-12$		
5	$4.02D-04$	$1.93D-07$	$1.78D-10$	$3.80D-13$	$2.12D-15$	$3.32D-17$	
6	$1.00D-04$	$1.21D-08$	$2.78D-12$	$1.49D-15$	$2.08D-18$	$8.21D-21$	$9.56D-23$
7	$2.51D-05$	$7.56D-10$	$4.34D-14$	$5.81D-18$	$2.03D-21$	$2.01D-24$	$5.89D-27$
8	$6.27D-06$	$4.72D-11$	$6.78D-16$	$2.27D-20$	$1.99D-24$	$4.91D-28$	$3.61D-31$
9	$1.57D-06$	$2.95D-12$	$1.06D-17$	$8.86D-23$	$1.94D-27$	$1.20D-31$	$5.35D-34$
10	$3.92D-07$	$1.85D-13$	$1.65D-19$	$3.46D-25$	$1.90D-30$	$6.92D-34$	$6.62D-34$

Remark. Note that all the above applies to sequences $\{A_m\}$ for which

$$A_m \sim A + \sum_{k=1}^{\infty} \alpha_k t_m^k \quad \text{as } m \to \infty,$$

where t_m are distinct and satisfy

$$|t_0| \geq |t_1| \geq \cdots ; \quad \lim_{m \to \infty} t_m = 0.$$

We have only to make the substitution $A(y_m) = A_m$ throughout.

Example 2.2.2 We now consider Example 1.1.1 on the method of Archimedes for approximating π. We use the notation of Example 1.1.1. Thus, $y = n^{-1}$, hence $t = y^2 = n^{-2}$, and $A(y) = a(t) = S_n$.

We start with the case of the inscribed regular polygon, for which $a(t)$ is infinitely differentiable everywhere and has the convergent Maclaurin expansion

$$a(t) = \pi + \sum_{k=1}^{\infty} \alpha_k t^k; \quad \alpha_k = (-1)^k |\alpha_k| = (-1)^k \frac{(2\pi)^{2k+1}}{2[(2k+1)!]}, \quad k = 1, 2, \ldots .$$

It can be shown that, for $t \leq 1/3^2$, the Maclaurin series of $a(t)$ and of its derivatives are alternating Leibnitz series so that $(-1)^r a^{(r)}(t) > 0$ for all $t \leq 1/3^2$.

Choosing $y_l = y_0 \omega^l$, with $y_0 = 1/4$ and $\omega = 1/2$, we have $t_l = 4^{-l-2}$ and $A(y_l) = a(t_l) = S_{4 \cdot 2^l} \equiv A_l, l = 0, 1, \ldots$. By using some trigonometric identities, it can be shown that, for the case of the inscribed polygon,

$$A_0 = 2, \quad A_{n+1} = \frac{\sqrt{2} A_n}{\sqrt{1 + \sqrt{1 - (A_n/2^{n+1})^2}}}, \quad n = 0, 1, \ldots .$$

Table 2.2.1 shows the relative errors $E_{n,i}^{(j)} = (\pi - A_n^{(j)})/\pi, 0 \leq n \leq 6$, that result from applying the polynomial Richardson extrapolation to $a(t)$. Note the sign pattern in $E_{n,i}^{(j)}$ that is consistent with $(-1)^r a^{(r)}(t) > 0$ for all $r \geq 0$ and $t \leq t_0$.

Table 2.2.2: *Polynomial Richardson extrapolation on the Archimedes method for approximating π by circumscribing regular polygons. Here $E_{n,c}^{(j)} = (\pi - A_n^{(j)})/\pi$*

j	$E_{0,c}^{(j)}$	$E_{1,c}^{(j)}$	$E_{2,c}^{(j)}$	$E_{3,c}^{(j)}$	$E_{4,c}^{(j)}$	$E_{5,c}^{(j)}$	$E_{6,c}^{(j)}$
0	$-2.73D-01$						
1	$-5.48D-02$	$1.80D-02$					
2	$-1.31D-02$	$8.59D-04$	$-2.86D-04$				
3	$-3.23D-03$	$5.05D-05$	$-3.36D-06$	$1.12D-06$			
4	$-8.04D-04$	$3.11D-06$	$-4.93D-08$	$3.29D-09$	$-1.10D-09$		
5	$-2.01D-04$	$1.94D-07$	$-7.59D-10$	$1.20D-11$	$-8.03D-13$	$2.68D-13$	
6	$-5.02D-05$	$1.21D-08$	$-1.18D-11$	$4.63D-14$	$-7.35D-16$	$4.90D-17$	$-1.63D-17$
7	$-1.26D-05$	$7.56D-10$	$-1.84D-13$	$1.80D-16$	$-7.07D-19$	$1.12D-20$	$-7.48D-22$
8	$-3.14D-06$	$4.73D-11$	$-2.88D-15$	$7.03D-19$	$-6.87D-22$	$2.70D-24$	$-4.28D-26$
9	$-7.84D-07$	$2.95D-12$	$-4.50D-17$	$2.75D-21$	$-6.71D-25$	$6.56D-28$	$-2.57D-30$
10	$-1.96D-07$	$1.85D-13$	$-7.03D-19$	$1.07D-23$	$-6.55D-28$	$1.60D-31$	$-1.53D-34$

Table 2.2.3: *Polynomial Richardson extrapolation on the Archimedes method for approximating π. Here $E_{n,i}^{(0)}$ and $E_{n,c}^{(0)}$ are exactly as in Table 2.2.1 and Table 2.2.2 respectively*

n	$E_{n,i}^{(0)}$	$E_{n,c}^{(0)}$
0	$3.63D - 01$	$-2.73D - 01$
1	$1.18D - 02$	$+1.80D - 02$
2	$4.45D - 05$	$-2.86D - 04$
3	$2.42D - 08$	$+1.12D - 06$
4	$2.14D - 12$	$-1.10D - 09$
5	$3.32D - 17$	$+2.68D - 13$
6	$9.56D - 23$	$-1.63D - 17$
7	$5.31D - 29$	$+2.49D - 22$
8	$4.19D - 34$	$-9.51D - 28$
9	$5.13D - 34$	$+8.92D - 34$

We have an analogous situation for the case of the circumscribing regular polygon. In this case too, $a(t)$ is infinitely differentiable for all small t and has a convergent Maclaurin expansion for $t \leq 1/3^2$:

$$a(t) = \pi + \sum_{k=1}^{\infty} \alpha_k t^k; \quad \alpha_k = \frac{(-1)^k 4^{k+1}(4^{k+1} - 1)B_{2k+2}\pi^{2k+1}}{(2k + 2)!} > 0, \quad k = 1, 2, \ldots .$$

From this expansion it is obvious that $a(t)$ and all its derivatives are positive for $t \leq 1/3^2$.

Choosing the t_l exactly as in the previous case, we now have

$$A_0 = 4, \quad A_{n+1} = \frac{2A_n}{1 + \sqrt{1 + (A_n/2^{n+2})^2}}, \quad n = 0, 1, \ldots .$$

Table 2.2.2 shows the relative errors $E_{n,c}^{(j)} = (\pi - A_n^{(j)})/\pi$ that result from applying the polynomial Richardson extrapolation to $a(t)$. Note the sign pattern in $E_{n,c}^{(j)}$ that is consistent with $a^{(r)}(t) > 0$ for all $r \geq 0$ and $t \leq t_0$.

Finally, in Table 2.2.3 we give the relative errors in the diagonal sequences $\{A_n^{(0)}\}_{n=0}^{\infty}$ for both the inscribed and circumscribing polygons. Note the remarkable rates of convergence.

In both cases, we are able to work with sequences $\{A_m\}_{m=0}^{\infty}$ whose computation involves only simple arithmetic operations and square roots.

Note that the results of Tables 2.2.1–2.2.3 have all been obtained in quadruple-precision arithmetic.

2.3 Application to Numerical Differentiation

The most immediate application of the polynomial Richardson extrapolation is to numerical differentiation. It was suggested by Rutishauser [245] about four decades ago. This topic is treated in almost all books on numerical analysis. See, for example, Ralston and Rabinowitz [235], Henrici [130], and Stoer and Bulirsch [326].

Two approaches to numerical differentiation are discussed in the literature: (i) polynomial interpolation followed by differentiation, and (ii) application of the polynomial Richardson extrapolation to a sequence of first-order divided differences.

Although the first approach is the most obvious for differentiation of numerical data, the second is recommended, and even preferred, for differentiation of functions that can be evaluated everywhere in a given interval.

We give a slightly generalized version of the extrapolation approach and show that the approximations produced by this approach can also be obtained by differentiating some suitable polynomials of interpolation to $f(x)$. Although this was known to be true for some simple special cases, the two approaches were not known to give identical results in general. In view of this fact, we reach the interesting conclusion that the extrapolation approach has no advantage over differentiation of interpolating polynomials, except for the simple and elegant algorithms that implement it. We show that these algorithms can also be obtained by differentiating the Neville–Aitken interpolation formula.

The material here is taken from Sidi [304], where additional problems are also discussed.

Let $f(x)$ be a given function that we assume to be in $C^\infty(I)$ for simplicity. Here, I is some interval. Assume that we wish to approximate $f'(a)$, where $a \in I$.

Let us first approximate $f'(a)$ by the first-order divided difference $\delta(h) = [f(a+h) - f(a)]/h$. By the fact that the Taylor series of $f(x)$ at a, whether convergent or not, is also its asymptotic expansion as $x \to a$, we have

$$\delta(h) \sim f'(a) + \sum_{k=1}^{\infty} \frac{f^{(k+1)}(a)}{(k+1)!} h^k \text{ as } h \to 0.$$

Therefore, we can apply the polynomial Richardson extrapolation to $\delta(h)$ with an arbitrary sequence of distinct h_m satisfying

$$|h_0| \geq |h_1| \geq \cdots ; \quad a + h_m \in I, \ m = 0, 1, \dots ; \quad \lim_{m\to\infty} h_m = 0.$$

(Note that we do not require the h_m to be of the same sign.) We obtain

$$A_0^{(j)} = \delta(h_j), \quad j = 0, 1, \dots ,$$

$$A_n^{(j)} = \frac{h_j A_{n-1}^{(j+1)} - h_{j+n} A_{n-1}^{(j)}}{h_j - h_{j+n}}, \quad j = 0, 1, \dots , \ n = 1, 2, \dots . \quad (2.3.1)$$

The following is the first main result of this section.

Theorem 2.3.1 *Let $x_m = a + h_m, m = 0, 1, \dots ,$ and denote by $Q_{n,j}(x)$ the polynomial of degree $n + 1$ that interpolates $f(x)$ at a and $x_j, x_{j+1}, \dots , x_{j+n}$. Then*

$$Q'_{n,j}(a) = A_n^{(j)} \text{ for all } j, n.$$

Hence

$$A_n^{(j)} - f'(a) = (-1)^n \frac{f^{(n+2)}(\xi_{n,j})}{(n+2)!} h_j h_{j+1} \cdots h_{j+n}, \text{ for some } \xi_{n,j} \in I. \quad (2.3.2)$$

Proof. First, we have $A_0^{(j)} = \delta(h_j) = Q'_{0,j}(a)$ for all j, as can easily be shown. Next, from the Neville–Aitken interpolation algorithm in (2.2.6), we have

$$Q_{n,j}(x) = \frac{(x - x_{j+n})Q_{n-1,j}(x) - (x - x_j)Q_{n-1,j+1}(x)}{x_j - x_{j+n}},$$

which, upon differentiating at $x = a$, gives

$$Q'_{n,j}(a) = \frac{h_j Q'_{n-1,j+1}(a) - h_{j+n} Q'_{n-1,j}(a)}{h_j - h_{j+n}}.$$

Comparing this with (2.3.1), and noting that $Q'_{n,j}(a)$ and $A_n^{(j)}$ satisfy the same recursion relation with the same initial values, we obtain the first result. The second is merely the error formula that results from differentiating the interpolation polynomial $Q_{n,j}(x)$ at a.   ∎

Special cases of the result of Theorem 2.3.1 have been known for equidistant x_i. See, for example, Henrici [130].

The second main result concerns the first-order centered difference $\delta_0(h) = [f(a + h) - f(a - h)]/(2h)$, for which we have

$$\delta_0(h) \sim f'(a) + \sum_{k=1}^{\infty} \frac{f^{(2k+1)}(a)}{(2k + 1)!} h^{2k} \quad \text{as } h \to 0.$$

We can now apply the polynomial Richardson extrapolation to $\delta_0(h)$ with an arbitrary sequence of distinct positive h_m that satisfy

$$h_0 > h_1 > \cdots ; \quad a \pm h_m \in I, \ m = 0, 1, \ldots ; \quad \lim_{m \to \infty} h_m = 0.$$

We obtain

$$B_0^{(j)} = \delta_0(h_j), \quad j = 0, 1, \ldots ,$$

$$B_n^{(j)} = \frac{h_j^2 B_{n-1}^{(j+1)} - h_{j+n}^2 B_{n-1}^{(j)}}{h_j^2 - h_{j+n}^2}, \quad j = 0, 1, \ldots , \ n = 1, 2, \ldots . \quad (2.3.3)$$

Theorem 2.3.2 *Let* $x_{\pm m} = a \pm h_m$, $m = 0, 1, \ldots ,$ *and denote by* $Q_{n,j}(x)$ *the polynomial of degree* $2n + 2$ *that interpolates* $f(x)$ *at the points* a *and* $x_{\pm j}$, $x_{\pm(j+1)}, \ldots , x_{\pm(j+n)}$. *Then*

$$Q'_{n,j}(a) = B_n^{(j)} \text{ for all } j, n.$$

Hence

$$B_n^{(j)} - f'(a) = (-1)^{n+1} \frac{f^{(2n+3)}(\xi_{n,j})}{(2n + 3)!} (h_j h_{j+1} \cdots h_{j+n})^2, \text{ for some } \xi_{n,j} \in I. \quad (2.3.4)$$

Proof. First, we have $B_0^{(j)} = \delta_0(h_j) = Q'_{0,j}(a)$ for all j, as can easily be shown. Next, the $Q_{n,j}(x)$ satisfy the following extension of the Neville–Aitken interpolation algorithm (which seems to be new):

$$Q_{n,j}(x) = \frac{(x - x_{j+n})(x - x_{-(j+n)})Q_{n-1,j}(x) - (x - x_j)(x - x_{-j})Q_{n-1,j+1}(x)}{h_j^2 - h_{j+n}^2}.$$

Upon differentiating this equality at $x = a$, we obtain

$$Q'_{n,j}(a) = \frac{h_j^2 Q'_{n-1,j+1}(a) - h_{j+n}^2 Q'_{n-1,j}(a)}{h_j^2 - h_{j+n}^2}.$$

Comparing this with (2.3.3), and noting that $Q'_{n,j}(a)$ and $A_n^{(j)}$ satisfy the same recursion relation with the same initial values, we obtain the first result. The second is merely the error formula that results from differentiating the interpolation polynomial $Q_{n,j}(x)$ at a. ∎

We can extend the preceding procedure to the approximation of the second derivative $f''(a)$. Let us use $\mu(h) = [f(a+h) - 2f(a) + f(a-h)]/h^2$, which satisfies

$$\mu(h) \sim f''(a) + 2 \sum_{k=1}^{\infty} \frac{f^{(2k+2)}(a)}{(2k+2)!} h^{2k} \quad \text{as } h \to 0.$$

We can now apply the polynomial Richardson extrapolation to $\mu(h)$ with an arbitrary sequence of distinct positive h_m that satisfy

$$h_0 > h_1 > \cdots; \quad a \pm h_m \in I, \ m = 0, 1, \ldots; \quad \lim_{m \to \infty} h_m = 0.$$

We obtain

$$C_0^{(j)} = \mu(h_j), \quad j = 0, 1, \ldots,$$

$$C_n^{(j)} = \frac{h_j^2 C_{n-1}^{(j+1)} - h_{j+n}^2 C_{n-1}^{(j)}}{h_j^2 - h_{j+n}^2}, \quad j = 0, 1, \ldots, \ n = 1, 2, \ldots. \quad (2.3.5)$$

Theorem 2.3.3 *Let $x_{\pm i}$ and $Q_{n,j}(x)$ be exactly as in Theorem 2.3.2. Then*

$$Q''_{n,j}(a) = C_n^{(j)} \ \text{for all } j, n.$$

Hence

$$C_n^{(j)} - f''(a) = (-1)^{n+1} 2 \frac{f^{(2n+4)}(\xi_{n,j})}{(2n+4)!} (h_j h_{j+1} \cdots h_{j+n})^2, \quad \text{for some } \xi_{n,j} \in I. \quad (2.3.6)$$

The proof can be carried out exactly as that of Theorem 2.3.2, and we leave it to the reader.

In practice, we implement the preceding extrapolation procedures by picking $h_m = h_0 \omega^m$ for some h_0 and some $\omega \in (0, 1)$, mostly $\omega = 1/2$. In this case, Theorem 1.5.4 guarantees that all three of $A_n^{(j)} - f'(a)$, $B_n^{(j)} - f'(a)$, and $C_n^{(j)} - f''(a)$ tend to zero faster than $e^{-\lambda n}$ as $n \to \infty$, for every $\lambda > 0$. Under the liberal growth condition that $\max_{x \in I} |f^{(k)}(x)| = O(e^{\beta k^\eta})$ as $k \to \infty$, for some $\eta < 2$ and β, they tend to zero as $n \to \infty$, like $\omega^{n^2/2}$ for $\delta(h)$, and like ω^{n^2} for $\delta_0(h)$ and $\mu(h)$. This can also be seen from (2.3.2), (2.3.4), and (2.3.6). [Note that the growth condition mentioned here covers the cases in which $\max_{x \in I} |f^{(k)}(x)| = O((\alpha k)!)$ as $k \to \infty$, for arbitrary α.]

The extrapolation processes, with h_l as in the preceding paragraph, are stable, as follows from Theorem 1.6.1 in the sense that initial errors in the $\delta(h_m)$, $\delta_0(h_m)$, and $\mu(h_m)$ are not magnified in the course of the process. Nevertheless, we should be aware

Table 2.3.1: *Errors in numerical differentiation via Richardson extrapolation of*
$f(x) = 2\sqrt{1+x}$ *at* $x = 0$. *Here* $h_0 = 1/4$ *and* $\omega = 1/2$, *and centered*
differences are being used

$8.03D - 03$					
$1.97D - 03$	$5.60D - 05$				
$4.89D - 04$	$3.38D - 06$	$1.30D - 07$			
$1.22D - 04$	$2.09D - 07$	$1.95D - 09$	$8.65D - 11$		
$3.05D - 05$	$1.30D - 08$	$3.01D - 11$	$3.33D - 13$	$4.66D - 15$	
$7.63D - 06$	$8.15D - 10$	$4.62D - 13$	$8.44D - 15$	$7.11D - 15$	$7.11D - 15$

of the danger of roundoff dominating the computation eventually. We mentioned in Example 1.1.2 that, as h tends to zero, both $f(a + h)$ and $f(a - h)$ become very close to $f(a)$ and hence to each other, so that, in finite-precision arithmetic, $\delta_0(h)$ has fewer correct significant figures than $f(a \pm h)$. Therefore, carrying out the computation of $\delta_0(h)$ beyond a certain threshold value of h is useless. Obviously, this threshold depends on the accuracy with which $f(a \pm h_m)$ can be computed. Our hope is to obtain a good approximation $B_n^{(0)}$ to $f'(a)$ while h_n is still sufficiently larger than the threshold value of h. This goal seems to be achieved in practice. All this applies to $\delta(h)$ and $\mu(h)$ as well. The following theorem concerns this subject in the context of the extrapolation of the first-order centered difference $\delta_0(h)$ for computing $f'(a)$. Its proof can be achieved by induction on n and is left to the reader.

Theorem 2.3.4 *Let* $h_m = h_0\omega^m, m = 0, 1, \ldots$, *for some* $\omega \in (0, 1)$, *in (2.3.3). Suppose that* $\bar{f}(x)$, *the computed value of* $f(x)$ *in* $\delta_0(h)$, *has relative error bounded by* η, *and denote by* $\bar{B}_n^{(j)}$ *the* $B_n^{(j)}$ *obtained from (2.3.3) with the initial conditions*

$$\bar{B}_0^{(j)} = \bar{\delta}(h_j) = [\bar{f}(a + h_j) - \bar{f}(a - h_j)]/(2h_j), \quad j = 0, 1, \ldots .$$

Then

$$|\bar{B}_n^{(j)} - B_n^{(j)}| \le K_n \|f\| \eta \, h_{j+n}^{-1},$$

where

$$K_n = \prod_{i=1}^{n} \frac{1 + \omega^{2i+1}}{1 - \omega^{2i}} < K_\infty = \prod_{i=1}^{\infty} \frac{1 + \omega^{2i+1}}{1 - \omega^{2i}} < \infty, \quad \|f\| = \max_{|x-a|\le h_0} |f(x)|.$$

Let us apply the method just described to the function $f(x) = 2\sqrt{1+x}$ with $\delta_0(h)$ and $a = 0$. We have $f'(0) = 1$. We pick $h_0 = 1/4$ and $\omega = 1/2$. We use double-precision arithmetic in our computations. The errors $|B_n^{(j)} - f'(0)|$, ordered as in Table 1.3.1, are given in Table 2.3.1. As this function is analytic in $(-1, +\infty)$, the convergence results mentioned above hold.

2.4 Application to Numerical Quadrature: Romberg Integration

In Example 1.1.3, we discussed the approximation of the integral $I[f] = \int_0^1 f(x)\,dx$ by the trapezoidal rule $T(h)$ and the midpoint rule $M(h)$. We mentioned there that if

$f \in C^\infty[0, 1]$, then the errors in these approximations have asymptotic expansions in powers of h^2, known as Euler–Maclaurin expansions. Thus, the polynomial Richardson extrapolation process can be applied to the numerical quadrature formulas $T(h)$ and $M(h)$ to obtain good approximations to $I[f]$. Letting $Q(h)$ stand for either $T(h)$ or $M(h)$, and picking a decreasing sequence $\{h_m\}_{m=0}^\infty$ from $\{1, 1/2, 1/3, \dots\}$, we compute the approximations $A_n^{(j)}$ to $I[f]$ as follows:

$$A_0^{(j)} = Q(h_j), \quad j = 0, 1, \dots,$$

$$A_n^{(j)} = \frac{h_j^2 A_{n-1}^{(j+1)} - h_{j+n}^2 A_{n-1}^{(j)}}{h_j^2 - h_{j+n}^2}, \quad j = 0, 1, \dots, \ n = 1, 2, \dots. \quad (2.4.1)$$

This scheme is known as *Romberg integration*. In the next theorem, we give an error expression for $A_n^{(j)}$ that is valid for arbitrary h_m and state convergence results for column sequences as well.

Theorem 2.4.1 *Let $f(x)$, $I[f]$, $Q(h)$, and $A_n^{(j)}$ be as in the preceding paragraph. Then the following hold:*

(i) *There exists a function $w(t) \in C^\infty[0, 1]$ such that $w(m^{-2}) = Q(m^{-1})$, $m = 1, 2, \dots$, for which*

$$A_n^{(j)} - I[f] = (-1)^n \frac{w^{(n+1)}(\hat{t}_{j,n})}{(n+1)!} \left(\prod_{i=0}^{n} h_{j+i} \right)^2, \quad \text{for some } \hat{t}_{j,n} \in (t_{j+n}, t_j), \quad (2.4.2)$$

where $t_m = h_m^2$ for each m. Thus, for arbitrary h_m, each column sequence $\{A_n^{(j)}\}_{j=0}^\infty$ converges, and there holds

$$A_n^{(j)} - I[f] = O((h_j h_{j+1} \cdots h_{j+n})^2) \quad \text{as } j \to \infty. \quad (2.4.3)$$

(ii) *In case $h_{m+1}/h_m \sim 1$ as $m \to \infty$, there holds*

$$A_n^{(j)} - I[f] \sim (-1)^n \frac{(\mu+1)_{n+1}}{(n+1)!} w_{n+1+\mu} h_j^{2(n+1+\mu)} \quad \text{as } j \to \infty, \quad (2.4.4)$$

where $w_k = e_k[f^{(2k-1)}(1) - f^{(2k-1)}(0)]$, $e_k = B_{2k}/(2k)!$ for $Q(h) = T(h)$ and $e_k = B_{2k}(\frac{1}{2})/(2k)!$ for $Q(h) = M(h)$, and $w_{n+1+\mu}$, with $\mu \geq 0$, is the first nonzero w_k with $k \geq n+1$. Thus, each column converges at least as quickly as the one preceding it. This holds when $h_m = 1/(m+1)$, $m = 0, 1, \dots$, in particular.

Proof. From Theorem D.4.1 in Appendix D, $Q(h)$ can be continued to a function $w(t) \in C^\infty[0, 1]$, such that $t = 1/m^2$ when $h = 1/m$ and $w(m^{-2}) = Q(m^{-1})$, $m = 1, 2, \dots$. From this and from (2.2.8), we obtain (2.4.2), and (2.4.3) follows from (2.4.2). Now, from the proof of Theorem D.4.1, it follows that $w(t) \sim I[f] + \sum_{k=1}^\infty w_k t^k$ as $t \to 0+$. From the fact that $w(t) \in C^\infty[0, 1]$, we also have that $w^{(n+1)}(t) \sim \sum_{k=n+1+\mu}^\infty k(k-1)\cdots(k-n)w_k t^{k-n-1}$ as $t \to 0+$. This implies that $w^{(n+1)}(t) \sim (\mu+1)_{n+1} w_{n+1+\mu} t^\mu$ as $t \to 0+$. Invoking this in (2.4.2), and realizing that $\hat{t}_{j,n} \sim t_j$ as $j \to \infty$, we finally obtain (2.4.4). ∎

If we compute the $A_n^{(j)}$ as in (2.4.1) by picking $h_m = h_0 \omega^m$, where $h_0 = 1$ and $\omega = 1/q$, where q is an integer greater than 1, and if $f \in C^{2s+2}[0, 1]$, then from Theorem 1.5.1 we have

$$A_n^{(j)} - I[f] = O(\omega^{2(p+1)j}) \text{ as } j \to \infty, \quad p = \min\{n, s\},$$

and from part (i) of Theorem 1.5.4 (with $d = 2$ there)

$$A_n^{(j)} - I[f] = O(\omega^{2(s+1)n}) \text{ as } n \to \infty.$$

If $f \in C^\infty[0, 1]$, then by part (ii) of Theorem 1.5.4

$$A_n^{(j)} - I[f] = O(e^{-\lambda n}) \text{ as } n \to \infty, \text{ for every } \lambda > 0.$$

The upper bound given in (1.5.19) now reads

$$|A_n^{(j)} - I[f]| \le |e_{n+1}| \left(\max_{x \in [0,1]} |f^{(2n+2)}(x)| \right) \left(\prod_{i=1}^n \frac{1 + \omega^{2i}}{1 - \omega^{2i}} \right) \omega^{2j(n+1)} \omega^{n^2+n},$$

where $e_k = B_{2k}/(2k)!$ for $Q(h) = T(h)$ and $e_k = B_{2k}(\frac{1}{2})/(2k)!$ for $Q(h) = M(h)$, and this is a very tight bound on $|A_n^{(j)} - I[f]|$. Part (iii) of Theorem 1.5.4 applies (with $d = 2$), and we practically have $|A_n^{(j)} - I[f]| = O(\omega^{n^2})$ as $n \to \infty$, again provided $f \in C^\infty[0, 1]$ and provided $\max_{x \in [0,1]} |f^{(k)}(x)| = O(e^{\beta k^\eta})$ as $k \to \infty$ for some $\eta < 2$ and β. [Here, we have also used the fact that $e_k = O((2\pi)^{-2k})$ as $k \to \infty$.] As mentioned before, this is a very liberal growth condition for $f^{(k)}(x)$. For analytic $f(x)$, we have a growth rate of $f^{(k)}(x) = O(k! e^{\beta k})$ as $k \to \infty$, which is much milder than the preceding growth condition. Even a large growth rate such as $f^{(k)}(x) = O((\alpha k)!)$ as $k \to \infty$ for some $\alpha > 0$ is accommodated by this growth condition. The extrapolation process is stable, as follows from Theorem 1.6.1, in the sense that initial errors in the $Q(h_m)$ are not magnified in the course of the process.

Romberg [240] was the first to propose the scheme in (2.4.1), with $\omega = 1/2$. A thorough analysis of this case was given in Bauer, Rutishauser, and Stiefel [20], where the following elegant expression for the error in case $f \in C^{2n+2}[0, 1]$ and $Q(h) = T(h)$ is also provided:

$$A_n^{(j)} - I[f] = \frac{4^{-j(n+1)} B_{2n+2}}{2^{n(n+1)}(2n + 2)!} f^{(2n+2)}(\xi) \text{ for some } \xi \in (0, 1).$$

Let us apply the Romberg integration with $Q(h) = T(h)$ and $\omega = 1/2$ to the integral $\int_0^1 f(x)\,dx$ when $f(x) = 1/(x + 1)$ for which $I[f] = \log 2$. We use double-precision arithmetic in our computations. The relative errors $|A_n^{(j)} - I[f]|/|I[f]|$, ordered as in Table 1.3.1, are presented in Table 2.4.1. As this function is analytic in $(-1, +\infty)$, the convergence results mentioned in the previous paragraph hold.

As we can easily see, when computing $Q(h_{k+1})$ with $h_m = \omega^m$, we are using all the integrand values of $Q(h_k)$, and this is a useful feature of the Romberg integration. On the other hand, the number of integrand values increases exponentially like $1/\omega^k$. Thus, when $\omega = 1/2$, $A_n^{(0)}$ that is computed from $Q(h_i)$, $0 \le i \le n$, requires 2^n integrand evaluations, so that increasing n by 1 results in doubling the number of integrand evaluations. To keep this number to a reasonable size, we should work with a sequence $\{h_m\}$ that tends to 0 at

Table 2.4.1: *Errors in Romberg integration for $\int_0^1 (x+1)^{-1}\,dx$. Here $h_0 = 1$ and $\omega = 1/2$, and the trapezoidal rule is being used*

$8.20D - 02$					
$2.19D - 02$	$1.87D - 03$				
$5.59D - 03$	$1.54D - 04$	$3.96D - 05$			
$1.41D - 03$	$1.06D - 05$	$1.04D - 06$	$4.29D - 07$		
$3.52D - 04$	$6.81D - 07$	$1.98D - 08$	$3.62D - 09$	$1.96D - 09$	
$8.80D - 05$	$4.29D - 08$	$3.29D - 10$	$1.94D - 11$	$5.30D - 12$	$3.39D - 12$

a rate more moderate than 2^{-m}. But for Romberg integration there is no sequence $\{h_m\}$ with $h_m = \omega^m$ and $\omega \in (1/2, 1)$. We can avoid this problem by using other types of $\{h_m\}$. Thus, to reduce the number of integrand values required to obtain $A_n^{(0)}$ with large n, two types of sequences $\{h_m\}$ have been used extensively in the literature: (i) $h_{m+1}/h_m \le \omega$ for some $\omega \in (0, 1)$, and (ii) $h_m = 1/(m+1)$. The extrapolation process is stable in the first case and unstable in the second. We do not pursue the subject further here, but we come back to it and analyze it in some detail later in Chapter 8. For more details and references, we refer the reader to Davis and Rabinowitz [63].

2.5 Rational Extrapolation

Bulirsch and Stoer [43] generalized the polynomial Richardson extrapolation by replacing the interpolating *polynomial* $p_{n,j}(t)$ of Section 2.2 by an interpolating *rational function* $q_{n,j}(t)$. The degree of the numerator of $q_{n,j}(t)$ is $\lfloor n/2 \rfloor$, the degree of its denominator is $\lfloor (n+1)/2 \rfloor$, and $q_{n,j}(t_l) = a(t_l)$, $j \le l \le j+n$. The approximations to A, which we now denote $T_n^{(j)}$, are obtained by setting $t = 0$ in $q_{n,j}(t)$, that is, $T_n^{(j)} = q_{n,j}(0)$. The resulting method is called *rational extrapolation*. Bulirsch and Stoer give the following elegant algorithm for computing the resulting $T_n^{(j)}$:

Algorithm 2.5.2

1. Set $T_{-1}^{(j)} = 0$, $T_0^{(j)} = a(t_j)$, $j = 0, 1, \ldots$.
2. For $j = 0, 1, \ldots$, and $n = 1, 2, \ldots$, compute $T_n^{(j)}$ recursively from

$$T_n^{(j)} = T_{n-1}^{(j+1)} + \frac{T_{n-1}^{(j+1)} - T_{n-1}^{(j)}}{\dfrac{t_j}{t_{j+n}}\left[1 - \dfrac{T_{n-1}^{(j+1)} - T_{n-1}^{(j)}}{T_{n-1}^{(j+1)} - T_{n-2}^{(j+1)}}\right] - 1}.$$

For a detailed derivation of this algorithm, see also Stoer and Bulirsch [326, pp. 67–71]. For more information on this method and its application to numerical integration and numerical solution of ordinary differential equations, we refer the reader to Bulirsch and Stoer [44], [45], [46].

Another approach that is essentially due to Wynn [369] and that produces the $T_{2s}^{(j)}$ can be derived through the Thiele continued fraction for rational interpolation; see Stoer and Bulirsch [326, pp. 63–67]. Let $R_{2s,j}(x)$ be the rational function in x with degree of numerator and denominator equal to s that interpolates $f(x)$ at the points $x_j, x_{j+1}, \ldots, x_{j+2s}$. It turns out that $\rho_{2s}^{(j)} \equiv \lim_{x \to \infty} R_{2s,j}(x)$ can be computed by the following recursive algorithm:

$$\rho_{-1}^{(j)} = 0, \quad \rho_0^{(j)} = f(x_j), \quad j = 0, 1, \ldots,$$

$$\rho_{k+1}^{(j)} = \rho_{k-1}^{(j+1)} + \frac{x_{j+k+1} - x_j}{\rho_k^{(j+1)} - \rho_k^{(j)}}, \quad j, k = 0, 1, \ldots . \tag{2.5.5}$$

The $\rho_k^{(j)}$ are called the *reciprocal differences* of $f(x)$. A determinantal expression for $\rho_{2s}^{(j)}$ is given in Nörlund [222, p. 419].

By making the substitution $t = x^{-1}$ in our problem, we see that $q_{2s,j}(x^{-1})$ is a rational function of x with degree of numerator and denominator equal to s, and it interpolates $a(x^{-1})$ at the points $t_j^{-1}, t_{j+1}^{-1}, \ldots, t_{j+2s}^{-1}$. In addition, $\lim_{x \to \infty} q_{2s,j}(x^{-1}) = T_{2s}^{(j)}$. Thus, the $T_{2s}^{(j)}$ can be computed via the following algorithm.

Algorithm 2.5.3

1. Set $r_{-1}^{(j)} = 0$, $r_0^{(j)} = a(t_j)$, $j = 0, 1, \ldots$.
2. Compute $r_n^{(j)}$ recursively from

$$r_{k+1}^{(j)} = r_{k-1}^{(j+1)} + \frac{t_{j+k+1}^{-1} - t_j^{-1}}{r_k^{(j+1)} - r_k^{(j)}}, \quad j, k = 0, 1, \ldots .$$

Of course, here $r_{2s}^{(j)} = T_{2s}^{(j)}$ for all j and s.

The following convergence theorem is stated in Gragg [105].

Theorem 2.5.4 *Let $a(t) \sim A + \sum_{i=1}^{\infty} \alpha_i t^i$ as $t \to 0+$, and let $T_n^{(j)}$ be as in Algorithm 2.5.2 with positive t_l that are chosen to satisfy $t_{l+1}/t_l \leq \omega$ for some $\omega \in (0, 1)$. Define*

$$H_r^{(m)} = \begin{vmatrix} \alpha_m & \alpha_{m+1} & \cdots & \alpha_{m+r-1} \\ \alpha_{m+1} & \alpha_{m+2} & \cdots & \alpha_{m+r} \\ \vdots & \vdots & & \vdots \\ \alpha_{m+r-1} & \alpha_{m+r} & \cdots & \alpha_{m+2r-2} \end{vmatrix}.$$

Assume that $H_r^{(m)} \neq 0$ for $m = 0, 1$ and all $r \geq 0$, and define

$$e_{2s} = H_{s+1}^{(0)}/H_s^{(0)} \quad and \quad e_{2s+1} = H_{s+1}^{(1)}/H_s^{(1)}.$$

Then

$$T_n^{(j)} - A = [e_{n+1} + O(t_j)](t_j t_{j+1} \cdots t_{j+n}) \quad as \ j \to \infty.$$

Letting $t = h^2$ and $t_j = h_j^2$, $j = 0, 1, \ldots$, Algorithm 2.5.3 can be applied to the trapezoidal or midpoint rule approximation $Q(h)$ of Section 2.4. For this application, see Brezinski [31]. See also Wuytack [367], where a different implementation of rational extrapolation is presented.

3

First Generalization of the Richardson
Extrapolation Process

3.1 Introduction

In Chapter 1, we considered the Richardson extrapolation process for a sequence $\{A_m\}$ derived from a function $A(y)$ that satisfies (1.1.1) or (1.1.2), through $A_m = A(y_m)$ with $y_m = y_0 \omega^m$, $m = 0, 1, \ldots$. In this chapter, we generalize somewhat certain aspects of the treatment of Chapter 1 to the case in which the function $A(y)$ has a rather general asymptotic behavior that also may be quite different from the ones in (1.1.1) or (1.1.2). In addition, the y_m are now arbitrary. Due to the generality of the asymptotic behavior of $A(y)$ and the arbitrariness of the y_m, and under suitable conditions, the approach of this chapter may serve as a "unifying" framework within which one can treat the various extrapolation methods that have appeared over the years. In particular, the convergence and stability results from this approach may be directly applicable to specific convergence acceleration methods in some cases. Unfortunately, we pay a price for the generality of the approach of this chapter: The problems of convergence and stability presented by it turn out to be very difficult mathematically, especially because of this generality. As a result, the number of the meaningful theorems that have been obtained and that pertain to convergence and stability has remained small.

Our treatment here closely follows that of Ford and Sidi [87] and of Sidi [290].

Let $A(y)$ be a function of the discrete or continuous variable y, defined for $y \in (0, b]$ for some $b > 0$. Assume that $A(y)$ has an expansion of the form

$$A(y) = A + \sum_{k=1}^{s} \alpha_k \phi_k(y) + O(\phi_{s+1}(y)) \quad \text{as } y \to 0+, \tag{3.1.1}$$

where A and the α_k are some scalars independent of y and $\{\phi_k(y)\}$ is an asymptotic sequence as $y \to 0+$, that is, it satisfies

$$\phi_{k+1}(y) = o(\phi_k(y)) \quad \text{as } y \to 0+, \quad k = 1, 2, \ldots . \tag{3.1.2}$$

Here, $A(y)$ and $\phi_k(y)$, $k = 1, 2, \ldots$, are assumed to be known for $y \in (0, b]$, but the α_k are not required to be known. The constant A that is in many cases $\lim_{y \to 0+} A(y)$ is what we are after. When $\lim_{y \to 0+} A(y)$ does not exist, A is the antilimit of $A(y)$ as $y \to 0+$, and in this case $\lim_{y \to 0+} \phi_i(y)$ does not exist at least for $i = 1$. If (3.1.1) is valid for every $s = 1, 2, \ldots$, then $A(y)$ has the bona fide asymptotic expansion

$$A(y) \sim A + \sum_{k=1}^{\infty} \alpha_k \phi_k(y) \text{ as } y \to 0+. \tag{3.1.3}$$

Note that the functions $A(y)$ treated in Chapter 1 are particular cases of the $A(y)$ treated in this chapter with $\phi_k(y) = y^{\sigma_k}$, $k = 1, 2, \ldots$. In the present case, it is not assumed that the $\phi_k(y)$ have any particular structure.

Definition 3.1.1 Let $A(y)$ be as described above. Pick a decreasing positive sequence $\{y_m\} \subset (0, b]$ such that $\lim_{m \to \infty} y_m = 0$. Then the approximation $A_n^{(j)}$ to A, whether A is the limit or antilimit of $A(y)$ for $y \to 0+$, is defined through the linear system

$$A(y_l) = A_n^{(j)} + \sum_{k=1}^{n} \bar{\alpha}_k \phi_k(y_l), \quad j \le l \le j + n, \tag{3.1.4}$$

$\bar{\alpha}_1, \ldots, \bar{\alpha}_n$ being the additional (auxiliary) unknowns. We call this process that generates the $A_n^{(j)}$ as in (3.1.4) the *first generalization of the Richardson extrapolation process.*

Comparing the equations (3.1.4) that define $A_n^{(j)}$ with the expansion of $A(y)$ for $y \to 0+$ given in (3.1.3), we realize that the former are obtained from the latter by truncating the asymptotic expansion at the term $\alpha_n \phi_n(y)$, replacing $\sim$ by $=$, A by $A_n^{(j)}$, and α_k by $\bar{\alpha}_k$, $k = 1, \ldots, n$, and finally collocating at $y = y_l$, $l = j, j + 1, \ldots, j + n$.

Note the analogy of Definition 3.1.1 to Theorem 1.4.5. In the next section, we show that, at least formally, this generalization of the Richardson extrapolation process does perform what is required of it, namely, that it eliminates $\phi_k(y)$, $k = 1, \ldots, n$, from the expansions in (3.1.1) or (3.1.3).

Before we go on, we would like to mention that the formal setting of the first generalization of the Richardson extrapolation process as given in (3.1.1)–(3.1.4) is not new. As far as is known to us, it first appeared in Hart et al. [125, p. 39]. It was considered in detail again by Schneider [259], who also gave the first recursive algorithm for computation of the $A_n^{(j)}$. We return to this in Section 3.3.

We would also like to mention that a great many convergence acceleration methods are defined directly or can be shown to be defined indirectly through a linear system of equations of the form (3.1.4). [The $\phi_k(y)$ in these equations now do not generally form asymptotic sequences, however.] Consequently, the analysis of this form can be thought of as a "unification" of the various acceleration methods, in terms of which their properties may be classified. As mentioned above, the number of meaningful mathematical results that follow from this "unification" is small. More will be said on this in Section 3.7 of this chapter.

Before we close this section, we would like to give an example of a function $A(y)$ of the type just discussed that arises in a nontrivial fashion from numerical integration.

Example 3.1.2 Trapezoidal Rule for Integrals with an Endpoint Singularity Consider the integral $I[G] = \int_0^1 G(x)\,dx$, where $G(x) = x^s \log x\, g(x)$ with $\Re s > -1$ and $g \in C^\infty[0, 1]$. Let $h = 1/n$, where n is a positive integer. Let us approximate $I[G]$ by the (modified) trapezoidal rule $T(h)$ given by

$$T(h) = h \left[\sum_{j=1}^{n-1} G(jh) + \frac{1}{2} G(1) \right]. \tag{3.1.5}$$

Then, by a result due to Navot [217] (see also Appendix D), we have the Euler–Maclaurin expansion

$$T(h) \sim I[G] + \sum_{i=1}^{\infty} a_i h^{2i} + \sum_{i=0}^{\infty} b_i h^{s+i+1} \rho_i(h) \quad \text{as } h \to 0, \tag{3.1.6}$$

where

$$a_i = \frac{B_{2i}}{(2i)!} G^{(2i-1)}(1), \quad i = 1, 2, \ldots,$$

$$b_i = \frac{g^{(i)}(0)}{i!}, \quad \rho_i(h) = \zeta(-s-i) \log h - \zeta'(-s-i), \quad i = 0, 1, \ldots. \tag{3.1.7}$$

Here B_k are the Bernoulli numbers, $\zeta(z)$ is the Riemann Zeta function, and $\zeta'(z) = \frac{d}{dz} \zeta(z)$.

Obviously, a_i and b_i are independent of h and depend only on $g(x)$, and $\rho_i(h)$ are independent of $g(x)$. Thus, $T(h)$ is analogous to a function $A(y)$ that satisfies (3.1.3) along with (3.1.2) in the following sense: $T(h) \leftrightarrow A(y)$, $h \leftrightarrow y$, and, in case $-1 < \Re s < 0$,

$$\phi_k(y) \leftrightarrow \begin{cases} h^{s+i+1} \rho_i(h), & i = \lfloor 2k/3 \rfloor, \ k = 1, 2, 4, 5, 7, 8, \ldots, \\ h^{2k/3}, & k = 3, 6, 9, \ldots, \end{cases} \tag{3.1.8}$$

Note that $\phi_k(y)$ are all known functions.

3.2 Algebraic Properties

Being the solution to the linear system in (3.1.4), with the help of Cramer's rule, $A_n^{(j)}$ can be expressed as the quotient of two determinants in the form

$$A_n^{(j)} = \frac{\begin{vmatrix} g_1(j) & g_2(j) & \cdots & g_n(j) & a(j) \\ g_1(j+1) & g_2(j+1) & \cdots & g_n(j+1) & a(j+1) \\ \vdots & \vdots & & \vdots & \vdots \\ g_1(j+n) & g_2(j+n) & \cdots & g_n(j+n) & a(j+n) \end{vmatrix}}{\begin{vmatrix} g_1(j) & g_2(j) & \cdots & g_n(j) & 1 \\ g_1(j+1) & g_2(j+1) & \cdots & g_n(j+1) & 1 \\ \vdots & \vdots & & \vdots & \vdots \\ g_1(j+n) & g_2(j+n) & \cdots & g_n(j+n) & 1 \end{vmatrix}}, \tag{3.2.1}$$

where, for convenience, we have defined

$$g_k(m) = \phi_k(y_m), \quad m = 0, 1, \ldots, \ k = 1, 2, \ldots, \quad \text{and } a(m) = A(y_m), \quad m = 0, 1, \ldots. \tag{3.2.2}$$

This notation is used interchangeably throughout this chapter. It seems that Levin [161] was the first to point to the determinantal representation of $A_n^{(j)}$ explicitly.

The following theorem was given by Schneider [259]. Our proof is different from that of Schneider, however.

Theorem 3.2.1 $A_n^{(j)}$ *can be expressed in the form*

$$A_n^{(j)} = \sum_{i=0}^{n} \gamma_{ni}^{(j)} A(y_{j+i}), \tag{3.2.3}$$

with the scalars $\gamma_{ni}^{(j)}$ determined by the linear system

$$\sum_{i=0}^{n} \gamma_{ni}^{(j)} = 1$$

$$\sum_{i=0}^{n} \gamma_{ni}^{(j)} \phi_k(y_{j+i}) = 0, \quad k = 1, 2, \ldots, n. \tag{3.2.4}$$

Proof. Denoting the cofactor of $a(j+i)$ in the numerator determinant of (3.2.1) by N_i, and expanding both the numerator and denominator determinants with respect to their last columns, we have

$$A_n^{(j)} = \frac{\sum_{i=0}^{n} N_i a(j+i)}{\sum_{i=0}^{n} N_i}, \tag{3.2.5}$$

from which (3.2.3) follows with

$$\gamma_{ni}^{(j)} = \frac{N_i}{\sum_{r=0}^{n} N_r}, \quad i = 0, 1, \ldots, n. \tag{3.2.6}$$

Thus, the first of the equations in (3.2.4) is satisfied. As for the rest of the equations in (3.2.4), we note that $\sum_{i=0}^{n} N_i g_k(j+i) = 0$ for $k = 1, \ldots, n$, because $[g_k(j), g_k(j+1), \ldots, g_k(j+n)]^T$, $k = 1, \ldots, n$, are the first n columns of the numerator determinant in (3.2.1) and N_i are the cofactors of its last column. ∎

The $\gamma_{ni}^{(j)}$ also turn out to be associated with a polynomial that has a form very similar to (3.2.1). This is the subject of Theorem 3.2.2 that was given originally in Sidi [290].

Theorem 3.2.2 *The $\gamma_{ni}^{(j)}$ satisfy*

$$\sum_{i=0}^{n} \gamma_{ni}^{(j)} z^i = \frac{H_n^{(j)}(z)}{H_n^{(j)}(1)}, \tag{3.2.7}$$

where $H_n^{(j)}(z)$ is a polynomial of degree at most n in z defined by

$$H_n^{(j)}(z) = \begin{vmatrix} g_1(j) & g_2(j) & \cdots & g_n(j) & 1 \\ g_1(j+1) & g_2(j+1) & \cdots & g_n(j+1) & z \\ \vdots & \vdots & & \vdots & \vdots \\ g_1(j+n) & g_2(j+n) & \cdots & g_n(j+n) & z^n \end{vmatrix}. \tag{3.2.8}$$

Proof. The proof of (3.2.7) and (3.2.8) follows from (3.2.6). We leave out the details. ∎

Note the similarity of the expression for $A_n^{(j)}$ given in (3.2.1) and that for $\sum_{i=0}^{n} \gamma_{ni}^{(j)} z^i$ given in (3.2.7) and (3.2.8).

The next result shows that, at least formally, the generalized Richardson extrapolation process that generates $A_n^{(j)}$ "eliminates" the $\phi_k(y)$ terms with $k = 1, 2, \ldots, n$, from the expansion in (3.1.1) or (3.1.3). We must emphasize though that this is *not* a convergence theorem by any means. It is a heuristic justification of the possible validity of (3.1.4) as a generalized Richardson extrapolation process.

Theorem 3.2.3 *Define*

$$R_s(y) = A(y) - A - \sum_{k=1}^{s} \alpha_k \phi_k(y). \tag{3.2.9}$$

Then, for all n, we have

$$A_n^{(j)} - A = \sum_{k=n+1}^{s} \alpha_k \left(\sum_{i=0}^{n} \gamma_{ni}^{(j)} \phi_k(y_{j+i}) \right) + \sum_{i=0}^{n} \gamma_{ni}^{(j)} R_s(y_{j+i}), \tag{3.2.10}$$

where the summation $\sum_{k=n+1}^{s}$ is taken to be zero for $n \geq s$. Consequently, when $A(y) = A + \sum_{k=0}^{s} \alpha_k \phi_k(y)$ for all possible y, we have $A_n^{(j)} = A$ for all $j \geq 0$ and all $n \geq s$.

Proof. The proof of (3.2.10) can be achieved by combining (3.2.9) with (3.2.3), and then invoking (3.2.4). The rest follows from (3.2.10) and from the fact that $R_s(y) \equiv 0$ when $A(y) = A + \sum_{k=0}^{s} \alpha_k \phi_k(y)$ for all possible y. ∎

As the $\phi_k(y)$ are not required to have any particular structure, the determinants given in (3.2.1) and (3.2.7), hence $A_n^{(j)}$ and $\sum_{i=0}^{n} \gamma_{ni}^{(j)} z^i$, cannot be expressed in simple terms. This makes their analysis rather difficult. It also does not enable us to devise algorithms as efficient as Algorithm 1.3.1, for example.

3.3 Recursive Algorithms for $A_n^{(j)}$

The simplest and most direct way to compute $A_n^{(j)}$ is by solving the linear system in (3.1.4). It is also possible to devise recursive algorithms for computing all the $A_n^{(j)}$ that can be determined from a given number of the $a(m) = A(y_m)$. In fact, there are two such algorithms in the literature: The first of these was presented by Schneider [259]. Schneider's algorithm was later rederived using different techniques by Håvie [129], and, after that, by Brezinski [37]. This algorithm has been known as the *E-algorithm*. The second one was given by Ford and Sidi [87], and we call it the *FS-algorithm*. As shown later, the FS-algorithm turns out to be much less expensive computationally than the E-algorithm. It also forms an integral part of the $W^{(m)}$-algorithm of Ford and Sidi [87] that is used in implementing a further generalization of the Richardson extrapolation process denoted GREP that is due to Sidi [272]. (GREP is the subject of Chapter 4, and the $W^{(m)}$-algorithm is considered in Chapter 7.) For these reasons, we start with the FS-algorithm.

3.3.1 The FS-Algorithm

We begin with some notation and definitions. We denote an arbitrary given sequence $\{b(l)\}_{l=0}^{\infty}$ by b. Thus, a stands for $\{a(l)\}_{l=0}^{\infty}$. Similarly, for each $k = 1, 2, \ldots$, g_k stands for $\{g_k(l)\}_{l=0}^{\infty}$. Finally, we denote the sequence $1, 1, 1, \ldots$, by I. For arbitrary sequences u_k, $k = 1, 2, \ldots$, and arbitrary integers $j \geq 0$ and $p \geq 1$, we define

$$|u_1(j)\, u_2(j)\, \cdots\, u_p(j)| = \begin{vmatrix} u_1(j) & u_2(j) & \cdots & u_p(j) \\ u_1(j+1) & u_2(j+1) & \cdots & u_p(j+1) \\ \vdots & \vdots & & \vdots \\ u_1(j+p-1) & u_2(j+p-1) & \cdots & u_p(j+p-1) \end{vmatrix}.$$

$$(3.3.1)$$

Let now

$$f_p^{(j)}(b) = |g_1(j)\, g_2(j)\, \cdots\, g_p(j)\, b(j)|. \tag{3.3.2}$$

With this notation, (3.2.1) can be rewritten as

$$A_n^{(j)} = \frac{f_n^{(j)}(a)}{f_n^{(j)}(I)}. \tag{3.3.3}$$

We next define

$$G_p^{(j)} = |g_1(j)\, g_2(j)\, \cdots\, g_p(j)|, \quad p \geq 1; \ G_0^{(j)} \equiv 1, \tag{3.3.4}$$

and, for an arbitrary sequence b, we let

$$\psi_p^{(j)}(b) = \frac{f_p^{(j)}(b)}{G_{p+1}^{(j)}}. \tag{3.3.5}$$

We can now reexpress $A_n^{(j)}$ as

$$A_n^{(j)} = \frac{\psi_n^{(j)}(a)}{\psi_n^{(j)}(I)}. \tag{3.3.6}$$

The FS-algorithm computes the $A_p^{(j)}$ indirectly through the $\psi_p^{(j)}(b)$ for various sequences b.

Because $G_{p+1}^{(j)} = f_p^{(j)}(g_{p+1})$, the determinants for $f_p^{(j)}(b)$ and $G_{p+1}^{(j)}$ differ only in their last columns. It is thus natural to seek a relation between these two quantities. This can be accomplished by means of the *Sylvester determinant identity* given in Theorem 3.3.1. For a proof of this theorem, see Gragg [106].

Theorem 3.3.1 *Let C be a square matrix, and let $C_{\rho\sigma}$ denote the matrix obtained by deleting row ρ and column σ of C. Also let $C_{\rho\rho';\sigma\sigma'}$ denote the matrix obtained by deleting* *rows* ρ *and* ρ' *and columns* σ *and* σ' *of C. Provided* $\rho < \rho'$ *and* $\sigma < \sigma'$,

$$\det C \det C_{\rho\rho';\sigma\sigma'} = \det C_{\rho\sigma} \det C_{\rho'\sigma'} - \det C_{\rho\sigma'} \det C_{\rho'\sigma}. \tag{3.3.7}$$

If C is a 2×2 *matrix, then (3.3.7) holds with* $C_{\rho\rho';\sigma\sigma'} = 1$.

Applying this theorem to the $(p+1) \times (p+1)$ determinant $f_p^{(j)}(b)$ in (3.3.2) with $\rho = 1$, $\sigma = p$, $\rho' = \sigma' = p+1$, and using (3.3.4), we obtain

$$f_p^{(j)}(b)G_{p-1}^{(j+1)} = f_{p-1}^{(j+1)}(b)G_p^{(j)} - f_{p-1}^{(j)}(b)G_p^{(j+1)}. \tag{3.3.8}$$

This is the desired relation. Upon invoking (3.3.5), and letting

$$D_p^{(j)} = \frac{G_{p+1}^{(j)}G_{p-1}^{(j+1)}}{G_p^{(j)}G_p^{(j+1)}}, \tag{3.3.9}$$

(3.3.8) becomes

$$\psi_p^{(j)}(b) = \frac{\psi_{p-1}^{(j+1)}(b) - \psi_{p-1}^{(j)}(b)}{D_p^{(j)}}. \tag{3.3.10}$$

From (3.3.6) and (3.3.10), we see that, once the $D_p^{(j)}$ are known, the $\psi_p^{(j)}(a)$ and $\psi_p^{(j)}(I)$ and hence $A_p^{(j)}$ can be computed recursively. Therefore, we aim at developing an efficient algorithm for determining the $D_p^{(j)}$. In the absence of detailed knowledge about the $g_k(l)$, which is the case we assume here, we can proceed by observing that, because $G_{p+1}^{(j)} = f_p^{(j)}(g_{p+1})$, (3.3.5) with $b = g_{p+1}$ reduces to

$$\psi_p^{(j)}(g_{p+1}) = 1. \tag{3.3.11}$$

Consequently, (3.3.10) becomes

$$D_p^{(j)} = \psi_{p-1}^{(j+1)}(g_{p+1}) - \psi_{p-1}^{(j)}(g_{p+1}), \tag{3.3.12}$$

which permits recursive evaluation of the $D_p^{(j)}$ through the quantities $\psi_p^{(j)}(g_k), k \geq p+1$. [Note that $\psi_p^{(j)}(g_k) = 0$ for $k = 1, 2, \ldots, p$.] Thus, (3.3.10) and (3.3.12) provide us with a recursive procedure for solving the general extrapolation problem of this chapter. We call this procedure the FS-algorithm. Its details are summarized in Algorithm 3.3.2.

Algorithm 3.3.2 (FS-algorithm)

1. For $j = 0, 1, 2, \ldots$, set

$$\psi_0^{(j)}(a) = \frac{a(j)}{g_1(j)}, \quad \psi_0^{(j)}(I) = \frac{1}{g_1(j)}, \quad \psi_0^{(j)}(g_k) = \frac{g_k(j)}{g_1(j)}, \quad k = 2, 3, \ldots.$$

2. For $j = 0, 1, \ldots$, and $p = 1, 2, \ldots$, compute $D_p^{(j)}$ by

$$D_p^{(j)} = \psi_{p-1}^{(j+1)}(g_{p+1}) - \psi_{p-1}^{(j)}(g_{p+1}),$$

Table 3.3.1: *The $\psi_p^{(j)}(a)$ and $\psi_p^{(j)}(I)$ arrays*

$\psi_0^{(0)}(a)$				$\psi_0^{(0)}(I)$			
$\psi_0^{(1)}(a)$	$\psi_1^{(0)}(a)$			$\psi_0^{(1)}(I)$	$\psi_1^{(0)}(I)$		
$\vdots$	$\vdots$	$\ddots$		$\vdots$	$\vdots$	$\ddots$	
$\psi_0^{(L)}(a)$	$\psi_1^{(L-1)}(a)$	$\cdots$	$\psi_L^{(0)}(a)$	$\psi_0^{(L)}(I)$	$\psi_1^{(L-1)}(I)$	$\cdots$	$\psi_L^{(0)}(I)$

and compute $\psi_p^{(j)}(b)$ with $b = a, I$, and $b = g_k$, $k = p+2, p+3, \ldots$, by the recursion

$$\psi_p^{(j)}(b) = \frac{\psi_{p-1}^{(j+1)}(b) - \psi_{p-1}^{(j)}(b)}{D_p^{(j)}},$$

and set

$$A_p^{(j)} = \frac{\psi_p^{(j)}(a)}{\psi_p^{(j)}(I)}.$$

The reader might find it helpful to see the relevant $\psi_p^{(j)}(b)$ and the $D_p^{(j)}$ arranged as in Tables 3.3.1 and 3.3.2, where we have set $\psi_{p,k}^{(j)} = \psi_p^{(j)}(g_k)$ for short. The flow of computation in these tables is exactly as that in Table 1.3.1.

Note that when $a(l) = A(y_l)$, $l = 0, 1, \ldots, L$, are given, then this algorithm enables us to compute all the $A_p^{(j)}$, $0 \le j + p \le L$, that are defined by these $a(l)$.

Remark. Judging from (3.3.5), one may be led to think incorrectly that the FS-algorithm requires knowledge of g_{L+1} for computing all the $A_p^{(j)}$, $0 \le j + p \le L$, in addition to $g_1, g_2, \ldots, g_L$ that are actually needed by (3.1.4). Really, the FS-algorithm employs g_{L+1} only for computing $D_L^{(0)}$, which is then used for determining $\psi_L^{(0)}(a)$ and $\psi_L^{(0)}(I)$, and $A_L^{(0)} = \psi_L^{(0)}(a)/\psi_L^{(0)}(I)$. As $A_L^{(0)}$ does not depend on g_{L+1}, in case it is not available, we can take g_{L+1} to be *any* sequence independent of $g_1, g_2, \ldots, g_L$, and I. Thus, when only $A(y_l)$, $l = 0, 1, \ldots, L$, are available, *no* specific knowledge of g_{L+1} is required in using the FS-algorithm, contrary to what is claimed in Brezinski and Redivo Zaglia [41, p. 62]. As suggested by Osada [226], one can avoid this altogether by computing the $A_n^{(0)}$, $n = 1, 2, \ldots$, from

$$A_n^{(0)} = \frac{\psi_{n-1}^{(1)}(a) - \psi_{n-1}^{(0)}(a)}{\psi_{n-1}^{(1)}(I) - \psi_{n-1}^{(0)}(I)}, \tag{3.3.13}$$

without having to compute $D_L^{(0)}$. This, of course, follows from (3.3.6) and (3.3.10).

3.3.2 The E-Algorithm

As we have seen, the FS-algorithm is a set of recursion relations that produces the quantities $\psi_p^{(j)}(b)$ and then gives $A_p^{(j)} = \psi_p^{(j)}(a)/\psi_p^{(j)}(I)$. The E-algorithm, on the other

Table 3.3.2: *The $\psi_p^{(j)}(g_k)$ and $D_p^{(j)}$ arrays*

$\psi_{0,2}^{(0)}$		$\psi_{0,3}^{(0)}$				$\psi_{0,L+1}^{(0)}$		
$\psi_{0,2}^{(1)}$ $D_1^{(0)}$		$\psi_{0,3}^{(1)}$ $\psi_{1,3}^{(0)}$				$\psi_{0,L+1}^{(1)}$ $\psi_{1,L+1}^{(0)}$		
$\psi_{0,2}^{(2)}$ $D_1^{(1)}$		$\psi_{0,3}^{(2)}$ $\psi_{1,3}^{(1)}$ $D_2^{(0)}$	$\cdots$			$\vdots$	$\vdots$	$\ddots$
$\vdots$ $\vdots$		$\vdots$ $\vdots$ $\vdots$				$\psi_{0,L+1}^{(L-1)}$	$\psi_{1,L+1}^{(L-2)}$ $\cdots$	$\psi_{L-1,L+1}^{(0)}$
$\psi_{0,2}^{(L)}$ $D_1^{(L-1)}$		$\psi_{0,3}^{(L)}$ $\psi_{1,3}^{(L-1)}$ $D_2^{(L-2)}$				$\psi_{0,L+1}^{(L)}$	$\psi_{1,L+1}^{(L-1)}$ $\cdots$	$\psi_{L-1,L+1}^{(1)}$ $D_L^{(0)}$

hand, is a different set of recursion relations that produces the quantities

$$\chi_p^{(j)}(b) = \frac{f_p^{(j)}(b)}{f_p^{(j)}(I)}, \tag{3.3.14}$$

from which we have $A_p^{(j)} = \chi_p^{(j)}(a)$. Starting from

$$\chi_p^{(j)}(b) = [f_p^{(j)}(b)G_{p-1}^{(j+1)}]/[f_p^{(j)}(I)G_{p-1}^{(j+1)}],$$

applying (3.3.8) to both the numerator and the denominator of this quotient, and realizing that $G_p^{(j)} = f_{p-1}^{(j)}(g_p)$, we obtain the recursion relation

$$\chi_p^{(j)}(b) = \frac{\chi_{p-1}^{(j+1)}(b)\chi_{p-1}^{(j)}(g_p) - \chi_{p-1}^{(j)}(b)\chi_{p-1}^{(j+1)}(g_p)}{\chi_{p-1}^{(j)}(g_p) - \chi_{p-1}^{(j+1)}(g_p)}, \tag{3.3.15}$$

where $b = a$ and $b = g_k$, $k = p + 1, p + 2, \ldots$. Here the initial conditions are $\chi_0^{(j)}(b) = b(j)$.

The E-algorithm, as given in (3.3.15), is not optimal costwise. The following form that can be found in Ford and Sidi [87] and that is almost identical to that suggested by Håvie [129] earlier, is less costly than (3.3.15):

$$\chi_p^{(j)}(b) = \frac{\chi_{p-1}^{(j+1)}(b) - w_p^{(j)}\chi_{p-1}^{(j)}(b)}{1 - w_p^{(j)}}, \tag{3.3.16}$$

where $w_p^{(j)} = \chi_{p-1}^{(j+1)}(g_p)/\chi_{p-1}^{(j)}(g_p)$.

Let us now compare the operation counts of the FS- and E-algorithms when $a(l)$, $l = 0, 1, \ldots, L$, are available. We note that most of the computational effort is spent in obtaining the $\psi_p^{(j)}(g_k)$ in the FS-algorithm and the $\chi_p^{(j)}(g_k)$ in the E-algorithm. The number of these quantities is $L^3/3 + O(L^2)$ for large L. Also, the division by $D_p^{(j)}$ in the FS-algorithm is carried out as multiplication by $1/D_p^{(j)}$, the latter quantity being computed only once. An analogous statement can be made about divisions in the E-algorithm by (3.3.15) and (3.3.16). Thus, we have the operation counts given in Table 3.3.3.

From Table 3.3.3 we see that the E-algorithm is about 50% more expensive than the FS-algorithm, even when it is implemented through (3.3.16).

Note that the E-algorithm was originally obtained by Schneider [259] by a technique different than that used here. The technique of this chapter was introduced by Brezinski [37]. The technique of Håvie [129] differs from both.

Table 3.3.3: *Operation counts of the FS- and E-algorithms*

Algorithm	No. of Multiplications	No. of Additions	No. of Divisions
FS	$L^3/3 + O(L^2)$	$L^3/3 + O(L^2)$	$O(L^2)$
E by (3.3.16)	$2L^3/3 + O(L^2)$	$L^3/3 + O(L^2)$	$O(L^2)$
E by (3.3.15)	$L^3 + O(L^2)$	$L^3/3 + O(L^2)$	$O(L^2)$

3.4 Numerical Assessment of Stability

Let us recall that, by Theorem 3.2.1, $A_n^{(j)}$ can be expressed as $A_n^{(j)} = \sum_{i=0}^n \gamma_{ni}^{(j)} A(y_{j+i})$, where the scalars $\gamma_{ni}^{(j)}$ satisfy $\sum_{i=0}^n \gamma_{ni}^{(j)} = 1$. As a result, the discussion of stability that we gave in Section 0.5 applies, and we conclude that the quantities $\Gamma_n^{(j)} = \sum_{i=0}^n |\gamma_{ni}^{(j)}|$ and $\Lambda_n^{(j)} = \sum_{i=0}^n |\gamma_{ni}^{(j)}| \, |A(y_{j+i})|$ control the propagation into $A_n^{(j)}$ of errors (roundoff and other) in the $A(y_i)$.

If we want to know $\Gamma_n^{(j)}$ and $\Lambda_n^{(j)}$, in general, we need to compute the relevant $\gamma_{ni}^{(j)}$. This can be done by solving the linear system in (3.2.4) numerically. It can also be accomplished by a simple extension of the FS-algorithm. Either way, the cost of computing the $\gamma_{ni}^{(j)}$ is high and it becomes even higher when we increase n. Thus, computation of $\Gamma_n^{(j)}$ and $\Lambda_n^{(j)}$ entails a large expense, in general.

If we are not interested in the exact $\Gamma_n^{(j)}$ but are satisfied with upper bounds for them, we can accomplish this simultaneously with the computation of the $A_n^{(j)}$ – for example, by the FS-algorithm, – and at almost no additional cost.

The following theorem gives a complete description of the computation of both the $\gamma_{ni}^{(j)}$ and the upper bounds on the $\Gamma_n^{(j)}$ via the FS-algorithm. We use the notation of the preceding section.

Theorem 3.4.1 *Set* $w_n^{(j)} = \psi_{n-1}^{(j)}(I)/\psi_{n-1}^{(j+1)}(I)$, *and define*

$$\lambda_n^{(j)} = \frac{1}{1 - w_n^{(j)}} \quad and \quad \mu_n^{(j)} = -\frac{w_n^{(j)}}{1 - w_n^{(j)}}, \tag{3.4.1}$$

for all $j = 0, 1, \ldots,$ *and* $n = 1, 2, \ldots.$

(i) *The* $\gamma_{ni}^{(j)}$ *can be computed recursively from*

$$\gamma_{ni}^{(j)} = \lambda_n^{(j)} \gamma_{n-1,i-1}^{(j+1)} + \mu_n^{(j)} \gamma_{n-1,i}^{(j)}, \quad i = 0, 1, \ldots, n, \tag{3.4.2}$$

where $\gamma_{00}^{(s)} = 1$ *for all* s, *and we have set* $\gamma_{ki}^{(s)} = 0$ *if* $i < 0$ *or* $i > k$.

(ii) *Consider now the recursion relation*

$$\tilde{\Gamma}_n^{(j)} = |\lambda_n^{(j)}| \tilde{\Gamma}_{n-1}^{(j+1)} + |\mu_n^{(j)}| \tilde{\Gamma}_{n-1}^{(j)}, \quad j = 0, 1, \ldots, \ n = 1, 2, \ldots, \tag{3.4.3}$$

with the initial conditions $\tilde{\Gamma}_0^{(j)} = \Gamma_0^{(j)} = 1, \ j = 0, 1, \ldots.$ *Then*

$$\Gamma_n^{(j)} \le \tilde{\Gamma}_n^{(j)}, \quad j = 0, 1, \ldots, \ n = 1, 2, \ldots, \tag{3.4.4}$$

with equality for $n = 1$.

Proof. We start by reexpressing $A_n^{(j)}$ in the form

$$A_n^{(j)} = \lambda_n^{(j)} A_{n-1}^{(j+1)} + \mu_n^{(j)} A_{n-1}^{(j)}, \tag{3.4.5}$$

which is obtained by substituting (3.3.10) with $b = a$ and $b = I$ in (3.3.6) and invoking (3.3.6) again. Next, combining (3.2.3) and (3.4.5), we obtain (3.4.2). Taking the moduli of both sides of (3.4.2) and summing over i, we obtain

$$\Gamma_n^{(j)} \leq |\lambda_n^{(j)}| \Gamma_{n-1}^{(j+1)} + |\mu_n^{(j)}| \Gamma_{n-1}^{(j)}. \tag{3.4.6}$$

The result in (3.4.4) now follows by subtracting (3.4.3) from (3.4.6) and using induction. ∎

From (3.4.1)–(3.4.3) and the fact that $w_n^{(j)}$ is given in terms of $\psi_{n-1}^{(j)}(I)$ and $\psi_{n-1}^{(j+1)}(I)$ that are already known, it is clear that the cost of computing the $\gamma_{ni}^{(j)}$, $0 \leq j + n \leq L$ is $O(L^3)$ arithmetic operations, while the cost of computing the $\tilde{\Gamma}_n^{(j)}$, $0 \leq j + n \leq L$, is $O(L^2)$ arithmetic operations.

Based on our numerical experience, we note that the $\tilde{\Gamma}_n^{(j)}$ tend to increase very quickly in some cases, thus creating the wrong impression that the corresponding extrapolation process is very unstable even when that is not the case. Therefore, it may be better to use (3.4.2) to compute the $\gamma_{ni}^{(j)}$ and then $\Gamma_n^{(j)}$ exactly.

A recursion relation for computing the $\gamma_{ni}^{(j)}$ within the context of the E-algorithm that is analogous to what we have in Theorem 3.4.1 was given by Håvie [129].

3.5 Analysis of Column Sequences

Because the functions $\phi_k(y)$ in the expansions (3.1.1) and (3.1.3) have no particular structure, the analysis of $A_n^{(j)}$ as defined by (3.1.4) is not a very well-defined task. Thus, we cannot expect to obtain convergence results that make sense without making reasonable assumptions about the $\phi_k(y)$ and the y_m. This probably is a major reason for the scarcity of meaningful results in this part of the theory of extrapolation methods. Furthermore, all the known results concern the column sequences $\{A_n^{(j)}\}_{j=0}^\infty$ for fixed n. Analysis of the diagonal sequences $\{A_n^{(j)}\}_{n=0}^\infty$ for fixed j seems to be much more difficult than that of the column sequences and most probably requires more complicated and detailed assumptions to be made on the $\phi_k(y)$ and the y_m.

All the results of the present section are obtained under the conditions that

$$\lim_{m \to \infty} \frac{g_k(m+1)}{g_k(m)} = \lim_{m \to \infty} \frac{\phi_k(y_{m+1})}{\phi_k(y_m)} = c_k \neq 1, \quad k = 1, 2, \dots, \tag{3.5.1}$$

and

$$c_k \neq c_q \text{ if } k \neq q, \tag{3.5.2}$$

in addition to that in (3.1.2). Also, the c_k can be complex numbers.

In view of these conditions, all the results of this section apply, in particular, to the case in which $\phi_k(y) = y^{\sigma_k}$, $k = 1, 2, \dots$, where $\sigma_k \neq 0$ for all k and $\Re\sigma_1 < \Re\sigma_2 < \cdots$, when $\lim_{m \to \infty} y_{m+1}/y_m = \omega$ with some fixed $\omega \in (0, 1)$. In this case, $c_k = \omega^{\sigma_k}$,

$k = 1, 2, \ldots$. With these $\phi_k(y)$, the function $A(y)$ is precisely that treated in Chapter 1. But this time the y_m do not necessarily satisfy $y_{m+1} = \omega y_m$, $m = 0, 1, 2, \ldots$, as assumed in Chapter 1; instead $y_{m+1} \sim \omega y_m$ as $m \to \infty$. Hence, this chapter provides additional results for the functions $A(y)$ of Chapter 1. These have been given as Theorem 2.1.1 in Chapter 2.

The source of all the results given in the next three subsections is Sidi [290]; those in the last two subsections are new. Some of these results are used in the analysis of other methods later in the book.

Before we go on, we would like to show that the conditions in (3.5.1 and (3.5.2) are satisfied in at least one realistic case, namely, that mentioned in Example 3.1.2.

Example 3.5.1 Let us consider the function $A(y)$ of Example 3.1.2. If we choose $\{y_m\}$ such that $\lim_{m\to\infty} y_{m+1}/y_m = \omega \in (0, 1)$, we realize that (3.5.1) and (3.5.2) are satisfied. Specifically, we have

$$c_k = \lim_{m \to \infty} \frac{\phi_k(y_{m+1})}{\phi_k(y_m)} = \begin{cases} \omega^{s+i+1}, & i = \lfloor 2k/3 \rfloor, \ k = 1, 2, 4, 5, 7, 8, \ldots, \\ \omega^{2k/3}, & k = 3, 6, 9, \ldots. \end{cases}$$

3.5.1 Convergence of the $\gamma_{ni}^{(j)}$

We start with a result on the polynomial $H_n^{(j)}(z)$ that we will use later.

Theorem 3.5.2 *For n fixed, the polynomial $H_n^{(j)}(z)$ satisfies*

$$\lim_{j \to \infty} \frac{H_n^{(j)}(z)}{\prod_{i=1}^n g_i(j)} = V(c_1, c_2, \ldots, c_n, z), \tag{3.5.3}$$

where $V(\xi_1, \xi_2, \ldots, \xi_k)$ is the Vandermonde determinant defined by

$$V(\xi_1, \xi_2, \ldots, \xi_k) = \begin{vmatrix} 1 & 1 & \cdots & 1 \\ \xi_1 & \xi_2 & \cdots & \xi_k \\ \vdots & \vdots & & \vdots \\ \xi_1^{k-1} & \xi_2^{k-1} & \cdots & \xi_k^{k-1} \end{vmatrix} = \prod_{1 \le i < j \le k} (\xi_j - \xi_i). \tag{3.5.4}$$

Proof. Dividing the ith column of $H_n^{(j)}(z)$ by $g_i(j)$, $i = 1, \ldots, n$, and letting

$$\tilde{g}_i^{(j)}(r) = \frac{g_i(j+r)}{g_i(j)}, \quad r = 0, 1, 2, \ldots, \tag{3.5.5}$$

we obtain

$$\frac{H_n^{(j)}(z)}{\prod_{i=1}^n g_i(j)} = \begin{vmatrix} 1 & \cdots & 1 & 1 \\ \tilde{g}_1^{(j)}(1) & \cdots & \tilde{g}_n^{(j)}(1) & z \\ \vdots & & \vdots & \vdots \\ \tilde{g}_1^{(j)}(n) & \cdots & \tilde{g}_n^{(j)}(n) & z^n \end{vmatrix}. \tag{3.5.6}$$

But

$$\tilde{g}_i^{(j)}(r) = \frac{g_i(j+r)}{g_i(j+r-1)} \frac{g_i(j+r-1)}{g_i(j+r-2)} \cdots \frac{g_i(j+1)}{g_i(j)}, \tag{3.5.7}$$

so that

$$\lim_{j\to\infty} \tilde{g}_i^{(j)}(r) = c_i^r, \quad r = 0, 1, \ldots, \tag{3.5.8}$$

from (3.5.1). The result in (3.5.3) now follows by taking the limit of both sides of (3.5.6). ∎

With the help of Theorems 3.2.2 and 3.5.2 and the conditions on the c_k given in (3.5.1) and (3.5.2), we obtain the following interesting result on the $\gamma_{ni}^{(j)}$.

Theorem 3.5.3 *The polynomial* $\sum_{i=0}^{n} \gamma_{ni}^{(j)} z^i$ *is such that*

$$\lim_{j\to\infty} \sum_{i=0}^{n} \gamma_{ni}^{(j)} z^i = U_n(z) = \prod_{i=1}^{n} \frac{z - c_i}{1 - c_i} \equiv \sum_{i=0}^{n} \rho_{ni} z^i. \tag{3.5.9}$$

Consequently, $\lim_{j\to\infty} \gamma_{ni}^{(j)} = \rho_{ni}$, $i = 0, 1, \ldots, n$, *as well.*

This result will be of use in the following analysis of $A_n^{(j)}$. Before we go on, we would like to draw attention to the similarity between (3.5.9) and (1.4.4).

3.5.2 Convergence and Stability of the $A_n^{(j)}$

Let us observe that from (3.3.14) and what we already know about the $\gamma_{ni}^{(j)}$, we have

$$\chi_n^{(j)}(b) = \sum_{i=0}^{n} \gamma_{ni}^{(j)} b(j+i) \tag{3.5.10}$$

for every sequence b. With this, we can now write (3.2.10) in the form

$$A_n^{(j)} - A = \sum_{k=n+1}^{s} \alpha_k \chi_n^{(j)}(g_k) + \chi_n^{(j)}(r_s), \tag{3.5.11}$$

where we have defined

$$r_s(i) = R_s(y_i), \quad i = 0, 1, \ldots. \tag{3.5.12}$$

The following theorem concerns the asymptotic behavior of the $\chi_n^{(j)}(g_k)$ and $\chi_n^{(j)}(r_s)$ as $j \to \infty$.

Theorem 3.5.4

(i) *For fixed* $n > 0$, *the sequence* $\{\chi_n^{(j)}(g_k)\}_{k=n+1}^{\infty}$ *is an asymptotic sequence as* $j \to \infty$; *that is,* $\lim_{j\to\infty} \chi_n^{(j)}(g_{k+1})/\chi_n^{(j)}(g_k) = 0$, *for all* $k \geq n+1$. *Actually, for each* $k \geq n+1$, *we have the asymptotic equality*

$$\chi_n^{(j)}(g_k) \sim U_n(c_k) g_k(j) \text{ as } j \to \infty, \tag{3.5.13}$$

with $U_n(z)$ *as defined in the previous theorem.*

(ii) *Similarly,* $\chi_n^{(j)}(r_s)$ *satisfies*

$$\chi_n^{(j)}(r_s) = O(\chi_n^{(j)}(g_{s+1})) \text{ as } j \to \infty. \tag{3.5.14}$$

Proof. From (3.5.10) and (3.5.5),

$$\chi_n^{(j)}(g_k) = \sum_{i=0}^{n} \gamma_{ni}^{(j)} g_k(j+i) = \left(\sum_{i=0}^{n} \gamma_{ni}^{(j)} \tilde{g}_k^{(j)}(i) \right) g_k(j). \tag{3.5.15}$$

Invoking (3.5.8) and Theorem 3.5.3 in (3.5.15), and using the fact that $U_n(c_k) \neq 0$ for $k \geq n+1$, the result in (3.5.13) follows. To prove (3.5.14), we start with

$$\left| \chi_n^{(j)}(r_s) \right| = \left| \sum_{i=0}^{n} \gamma_{ni}^{(j)} r_s(j+i) \right| \leq \sum_{i=0}^{n} |\gamma_{ni}^{(j)}| \, |R_s(y_{j+i})| . \tag{3.5.16}$$

From the fact that the $\gamma_{ni}^{(j)}$ are bounded in j by Theorem 3.5.3 and from $R_s(y) = O(\phi_{s+1}(y))$ as $y \to 0+$ and $\phi_{s+1}(y_{j+i}) \sim c_{s+1}^i \phi_{s+1}(y_j)$ as $j \to \infty$ that follows from (3.5.8), we first obtain $\chi_n^{(j)}(r_s) = O(g_{s+1}(j))$ as $j \to \infty$. The result now follows from the fact that $g_{s+1}(j) = O(\chi_n^{(j)}(r_s))$ as $j \to \infty$, which is a consequence of the asymptotic equality in (3.5.13) with $k = s+1$. ∎

The next theorem concerns the behavior of $A_n^{(j)}$ for $j \to \infty$ and is best possible asymptotically.

Theorem 3.5.5

(i) *Suppose that $A(y)$ satisfies (3.1.3). Then $A_n^{(j)} - A$ has the bona fide asymptotic expansion*

$$A_n^{(j)} - A \sim \sum_{k=n+1}^{\infty} \alpha_k \chi_n^{(j)}(g_k) \text{ as } j \to \infty,$$

$$= O(g_{n+1}(j)) \text{ as } j \to \infty. \tag{3.5.17}$$

Consequently, if $\alpha_{n+\mu}$ is the first nonzero α_{n+i} with $i \geq 1$ in (3.1.3), then we have the asymptotic equality

$$A_n^{(j)} - A \sim \alpha_{n+\mu} \chi_n^{(j)}(g_{n+\mu}) \sim \left(\prod_{i=1}^{n} \frac{c_{n+\mu} - c_i}{1 - c_i} \right) \alpha_{n+\mu} \phi_{n+\mu}(y_j) \text{ as } j \to \infty.$$

$$\tag{3.5.18}$$

(ii) *In case $A(y)$ satisfies (3.1.1) with s a finite largest possible integer, and $\alpha_{n+\mu}$ is the first nonzero α_{n+i} with $i \geq 1$ there, then (3.5.18) holds. If $\alpha_{n+1} = \alpha_{n+2} = \cdots = \alpha_s = 0$ or $n \geq s$, we have*

$$A_n^{(j)} - A = O(\phi_{s+1}(y_j)) \text{ as } j \to \infty. \tag{3.5.19}$$

These results are valid whether $\lim_{y \to 0+} A(y)$ exists or not.

Proof. For (3.5.17) to have a meaning as an asymptotic expansion and to be valid, first $\{\chi_n^{(j)}(g_k)\}_{k=n+1}^{\infty}$ must be an asymptotic sequence as $j \to \infty$, and then $A_n^{(j)} - A - \sum_{k=n+1}^{s} \alpha_k \chi_n^{(j)}(g_k) = O(\chi_n^{(j)}(g_{s+1}))$ as $j \to \infty$ must hold for each $s \geq n+1$. Now, both of these conditions hold by Theorem 3.5.4. Therefore, (3.5.17) is valid as an asymptotic expansion. The asymptotic equality in (3.5.18) is a direct consequence of (3.5.17). This completes the proof of part (i) of the theorem. The proof of part (ii) can be achieved similarly and is left to the reader. ∎

Remarks.

1. Comparing (3.5.17) with (3.5.11), one may be led to believe erroneously that (3.5.17) follows from (3.5.11) in a trivial way by letting $s \to \infty$ in the latter. This is far from being the case as is clear from the proof of Theorem 3.5.5, which, in turn, depends entirely on Theorem 3.5.4.
2. By imposing the additional conditions that $|\alpha_k| < \lambda^k$ for all k and for some $\lambda > 0$, and that $A(y)$ has the convergent expansion $A(y) = A + \sum_{k=1}^{\infty} \alpha_k \phi_k(y)$, and that this expansion converges absolutely and uniformly in y, Wimp [366, pp. 188–189] has proved the result of Theorem 3.5.5. Clearly, these conditions are very restrictive. Brezinski [37] stated an acceleration result under even more restrictive conditions that also impose limitations on the $A_n^{(j)}$ themselves, when we are actually investigating the properties of the latter. Needless to say, the result of [37] follows in a trivial way from Theorem 3.5.5 that has been obtained under the smallest number of conditions on $A(y)$ only.

We now consider the stability of the column sequences $\{A_n^{(j)}\}_{j=0}^{\infty}$. From the discussion on stability in Section 0.5, the relevant quantity to be analyzed in this connection is

$$\Gamma_n^{(j)} = \sum_{i=0}^{n} |\gamma_{ni}^{(j)}|. \tag{3.5.20}$$

The following theorem on stability is a direct consequence of Theorem 3.5.3.

Theorem 3.5.6 *The $\gamma_{ni}^{(j)}$ satisfy*

$$\lim_{j \to \infty} \sum_{i=0}^{n} |\gamma_{ni}^{(j)}| = \sum_{i=0}^{n} |\rho_{ni}| \leq \prod_{i=1}^{n} \frac{1 + |c_i|}{|1 - c_i|}, \tag{3.5.21}$$

with ρ_{ni} as defined by (3.5.9). Consequently, the extrapolation process that generates the sequence $\{A_n^{(j)}\}_{j=0}^{\infty}$ is stable in the sense that

$$\sup_{j} \Gamma_n^{(j)} < \infty. \tag{3.5.22}$$

In case the c_k all have the same phase, the inequality in (3.5.21) becomes an equality. In case the c_k are real and negative, (3.5.21) becomes $\lim_{j \to \infty} \Gamma_n^{(j)} = 1$.

Proof. The result in (3.5.21) follows from Theorems 3.5.3 and 1.4.3, and that in (3.5.22) follows directly from (3.5.21). ∎

Remark. As is seen from (3.5.21), the size of the bound on $\Gamma_n^{(j)} = \sum_{i=0}^{n} |\gamma_{ni}^{(j)}|$ is proportional to $D = \prod_{i=1}^{n} |1 - c_i|^{-1}$. This implies that $\Gamma_n^{(j)}$ will be small provided the c_i are away from 1. We also see from (3.5.18) that the coefficient of $\phi_{n+\mu}(y_j)$ there is proportional to D. From this, we conclude that the more stable the extrapolation process, the better the quality of the (theoretical) error $A_n^{(j)} - A$, which is somewhat surprising.

3.5.3 *Convergence of the* $\bar{\alpha}_k$

As mentioned previously, the $\bar{\alpha}_k \equiv \alpha_{nk}^{(j)}$ in the equations in (3.1.4) turn out to be approximations to the corresponding α_k, $k = 1, \ldots, n$. In fact, $\alpha_{nk}^{(j)} \to \alpha_k$ for $j \to \infty$, as we show in the next theorem that is best possible asymptotically.

Theorem 3.5.7 *Assume that $A(y)$ is as in (3.1.3). Then, for $k = 1, \ldots, n$, with n fixed, we have $\lim_{j\to\infty} \alpha_{nk}^{(j)} = \alpha_k$. In fact,*

$$\alpha_{nk}^{(j)} - \alpha_k \sim \alpha_{n+\mu} \left(\frac{c_{n+\mu} - 1}{c_k - 1} \prod_{\substack{i=1 \\ i \neq k}}^{n} \frac{c_{n+\mu} - c_i}{c_k - c_i} \right) \frac{\phi_{n+\mu}(y_j)}{\phi_k(y_j)} \quad as \ j \to \infty, \quad (3.5.23)$$

where $\alpha_{n+\mu}$ is the first nonzero α_{n+i} with $i \geq 1$.

Proof. By Cramer's rule the solution of the linear system in (3.1.4) for $\bar{\alpha}_k$ is given by

$$\alpha_{nk}^{(j)} = \frac{1}{f_n^{(j)}(I)} |g_1(j) \ \cdots \ g_{k-1}(j) \, a(j) \, g_{k+1}(j) \ \cdots \ g_n(j) \, I(j)| . \quad (3.5.24)$$

Now, under the given conditions, the expansion

$$a(j) = A + \sum_{k=1}^{n} \alpha_k g_k(j) + \alpha_{n+\mu} \hat{g}_{n+\mu}(j), \quad (3.5.25)$$

where $\hat{g}_{n+\mu}(j) = g_{n+\mu}(j)[1 + \epsilon_{n+\mu}(j)]$ and $\epsilon_{n+\mu}(j) = o(1)$ as $j \to \infty$, is valid. Substituting (3.5.25) in (3.5.24), and expanding the numerator determinant there with respect to its kth column, we obtain

$$\alpha_{nk}^{(j)} - \alpha_k = \frac{V_{nk}^{(j)}}{f_n^{(j)}(I)},$$

$$V_{nk}^{(j)} = \alpha_{n+\mu} |g_1(j) \ \cdots \ g_{k-1}(j) \, \hat{g}_{n+\mu}(j) \, g_{k+1}(j) \ \cdots \ g_n(j) \, I(j)|. \quad (3.5.26)$$

The rest can be completed by dividing the ith column of $V_{nk}^{(j)}$ by $g_i(j)$, $i = 1, \ldots, n$, $i \neq k$, and the kth column by $g_{n+\mu}(j)$, and taking the limit for $j \to \infty$. Following the steps of the proof of Theorem 3.5.2, we have

$$\lim_{j\to\infty} \frac{V_{nk}^{(j)}}{g_{n+\mu}(j) \prod_{\substack{i=1 \\ i \neq k}}^{n} g_i(j)} = \alpha_{n+\mu} V(c_1, \ldots, c_{k-1}, c_{n+\mu}, c_{k+1}, \ldots, c_n, 1). \quad (3.5.27)$$

Combining this with Theorem 3.5.2 on $f_n^{(j)}(I) = H_n^{(j)}(1)$ in (3.5.26), we obtain (3.5.23). ∎

Even though we have convergence of $\alpha_{nk}^{(j)}$ to α_k for $j \to \infty$, in finite-precision arithmetic the computed $\alpha_{nk}^{(j)}$ is a poor approximation to α_k in most cases. The precise reason for this is given in the next theorem that is best possible asymptotically.

Theorem 3.5.8 *If we write*

$$\alpha_{nk}^{(j)} = \sum_{i=0}^{n} \delta_{nki}^{(j)} a(j+i), \tag{3.5.28}$$

which is true from (3.1.4), then the $\delta_{nki}^{(j)}$ satisfy

$$\lim_{j\to\infty}\left[\phi_k(y_j)\sum_{i=0}^{n}\delta_{nki}^{(j)}z^i\right] = \frac{z-1}{c_k-1}\prod_{\substack{i=1\\i\neq k}}^{n}\frac{z-c_i}{c_k-c_i} \equiv \sum_{i=0}^{n}\tilde{\delta}_{nki}z^i, \tag{3.5.29}$$

from which we have

$$\sum_{i=0}^{n}|\delta_{nki}^{(j)}| \sim \left(\sum_{i=0}^{n}|\tilde{\delta}_{nki}|\right)\frac{1}{|\phi_k(y_j)|} \quad \text{as } j \to \infty. \tag{3.5.30}$$

Thus, if $\lim_{y\to 0+}\phi_k(y) = 0$, then the computation of $\alpha_{nk}^{(j)}$ from (3.1.4) is unstable for large j.

Proof. Let us denote the matrix of coefficients of the system in (3.1.4) by B. Then $\delta_{nki}^{(j)}$ are the elements of the row of B^{-1} that is appropriate for $\alpha_{nk}^{(j)}$. In fact, $\delta_{nki}^{(j)}$ can be identified by expanding the numerator determinant in (3.5.24) with respect to its kth column. We have

$$\sum_{i=0}^{n}\delta_{nki}^{(j)}z^i = \frac{z^{-j}}{f_n^{(j)}(I)}|g_1(j) \cdots g_{k-1}(j)\, Z(j)\, g_{k+1}(j) \cdots g_n(j)\, I(j)|, \quad (3.5.31)$$

where $Z(l) = z^l$, $l = 0, 1, \ldots$. Proceeding as in the proof of Theorem 3.5.2, we obtain

$$\sum_{i=0}^{n}\delta_{nki}^{(j)}z^i \sim \frac{V(c_1,\ldots,c_{k-1},z,c_{k+1},\ldots,c_n,1)}{V(c_1,\ldots,c_n,1)}\frac{1}{g_k(j)} \quad \text{as } j \to \infty. \quad (3.5.32)$$

The results in (3.5.29) and (3.5.30) now follow. We leave the details to the reader. ∎

By the condition in (3.1.2) and Theorems 3.5.7 and 3.5.8, it is clear that both the theoretical quality of $\alpha_{nk}^{(j)}$ and the quality of the computed $\alpha_{nk}^{(j)}$ decrease for larger k.

3.5.4 Conditioning of the System (3.1.4)

The stability results of Theorems 3.5.6 and 3.5.8 are somewhat related to the conditioning of the linear system (3.1.4) that defines the first generalization of the Richardson extrapolation process. In this section, we study the extent of this relation under the conditions of the preceding section.

As in the proof of Theorem 3.5.8, let us denote the matrix of coefficients of the linear system in (3.1.4) by B. Then the l_∞-norm of B is given by

$$\| B \|_\infty = \max_{0 \le r \le n} \left\{ 1 + \sum_{i=1}^{n} |g_i(j+r)| \right\}. \tag{3.5.33}$$

The row of B^{-1} that gives us $A_n^{(j)}$ is nothing but $(\gamma_{n0}^{(j)}, \ldots, \gamma_{nn}^{(j)})$, and the row that gives $\alpha_{nk}^{(j)}$ is similarly $(\delta_{nk0}^{(j)}, \ldots, \delta_{nkn}^{(j)})$, so that

$$\| B^{-1} \|_\infty = \max \left\{ \sum_{i=0}^{n} |\gamma_{ni}^{(j)}|, \sum_{i=0}^{n} |\delta_{nki}^{(j)}|, \ k = 1, \ldots, n \right\}. \tag{3.5.34}$$

From Theorems 3.5.6 and 3.5.8 we, therefore, have

$$\| B^{-1} \|_\infty \sim \max\{M_0, M_n/|g_n(j)|\} \ \text{as} \ j \to \infty, \tag{3.5.35}$$

with

$$M_0 = \sum_{i=0}^{n} |\rho_{ni}| \ \text{and} \ M_n = \sum_{i=0}^{n} |\tilde{\delta}_{nni}|, \tag{3.5.36}$$

and these are best possible asymptotically.

From (3.5.33) and (3.5.35), we can now determine how the condition number of B behaves for $j \to \infty$. In particular, we have the following result.

Theorem 3.5.9 *Denote by $\kappa_\infty(B)$ the l_∞ condition number of the matrix B of the linear system in (3.1.4); that is, $\kappa_\infty(B) = \|B\|_\infty \|B^{-1}\|_\infty$. In general, $\kappa_\infty(B) \to \infty$ as $j \to \infty$ at least as quickly as $g_1(j)/g_n(j)$. In particular, when $\lim_{y \to 0+} \phi_1(y) = 0$, we have*

$$\kappa_\infty(B) \sim \frac{M_n}{|\phi_n(y_j)|} \ \text{as} \ j \to \infty. \tag{3.5.37}$$

Remarks. It must be noted that $\kappa_\infty(B)$, the condition number of the matrix B, does not affect the numerical solution for the unknowns $A_n^{(j)}$ and $\alpha_{nk}^{(j)} = \bar{\alpha}_k$, $k = 1, \ldots, n$, uniformly. In fact, it has *no* effect on the computed $A_n^{(j)}$. It is $\Gamma_n^{(j)} = \sum_{i=0}^{n} |\gamma_{ni}^{(j)}|$ that controls the influence of errors in the $a(m) = A(y_m)$ (including roundoff errors) on $A_n^{(j)}$, and that $\Gamma_n^{(j)} \sim M_0 < \infty$ as $j \to \infty$, independently of $\kappa_\infty(B)$. Similarly, the stability properties of $\alpha_{nk}^{(j)}$, $1 \le k \le n-1$, are determined by $\sum_{i=0}^{n} |\delta_{nki}^{(j)}|$ and not by $\kappa_\infty(B)$. The stability properties only of $\alpha_{nn}^{(j)}$ appear to be controlled by $\kappa_\infty(B)$.

3.5.5 Conditioning of (3.1.4) for the Richardson Extrapolation Process on Diagonal Sequences

So far, we have analyzed the conditioning of the linear system in (3.1.4) for column sequences under the conditions in (3.5.1) and (3.5.2). Along the way to the main result in Theorem 3.5.9, we have also developed some of the main ingredients for treatment of the conditioning related to *diagonal* sequences for the Richardson extrapolation process of Chapter 1, to which we now turn. In this case, we have $\phi_k(y) = y^{\sigma_k}$, $k = 1, 2, \ldots,$

and $y_{m+1} = \omega y_m$, $m = 0, 1, \ldots$, so that $g_k(n+1)/g_k(n) = \omega^{\sigma_k} = c_k$, $k = 1, 2, \ldots$.
Going over the proof of Theorem 3.5.8, we realize that (3.5.29) can now be replaced by

$$y_j^{\sigma_k} \sum_{i=0}^{n} \delta_{nki}^{(j)} z^i = \frac{z-1}{c_k - 1} \prod_{\substack{i=1 \\ i \neq k}}^{n} \frac{z - c_i}{c_k - c_i} = \sum_{i=0}^{n} \tilde{\delta}_{nki} z^i \qquad (3.5.38)$$

that is valid for every j and n. Obviously, (3.5.33) and (3.5.34) are always valid. From (3.5.33), (3.5.34), and (3.5.38), we can write

$$\| B \|_\infty > |y_j^{\sigma_n}| \quad \text{and} \quad \| B^{-1} \|_\infty \geq \sum_{i=0}^{n} |\delta_{nni}^{(j)}| = |y_j^{-\sigma_n}| \sum_{i=0}^{n} |\tilde{\delta}_{nni}|, \qquad (3.5.39)$$

so that

$$\kappa_\infty(B) = \| B \|_\infty \| B^{-1} \|_\infty > \sum_{i=0}^{n} |\tilde{\delta}_{nni}|. \qquad (3.5.40)$$

The following lemma that complements Lemma 1.4.4 will be useful in the sequel.

Lemma 3.5.10 *Let* $Q(z) = \sum_{i=0}^{n} a_i z^i$, $a_n = 1$. *Denote the zeros of* $Q(z)$ *that are not on the unit circle* $K = \{z : |z| = 1\}$ *by* $z_1, z_2, \ldots, z_p$, *so that* $Q(z) = R(z) \prod_{i=1}^{p} (z - z_i)$ *for some polynomial* $R(z)$ *of degree* $n - p$ *with all its zeros on* K. *Then*

$$\sum_{i=0}^{n} |a_i| \geq \left(\max_{|z|=1} |R(z)| \right) \prod_{i=1}^{p} |1 - |z_i|| > 0. \qquad (3.5.41)$$

Proof. We start with

$$\sum_{i=0}^{n} |a_i| \geq \left| \sum_{i=0}^{n} a_i z^i \right| = |R(z)| \prod_{i=1}^{p} |z - z_i| \quad \text{for every } z \in K.$$

Next, we have

$$\sum_{i=0}^{n} |a_i| \geq |R(z)| \prod_{i=1}^{p} ||z| - |z_i|| = |R(z)| \prod_{i=1}^{p} |1 - |z_i|| \quad \text{for every } z \in K.$$

The result follows by maximizing both sides of this inequality on K. ∎

Applying Lemma 3.5.10 to the polynomial

$$\sum_{i=0}^{n} \tilde{\delta}_{nni} z^i = \frac{z-1}{c_n - 1} \prod_{i=1}^{n-1} \frac{z - c_i}{c_n - c_i},$$

and using the fact that $\lim_{m \to \infty} c_m = 0$, we have

$$\sum_{i=0}^{n} |\tilde{\delta}_{nni}| \geq L_1 \frac{\prod_{i=r}^{n-1} |1 - |c_i||}{\prod_{i=1}^{n-1} |c_n - c_i|}$$

for all n, where L_1 is some positive constant and r is some positive integer for which $|c_r| < 1$, so that $|c_i| < 1$ for $i \geq r$ as well.

Now, from the fact that $\prod_{i=1}^{n-1} |c_n - c_i| \leq |\prod_{i=1}^{n-1} c_i| \prod_{i=1}^{n-1} (1 + |c_n/c_i|)$ and from (1.5.20), which holds under the condition in (1.5.12), we have $\prod_{i=1}^{n-1} |c_n - c_i| \leq L_2 |\prod_{i=1}^{n-1} c_i|$ for all n, where L_2 is some positive constant. Similarly, $\prod_{i=r}^{n-1} |1 - |c_i|| > L_3$ for all n, where L_3 is some positive constant, since the infinite product $\prod_{i=1}^{\infty} |1 - c_i|$ converges, as shown in the proof of Theorem 1.5.4. Combining all these, we have

$$\sum_{i=0}^{n} |\tilde{\delta}_{nni}| \geq L \left(\prod_{i=1}^{n-1} |c_i| \right)^{-1} \quad \text{for all } n, \text{ some constant } L > 0. \qquad (3.5.42)$$

We have thus obtained the following result.

Theorem 3.5.11 *When $\phi_k(y) = y^{\sigma_k}$, $k = 1, 2, \ldots$, and $y_{m+1} = \omega y_m$, $m = 0, 1, \ldots$, and the σ_k satisfy the condition in (1.5.12) of Theorem 1.5.4, the condition number of the matrix B of the linear system (3.1.4) satisfies*

$$\kappa_\infty(B) > L \left(\prod_{i=1}^{n-1} |c_i| \right)^{-1} \geq L\omega^{-dn^2/2+gn} \quad \text{for some constants } L > 0 \text{ and } g. \quad (3.5.43)$$

Thus, $\kappa_\infty(B) \to \infty$ as $n \to \infty$ practically at the rate of $\omega^{-dn^2/2}$. ∎

Note again that, if Gaussian elimination is used for solving (3.1.4) in finite-precision arithmetic, the condition number $\kappa_\infty(B)$ has no effect whatsoever on the computed value of $A_n^{(j)}$ for any value of n, even though $\kappa_\infty(B) \to \infty$ for $n \to \infty$ very strongly.

3.6 Further Results for Column Sequences

The convergence and stability analysis for the sequences $\{A_n^{(j)}\}_{j=0}^{\infty}$ of the preceding section was carried out under the assumptions in (3.1.2) on the $\phi_k(y)$ and (3.5.1) and (3.5.2) on the $g_k(m)$. It is seen from the proofs of the theorems of Section 3.5 that (3.1.2) can be relaxed somewhat without changing any of the results there. Thus, we can assume that

$$\phi_{k+1}(y) = O(\phi_k(y)) \quad \text{as } y \to 0+ \quad \text{for all } k, \quad \text{and}$$

$$\phi_{k+1}(y) = o(\phi_k(y)) \quad \text{as } y \to 0+ \quad \text{for infinitely many } k.$$

Of course, in this case we take the asymptotic expansion in (3.1.3) to mean that (3.1.1) holds for each integer s. This is a special case of a more general situation treated by Sidi [297]. The results of [297] are briefly mentioned in Subsection 14.1.4.

Matos and Prévost [207] have considered a set of conditions different from those in (3.5.1) and (3.5.2). We bring their main result (Theorem 3 in [207]) here without proof.

Theorem 3.6.1 *Let the functions $\phi_k(y)$ satisfy (3.1.2) and let $\phi_1(y) = o(1)$ as $y \to 0+$ as well. Let the $g_k(m) = \phi_k(y_m)$ satisfy for all $i \geq 0$ and $p \geq 0$*

$$|g_{i+p}(j) \cdots g_{i+1}(j) g_i(j)| \geq 0, \quad \text{for } j \geq J, \text{ some } J, \qquad (3.6.1)$$

where $g_0(j) = 1$, $j = 0, 1, \ldots$ *. Then, for fixed* n, $\{\chi_n^{(j)}(g_k)\}_{k=n+1}^{\infty}$ *is an asymptotic sequence as* $j \to \infty$; *that is,* $\lim_{j\to\infty} \chi_n^{(j)}(g_{k+1})/\chi_n^{(j)}(g_k) = 0$ *for all* $k \geq n+1$.

Two remarks on the conditions of this theorem are now in order. We first realize from (3.6.1) that $g_k(m)$ are all real. Next, the fact that $\phi_1(y) = o(1)$ as $y \to 0+$ implies that $\lim_{y\to 0+} A(y) = A$ is assumed.

These authors also state that, provided $\alpha_{n+1} \neq 0$, the following convergence acceleration result holds in addition to Theorem 3.6.1:

$$\lim_{j\to\infty} \frac{A_{n+1}^{(j)} - A}{A_n^{(j)} - A} = 0. \tag{3.6.2}$$

For this to be true, we must also have for each s that

$$A_n^{(j)} - A = \sum_{k=n+1}^{s} \alpha_k \chi_n^{(j)}(g_k) + O(\chi_n^{(j)}(g_{s+1})) \text{ as } j \to \infty. \tag{3.6.3}$$

However, careful reading of the proof in [207] reveals that this has not been shown. We have not been able to show the truth of this under the conditions given in Theorem 3.6.1 either. It is easy to see that (3.6.3) holds in case $A(y) = A + \sum_{k=1}^{N} \alpha_k \phi_k(y)$ for some finite N, but this, of course, restricts the scope of (3.6.3) considerably. The most common cases that arise in practice involve expansions such as (3.1.1) or truly divergent asymptotic expansions such as (3.1.3). By comparison, all the results of the preceding section have been obtained for $A(y)$ as in (3.1.1) and (3.1.3). Finally, unlike that in (3.5.18), the result in (3.6.2) gives no information about the rates of convergence and acceleration of the extrapolation process.

Here are some examples of the $g_k(m)$ that have been given in [207] and that are covered by Theorem 3.6.1.

- $g_1(m) = g(m)$ and $g_k(m) = (-1)^k \Delta^k g(m)$, $k = 2, 3, \ldots$, where $\{g(m)\}$ is a logarithmic totally monotonic sequence. [That is, $\lim_{m\to\infty} g(m+1)/g(m) = 1$ and $(-1)^k \Delta^k g(m) \geq 0$, $k = 0, 1, 2, \ldots$.]
- $g_k(m) = x_m^{\alpha_k}$ with $1 > x_1 > x_2 > \cdots > 0$ and $0 < \alpha_1 < \alpha_2 < \cdots$.
- $g_k(m) = c_k^m$ with $1 > c_1 > c_2 > \cdots > 0$.
- $g_k(m) = 1 \big/ \big[(m+1)^{\alpha_k}(\log(m+2))^{\beta_k}\big]$ with $0 < \alpha_1 \leq \alpha_2 \leq \cdots$, and $\beta_k < \beta_{k+1}$ if $\alpha_k = \alpha_{k+1}$.

Of these, the third example is covered in full detail in Chapter 1 (see Section 1.9), where results for both column and diagonal sequences are presented. Furthermore, the treatment in Chapter 1 does not assume the c_k to be real, in general.

3.7 Further Remarks on (3.1.4): "Related" Convergence Acceleration Methods

Let us replace $A(y_l)$ by A_l and $\phi_k(y_l)$ by $g_k(l)$ in the linear system (3.1.4). This system then becomes

$$A_l = A_n^{(j)} + \sum_{k=0}^{n} \bar{\alpha}_k g_k(l), \quad j \leq l \leq j+n. \tag{3.7.1}$$

Now let us take $\{g_k(m)\}_{m=0}^{\infty}$, $k = 1, 2, \ldots$, to be arbitrary sequences, not necessarily related to any functions $\phi_k(y)$, $k = 1, 2, \ldots$, that satisfy (3.1.2).

We mentioned previously that, with appropriate $g_k(m)$, many known convergence acceleration methods for a given infinite sequence $\{A_m\}$ are either directly defined or can be shown to be defined through linear systems of the form (3.7.1). Also, as observed by Brezinski [37], various other methods can be cast into the form of (3.7.1). Then, what differentiates between the various methods is the $g_k(m)$ that accompany them. Again, $A_n^{(j)}$ are taken to be approximations to the limit or antilimit of $\{A_m\}$. In this sense, the formalism represented by (3.7.1) may serve as a "unifying" framework for (defining) these convergence acceleration methods. This formalism may also be used to unify the convergence studies for the different methods, provided their corresponding $g_k(m)$ satisfy

$$g_{k+1}(m) = o(g_k(m)) \text{ as } m \to \infty, \ k = 1, 2, \ldots, \tag{3.7.2}$$

and also

$$A_m \sim A + \sum_{k=1}^{\infty} \alpha_k g_k(m) \text{ as } m \to \infty. \tag{3.7.3}$$

[Note that (3.7.3) has a meaning only when (3.7.2) is satisfied.] The results of Sections 3.5 and 3.6, for example, may be used for this purpose. However, this seems to be of a limited scope because (3.7.2) and hence (3.7.3) are not satisfied in all cases of practical importance. We illustrate this limitation at the end of this section with the transformation of Shanks.

Here is a short list of convergence acceleration methods that are defined through the linear system in (3.7.1) with their corresponding $g_k(m)$. This list is taken in part from Brezinski [37]. The FS- and E-algorithms may serve as implementations for all the convergence acceleration methods that fall within this formalism, but they are much less efficient than the algorithms that have been designed specially for these methods.

- The Richardson extrapolation process of Chapter 1
  $g_k(m) = c_k^m$, where $c_k \neq 0, 1$, are known constants. See Section 1.9 and Theorem 1.4.5. It can be implemented by Algorithm 1.3.1 with ω^{σ_n} there replaced by c_n.
- The polynomial Richardson extrapolation process
  $g_k(m) = x_m^k$, where $\{x_i\}$ is a given sequence. It can be implemented by Algorithm 2.2.1 of Bulirsch and Stoer [43].
- The G-transformation of Gray, Atchison, and McWilliams [112]
  $g_k(m) = x_{m+k-1}$, where $\{x_i\}$ is a given sequence. It can be implemented by the rs-algorithm of Pye and Atchison [233]. We have also developed a new procedure called the FS/qd-algorithm. Both the rs- and FS/qd-algorithms are presented in Chapter 21.
- The Shanks [264] transformation
  $g_k(m) = \Delta A_{m+k-1}$. It can be implemented by the ε-algorithm of Wynn [368]. See Section 0.4.
- Levin [161] transformations
  $g_k(m) = R_m/(m+1)^{k-1}$, where $\{R_i\}$ is a given sequence. It can be implemented by an algorithm due to Longman [178] and Fessler, Ford, and Smith [83] and also by the W-algorithm of Sidi [278]. Levin gives three different sets of R_m for three methods,

namely, $R_m = a_m$ for the t-transformation, $R_m = (m+1)a_m$ for the u-transformation, and $R_m = a_m a_{m+1}/(a_{m+1} - a_m)$ for the v-transformation, where $a_0 = A_0$ and $a_m = \Delta A_{m-1}$ for $m \geq 1$.

- The transformation of Wimp [361]
  $g_k(m) = (\Delta A_m)^k$. It can be implemented by the algorithm of Bulirsch and Stoer [43]. (This transformation was rediscovered later by Germain-Bonne [96].)
- The Thiele rational extrapolation
  $g_k(m) = x_m^k$, $k = 1, 2, \ldots, p$, and $g_{p+k}(m) = A_m x_m^k$, $k = 1, 2, \ldots, p$, with $n = 2p$, where $\{x_i\}$ is a given sequence. It can be implemented by a recursive algorithm that involves inverted differences. By taking $x_i = 1/i$, $i = 1, 2, \ldots$, we obtain the extrapolation method that has been known by the name ρ-algorithm and that is due to Wynn [369]. ρ-algorithm is also the name of the recursive implementation of the method.

Brezinski [37] also shows that the solution for $A_n^{(j)}$ of the generalized rational extrapolation problem

$$A_l = \frac{A_n^{(j)} + \sum_{i=1}^p \alpha_i f_i(l)}{1 + \sum_{i=1}^q \beta_i h_i(l)}, \quad l = j, j+1, \ldots, j+n, \text{ with } n = p+q,$$

can be cast into the form of (3.7.1) by identifying $g_k(m) = f_k(m)$, $k = 1, 2, \ldots, p$, and $g_{p+k}(m) = A_m h_k(m)$, $k = 1, 2, \ldots, q$. Here $\{f_i(m)\}$ and $\{g_i(m)\}$ are known sequences.

We mentioned that (3.7.2) is not satisfied in all cases of practical importance. We now want to show that the $g_k(m)$ associated with the transformation of Shanks generally do not satisfy (3.7.2). To illustrate this, let us consider applying this transformation to the sequence $\{A_m\}$, where $A_m = \sum_{k=0}^m a_k$, $m = 0, 1, \ldots$, with $a_k = (-1)^k/(k+1)$. (The Shanks transformation is extremely effective on the sequence $\{A_m\}$. In fact, it is one of the best acceleration methods on this sequence.) It is easy to show that $\lim_{m \to \infty} g_{k+1}(m)/g_k(m) = -1 \neq 0$, so that (3.7.2) fails to be satisfied. This observation is not limited only to the case $a_k = (-1)^k/(k+1)$, but it applies to the general case in which $a_k \sim z^k \sum_{i=0}^{\infty} \alpha_i k^{\beta-i}$ as $k \to \infty$, with $|z| \leq 1$ but $z \neq 1$, as well. (The Shanks transformation is extremely effective in this general case too.) A similar statement can be made about the G-transformation. This shows very clearly that the preceding formalism is of limited use in the analysis of the various convergence acceleration methods mentioned above. Indeed, there is no unifying framework within which convergence acceleration methods can be analyzed in a serious manner. In reality, we need different approaches and techniques for analyzing the different convergence acceleration methods and for the different classes of sequences.

3.8 Epilogue: What Is the E-Algorithm? What Is It Not?

There has been a great confusion in the literature about what the E-algorithm is and what it is not. We believe this confusion can now be removed in view of the contents of this chapter.

It has been written in different places that the E-algorithm is "the most general extrapolation method actually known," or that it "covers most of the other algorithms," or that it "includes most of the convergence acceleration algorithms actually known," or that "most algorithms for convergence acceleration turn out to be special cases of it." These statements are false and misleading.

First, the E-algorithm is *not* a sequence transformation (equivalently, a convergence acceleration method), it is only a computational procedure (one of many) for solving linear systems of the form (3.7.1) for the unknown $A_n^{(j)}$. The E-algorithm has nothing to do with the fact that most sequence transformations can be defined via linear systems of the form given in (3.7.1).

Next, the formalism represented by (3.7.1) is *not* a convergence acceleration method by itself. It remains a formalism as long as the $g_k(m)$ are not specified. It becomes a convergence acceleration method only after the accompanying $g_k(m)$ are specified. As soon as the $g_k(m)$ are specified, the idea of using extrapolation suggests itself, and the source of this appears to be Hart et al. [125, p. 39].

The derivation of the appropriate $g_k(m)$ – for example, through an asymptotic analysis of $\{A_m\}$ – is much more important than the mere observation that a convergence acceleration method falls within the formalism of (3.7.1).

Finally, as we have already noted, very little has been learned about existing extrapolation methods through this formalism. Much more can be learned by studying the individual methods. This is precisely what we do in the remainder of this book.

4

GREP: Further Generalization of the Richardson Extrapolation Process

4.1 The Set $\mathbf{F}^{(m)}$

The first generalization of the Richardson extrapolation process that we defined and studied in Chapter 3 is essentially based on the asymptotic expansion in (3.1.3), where $\{\phi_k(y)\}$ is an asymptotic sequence in the sense of (3.1.2). In general, the $\phi_k(y)$ considered in Chapter 3 are assumed to have *no* particular structure. The Richardson extrapolation process that we studied in Chapter 1, on the other hand, is based on the asymptotic expansion in (1.1.2) that has a simple *well-defined* structure. In this chapter, we describe a further generalization of the Richardson extrapolation process due to Sidi [272], and denoted there GREP for short, that is based on an asymptotic expansion of $A(y)$ that possesses both of these features. The exact form of this expansion is presented in Definition 4.1.1. Following that, GREP is defined in the next section.

Definition 4.1.1 We shall say that a function $A(y)$, defined for $y \in (0, b]$, for some $b > 0$, where y is a discrete or continuous variable, belongs to the set $\mathbf{F}^{(m)}$ for some positive integer m, if there exist functions $\phi_k(y)$ and $\beta_k(y)$, $k = 1, 2, \ldots, m$, and a constant A, such that

$$A(y) = A + \sum_{k=1}^{m} \phi_k(y)\beta_k(y), \tag{4.1.1}$$

where the functions $\phi_k(y)$ are defined for $y \in (0, b]$ and $\beta_k(\xi)$, as functions of the continuous variable ξ, are continuous in $[0, \hat{\xi}]$ for some $\hat{\xi} \leq b$, and for some constants $r_k > 0$, have Poincaré-type asymptotic expansions of the form

$$\beta_k(\xi) \sim \sum_{i=0}^{\infty} \beta_{ki}\xi^{ir_k} \quad \text{as } \xi \to 0+, \quad k = 1, \ldots, m. \tag{4.1.2}$$

If, in addition, $B_k(t) \equiv \beta_k(t^{1/r_k})$, as a function of the continuous variable t, is infinitely differentiable in $[0, \hat{\xi}^{r_k}]$, $k = 1, \ldots, m$, we say that $A(y)$ belongs to the set $\mathbf{F}_\infty^{(m)}$. [Thus, $\mathbf{F}_\infty^{(m)} \subset \mathbf{F}^{(m)}$.]

The reader may be wondering why we are satisfied with the requirement that the functions $\beta_k(\xi)$ in Definition 4.1.1 be continuous on $[0, \hat{\xi}]$ with $\hat{\xi} \leq b$, and not necessarily on the larger interval $[0, b]$ where $A(y)$ and the $\phi_k(y)$ are defined. The reason is that it is quite easy to construct practical examples in which $A(y)$ and the $\phi_k(y)$ are defined in some interval $[0, b]$ but the $\beta_k(\xi)$ are continuous only in a smaller interval $[0, \hat{\xi}], \hat{\xi} < b$. Definition 4.1.1 accommodates this general case.

We assume that A in (4.1.1) is either the limit or the antilimit of $A(y)$ as $y \rightarrow 0+$. We also assume that $A(y)$ and $\phi_k(y),\ k = 1, \ldots, m$, are all known (or computable) for all possible values that y is allowed to assume in $(0, b]$ and that $r_k,\ k = 1, \ldots, m$, are known as well. We do not assume that the constants β_{ki} are known. We do not assume the functions $\phi_k(y)$ to have any particular structure. Nor do we assume that they satisfy $\phi_{k+1}(y) = o(\phi_k(y))$ as $y \rightarrow 0+$, cf. (3.1.2). Finally, we are interested in finding (or approximating) A, whether it is the limit or the antilimit of $A(y)$ as $y \rightarrow 0+$.

Obviously, when $\lim_{y \rightarrow 0+} \phi_k(y) = 0,\ k = 1, \ldots, m$, we also have $\lim_{y \rightarrow 0+} A(y) = A$. Otherwise, the existence of $\lim_{y \rightarrow 0+} A(y)$ is not guaranteed. In case $\lim_{y \rightarrow 0+} A(y)$ does not exist, it is clear that $\lim_{y \rightarrow 0+} \phi_k(y)$ does not exist for at least one value of k.

It is worth emphasizing that the function $A(y)$ above is such that $A(y) - A$ is the sum of m terms; each one of these terms is the product of a function $\phi_k(y)$ that may have *arbitrary* behavior for $y \rightarrow 0+$ and another, $\beta_k(y)$, that has a *well-defined* (and smooth) behavior for $y \rightarrow 0+$ as a function of y^{r_k}.

Obviously, when $A(y) \in \mathbf{F}^{(m)}$ with A as its limit (or antilimit), $aA(y) + b \in \mathbf{F}^{(m)}$ as well, with $aA + b$ as its limit (or antilimit).

In addition, we have some kind of a "closure" property among the functions in the sets $\mathbf{F}^{(m)}$. This is described in Proposition 4.1.2, whose verification we leave to the reader.

Proposition 4.1.2 *Let $A_1(y) \in \mathbf{F}^{(m_1)}$ with limit or antilimit A_1 and $A_2(y) \in \mathbf{F}^{(m_2)}$ with limit or antilimit A_2. Then $A_1(y) + A_2(y) \in \mathbf{F}^{(m)}$ with limit or antilimit $A_1 + A_2$ and with some $m \leq m_1 + m_2$.*

Another simple observation is contained in Proposition 4.1.3 that is given next.

Proposition 4.1.3 *If $A(y) \in \mathbf{F}^{(m)}$, then $A(y) \in \mathbf{F}^{(m')}$ for $m' > m$ as well.*

This observation raises the following natural question: What is the smallest possible value of m that is appropriate for the particular function $A(y)$, and how can it be determined? As we show later, this question is relevant and has important practical consequences.

We now pause to give a few simple examples of functions $A(y)$ in $\mathbf{F}^{(1)}$ and $\mathbf{F}^{(2)}$ that arise in natural ways. These examples are generalized in the next few chapters. They and the examples that we provide later show that the classes $\mathbf{F}^{(m)}$ and $\mathbf{F}^{(m)}_{\infty}$ are very rich in the sense that many sequences that we are likely to encounter in applied work are associated with functions $A(y)$ that belong to the classes $\mathbf{F}^{(m)}$ or $\mathbf{F}^{(m)}_{\infty}$.

The functions $A(y)$ in Examples 4.1.5–4.1.9 below are in $\mathbf{F}^{(1)}_{\infty}$ and $\mathbf{F}^{(2)}_{\infty}$. Verification of this requires special techniques and is considered later.

Example 4.1.4 Trapezoidal Rule for Integrals with an Endpoint Singularity Consider the integral $I[G] = \int_0^1 G(x)\,dx$, where $G(x) = x^s g(x)$ with $\Re s > -1$ but s not an integer and $g \in C^\infty[0, 1]$. Thus, the integrand $G(x)$ has an algebraic singularity at the left endpoint $x = 0$. Let $h = 1/n$, where n is a positive integer. Let us approximate $I[G]$ by the (modified) trapezoidal rule $T(h)$ or by the midpoint rule $M(h)$, where these are given as

$$T(h) = h\left[\sum_{j=1}^{n-1} G(jh) + \frac{1}{2}G(1)\right] \text{ and } M(h) = h\sum_{j=1}^{n} G((j-1/2)h). \quad (4.1.3)$$

If we now let $Q(h)$ stand for either $T(h)$ or $M(h)$, we have the asymptotic expansion

$$Q(h) \sim I[G] + \sum_{i=1}^{\infty} a_i h^{2i} + \sum_{i=0}^{\infty} b_i h^{s+i+1} \text{ as } h \to 0, \quad (4.1.4)$$

where a_i and b_i are constants independent of h. For $T(h)$, these constants are given by

$$a_i = \frac{B_{2i}}{(2i)!}G^{(2i-1)}(1), \ i = 1, 2, \ldots; \ b_i = \frac{\zeta(-s-i)}{i!}g^{(i)}(0), \ i = 0, 1, \ldots, \quad (4.1.5)$$

where B_i are the Bernoulli numbers and $\zeta(z)$ is the Riemann Zeta function, defined by $\zeta(z) = \sum_{k=1}^{\infty} k^{-z}$ for $\Re z > 1$ and then continued analytically to the complex plane. Similarly, for $M(h)$, a_i and b_i are given by

$$a_i = \frac{D_{2i}}{(2i)!}G^{(2i-1)}(1), \ i = 1, 2, \ldots; \ b_i = \frac{\zeta(-s-i, 1/2)}{i!}g^{(i)}(0), \ i = 0, 1, \ldots,$$
$$(4.1.6)$$

where $D_{2i} = -(1 - 2^{1-2i})B_{2i}$, $i = 1, 2, \ldots$, and $\zeta(z, \epsilon)$ is the generalized Zeta function defined by $\zeta(z, \epsilon) = \sum_{k=0}^{\infty}(k+\epsilon)^{-z}$ for $\Re z > 1$ and then continued analytically to the complex plane. Obviously, $Q(h)$ is analogous to a function $A(y) \in \mathbf{F}^{(2)}$ in the following sense: $Q(h) \leftrightarrow A(y)$, $h \leftrightarrow y$, $h^2 \leftrightarrow \phi_1(y)$, $r_1 = 2$, $h^{s+1} \leftrightarrow \phi_2(y)$, $r_2 = 1$, and $I[G] \leftrightarrow A$. The variable y is, of course, discrete and takes on the values $1, 1/2, 1/3, \ldots$.

The asymptotic expansions for $T(h)$ and $M(h)$ described above are generalizations of the Euler–Maclaurin expansions for regular integrands discussed in Example 1.1.2 and are obtained as special cases of that derived by Navot [216] (see Appendix D) for the offset trapezoidal rule.

Example 4.1.5 Approximation of an Infinite-Range Integral Consider the infinite-range integral $I[f] = \int_0^\infty (\sin t/t)\,dt$. Assume it is being approximated by the finite integral $F(x) = \int_0^x (\sin t/t)\,dt$ for sufficiently large x. By repeated integration by parts of $\int_x^\infty (\sin t/t)\,dt$, it can be shown that $F(x)$ has the asymptotic expansion

$$F(x) \sim I[f] - \frac{\cos x}{x}\sum_{i=0}^{\infty}(-1)^i\frac{(2i)!}{x^{2i}} - \frac{\sin x}{x^2}\sum_{i=0}^{\infty}(-1)^i\frac{(2i+1)!}{x^{2i}} \text{ as } x \to \infty. \quad (4.1.7)$$

We now have that $F(x)$ is analogous to a function $A(y) \in \mathbf{F}^{(2)}$ in the following sense: $F(x) \leftrightarrow A(y)$, $x^{-1} \leftrightarrow y$, $\cos x/x \leftrightarrow \phi_1(y)$, $r_1 = 2$, $\sin x/x^2 \leftrightarrow \phi_2(y)$, $r_2 = 2$, and $F(\infty) = I[f] \leftrightarrow A$. The variable y is continuous in this case.

Example 4.1.6 Approximation of an Infinite-Range Integral Continued Let us allow the variable x in the previous example to assume only the discrete values $x_s = s\pi$, $s = 1, 2, \ldots$. Then (4.1.7) reduces to

$$F(x_s) \sim I[f] - \frac{\cos x_s}{x_s} \sum_{i=0}^{\infty} (-1)^i \frac{(2i)!}{x_s^{2i}} \quad \text{as } s \to \infty. \tag{4.1.8}$$

In this case, $F(x)$ is analogous to a function $A(y) \in \mathbf{F}^{(1)}$ in the following sense: $F(x) \leftrightarrow A(y)$, $x^{-1} \leftrightarrow y$, $\cos x / x \leftrightarrow \phi_1(y)$, $r_1 = 2$, and $F(\infty) = I[f] \leftrightarrow A$. The variable y assumes only the discrete values $(i\pi)^{-1}$, $i = 1, 2, \ldots$.

Example 4.1.7 Summation of the Riemann Zeta Function Series Let us go back to Example 1.1.4, where we considered the summation of the infinite series $\sum_{k=1}^{\infty} k^{-z}$ that converges to and defines the Riemann Zeta function $\zeta(z)$ for $\Re z > 1$. We saw there that the partial sum $A_n = \sum_{k=1}^{n} k^{-z}$ has the asymptotic expansion given in (1.1.18). This expansion can be rewritten in the form

$$A_n \sim \zeta(z) + n^{-z+1} \sum_{i=0}^{\infty} \frac{\mu_i}{n^i} \quad \text{as } n \to \infty, \tag{4.1.9}$$

where μ_i depend only on z. From this, we see that A_n is analogous to a function $A(y) \in \mathbf{F}^{(1)}$ in the following sense: $A_n \leftrightarrow A(y)$, $n^{-1} \leftrightarrow y$, $n^{-z+1} \leftrightarrow \phi_1(y)$, $r_1 = 1$, and $\zeta(z) \leftrightarrow A$. The variable y is discrete and takes on the values $1, 1/2, 1/3, \ldots$.

Recall that $\zeta(z)$ is the limit of $\{A_n\}$ when $\Re z > 1$ and its antilimit otherwise, provided $z \neq 1, 0, -1, -2, \ldots$.

Example 4.1.8 Summation of the Logarithmic Function Series Consider the infinite power series $\sum_{k=1}^{\infty} z^k / k$ that converges to $\log(1 - z)^{-1}$ when $|z| \leq 1$ but $z \neq 1$. We show below that, as long as $z \notin [1, +\infty)$, the partial sum $A_n = \sum_{k=1}^{n} z^k / k$ is analogous to a function $A(y) \in \mathbf{F}^{(1)}$, whether $\sum_{k=1}^{\infty} z^k / k$ converges or not. We also provide the precise description of the relevant antilimit when $\sum_{k=1}^{\infty} z^k / k$ diverges.

Invoking $\int_0^{\infty} e^{-kt} \, dt = 1/k$, $k > 0$, in $A_n = \sum_{k=1}^{n} z^k / k$ with $z \notin [1, +\infty)$, changing the order of the integration and summation, and summing the resulting geometric series $\sum_{k=1}^{n} (ze^{-t})^k$, we obtain

$$\sum_{k=1}^{n} \frac{z^k}{k} = \int_0^{\infty} \frac{ze^{-t}}{1 - ze^{-t}} \, dt - \int_0^{\infty} \frac{(ze^{-t})^{n+1}}{1 - ze^{-t}} \, dt. \tag{4.1.10}$$

Now, the first integral in (4.1.10) is $\log(1 - z)^{-1}$ with its branch cut along the real interval $[1, +\infty)$. We rewrite the second integral in the form $z^{n+1} \int_0^{\infty} e^{-nt} (e^t - z)^{-1} \, dt$ and apply Watson's lemma (see, for example, Olver [223]; see also Appendix B). Combining all this in (4.1.10), we have the asymptotic expansion

$$A_n \sim \log(1 - z)^{-1} - \frac{z^{n+1}}{n} \sum_{i=0}^{\infty} \frac{\mu_i}{n^i} \quad \text{as } n \to \infty, \tag{4.1.11}$$

where $\mu_i = (d/dt)^i (e^t - z)^{-1} \, |_{t=0}$, $i = 0, 1, \ldots$. In particular, $\mu_0 = 1/(1 - z)$ and $\mu_1 = -1/(1 - z)^2$.

Thus, we have that A_n is analogous to a function $A(y) \in \mathbf{F}^{(1)}$ in the following sense: $A_n \leftrightarrow A(y), n^{-1} \leftrightarrow y, z^n/n \leftrightarrow \phi_1(y), r_1 = 1,$ and $\log(1-z)^{-1} \leftrightarrow A$, with $\log(1-z)^{-1}$ having its branch cut along $[1, +\infty)$. When $|z| \leq 1$ but $z \neq 1$, $\sum_{k=1}^{\infty} z^k/k$ converges and $\log(1-z)^{-1}$ is $\lim_{n\to\infty} A_n$. When $|z| > 1$ but $z \notin [1, +\infty)$, $\sum_{k=1}^{\infty} z^k/k$ diverges, but $\log(1-z)^{-1}$ serves as the antilimit of $\{A_n\}$. (See Example 0.2.1.) The variable y is discrete and takes on the values $1, 1/2, 1/3, \ldots$.

Example 4.1.9 Summation of a Fourier Cosine Series Consider the convergent Fourier cosine series $\sum_{k=1}^{\infty} \cos k\theta/k$ whose sum is $\log(2 - 2\cos\theta)^{-1/2}$ when $\theta \neq 2k\pi$, $k = 0, \pm 1, \pm 2, \ldots$. This series can be obtained by letting $z = e^{i\theta}$ in $\sum_{k=1}^{\infty} z^k/k$ that we treated in the preceding example and then taking the real part of the latter. By doing the same in (4.1.11), we see that the partial sum $A_n = \sum_{k=1}^{n} \cos k\theta/k$ has the asymptotic expansion

$$A_n \sim \log(2 - 2\cos\theta)^{-1/2} + \frac{\cos n\theta}{n} \sum_{i=0}^{\infty} \frac{\gamma_i}{n^i} + \frac{\sin n\theta}{n} \sum_{i=0}^{\infty} \frac{\delta_i}{n^i} \quad \text{as } n \to \infty, \quad (4.1.12)$$

for some γ_i and δ_i that depend only on θ.

Thus, A_n is analogous to a function $A(y) \in \mathbf{F}^{(2)}$ in the following sense: $A_n \leftrightarrow A(y)$, $n^{-1} \leftrightarrow y$, $\cos n\theta/n \leftrightarrow \phi_1(y)$, $r_1 = 1$, $\sin n\theta/n \leftrightarrow \phi_2(y)$, $r_2 = 1$, and $\log(2 - 2\cos\theta)^{-1/2} \leftrightarrow A$. The variable y is discrete and takes on the values $1, 1/2, 1/3, \ldots$.

4.2 Definition of the Extrapolation Method GREP

Definition 4.2.1 Let $A(y)$ belong to $\mathbf{F}^{(m)}$ with the notation of Definition 4.1.1. Pick a decreasing positive sequence $\{y_l\} \subset (0, b]$ such that $\lim_{l\to\infty} y_l = 0$. Let $n \equiv (n_1, n_2, \ldots, n_m)$, where $n_1, \ldots, n_m$ are nonnegative integers. Then, the approximation $A_n^{(m,j)}$ to A, whether A is the limit or the antilimit of $A(y)$ as $y \to 0+$, is defined through the linear system

$$A(y_l) = A_n^{(m,j)} + \sum_{k=1}^{m} \phi_k(y_l) \sum_{i=0}^{n_k-1} \bar{\beta}_{ki} y_l^{ir_k}, \quad j \leq l \leq j + N; \quad N = \sum_{k=1}^{m} n_k, \quad (4.2.1)$$

$\bar{\beta}_{ki}$ being the additional (auxiliary) N unknowns. In (4.2.1), $\sum_{i=0}^{-1} c_i \equiv 0$ so that $A_{(0,\ldots,0)}^{(m,j)} = A(y_j)$ for all j. This generalization of the Richardson extrapolation process that generates the $A_n^{(m,j)}$ is denoted $GREP^{(m)}$. When there is no room for confusion, we will write GREP instead of $GREP^{(m)}$ for short.

Comparing the equations in (4.2.1) with the expansion of $A(y)$ that is given in (4.1.1) with (4.1.2), we realize that the former are obtained from the latter by substituting in (4.1.1) the asymptotic expansion of $\beta_k(y)$ given in (4.1.2), truncating the latter at the term $\beta_{k,n_k-1} y^{(n_k-1)r_k}$, $k = 1, \ldots, m$, and finally collocating at $y = y_l$, $l = j, j+1, \ldots,$ $j + N$, where $N = \sum_{k=1}^{m} n_k$.

The following theorem states that $A_n^{(m,j)}$ is expressible in the form (0.3.2) with (0.3.3).

Theorem 4.2.2 $A_n^{(m,j)}$ *can be expressed in the form*

$$A_n^{(m,j)} = \sum_{i=0}^{N} \gamma_{ni}^{(m,j)} A(y_{j+i}) \qquad (4.2.2)$$

with the $\gamma_{ni}^{(m,j)}$ determined by the linear system

$$\sum_{i=0}^{N} \gamma_{ni}^{(m,j)} = 1$$

$$\sum_{i=0}^{N} \gamma_{ni}^{(m,j)} \phi_k(y_{j+i}) y_{j+i}^{sr_k} = 0, \ s = 0, 1, \ldots, n_k - 1, \ k = 1, \ldots, m. \quad (4.2.3)$$

The proof of Theorem 4.2.2 follows from a simple analysis of the linear system in (4.2.1) and is identical to that of Theorem 3.2.1. Note that if we denote by M the matrix of coefficients of (4.2.1), such that $A_n^{(m,j)}$ is the first element of the vector of unknowns, then the vector $[\gamma_0, \gamma_1, \ldots, \gamma_N]$, where $\gamma_i \equiv \gamma_{ni}^{(m,j)}$ for short, is the first row of the matrix M^{-1}; this, obviously, is entirely consistent with (4.2.3).

The next theorem shows that the extrapolation process GREP that generates $A_n^{(m,j)}$ "eliminates" the first n_k terms y^{ir_k}, $i = 0, 1, \ldots, n_k - 1$, from the asymptotic expansion of $\beta_k(y)$ given in (4.1.2), for $k = 1, 2, \ldots, m$. Just as Theorem 3.2.3, the next theorem too is not a convergence theorem but a heuristic justification of the possible validity of (4.2.1) as an extrapolation process.

Theorem 4.2.3 *Let n and N be as before, and define*

$$R_n(y) = A(y) - A - \sum_{k=1}^{m} \phi_k(y) \sum_{s=0}^{n_k-1} \beta_{ks} y^{sr_k}. \qquad (4.2.4)$$

Then

$$A_n^{(m,j)} - A = \sum_{i=0}^{N} \gamma_{ni}^{(m,j)} R_n(y_{j+i}). \qquad (4.2.5)$$

In particular, when $A(y) = A + \sum_{k=1}^{m} \phi_k(y) \sum_{s=0}^{n_k-1} \beta_{ks} y^{sr_k}$ for all possible y, we have $A_{n'}^{(m,j)} = A$, for all $j \geq 0$ and $n' = (n'_1, \ldots, n'_m)$ such that $n'_k \geq n_k$, $k = 1, \ldots, m$.

The proof of Theorem 4.2.3 can be achieved by invoking Theorem 4.2.2. We leave the details to the reader.

Before we close this section, we would like to mention that important examples of GREP are the Levin transformations, the Levin–Sidi D- and d-transformations and the Sidi $\bar{D}$-, $\tilde{D}$-, W-, and mW-transformations, which are considered in great detail in the next chapters.

4.3 General Remarks on $\mathbf{F}^{(m)}$ and GREP

1. We first note that the expansion in (4.1.1) with (4.1.2) is a true generalization of the expansion $A(y) \sim A + \sum_{i=1}^{\infty} \alpha_i y^{ir}$ as $y \to 0+$ that forms the basis of the polynomial Richardson extrapolation process in the sense that the former reduces to the latter when we let $m = 1$, $\phi_1(y) = y^{r_1}$, and $r_1 = r$ in (4.1.1) and (4.1.2). Thus, $\mathrm{GREP}^{(1)}$ is a true generalization of the polynomial Richardson extrapolation process.

   Next, the expansion in (4.1.1) with (4.1.2) can be viewed as a generalization also of the expansion in (3.1.3) in the following sense: (i) The unknown *constants* α_k in (3.1.3) are now being replaced by some unknown *smooth functions* $\beta_k(y)$ that possess asymptotic expansions of *known forms* for $y \to 0+$. (ii) In addition, unlike the $\phi_k(y)$ in (3.1.3), the $\phi_k(y)$ in (4.1.1) are not required to satisfy the condition in (3.1.2). That is, the $\phi_k(y)$ in (4.1.1) may have growth rates for $y \to 0+$ that may be independent of each other; they may even have the same growth rates. Thus, in this sense, $\mathrm{GREP}^{(m)}$ is a generalization of the Richardson extrapolation process *beyond* its first generalization.

   Finally, as opposed to the $A(y)$ in Chapter 3 that are represented by one asymptotic expansion as $y \to 0+$, the $A(y)$ in this chapter are represented by sums of m asymptotic expansions as $y \to 0+$. As the union $\cup_{m=1}^{\bar{m}} \mathbf{F}^{(m)}$ is a very large set that further expands with $\bar{m}$, we realize that GREP is a comprehensive extrapolation procedure with a very large scope.

2. Because successful application of GREP to find the limit or antilimit of a function $A(y) \in \mathbf{F}^{(m)}$ requires as input the integer m together with the functions $\phi_k(y)$ and the numbers r_k, $k = 1, \dots, m$, one may wonder how these can be determined. As seen from the examples in Section 4.1, this problem is far from trivial, and any general technique for solving it is of utmost importance.

   By Definition 4.1.1, the information about m, the $\phi_k(y)$, and r_k is contained in the asymptotic expansion of $A(y)$ for $y \to 0+$. Therefore, we need to analyze $A(y)$ asymptotically for $y \to 0+$ to obtain this information. As we do not want to have to go through such an analysis in each case, it is very crucial that we have theorems of a general nature that provide us with the appropriate m, $\phi_k(y)$, and r_k for large classes of $A(y)$. (Example 4.1.4 actually contains one such theorem. Being specific, Examples 4.1.5–4.1.9 do not.)

   Fortunately, powerful theorems concerning the asymptotic behavior of $A(y)$ in many cases of practical interest can be proved. From these theorems, it follows that sets of convenient $\phi_k(y)$ can be obtained directly from $A(y)$ in simple ways. The fact that the $\phi_k(y)$ are readily available makes GREP a very useful tool for computing limits and antilimits. (For $m > 1$, GREP is usually the only effective tool.)

3. The fact that there are only a *finite number* of structureless functions, namely, the m functions $\phi_k(y)$, and that the $\beta_k(y)$ are essentially *polynomial* in nature, enables us to design algorithms for implementing GREP, such as the W- and $W^{(m)}$-algorithms, that are extremely efficient, unlike those for the first generalization. The existence of such efficient algorithms makes GREP especially attractive to the user. (The subject of algorithms for GREP is treated in detail in Chapter 7.)

4. Now, the true nature of $A(y)$ as $y \to 0+$ is actually contained in the functions $\phi_k(y)$. Viewed in this light, the terms *shape function* and *form factor* commonly used in

nuclear physics, become appropriate for the $\phi_k(y)$ as well. We use this terminology throughout.

Weniger [353] calls the $\phi_k(y)$ "remainder estimates." When $m > 1$, this terminology is not appropriate, because, in general, none of the $\phi_k(y)$ by itself can be considered an estimate of the remainder $A(y) - A$. This should be obvious from Examples 4.1.5 and 4.1.9.

5. We note that the $\phi_k(y)$ are not unique in the sense that they can be replaced by some other functions $\tilde{\phi}_k(y)$, at the same time preserving the form of the expansion in (4.1.1) and (4.1.2). In other words, we can have $\sum_{k=1}^m \phi_k(y)\beta_k(y) = \sum_{k=1}^m \tilde{\phi}_k(y)\tilde{\beta}_k(y)$, with $\tilde{\beta}_k(\xi)$ having asymptotic expansions of the form similar to that given in (4.1.2).

As the $\phi_k(y)$ are not unique, we can now aim at obtaining $\phi_k(y)$ that are as simple and as user-friendly as possible for use with GREP. Practical examples of this strategy, related to the acceleration of convergence of infinite-range integrals and infinite series and sequences, are considered in Chapters 5, 6, and 11.

4.4 A Convergence Theory for GREP

From our convention that n in $A_n^{(m,j)}$ stands for the vector of integers $(n_1, n_2, \ldots, n_m)$, it is clear that only for $m = 1$ can we arrange the $A_n^{(m,j)} = A_{n_1}^{(1,j)}$ in a two-dimensional array as in Table 1.3.1, but for $m = 2, 3, \ldots$, multidimensional arrays are needed. Note that the dimension of the array for a given value of m is $m + 1$. As a result of the multidimensionality of the arrays involved, we realize that there can be infinitely many sequences of the $A_{(n_1,\ldots,n_m)}^{(m,j)}$ that can be studied and also used in practice.

Because of the multitude of choices, we may be confused about which sequences we should use and study. This ultimately ties in with the question of the order in which the functions $\phi_k(y)y^{ir_k}$ should be "eliminated" from the expansion in (4.1.1) and (4.1.2).

Recall that in the first generalization of the Richardson extrapolation process we "eliminated" the $\phi_k(y)$ from (3.1.3) in the order $\phi_1(y), \phi_2(y), \ldots$, because they satisfied (3.1.2). In GREP, however, we do not assume that the $\phi_k(y)$ satisfy a relation analogous to (3.1.2). Therefore, it seems it is not clear a priori in what order the functions $\phi_k(y)y^{ir_k}$ should be "eliminated" from (4.1.1) and (4.1.2) for an arbitrary $A(y)$.

Fortunately, there are certain orders of elimination that are universal and give rise to good sequences of $A_n^{(m,j)}$. We restrict our attention to two types of such sequences because of their analogy to the column and diagonal sequences of Chapters 1–3.

1. $\{A_n^{(m,j)}\}_{j=0}^\infty$ with $n = (n_1, \ldots, n_m)$ fixed. Such sequences are analogous to the column sequences of Chapters 1 and 3. In connection with these sequences, the limiting process in which $j \to \infty$ and $n = (n_1, \ldots, n_m)$ is being held fixed has been denoted Process I.

2. $\{A_{q+(\nu,\nu,\ldots,\nu)}^{(m,j)}\}_{\nu=0}^\infty$ with j and $q = (q_1, \ldots, q_m)$ fixed. In particular, the sequence $\{A_{(\nu,\nu,\ldots,\nu)}^{(m,j)}\}_{\nu=0}^\infty$ with j fixed is of this type. Such sequences are analogous to the diagonal sequences of Chapters 1 and 3. In connection with these sequences, the limiting process in which $n_k \to \infty$, $k = 1, \ldots, m$, simultaneously and j is being held fixed has been denoted Process II.

Numerical experience and theoretical results indicate that Process II is the more effective of the two. In view of this, in practice we look at the sequences $\{A^{(m,0)}_{(\nu,\nu,\dots,\nu)}\}^{\infty}_{\nu=0}$ as these seem to give the best accuracy for a given number of the $A(y_i)$. The theory we propose next supports this observation well. It also provides strong justification of the practical relevance of Process I and Process II. Finally, it is directly applicable to certain GREPs used in the summation of some oscillatory infinite-range integrals and infinite series, as we will see later in the book.

Throughout the rest of this section, we take $A(y)$ and $A^{(m,j)}_n$ to be exactly as in Definition 4.1.1 and Definition 4.2.1, respectively, with the same notation. We also define

$$\Gamma^{(m,j)}_n = \sum_{i=0}^{N}\left|\gamma^{(m,j)}_{ni}\right|, \tag{4.4.1}$$

where the $\gamma^{(m,j)}_{ni}$ are as defined in Theorem 4.2.2. Thus, $\Gamma^{(m,j)}_n$ is the quantity that controls the propagation of the errors in the $A(y_i)$ into $A^{(m,j)}_n$. Finally, we let Π_ν denote the set of all polynomials $u(t)$ of degree ν or less in t.

The following preliminary lemma is the starting point of our analysis.

Lemma 4.4.1 *The error in $A^{(m,j)}_n$ satisfies*

$$A^{(m,j)}_n - A = \sum_{i=0}^{N}\gamma^{(m,j)}_{ni}\sum_{k=1}^{m}\phi_k(y_{j+i})[\beta_k(y_{j+i}) - u_k(y^{r_k}_{j+i})], \tag{4.4.2}$$

where, for each $k = 1,\dots,m$, $u_k(t)$ is in Π_{n_k-1} and arbitrary.

Proof. The proof can be achieved by invoking Theorem 4.2.2, just as that of Theorem 4.2.3. ∎

A convergence study of Process I and Process II can now be made by specializing the polynomials $u_k(t)$ in Lemma 4.4.1.

4.4.1 Study of Process I

Theorem 4.4.2 *Provided $\sup_j \Gamma^{(m,j)}_n = \Omega^{(m)}_n < \infty$ for fixed $n_1,\dots,n_m$, we have*

$$|A^{(m,j)}_n - A| \le \sum_{k=1}^{m}\left(\max_{0\le i\le N}\left|\phi_k(y_{j+i})\right|\right)O(y^{n_k r_k}_j)\ \text{ as } j\to\infty. \tag{4.4.3}$$

If, in addition, $\phi_k(y_{j+i}) = O(\phi_k(y_j))$ as $j\to\infty$, then

$$|A^{(m,j)}_n - A| \le \sum_{k=1}^{m}|\phi_k(y_j)|O(y^{n_k r_k}_j)\ \text{ as } j\to\infty, \tag{4.4.4}$$

whether $\lim_{y\to 0+} A(y)$ exists or not.

Proof. Let us pick $u_k(t) = \sum_{i=0}^{n_k-1}\beta_{ki}t^i$ in Lemma 4.4.1, with β_{ki} as in (4.1.2). The result in (4.4.3) follows from the fact that $\beta_k(y) - u_k(y^{r_k}) = O(y^{n_k r_k})$ as $y\to 0+$, which is a consequence of (4.1.2), and from the fact that $y_j > y_{j+1} > \cdots$, with $\lim_{j\to\infty} y_j = 0$. (4.4.4) is a direct consequence of (4.4.3). We leave the details to the reader. ∎

Comparing (4.4.4) with $A(y_j) - A = \sum_{k=1}^{m} \phi_k(y_j)O(1)$ as $j \to \infty$, that follows from (4.1.1) with (4.1.2), we realize that, generally speaking, $\{A_n^{(m,j)}\}_{j=0}^{\infty}$ converges to A more quickly than $\{A(y_i)\}$ when the latter converges. Depending on the growth rates of the $\phi_k(y_i)$, $\{A_n^{(m,j)}\}_{j=0}^{\infty}$ may converge to A even when $\{A(y_i)\}$ does not. Also, more refined results similar to those in (1.5.5) and (3.5.18) in case some or all of β_{kn_k} are zero can easily be written down. We leave this to the reader. Finally, by imposing the condition $\sup_j \Gamma_n^{(m,j)} < \infty$, we are actually assuming in Theorem 4.4.2 that Process I is a stable extrapolation method with the given $\phi_k(y)$ and the y_l.

4.4.2 Study of Process II

Theorem 4.4.3 *Let the integer j be fixed and assume that $\hat{\xi} \geq y_j$ in Definition 4.1.1. Then, provided $\sup_n \Gamma_n^{(m,j)} = \Omega^{(m,j)} < \infty$, there holds*

$$|A_n^{(m,j)} - A| \leq \Omega^{(m,j)} \sum_{k=1}^{m} \left(\max_{y\in I_{j,N}} |\phi_k(y)| \right) E_{k,n_k}^{(j)}, \qquad (4.4.5)$$

where $I_{j,N} = [y_{j+N}, y_j]$ and

$$E_{k,v}^{(j)} = \min_{u\in\Pi_{v-1}} \max_{y\in[0,y_j]} |\beta_k(y) - u(y^{r_k})|, \quad k = 1, \dots, m. \qquad (4.4.6)$$

If, in addition, $A(y) \in \mathbf{F}_\infty^{(m)}$ with $B_k(t) \in C^\infty[0, \hat{\xi}^{r_k}]$, $k = 1, \dots, m$, with the same $\hat{\xi}$, and $\max_{y\in I_{j,N}} |\phi_k(y)| = O(N^{\alpha_k})$ as $N \to \infty$ for some α_k, $k = 1, \dots, m$, then

$$A_n^{(m,j)} - A = O(N^\alpha \hat{n}^{-\mu}) \ as \ n_1, \dots, n_m \to \infty, \ for \ every \ \mu > 0, \qquad (4.4.7)$$

where $\alpha = \max\{\alpha_1, \dots, \alpha_m\}$ and $\hat{n} = \min\{n_1, \dots, n_m\}$. Thus,

$$A_n^{(m,j)} - A = O(\hat{n}^{-\mu}) \ as \ n_1, \dots, n_m \to \infty, \ for \ every \ \mu > 0, \qquad (4.4.8)$$

holds (i) with no extra condition on the n_k when $\alpha \leq 0$, and (ii) provided $N = O(\hat{n}^\rho)$ as $\hat{n} \to \infty$ for some $\rho > 0$, when $\alpha > 0$.

Proof. For each k, let us pick $u_k(t)$ in Lemma 4.4.1 to be the best polynomial approximation of degree at most $n_k - 1$ to the function $B_k(t) \equiv \beta_k(t^{1/r_k})$ on $[0, t_j]$, where $t_j = y_j^{r_k}$, in the maximum norm. Thus, (4.4.6) is equivalent to

$$E_{k,v}^{(j)} = \min_{u\in\Pi_{v-1}} \max_{t\in[0,t_j]} |B_k(t) - u(t)| = \max_{t\in[0,t_j]} |B_k(t) - u_k(t)|. \qquad (4.4.9)$$

The result in (4.4.5) now follows by taking the modulus of both sides of (4.4.2) and manipulating its right-hand side appropriately. When $A(y) \in \mathbf{F}_\infty^{(m)}$, each $B_k(t)$ is infinitely differentiable on $[0, t_j]$ because $t_j = y_j^{r_k} \leq \hat{\xi}^{r_k}$. Therefore, from a standard result in polynomial approximation theory (see, for example, Cheney [47]; see also Appendix F), we have that $E_{k,\nu}^{(j)} = O(\nu^{-\mu})$ as $\nu \to \infty$, for every $\mu > 0$. The result in (4.4.7) now follows. The remaining part of the theorem follows from (4.4.7) and its proof is left to the reader. ∎

As is clear from (4.4.8), under the conditions stated following (4.4.8), Process II converges to A, whether $\{A(y_i)\}$ does or not. Recalling the remark following the proof of Theorem 4.4.2 on Process I, we realize that Process II has convergence properties superior to those of Process I. Finally, by imposing the condition $\sup_n \Gamma_n^{(m,j)} < \infty$, we are actually assuming in Theorem 4.4.3 that Process II is a stable extrapolation method with the given $\phi_k(y)$ and the y_l.

Note that the condition that $N = O(\hat{n}^\rho)$ as $\hat{n} \to \infty$ is satisfied in the case $n_k = q_k + \nu$, $k = 1, \ldots, m$, where q_k are fixed nonnegative integers. In this case, $\hat{n} = O(\nu)$ as $\nu \to \infty$ and, consequently, (4.4.8) now reads

$$A_{q+(\nu,\ldots,\nu)}^{(m,j)} - A = O(\nu^{-\mu}) \text{ as } \nu \to \infty, \text{ for every } \mu > 0, \qquad (4.4.10)$$

both when $\alpha \leq 0$ and when $\alpha > 0$.

Also, in many cases of interest, $A(y)$, in addition to being in $\mathbf{F}_\infty^{(m)}$, is such that its associated $\beta_k(y)$ satisfy $E_{k,\nu}^{(j)} = O(\exp(-\lambda_k \nu^{\delta_k}))$ as $\nu \to \infty$ for some $\lambda_k > 0$ and $\delta_k > 0$, $k = 1, \ldots, m$. (In particular, $\delta_k \geq 1$ if $B_k(t) \equiv \beta_k(t^{1/r_k})$ is not only in $C^\infty[0, b^{r_k}]$ but is analytic on $[0, b^{r_k}]$ as well. Otherwise, $0 < \delta_k < 1$. For examples of the latter case, we refer the reader to the papers of Németh [218], Miller [212], and Boyd [29], [30].) This improves the results in Theorem 4.4.3 considerably. In particular, (4.4.10) becomes

$$A_{q+(\nu,\ldots,\nu)}^{(m,j)} - A = O(\exp(-\lambda \nu^\delta)) \text{ as } \nu \to \infty, \qquad (4.4.11)$$

where $\delta = \min\{\delta_1, \ldots, \delta_m\}$ and $\lambda = \min\{\lambda_k : \delta_k = \delta\} - \epsilon$ for arbitrary $\epsilon > 0$.

4.4.3 Further Remarks on Convergence Theory

Our treatment of the convergence analysis was carried out under the stability conditions that $\sup_j \Gamma_n^{(m,j)} < \infty$ for Process I and $\sup_n \Gamma_n^{(m,j)} < \infty$ for Process II. Although these conditions can be fulfilled by clever choices of the $\{y_l\}$ and verified rigorously as well, they do not have to hold in general. Interestingly, however, even when they do not have to hold, Process I and Process II may converge, as has been observed numerically in many cases and as can be shown rigorously in some cases.

Furthermore, we treated the convergence of Process I and Process II for arbitrary m and for arbitrary $\{y_l\}$. We can obtain much stronger and interesting results for $m = 1$ and for certain choices of $\{y_l\}$ that have been used extensively; for example, in the literature of numerical quadrature for one- and multi-dimensional integrals. We intend to come back to GREP$^{(1)}$ in Chapters 8–10, where we present an extensive analysis for it along with a new set of mathematical techniques developed recently.

4.4.4 Remarks on Convergence of the $\bar{\beta}_{ki}$

So far we have been concerned only with the convergence of the $A_n^{(m,j)}$ and mentioned nothing about the $\bar{\beta}_{ki}$, the remaining unknowns in (4.2.1). Just as the $\bar{\alpha}_k$ in the first generalization of the Richardson extrapolation process tend to the corresponding α_k under certain conditions, the $\bar{\beta}_{ki}$ in GREP tend to the corresponding β_{ki}, again under certain conditions. Even though their convergence may be quick theoretically, they themselves are prone to severe roundoff in finite-precision arithmetic. Therefore, the $\bar{\beta}_{ki}$ have very poor accuracy in finite-precision arithmetic. For this reason, we do not pursue the theoretical study of the $\bar{\beta}_{ki}$ any further. We refer the reader to Section 4 of Levin and Sidi [165] for some numerical examples that also provide a few computed values of the $\bar{\beta}_{ki}$.

4.4.5 Knowing the Minimal m Pays

Let us go back to the error bound on $A_n^{(m,j)}$ given in (4.4.5) and (4.4.6) of Theorem 4.4.3. Of course, this bound is valid whether m is minimal or not. In addition, in many cases of interest, the factors $\Omega^{(m,j)}$ and $\sum_{k=1}^{m}\left(\max_{y \in I_{j,N}} |\phi_k(y)|\right) E_{k,n_k}^{(j)}$ vary slowly with m. [Actually, $\Omega^{(m,j)}$ is smaller for smaller m in most cases.] In other words, the quality of the approximation $A_n^{(m,j)}$ may be nearly independent of the value of m. The cost of computing $A_n^{(m,j)}$, on the other hand, increases linearly with m. For example, computation of the approximations $A_{(\nu,\ldots,\nu)}^{(m,0)}$, $\nu = 0, 1, \ldots , \nu_{\max}$, that are associated with Process II entails a cost of $m\nu_{\max} + 1$ evaluations of $A(y)$. Employing the minimal value of m, therefore, reduces this cost considerably. This reduction is significant, especially when the minimal m is a small integer, such as 1 or 2. Such cases do occur most frequently in applications, as we will see later.

This discussion shows clearly that there is great importance to knowing the minimal value of m for which $A(y) \in \mathbf{F}^{(m)}$. In the two examples of GREP that we discuss in the next two chapters, we present simple heuristic ways to determine the minimal m for various functions $A(y)$ that are related to some infinite-range integrals and infinite series of varying degrees of complexity.

4.5 Remarks on Stability of GREP

As mentioned earlier, $\Gamma_n^{(m,j)}$ defined in (4.4.1) is the quantity that controls the propagation of errors in the $A(y_l)$ into $A_n^{(m,j)}$, the approximation to A. By (4.2.3), we have obviously that $\Gamma_n^{(m,j)} \geq 1$ for all j and n. Thus, the closer $\Gamma_n^{(m,j)}$ is to unity, the better the numerical stability of $A_n^{(m,j)}$. It is also clear that $\Gamma_n^{(m,j)}$ is a function of the $\phi_k(y)$ and the y_l. As $\phi_k(y)$ are given and thus not under the user's control, the behavior of $\Gamma_n^{(m,j)}$ ultimately depends on the y_l that can be picked at will. Thus, there is great value to knowing how to pick the y_l appropriately.

Although it is impossible to analyze the behavior of $\Gamma_n^{(m,j)}$ in the most general case, we can nevertheless state a few practical conclusions and rules of thumb that have been derived from numerous applications.

First, the smaller $\Gamma_n^{(m,j)}$ is, the more accurate an approximation $A_n^{(m,j)}$ is to A theoretically as well. Next, when the sequences of the $\Gamma_n^{(m,j)}$ increase mildly or remain bounded in Process I or Process II, the corresponding sequences of the $A_n^{(m,j)}$ converge to A quickly.

In practice, the growth of the $\Gamma_n^{(m,j)}$ both in Process I and in Process II can be reduced very effectively by picking the y_l such that, for each k, the sequences $\{\phi_k(y_l)\}_{l=0}^\infty$ are quickly varying. Quick variation can come about if, for example, $\phi_k(y_l) = C_k(l)\exp(u_k(l))$, where $C_k(l)$ and $u_k(l)$ are slowly varying functions of l. A function $H(l)$ is slowly varying in l if $\lim_{l\to\infty} H(l+1)/H(l) = 1$. Thus, $H(l)$ can be slowly varying if, for instance, it is monotonic and behaves like κl^α as $l \to \infty$, for some $\alpha \neq 0$ and $\kappa \neq 0$ that can be complex in general. The case in which $u_k(l) \sim \kappa l^\alpha$ as $l \to \infty$, where $\alpha = 1$ and $\kappa = i\pi$, produces one of the best types of quick variation in that now $\phi_k(y_l) \sim C_k(l)(-1)^l$ as $l \to \infty$, and hence $\lim_{l\to\infty} \phi_k(y_{l+1})/\phi_k(y_l) = -1$.

To illustrate these points, let us look at the following two cases:

(i) $m = 1$ and $\phi_1(y) = y^\delta$ for some $\delta \neq 0, -1, -2, \dots$. Quick variation in this case is achieved by picking $y_l = y_0\omega^l$ for some $\omega \in (0, 1)$. With this choice of the y_l, we have $\phi_1(y_l) = y_0^\delta e^{(\log\omega)\delta l}$ for all l, and both Process I and Process II are stable. By the stability theory of Chapter 1 with these y_l we actually have $\Gamma_{(v)}^{(1,j)} = \prod_{i=1}^v \frac{1+|c_i|}{|1-c_i|}$ for all j and v, where $c_i = \omega^{\delta+i-1}$, and hence $\Gamma_{(v)}^{(1,j)}$ are bounded both in j and in v. The choice $y_l = y_0/(l+1)$, on the other hand, leads to extremely unstable extrapolation processes, as follows from Theorem 2.1.2.

(ii) $m = 1$ and $\phi_1(y) = e^{i\pi/y}y^\delta$ for some δ. Best results can be achieved by picking $y_l = 1/(l+1)$, as this gives $\phi_1(y_l) \sim K l^{-\delta}(-1)^l$ as $l \to \infty$. As shown later, for this choice of the y_l, we actually have $\Gamma_{(v)}^{(1,j)} = 1$ for all j and v, when δ is real. Hence, both Process I and Process II are stable.

We illustrate all these points with numerical examples and also with ample theoretical developments in the next chapters.

4.6 Extensions of GREP

Before closing this chapter, we mention that GREP can be extended to cover those functions $A(y)$ that are as in (4.1.1), for which the functions $\beta_k(y)$ have asymptotic expansions of the form

$$\beta_k(y) \sim \sum_{i=0}^\infty \beta_{ki} y^{\tau_{ki}} \text{ as } y \to 0+, \tag{4.6.1}$$

where the τ_{ki} are known constants that satisfy

$$\tau_{ki} \neq 0, \ i = 0, 1, \dots, \ \Re\tau_{k0} < \Re\tau_{k1} < \cdots, \text{ and } \lim_{i\to\infty} \Re\tau_{ki} = +\infty, \tag{4.6.2}$$

or of the more general form

$$\beta_k(y) \sim \sum_{i=0}^\infty \beta_{ki} u_{ki}(y) \text{ as } y \to 0+, \tag{4.6.3}$$

where $u_{ki}(y)$ are known functions that satisfy

$$u_{k,i+1}(y) = o(u_{ki}(y)) \text{ as } y \to 0+, \ i = 0, 1, \dots. \tag{4.6.4}$$

The approximations $A_n^{(m,j)}$, where $n = (n_1, \ldots, n_m)$, are now defined by the linear equations

$$A(y_l) = A_n^{(m,j)} + \sum_{k=1}^{m} \phi_k(y_l) \sum_{i=0}^{n_k-1} \bar{\beta}_{ki} u_{ki}(y_l), \;\; j \le l \le j + N; \; N = \sum_{k=1}^{m} n_k. \quad (4.6.5)$$

These extensions of GREP were originally suggested by Sidi [287]. It would be reasonable to expect such extensions of GREP to be effective if and when needed.

Note that the Richardson extrapolation process of Chapter 1 is an extended GREP$^{(1)}$ with $\phi_1(y) = 1$ and $\tau_{1i} = \sigma_{i+1}$, $i = 0, 1, \ldots$, in (4.6.1).

Another example of such transformations is one in which

$$u_{ki}(y) \sim \sum_{s=i}^{\infty} \delta_{kis} y^{sr_k} \;\; \text{as } y \to 0+, \;\; i = 0, 1, \ldots, \quad (4.6.6)$$

with r_k as before. An interesting special case of this is $u_{ki}(y) = 1/(y^{-r_k})_i$, where $(z)_i = z(z+1)\cdots(z+i-1)$ is the Pochhammer symbol. (Note that if $\beta(y) \sim \sum_{i=0}^{\infty} \beta_{ki} y^{ir_k}$ as $y \to 0+$, then $\beta(y) \sim \sum_{i=0}^{\infty} \beta'_{ki}/(y^{-r_k})_i$ as $y \to 0+$ as well.) Such an extrapolation method with $m = 1$ was proposed earlier by Sidi and has been called the $\mathcal{S}$-transformation. This method turns out to be very effective in the summation of strongly divergent series. We come back to it later.

5

The D-Transformation: A GREP for Infinite-Range Integrals

5.1 The Class $\mathbf{B}^{(m)}$ and Related Asymptotic Expansions

In many applications, we may need to compute numerically infinite-range integrals of the form

$$I[f] = \int_0^\infty f(t)\,dt. \tag{5.1.1}$$

A direct way to achieve this is by truncating the infinite range and taking the (numerically computed) finite-range integral

$$F(x) = \int_0^x f(t)\,dt \tag{5.1.2}$$

for some sufficiently large x as an approximation to $I[f]$. In many cases, however, $f(x)$ decays very slowly as $x \to \infty$, and this causes $F(x)$ to converge to $I[f]$ very slowly as $x \to \infty$. The decay of $f(x)$ may be so slow that we hardly notice the convergence of $F(x)$ numerically. Even worse is the case in which $f(x)$ decays slowly and oscillates an infinite number of times as $x \to \infty$. Thus, this approach of approximating $I[f]$ by $F(x)$ for some large x is of limited use at best.

Commonly occurring examples of infinite-range integrals are Fourier cosine and sine transforms $\int_0^\infty \cos \omega t\, g(t)\,dt$ and $\int_0^\infty \sin \omega t\, g(t)\,dt$ and Hankel transforms $\int_0^\infty J_\nu(\omega t) g(t)\,dt$, where $J_\nu(x)$ is the Bessel function of the first kind of order ν. Now the kernels $\cos x$, $\sin x$, and $J_\nu(x)$ of these transforms satisfy linear homogeneous ordinary differential equations (of order 2) whose coefficients have asymptotic expansions in x^{-1} for $x \to \infty$. In many cases, the functions $g(x)$ also satisfy differential equations of the same nature, and this puts the integrands $\cos \omega t\, g(t)$, $\sin \omega t\, g(t)$, and $J_\nu(\omega t) g(t)$ in some function classes that we denote $\mathbf{B}^{(m)}$, where $m = 1, 2, \ldots$. The precise description of $\mathbf{B}^{(m)}$ is given in Definition 5.1.2.

As we show later, when the integrand $f(x)$ is in $\mathbf{B}^{(m)}$ for some m, $F(x)$ is analogous to a function $A(y)$ in $\mathbf{F}^{(m)}$ with the same m, where $y = x^{-1}$. Consequently, $\mathrm{GREP}^{(m)}$ can be applied to obtain good approximations to $A \leftrightarrow I[f]$ at a small cost. The resulting $\mathrm{GREP}^{(m)}$ for this case is now called the $D^{(m)}$-transformation.

Note that, in case the integral to be computed is $\int_a^\infty f(t)\,dt$ with $a \neq 0$, we apply the $D^{(m)}$-transformation to $\int_0^\infty \tilde{f}(t)\,dt$, where $\tilde{f}(x) = f(a + x)$.

95

5.1.1 Description of the Class $\mathbf{A}^{(\gamma)}$

Before we embark on the definition of the class $\mathbf{B}^{(m)}$, we need to define another important class of functions, which we denote $\mathbf{A}^{(\gamma)}$, that will serve us throughout this chapter and the rest of this work.

Definition 5.1.1 A function $\alpha(x)$ belongs to the set $\mathbf{A}^{(\gamma)}$ if it is infinitely differentiable for all large $x > 0$ and has a Poincaré-type asymptotic expansion of the form

$$\alpha(x) \sim \sum_{i=0}^{\infty} \alpha_i x^{\gamma-i} \quad \text{as } x \to \infty, \tag{5.1.3}$$

and its derivatives have Poincaré-type asymptotic expansions obtained by differentiating that in (5.1.3) formally term by term. If, in addition, $\alpha_0 \neq 0$ in (5.1.3), then $\alpha(x)$ is said to belong to $\mathbf{A}^{(\gamma)}$ strictly. Here γ is complex in general.

Remarks.

1. $\mathbf{A}^{(\gamma)} \supset \mathbf{A}^{(\gamma-1)} \supset \mathbf{A}^{(\gamma-2)} \supset \cdots$, so that if $\alpha \in \mathbf{A}^{(\gamma)}$, then, for any positive integer k, $\alpha \in \mathbf{A}^{(\gamma+k)}$ but not strictly. Conversely, if $\alpha \in \mathbf{A}^{(\delta)}$ but not strictly, then $\alpha \in \mathbf{A}^{(\delta-k)}$ strictly for a unique positive integer k.
2. If $\alpha \in \mathbf{A}^{(\gamma)}$ strictly, then $\alpha \notin \mathbf{A}^{(\gamma-1)}$.
3. If $\alpha \in \mathbf{A}^{(\gamma)}$ strictly, and $\beta(x) = \alpha(cx + d)$ for some arbitrary constants $c > 0$ and d, then $\beta \in \mathbf{A}^{(\gamma)}$ strictly as well.
4. If $\alpha, \beta \in \mathbf{A}^{(\gamma)}$, then $\alpha \pm \beta \in \mathbf{A}^{(\gamma)}$ as well. (This implies that the zero function is included in $\mathbf{A}^{(\gamma)}$.) If $\alpha \in \mathbf{A}^{(\gamma)}$ and $\beta \in \mathbf{A}^{(\gamma+k)}$ strictly for some positive integer k, then $\alpha \pm \beta \in \mathbf{A}^{(\gamma+k)}$ strictly.
5. If $\alpha \in \mathbf{A}^{(\gamma)}$ and $\beta \in \mathbf{A}^{(\delta)}$, then $\alpha\beta \in \mathbf{A}^{(\gamma+\delta)}$; if, in addition, $\beta \in \mathbf{A}^{(\delta)}$ strictly, then $\alpha/\beta \in \mathbf{A}^{(\gamma-\delta)}$.
6. If $\alpha \in \mathbf{A}^{(\gamma)}$ strictly, such that $\alpha(x) > 0$ for all large x, and we define $\theta(x) = [\alpha(x)]^{\xi}$, then $\theta \in \mathbf{A}^{(\gamma\xi)}$ strictly.
7. If $\alpha \in \mathbf{A}^{(\gamma)}$ strictly and $\beta \in \mathbf{A}^{(k)}$ strictly for some positive integer k, such that $\beta(x) > 0$ for all large $x > 0$, and we define $\theta(x) = \alpha(\beta(x))$, then $\theta \in \mathbf{A}^{(k\gamma)}$ strictly. Similarly, if $\mu(x^{-1}) \in \mathbf{A}^{(-\delta)}$ strictly so that $\mu(t) \sim \sum_{i=0}^{\infty} \mu_i t^{\delta+i}$ as $t \to 0+$, $\mu_0 \neq 0$, and if $\beta \in \mathbf{A}^{(-k)}$ strictly for some positive integer k, such that $\beta(x) > 0$ for all large $x > 0$, and we define $\psi(x) = \mu(\beta(x))$, then $\psi \in \mathbf{A}^{(-k\delta)}$ strictly.
8. If $\alpha \in \mathbf{A}^{(\gamma)}$ (strictly) and $\gamma \neq 0$, then $\alpha' \in \mathbf{A}^{(\gamma-1)}$ (strictly). If $\alpha \in \mathbf{A}^{(0)}$, then $\alpha' \in \mathbf{A}^{(-2)}$.
9. Let $\alpha(x)$ be in $\mathbf{A}^{(\gamma)}$ and satisfy (5.1.3). If we define the function $\hat{\alpha}(y)$ by $\hat{\alpha}(y) = y^{\gamma}\alpha(y^{-1})$, then $\hat{\alpha}(y)$ is infinitely differentiable for $0 \leq y < c$ for some $c > 0$, and $\hat{\alpha}^{(i)}(0)/i! = \alpha_i$, $i = 0, 1, \ldots$. In other words, the series $\sum_{i=0}^{\infty} \alpha_i y^i$, whether convergent or not, is the Maclaurin series of $\hat{\alpha}(y)$. The next two remarks are immediate consequences of this.
10. If $\alpha \in \mathbf{A}^{(0)}$, then it is infinitely differentiable for all large $x > 0$ up to and including $x = \infty$, although it is not necessarily analytic at $x = \infty$.

11. If $x^{-\gamma}\alpha(x)$ is infinitely differentiable for all large $x > 0$ up to and including $x = \infty$, and thus has an infinite Taylor series expansion in powers of x^{-1}, then $\alpha \in \mathbf{A}^{(\gamma)}$. This is true whether the Taylor series converges or not. Furthermore, the asymptotic expansion of $\alpha(x)$ as $x \to \infty$ is x^{γ} times the Taylor series.

All these are simple consequences of Definition 5.1.1; we leave their verification to the reader.

Here are a few simple examples of functions in $\mathbf{A}^{(\gamma)}$ for some values of γ:

- The function $\sqrt{x^3 + x}$ is in $\mathbf{A}^{(3/2)}$ since it has the (convergent) expansion

$$\sqrt{x^3 + x} = x^{3/2} \sum_{i=0}^{\infty} \binom{1/2}{i} \frac{1}{x^{2i}}, \quad x > 1,$$

that is also its asymptotic expansion as $x \to \infty$.
- The function $\sin(1/\sqrt{x})$ is in $\mathbf{A}^{(-1/2)}$ since it has the (convergent) expansion

$$\sin\left(\frac{1}{\sqrt{x}}\right) = x^{-1/2} \sum_{i=0}^{\infty} \frac{(-1)^i}{(2i+1)!} \frac{1}{x^i}, \quad \text{all } x > 0,$$

that is also its asymptotic expansion as $x \to \infty$.
- Any rational function $R(x)$ whose numerator and denominator polynomials have degrees exactly m and n, respectively, is in $\mathbf{A}^{(m-n)}$ strictly. In addition, its asymptotic expansion is convergent for $x > a$ with some $a > 0$.
- If $\alpha(x) = \int_0^\infty e^{-xt} t^{-\gamma-1} g(t)\, dt$, where $\Re\gamma < 0$ and $g(t)$ is in $C^\infty[0, \infty)$ and satisfies $g(t) \sim \sum_{i=0}^\infty g_i t^i$ as $t \to 0+$, with $g_0 = g(0) \neq 0$, and $g(t) = O(e^{ct})$ as $t \to \infty$, for some constant c, then $\alpha(x)$ is in $\mathbf{A}^{(\gamma)}$ strictly. This can be shown by using Watson's lemma, which gives the asymptotic expansion

$$\alpha(x) \sim \sum_{i=0}^{\infty} g_i \Gamma(-\gamma + i) x^{\gamma - i} \quad \text{as } x \to \infty,$$

and the fact that $\alpha'(x) = -\int_0^\infty e^{-xt} t^{-\gamma} g(t)\, dt$, a known property of Laplace transforms. Here $\Gamma(z)$ is the Gamma function, as usual. Let us take $\gamma = -1$ and $g(t) = 1/(1+t)$ as an example. Then $\alpha(x) \sim x^{-1} \sum_{i=0}^\infty (-1)^i i!\, x^{-i}$ as $x \to \infty$. Note that, for this $g(t)$, $\alpha(x) = e^x E_1(x)$, where $E_1(x) = \int_x^\infty t^{-1} e^{-t}\, dt$ is the *exponential integral*, and the asymptotic expansion of $\alpha(x)$ is a divergent series for all $x \neq \infty$.
- The function $e^x K_0(x)$, where $K_0(x)$ is the modified Bessel function of order 0 of the second kind, is in $\mathbf{A}^{(-1/2)}$ strictly. This can be shown by applying the previous result to the integral representation of $K_0(x)$, namely, $K_0(x) = \int_1^\infty e^{-xt}(t^2 - 1)^{-1/2}\, dt$, following a suitable transformation of variable. Indeed, it has the asymptotic expansion

$$e^x K_0(x) \sim \sqrt{\frac{\pi}{2x}} \left\{ 1 - \frac{1}{8x} + \frac{9}{2!\,(8x)^2} + \cdots \right\} \quad \text{as } x \to \infty.$$

Before going on, we would like to note that, by the way $\mathbf{A}^{(\gamma)}$ is defined, there may be any number of functions in $\mathbf{A}^{(\gamma)}$ having the same asymptotic expansion. In certain places it will be convenient to work with subsets $\mathbf{X}^{(\gamma)}$ of $\mathbf{A}^{(\gamma)}$ that are defined for all γ collectively as follows:

(i) A function α belongs to $\mathbf{X}^{(\gamma)}$ if either $\alpha \equiv 0$ or $\alpha \in \mathbf{A}^{(\gamma-k)}$ strictly for some non-negative integer k.

(ii) $\mathbf{X}^{(\gamma)}$ is closed under addition and multiplication by scalars.

(iii) If $\alpha \in \mathbf{X}^{(\gamma)}$ and $\beta \in \mathbf{X}^{(\delta)}$, then $\alpha\beta \in \mathbf{X}^{(\gamma+\delta)}$; if, in addition, $\beta \in \mathbf{A}^{(\delta)}$ strictly, then $\alpha/\beta \in \mathbf{X}^{(\gamma-\delta)}$.

(iv) If $\alpha \in \mathbf{X}^{(\gamma)}$, then $\alpha' \in \mathbf{X}^{(\gamma-1)}$.

It is obvious that no two functions in $\mathbf{X}^{(\gamma)}$ have the same asymptotic expansion, since if $\alpha, \beta \in \mathbf{X}^{(\gamma)}$, then either $\alpha \equiv \beta$ or $\alpha - \beta \in \mathbf{A}^{(\gamma-k)}$ strictly for some nonnegative integer k. Thus, $\mathbf{X}^{(\gamma)}$ does not contain functions $\alpha(x) \not\equiv 0$ that satisfy $\alpha(x) = O(x^{-\mu})$ as $x \to \infty$ for every $\mu > 0$, such as $\exp(-cx^s)$ with $c, s > 0$.

Functions $\alpha(x)$ that are given as sums of series $\sum_{i=0}^{\infty} \alpha_i x^{\gamma-i}$ that converge for all large x form such a subset; obviously, such functions are of the form $\alpha(x) = x^\gamma R(x)$ with $R(x)$ analytic at infinity. Thus, $R(x)$ can be rational functions that are bounded at infinity, for example.

(Concerning the uniqueness of $\alpha(x)$, see the last paragraph of Section A.2 of Appendix A.)

5.1.2 Description of the Class $\mathbf{B}^{(m)}$

Definition 5.1.2 A function $f(x)$ that is infinitely differentiable for all large x belongs to the set $\mathbf{B}^{(m)}$ if it satisfies a linear homogeneous ordinary differential equation of order m of the form

$$f(x) = \sum_{k=1}^{m} p_k(x) f^{(k)}(x), \qquad (5.1.4)$$

where $p_k \in \mathbf{A}^{(k)}$, $k = 1, \ldots, m$, such that $p_k \in \mathbf{A}^{(i_k)}$ strictly for some integer $i_k \leq k$.

The following simple result is a consequence of this definition.

Proposition 5.1.3 *If $f \in \mathbf{B}^{(m)}$, then $f \in \mathbf{B}^{(m')}$ for every $m' > m$.*

Proof. It is enough to consider the case $m' = m + 1$. Let (5.1.4) be the ordinary differential equation satisfied by $f(x)$. Applying to both sides of (5.1.4) the differential operator $[1 + \mu(x)d/dx]$, where $\mu(x)$ is an arbitrary function in $\mathbf{A}^{(1)}$, we have $f(x) = \sum_{k=1}^{m+1} q_k(x) f^{(k)}(x)$ with $q_1 = p_1 + \mu p_1' - \mu$, $q_k = p_k + \mu p_k' + \mu p_{k-1}$, $k = 2, \ldots, m$, and $q_{m+1} = \mu p_m$. From the fact that $\mu \in \mathbf{A}^{(1)}$ and $p_k \in \mathbf{A}^{(k)}$, $k = 1, \ldots, m$, it follows that $q_k \in \mathbf{A}^{(k)}$, $k = 1, \ldots, m+1$. ∎

We observe from the proof of Proposition 5.1.3 that if $f \in \mathbf{B}^{(m)}$, then, for any $m' > m$, there are infinitely many differential equations of the form $f = \sum_{k=1}^{m'} q_k f^{(k)}$ with $q_k \in \mathbf{A}^{(k)}$. An interesting question concerning the situation in which $f \in \mathbf{B}^{(m)}$ with *minimal* m is whether the differential equation $f = \sum_{k=1}^{m} p_k f^{(k)}$ with $p_k \in \mathbf{A}^{(k)}$ is unique. By restricting the p_k such that $p_k \in \mathbf{X}^{(k)}$ with $\mathbf{X}^{(\gamma)}$ as defined above, we can actually show that it is; this is the subject of Proposition 5.1.5 below. In general, we assume that this differential equation is unique for minimal m, and we invoke this assumption later in Theorem 5.6.4.

Knowing the minimal m is important for computational economy when using the D-transformation since the cost the latter increases with increasing m, as will become clear shortly.

We start with the following auxiliary result.

Proposition 5.1.4 *Let $f(x)$ be infinitely differentiable for all large x and satisfy an ordinary differential equation of order m of the form $f(x) = \sum_{k=1}^{m} p_k(x) f^{(k)}(x)$ with $p_k \in \mathbf{A}^{(\nu_k)}$ for some integers ν_k. If m is smallest possible, then $f^{(k)}(x)$, $k = 0, 1, \dots, m-1$, are independent in the sense that there do not exist functions $\nu_k(x)$, $k = 0, 1, \dots, m-1$, not all identically zero, and $\nu_k \in \mathbf{A}^{(\tau_k)}$ with τ_k integers, such that $\sum_{k=0}^{m-1} \nu_k f^{(k)} = 0$. In addition, $f^{(i)}(x)$, $i = m, m+1, \dots$, can all be expressed in the form $f^{(i)} = \sum_{k=0}^{m-1} w_{ik} f^{(k)}$, where $w_{ik} \in \mathbf{A}^{(\mu_{ik})}$ for some integers μ_{ik}. This applies, in particular, when $f \in \mathbf{B}^{(m)}$.*

Proof. Suppose, to the contrary, that $\sum_{k=0}^{s} \nu_k f^{(k)} = 0$ for some $s \le m-1$ and $\nu_k \in \mathbf{A}^{(\nu_k)}$ with ν_k integers. If $\nu_0 \not\equiv 0$, then we have $f = \sum_{k=1}^{s} \tilde{p}_k f^{(k)}$, where $\tilde{p}_k = -\nu_k/\nu_0 \in \mathbf{A}^{(\nu_k')}$, ν_k' an integer, contradicting the assumption that m is minimal. If $\nu_0 \equiv 0$, differentiating the equality $\sum_{k=1}^{s} \nu_k f^{(k)} = 0$ $m-s$ times, we have $\sum_{k=1}^{m-1} w_k f^{(k)} + \nu_s f^{(m)} = 0$. The functions w_k are obviously in $\mathbf{A}^{(\mu_k)}$ for some integers μ_k. Solving this last equation for $f^{(m)}$, and substituting in $f = \sum_{k=1}^{m} p_k f^{(k)}$, we obtain the differential equation $f = \sum_{k=1}^{m-1} \tilde{p}_k f^{(k)}$, where $\tilde{p}_k = p_k - p_m w_k/\nu_s \in \mathbf{A}^{(\lambda_k)}$ for some integers λ_k. This too contradicts the assumption that m is minimal. We leave the rest of the proof to the reader. ∎

Proposition 5.1.5 *If $f(x)$ satisfies an ordinary differential equation of the form $f(x) = \sum_{k=1}^{m} p_k(x) f^{(k)}(x)$ with $p_k \in \mathbf{X}^{(\nu_k)}$ for some integers ν_k, and if m is smallest possible, then the functions $p_k(x)$ in this differential equation are unique. This applies, in particular, when $f \in \mathbf{B}^{(m)}$.*

Proof. Suppose, to the contrary, that $f(x)$ satisfies also the differential equation $f = \sum_{k=1}^{m} q_k f^{(k)}$ with $q_k \in \mathbf{X}^{(\sigma_k)}$ for some integers σ_k, such that $p_k(x) \not\equiv q_k(x)$ for at least one value of k. Eliminating $f^{(m)}$ from both differential equations, we obtain $\sum_{k=0}^{m-1} \nu_k f^{(k)} = 0$, where $\nu_0 = q_m - p_m$ and $\nu_k = -(p_k q_m - q_k p_m)$, $k = 1, \dots, m-1$, $\nu_k \in \mathbf{X}^{(\lambda_k)}$ for some integers λ_k, and $\nu_k(x) \not\equiv 0$ for at least one value of k. Since m is minimal, this is impossible by Proposition 5.1.4. Therefore, we must have $p_k \equiv q_k$, $1 \le k \le m$. ∎

The following proposition, whose proof we leave to the reader, concerns the derivatives of $f(x)$ when $f \in \mathbf{B}^{(m)}$ in particular.

Proposition 5.1.6 *If $f(x)$ satisfies an ordinary differential equation of the form $f(x) = \sum_{k=1}^{m} p_k(x) f^{(k)}(x)$ with $p_k \in \mathbf{A}^{(\nu_k)}$ for some integers ν_k, then $f'(x)$ satisfies an ordinary* *differential equation of the same form, namely,* $f'(x) = \sum_{k=1}^{m} q_k(x) f^{(k+1)}(x)$ *with* $q_k \in$ $\mathbf{A}^{(\mu_k)}$ *for some integers* μ_k, *provided* $[1 - p_1'(x)] \in \mathbf{A}^{(\tau)}$ *strictly for some integer* τ. *In particular, if* $f \in \mathbf{B}^{(m)}$, *then* $f' \in \mathbf{B}^{(m)}$ *as well, provided* $\lim_{x \to \infty} x^{-1} p_1(x) \neq 1$.

Let us now give a few examples of functions in the classes $\mathbf{B}^{(1)}$ and $\mathbf{B}^{(2)}$. In these examples, we make free use of the remarks following Definition 5.1.1.

Example 5.1.7 The Bessel function of the first kind $J_\nu(x)$ is in $\mathbf{B}^{(2)}$ since it satisfies $y = \frac{x}{\nu^2 - x^2} y' + \frac{x^2}{\nu^2 - x^2} y''$ so that $p_1(x) = x/(\nu^2 - x^2) \in \mathbf{A}^{(-1)}$ and $p_2(x) = x^2/(\nu^2 - x^2) \in \mathbf{A}^{(0)}$. The same applies to the Bessel function of the second kind $Y_\nu(x)$ and to all linear combinations $b J_\nu(x) + c Y_\nu(x)$.

Example 5.1.8 The function $\cos x / x$ is in $\mathbf{B}^{(2)}$ since it satisfies $y = -\frac{2}{x} y' - y''$ so that $p_1(x) = -2/x \in \mathbf{A}^{(-1)}$ and $p_2(x) = -1 \in \mathbf{A}^{(0)}$. The same applies to the function $\sin x / x$ and to all linear combinations $B \cos x / x + C \sin x / x$.

Example 5.1.9 The function $f(x) = \log(1 + x)/(1 + x^2)$ is in $\mathbf{B}^{(2)}$ since it satisfies $y = p_1(x) y' + p_2(x) y''$, where $p_1(x) = -(5x^2 + 4x + 1)/(4x + 2) \in \mathbf{A}^{(1)}$ strictly and $p_2(x) = -(x^2 + 1)(x + 1)/(4x + 2) \in \mathbf{A}^{(2)}$ strictly.

Example 5.1.10 A function $f(x) \in \mathbf{A}^{(\gamma)}$ strictly, for arbitrary $\gamma \neq 0$, is in $\mathbf{B}^{(1)}$ since it satisfies $y = \frac{f(x)}{f'(x)} y'$ so that $p_1(x) = f(x)/f'(x) \in \mathbf{A}^{(1)}$ strictly. That $p_1 \in \mathbf{A}^{(1)}$ strictly follows from the fact that $f' \in \mathbf{A}^{(\gamma-1)}$ strictly.

Example 5.1.11 A function $f(x) = e^{\theta(x)} h(x)$, where $\theta \in \mathbf{A}^{(s)}$ strictly for some positive integer s and $h \in \mathbf{A}^{(\gamma)}$ for arbitrary γ, is in $\mathbf{B}^{(1)}$ since it satisfies $y = p_1(x) y'$ with $p_1(x) = 1/[\theta'(x) + h'(x)/h(x)] \in \mathbf{A}^{(-s+1)}$ and $-s + 1 \leq 0$. That $p_1 \in \mathbf{A}^{(-s+1)}$ can be seen as follows: Now $\theta' \in \mathbf{A}^{(s-1)}$ strictly and $h'/h \in \mathbf{A}^{(-1)} \subset \mathbf{A}^{(s-1)}$ because $s - 1$ is a nonnegative integer. Therefore, $\theta' + h'/h \in \mathbf{A}^{(s-1)}$ strictly. Consequently, $p_1 = (\theta' + h'/h)^{-1} \in \mathbf{A}^{(-s+1)}$ strictly.

5.1.3 Asymptotic Expansion of $F(x)$ When $f(x) \in \mathbf{B}^{(m)}$

We now state a general theorem due to Levin and Sidi [165] concerning the asymptotic behavior of $F(x) = \int_0^x f(t) \, dt$ as $x \to \infty$ when $f \in \mathbf{B}^{(m)}$ for some m and is integrable at infinity. The proof of this theorem requires a good understanding of asymptotic expansions and is quite involved. We refer the reader to Section 5.6 for a detailed constructive proof.

Theorem 5.1.12 *Let* $f(x)$ *be a function in* $\mathbf{B}^{(m)}$ *that is also integrable at infinity. Assume, in addition, that*

$$\lim_{x \to \infty} p_k^{(j-1)}(x) f^{(k-j)}(x) = 0, \quad k = j, j+1, \ldots, m, \quad j = 1, 2, \ldots, m, \quad (5.1.5)$$

and that

$$\sum_{k=1}^{m} l(l-1)\cdots(l-k+1)\bar{p}_k \neq 1, \quad l = \pm 1, 2, 3, \ldots, \tag{5.1.6}$$

where

$$\bar{p}_k = \lim_{x \to \infty} x^{-k} p_k(x), \quad k = 1, \ldots, m. \tag{5.1.7}$$

Then

$$F(x) = I[f] + \sum_{k=0}^{m-1} x^{\rho_k} f^{(k)}(x) g_k(x) \tag{5.1.8}$$

*for some integers $\rho_k \leq k+1$ and functions $g_k \in$ **A**$^{(0)}$, $k = 0, 1, \ldots, m-1$. Actually, if $p_k \in$ **A**$^{(i_k)}$ strictly for some integer $i_k \leq k$, $k = 1, \ldots, m$, then*

$$\rho_k \leq \bar{\rho}_k \equiv \max\{i_{k+1}, i_{k+2} - 1, \ldots, i_m - m + k + 1\} \leq k + 1, \quad k = 0, 1, \ldots, m-1. \tag{5.1.9}$$

*Equality holds in (5.1.9) when the integers whose maximum is being considered are distinct. Finally, being in **A**$^{(0)}$, the functions $g_k(x)$ have asymptotic expansions of the form*

$$g_k(x) \sim \sum_{i=0}^{\infty} g_{ki} x^{-i} \quad as \; x \to \infty. \tag{5.1.10}$$

Remarks.

1. By (5.1.7), $\bar{p}_k \neq 0$ if and only if $p_k \in$ **A**$^{(k)}$ strictly. Thus, whenever $p_k \in$ **A**$^{(i_k)}$ with $i_k < k$, we have $\bar{p}_k = 0$. This implies that whenever $i_k < k$, $k = 1, \ldots, m$, we have $\bar{p}_k = 0$, $k = 1, \ldots, m$, and the condition in (5.1.6) is automatically satisfied as the left-hand side of the inequality there is zero for all values of l.
2. It follows from (5.1.9) that $\rho_{m-1} = i_m$ always.
3. Similarly, for $m = 1$ we have $\rho_0 = i_1$ precisely.
4. For numerous examples we have treated, equality seems to hold in (5.1.9) for all $k = 1, \ldots, m$.
5. The integers ρ_k and the functions $g_k(x)$ in (5.1.8) depend only on the functions $p_k(x)$ in the ordinary differential equation in (5.1.4). This being the case, they are the same for all solutions $f(x)$ of (5.1.4) that are integrable at infinity and that satisfy (5.1.5).
6. From (5.1.5) and (5.1.9), we also have that $\lim_{x \to \infty} x^{\bar{\rho}_k} f^{(k)}(x) = 0$, $k = 0, 1, \ldots, m-1$. Thus, $\lim_{x \to \infty} x^{\rho_k} f^{(k)}(x) = 0$, $k = 0, 1, \ldots, m-1$, as well.
7. Finally, Theorem 5.1.12 says that the function $G(x) \equiv I[f] - F(x) = \int_x^{\infty} f(t)\,dt$ is in **B**$^{(m)}$ if $f \in$ **B**$^{(m)}$ too. This follows from the fact that $G^{(k)}(x) = -f^{(k-1)}(x)$, $k = 1, 2, \ldots$.

By making the analogy $F(x) \leftrightarrow A(y)$, $x^{-1} \leftrightarrow y$, $x^{\rho_{k-1}} f^{(k-1)}(x) \leftrightarrow \phi_k(y)$ and $r_k = 1$, $k = 1, \ldots, m$, and $I[f] \leftrightarrow A$, we realize that $A(y)$ is in **F**$^{(m)}$. Actually, $A(y)$ is even in $\mathbf{F}_\infty^{(m)}$ because of the differentiability conditions imposed on $f(x)$ and the $p_k(x)$. Finally, the variable y is continuous for this case.

All the conditions of Theorem 5.1.12 are satisfied by Examples 5.1.7–5.1.9. They are satisfied by Example 5.1.10, provided $\Re\gamma < -1$ so that $f(x)$ becomes integrable at infinity. Similarly, they are satisfied by Example 5.1.11 provided $\lim_{x\to\infty}\Re\theta(x) = -\infty$ so that $f(x)$ becomes integrable at infinity. We leave the verification of these claims to the reader.

The numerous examples we have studied seem to indicate that the requirement that $f(x)$ be in $\mathbf{B}^{(m)}$ for some m is the most crucial of the conditions in Theorem 5.1.12. The rest of the conditions, namely, (5.1.5)–(5.1.7), seem to be satisfied automatically. Therefore, to decide whether $A(y) \equiv F(x)$, where $y = x^{-1}$, is in $\mathbf{F}^{(m)}$ for some m, it is practically sufficient to check whether $f(x)$ is in $\mathbf{B}^{(m)}$. Later in this chapter, we provide some simple ways to check this point.

Finally, even though Theorem 5.1.12 is stated for functions $f \in \mathbf{B}^{(m)}$ that are integrable at infinity, $F(x)$ may satisfy (5.1.8)–(5.1.10) also when $f \in \mathbf{B}^{(m)}$ but is not integrable at infinity, at least in some cases. In such a case, the constant $I[f]$ in (5.1.8) will be the antilimit of $F(x)$ as $x \to \infty$. In Theorem 5.7.3 at the end of this chapter, we show that (5.1.8)–(5.1.10) hold (i) for *all* functions $f(x)$ in $\mathbf{B}^{(1)}$ that are integrable at infinity and (ii) for a large subset of functions in $\mathbf{B}^{(1)}$ that are not integrable there but grow at most like a power of x as $x \to \infty$.

We now demonstrate the result of Theorem 5.1.12 with the two functions $f(x) = J_0(x)$ and $f(x) = \sin x/x$ that were shown to be in $\mathbf{B}^{(2)}$ in Examples 5.1.7 and 5.1.8, respectively.

Example 5.1.13 Let $f(x) = J_0(x)$. From Longman [172], we have the asymptotic expansion

$$F(x) \sim I[f] - J_0(x)\sum_{i=0}^{\infty}(-1)^i\frac{[(2i+1)!!]^2}{2i+1}\frac{1}{x^{2i+1}}$$

$$+ J_1(x)\sum_{i=0}^{\infty}(-1)^i\left[\frac{(2i+1)!!}{2i+1}\right]^2\frac{1}{x^{2i}} \quad \text{as } x \to \infty, \qquad (5.1.11)$$

completely in accordance with Theorem 5.1.12, since $J_1(x) = -J_0'(x)$.

Example 5.1.14 Let $f(x) = \sin x/x$. We already have an asymptotic expansion for $F(x)$ that is given in (4.1.7) in Example 4.1.5. Rearranging this expansion, we also have

$$F(x) \sim I[f] - \frac{\sin x}{x}\sum_{i=0}^{\infty}(-1)^i\frac{(2i)!(2i+2)}{x^{2i+1}}$$

$$- \left(\frac{\sin x}{x}\right)'\sum_{i=0}^{\infty}(-1)^i\frac{(2i)!}{x^{2i}} \quad \text{as } x \to \infty, \qquad (5.1.12)$$

completely in accordance with Theorem 5.1.12.

5.1.4 Remarks on the Asymptotic Expansion of $F(x)$ and a Simplification

An interesting and potentially useful feature of Theorem 5.1.12 is the *simplicity* of the asymptotic expansion of $F(x)$ given in (5.1.8)–(5.1.10). This is noteworthy especially because the function $f(x)$ may itself have a very complicated asymptotic behavior as $x \to \infty$, and we would normally expect this behavior to show in that of $F(x)$ explicitly. For instance, $f(x)$ may be oscillatory or monotonic or it may be a combination of oscillatory and monotonic functions. It is clear that the asymptotic expansion of $F(x)$, as given in (5.1.8)–(5.1.10), contains no reference to any of this; it is expressed solely in terms of $f(x)$ and the first $m-1$ derivatives of it.

We noted earlier that $F(x)$ is analogous to a function $A(y)$ in $\mathbf{F}^{(m)}$, with $\phi_k(y) \leftrightarrow x^{\rho_k-1} f^{(k-1)}(x)$, $k = 1, \ldots, m$. Thus, it seems that knowledge of the integers ρ_k may be necessary to proceed with $\mathrm{GREP}^{(m)}$ to find approximations to $I[f]$. Now the ρ_k are determined by the integers i_k, and the i_k are given by the differential equation that $f(x)$ satisfies. As we do not expect, nor do we intend, to know what this differential equation is in general, determining the ρ_k exactly may not be possible in all cases. This could cause us to conclude that the $\phi_k(y)$ may not be known completely, and hence that $\mathrm{GREP}^{(m)}$ may not be applicable for approximating $I[f]$. Really, we do *not* need to know the ρ_k exactly; we can replace each ρ_k by its known upper bound $k+1$, and rewrite (5.1.8) in the form

$$F(x) = I[f] + \sum_{k=0}^{m-1} x^{k+1} f^{(k)}(x) h_k(x), \qquad (5.1.13)$$

where $h_k(x) = x^{\rho_k-k-1} g_k(x)$, hence $h_k \in \mathbf{A}^{(\rho_k-k-1)} \subseteq \mathbf{A}^{(0)}$ for each k. Note that, if $\rho_k = k+1$, then $h_k(x) = g_k(x)$, while if $\rho_k < k+1$, then from (5.1.10) we have

$$h_k(x) \sim \sum_{i=0}^{\infty} h_{ki} x^{-i} \equiv 0 \cdot x^0 + 0 \cdot x^{-1} + \cdots + 0 \cdot x^{\rho_k-k}$$

$$+ g_{k0} x^{\rho_k-k-1} + g_{k1} x^{\rho_k-k-2} + \cdots \text{ as } x \to \infty. \quad (5.1.14)$$

Now that we have established the validity of (5.1.13) with $h_k \in \mathbf{A}^{(0)}$ for each k, we have also determined a set of simple and readily known form factors (or shape functions) $\phi_k(y)$, namely, $\phi_k(y) \leftrightarrow x^k f^{(k-1)}(x)$, $k = 1, \ldots, m$.

5.2 Definition of the $D^{(m)}$-Transformation

We have seen that the functions $h_k(x)$ in (5.1.13) are all in $\mathbf{A}^{(0)}$. Reexpanding them in negative powers of $x + \alpha$ for some fixed α, which is legitimate, we obtain the following asymptotic expansion for $F(x)$:

$$F(x) \sim I[f] + \sum_{k=0}^{m-1} x^{k+1} f^{(k)}(x) \sum_{i=0}^{\infty} \frac{\tilde{h}_{ki}}{(x+\alpha)^i} \text{ as } x \to \infty.$$

Based on this asymptotic expansion of $F(x)$, we now define the Levin–Sidi $D^{(m)}$-transformation for approximating the infinite-range integral $I[f] = \int_0^\infty f(t)\, dt$. As mentioned earlier, the $D^{(m)}$-transformation is a $\mathrm{GREP}^{(m)}$.

Definition 5.2.1 Pick an increasing positive sequence $\{x_l\}$ such that $\lim_{l\to\infty} x_l = \infty$. Let $n \equiv (n_1, n_2, \ldots, n_m)$, where $n_1, \ldots, n_m$ are nonnegative integers. Then the approximation $D_n^{(m,j)}$ to $I[f]$ is defined through the linear system

$$F(x_l) = D_n^{(m,j)} + \sum_{k=1}^{m} x_l^k f^{(k-1)}(x_l) \sum_{i=0}^{n_k-1} \frac{\bar{\beta}_{ki}}{(x_l + \alpha)^i}, \quad j \le l \le j + N; \ N = \sum_{k=1}^{m} n_k,$$

(5.2.1)

$\alpha > -x_0$ being a parameter at our disposal and $\bar{\beta}_{ki}$ being the additional (auxiliary) N unknowns. In (5.2.1), $\sum_{i=0}^{-1} c_i \equiv 0$ so that $D_{(0,\ldots,0)}^{(m,j)} = F(x_j)$ for all j. We call this GREP that generates the $D_n^{(m,j)}$ the $D^{(m)}$-transformation. When there is no room for confusion, we call it the D-transformation for short. [Of course, in case the $\bar{\rho}_k$ are known, the factors $x_l^k f^{(k-1)}(x_l)$ in (5.2.1) can be replaced by $x_l^{\bar{\rho}_{k-1}} f^{(k-1)}(x_l)$.]

We mention at this point that the P-transformation of Levin [162] for the Bromwich integral (see Appendix B) is a $D^{(1)}$-transformation with $x_l = l + 1, l = 0, 1, \ldots$.

Remarks.

1. A good choice of the parameter α appears to be $\alpha = 0$. We adopt this choice in the sequel. We have adopted it in all our numerical experiments as well.
2. We observe from (5.2.1) that the input needed for the $D^{(m)}$-transformation is the integer m, the function $f(x)$ and its first $m-1$ derivatives, and the finite integrals $F(x_l) = \int_0^{x_l} f(t)\,dt, \ l = 0, 1, \ldots$. We dwell on how to determine m later in this chapter. Suffice it to say that there is no harm done if m is overestimated. However, acceleration of convergence cannot be expected in every case if m is underestimated. As for the derivatives of $f(x)$, we may assume that they can somehow be obtained analytically if $f(x)$ is given analytically as well. In difficult cases, this may be achieved by symbolic computation, for example. In case m is small, we can compute the derivatives $f^{(i)}(x), \ i = 1, \ldots, m-1$, also numerically with sufficient accuracy. We can use the polynomial Richardson extrapolation process of Chapter 2 very effectively for this purpose, as discussed in Section 2.3. Finally, $F(x_l)$ can be evaluated in the form $F(x_l) = \sum_{i=0}^{l} \nabla F_i$, where $\nabla F_i = \int_{x_{i-1}}^{x_i} f(t)\,dt$, with $x_{-1} = 0$. The finite-range integrals ∇F_i can be computed numerically to very high accuracy by using, for example, a low-order Gaussian quadrature formula.
3. Because we have the freedom to choose the x_l as we wish, we can make choices that induce better convergence acceleration to $I[f]$ and/or better numerical stability. This is one of the important advantages of the D-transformation over other existing methods that we discuss later in this book.
4. The way the $D^{(m)}$-transformation is defined depends only on the integrand $f(x)$ and is totally independent of whether or not $f(x)$ is in $\mathbf{B}^{(m)}$ and/or satisfies Theorem 5.1.12. This implies that the D-transformation can be applied to any integral

$I[f]$, whether $f(x)$ is in $\mathbf{B}^{(m)}$ or not. Whether this application is successful depends on the asymptotic behavior of $f(x)$. If $f(x) \in \mathbf{B}^{(m)}$ for some m, then the $D^{(m)}$-transformation will produce good results. It may produce good results with some m even when $f(x) \in \mathbf{B}^{(q)}$ for some $q > m$ but $f(x) \notin \mathbf{B}^{(m)}$, at least in some cases of interest.

5. Finally, the definition of $D_n^{(m,j)}$ via the linear system in (5.2.1) may look very complicated at first and may create the impression that the $D^{(m)}$-transformation is difficult and/or expensive to apply. This impression is wrong, however. The $D^{(m)}$-transformation can be implemented very efficiently by the W-algorithm (see Section 7.2) when $m = 1$ and by the $W^{(m)}$-algorithm (see Section 7.3) when $m \geq 2$.

5.2.1 Kernel of the $D^{(m)}$-Transformation

From Definition 5.2.1, it is clear that the kernel of the $D^{(m)}$-transformation (with $\alpha = 0$) is all integrals $F(x) = \int_0^x f(t)\,dt$, such that

$$\int_x^\infty f(t)\,dt = \sum_{k=1}^m f^{(k-1)}(x) \sum_{i=0}^{n_k-1} \alpha_{ki} x^{k-i}, \quad \text{some finite } n_k. \qquad (5.2.2)$$

For these integrals there holds $D_n^{(m,j)} = I[f]$ for all j, when $n = (n_1, \ldots, n_m)$. From the proof of Theorem 5.1.12 in Section 5.6, it can be seen that (5.2.2) holds, for example, when $f(x)$ satisfies $f(x) = \sum_{k=1}^m p_k(x) f^{(k)}(x)$, $p_k(x)$ being a polynomial of degree at most k for each k. In this case, (5.2.2) holds with $n_k - 1 = k$.

Another interesting example of (5.2.2) occurs when $f(x) = J_{2k+1}(x)$, $k = 0, 1, \ldots$. Here, $J_\nu(x)$ is the Bessel function of the first kind of order ν, as before. For instance, when $k = 1$, we have

$$\int_x^\infty J_3(t)\,dt = J_3(x)\left(\frac{1}{x} + \frac{24}{x^3}\right) + J_3'(x)\left(1 + \frac{8}{x^2}\right).$$

For details, see Levin and Sidi [165].

5.3 A Simplification of the $D^{(m)}$-Transformation: The $sD^{(m)}$-Transformation

In Section 4.3, we mentioned that the form factors $\phi_k(y)$ that accompany GREP are not unique and that they can be replaced by some other functions $\tilde{\phi}_k(y)$, at the same time preserving the *form* of the expansion in (4.1.1) and (4.1.2). This observation enables us to simplify the D-transformation considerably in some cases of practical interest, such as Fourier cosine and sine transforms and Hankel transforms that were mentioned in Section 5.1.

Let us assume that the integrand $f(x)$ can be written as $f(x) = u(x) Q(x)$, where $Q(x)$ is a simple function and $u(x)$ is strictly in $\mathbf{A}^{(\gamma)}$ for some γ and may be complicated. By $Q(x)$ being simple we actually mean that its derivatives are easier to obtain than those of $f(x)$ itself. Then, from (5.1.13) we have

$$F(x) - I[f] = \sum_{s=0}^{m-1} x^{\rho_s} f^{(s)}(x) h_s(x)$$

$$= \sum_{s=0}^{m-1} x^{\rho_s} \left[\sum_{k=0}^{s} \binom{s}{k} u^{(s-k)}(x) Q^{(k)}(x) \right] h_s(x)$$

$$= \sum_{k=0}^{m-1} \left[\sum_{s=k}^{m-1} \binom{s}{k} u^{(s-k)}(x) h_s(x) x^{\rho_s} \right] Q^{(k)}(x). \qquad (5.3.1)$$

Because $h_s \in \mathbf{A}^{(0)}$ for all s and $u^{(s-k)} \in \mathbf{A}^{(\gamma-s+k)}$, we see that $u^{(s-k)}(x) h_s(x) x^{\rho_s} \in \mathbf{A}^{(\gamma+k+\rho_s-s)}$ so that the term inside the brackets that multiplies $Q^{(k)}(x)$ is in $\mathbf{A}^{(\gamma+\rho'_k)}$, where ρ'_k, just as ρ_k, are integers satisfying $\rho'_k \leq \bar{\rho}_k$ and hence $\rho'_k \leq k+1$ for each k, as can be shown by invoking (5.1.9). By the fact that $u \in \mathbf{A}^{(\gamma)}$ strictly, this term can be written as $x^{\rho'_k} u(x) \tilde{h}_k(x)$ for some $\tilde{h}_k \in \mathbf{A}^{(0)}$. We have thus obtained

$$F(x) = I[f] + \sum_{k=0}^{m-1} x^{\rho'_k} [u(x) Q^{(k)}(x)] \tilde{h}_k(x). \qquad (5.3.2)$$

Consequently, the definition of the $D^{(m)}$-transformation can be modified by replacing the form factors $x_l^k f^{(k-1)}(x_l)$ in (5.2.1) by $x_l^k u(x_l) Q^{(k-1)}(x_l)$ [or by $x_l^{\bar{\rho}_{k-1}} u(x_l) Q^{(k-1)}(x_l)$ when the $\bar{\rho}_k$ are known]; everything else stays the same.

For this simplified $D^{(m)}$-transformation, we do not need to compute derivatives of $f(x)$; we need only those of $Q(x)$, which are easier to obtain. We denote this new version of the $D^{(m)}$-transformation the $sD^{(m)}$-transformation.

We use this approach later to compute Fourier and Hankel transforms, for which we derive further simplifications and modifications of the D-transformation.

5.4 How to Determine m

As mentioned earlier, to apply the $D^{(m)}$-transformation to a given integral $\int_0^\infty f(t)\, dt$, we need to have a value for the integer m, for which $f \in \mathbf{B}^{(m)}$. In this section, we deal with the question of how to determine the smallest value of m or an upper bound for it in a simple manner.

5.4.1 By Trial and Error

The simplest approach to this problem is trial and error. We start by applying the $D^{(1)}$-transformation. If this is successful, then we accept $m = 1$ and stop. If not, we apply the $D^{(2)}$-transformation. If this is successful, then we accept $m = 2$ and stop. Otherwise, we try the $D^{(3)}$-transformation, and so on. Our hope is that for some (smallest) value of m the $D^{(m)}$-transformation will perform well. Of course, this will be the case when $f \in \mathbf{B}^{(m)}$ for this m. We also note that the $D^{(m)}$-transformation may perform well for some m even when $f \notin \mathbf{B}^{(s)}$ for any s, at least for some cases, as was mentioned earlier.

The trial-and-error approach has almost no extra cost involved as the integrals $F(x_l)$ that we need as input need to be computed once only, and they can be used again for each value of m. In addition, the computational effort due to implementing the $D^{(m)}$-transformation for different values of m is small when the W- and $W^{(m)}$-algorithms of Chapter 7 are used for this purpose.

5.4.2 Upper Bounds on m

We now use a heuristic approach that allows us to determine by inspection of f some values for the integer m for which $f \in \mathbf{B}^{(m)}$. Only this time, we relax somewhat the conditions in Definition 5.1.2: we assume that $f(x)$ satisfies a differential equation of the form (5.1.4) with $p_k \in \mathbf{A}^{(v_k)}$, where v_k is an integer not necessarily less than or equal to k, $k = 1, 2, \ldots, m$. [Our experience shows that, if f belongs to this new set $\mathbf{B}^{(m)}$ and is integrable at infinity, then it belongs to the set $\mathbf{B}^{(m)}$ of Definition 5.1.2 as well.] The idea in the present approach is that the integrand $f(x)$ is viewed as a product or as a sum of simpler functions, and it is assumed that we know to which classes $\mathbf{B}^{(s)}$ these simpler functions belong.

The following results, which we denote Heuristics 5.4.1–5.4.3, pertain precisely to this subject. The demonstrations of these results are based on the assumption that certain linear systems, whose entries are in $\mathbf{A}^{(\gamma)}$ for various integer values of γ, are invertible. For this reason, we call these results heuristics, and not lemmas or theorems, and refer to their demonstrations as "proofs".

Heuristic 5.4.1 *Let $g \in \mathbf{B}^{(r)}$ and $h \in \mathbf{B}^{(s)}$ and assume that g and h satisfy different ordinary differential equations of the form described in Definition 5.1.2. Then*

(i) $gh \in \mathbf{B}^{(m)}$ *with $m \le rs$, and*
(ii) $g + h \in \mathbf{B}^{(m)}$ *with $m \le r + s$.*

"Proof". Let g and h satisfy

$$g = \sum_{k=1}^{r} u_k g^{(k)} \quad \text{and} \quad h = \sum_{k=1}^{s} v_k h^{(k)}. \tag{5.4.1}$$

To prove part (i), we need to show that gh satisfies an ordinary differential equation of the form

$$gh = \sum_{k=1}^{rs} p_k (gh)^{(k)}, \tag{5.4.2}$$

where $p_k \in \mathbf{A}^{(v_k)}$, $k = 1, \ldots, rs$, for some integers v_k, and recall that, by Proposition 5.1.3, the actual order of the ordinary differential equation satisfied by gh may possibly be less than rs too.

Let us multiply the two equations in (5.4.1). We obtain

$$gh = \sum_{k=1}^{r} \sum_{l=1}^{s} u_k v_l g^{(k)} h^{(l)}. \tag{5.4.3}$$

We now want to demonstrate that the rs products $g^{(k)} h^{(l)}$, $1 \le k \le r$, $1 \le l \le s$, can be expressed as combinations of the rs derivatives $(gh)^{(k)}$, $1 \le k \le rs$, under a mild condition. We have

$$(gh)^{(j)} = \sum_{i=0}^{j} \binom{j}{i} g^{(i)} h^{(j-i)}, \quad j = 1, 2, \ldots, rs. \tag{5.4.4}$$

Let us now use the differential equations in (5.4.1) to express $g^{(0)} = g$ and $h^{(0)} = h$ in (5.4.4) as combinations of $g^{(i)}$, $1 \le i \le r$, and $h^{(i)}$, $1 \le i \le s$, respectively. Next, let us express $g^{(i)}$ with $i > r$ and $h^{(i)}$ with $i > s$ as combinations of $g^{(i)}$, $1 \le i \le r$, and $h^{(i)}$, $1 \le i \le s$, respectively, as well. That this is possible can be shown by differentiating the equations in (5.4.1) as many times as is necessary. For instance, if $g^{(r+1)}$ is required, we can obtain it from

$$g' = \left[\sum_{k=1}^{r} u_k g^{(k)} \right]' = \sum_{k=1}^{r} u_k' g^{(k)} + \sum_{k=1}^{r} u_k g^{(k+1)}, \tag{5.4.5}$$

in the form

$$g^{(r+1)} = \left[(1 - u_1') g' - \sum_{k=2}^{r} (u_k' + u_{k-1}) g^{(k)} \right] / u_r. \tag{5.4.6}$$

To obtain $g^{(r+2)}$ as a combination of $g^{(i)}$, $1 \le i \le r$, we differentiate (5.4.5) and invoke (5.4.6) as well. We continue similarly for $g^{(r+3)}$, $g^{(r+4)}$, ... As a result, (5.4.4) can be rewritten in the form

$$(gh)^{(j)} = \sum_{k=1}^{r} \sum_{l=1}^{s} w_{jkl} g^{(k)} h^{(l)}, \quad j = 1, 2, \ldots, rs, \tag{5.4.7}$$

where $w_{jkl} \in \mathbf{A}^{(\mu_{jkl})}$ for some integers μ_{jkl}. Now (5.4.7) is a linear system of rs equations for the rs unknowns $g^{(k)} h^{(l)}$, $1 \le k \le r$, $1 \le l \le s$. Assuming that the matrix of this system is nonsingular for all large x, we can solve by Cramer's rule for the products $g^{(k)} h^{(l)}$ in terms of the $(gh)^{(i)}$, $i = 1, 2, \ldots, rs$. Substituting this solution in (5.4.3), the proof of part (i) is achieved.

To prove part (ii), we proceed similarly. What we need to show is that $g + h$ satisfies an ordinary differential equation of the form

$$g + h = \sum_{k=1}^{r+s} p_k (g + h)^{(k)}, \tag{5.4.8}$$

where $p_k \in \mathbf{A}^{(\nu_k)}$, $k = 1, \ldots, r + s$, for some integers ν_k, and again recall that, by Proposition 5.1.3, the order of the differential equation satisfied by $g + h$ may possibly be less than $r + s$ too.

Adding the two equations in (5.4.1) we obtain

$$g + h = \sum_{k=1}^{r} u_k g^{(k)} + \sum_{l=1}^{s} v_l h^{(l)}. \tag{5.4.9}$$

Next, we have

$$(g + h)^{(j)} = g^{(j)} + h^{(j)}, \quad j = 1, 2, \ldots, r + s, \tag{5.4.10}$$

which, by the argument given above, can be reexpressed in the form

$$(g+h)^{(j)} = \sum_{k=1}^{r} w_{g;jk} g^{(k)} + \sum_{l=1}^{s} w_{h;jl} h^{(l)}, \quad j = 1, 2, \dots, r+s, \quad (5.4.11)$$

where $w_{g;jk} \in \mathbf{A}^{(\mu_{g;jk})}$ and $w_{h;jl} \in \mathbf{A}^{(\mu_{h;jl})}$ for some integers $\mu_{g;jk}$ and $\mu_{h;jl}$. We observe that (5.4.11) is a linear system of $r + s$ equations for the $r + s$ unknowns $g^{(k)}$, $1 \leq k \leq r$, and $h^{(l)}$, $1 \leq l \leq s$. The proof can now be completed as that of part (i). ∎

Heuristic 5.4.2 *Let $g \in \mathbf{B}^{(r)}$ and $h \in \mathbf{B}^{(r)}$ and assume that g and h satisfy the same ordinary differential equation of the form described in Definition 5.1.2. Then*

(i) *$gh \in \mathbf{B}^{(m)}$ with $m \leq r(r+1)/2$, and*
(ii) *$g + h \in \mathbf{B}^{(m)}$ with $m \leq r$.*

"Proof". The proof of part (i) is almost the same as that of part (i) of Heuristic 5.4.1, the difference being that the set $\{g^{(k)} h^{(l)} : 1 \leq k \leq r, \ 1 \leq l \leq s\}$ in Heuristic 5.4.1 is now replaced by the smaller set $\{g^{(k)} h^{(l)} + g^{(l)} h^{(k)} : 1 \leq k \leq l \leq r\}$ that contains $r(r+1)/2$ functions. As for part (ii), its proof follows immediately from the fact that $g + h$ satisfies the same ordinary differential equation that g and h satisfy separately. We leave the details to the reader. ∎

We now give a generalization of part (i) of the previous heuristic. Even though its proof is similar to that of the latter, it is quite involved. Therefore, we leave its details to the interested reader.

Heuristic 5.4.3 *Let $g_i \in \mathbf{B}^{(r)}$, $i = 1, \dots, \mu$, and assume that they all satisfy the same ordinary differential equation of the form described in Definition 5.1.2. Define $f = \prod_{i=1}^{\mu} g_i$. Then $f \in \mathbf{B}^{(m)}$ with $m \leq \binom{r+\mu-1}{\mu}$. In particular, if $g \in \mathbf{B}^{(r)}$, then $(g)^{\mu} \in \mathbf{B}^{(m)}$ with $m \leq \binom{r+\mu-1}{\mu}$.*

The preceding results are important because most integrands occurring in practical applications are products or sums of functions that are in the classes $\mathbf{B}^{(r)}$ for low values of r, such as $r = 1$ and $r = 2$.

Let us now apply these results to a few examples.

Example 5.4.4 Consider the function $f(x) = \sum_{i=1}^{m} f_i(x)$, where $f_i \in \mathbf{B}^{(1)}$ for each i. By part (ii) of Heuristic 5.4.1, we have that $f \in \mathbf{B}^{(m')}$ for some $m' \leq m$. This occurs, in particular, in the following two cases, among many others: (i) when $f_i \in \mathbf{A}^{(\gamma_i)}$ for some distinct $\gamma_i \neq 0, 1, \dots$, since $h_i \in \mathbf{B}^{(1)}$, by Example 5.1.10, and (ii) when $f_i(x) = e^{\lambda_i x} h_i(x)$, such that λ_i are distinct and nonzero and $h_i \in \mathbf{A}^{(\gamma_i)}$ for arbitrary γ_i that are not necessarily distinct, since $f_i \in \mathbf{B}^{(1)}$, by Example 5.1.11. In both cases, it can be shown that $f \in \mathbf{B}^{(m)}$ with $\mathbf{B}^{(m)}$ exactly as in Definition 5.1.2. We do not prove this here but refer the reader to the analogous proofs of Theorem 6.8.3 [for case (i)] and of Theorem 6.8.7 [for case (ii)] in the next chapter.

Example 5.4.5 Consider the function $f(x) = g(x)\mathcal{C}_\nu(x)$, where $\mathcal{C}_\nu(x) = bJ_\nu(x) + cY_\nu(x)$ is an arbitrary solution of the Bessel equation of order ν and $g \in \mathbf{A}^{(\gamma)}$ for some γ. By Example 5.1.10, $g \in \mathbf{B}^{(1)}$; by Example 5.1.7, $\mathcal{C}_\nu \in \mathbf{B}^{(2)}$. Applying part (i) of Heuristic 5.4.1, we conclude that $f \in \mathbf{B}^{(2)}$. Indeed, by the fact that $\mathcal{C}_\nu(x) = f(x)/g(x)$ satisfies the Bessel equation of order ν, after some manipulation of the latter, we obtain the ordinary differential equation $f = p_1 f' + p_2 f''$ with

$$p_1(x) = \frac{2x^2 g'(x)/g(x) - x}{w(x)} \quad \text{and} \quad p_2(x) = -\frac{x^2}{w(x)},$$

where

$$w(x) = x^2\left[\left(\frac{g'(x)}{g(x)}\right)^2 - \left(\frac{g'(x)}{g(x)}\right)'\right] - x\frac{g'(x)}{g(x)} + x^2 - \nu^2.$$

Consequently, $p_1 \in \mathbf{A}^{(-1)}$ and $p_2 \in \mathbf{A}^{(0)}$. That is, $f \in \mathbf{B}^{(2)}$ with $\mathbf{B}^{(m)}$ precisely as in Definition 5.1.2. Finally, provided also that $\Re\gamma < 1/2$ so that $f(x)$ is integrable at infinity, Theorem 5.1.12 applies as all of its conditions are satisfied, and we also have $\rho_0 \le \max\{i_1, i_2 - 1\} = -1$ and $\rho_1 = i_2 = 0$ in Theorem 5.1.12.

Example 5.4.6 Consider the function $f(x) = g(x)h(x)\mathcal{C}_\nu(x)$, where $g(x)$ and $\mathcal{C}_\nu(x)$ are exactly as in the preceding example and $h(x) = B\cos\alpha x + C\sin\alpha x$. From the preceding example, we have that $g(x)\mathcal{C}_\nu(x)$ is in $\mathbf{B}^{(2)}$. Similarly, the function $h(x)$ is in $\mathbf{B}^{(2)}$ as it satisfies the ordinary differential equation $y'' + \alpha^2 y = 0$. By part (i) of Heuristic 5.4.1, we therefore conclude that $f \in \mathbf{B}^{(4)}$. It can be shown by using different techniques that, when $\alpha = 1$, $f \in \mathbf{B}^{(3)}$, and this too is in agreement with part (i) of Heuristic 5.4.1.

Example 5.4.7 The function $f(x) = g(x)\prod_{i=1}^{q}\mathcal{C}_{\nu_i}(\alpha_i x)$, where $\mathcal{C}_{\nu_i}(z)$ are as in the previous examples and $g \in \mathbf{A}^{(\gamma)}$ for some γ, is in $\mathbf{B}^{(2^q)}$. This follows by repeated application of part (i) of Heuristic 5.4.1. When $\alpha_1, \ldots, \alpha_q$ are not all distinct, then $f \in \mathbf{B}^{(m)}$ with $m < 2^q$, as shown by different techniques later.

Example 5.4.8 The function $f(x) = (\sin x/x)^2$ is in $\mathbf{B}^{(3)}$. This follows from Example 5.1.8, which says that $\sin x/x$ is in $\mathbf{B}^{(2)}$ and from part (i) of Heuristic 5.4.2. Indeed, $f(x)$ satisfies the ordinary differential equation $y = \sum_{k=1}^{3} p_k(x)y^{(k)}$ with

$$p_1(x) = -\frac{2x^2 + 3}{4x} \in \mathbf{A}^{(1)}, \quad p_2(x) = -\frac{3}{4} \in \mathbf{A}^{(0)}, \quad \text{and} \quad p_3(x) = -\frac{x}{8} \in \mathbf{A}^{(1)}.$$

Finally, Theorem 5.1.12 applies, as all its conditions are satisfied, and we also have $\rho_0 \le 1$, $\rho_1 \le 0$, and $\rho_2 = 1$. Another way to see that $f \in \mathbf{B}^{(3)}$ is as follows: We can write $f(x) = (1 - \cos 2x)/(2x^2) = \frac{1}{2}x^{-2} - \frac{1}{2}x^{-2}\cos 2x$. Now, being in $\mathbf{A}^{(-2)}$, the function $\frac{1}{2}x^{-2}$ is in $\mathbf{B}^{(1)}$. (See Example 5.1.10.) Next, since $\frac{1}{2}x^{-2} \in \mathbf{B}^{(1)}$ and $\cos 2x \in \mathbf{B}^{(2)}$, their product is also in $\mathbf{B}^{(2)}$ by part (i) of Heuristic 5.4.1. Therefore, by part (ii) of Heuristic 5.4.1, $f \in \mathbf{B}^{(3)}$.

Example 5.4.9 Let $g \in \mathbf{B}^{(r)}$ and $h(x) = \sum_{s=0}^{q} u_s(x)(\log x)^s$, with $u_s \in \mathbf{A}^{(\gamma)}$ for every s. Here some or all of the $u_s(x)$, $0 \le s \le q - 1$, can be identically zero, but $u_q(x) \not\equiv 0$.

Then $gh \in \mathbf{B}^{(q+1)r}$. To show this, we start with

$$h^{(k)}(x) = \sum_{s=0}^{q} w_{ks}(x)(\log x)^s, \quad w_{ks} \in \mathbf{A}^{(\gamma-k)}, \quad \text{for all } k, s.$$

Treating $q + 1$ of these equalities, with $k = 1, \ldots, q + 1$, as a linear system of equations for the "unknowns" $(\log x)^s$, $s = 0, 1, \ldots, q$, and invoking Cramer's rule, we can show that $(\log x)^s = \sum_{k=1}^{q+1} \mu_{sk}(x)h^{(k)}(x)$, $\mu_{sk} \in \mathbf{A}^{(\nu_{sk}-\gamma)}$ for some integers ν_{sk}. Substituting these in $h(x) = \sum_{s=0}^{q} u_s(x)(\log x)^s$, we get $h(x) = \sum_{k=1}^{q+1} e_k(x)h^{(k)}(x)$, $e_k \in \mathbf{A}^{(\sigma_k)}$ for some integers σ_k. Thus, $h \in \mathbf{B}^{(q+1)}$ in the relaxed sense. Invoking now part (i) of Heuristic 5.4.1, the result follows.

Important Remark. When we know that $f = g + h$ with $g \in \mathbf{B}^{(r)}$ and $h \in \mathbf{B}^{(s)}$, and we can compute g and h separately, we should go ahead and compute $\int_0^\infty g(t)\,dt$ and $\int_0^\infty h(t)\,dt$ by the $D^{(r)}$- and $D^{(s)}$-transformations, respectively, instead of computing $\int_0^\infty f(t)\,dt$ by the $D^{(r+s)}$-transformation. The reason for this is that less computing is needed for the $D^{(r)}$- and $D^{(s)}$-transformations than for the $D^{(r+s)}$-transformation.

5.5 Numerical Examples

We now apply the D-transformation (with $\alpha = 0$ in Definition 5.2.1) to two infinite-range integrals of different nature. The numerical results that we present have been obtained using double-precision arithmetic (approximately 16 decimal digits). For more examples, we refer the reader to Levin and Sidi [165] and Sidi [272].

Example 5.5.1 Consider the integral

$$I[f] = \int_0^\infty \frac{\log(1+t)}{1+t^2}\,dt = \frac{\pi}{4}\log 2 + G = 1.4603621167531195\cdots,$$

where G is Catalan's constant. As we saw in Example 5.1.9, $f(x) = \log(1+x)/(1+x^2) \in \mathbf{B}^{(2)}$. Also, $f(x)$ satisfies all the conditions of Theorem 5.1.12. We applied the $D^{(2)}$-transformation to this integral with $x_l = e^{0.4l}$, $l = 0, 1, \ldots$. With this choice of the x_l, the approximations to $I[f]$ produced by the $D^{(2)}$-transformation enjoy a great amount of stability. Indeed, we have that both $x_l f(x_l)$ and $x_l^2 f'(x_l)$ are $O(x_l^{-1}\log x_l) = O(le^{-0.4l})$ as $l \to \infty$. [That is, $\phi_k(y_l) = C_k(l)\exp(u_k(l))$, with $C_k(l) = O(l)$ as $l \to \infty$, and $u_k(l) = -0.4l$, and both $C_k(l)$ and $u_k(l)$ vary slowly with l. See the discussion on stability of GREP in Section 4.5.]

The relative errors $|F(x_{2\nu}) - I[f]|/|I[f]|$ and $|\breve{D}_{(\nu,\nu)}^{(0,2)} - I[f]|/|I[f]|$, where $\breve{D}_{(\nu,\nu)}^{(0,2)}$ is the computed $D_{(\nu,\nu)}^{(0,2)}$, are given in Table 5.5.1. Note that $F(x_{2\nu})$ is the best of all the $F(x_l)$ that are used in computing $D_{(\nu,\nu)}^{(0,2)}$. Observe also that the $\breve{D}_{(\nu,\nu)}^{(0,2)}$ retain their accuracy with increasing ν, which is caused by the good stability properties of the extrapolation process in this example.

Example 5.5.2 Consider the integral

$$I[f] = \int_0^\infty \left(\frac{\sin t}{t}\right)^2 dt = \frac{\pi}{2}.$$

Table 5.5.1: *Numerical results for Example 5.5.1*

ν	$\|F(x_{2\nu}) - I[f]\|/\|I[f]\|$	$\|\check{D}^{(0,2)}_{(\nu,\nu)} - I[f]\|/\|I[f]\|$
0	$8.14D - 01$	$8.14D - 01$
1	$5.88D - 01$	$6.13D - 01$
2	$3.69D - 01$	$5.62D - 03$
3	$2.13D - 01$	$5.01D - 04$
4	$1.18D - 01$	$2.19D - 05$
5	$6.28D - 02$	$4.81D - 07$
6	$3.27D - 02$	$1.41D - 09$
7	$1.67D - 02$	$1.69D - 12$
8	$8.42D - 03$	$4.59D - 14$
9	$4.19D - 03$	$4.90D - 14$
10	$2.07D - 03$	$8.59D - 14$

As we saw in Example 5.4.8, $f(x) = (\sin x/x)^2 \in \mathbf{B}^{(3)}$ and satisfies all the conditions of Theorem 5.1.12. We applied the $D^{(3)}$-transformation to this integral with $x_l = \frac{3}{2}(l+1)$, $l = 0, 1, \ldots$. The relative errors $\|F(x_{3\nu}) - I[f]\|/\|I[f]\|$ and $\|\check{D}^{(0,3)}_{(\nu,\nu,\nu)} - I[f]\|/\|I[f]\|$, where $\check{D}^{(0,3)}_{(\nu,\nu,\nu)}$ is the computed $D^{(0,3)}_{(\nu,\nu,\nu)}$, are given in Table 5.5.2. Note that $F(x_{3\nu})$ is the best of all the $F(x_l)$ that are used in computing $D^{(0,3)}_{(\nu,\nu,\nu)}$. Note also the loss of accuracy in the $\check{D}^{(0,3)}_{(\nu,\nu,\nu)}$ with increasing ν that is caused by the instability of the extrapolation process in this example. Nevertheless, we are able to obtain approximations with as many as 12 correct decimal digits.

5.6 Proof of Theorem 5.1.12

In this section, we give a complete proof of Theorem 5.1.12 by actually constructing the asymptotic expansion of the integral $\int_x^\infty f(t)\,dt$ for $x \to \infty$. We start with the following simple lemma whose proof can be achieved by integration by parts.

Lemma 5.6.1 *Let $P(x)$ be differentiable a sufficient number of times on $(0, \infty)$. Then, provided*

$$\lim_{x \to \infty} P^{(j-1)}(x)f^{(k-j)}(x) = 0, \quad j = 1, \ldots, k, \tag{5.6.1}$$

and provided $P(x)f^{(k)}(x)$ is integrable at infinity, we have

$$\int_x^\infty P(t)f^{(k)}(t)\,dt = \sum_{j=1}^{k}(-1)^j P^{(j-1)}(x)f^{(k-j)}(x) + (-1)^k \int_x^\infty P^{(k)}(t)f(t)\,dt. \tag{5.6.2}$$

Using Lemma 5.6.1 and the ordinary differential equation (5.1.4) that is satisfied by $f(x)$, we can now state the following lemma.

Lemma 5.6.2 *Let $Q(x)$ be differentiable a sufficient number of times on $(0, \infty)$. Define*

$$\hat{Q}_k(x) = \sum_{j=k+1}^{m}(-1)^{j+k}[Q(x)p_j(x)]^{(j-k-1)}, \quad k = 0, 1, \ldots, m-1, \tag{5.6.3}$$

Table 5.5.2: *Numerical results for Example 5.5.2*

ν	$|F(x_{3\nu}) - I[f]|/|I[f]|$	$|\check{D}^{(0,3)}_{(\nu,\nu,\nu)} - I[f]|/|I[f]|$
0	$2.45D-01$	$2.45D-01$
1	$5.02D-02$	$4.47D-02$
2	$3.16D-02$	$4.30D-03$
3	$2.05D-02$	$1.08D-04$
4	$1.67D-02$	$2.82D-07$
5	$1.31D-02$	$1.60D-07$
6	$1.12D-02$	$4.07D-09$
7	$9.65D-03$	$2.65D-13$
8	$8.44D-03$	$6.39D-12$
9	$7.65D-03$	$5.93D-11$
10	$6.78D-03$	$6.81D-11$

and

$$\hat{Q}_{-1}(x) = \sum_{k=1}^{m}(-1)^k[Q(x)p_k(x)]^{(k)}.$$ (5.6.4)

Then, provided

$$\lim_{x\to\infty}\hat{Q}_k(x)f^{(k)}(x) = 0, \quad k=0,1,\dots,m-1,$$ (5.6.5)

and provided $Q(x)f(x)$ is integrable at infinity, we have

$$\int_x^\infty Q(t)f(t)\,dt = \sum_{k=0}^{m-1}\hat{Q}_k(x)f^{(k)}(x) + \int_x^\infty \hat{Q}_{-1}(t)f(t)\,dt.$$ (5.6.6)

Proof. By (5.1.4), we first have

$$\int_x^\infty Q(t)f(t)\,dt = \sum_{k=1}^{m}\int_x^\infty[Q(t)p_k(t)]f^{(k)}(t)\,dt.$$ (5.6.7)

The result follows by applying Lemma 5.6.1 to each integral on the right-hand side of (5.6.7). We leave the details to the reader. ∎

By imposing additional conditions on the function $Q(x)$ in Lemma 5.6.2, we obtain the following key result.

Lemma 5.6.3 *In Lemma 5.6.2, let $Q \in \mathbf{A}^{(-l-1)}$ strictly for some integer l, $l = \pm 1,2,3,\dots$. Let also*

$$\alpha_l = \sum_{k=1}^{m}l(l-1)\cdots(l-k+1)\bar{p}_k,$$ (5.6.8)

so that $\alpha_l \neq 1$ by (5.1.6). Define

$$\tilde{Q}_k(x) = \frac{\hat{Q}_k(x)}{1 - \alpha_l}, \quad k = 0, 1, \dots, m - 1,$$

$$\tilde{Q}_{-1}(x) = \frac{\hat{Q}_{-1}(x) - \alpha_l Q(x)}{1 - \alpha_l}, \tag{5.6.9}$$

Then

$$\int_x^\infty Q(t) f(t) \, dt = \sum_{k=0}^{m-1} \tilde{Q}_k(x) f^{(k)}(x) + \int_x^\infty \tilde{Q}_{-1}(t) f(t) \, dt. \tag{5.6.10}$$

We also have that $\tilde{Q}_k \in \mathbf{A}^{(\bar{\rho}_k - l - 1)}$, $k = 0, 1, \dots, m - 1$, and that $\tilde{Q}_{-1} \in \mathbf{A}^{(-l-2)}$. Thus, $\tilde{Q}_{-1}(x)/Q(x) = O(x^{-1})$ as $x \to \infty$. This result is improved to $\tilde{Q}_{-1} \in \mathbf{A}^{(-2)}$ when $Q(x)$ is constant for all x.

Proof. By the assumptions that $Q \in \mathbf{A}^{(-l-1)}$ and $p_k \in \mathbf{A}^{(k)}$, we have $Qp_k \in \mathbf{A}^{(k-l-1)}$ so that $(Qp_k)^{(k)} \in \mathbf{A}^{(-l-1)}$, $k = 1, \dots, m$. Consequently, by (5.6.4), $\hat{Q}_{-1} \in \mathbf{A}^{(-l-1)}$ as well. Being in $\mathbf{A}^{(-l-1)}$, $Q(x)$ is of the form $Q(x) = \bar{q} x^{-l-1} + R(x)$ for some nonzero constant $\bar{q}$ and some $R \in \mathbf{A}^{(-l-2)}$. Thus, $\hat{Q}_{-1}(x) = \alpha_l \bar{q} x^{-l-1} + S(x)$ with α_l as in (5.6.8) and some $S \in \mathbf{A}^{(-l-2)}$. This can be rewritten as $\hat{Q}_{-1}(x) = \alpha_l Q(x) + T(x)$ for some $T \in \mathbf{A}^{(-l-2)}$ whether $\alpha_l \neq 0$ or not. As a result, we have

$$\int_x^\infty \hat{Q}_{-1}(t) f(t) \, dt = \alpha_l \int_x^\infty Q(t) f(t) \, dt + \int_x^\infty T(t) f(t) \, dt.$$

Substituting this in (5.6.6), and solving for $\int_x^\infty Q(t) f(t) \, dt$, we obtain (5.6.10) with (5.6.9). Also, $\tilde{Q}_{-1}(x) = T(x)/(1 - \alpha_l) = [S(x) - \alpha_l R(x)]/(1 - \alpha_l)$.

We can similarly show that $\tilde{Q}_k \in \mathbf{A}^{(\bar{\rho}_k - l - 1)}$, $k = 0, 1, \dots, m - 1$. For this, we combine (5.6.3) and the fact that $p_k \in \mathbf{A}^{(i_k)}$, $i = 1, \dots, m$, in (5.6.9), and then invoke (5.1.9). We leave the details to the reader. ∎

What we have achieved through Lemma 5.6.3 is that the new integral $\int_x^\infty \tilde{Q}_{-1}(t) f(t) \, dt$ converges to zero as $x \to \infty$ more quickly than the original integral $\int_x^\infty Q(t) f(t) \, dt$. This is an important step in the derivation of the result in Theorem 5.1.12.

We now apply Lemma 5.6.3 to the integral $\int_x^\infty f(t) \, dt$, that is, we apply it with $Q(x) = 1$. We can easily verify that the integrability of $f(x)$ at infinity and the conditions in (5.1.5)–(5.1.7) guarantee that Lemma 5.6.3 applies with $l = -1$. We obtain

$$\int_x^\infty f(t) \, dt = \sum_{k=0}^{m-1} b_{1,k}(x) f^{(k)}(x) + \int_x^\infty b_1(t) f(t) \, dt, \tag{5.6.11}$$

with $b_{1,k} \in \mathbf{A}^{(\bar{\rho}_k)}$, $k = 0, 1, \dots, m - 1$, and $b_1 \in \mathbf{A}^{(-\sigma_1 - 1)}$ strictly for some integer $\sigma_1 \geq 1$.

Now let us apply Lemma 5.6.3 to the integral $\int_x^\infty b_1(t)f(t)\,dt$. This results in

$$\int_x^\infty b_1(t)f(t)\,dt = \sum_{k=0}^{m-1} b_{2,k}(x)f^{(k)}(x) + \int_x^\infty b_2(t)f(t)\,dt, \qquad (5.6.12)$$

with $b_{2,k} \in \mathbf{A}^{(\bar{\rho}_k - \sigma_1 - 1)}$, $k = 0, 1, \ldots, m-1$, and $b_2 \in \mathbf{A}^{(-\sigma_2 - 1)}$ strictly for some integer $\sigma_2 \geq \sigma_1 + 1 \geq 2$.

Next, apply Lemma 5.6.3 to the integral $\int_x^\infty b_2(t)f(t)\,dt$. This results in

$$\int_x^\infty b_2(t)f(t)\,dt = \sum_{k=0}^{m-1} b_{3,k}(x)f^{(k)}(x) + \int_x^\infty b_3(t)f(t)\,dt, \qquad (5.6.13)$$

with $b_{3,k} \in \mathbf{A}^{(\bar{\rho}_k - \sigma_2 - 1)}$, $k = 0, 1, \ldots, m-1$, and $b_3 \in \mathbf{A}^{(-\sigma_3 - 1)}$ strictly for some integer $\sigma_3 \geq \sigma_2 + 1 \geq 3$.

Continuing this way, and combining the results in (5.6.11), (5.6.12), (5.6.13), and so on, we obtain

$$\int_x^\infty f(t)\,dt = \sum_{k=0}^{m-1} \beta_{[s;k]}(x)f^{(k)}(x) + \int_x^\infty b_s(t)f(t)\,dt, \qquad (5.6.14)$$

where s is an arbitrary positive integer and

$$\beta_{[s;k]}(x) = \sum_{i=1}^{s} b_{i,k}(x), \quad k = 0, 1, \ldots, m-1. \qquad (5.6.15)$$

Here $b_{i,k} \in \mathbf{A}^{(\bar{\rho}_k - \sigma_{i-1} - 1)}$ for $0 \leq k \leq m-1$ and $i \geq 1$, and $b_s \in \mathbf{A}^{(-\sigma_s - 1)}$ strictly for some integer $\sigma_s \geq \sigma_{s-1} + 1 \geq s$.

Now, from one of the remarks following the statement of Theorem 5.1.12, we have $\lim_{x \to \infty} x^{\bar{\rho}_0} f(x) = 0$. Thus

$$\int_x^\infty b_s(t)f(t)\,dt = o(x^{-\sigma_s - \bar{\rho}_0}) \text{ as } x \to \infty, \qquad (5.6.16)$$

Next, using the fact that $\beta_{[s+1;k]}(x) = \beta_{[s;k]}(x) + b_{s+1,k}(x)$ and $b_{s+1,k} \in \mathbf{A}^{(\bar{\rho}_k - \sigma_s - 1)}$, we can expand $\beta_{[s;k]}(x)$ to obtain

$$\beta_{[s;k]}(x) = \sum_{i=0}^{\sigma_s} \beta'_{ki} x^{\bar{\rho}_k - i} + O(x^{\bar{\rho}_k - \sigma_s - 1}) \text{ as } x \to \infty. \qquad (5.6.17)$$

Note that β'_{ki}, $0 \leq i \leq \sigma_s$, remain unchanged in the expansion of $\beta_{[s';k]}(x)$ for all $s' > s$.

Combining (5.6.16) and (5.6.17) in (5.6.14), we see that $\int_x^\infty f(t)\,dt$ has a genuine asymptotic expansion given by

$$\int_x^\infty f(t)\,dt \sim \sum_{k=0}^{m-1} x^{\bar{\rho}_k} f^{(k)}(x) \sum_{i=0}^{\infty} \beta'_{ki} x^{-i} \text{ as } x \to \infty. \qquad (5.6.18)$$

The proof of Theorem 5.1.12 can now be completed easily.

We end this section by showing that the functions $x^{\rho_k} g_k(x) \equiv \beta_k(x)$, $k = 0, 1, \ldots, m-1$, can also be determined from a system of linear first-order differential equations that are expressed solely in terms of the $p_k(x)$. This theorem also produces the relation between the ρ_k and the $\bar{\rho}_k$ that is given in (5.1.9).

Theorem 5.6.4 *Let $f \in \mathbf{B}^{(m)}$ with minimal m and satisfy (5.1.4). Then, the functions $\beta_k(x)$ above satisfy the first-order linear system of differential equations*

$$p_k(x)\beta_0'(x) + \beta_k'(x) + \beta_{k-1}(x) + p_k(x) = 0, \quad k = 1, \ldots, m; \quad \beta_m(x) \equiv 0. \quad (5.6.19)$$

In particular, $\beta_0(x)$ is a solution of the mth-order differential equation

$$\beta_0 + \sum_{k=1}^{m}(-1)^{k-1}(p_k\beta_0')^{(k-1)} + \sum_{k=1}^{m}(-1)^{k-1}p_k^{(k-1)} = 0. \quad (5.6.20)$$

Once $\beta_0(x)$ has been determined, the rest of the $\beta_k(x)$ can be obtained from (5.6.19) in the order $k = m - 1, m - 2, \ldots, 1$.

Proof. Differentiating the already known relation

$$\int_x^\infty f(t)\,dt = \sum_{k=0}^{m-1}\beta_k(x)f^{(k)}(x), \quad (5.6.21)$$

we obtain

$$f = \sum_{k=1}^{m-1}\left(-\frac{\beta_k' + \beta_{k-1}}{\beta_0' + 1}\right)f^{(k)} + \left(-\frac{\beta_{m-1}}{\beta_0' + 1}\right)f^{(m)}. \quad (5.6.22)$$

Since $f \in \mathbf{B}^{(m)}$ and m is minimal, the $p_k(x)$ are unique by our assumption following Proposition 5.1.3. We can therefore identify

$$p_k = -\frac{\beta_k' + \beta_{k-1}}{\beta_0' + 1}, \quad k = 1, \ldots, m-1, \quad \text{and} \quad p_m = -\frac{\beta_{m-1}}{\beta_0' + 1}, \quad (5.6.23)$$

from which (5.6.19) follows. Applying the differential operator $(-1)^{k-1}\frac{d^{k-1}}{dx^{k-1}}$ to the kth equation in (5.6.19) and summing over k, we obtain the differential equation given in (5.6.20). The rest is immediate. ∎

It is important to note that the differential equation in (5.6.20) actually has a solution for $\beta_0(x)$ that has an asymptotic expansion of the form $\sum_{i=0}^{\infty}\delta_i x^{1-i}$ for $x \to \infty$. It can be shown that, under the condition given in (5.1.6), the coefficients δ_i of this expansion are uniquely determined from (5.6.20). It can further be shown that, with α_l as defined in (5.6.8),

$$\beta_0'(x) = \frac{\alpha_{-1}}{1 - \alpha_{-1}} + O(x^{-1}) \quad \text{as } x \to \infty,$$

so that $\beta_0'(x) + 1 = 1/(1 - \alpha_{-1}) + O(x^{-1})$ and hence $(\beta_0' + 1) \in \mathbf{A}^{(0)}$ strictly. Using this fact in the equation $\beta_{m-1} = -(\beta_0' + 1)p_m$ that follows from the mth of the equations in (5.6.19), we see that $\beta_{m-1} \in \mathbf{A}^{(i_m)}$ strictly. Using these two facts about $(\beta_0' + 1)$ and β_{m-1} in the equation $\beta_{m-2} = -(\beta_0' + 1)p_{m-1} - \beta_{m-1}'$ that follows from the $(m-1)$st of the equations in (5.6.19), we see that $\beta_{m-2} \in \mathbf{A}^{(\bar{\rho}_{m-2})}$. Continuing this way we can show that

$\beta_k = -(\beta_0' + 1)p_{k+1} - \beta_{k+1}' \in \mathbf{A}^{(\bar{\rho}_k)}, \ k = m - 3, \ldots, 1, 0$. Also, once $\beta_0(x)$ has been determined, the equations from which the rest of the $\beta_k(x)$ are determined are algebraic (as opposed to differential) equations.

5.7 Characterization and Integral Properties of Functions in $\mathbf{B}^{(1)}$

5.7.1 Integral Properties of Functions in $\mathbf{A}^{(\gamma)}$

From Heuristic 5.4.1 and Example 5.4.4, it is clear that functions in $\mathbf{B}^{(1)}$ are important building blocks for functions in $\mathbf{B}^{(m)}$ with arbitrary m. This provides ample justification for studying them in detail. Therefore, we close this chapter by proving a characterization theorem for functions in $\mathbf{B}^{(1)}$ and also a theorem on their integral properties. For this, we need the following result that provides the integral properties of functions in $\mathbf{A}^{(\gamma)}$ and thus extends the list of remarks following Definition 5.1.1.

Theorem 5.7.1 *Let* $g \in \mathbf{A}^{(\gamma)}$ *strictly for some* γ *with* $g(x) \sim \sum_{i=0}^{\infty} g_i x^{\gamma - i}$ *as* $x \to \infty$, *and define* $G(x) = \int_a^x g(t)\,dt$, *where* $a > 0$ *without loss of generality. Then*

$$G(x) = b + c \log x + \tilde{G}(x), \tag{5.7.1}$$

where b *and* c *are constants and* $\tilde{G} \in \mathbf{A}^{(\gamma+1)}$ *strictly if* $\gamma \neq -1$, *while* $\tilde{G} \in \mathbf{A}^{(-1)}$ *if* $\gamma = -1$. *When* $\gamma + 1 \neq 0, 1, 2, \ldots$, b *is simply the value of* $\int_0^{\infty} g(t)\,dt$ *or of its Hadamard finite part and* $c = 0$. *When* $\gamma + 1 = k \in \{0, 1, 2, \ldots\}$, $c = g_k$. *Explicit expressions for* b *and* $\tilde{G}(x)$ *are given in the following proof.*

Proof. Let us start by noting that $\int_a^{\infty} g(t)\,dt$ converges only when $\Re\gamma + 1 < 0$. Let N be an arbitrary positive integer greater than $\Re\gamma + 1$, and define $\hat{g}(x) = g(x) - \sum_{i=0}^{N-1} g_i x^{\gamma - i}$. Thus, $\hat{g} \in \mathbf{A}^{(\gamma - N)}$ so that $\hat{g}(x)$ is $O(x^{\gamma - N})$ as $x \to \infty$ and is integrable at infinity by the fact that $\Re\gamma - N + 1 < 0$. That is, $U_N(x) = \int_x^{\infty} \hat{g}(t)\,dt$ exists for all $x \geq a$. Therefore, we can write

$$G(x) = \int_a^x \left(\sum_{i=0}^{N-1} g_i t^{\gamma - i} \right) dt + U_N(a) - U_N(x). \tag{5.7.2}$$

Integrating the asymptotic expansion $\hat{g}(x) \sim \sum_{i=N}^{\infty} g_i x^{\gamma - i}$ as $x \to \infty$ term by term, which is justified, we obtain

$$U_N(x) \sim -\sum_{i=N}^{\infty} \frac{g_i}{\gamma - i + 1} x^{\gamma - i + 1} \quad \text{as } x \to \infty. \tag{5.7.3}$$

Carrying out the integration on the right-hand side of (5.7.2), the result follows with

$$b = U_N(a) - \sum_{\substack{i=0 \\ \gamma - i \neq -1}}^{N-1} \frac{g_i}{\gamma - i + 1} a^{\gamma - i + 1} - c \log a \tag{5.7.4}$$

and

$$\tilde{G}(x) = -U_N(x) + \sum_{\substack{i=0 \\ \gamma-i\neq-1}}^{N-1} \frac{g_i}{\gamma-i+1} x^{\gamma-i+1}. \tag{5.7.5}$$

It can easily be verified from (5.7.4) and (5.7.5) that, despite their appearance, both b and $\tilde{G}(x)$ are independent of N. Substituting (5.7.3) in (5.7.5), we see that $\tilde{G}(x)$ has the asymptotic expansion

$$\tilde{G}(x) \sim \sum_{\substack{i=0 \\ \gamma-i\neq-1}}^{\infty} \frac{g_i}{\gamma-i+1} x^{\gamma-i+1} \quad \text{as } x \to \infty. \tag{5.7.6}$$

We leave the rest of the proof to the reader. ∎

5.7.2 A Characterization Theorem for Functions in $\mathbf{B}^{(1)}$

The following result is a characterization theorem for functions in $\mathbf{B}^{(1)}$.

Theorem 5.7.2 *A function $f(x)$ is in $\mathbf{B}^{(1)}$ if and only if either (i) $f \in \mathbf{A}^{(\gamma)}$ for some arbitrary $\gamma \neq 0$, or (ii) $f(x) = e^{\theta(x)}h(x)$, where $\theta \in \mathbf{A}^{(s)}$ strictly for some positive integer s and $h \in \mathbf{A}^{(\gamma)}$ for some arbitrary γ. If we write $f(x) = p(x)f'(x)$, then $p \in \mathbf{A}^{(\rho)}$ strictly, with $\rho = 1$ in case (i) and $\rho = -s + 1 \leq 0$ in case (ii).*

Proof. The proof of sufficiency is already contained in Examples 5.1.10 and 5.1.11. Therefore, we need be concerned only with necessity. From the fact that $f(x) = p(x)f'(x)$, we have first $f(x) = K \exp[G(x)]$, where K is some constant and $G(x) = \int_r^x g(t)\,dt$ with $g(x) = 1/p(x)$ and with r sufficiently large so that $p(x) \neq 0$ for $x \geq r$ by the fact that $p \in \mathbf{A}^{(\rho)}$ strictly. Now because ρ is an integer ≤ 1, we have that $g = 1/p \in \mathbf{A}^{(k)}$ strictly, with $k = -\rho$. Thus, k can assume only one of the values $-1, 0, 1, 2, \ldots$. From the preceding theorem, we have that $G(x)$ is of the form given in (5.7.1) with $c = g_{k+1}$ and $\tilde{G}(x)$ as in (5.7.5) with γ there replaced by k. Here, the g_i are defined via $g(x) \sim \sum_{i=0}^{\infty} g_i x^{k-i}$ as $x \to \infty$. Therefore, we can write $f(x) = K e^b x^c \exp[\tilde{G}(x)]$. Let us recall that when $k \neq -1$, $\tilde{G} \in \mathbf{A}^{(k+1)}$ strictly. The proof of case (ii) can now be completed by identifying $\theta(x) = \tilde{G}(x)$ and $h(x) = K e^b x^c$ when $k \in \{0, 1, \ldots\}$. In particular, $s = k + 1$ and $\gamma = c$. The proof of case (i) is completed by recalling that when $k = -1$, $\tilde{G} \in \mathbf{A}^{(-1)}$ so that $\exp[\tilde{G}(x)] \in \mathbf{A}^{(0)}$ strictly, as a result of which, $f \in \mathbf{A}^{(\gamma)}$ strictly with $\gamma = c$. In addition, $\gamma \neq 0$ necessarily because $c = g_0 \neq 0$. ∎

5.7.3 Asymptotic Expansion of $F(x)$ When $f(x) \in \mathbf{B}^{(1)}$

Theorems 5.7.1 and 5.7.2 will now be used in the study of integral properties of functions $f(x)$ in $\mathbf{B}^{(1)}$ that satisfy $f(x) = O(x^\lambda)$ as $x \to \infty$ for some λ. More specifically, we are concerned with the following two cases in the notation of Theorem 5.7.2.

(i) $f \in \mathbf{A}^{(\gamma)}$ strictly for some $\gamma \neq -1, 0, 1, 2, \dots$. In this case, $f(x) = p(x)f'(x)$ with $p \in \mathbf{A}^{(\rho)}$ strictly, $\rho = 1$.
(ii) $f(x) = e^{\theta(x)}h(x)$, where $h \in \mathbf{A}^{(\gamma)}$ strictly for some γ and $\theta \in \mathbf{A}^{(s)}$ strictly for some positive integer s, and that either (a) $\lim_{x\to\infty} \Re\theta(x) = -\infty$, or (b) $\lim_{x\to\infty} \Re\theta(x)$ is finite. In this case, $f(x) = p(x)f'(x)$ with $p \in \mathbf{A}^{(\rho)}$ strictly, $\rho = -s + 1 \leq 0$.

We already know that in case (i) $f(x)$ is integrable at infinity only when $\Re\gamma < -1$. In case (ii) $f(x)$ is integrable at infinity (a) for all γ when $\lim_{x\to\infty} \Re\theta(x) = -\infty$ and (b) for $\Re\gamma < s - 1$ when $\lim_{x\to\infty} \Re\theta(x)$ is finite; otherwise, $f(x)$ is not integrable at infinity. The validity of this assertion in case (i) and in case (ii-a) is obvious; for case (ii-b), it follows from Theorem 5.7.3 that we give next. Finally, in case (ii-a) $|f(x)| \sim C_1 x^{\Re\gamma} e^{\Re\theta(x)}$ as $x \to \infty$, and in case (ii-b) $|f(x)| \sim C_2 x^{\Re\gamma}$ as $x \to \infty$.

Theorem 5.7.3 *Let* $f \in \mathbf{B}^{(1)}$ *be as in the preceding paragraph. Then there exist a constant* $I[f]$ *and a function* $g \in \mathbf{A}^{(0)}$ *strictly such that*

$$F(x) = \int_0^x f(t)\,dt = I[f] + x^\rho f(x)g(x), \tag{5.7.7}$$

whether $f(x)$ *is integrable at infinity or not.*

Remark. In case $f(x)$ is integrable at infinity, this theorem is simply a special case of Theorem 5.1.12, and $I[f] = \int_0^\infty f(t)\,dt$, as can easily be verified. The cases in which $f(x)$ is not integrable at infinity were originally treated by Sidi [286], [300].

Proof. In case (i), Theorem 5.7.1 applies and we have $F(x) = b + \tilde{F}(x)$ for some constant b and a function $\tilde{F} \in \mathbf{A}^{(\gamma+1)}$ strictly. Since $xf(x)$ is in $\mathbf{A}^{(\gamma+1)}$ strictly as well, $g(x) \equiv \tilde{F}(x)/[xf(x)]$ is in $\mathbf{A}^{(0)}$ strictly. The result in (5.7.7) now follows with $I[f] = b$.
   For case (ii), we proceed by integrating $\int_0^x f(t)\,dt = \int_0^x p(t)f'(t)\,dt$ by parts:

$$\int_0^x f(t)\,dt = p(t)f(t)\big|_{t=0}^{t=x} - \int_0^x p'(t)f(t)\,dt.$$

Defining next

$$u_0(x) = 1; \quad v_{i+1}(x) = p(x)u_i(x), \; u_{i+1}(x) = -v'_{i+1}(x), \quad i = 0, 1, \dots, \tag{5.7.8}$$

we have, again by integration by parts,

$$\int_0^x u_i(t)f(t)\,dt = v_{i+1}(t)f(t)\big|_{t=0}^{t=x} + \int_0^x u_{i+1}(t)f(t)\,dt, \quad i = 0, 1, \dots . \tag{5.7.9}$$

Summing all these equalities, we obtain

$$\int_0^x f(t)\,dt = \sum_{i=1}^N v_i(t)f(t)\big|_{t=0}^{t=x} + \int_0^x u_N(t)f(t)\,dt, \tag{5.7.10}$$

where N is as large as we wish. Now by the fact that $p \in \mathbf{A}^{(\rho)}$ strictly, we have $v_i \in \mathbf{A}^{(\tau_i+1)}$ and $u_i \in \mathbf{A}^{(\tau_i)}$, where $\tau_i = i(\rho - 1)$, $i = 1, 2, \ldots$. In addition, by the fact that $v_1(x) = p(x)$, $v_1 \in \mathbf{A}^{(\rho)}$ strictly. Let us pick $N > (1 + \Re\gamma)/(1 - \rho)$ so that $\tau_N + \Re\gamma < -1$. Then the integral $\int_0^\infty u_N(t) f(t) \, dt$ converges, and we can rewrite (5.7.10) as in

$$\int_0^x f(t) \, dt = b + \left[\sum_{i=1}^N v_i(x) \right] f(x) - \int_x^\infty u_N(t) f(t) \, dt, \qquad (5.7.11)$$

where

$$b = -\lim_{x \to 0} \left[\sum_{i=1}^N v_i(x) \right] f(x) + \int_0^\infty u_N(t) f(t) \, dt. \qquad (5.7.12)$$

Realizing that $\{v_i(x)\}_{i=0}^\infty$ is an asymptotic sequence as $x \to \infty$ and expanding the functions $v_i(x)$, we see that

$$\sum_{i=1}^N v_i(x) = \sum_{i=0}^{\nu_r} \beta_i x^{\rho-i} + O(v_r(x)) \text{ as } x \to \infty, \quad \beta_0 \neq 0, \ r \leq N, \quad (5.7.13)$$

where the integers ν_r are defined via $\rho - \nu_r - 1 = \tau_r + 1$. Here $\beta_0, \beta_1, \ldots, \beta_{\nu_r}$ are obtained by expanding only $v_1(x), \ldots, v_{r-1}(x)$ and are not affected by $v_i(x)$, $i \geq r$. In addition, by the fact that $|f(x)| \sim C_1 x^{\Re\gamma} \exp[\Re\theta(x)]$ as $x \to \infty$ in case (ii-a) and $|f(x)| \sim C_2 x^{\Re\gamma}$ as $x \to \infty$ in case (ii-b), we also have that

$$\int_x^\infty u_N(t) f(t) \, dt = O(x^{\tau_N+1} f(x)) \text{ as } x \to \infty. \qquad (5.7.14)$$

Combining (5.7.13) and (5.7.14) in (5.7.11), and keeping in mind that N is arbitrary, we thus obtain

$$\int_0^x f(t) \, dt = I[f] + x^\rho f(x) \left[\sum_{i=0}^{\nu_r} \beta_i x^{-i} + O(x^{-\nu_r-1}) \right] \text{ as } x \to \infty, \quad (5.7.15)$$

with $I[f] = b$. This completes the proof. (Note that $I[f]$ is independent of N despite its appearance.) ∎

Remarks.

1. When $f(x)$ is not integrable at infinity, we can say the following about $I[f]$: In case (i), $I[f]$ is the Hadamard finite part of the (divergent) integral $\int_0^\infty f(t) \, dt$, while in case (ii), $I[f]$ is the Abel sum of $\int_0^\infty f(t) \, dt$ that is defined as $\lim_{\epsilon \to 0+} \int_0^\infty e^{-\epsilon t} f(t) \, dt$. We have already shown the former in Theorem 5.7.1. The proof of the latter can be achieved by applying Theorem 5.7.3 to $e^{-\epsilon t} f(t)$ and letting $\epsilon \to 0+$. We leave the details to the interested reader.

2. Since the asymptotic expansion of $F(x)$ as $x \to \infty$ is of one and the same form whether $\int_0^\infty f(t) \, dt$ converges or not, the $D^{(1)}$-transformation can be applied to approximate $I[f]$ in all cases considered above.

6

The d-Transformation: A GREP for Infinite Series and Sequences

6.1 The Class $\mathbf{b}^{(m)}$ and Related Asymptotic Expansions

The summation of infinite series of the form

$$S(\{a_k\}) = \sum_{k=1}^{\infty} a_k \tag{6.1.1}$$

is a very common problem in many branches of science and engineering. A direct way to achieve this is by computing the sequence $\{A_n\}$ of its partial sums, namely,

$$A_n = \sum_{k=1}^{n} a_k, \quad n = 1, 2, \dots, \tag{6.1.2}$$

hoping that A_n, for n not too large, approximates the sum $S(\{a_k\})$ sufficiently well. In many cases, however, the terms a_n decay very slowly as $n \to \infty$, and this causes A_n to converge to this sum very slowly. In many instances, it may even be practically impossible to notice the convergence of $\{A_n\}$ to $S(\{a_k\})$ numerically. Thus, use of the sequence of partial sums $\{A_n\}$ to approximate $S(\{a_k\})$ may be of limited benefit in most cases of practical interest.

Infinite series that occur most commonly in applications are power series $\sum_{n=0}^{\infty} c_n z^n$, Fourier cosine and sine series $\sum_{n=0}^{\infty} c_n \cos nx$ and $\sum_{n=1}^{\infty} c_n \sin nx$, series of orthogonal polynomials such as Fourier–Legendre series $\sum_{n=0}^{\infty} c_n P_n(x)$, and series of other special functions. Now the powers z^n satisfy the homogeneous two-term recursion relation $z^{n+1} = z \cdot z^n$. Similarly, the functions $\cos nx$ and $\sin nx$ satisfy the homogeneous three-term recursion relation $f_{n+1} = 2(\cos x) f_n - f_{n-1}$. Both of these recursions involve coefficients that are constant in n. More generally, the Legendre polynomials $P_n(x)$, as well as many other special functions, satisfy linear homogeneous (three-term) recursion relations [or, equivalently, they satisfy linear homogeneous difference equations (of order 2)], whose coefficients have asymptotic expansions in n^{-1} for $n \to \infty$. In many cases, the coefficients c_n in these series as well satisfy difference equations of a similar nature, and this puts the terms $c_n z^n$, $c_n \cos nx$, $c_n \sin nx$, $c_n P_n(x)$, etc., in some sequence classes that we shall denote $\mathbf{b}^{(m)}$, where $m = 1, 2, \dots$.

As will be seen shortly, the class $\mathbf{b}^{(m)}$ is a discrete counterpart of the class $\mathbf{B}^{(m)}$ that was defined in Chapter 5 on the D-transformation for infinite-range integrals. In

fact, all the developments of this chapter parallel those of Chapter 5. In particular, the
d-transformation for the infinite series we develop here is a genuine discrete analogue
of the *D*-transformation.

6.1.1 Description of the Class $\mathbf{A}_0^{(\gamma)}$

Before going on to the definition of the class $\mathbf{b}^{(m)}$, we need to define another class of
functions we denote $\mathbf{A}_0^{(\gamma)}$.

Definition 6.1.1 A function $\alpha(x)$ defined for all large $x > 0$ is in the set $\mathbf{A}_0^{(\gamma)}$ if it has a
Poincaré-type asymptotic expansion of the form

$$\alpha(x) \sim \sum_{i=0}^{\infty} \alpha_i x^{\gamma - i} \quad \text{as } x \to \infty. \tag{6.1.3}$$

If, in addition, $\alpha_0 \neq 0$ in (6.1.3), then $\alpha(x)$ is said to belong to $\mathbf{A}_0^{(\gamma)}$ strictly. Here γ is
complex in general.

Comparing Definition 6.1.1 with Definition 5.1.1, we realize that $\mathbf{A}^{(\gamma)} \subset \mathbf{A}_0^{(\gamma)}$, and that
Remarks 1–7 following Definition 5.1.1 that apply to the sets $\mathbf{A}^{(\gamma)}$ apply to the sets $\mathbf{A}_0^{(\gamma)}$ as
well. Remarks 8–11 are irrelevant to the sets $\mathbf{A}_0^{(\gamma)}$, as functions in $\mathbf{A}_0^{(\gamma)}$ are not required to
have any differentiability properties. There is a discrete analogue of Remark 8, however.
For the sake of completeness, we provide all these as Remarks 1–8 here.

Remarks.

1. $\mathbf{A}_0^{(\gamma)} \supset \mathbf{A}_0^{(\gamma-1)} \supset \mathbf{A}_0^{(\gamma-2)} \supset \cdots$, so that if $\alpha \in \mathbf{A}_0^{(\gamma)}$, then, for any positive integer k,
   $\alpha \in \mathbf{A}_0^{(\gamma+k)}$ but not strictly. Conversely, if $\alpha \in \mathbf{A}_0^{(\delta)}$ but not strictly, then $\alpha \in \mathbf{A}_0^{(\delta-k)}$
   strictly for a unique positive integer k.
2. If $\alpha \in \mathbf{A}_0^{(\gamma)}$ strictly, then $\alpha \notin \mathbf{A}_0^{(\gamma-1)}$.
3. If $\alpha \in \mathbf{A}_0^{(\gamma)}$ strictly, and $\beta(x) = \alpha(cx + d)$ for some arbitrary constants $c > 0$ and d,
   then $\beta \in \mathbf{A}_0^{(\gamma)}$ strictly as well.
4. If $\alpha, \beta \in \mathbf{A}_0^{(\gamma)}$, then $\alpha \pm \beta \in \mathbf{A}_0^{(\gamma)}$ as well. (This implies that the zero function is
   included in $\mathbf{A}_0^{(\gamma)}$.) If $\alpha \in \mathbf{A}_0^{(\gamma)}$ and $\beta \in \mathbf{A}_0^{(\gamma+k)}$ strictly for some positive integer k, then
   $\alpha \pm \beta \in \mathbf{A}_0^{(\gamma+k)}$ strictly.
5. If $\alpha \in \mathbf{A}_0^{(\gamma)}$ and $\beta \in \mathbf{A}_0^{(\delta)}$, then $\alpha\beta \in \mathbf{A}_0^{(\gamma+\delta)}$; if, in addition, $\beta \in \mathbf{A}_0^{(\delta)}$ strictly, then
   $\alpha/\beta \in \mathbf{A}_0^{(\gamma-\delta)}$.
6. If $\alpha \in \mathbf{A}_0^{(\gamma)}$ strictly, such that $\alpha(x) > 0$ for all large x, and we define $\theta(x) = [\alpha(x)]^\xi$,
   then $\theta \in \mathbf{A}_0^{(\gamma\xi)}$ strictly.
7. If $\alpha \in \mathbf{A}_0^{(\gamma)}$ strictly and $\beta \in \mathbf{A}_0^{(k)}$ strictly for some positive integer k, such that $\beta(x) > 0$
   for all large $x > 0$, and we define $\theta(x) = \alpha(\beta(x))$, then $\theta \in \mathbf{A}_0^{(k\gamma)}$ strictly. Similarly,
   if $\mu(t) \sim \sum_{i=0}^{\infty} \mu_i t^{\delta+i}$ as $t \to 0+$, $\mu_0 \neq 0$, and if $\beta \in \mathbf{A}_0^{(-k)}$ strictly for some positive
   integer k, such that $\beta(x) > 0$ for all large $x > 0$, and we define $\psi(x) = \mu(\beta(x))$, then
   $\psi \in \mathbf{A}_0^{(-k\delta)}$ strictly.

8. If $\alpha \in \mathbf{A}_0^{(\gamma)}$ (strictly), and $\beta(x) = \alpha(x+d) - \alpha(x)$ for an arbitrary constant $d \neq 0$, then $\beta \in \mathbf{A}_0^{(\gamma-1)}$ (strictly) when $\gamma \neq 0$. If $\alpha \in \mathbf{A}_0^{(0)}$, then $\beta \in \mathbf{A}_0^{(-2)}$.

Before going on, we define subsets $\mathbf{X}_0^{(\gamma)}$ of $\mathbf{A}_0^{(\gamma)}$ as follows:

(i) A function α belongs to $\mathbf{X}_0^{(\gamma)}$ if either $\alpha \equiv 0$ or $\alpha \in \mathbf{A}_0^{(\gamma-k)}$ strictly for some non-negative integer k.

(ii) $\mathbf{X}_0^{(\gamma)}$ is closed under addition and multiplication by scalars.

(iii) If $\alpha \in \mathbf{X}_0^{(\gamma)}$ and $\beta \in \mathbf{X}_0^{(\delta)}$, then $\alpha\beta \in \mathbf{X}_0^{(\gamma+\delta)}$; if, in addition, $\beta \in \mathbf{A}_0^{(\delta)}$ strictly, then $\alpha/\beta \in \mathbf{X}_0^{(\gamma-\delta)}$.

It is obvious that no two functions in $\mathbf{X}_0^{(\gamma)}$ have the same asymptotic expansion, since if $\alpha, \beta \in \mathbf{X}_0^{(\gamma)}$, then either $\alpha \equiv \beta$ or $\alpha - \beta \in \mathbf{A}_0^{(\gamma-k)}$ strictly for some nonnegative integer k. Thus, $\mathbf{X}_0^{(\gamma)}$ does not contain functions $\alpha(x) \not\equiv 0$ that satisfy $\alpha(x) = O(x^{-\mu})$ as $x \to \infty$ for every $\mu > 0$, such as $\exp(-cx^s)$ with $c, s > 0$. Needless to say, the subsets $\mathbf{X}_0^{(\gamma)}$ defined here are analogous to the subsets $\mathbf{X}^{(\gamma)}$ defined in Chapter 5. Furthermore, $\mathbf{X}^{(\gamma)} \subset \mathbf{X}_0^{(\gamma)}$.

6.1.2 Description of the Class $\mathbf{b}^{(m)}$

Definition 6.1.2 A sequence $\{a_n\}$ belongs to the set $\mathbf{b}^{(m)}$ if it satisfies a linear homogeneous difference equation of order m of the form

$$a_n = \sum_{k=1}^{m} p_k(n) \Delta^k a_n, \tag{6.1.4}$$

where $p_k \in \mathbf{A}_0^{(k)}$, $k = 1, \ldots, m$, such that $p_k \in \mathbf{A}_0^{(i_k)}$ strictly for some integer $i_k \leq k$. Here $\Delta^0 a_n = a_n$, $\Delta^1 a_n = \Delta a_n = a_{n+1} - a_n$, and $\Delta^k a_n = \Delta(\Delta^{k-1}a_n)$, $k = 2, 3, \ldots$.

By recalling that

$$\Delta^k a_n = \sum_{i=0}^{k} (-1)^{k-i} \binom{k}{i} a_{n+i}, \tag{6.1.5}$$

it is easy to see that the difference equation in (6.1.4) can be expressed as an equivalent $(m+1)$-term recursion relation of the form

$$a_{n+m} = \sum_{i=0}^{m-1} u_i(n) a_{n+i}, \tag{6.1.6}$$

with $u_i \in \mathbf{A}_0^{(v_i)}$, where v_i are some integers. [Note, however, that not every sequence $\{a_n\}$ that satisfies such a recursion relation is in $\mathbf{b}^{(m)}$.] Conversely, a recursion relation of the form (6.1.6) can be rewritten as a difference equation of the form $\sum_{k=0}^{m} v_k(n)\Delta^k a_n = 0$

with $v_k \in \mathbf{A}_0^{(\mu_k)}$ for some integers μ_k by using

$$a_{n+i} = \sum_{k=0}^{i} \binom{i}{k} \Delta^k a_n. \tag{6.1.7}$$

This fact is used in the examples below.

The following results are consequences of Definition 6.1.2 and form discrete analogues of Propositions 5.1.3–5.1.6.

Proposition 6.1.3 *If* $\{a_n\} \in \mathbf{b}^{(m)}$, *then* $\{a_n\} \in \mathbf{b}^{(m')}$ *for every* $m' > m$.

Proof. It is enough to consider $m' = m + 1$. Let (6.1.4) be the difference equation satisfied by $\{a_n\}$. Applying to both sides of (6.1.4) the difference operator $[1 + \mu(n)\Delta]$, where $\mu(n)$ is an arbitrary function in $\mathbf{A}_0^{(1)}$, and using the fact that

$$\Delta(u_n v_n) = u_{n+1}\Delta v_n + (\Delta u_n)v_n, \tag{6.1.8}$$

we have $a_n = \sum_{k=1}^{m+1} q_k(n)\Delta^k a_n$ with $q_1(n) = p_1(n) + \mu(n)\Delta p_1(n) - \mu(n)$, $q_k(n) = p_k(n) + \mu(n)\Delta p_k(n) + \mu(n)p_{k-1}(n+1)$, $k = 2, \ldots, m$, and $q_{m+1}(n) = \mu(n)p_m(n+1)$. From the fact that $\mu \in \mathbf{A}_0^{(1)}$ and $p_k \in \mathbf{A}_0^{(k)}$, $k = 1, \ldots, m$, it follows that $q_k \in \mathbf{A}_0^{(k)}$, $k = 1, \ldots, m+1$. ∎

We observe from the proof of Proposition 6.1.3 that if $\{a_n\} \in \mathbf{b}^{(m)}$, then, for any $m' > m$, there are infinitely many difference equations of the form $a_n = \sum_{k=1}^{m'} q_k(n)\Delta^k a_n$ with $q_k \in \mathbf{A}_0^{(k)}$. Analogously to what we did in Chapter 5, we ask concerning the situation in which $\{a_n\} \in \mathbf{b}^{(m)}$ with *minimal* m whether the difference equation $a_n = \sum_{k=1}^{m} p_k(n)\Delta^k a_n$ with $p_k \in \mathbf{A}_0^{(k)}$ is unique. We assume that it is in general; we can prove that it is when $p_k(n)$ are restricted to the subsets $\mathbf{X}_0^{(k)}$. This assumption can be invoked to prove a result analogous to Theorem 5.6.4.

Since the cost of the d-transformation increases with increasing m, knowing the minimal m is important for computational economy.

Concerning the minimal m, we can prove the following results, whose proofs we leave out as they are analogous to those of Propositions 5.1.4 and 5.1.5.

Proposition 6.1.4 *If* $\{a_n\}$ *satisfies a difference equation of order* m *of the form* $a_n = \sum_{k=1}^{m} p_k(n)\Delta^k a_n$ *with* $p_k \in \mathbf{A}_0^{(v_k)}$ *for some integers* v_k, *and if* m *is smallest possible, then* $\Delta^k a_n$, $k = 0, 1, \ldots, m - 1$, *are independent in the sense that there do not exist functions* $v_k(n)$, $k = 0, 1, \ldots, m - 1$, *not all identically zero and* $v_k \in \mathbf{A}_0^{(\tau_k)}$ *with* τ_k *integers, such that* $\sum_{k=0}^{m-1} v_k(n)\Delta^k a_n = 0$. *In addition,* $\Delta^i a_n$, $i = m, m + 1, \ldots$, *can all be expressed in the form* $\Delta^i a_n = \sum_{k=0}^{m-1} w_{ik}(n)\Delta^k a_n$, *where* $w_{ik} \in \mathbf{A}_0^{(\mu_{ik})}$ *for some integers* μ_{ik}. *This applies, in particular, when* $\{a_n\} \in \mathbf{b}^{(m)}$.

Proposition 6.1.5 *If* $\{a_n\}$ *satisfies a difference equation of order* m *of the form* $a_n = \sum_{k=1}^{m} p_k(n)\Delta^k a_n$ *with* $p_k \in \mathbf{X}_0^{(v_k)}$ *for some integers* v_k, *and if* m *is smallest possible,* *then* $p_k(n)$ *in this difference equation are unique. This applies, in particular, when* $\{a_n\} \in \mathbf{b}^{(m)}$.

The following proposition concerns the sequence $\{\Delta a_n\}$ when $\{a_n\} \in \mathbf{b}^{(m)}$ in particular.

Proposition 6.1.6 *If* $\{a_n\}$ *satisfies a difference equation of order* m *of the form* $a_n = \sum_{k=1}^m p_k(n)\Delta^k a_n$ *with* $p_k \in \mathbf{A}_0^{(v_k)}$ *for some integers* v_k, *then* $\{\Delta a_n\}$ *satisfies a difference equation of the same form, namely,* $\Delta a_n = \sum_{k=1}^m q_k(n)\Delta^{k+1} a_n$ *with* $q_k \in \mathbf{A}_0^{(\mu_k)}$ *for some integers* μ_k, *provided* $[1 - \Delta p_1(n)] \in \mathbf{A}_0^{(\tau)}$ *strictly for some integer* τ. *In particular, if* $\{a_n\} \in \mathbf{b}^{(m)}$, *then* $\{\Delta a_n\} \in \mathbf{b}^{(m)}$ *as well, provided* $\lim_{n\to\infty} n^{-1} p_1(n) \neq 1$.

Following are a few examples of sequences $\{a_n\}$ in the classes $\mathbf{b}^{(1)}$ and $\mathbf{b}^{(2)}$.

Example 6.1.7 The sequence $\{a_n\}$ with $a_n = u_n/n$, where $u_n = B\cos n\theta + C\sin n\theta$ with arbitrary constants B and C, is in $\mathbf{b}^{(2)}$ when $\theta \neq 2k\pi$, $k = 0, \pm 1, \pm 2, \dots$, since it satisfies the difference equation $a_n = p_1(n)\Delta a_n + p_2(n)\Delta^2 a_n$, where

$$p_1(n) = \frac{(\xi - 1)n + \xi - 2}{(1 - \xi)(n + 1)} \quad \text{and} \quad p_2(n) = \frac{n + 2}{2(\xi - 1)(n + 1)}, \quad \text{with } \xi = \cos\theta \neq 1.$$

Thus, $p_1 \in \mathbf{A}_0^{(0)}$ and $p_2 \in \mathbf{A}_0^{(0)}$, and both strictly. Also note that, as $p_1(n)$ and $p_2(n)$ are rational functions in n, they are both in $\mathbf{A}^{(0)}$ strictly. To derive this difference equation, we start with the known three-term recursion relation that is satisfied by the u_n, namely, $u_{n+2} = 2\xi u_{n+1} - u_n$. We next substitute $u_k = ka_k$ in this recursion, to obtain $(n + 2)a_{n+2} - 2\xi(n + 1)a_{n+1} + na_n = 0$, and finally use (6.1.7). From the recursion relation for the u_n, we can also deduce that $\{u_n\} \in \mathbf{b}^{(2)}$.

Example 6.1.8 The sequence $\{a_n\}$ with $a_n = P_n(x)/(n + 1)$ is in $\mathbf{b}^{(2)}$ because it satisfies the difference equation $a_n = p_1(n)\Delta a_n + p_2(n)\Delta^2 a_n$, where

$$p_1(n) = -\frac{2n^2(1 - x) + n(10 - 7x) + 12 - 6x}{(2n^2 + 7n)(1 - x) + 7 - 6x},$$

$$p_2(n) = -\frac{n^2 + 5n + 6}{(2n^2 + 7n)(1 - x) + 7 - 6x}.$$

Obviously, $p_1 \in \mathbf{A}_0^{(0)}$ and $p_2 \in \mathbf{A}_0^{(0)}$ strictly provided $x \neq 1$. (When $x = 1$, $p_1 \in \mathbf{A}_0^{(1)}$ and $p_2 \in \mathbf{A}_0^{(2)}$.) Also note that, as both $p_1(n)$ and $p_2(n)$ are rational functions of n, they are in the corresponding sets $\mathbf{A}^{(\gamma)}$ strictly as well. This difference equation can be derived from the known recursion relation for Legendre polynomials $P_n(x)$, namely,

$$(n + 2)P_{n+2}(x) - (2n + 3)x P_{n+1}(x) + (n + 1)P_n(x) = 0,$$

first by substituting $P_k(x) = (k + 1)a_k$ to obtain

$$(n + 2)(n + 3)a_{n+2} - (n + 2)(2n + 3)x a_{n+1} + (n + 1)^2 a_n = 0,$$

and next by using (6.1.7). From the recursion relation for the $P_n(x)$, we can also deduce that $\{P_n(x)\} \in \mathbf{b}^{(2)}$.

Example 6.1.9 The sequence $\{a_n\}$ with $a_n = H_n/[n(n+1)]$ and $H_n = \sum_{k=1}^{n} 1/k$ is in $\mathbf{b}^{(2)}$. To see this, observe that $\{a_n\}$ satisfies the difference equation $a_n = p_1(n)\Delta a_n + p_2(n)\Delta^2 a_n$, where

$$p_1(n) = -\frac{(n+2)(5n+9)}{2(2n+3)} \quad \text{and} \quad p_2(n) = -\frac{(n+2)^2(n+3)}{2(2n+3)}.$$

Obviously, $p_1 \in \mathbf{A}^{(1)}$ and $p_2 \in \mathbf{A}^{(2)}$ strictly, because both are rational functions of n. This difference equation can be derived as follows: First, we have $\Delta[n(n+1)a_n] = \Delta H_n = (n+1)^{-1}$. Next, we have $\Delta\{(n+1)\Delta[n(n+1)a_n]\} = 0$. Finally, we apply (6.1.7). [Note that, because $H_n \sim \log n + \sum_{i=0}^{\infty} c_i/n^i$ as $n \to \infty$, for some constants c_i, we have $a_n = \beta(n)\log n + \gamma(n)$ for some $\beta, \gamma \in \mathbf{A}^{(-2)}$.]

Example 6.1.10 The sequence $\{a_n\}$ with $a_n = h(n)$, where $h \in \mathbf{A}_0^{(\gamma)}$ for arbitrary $\gamma \neq 0$, is in $\mathbf{b}^{(1)}$. To see this, observe that $\{a_n\}$ satisfies $a_n = p_1(n)\Delta a_n$ with

$$p_1(n) = \frac{a_n}{\Delta a_n} = \left[\frac{h(n+1)}{h(n)} - 1\right]^{-1} \sim \gamma^{-1}n + \sum_{i=0}^{\infty} e_i n^{-i} \quad \text{as } n \to \infty,$$

so that $p_1 \in \mathbf{A}_0^{(1)}$ strictly.

As a special case, consider $h(n) = n^{-2}$. Then $p_1(n) = -(n+1)^2/(2n+1)$ exactly. Thus, in this case p_1 is in $\mathbf{A}^{(1)}$ strictly, as well as being in $\mathbf{A}_0^{(1)}$ strictly.

Example 6.1.11 The sequence $\{a_n\}$ with $a_n = \zeta^n h(n)$, where $\zeta \neq 1$ and $h \in \mathbf{A}_0^{(\gamma)}$ for arbitrary γ, is in $\mathbf{b}^{(1)}$. To see this, observe that $\{a_n\}$ satisfies $a_n = p_1(n)\Delta a_n$ with

$$p_1(n) = \frac{a_n}{\Delta a_n} = \left[\zeta\frac{h(n+1)}{h(n)} - 1\right]^{-1} \sim (\zeta-1)^{-1} + \sum_{i=1}^{\infty} e_i n^{-i} \quad \text{as } n \to \infty,$$

so that $p_1 \in \mathbf{A}_0^{(0)}$ strictly.

As a special case, consider $h(n) = n^{-1}$. Then $p_1(n) = (n+1)/[(\zeta-1)n-1]$ exactly. Thus, in this case p_1 is in $\mathbf{A}^{(0)}$ strictly, as well as being in $\mathbf{A}_0^{(0)}$ strictly.

6.1.3 Asymptotic Expansion of A_n When $\{a_n\} \in \mathbf{b}^{(m)}$

We now state a general theorem due to Levin and Sidi [165] concerning the asymptotic behavior of the partial sum A_n as $n \to \infty$ when $\{a_n\} \in \mathbf{b}^{(m)}$ for some m and $\sum_{k=1}^{\infty} a_k$ converges. This theorem is the discrete analogue of Theorem 5.1.12 for infinite-range integrals. Its proof is analogous to that of Theorem 5.1.12 given in Section 5.6. It can be achieved by replacing integration by parts by summation by parts and derivatives by forward differences. See also Levin and Sidi [165] for a sketch. A complete proof for the case $m = 1$ is given in Sidi [270], and this proof is now contained in the proof of Theorem 6.6.6 in this chapter. By imposing the assumption about the uniqueness of the difference equation $a_n = \sum_{k=1}^{m} p_k(n)\Delta^k a_n$ when m is minimal, a result analogous to Theorem 5.6.4 can be proved concerning A_n as well.

Theorem 6.1.12 *Let the sequence* $\{a_n\}$ *be in* $\mathbf{b}^{(m)}$ *and let* $\sum_{k=1}^{\infty} a_k$ *be a convergent series. Assume, in addition, that*

$$\lim_{n \to \infty} \left(\Delta^{j-1} p_k(n) \right) \left(\Delta^{k-j} a_n \right) = 0, \quad k = j, j+1, \ldots, m, \quad j = 1, 2, \ldots, m, \quad (6.1.9)$$

and that

$$\sum_{k=1}^{m} l(l-1) \cdots (l-k+1) \bar{p}_k \neq 1, \quad l = \pm 1, 2, 3, \ldots, \quad (6.1.10)$$

where

$$\bar{p}_k = \lim_{n \to \infty} n^{-k} p_k(n), \quad k = 1, \ldots, m. \quad (6.1.11)$$

Then

$$A_{n-1} = S(\{a_k\}) + \sum_{k=0}^{m-1} n^{\rho_k} (\Delta^k a_n) g_k(n) \quad (6.1.12)$$

for some integers $\rho_k \leq k+1$, *and functions* $g_k \in \mathbf{A}_0^{(0)}$, $k = 0, 1, \ldots, m-1$. *Actually, if* $p_k \in \mathbf{A}_0^{(i_k)}$ *strictly for some integer* $i_k \leq k$, $k = 1, \ldots, m$, *then*

$$\rho_k \leq \bar{\rho}_k \equiv \max\{i_{k+1}, i_{k+2} - 1, \ldots, i_m - m + k + 1\} \leq k + 1, \quad k = 0, 1, \ldots, m-1. \quad (6.1.13)$$

Equality holds in (6.1.13) when the integers whose maximum is being considered are distinct. Finally, being in $\mathbf{A}_0^{(0)}$, *the functions* $g_k(n)$ *have asymptotic expansions of the form*

$$g_k(n) \sim \sum_{i=0}^{\infty} g_{ki} n^{-i} \quad \text{as } n \to \infty. \quad (6.1.14)$$

Remarks.

1. By (6.1.11), $\bar{p}_k \neq 0$ if and only if $p_k \in \mathbf{A}_0^{(k)}$ strictly. Thus, whenever $p_k \in \mathbf{A}_0^{(i_k)}$ with $i_k < k$, we have $\bar{p}_k = 0$. This implies that whenever $i_k < k$, $k = 1, \ldots, m$, we have $\bar{p}_k = 0$, $k = 1, \ldots, m$, and the condition in (6.1.10) is automatically satisfied.
2. It follows from (6.1.13) that $\rho_{m-1} = i_m$ always.
3. Similarly, for $m = 1$ we have $\rho_0 = i_1$ precisely.
4. For numerous examples we have treated, equality seems to hold in (6.1.13) for all $k = 1, \ldots, m$.
5. The integers ρ_k and the functions $g_k(n)$ in (6.1.12) depend only on the functions $p_k(n)$ in the difference equation in (6.1.4). This being the case, they are the same for all solutions a_n of (6.1.4) that satisfy (6.1.9) and for which $\sum_{k=1}^{\infty} a_k$ converges.
6. From (6.1.9) and (6.1.13), we also have that $\lim_{n \to \infty} n^{\bar{\rho}_k} \Delta^k a_n = 0$, $k = 0, 1, \ldots, m-1$.
7. Finally, Theorem 6.1.12 says that the sequence $\{G_n = S(\{a_k\}) - A_{n-1}\}$ of the remainders of $\sum_{k=1}^{\infty} a_k$ is in $\mathbf{b}^{(m)}$ if $\{a_n\} \in \mathbf{b}^{(m)}$ too. This follows from the fact that $\Delta^k G_n = -\Delta^{k-1} a_n$, $k = 1, 2, \ldots$.

By making the analogy $A_{n-1} \leftrightarrow A(y)$, $n^{-1} \leftrightarrow y$, $n^{\rho_k-1}(\Delta^{k-1}a_n) \leftrightarrow \phi_k(y)$ and $r_k = 1$, $k = 1, \ldots, m$, and $S(\{a_k\}) \leftrightarrow A$, we realize that $A(y)$ is in $\mathbf{F}^{(m)}$. Finally, the variable y is discrete in this case and assumes the values $1, 1/2, 1/3, \ldots$.

All the conditions of Theorem 6.1.12 are satisfied by Examples 6.1.7 and 6.1.8. They are satisfied by Examples 6.1.10 and 6.1.11 provided the corresponding series $\sum_{k=1}^{\infty} a_k$ converge. The series of Example 6.1.10 converges provided $\Re\gamma < -1$. The series of Example 6.1.11 converges provided either (i) $|z| < 1$ or (ii) $|z| = 1$, $z \neq 1$, and $\Re\gamma < 0$.

The many examples we have studied seem to indicate that the requirement that $\{a_n\} \in \mathbf{b}^{(m)}$ for some m is the most crucial of the conditions in Theorem 6.1.12. The rest of the conditions, namely, (6.1.9)–(6.1.11), appear to be satisfied automatically. Therefore, to decide whether $A(y) \equiv A_{n-1}$, where $y = n^{-1}$, is in $\mathbf{F}^{(m)}$ for some m, it is practically sufficient to check whether $\{a_n\}$ is in $\mathbf{b}^{(m)}$. Later in this chapter, we provide some simple ways to check this point.

Finally, even though Theorem 6.1.12 is stated for sequences $\{a_n\} \in \mathbf{b}^{(m)}$ for which $\sum_{k=1}^{\infty} a_k$ converges, A_n may satisfy (6.1.10)–(6.1.12) also when $\{a_n\} \in \mathbf{b}^{(m)}$ without $\sum_{k=1}^{\infty} a_k$ being convergent, at least in some cases. In such a case, the constant $S(\{a_k\})$ in (6.1.12) will be the antilimit of $\{A_n\}$. In Theorem 6.6.6, we show that (6.1.10)–(6.1.12) hold for *all* $\{a_n\} \in \mathbf{b}^{(1)}$ for which $\sum_{k=1}^{\infty} a_k$ converge and for a large subset of sequences $\{a_n\} \in \mathbf{b}^{(1)}$ for which $\sum_{k=1}^{\infty} a_k$ do not converge but a_n grow at most like a power of n as $n \to \infty$. As a matter of fact, Theorem 6.6.6 is valid for a class of sequences denoted $\tilde{\mathbf{b}}^{(m)}$ that *includes* $\mathbf{b}^{(1)}$.

We now demonstrate the result of Theorem 6.1.12 via some examples that were treated earlier.

Example 6.1.13 Consider the sequence $\{n^{-z}\}_{n=1}^{\infty}$ with $\Re z > -1$ that was treated in Example 4.1.7, and, prior to that, in Example 1.1.4. From Example 6.1.10, we know that this sequence is in $\mathbf{b}^{(1)}$. The asymptotic expansion in (4.1.9) can be rewritten in the form

$$A_{n-1} \sim \zeta(z) + na_n \sum_{i=0}^{\infty} g_{0i}n^{-i} \quad \text{as } n \to \infty,$$

for some constants g_{0i}, with $g_{00} = (1-z)^{-1} \neq 0$, completely in accordance with Theorem 6.1.12. This expansion is valid also when $z \neq 1, 0, -1, -2, \ldots$, as shown in Example 4.1.7.

Example 6.1.14 Consider the sequence $\{z^n/n\}_{n=1}^{\infty}$ with $|z| \leq 1$ and $z \neq 1$ that was treated in Example 4.1.8. From Example 6.1.11, we know that this sequence is in $\mathbf{b}^{(1)}$. The asymptotic expansion in (4.1.11) is actually

$$A_{n-1} \sim \log(1-z)^{-1} + a_n \sum_{i=0}^{\infty} g_{0i}n^{-i} \quad \text{as } n \to \infty,$$

for some constants g_{0i} with $g_{00} = (z-1)^{-1}$, completely in accordance with Theorem 6.1.12. Furthermore, this expansion is valid also when $\sum_{k=1}^{\infty} a_k$ diverges with z not on the branch cut of $\log(1-z)$, that is, also when $|z| \geq 1$ but $z \notin [1, +\infty)$, as shown in Example 4.1.8.

Example 6.1.15 Consider the sequence $\{\cos n\theta / n\}_{n=1}^{\infty}$ with $\theta \neq 2\pi k, \ k = 0, \pm 1,$ $\pm 2, \ldots$, that was treated in Example 4.1.9. From Example 6.1.7, we know that this sequence is in $\mathbf{b}^{(2)}$. Substituting

$$\frac{\sin n\theta}{n} = \left[-(1 + n^{-1}) \csc \theta + \cot \theta \right] \frac{\cos n\theta}{n} - \left[(1 + n^{-1}) \csc \theta \right] \Delta \left(\frac{\cos n\theta}{n} \right)$$

in the asymptotic expansion of (4.1.12), we obtain

$$A_{n-1} \sim -\log \left| 2 \sin \frac{\theta}{2} \right| + a_n \sum_{i=0}^{\infty} g_{0i} n^{-i} + \Delta a_n \sum_{i=0}^{\infty} g_{1i} n^{-i} \quad \text{as } n \to \infty,$$

for some constants g_{0i} and g_{1i} that depend only on θ, completely in accordance with Theorem 6.1.12.

6.1.4 Remarks on the Asymptotic Expansion of A_n and a Simplification

A very useful feature of Theorem 6.1.12 is the *simplicity* of the asymptotic expansion of A_{n-1} given in (6.1.12)–(6.1.14). This is worth noting especially as the term a_n may itself have a very complicated behavior as $n \to \infty$, and this behavior does not show explicitly in (6.1.12). It is present there *implicitly* through $\Delta^k a_n, \ k = 0, 1, \ldots, m - 1$, or equivalently, through $a_{n+k}, \ k = 0, 1, \ldots, m - 1$.

As noted earlier, A_{n-1} is analogous to an $A(y)$ in $\mathbf{F}^{(m)}$ with $\phi_k(y) \leftrightarrow n^{\rho_k - 1} \Delta^{k-1} a_n$, $k = 1, \ldots, m$. This may give the impression that we have to know the integers ρ_k to proceed with GREP$^{(m)}$ to find approximations to $S(\{a_k\})$. As we have seen, ρ_k depend on the difference equation in (6.1.4), which we do not expect or intend to know in general. This lack of precise knowledge of the ρ_k could lead us to conclude that we do not know the $\phi_k(y)$ precisely, and hence cannot apply GREP$^{(m)}$ in all cases. Really, we do *not* need to know the ρ_k exactly; we can replace each ρ_k by its known upper bound $k + 1$, and rewrite (6.1.12) in the form

$$A_{n-1} = S(\{a_k\}) + \sum_{k=0}^{m-1} n^{k+1} (\Delta^k a_n) h_k(n), \quad (6.1.15)$$

where $h_k(n) = n^{\rho_k - k - 1} g_k(n)$, hence $h_k \in \mathbf{A}_0^{(\rho_k - k - 1)} \subseteq \mathbf{A}_0^{(0)}$ for each k. Note that, when $\rho_k = k + 1$, we have $h_k(n) = g_k(n)$, and when $\rho_k < k + 1$, we have

$$h_k(n) \sim \sum_{i=0}^{\infty} h_{ki} n^{-i} \equiv 0 \cdot n^0 + 0 \cdot n^{-1} + \cdots + 0 \cdot n^{\rho_k - k}$$

$$+ g_{k0} n^{\rho_k - k - 1} + g_{k1} n^{\rho_k - k - 2} + \cdots \quad \text{as } n \to \infty. \quad (6.1.16)$$

Now that we have established the validity of (6.1.15) with $h_k \in \mathbf{A}_0^{(0)}$ for each k, we have also derived a new set of form factors or shape functions $\phi_k(y)$, namely, $\phi_k(y) \leftrightarrow n^k \Delta^{k-1} a_n, k = 1, \ldots, m$. Furthermore, these $\phi_k(y)$ are immediately available and simply expressible in terms of the series elements a_n. [Compare them with the $\phi_k(y)$ of Chapter 5 that are expressed in terms of the integrand $f(x)$ and its derivatives.]

Finally, before we turn to the definition of the d-transformation, we make one minor change in (6.1.15) by adding the term a_n to both sides. This results in

$$A_n = S(\{a_k\}) + n[h_0(n) + n^{-1}]a_n + \sum_{k=1}^{m-1} n^{k+1}(\Delta^k a_n)h_k(n). \qquad (6.1.17)$$

As $h_0(n) + n^{-1}$, just as $h_0(n)$, is in $\mathbf{A}_0^{(0)}$, the asymptotic expansion of A_n in (6.1.17) is of the same form as that of A_{n-1} in (6.1.15). Thus, we have

$$A_n \sim S(\{a_k\}) + \sum_{k=0}^{m-1} n^{k+1}(\Delta^k a_n) \sum_{i=0}^{\infty} \frac{h_{ki}}{n^i} \quad \text{as } n \to \infty. \qquad (6.1.18)$$

6.2 Definition of the $d^{(m)}$-Transformation

Let us reexpand the functions $h_k(n)$ in negative powers of $n + \alpha$ for some fixed α, which is legitimate. The asymptotic expansion in (6.1.18) then assumes the form

$$A_n \sim S(\{a_k\}) + \sum_{k=0}^{m-1} n^{k+1}(\Delta^k a_n) \sum_{i=0}^{\infty} \frac{\tilde{h}_{ki}}{(n+\alpha)^i} \quad \text{as } n \to \infty.$$

Based on this asymptotic expansion of A_n, we now give the definition of the Levin–Sidi d-transformation for approximating the sum $S(\{a_k\})$ of the infinite series $\sum_{k=1}^{\infty} a_k$. As mentioned earlier, the $d^{(m)}$-transformation is a GREP$^{(m)}$.

Definition 6.2.1 Pick a sequence of integers $\{R_l\}_{l=0}^{\infty}$, $1 \le R_0 < R_1 < R_2 < \cdots$. Let $n \equiv (n_1, \ldots, n_m)$, where $n_1, \ldots, n_m$ are nonnegative integers. Then the approximation $d_n^{(m,j)}$ to $S(\{a_k\})$ is defined through the linear system

$$A_{R_l} = d_n^{(m,j)} + \sum_{k=1}^{m} R_l^k(\Delta^{k-1} a_{R_l}) \sum_{i=0}^{n_k-1} \frac{\bar{\beta}_{ki}}{(R_l+\alpha)^i}, \quad j \le l \le j+N; \quad N = \sum_{k=1}^{m} n_k, \qquad (6.2.1)$$

$\alpha > -R_0$ being a parameter at our disposal and $\bar{\beta}_{ki}$ being the additional (auxiliary) N unknowns. In (6.2.1), $\sum_{i=0}^{-1} c_i \equiv 0$ so that $d_{(0,\ldots,0)}^{(m,j)} = A_j$ for all j. We call this GREP that generates the $d_n^{(m,j)}$ the $d^{(m)}$-transformation. When there is no room for confusion, we call it the d-transformation for short. [Of course, if the $\bar{\rho}_k$ are known, we can replace the factors $R_l^k(\Delta^{k-1} a_{R_l})$ in (6.2.1) by $R_l^{\bar{\rho}_{k-1}}(\Delta^{k-1} a_{R_l})$.]

Remarks.

1. A good choice of the parameter α appears to be $\alpha = 0$. We adopt this choice in the sequel. We have adopted it in all our numerical experiments as well.
2. From (6.2.1), it is clear that the input needed for the $d^{(m)}$-transformation is the integer m, integers R_l, and the sequence elements a_i, $1 \le i \le R_{j+N}$. We consider the issue of determining m later in this chapter. We note only that no harm is done if m is overestimated, but no acceleration of convergence should be expected in general if m is underestimated.

3. Because we have the freedom to pick the integers R_l as we wish, we can pick them to induce better convergence acceleration to $S(\{a_k\})$ and/or better numerical stability. This is one of the important advantages of the d-transformation over other convergence acceleration methods for infinite series and sequences that we discuss later.

4. The way the $d^{(m)}$-transformation is defined depends only on the sequence $\{a_k\}$ and is totally independent of whether or not this sequence is in $\mathbf{b}^{(m)}$ and/or satisfies Theorem 6.1.12. Therefore, the d-transformation can be applied to any infinite series $\sum_{k=1}^{\infty} a_k$, whether $\{a_k\} \in \mathbf{b}^{(m)}$ or not. Whether this application will produce good approximations to the sum of the series depends on the asymptotic behavior of a_k. If $\{a_k\} \in \mathbf{b}^{(m)}$ for some m, then the $d^{(m)}$-transformation will produce good results. It may produce good results with some m even when $\{a_k\} \in \mathbf{b}^{(q)}$ for some $q > m$ but $\{a_k\} \notin \mathbf{b}^{(m)}$, at least in some cases of interest.

5. Despite its somewhat complicated appearance in Definition 6.2.1, the $d^{(m)}$-transformation can be implemented very efficiently by the W-algorithm (see Section 7.2) when $m = 1$ and by the $\mathrm{W}^{(m)}$-algorithm (see Section 7.3) when $m \geq 2$. We present the implementation of the $d^{(1)}$-transformation via the W-algorithm in he next section.

6.2.1 Kernel of the $d^{(m)}$-Transformation

From Definition 6.2.1, it is clear that the kernel of the $d^{(m)}$-transformation (with $\alpha = 0$) is all sequences $\{A_r = \sum_{k=1}^{r} a_k\}$, such that

$$\sum_{k=r+1}^{\infty} a_k = \sum_{k=1}^{m} (\Delta^{k-1} a_r) \sum_{i=0}^{n_k-1} \alpha_{ki} r^{k-i}, \quad \text{some finite } n_k. \tag{6.2.2}$$

For these sequences, there holds $d_n^{(m,j)} = S(\{a_k\})$ for all j, when $n = (n_1, \dots, n_m)$. This holds, for example, when $\{a_k\}$ satisfies $a_r = \sum_{k=1}^{m} p_k(r) \Delta^k a_r$, $p_k(r)$ being a polynomial in r of degree at most k for each k. In this case, (6.2.2) holds with $n_k - 1 = k$.

6.2.2 The $d^{(m)}$-Transformation for Infinite Sequences

So far, we have defined the $d^{(m)}$-transformation for infinite series $\sum_{k=1}^{\infty} a_k$. Noting that $a_n = \Delta A_{n-1} = A_n - A_{n-1}$, $n = 1, 2, \dots$, where $A_0 \equiv 0$, we can reformulate the $d^{(m)}$-transformation for arbitrary infinite sequences $\{A_k\}_{k=1}^{\infty}$, whether (6.1.15) is satisfied or not.

Definition 6.2.2 Let $n \equiv (n_1, \dots, n_m)$ and the R_l be as in Definition 6.2.1. Let $\{A_k\}$ be a sequence with limit or antilimit A. Then, the approximation $d_n^{(m,j)}$ to A along with the additional (auxiliary) unknowns $\bar{\beta}_{ki}$, $0 \leq i \leq n_k - 1$, $1 \leq k \leq m$, is defined through the linear system

$$A_{R_l} = d_n^{(m,j)} + \sum_{k=1}^{m} R_l^k (\Delta^k A_{R_l-1}) \sum_{i=0}^{n_k-1} \frac{\bar{\beta}_{ki}}{(R_l + \alpha)^i}, \quad j \leq l \leq j + N; \ N = \sum_{k=1}^{m} n_k. \tag{6.2.3}$$

In (6.2.3), $\sum_{i=0}^{-1} c_i \equiv 0$ so that $d_{(0,\dots,0)}^{(m,j)} = A_j$ for all j. Also, $A_0 \equiv 0$.

In this form, the $d^{(m)}$-transformation is a truly universal extrapolation method for infinite sequences.

6.2.3 The Factorial $d^{(m)}$-Transformation

By rewriting the asymptotic expansions of the functions $h_k(n)$ in (6.1.18) in different forms, as explained in Section 4.6 on extensions of GREP, we obtain other forms of the d-transformation. For example, we can write an arbitrary asymptotic series $\sum_{i=0}^{\infty} \gamma_i/n^i$ as $n \to \infty$ also in the form $\sum_{i=0}^{\infty} \hat{\gamma}_i/(n)_i$ as $n \to \infty$, where $(n)_0 = 1$ and $(n)_i = \prod_{s=0}^{i-1}(n+s)$ for $i \geq 1$. Here $\hat{\gamma}_i = \gamma_i$, for $0 \leq i \leq 2$, $\hat{\gamma}_3 = \gamma_2 + \gamma_3$, and so on. For each i, $\hat{\gamma}_i$ is uniquely determined by $\gamma_0, \gamma_1, \dots, \gamma_i$.

If we now rewrite the asymptotic expansions $\sum_{i=0}^{\infty} h_{ki}/n^i$ as $n \to \infty$ in the form $\sum_{i=0}^{\infty} \hat{h}_{ki}/(n)_i$ as $n \to \infty$, and proceed as before, we can define the *factorial $d^{(m)}$-transformation* for infinite series via the linear equations

$$A_{R_l} = d_n^{(m,j)} + \sum_{k=1}^{m} R_l^k(\Delta^{k-1} a_{R_l}) \sum_{i=0}^{n_k-1} \frac{\bar{\beta}_{ki}}{(R_l+\alpha)_i}, \quad j \leq l \leq j+N; \quad N = \sum_{k=1}^{m} n_k, \tag{6.2.4}$$

and that for infinite sequences via linear the equations

$$A_{R_l} = d_n^{(m,j)} + \sum_{k=1}^{m} R_l^k(\Delta^k A_{R_l-1}) \sum_{i=0}^{n_k-1} \frac{\bar{\beta}_{ki}}{(R_l+\alpha)_i}, \quad j \leq l \leq j+N; \quad N = \sum_{k=1}^{m} n_k. \tag{6.2.5}$$

6.3 Special Cases with $m = 1$
6.3.1 The $d^{(1)}$-Transformation

Let us replace the R_l^k in the equations in (6.2.1) by $R_l^{\rho_k}$, and take $\alpha = 0$ for simplicity. When $m = 1$, these equations assume the form

$$A_{R_l} = d_n^{(1,j)} + \omega_{R_l} \sum_{i=0}^{n-1} \frac{\bar{\beta}_i}{R_l^i}, \quad j \leq l \leq j+n; \quad \omega_r = r^\rho a_r, \tag{6.3.1}$$

where n now is a positive integer and ρ stands for ρ_1. These equations can be solved for $d_n^{(1,j)}$ (with arbitrary R_l) very simply and efficiently via the W-algorithm of [278] as follows:

$$M_0^{(j)} = \frac{A_{R_j}}{\omega_{R_j}}, \quad N_0^{(j)} = \frac{1}{\omega_{R_j}}, \quad j \geq 0; \quad \omega_r = r^\rho a_r,$$

$$M_n^{(j)} = \frac{M_{n-1}^{(j+1)} - M_{n-1}^{(j)}}{R_{j+n}^{-1} - R_j^{-1}}, \quad N_n^{(j)} = \frac{N_{n-1}^{(j+1)} - N_{n-1}^{(j)}}{R_{j+n}^{-1} - R_j^{-1}}, \quad j \geq 0, \ n \geq 1.$$

$$d_n^{(1,j)} = \frac{M_n^{(j)}}{N_n^{(j)}}, \quad j, n \geq 0.$$

6.3.2 The Levin $\mathcal{L}$-Transformation

Choosing $R_l = l + 1$ in (6.3.1), we obtain

$$A_r = d_n^{(1,j)} + \omega_r \sum_{i=0}^{n-1} \frac{\bar{\beta}_i}{r^i}, \quad J \le r \le J + n; \quad \omega_r = r^\rho a_r, \quad J = j + 1, \quad (6.3.2)$$

the resulting $d^{(1)}$-transformation being nothing but the famous t- and u-transformations of Levin [161], with $\rho = 0$ and $\rho = 1$, respectively. Let us denote $d_n^{(1,j)}$ in (6.3.2) by $\mathcal{L}_n^{(j)}$. Then $\mathcal{L}_n^{(j)}$ has the following known closed form that was given in [161]:

$$\mathcal{L}_n^{(j)} = \frac{\Delta^n \left(J^{n-1} A_J / \omega_J \right)}{\Delta^n \left(J^{n-1} / \omega_J \right)} = \frac{\sum_{i=0}^n (-1)^i \binom{n}{i} (J+i)^{n-1} A_{J+i} / \omega_{J+i}}{\sum_{i=0}^n (-1)^i \binom{n}{i} (J+i)^{n-1} / \omega_{J+i}}; \quad J = j + 1.$$

$$(6.3.3)$$

The comparative study of Smith and Ford [317], [318] has shown that the Levin transformations are extremely efficient for summing a large class of infinite series $\sum_{k=1}^\infty a_k$ with $\{a_n\}_{n=1}^\infty \in \mathbf{b}^{(1)}$. We return to these transformations in Chapters 12 and 19.

6.3.3 The Sidi $\mathcal{S}$-Transformation

Letting $m = 1$ and $R_l = l + 1$, and replacing R_l^k by $R_l^{\rho_k}$, the equations in (6.2.4) assume the form

$$A_r = d_n^{(1,j)} + \omega_r \sum_{i=0}^{n-1} \frac{\bar{\beta}_i}{(r)_i}, \quad J \le r \le J + n; \quad \omega_r = r^\rho a_r, \quad J = j + 1. \quad (6.3.4)$$

The resulting factorial $d^{(1)}$-transformation is the $\mathcal{S}$-transformation of Sidi. Let us denote $d_n^{(1,j)}$ in (6.3.4) by $\mathcal{S}_n^{(j)}$. Then $\mathcal{S}_n^{(j)}$ has the following known closed form that was given in [277]:

$$\mathcal{S}_n^{(j)} = \frac{\Delta^n \left((J)_{n-1} A_J / \omega_J \right)}{\Delta^n \left((J)_{n-1} / \omega_J \right)} = \frac{\sum_{i=0}^n (-1)^i \binom{n}{i} (J+i)_{n-1} A_{J+i} / \omega_{J+i}}{\sum_{i=0}^n (-1)^i \binom{n}{i} (J+i)_{n-1} / \omega_{J+i}}; \quad J = j + 1.$$

$$(6.3.5)$$

The $\mathcal{S}$-transformation was first used for summing infinite power series in the M.Sc. thesis of Shelef [265] that was done under the supervision of the author. The comparative study of Grotendorst [116] has shown that it is one of the most effective methods for summing a large class of everywhere-divergent power series. See also Weniger [353], who called the method the $\mathcal{S}$-transformation. We return to it in Chapters 12 and 19.

6.4 How to Determine m

As in the case of the D-transformation for infinite-range integrals, in applying the d-transformation to a given infinite series $\sum_{k=1}^\infty a_k$, we must first assign a value to the integer m, for which $\{a_n\} \in \mathbf{b}^{(m)}$. In this section, we deal with the question of how to determine the smallest value of m or an upper bound for it in a simple manner. Our approach here parallels that of Section 5.4.

6.4.1 By Trial and Error

The simplest approach to this problem is via trial and error. We start by applying the $d^{(1)}$-transformation. If this is successful, then we accept $m = 1$ and stop. If not, we apply the $d^{(2)}$-transformation. If this is successful, then we accept $m = 2$ and stop. Otherwise, we try the $d^{(3)}$-transformation, and so on. We hope that for some (smallest) value of m the $d^{(m)}$-transformation will perform well. Of course, this will be the case when $\{a_n\} \in \mathbf{b}^{(m)}$ for this m. We also note that the $d^{(m)}$-transformation may perform well for some m even when $\{a_n\} \notin \mathbf{b}^{(s)}$ for any s, at least for some cases, as mentioned previously.

The trial-and-error approach has almost no extra cost involved as the partial sums A_{R_l} that we need as input need to be computed once only, and they can be used again for each value of m. In addition, the computational effort due to implementing the $d^{(m)}$-transformation for different values of m is small when the W- and W$^{(m)}$-algorithms of Chapter 7 are used for this purpose.

6.4.2 Upper Bounds on m

We now use a heuristic approach that allows us to determine by inspection of a_n some values for the integer m for which $\{a_n\} \in \mathbf{b}^{(m)}$. Only this time we relax somewhat the conditions in Definition 6.1.2: we assume that a_n satisfies a difference equation of the form (6.1.4) with $p_k \in \mathbf{A}_0^{(v_k)}$, where v_k is an integer not necessarily less than or equal to k, $k = 1, 2, \dots, m$. [Our experience shows that if $\{a_n\}$ belongs to this new set $\mathbf{b}^{(m)}$ and $\sum_{k=1}^{\infty} a_k$ converges, then $\{a_n\}$ belongs to the set $\mathbf{b}^{(m)}$ of Definition 6.1.2 as well.] The idea in the present approach is that the general term a_n is viewed as a product or as a sum of simpler terms, and it is assumed that we know to which classes $\mathbf{b}^{(s)}$ the sequences of these simpler terms belong.

The following results, which we denote Heuristics 6.4.1–6.4.3, pertain precisely to this subject. The demonstrations of these results are based on the assumption that certain linear systems, whose entries are in $\mathbf{A}_0^{(\gamma)}$ for various integer values of γ, are invertible. For this reason, we call these results heuristics and not lemmas or theorems, and we refer to their demonstrations as "proofs".

Heuristic 6.4.1 *Let $\{g_n\} \in \mathbf{b}^{(r)}$ and $\{h_n\} \in \mathbf{b}^{(s)}$ and assume that $\{g_n\}$ and $\{h_n\}$ satisfy different difference equations of the form described in Definition 6.1.2. Then*

(i) $\{g_n h_n\} \in \mathbf{b}^{(m)}$ *with $m \leq rs$, and*
(ii) $\{g_n + h_n\} \in \mathbf{b}^{(m)}$ *with $m \leq r + s$.*

Heuristic 6.4.2 *Let $\{g_n\} \in \mathbf{b}^{(r)}$ and $\{h_n\} \in \mathbf{b}^{(r)}$ and assume that $\{g_n\}$ and $\{h_n\}$ satisfy the same difference equation of the form described in Definition 6.1.2. Then*

(i) $\{g_n h_n\} \in \mathbf{b}^{(m)}$ *with $m \leq r(r + 1)/2$, and*
(ii) $\{g_n + h_n\} \in \mathbf{b}^{(m)}$ *with $m \leq r$.*

Heuristic 6.4.3 *Let* $\{g_n^{(i)}\}_{n=1}^\infty \in \mathbf{b}^{(r)}$, $i = 1, \dots, \mu$, *and assume that all* μ *sequences satisfy the same difference equation of the form described in Definition 6.1.2. Let* $a_n = \prod_{i=1}^\mu g_n^{(i)}$ *for all n. Then* $\{a_n\} \in \mathbf{b}^{(m)}$ *with* $m \leq \binom{r+\mu-1}{\mu}$. *In particular, if* $\{g_n\} \in \mathbf{b}^{(r)}$, *then* $\{(g_n)^\mu\} \in \mathbf{b}^{(m)}$ *with* $m \leq \binom{r+\mu-1}{\mu}$.

The "proofs" of these are analogous to those of Heuristics 5.4.1–5.4.3. They can be achieved by using (6.1.8) and by replacing (5.4.4) with

$$\Delta^j(g_n h_n) = \sum_{i=0}^j \binom{j}{i}(\Delta^{j-i}g_{n+i})(\Delta^i h_n) = \sum_{i=0}^j \binom{j}{i}\sum_{s=0}^i \binom{i}{s}(\Delta^{j-s}g_n)(\Delta^i h_n). \quad (6.4.1)$$

We leave the details to the reader.

These results are important because, in most instances of practical interest, a_n is a product or a sum of terms $g_n, h_n, \dots$, such that $\{g_n\}, \{h_n\}, \dots$, are in the classes $\mathbf{b}^{(r)}$ for low values of r, such as $r = 1$ and $r = 2$.

We now apply these results to a few examples.

Example 6.4.4 Consider $a_n = \sum_{i=1}^m a_n^{(i)}$, where $\{a_n^{(i)}\} \in \mathbf{b}^{(1)}$ for each i. By part (ii) of Heuristic 6.4.1, we conclude that $\{a_n\} \in \mathbf{b}^{(m')}$ for some $m' \leq m$. This occurs, in particular, when $a_n = \sum_{i=1}^m h_i(n)$, where $h_i \in \mathbf{A}_0^{(\gamma_i)}$ for some distinct $\gamma_i \neq 0$, since $\{h_i(n)\} \in \mathbf{b}^{(1)}$ for each i, by Example 6.1.10. It also occurs when $a_n = \sum_{i=1}^m \zeta_i^n h_i(n)$, where $\zeta_i \neq 1$ and are distinct and $h_i \in \mathbf{A}_0^{(\gamma_i)}$ for some arbitrary γ_i not necessarily distinct, since $\{\zeta_i^n h_i(n)\} \in \mathbf{b}^{(1)}$ for each i, by Example 6.1.11. Such sequences $\{a_n\}$ are very common and are considered again and in greater detail in Section 6.8, where we prove that they are in $\mathbf{b}^{(m)}$ with $\mathbf{b}^{(m)}$ exactly as in Definition 6.1.2.

Example 6.4.5 Consider $a_n = g(n)u_n$ where $g \in \mathbf{A}_0^{(\gamma)}$ for some γ and $u_n = B\cos n\theta + C\sin n\theta$ with B and C constants. From Example 6.1.7, we already know that $\{u_n\} \in \mathbf{b}^{(2)}$. From Example 6.1.10, we also know that $\{g(n)\} \in \mathbf{b}^{(1)}$ since $g \in \mathbf{A}_0^{(\gamma)}$. By part (i) of Heuristic 6.4.1, we therefore conclude that $\{a_n\} \in \mathbf{b}^{(2)}$. Indeed, using the technique of Example 6.1.7, we can show that $\{a_n\}$ satisfies the difference equation $a_n = p_1(n)\Delta a_n + p_2(n)\Delta^2 a_n$ with

$$p_1(n) = \frac{2g(n)[g(n+1) - \xi g(n+2)]}{w(n)} \quad \text{and} \quad p_2(n) = -\frac{g(n)g(n+1)}{w(n)},$$

where

$$w(n) = g(n)g(n+1) - 2\xi g(n)g(n+2) + g(n+1)g(n+2) \quad \text{and} \quad \xi = \cos\theta.$$

Thus, $p_1, p_2 \in \mathbf{A}_0^{(0)}$ strictly when $\xi \neq 1$, and, therefore, $\{a_n\} \in \mathbf{b}^{(2)}$ with $\mathbf{b}^{(m)}$ exactly as in Definition 6.1.2.

Example 6.4.6 Consider $a_n = g(n)P_n(x)$, where $P_n(x)$ is the Legendre polynomial of degree n and $g \in \mathbf{A}_0^{(\gamma)}$ for some γ. We already know that $\{P_n(x)\} \in \mathbf{b}^{(2)}$. Also,

$\{g(n)\} \in \mathbf{b}^{(1)}$ because $g \in \mathbf{A}_0^{(\gamma)}$, by Example 6.1.10. By part (i) of Heuristic 6.4.1, we conclude that $\{a_n\} \in \mathbf{b}^{(2)}$. Indeed, using the technique of Example 6.1.8, we can show that $a_n = p_1(n)\Delta a_n + p_2(n)\Delta^2 a_n$, with

$$p_1(n) = \frac{[x(2n+3)g(n+2) - 2(n+2)g(n+1)]g(n)}{w(n)},$$

$$p_2(n) = -\frac{(n+2)g(n)g(n+1)}{w(n)},$$

where

$$w(n) = (n+2)g(n)g(n+1) - x(2n+3)g(n)g(n+2) + (n+1)g(n+1)g(n+2).$$

Also, when $x \neq 1$, we have that $p_1, p_2 \in \mathbf{A}_0^{(0)}$ strictly because $w \in \mathbf{A}_0^{(2\gamma+1)}$ strictly, and, therefore, $\{a_n\} \in \mathbf{b}^{(2)}$ with $\mathbf{b}^{(m)}$ exactly as in Definition 6.1.2. [Recall that Example 6.1.8 is a special case with $g(n) = 1/(n+1)$.]

Example 6.4.7 Consider $a_n = g(n)u_n P_n(x)$, where $g(n)$, u_n, and $P_n(x)$ are as in Examples 6.4.5 and 6.4.6. We already have that $\{g(n)\} \in \mathbf{b}^{(1)}$, $\{u_n\} \in \mathbf{b}^{(2)}$, and $\{P_n(x)\} \in \mathbf{b}^{(2)}$. Thus, from part (i) of Heuristic 6.4.1, we conclude that $\{a_n\} \in \mathbf{b}^{(4)}$. It can be shown by different techniques that, when $x = \cos\theta$, $\{a_n\} \in \mathbf{b}^{(3)}$, again in agreement with part (i) of Heuristic 6.4.1.

Example 6.4.8 Consider $a_n = (\sin n\theta/n)^2$. Then $\{a_n\} \in \mathbf{b}^{(3)}$. This follows from the fact that $\{\sin n\theta/n\} \in \mathbf{b}^{(2)}$ and from part (i) of Heuristic 6.4.2. Another way to see this is as follows: First, write $a_n = (1 - \cos 2n\theta)/(2n^2) = \frac{1}{2}n^{-2} - \frac{1}{2}n^{-2}\cos 2n\theta$. Now $\{\frac{1}{2}n^{-2}\} \in \mathbf{b}^{(1)}$ since $\frac{1}{2}x^{-2} \in \mathbf{A}^{(-2)}$. Next, because $\{\frac{1}{2}n^{-2}\} \in \mathbf{b}^{(1)}$ and $\{\cos 2n\theta\} \in \mathbf{b}^{(2)}$, their product is also in $\mathbf{b}^{(2)}$ by part (i) of Heuristic 6.4.1. Therefore, by part (ii) of Heuristic 6.4.1, $\{a_n\} \in \mathbf{b}^{(3)}$.

Example 6.4.9 Let $\{g_n\} \in \mathbf{b}^{(r)}$ and $h_n = \sum_{s=0}^q u_s(n)(\log n)^s$, with $u_s \in \mathbf{A}_0^{(\gamma)}$ for every s. Here, some or all of the $u_s(n)$, $0 \leq s \leq q-1$, can be identically zero, but $u_q(n) \not\equiv 0$. Then $\{g_n h_n\} \in \mathbf{b}^{(q+1)r}$. To show this we start with the equalities

$$\Delta^k h_n = \sum_{s=0}^q w_{ks}(n)(\log n)^s, \quad w_{ks} \in \mathbf{A}_0^{(\gamma-k)}, \quad \text{for all } k, s.$$

Treating $q + 1$ of these equalities with $k = 1, \ldots, q + 1$ as a linear system of equations for the "unknowns" $(\log n)^s$, $s = 0, 1, \ldots, q$, and invoking Cramer's rule, we can show that $(\log n)^s = \sum_{k=1}^{q+1} \mu_{sk}(n)\Delta^k h_n$, $\mu_{sk} \in \mathbf{A}_0^{(\nu_{sk}-\gamma)}$ for some integers ν_{sk}. Substituting these in $h_n = \sum_{s=0}^q u_s(n)(\log n)^s$, we get $h_n = \sum_{k=1}^{q+1} e_k(n)\Delta^k h_n$, $e_k \in \mathbf{A}_0^{(\sigma_k)}$ for some integers σ_k. Thus, $\{h_n\} \in \mathbf{b}^{(q+1)}$ in the relaxed sense. Invoking now part (i) of Heuristic 6.4.1, the result follows. We mention that such sequences (and sums of them) arise from the trapezoidal rule approximation of simple and multidimensional integrals with corner and/or edge and/or surface singularities. Thus, the result

Table 6.5.1: *Relative floating-point errors* $\bar{E}_\nu^{(0)} = |\bar{d}_\nu^{(1,0)} - S|/|S|$ *and* $\Gamma_\nu^{(1,0)}$ *for the Riemann Zeta function series* $\sum_{k=1}^\infty k^{-2}$ *via the* $d^{(1)}$*-transformation, with* $R_l = l + 1$. *Here* $S = \pi^2/6$, $\bar{d}_\nu^{(1,0)}$ *are the computed* $d_\nu^{(1,0)}$, *and* R_ν *is the number of terms used in computing* $d_\nu^{(1,0)}$. $\bar{d}_\nu^{(1,0)}$(d) *and* $\bar{d}_\nu^{(1,0)}$(q) *are computed, respectively, in double precision (approximately 16 decimal digits) and in quadruple precision (approximately 35 decimal digits)*

ν	R_ν	$\bar{E}_\nu^{(0)}$(d)	$\bar{E}_\nu^{(0)}$(q)	$\Gamma_\nu^{(1,0)}$
0	1	$3.92D - 01$	$3.92D - 01$	$1.00D + 00$
2	3	$1.21D - 02$	$1.21D - 02$	$9.00D + 00$
4	5	$1.90D - 05$	$1.90D - 05$	$9.17D + 01$
6	7	$6.80D - 07$	$6.80D - 07$	$1.01D + 03$
8	9	$1.56D - 08$	$1.56D - 08$	$1.15D + 04$
10	11	$1.85D - 10$	$1.83D - 10$	$1.35D + 05$
12	13	$1.09D - 11$	$6.38D - 13$	$1.60D + 06$
14	15	$2.11D - 10$	$2.38D - 14$	$1.92D + 07$
16	17	$7.99D - 09$	$6.18D - 16$	$2.33D + 08$
18	19	$6.10D - 08$	$7.78D - 18$	$2.85D + 09$
20	21	$1.06D - 07$	$3.05D - 20$	$3.50D + 10$
22	23	$1.24D - 05$	$1.03D - 21$	$4.31D + 11$
24	25	$3.10D - 04$	$1.62D - 22$	$5.33D + 12$
26	27	$3.54D - 03$	$4.33D - 21$	$6.62D + 13$
28	29	$1.80D - 02$	$5.44D - 20$	$8.24D + 14$
30	31	$1.15D - 01$	$4.74D - 19$	$1.03D + 16$

obtained here implies that the d-transformation can be used successfully to accelerate the convergence of sequences of trapezoidal rule approximations. We come back to this in Chapter 25.

Important Remark. When we know that $a_n = g_n + h_n$ with $\{g_n\} \in \mathbf{b}^{(r)}$ and $\{h_n\} \in \mathbf{b}^{(s)}$, and we can compute g_n and h_n separately, we should go ahead and compute $\sum_{n=1}^\infty g_n$ and $\sum_{n=1}^\infty h_n$ by the $d^{(r)}$- and $d^{(s)}$-transformations, respectively, instead of computing $\sum_{n=1}^\infty a_n$ by the $d^{(r+s)}$-transformation. The reason for this is that, for a given required level of accuracy, fewer terms are needed for the $d^{(r)}$- and $d^{(s)}$-transformations than for the $d^{(r+s)}$-transformation.

6.5 Numerical Examples

We now illustrate the use of the $d^{(m)}$-transformation (with $\alpha = 0$ in Definition 6.2.1) in the summation of a few infinite series of varying complexity. For more examples, we refer the reader to Levin and Sidi [165], Sidi and Levin [312], and Sidi [294], [295].

Example 6.5.1 Consider the Riemann Zeta function series $\sum_{n=1}^\infty n^{-z}$, $\Re z > 1$. We saw earlier that, for $\Re z > 1$, this series converges and its sum is $\zeta(z)$. Also, as we saw in Example 6.1.13, $\{n^{-z}\} \in \mathbf{b}^{(1)}$ and satisfies all the conditions of Theorem 6.1.12.

Table 6.5.2: *Relative floating-point errors* $\bar{E}_\nu^{(0)} = |\bar{d}_\nu^{(1,0)} - S|/|S|$ *and* $\Gamma_\nu^{(1,0)}$ *for the Riemann Zeta function series* $\sum_{k=1}^\infty k^{-2}$ *via the* $d^{(1)}$*-transformation, with* R_l *as in (6.5.1) and* $\sigma = 1.3$ *there. Here* $S = \pi^2/6$, $\bar{d}_\nu^{(1,0)}$ *are the computed* $d_\nu^{(1,0)}$, *and* R_ν *is the number of terms used in computing* $d_\nu^{(1,0)}$. $\bar{d}_\nu^{(1,0)}(d)$ *and* $\bar{d}_\nu^{(1,0)}(q)$ *are computed, respectively, in double precision (approximately 16 decimal digits) and in quadruple precision (approximately 35 decimal digits)*

ν	R_ν	$\bar{E}_\nu^{(0)}(d)$	$\bar{E}_\nu^{(0)}(q)$	$\Gamma_\nu^{(1,0)}$
0	1	$3.92D-01$	$3.92D-01$	$1.00D+00$
2	3	$1.21D-02$	$1.21D-02$	$9.00D+00$
4	5	$1.90D-05$	$1.90D-05$	$9.17D+01$
6	7	$6.80D-07$	$6.80D-07$	$1.01D+03$
8	11	$1.14D-08$	$1.14D-08$	$3.04D+03$
10	18	$6.58D-11$	$6.59D-11$	$3.75D+03$
12	29	$1.58D-13$	$1.20D-13$	$3.36D+03$
14	48	$1.55D-15$	$4.05D-17$	$3.24D+03$
16	80	$7.11D-15$	$2.35D-19$	$2.76D+03$
18	135	$5.46D-14$	$1.43D-22$	$2.32D+03$
20	227	$8.22D-14$	$2.80D-26$	$2.09D+03$
22	383	$1.91D-13$	$2.02D-30$	$1.97D+03$
24	646	$1.00D-13$	$4.43D-32$	$1.90D+03$
26	1090	$4.21D-14$	$7.24D-32$	$1.86D+03$
28	1842	$6.07D-14$	$3.27D-31$	$1.82D+03$
30	3112	$1.24D-13$	$2.52D-31$	$1.79D+03$

In Tables 6.5.1 and 6.5.2, we present the numerical results obtained by applying the $d^{(1)}$-transformation to this series with $z = 2$, for which we have $\zeta(2) = \pi^2/6$. We have done the computations once by choosing $R_l = l + 1$ and once by choosing

$$R_0 = 1, \quad R_l = \begin{cases} R_{l-1} + 1 \text{ if } \lfloor \sigma R_{l-1} \rfloor = R_{l-1} \\ \lfloor \sigma R_{l-1} \rfloor \text{ otherwise} \end{cases}, \quad l = 1, 2, \ldots; \quad \text{for some } \sigma > 1,$$

(6.5.1)

with $\sigma = 1.3$. The former choice of the R_l gives rise to the Levin u-transformation, as we mentioned previously. The latter choice is quite different and induces a great amount of numerical stability. [The choice of the R_l as in (6.5.1) is called *geometric progression sampling* (GPS) and is discussed in detail in Chapter 10.]

Note that, in both tables, we have given floating-point arithmetic results in quadruple precision, as well as in double precision. These show that, with the first choice of the R_l, the maximum accuracy that can be attained is 11 digits in double precision and 22 digits in quadruple precision. Adding more terms to the process does not improve the accuracy; to the contrary, the accuracy dwindles quite quickly. With the R_l as in (6.5.1), on the other hand, we are able to improve the accuracy to almost machine precision.

Example 6.5.2 Consider the Fourier cosine series $\sum_{n=1}^\infty [\cos(2n-1)\theta]/(2n-1)$. This series converges to the function $f(\theta) = -\frac{1}{2} \log |\tan(\theta/2)|$ for every real θ except at

Table 6.5.3: *Relative floating-point errors $\bar{E}_\nu^{(0)} = |\bar{d}_{(\nu,\nu)}^{(2,0)} - S|/|S|$ for the Fourier cosine series of Example 6.5.2 via the $d^{(2)}$-transformation, with $R_l = l + 1$ in the third and fifth columns and with $R_l = 2(l+1)$ in the seventh column. Here $S = f(\theta)$, $\bar{d}_{(\nu,\nu)}^{(2,0)}$ are the computed $d_{(\nu,\nu)}^{(2,0)}$, and $R_{2\nu}$ is the number of terms used in computing $d_{(\nu,\nu)}^{(2,0)}$. $\bar{d}_{(\nu,\nu)}^{(2,0)}$ are computed in double precision (approximately 16 decimal digits)*

ν	$R_{2\nu}$	$\bar{E}_\nu^{(0)}(\theta=\pi/3)$	$R_{2\nu}$	$\bar{E}_\nu^{(0)}(\theta=\pi/6)$	$R_{2\nu}$	$\bar{E}_\nu^{(0)}(\theta=\pi/6)$
0	1	$8.21D-01$	1	$3.15D-01$	2	$3.15D-01$
1	3	$1.56D+00$	3	$1.39D-01$	6	$1.27D-01$
2	5	$6.94D-03$	5	$7.49D-03$	10	$2.01D-03$
3	7	$1.71D-05$	7	$2.77D-02$	14	$1.14D-04$
4	9	$2.84D-07$	9	$3.34D-03$	18	$6.76D-06$
5	11	$5.56D-08$	11	$2.30D-05$	22	$2.27D-07$
6	13	$3.44D-09$	13	$9.65D-06$	26	$4.12D-09$
7	15	$6.44D-11$	15	$4.78D-07$	30	$4.28D-11$
8	17	$9.51D-13$	17	$1.33D-07$	34	$3.97D-13$
9	19	$1.01D-13$	19	$4.71D-08$	38	$2.02D-14$
10	21	$3.68D-14$	21	$5.69D-08$	42	$1.35D-15$
11	23	$9.50D-15$	23	$4.62D-10$	46	$1.18D-15$
12	25	$2.63D-15$	25	$4.50D-11$	50	$1.18D-15$
13	27	$6.06D-16$	27	$2.17D-11$	54	$5.06D-16$
14	29	$8.09D-16$	29	$5.32D-11$	58	$9.10D-15$
15	31	$6.27D-15$	31	$3.57D-11$	62	$9.39D-14$

$\theta = k\pi$, $k = 0, \pm1, \pm2, \ldots$, where $f(\theta)$ has singularities. It is easy to show that $\{[\cos(2n-1)\theta]/(2n-1)\} \in \mathbf{b}^{(2)}$. Consequently, the $d^{(2)}$-transformation is very effective. It turns out that the choice $R_l = \kappa(l+1)$ with some positive integer κ is suitable. When θ is away from the points of singularity, $\kappa = 1$ is sufficient. As θ approaches a point of singularity, κ should be increased. [The choice of the R_l as given here is called *arithmetic progression sampling* (APS) and is discussed in detail in Chapter 10.]

Table 6.5.3 presents the results obtained for $\theta = \pi/3$ and $\theta = \pi/6$ in double-precision arithmetic. While machine precision is reached for $\theta = \pi/3$ with $R_l = l+1$, only 11-digit accuracy is achieved for $\theta = \pi/6$ with the same choice of the R_l. Machine precision is achieved for $\theta = \pi/6$ with $R_l = 2(l+1)$. This phenomenon is analyzed and explained in detail in later chapters.

Example 6.5.3 Consider the Legendre series $\sum_{n=1}^\infty \frac{2n+1}{n(n+1)} P_n(x)$. This series converges to the function $f(x) = -\log[(1-x)/2] - 1$ for $-1 \le x < 1$, and it diverges for all other x. The function $f(x)$ has a branch cut along $[1, +\infty)$, whereas it is well-defined for $x < -1$ even though the series diverges for such x. It is easy to show that $\{\frac{2n+1}{n(n+1)} P_n(x)\} \in \mathbf{b}^{(2)}$. Consequently, the $d^{(2)}$-transformation is very effective in this case too. Again, the choice $R_l = \kappa(l+1)$ with some positive integer κ is suitable. When $x < 1$ and x is away from 1, the branch point of $f(x)$, $\kappa = 1$ is sufficient. As x approaches 1, κ should be increased.

Table 6.5.4 presents the results obtained for $x = 0.3$ with $R_l = l+1$, for $x = 0.9$ with $R_l = 3(l+1)$, and for $x = -1.5$ with $R_l = l+1$, in double-precision arithmetic.

Table 6.5.4: *Relative floating-point errors $\bar{E}_\nu^{(0)} = |\bar{d}_{(\nu,\nu)}^{(2,0)} - S|/|S|$ for the Legendre series of Example 6.5.3 via the $d^{(2)}$-transformation, with $R_l = l + 1$ in the third and seventh columns and with $R_l = 3(l + 1)$ in the fifth column. Here $S = f(x)$, $\bar{d}_{(\nu,\nu)}^{(2,0)}$ are the computed $d_{(\nu,\nu)}^{(2,0)}$, and $R_{2\nu}$ is the number of terms used in computing $d_{(\nu,\nu)}^{(2,0)}$. $\bar{d}_{(\nu,\nu)}^{(2,0)}$ are computed in double precision (approximately 16 decimal digits)*

ν	$R_{2\nu}$	$\bar{E}_\nu^{(0)}(x = 0.3)$	$R_{2\nu}$	$\bar{E}_\nu^{(0)}(x = 0.9)$	$R_{2\nu}$	$\bar{E}_\nu^{(0)}(x = -1.5)$
0	1	$4.00D - 01$	3	$2.26D - 01$	1	$1.03D + 00$
1	3	$2.39D - 01$	9	$1.23D - 01$	3	$1.07D - 01$
2	5	$5.42D - 02$	15	$2.08D - 02$	5	$5.77D - 04$
3	7	$1.45D - 03$	21	$8.43D - 04$	7	$5.73D - 04$
4	9	$1.67D - 04$	27	$9.47D - 05$	9	$1.35D - 06$
5	11	$8.91D - 06$	33	$2.76D - 07$	11	$2.49D - 09$
6	13	$7.80D - 07$	39	$1.10D - 06$	13	$2.55D - 10$
7	15	$2.72D - 07$	45	$1.45D - 07$	15	$3.28D - 13$
8	17	$4.26D - 08$	51	$3.39D - 09$	17	$3.52D - 13$
9	19	$3.22D - 09$	57	$7.16D - 09$	19	$4.67D - 13$
10	21	$2.99D - 10$	63	$2.99D - 10$	21	$6.20D - 13$
11	23	$1.58D - 11$	69	$4.72D - 12$	23	$2.97D - 13$
12	25	$2.59D - 13$	75	$2.28D - 13$	25	$2.27D - 12$
13	27	$4.14D - 14$	81	$1.91D - 14$	27	$1.86D - 11$
14	29	$3.11D - 15$	87	$4.66D - 15$	29	$3.42D - 11$
15	31	$3.76D - 15$	93	$4.89D - 15$	31	$1.13D - 11$

Note that almost machine precision is reached when $x = 0.3$ and $x = 0.9$ for which the series converges. Even though the series diverges for $x = -1.5$, the $d^{(2)}$-transformation produces $f(x)$ to almost 13-digit accuracy. This suggests that the $d^{(m)}$-transformation may be a useful tool for analytic continuation. (Note that the accuracy for $x = -1.5$ decreases as we increase the number of terms of the series used in extrapolation. This is because the partial sums A_n are unbounded as $n \to \infty$.)

6.6 A Further Class of Sequences in $\mathbf{b}^{(m)}$: The Class $\tilde{\mathbf{b}}^{(m)}$

In the preceding sections, we presented examples of sequences in $\mathbf{b}^{(m)}$ and showed how we construct others via Heuristics 6.4.1 and 6.4.2. In this section, we would like to derive a very general class of sequences in $\mathbf{b}^{(m)}$ for arbitrary m. For $m = 1$, this class will turn out to be *all* of $\mathbf{b}^{(1)}$. For $m \geq 2$, we will see that, even though it is not all of $\mathbf{b}^{(m)}$, it is quite large nevertheless. We derive this class by extending $\mathbf{A}_0^{(\gamma)}$ and $\mathbf{b}^{(1)}$ in an appropriate fashion.

We note that the contents of this section are related to the paper by Wimp [362].

6.6.1 The Function Class $\tilde{\mathbf{A}}_0^{(\gamma,m)}$ and Its Summation Properties

We start by generalizing the function class $\mathbf{A}_0^{(\gamma)}$. In addition to being of interest in itself, this generalization will also help us keep the notation simple in the sequel.

Definition 6.6.1 A function $\alpha(x)$ defined for all large x is in the set $\tilde{\mathbf{A}}_0^{(\gamma,m)}$, m a positive integer, if it has a Poincaré-type asymptotic expansion of the form

$$\alpha(x) \sim \sum_{i=0}^{\infty} \alpha_i x^{\gamma-i/m} \quad \text{as } x \to \infty. \tag{6.6.1}$$

In addition, if $\alpha_0 \neq 0$ in (6.6.1), then $\alpha(x)$ is said to belong to $\tilde{\mathbf{A}}_0^{(\gamma,m)}$ strictly. Here γ is complex in general.

With this definition, we obviously have $\tilde{\mathbf{A}}_0^{(\gamma,1)} = \mathbf{A}_0^{(\gamma)}$. Furthermore, all the remarks on the sets $\mathbf{A}_0^{(\gamma)}$ following Definition 6.1.1 apply to the sets $\tilde{\mathbf{A}}_0^{(\gamma,m)}$ with suitable and obvious modifications, which we leave to the reader.

It is interesting to note that, if $\alpha \in \tilde{\mathbf{A}}_0^{(\gamma,m)}$, then $\alpha(x) = \sum_{s=0}^{m-1} q_s(x)$, where $q_s \in \mathbf{A}_0^{(\gamma-s/m)}$ and $q_s(x) \sim \sum_{i=0}^{\infty} \alpha_{s+mi} x^{\gamma-s/m-i}$ as $x \to \infty$, $s = 0, 1, \ldots, m-1$. In other words, $\alpha(x)$ is the sum of m functions $q_s(x)$, each in a class $\mathbf{A}_0^{(\gamma_s)}$. Thus, in view of Example 6.4.4 and by Theorem 6.8.3 of Section 6.8, we may conclude rigorously that if a sequence $\{a_n\}$ is such that $a_n = \alpha(n)$, where $\alpha \in \tilde{\mathbf{A}}_0^{(\gamma,m)}$ for some $\gamma \neq s/m$, $s = 0, 1, \ldots, m-1$, then $\{a_n\} \in \mathbf{b}^{(m)}$. Based on this, we realize that the class $\tilde{\mathbf{A}}_0^{(\gamma,m)}$ may become useful in constructing sequences in $\mathbf{b}^{(m)}$, hence deserves some attention.

We start with the following result on the summation properties of functions in $\tilde{\mathbf{A}}_0^{(\gamma,m)}$. This result extends the list of remarks that succeeds Definition 6.1.1.

Theorem 6.6.2 *Let $g \in \tilde{\mathbf{A}}_0^{(\gamma,m)}$ strictly for some γ with $g(x) \sim \sum_{i=0}^{\infty} g_i x^{\gamma-i/m}$ as $x \to \infty$, and define $G(n) = \sum_{r=1}^{n-1} g(r)$. Then*

$$G(n) = b + c \log n + \tilde{G}(n), \tag{6.6.2}$$

where b and c are constants and $\tilde{G} \in \tilde{\mathbf{A}}_0^{(\gamma+1,m)}$. If $\gamma \neq -1$, then $\tilde{G} \in \tilde{\mathbf{A}}_0^{(\gamma+1,m)}$ strictly, while $\tilde{G} \in \tilde{\mathbf{A}}_0^{(-1/m,m)}$ if $\gamma = -1$. If $\gamma + 1 \neq i/m$, $i = 0, 1, \ldots$, then b is the limit or antilimit of $G(n)$ as $n \to \infty$ and $c = 0$. When $\gamma + 1 = k/m$ for some integer $k \geq 0$, $c = g_k$. Finally,

$$\tilde{G}(n) = \sum_{\substack{i=0 \\ \gamma-i/m\neq-1}}^{m-1} \frac{g_i}{\gamma-i/m+1} n^{\gamma-i/m+1} + O(n^\gamma) \quad \text{as } n \to \infty. \tag{6.6.3}$$

Proof. Let N be an arbitrary integer greater than $(\Re\gamma + 1)m$, and define $\hat{g}(x) = g(x) - \sum_{i=0}^{N-1} g_i x^{\gamma-i/m}$. Thus, $\hat{g} \in \tilde{\mathbf{A}}_0^{(\gamma-N/m,m)}$, and $\sum_{r=n}^{\infty} \hat{g}(r)$ converges since $\hat{g}(x) = O(x^{\gamma-N/m})$ as $x \to \infty$ and $\Re\gamma - N/m < -1$. In fact, $U_N(n) = \sum_{r=n}^{\infty} \hat{g}(r) = O(n^{\gamma-N/m+1})$ as $n \to \infty$. Consequently,

$$G(n) = \sum_{r=1}^{n-1}\left(\sum_{i=0}^{N-1} g_i r^{\gamma-i/m}\right) + U_N(1) - U_N(n)$$

$$= \sum_{i=0}^{N-1} g_i \left(\sum_{r=1}^{n-1} r^{\gamma-i/m}\right) + U_N(1) + O(n^{\gamma-N/m+1}) \quad \text{as } n \to \infty. \tag{6.6.4}$$

From Example 1.1.4 on the Zeta function series, we know that for $\gamma - i/m \neq -1$,

$$\sum_{r=1}^{n-1} r^{\gamma - i/m} = \zeta(-\gamma + i/m) + T_i(n), \quad T_i \in \mathbf{A}^{(\gamma - i/m+1)} \text{ strictly,} \qquad (6.6.5)$$

while if $\gamma - k/m = -1$ for some nonnegative integer k,

$$\sum_{r=1}^{n-1} r^{\gamma - k/m} = \sum_{r=1}^{n-1} r^{-1} = \log n + C + \tilde{T}(n), \quad \tilde{T} \in \mathbf{A}^{(-1)} \text{ strictly,} \qquad (6.6.6)$$

where C is Euler's constant. The result in (6.6.2) can now be obtained by substituting (6.6.5) and (6.6.6) in (6.6.4), and recalling that $g_0 \neq 0$ and that N is arbitrary. We also realize that if $\gamma - i/m \neq -1$, $i = 0, 1, \ldots$, then $b = U_N(1) + \sum_{i=0}^{N-1} g_i \zeta(-\gamma + i/m)$ and b is independent of N, and $c = 0$. If $\gamma - k/m = -1$ for a nonnegative integer k, then $b = U_N(1) + g_k C + \sum_{\substack{i=0 \\ i \neq k}}^{N-1} g_i \zeta(-\gamma + i/m)$ and b is again independent of N, and $c = g_k$. Finally, (6.6.3) follows from the fact that $T_i(n) = \frac{1}{\gamma - i/m+1} n^{\gamma - i/m+1} + O(n^{\gamma - i/m})$ as $n \to \infty$ whenever $\gamma - i/m \neq -1$.   ∎

6.6.2 The Sequence Class $\tilde{\mathbf{b}}^{(m)}$ and a Characterization Theorem

With the classes $\tilde{\mathbf{A}}_0^{(\gamma, m)}$ already defined, we now go on to define the sequence class $\tilde{\mathbf{b}}^{(m)}$ as a generalization of the class $\mathbf{b}^{(1)}$.

Definition 6.6.3 A sequence $\{a_n\}$ belongs to the set $\tilde{\mathbf{b}}^{(m)}$ if it satisfies a linear homogeneous difference equation of first order of the form $a_n = p(n)\Delta a_n$ with $p \in \tilde{\mathbf{A}}_0^{(q/m, m)}$ for some integer $q \leq m$.

Note the analogy between the classes $\tilde{\mathbf{b}}^{(m)}$ and $\mathbf{b}^{(1)}$. Also note that $\tilde{\mathbf{b}}^{(1)}$ is simply $\mathbf{b}^{(1)}$.

Now the difference equation in Definition 6.6.3 can also be expressed as a two-term recursion relation of the form $a_{n+1} = c(n)a_n$ with $c(n) = 1 + 1/p(n)$. We make use of this recursion relation to explore the nature of the sequences in $\tilde{\mathbf{b}}^{(m)}$. In this respect, the following theorem that is closely related to the theory of Birkhoff and Trjitzinsky [25] on general linear difference equations is of major importance.

Theorem 6.6.4

(i) *Let $a_{n+1} = c(n)a_n$ such that $c \in \tilde{\mathbf{A}}_0^{(\mu, m)}$ strictly with μ in general complex. Then a_n is of the form*

$$a_n = [(n-1)!]^\mu \exp[Q(n)] n^\gamma w(n), \qquad (6.6.7)$$

where

$$Q(n) = \sum_{i=0}^{m-1} \theta_i n^{1-i/m} \quad and \quad w \in \tilde{\mathbf{A}}_0^{(0, m)} \text{ strictly.} \qquad (6.6.8)$$

Given that $c(n) \sim \sum_{i=0}^{\infty} c_i n^{\mu - i/m}$ *as* $n \to \infty$, *we have*

$$e^{\theta_0} = c_0; \quad \theta_i = \frac{\epsilon_i}{1 - i/m}, \quad i = 1, \ldots, m-1; \quad \gamma = \epsilon_m, \tag{6.6.9}$$

where the ϵ_i *are defined via*

$$\sum_{s=1}^{m} \frac{(-1)^{s+1}}{s} \left(\sum_{i=1}^{m} \frac{c_i}{c_0} z^i \right)^s = \sum_{i=1}^{m} \epsilon_i z^i + O(z^{m+1}) \text{ as } z \to 0. \tag{6.6.10}$$

(ii) *The converse is also true, that is, if* a_n *is as in (6.6.7) and (6.6.8), then* $a_{n+1} = c(n)a_n$ *with* $c \in \tilde{\mathbf{A}}_0^{(\mu,m)}$ *strictly.*

(iii) *Finally, (a)* $\theta_1 = \cdots = \theta_{m-1} = 0$ *if and only if* $c_1 = \cdots = c_{m-1} = 0$, *and (b)* $\theta_1 = \cdots = \theta_{r-1} = 0$ *and* $\theta_r \neq 0$ *if and only if* $c_1 = \cdots = c_{r-1} = 0$ *and* $c_r \neq 0$, $r \in \{1, \ldots, m-1\}$.

Remark. When $m = 1$, we have $a_n = [(n-1)!]^{\mu} \zeta^n n^{\gamma} w(n)$, with $\zeta = c_0$, $\gamma = c_1/c_0$, and $w(n) \in \mathbf{A}_0^{(0)}$ strictly, as follows easily from (6.6.7)–(6.6.10) above.

Proof. We start with the fact that $a_n = a_1 \left(\prod_{r=1}^{n-1} c(r) \right)$. Next, writing $c(x) = c_0 x^{\mu} u(x)$, where $u \in \tilde{\mathbf{A}}_0^{(0,m)}$ and $u(x) \equiv 1 + v(x) \sim 1 + \sum_{i=1}^{\infty} (c_i/c_0) x^{-i/m}$ as $x \to \infty$, we obtain

$$a_n = a_1 c_0^{n-1} [(n-1)!]^{\mu} \left(\prod_{r=1}^{n-1} u(r) \right). \tag{6.6.11}$$

Now

$$\prod_{r=1}^{n-1} u(r) = \exp \left(\sum_{r=1}^{n-1} \log u(r) \right) = \exp \left(\sum_{r=1}^{n-1} \log[1 + v(r)] \right). \tag{6.6.12}$$

By the fact that $\log(1 + z) = \sum_{s=1}^{\infty} \frac{(-1)^{s+1}}{s} z^s$ for $|z| < 1$ and $v(x) = O(x^{-1/m}) = o(1)$ as $x \to \infty$, it follows that $\log u(x) = \sum_{s=1}^{\infty} \frac{(-1)^{s+1}}{s} [v(x)]^s$ for all large x. Since this (convergent) series also gives the asymptotic expansion of $\log u(x)$ as $x \to \infty$, we see that $\log u(x) \in \tilde{\mathbf{A}}_0^{(-1/m,m)}$. Actually, $\log u(x) \sim \sum_{i=1}^{\infty} \epsilon_i x^{-i/m}$ as $x \to \infty$, where $\epsilon_1, \ldots, \epsilon_m$ are defined exclusively by $c_1/c_0, \ldots, c_m/c_0$ as in (6.6.10). Applying now Theorem 6.6.2 to the sum $\sum_{r=1}^{n-1} \log u(r)$, we obtain

$$\sum_{r=1}^{n-1} \log u(r) = b + \epsilon_m \log n + T(n), \quad T \in \tilde{\mathbf{A}}_0^{(1-1/m,m)}, \tag{6.6.13}$$

where $T(n) = \sum_{i=1}^{m-1} \theta_i n^{1-i/m} + O(1)$ as $n \to \infty$ with $\theta_1, \ldots, \theta_{m-1}$ as in (6.6.9), and b is a constant. The result now follows by combining (6.6.11)–(6.6.13) and defining θ_0 and γ as in (6.6.9). This completes the proof of part (i). Also, by (6.6.9) and (6.6.10), it is clear that $c_1 = \cdots = c_{m-1} = 0$ forces $\epsilon_1 = \cdots = \epsilon_{m-1} = 0$, which in turn forces $\theta_1 = \cdots = \theta_{m-1} = 0$. Next, if $c_1 \neq 0$, then $\epsilon_1 = c_1/c_0 \neq 0$ and hence $\theta_1 = \epsilon_1/(1 - 1/m) \neq 0$. Finally, if $c_1 = \cdots = c_{r-1} = 0$ and $c_r \neq 0$, then $\epsilon_1 = \cdots = \epsilon_{r-1} = 0$ and $\epsilon_r = c_r/c_0 \neq 0$ and hence $\theta_r = \epsilon_r/(1 - r/m) \neq 0$, $r \in \{1, \ldots, m-1\}$. This proves half of part (iii) that assumes the conditions of part (i).

For the proof of part (ii), we begin by forming $c(n) = a_{n+1}/a_n$. We obtain

$$c(n) = n^{\mu}(1 + n^{-1})^{\gamma} \, \frac{w(n+1)}{w(n)} \, \exp\left[\Delta Q(n)\right]; \quad \Delta Q(n) = Q(n+1) - Q(n).$$

$$(6.6.14)$$

Now $X(n) \equiv (1 + n^{-1})^{\gamma}$, $Y(n) \equiv w(n+1)/w(n)$, and $Z(n) \equiv \exp\left[\Delta Q(n)\right]$ are each in $\tilde{\mathbf{A}}_0^{(0,m)}$ strictly. For $X(n)$ and $Y(n)$, this assertion can be verified in a straightforward manner. For $Z(n)$, its truth follows from the fact that $\Delta Q(n) = Q(n+1) - Q(n)$ is either a constant or is in $\tilde{\mathbf{A}}_0^{(0,m)}$. This completes the proof of part (ii). A more careful study is needed to prove the second half of part (iii) that assumes the conditions of part (ii). First, we have $X(n) = 1 + O(n^{-1})$ as $n \to \infty$. Next, $Y(n) = 1 + O(n^{-1-1/m})$ as $n \to \infty$. This is a result of the not so obvious fact that since $w(n) = \sum_{i=0}^{m} w_i n^{-i/m} + O(n^{-1-1/m})$ as $n \to \infty$, then $w(n+1) = \sum_{i=0}^{m} w_i n^{-i/m} + O(n^{-1-1/m})$ as $n \to \infty$ as well. As for $Z(n)$, we have two different cases: When $\theta_1 = \cdots = \theta_{m-1} = 0$, we have $\Delta Q(n) = \theta_0$, from which $Z(n) = e^{\theta_0}$. When $\theta_1 = \cdots = \theta_{r-1} = 0$ and $\theta_r \neq 0$, $r \in \{1, \ldots, m-1\}$, we have

$$\Delta Q(n) = \theta_0 + \sum_{i=r}^{m-1} \theta_i (1 - i/m) n^{-i/m} [1 + O(n^{-1})]$$

$$= \theta_0 + \sum_{i=r}^{m-1} \theta_i (1 - i/m) n^{-i/m} + O(n^{-1-r/m}) \text{ as } n \to \infty.$$

Exponentiating $\Delta Q(n)$, we obtain

$$Z(n) = e^{\theta_0}[1 + \theta_r(1 - r/m)n^{-r/m} + O(n^{-(r+1)/m})] \text{ as } n \to \infty.$$

Combining everything in (6.6.14), the second half of part (iii) now follows. ∎

The next theorem gives necessary and sufficient conditions for a sequence $\{a_n\}$ to be in $\tilde{\mathbf{b}}^{(m)}$. In this sense, it is a characterization theorem for sequences in $\tilde{\mathbf{b}}^{(m)}$. Theorem 6.6.4 becomes useful in the proof.

Theorem 6.6.5 *A sequence $\{a_n\}$ is in $\tilde{\mathbf{b}}^{(m)}$ if and only if its members satisfy $a_{n+1} = c(n)a_n$ with $c \in \tilde{\mathbf{A}}_0^{(s/m,m)}$ for an arbitrary integer s and $c(n) \neq 1 + O(n^{-1-1/m})$ as $n \to \infty$. Specifically, if a_n is as in (6.6.7) with (6.6.8) and with $\mu = s/m$, then $a_n = p(n)\Delta a_n$ with $p \in \tilde{\mathbf{A}}_0^{(\sigma,m)}$ strictly, where $\sigma = q/m$ and q is an integer $\leq m$. In particular, (i) $\sigma = 1$ when $\mu = 0$, $Q(n) \equiv 0$, and $\gamma \neq 0$, (ii) $\sigma = r/m$ when $\mu = 0$, $Q(n) = \sum_{i=r}^{m-1} \theta_i n^{1-i/m}$ with $\theta_r \neq 0$, for $r \in \{0, 1, \ldots, m-1\}$, (iii) $\sigma = 0$ when $\mu < 0$, and (iv) $\sigma = -\mu = -s/m$ when $\mu > 0$.*

Proof. The first part can be proved by analyzing $p(n) = [c(n) - 1]^{-1}$ when $c(n)$ is given, and $c(n) = 1 + 1/p(n)$ when $p(n)$ is given. The second part can be proved by invoking Theorem 6.6.4. We leave the details to the reader.   ∎

6.6.3 *Asymptotic Expansion of* A_n *When* $\{a_n\} \in \tilde{\mathbf{b}}^{(m)}$

With Theorems 6.6.2–6.6.5 available, we go on to the study of the partial sums $A_n = \sum_{k=1}^{n} a_k$ when $\{a_n\} \in \tilde{\mathbf{b}}^{(m)}$ with $a_n = O(n^\lambda)$ as $n \to \infty$ for some λ. [At this point it is worth mentioning that Theorem 6.6.4 remains valid when $Q(n)$ there is replaced by $\theta(n), \theta \in \tilde{\mathbf{A}}_0^{(1,m)}$, and we use this fact here.] More specifically, we are concerned with the following cases in the notation of Theorem 6.6.5:

(i) $a_n = h(n) \in \tilde{\mathbf{A}}_0^{(\gamma,m)}$ strictly for some $\gamma \neq -1 + i/m, \ i = 0, 1, \dots$. In this case, $a_n = p(n)\Delta a_n$ with $p \in \tilde{\mathbf{A}}_0^{(\sigma,m)}$ strictly, $\sigma = 1$.

(ii) $a_n = e^{\theta(n)} h(n)$, where $h \in \tilde{\mathbf{A}}_0^{(\gamma,m)}$ strictly for arbitrary γ and $\theta \in \tilde{\mathbf{A}}_0^{(1-r/m,m)}$ strictly for some $r \in \{0, 1, \dots, m-1\}$, and either (a) $\lim_{n\to\infty} \Re\theta(n) = -\infty$, or (b) $\lim_{n\to\infty} \Re\theta(n)$ is finite. In this case, $a_n = p(n)\Delta a_n$ with $p \in \tilde{\mathbf{A}}_0^{(\sigma,m)}$ strictly, $\sigma = r/m$, thus $0 \leq \sigma < 1$.

(iii) $a_n = [(n-1)!]^\mu e^{\theta(n)} h(n)$, where $h \in \tilde{\mathbf{A}}_0^{(\gamma,m)}$ strictly for arbitrary γ, $\theta \in \tilde{\mathbf{A}}_0^{(1,m)}$ and is arbitrary, and $\mu = s/m$ for an arbitrary integer $s < 0$. In this case, $a_n = p(n)\Delta a_n$ with $p \in \tilde{\mathbf{A}}_0^{(\sigma,m)}$ strictly, $\sigma = 0$.

We already know that, in case (i), $\sum_{k=1}^{\infty} a_k$ converges only when $\Re\gamma < -1$. In case (ii), $\sum_{k=1}^{\infty} a_k$ converges (a) for all γ when $\lim_{n\to\infty} \Re\theta(n) = -\infty$ and (b) for $\Re\gamma < -r/m$ when $\lim_{n\to\infty} \Re\theta(n)$ is finite. In case (iii), convergence takes place always. In all other cases, $\sum_{k=1}^{\infty} a_k$ diverges. The validity of this assertion in cases (i), (ii-a), and (iii) is obvious, for case (ii-b) it follows from Theorem 6.6.6 below. Finally, in case (ii-a) $|a_n| \sim C_1 n^{\Re\gamma} e^{\Re\theta(n)}$ as $n \to \infty$, whereas in case (ii-b) $|a_n| \sim C_2 n^{\Re\gamma}$ as $n \to \infty$. A similar relation holds for case (iii).

Theorem 6.6.6 *Let* $\{a_n\} \in \tilde{\mathbf{b}}^{(m)}$ *be as in the previous paragraph. Then there exist a constant* $S(\{a_k\})$ *and a function* $g \in \tilde{\mathbf{A}}_0^{(0,m)}$ *strictly such that*

$$A_{n-1} = S(\{a_k\}) + n^\sigma a_n \, g(n), \qquad (6.6.15)$$

whether $\sum_{k=1}^{\infty} a_k$ *converges or not.*

Remark. In case $m = 1$ and $\sum_{k=1}^{\infty} a_k$ converges, this theorem is a special case of Theorem 6.1.12 as can easily be verified.

Proof. We start with the proof of case (i) as it is the simplest. In this case, Theorem 6.6.2 applies, and we have $A_{n-1} = b + V(n)$, where $V \in \tilde{\mathbf{A}}_0^{(\gamma+1,m)}$ strictly. Because $na_n = nh(n) \in \tilde{\mathbf{A}}_0^{(\gamma+1,m)}$ strictly as well, $g(n) \equiv V(n)/(na_n)$ is in $\tilde{\mathbf{A}}_0^{(0,m)}$ strictly. The result in (6.6.15) now follows by identifying $S(\{a_k\}) = b$.

For cases (ii) and (iii), we proceed by applying "summation by parts", namely,

$$\sum_{k=r}^{s} x_k \Delta y_k = x_s y_{s+1} - x_{r-1} y_r - \sum_{k=r}^{s} (\Delta x_{k-1}) y_k, \qquad (6.6.16)$$

to $A_{n-1} = \sum_{k=1}^{n-1} a_k = \sum_{k=1}^{n-1} p(k) \Delta a_k$:

$$\sum_{k=1}^{n-1} a_k = p(n-1)a_n - \sum_{k=1}^{n-1} [\Delta p(k-1)] a_k, \qquad (6.6.17)$$

where we have defined $p(k) = 0$ for $k \leq 0$. (Hence $a_k = 0$ for $k \leq 0$ too.) We can now do the same with the series $\sum_{k=1}^{n-1} [\Delta p(k-1)] a_k$ and repeat as many times as we wish. This procedure can be expressed in a simple way by defining

$$u_0(n) = 1; \quad v_{i+1}(n) = p(n-1)u_i(n-1), \quad u_{i+1}(n) = -\Delta v_{i+1}(n), \quad i = 0, 1, \ldots. \qquad (6.6.18)$$

With these definitions, we have by summation by parts

$$\sum_{k=1}^{n-1} u_i(k) a_k = v_{i+1}(n) a_n + \sum_{k=1}^{n-1} u_{i+1}(k) a_k, \quad i = 0, 1, \ldots. \qquad (6.6.19)$$

Summing all these equations, we obtain

$$A_{n-1} = \left[\sum_{i=1}^{N} v_i(n) \right] a_n + \sum_{k=1}^{n-1} u_N(k) a_k, \qquad (6.6.20)$$

for any positive integer N. Now, by the fact that $p \in \tilde{\mathbf{A}}_0^{(\sigma,m)}$ strictly, we have $u_i \in \tilde{\mathbf{A}}_0^{(\tau_i,m)}$ and $v_i \in \tilde{\mathbf{A}}_0^{(\tau_i+1,m)}$, where $\tau_i = i(\sigma-1)$, $i = 1, 2, \ldots$. In addition, by the fact that $v_1(n) = p(n-1)$, $v_1 \in \tilde{\mathbf{A}}_0^{(\sigma,m)}$ strictly. Let us pick $N > (1 + \Re\gamma)/(1-\sigma)$ so that $\tau_N + \Re\gamma < -1$. Then, the infinite series $\sum_{k=1}^{\infty} u_N(k) a_k$ converges, and we can write

$$A_{n-1} = \sum_{k=1}^{\infty} u_N(k) a_k + \left[\sum_{i=1}^{N} v_i(n) \right] a_n - \sum_{k=n}^{\infty} u_N(k) a_k. \qquad (6.6.21)$$

Realizing that $\{v_i(n)\}_{i=1}^{\infty}$ is an asymptotic sequence as $n \to \infty$ and expanding the functions $v_i(n)$, we see that

$$\sum_{i=1}^{N} v_i(n) = \sum_{i=0}^{v_r} \beta_i n^{\sigma-i/m} + O(v_r(n)) \quad \text{as } n \to \infty, \quad \beta_0 \neq 0, \ r \leq N, \qquad (6.6.22)$$

where the integer v_r is defined via $\sigma - (v_r + 1)/m = \tau_r + 1$. Here $\beta_0, \beta_1, \ldots, \beta_{v_r}$ are obtained by expanding only $v_1(n), \ldots, v_{r-1}(n)$, and are not affected by $v_i(n)$, $i \geq r$. In addition, by the asymptotic behavior of $|a_n|$ mentioned above, we also have

$$\sum_{k=n}^{\infty} u_N(k) a_k = O(n^{\tau_N+1} a_n) \quad \text{as } n \to \infty. \qquad (6.6.23)$$

Combining (6.6.20)–(6.6.23), we obtain

$$A_{n-1} = S(\{a_k\}) + n^\sigma \left[\sum_{i=0}^{\nu_r} \beta_i n^{-i/m} + O(n^{-(\nu_r+1)/m}) \right] a_n \quad \text{as } n \to \infty, \quad (6.6.24)$$

with $S(\{a_k\}) = \sum_{k=1}^{\infty} u_N(k) a_k$. This completes the proof. [Note that $S(\{a_k\})$ is independent of N despite its appearance.]   ∎

Remarks.

1. When $\sum_{k=1}^{\infty} a_k$ does not converge, we can say the following about $S(\{a_k\})$: In case (i), $S(\{a_k\})$ is the analytic continuation of the sum of $\sum_{k=1}^{\infty} a_k$ in γ at least when $a_n = n^\gamma w(n)$ with $w(n)$ independent of γ. In case (ii), $S(\{a_k\})$ is the Abelian mean defined by $\lim_{\epsilon \to 0+} \sum_{k=1}^{\infty} e^{-\epsilon k^{1/m}} a_k$. It is also related to generalized functions as we will see when we consider the problem of the summation of Fourier series and their generalizations. [When $\sum_{k=1}^{\infty} a_k$ converges, we have $S(\{a_k\}) = \sum_{k=1}^{\infty} a_k$.]
2. Because the asymptotic expansion of A_n as $n \to \infty$ is of one and the same form whether $\sum_{k=1}^{\infty} a_k$ converges or not, the $\tilde{d}^{(m)}$-transformation that we define next can be applied to approximate $S(\{a_k\})$ in all cases considered above. We will also see soon that the $d^{(m)}$-transformation can be applied effectively as $\{a_k\} \in \tilde{\mathbf{b}}^{(m)}$ implies at least heuristically that $\{a_k\} \in \mathbf{b}^{(m)}$.

6.6.4 The $\tilde{d}^{(m)}$-Transformation

Adding a_n to both sides of (6.6.15), and recalling that $\sigma = q/m$ for $q \in \{0, 1, \dots, m\}$ and $w \in \tilde{\mathbf{A}}_0^{(0,m)}$, we observe that

$$A_n \sim S(\{a_k\}) + n^\sigma a_n \sum_{i=0}^{\infty} \beta_i n^{-i/m} \quad \text{as } n \to \infty, \quad (6.6.25)$$

with all β_i exactly as in (6.6.24) except β_q, to which we have added 1. This means that A_n is analogous to a function $A(y) \in \mathbf{F}^{(1)}$ in the following sense: $A_n \leftrightarrow A(y)$, $n^{-1} \leftrightarrow y$, $n^\sigma a_n \leftrightarrow \phi_1(y)$, $r_1 = 1/m$, and $S(\{a_k\}) \leftrightarrow A$. The variable y is discrete and assumes the values $1, 1/2, 1/3, \dots$. Thus, we can apply GREP$^{(1)}$ to $A(y)$ to obtain good approximations to A. In case we do not wish to bother with the exact value of σ, we can simply replace it by 1, its maximum possible value, retaining the form of (6.6.25) at the same time. That is to say, we now have $na_n \leftrightarrow \phi_1(y)$. As we recall, the W-algorithm can be used to implement GREP$^{(1)}$ very efficiently. (Needless to say, if we know the exact value of σ, especially $\sigma = 0$, we should use it.)

For the sake of completeness, here are the equations that define GREP$^{(1)}$ for the problem at hand:

$$A_{R_l} = \tilde{d}_n^{(m,j)} + R_l^{\hat{\sigma}} a_{R_l} \sum_{i=0}^{n-1} \frac{\bar{\beta}_i}{(R_l + \alpha)^{i/m}}, \quad j \leq l \leq j + n, \quad (6.6.26)$$

where $\hat{\sigma} = \sigma$ when σ is known or $\hat{\sigma} = 1$ otherwise. Again, $\alpha > -R_0$ and a good choice is $\alpha = 0$. We call this GREP$^{(1)}$ the $\tilde{d}^{(m)}$-transformation.

Again, for the sake of completeness, we also give the implementation of this new transformation (with $\alpha = 0$) via the W-algorithm:

$$M_0^{(j)} = \frac{A_{R_j}}{\omega_{R_j}}, \quad N_0^{(j)} = \frac{1}{\omega_{R_j}}, \quad j \geq 0; \quad \omega_r = r^{\hat{\sigma}}(\Delta A_{r-1}),$$

$$M_n^{(j)} = \frac{M_{n-1}^{(j+1)} - M_{n-1}^{(j)}}{R_{j+n}^{-1/m} - R_j^{-1/m}}, \quad N_n^{(j)} = \frac{N_{n-1}^{(j+1)} - N_{n-1}^{(j)}}{R_{j+n}^{-1/m} - R_j^{-1/m}}, \quad j \geq 0, \, n \geq 1,$$

$$\tilde{d}_n^{(m,j)} = \frac{M_n^{(j)}}{N_n^{(j)}}, \quad j, n \geq 0.$$

6.6.5 Does $\{a_n\} \in \tilde{\mathbf{b}}^{(m)}$ Imply $\{a_n\} \in \mathbf{b}^{(m)}$? A Heuristic Approach

Following Definition 6.6.1, we concluded in a rigorous manner that if $a_n = \alpha(n)$, $\alpha \in \tilde{\mathbf{A}}_0^{(\gamma,m)}$, then, subject to some condition on γ, $\{a_n\} \in \mathbf{b}^{(m)}$. Subject to a similar condition on γ, we know from Theorem 6.6.5 that this sequence is in $\tilde{\mathbf{b}}^{(m)}$ as well. In view of this, we now ask whether *every* sequence in $\tilde{\mathbf{b}}^{(m)}$ is also in $\mathbf{b}^{(m)}$. In the sequel, we show heuristically that this is so. We do this again by relaxing the conditions on the set $\mathbf{b}^{(m)}$, exactly as was done in Subsection 6.4.2.

We start with the fact that $a_{n+1} = c(n)a_n$, where $c \in \tilde{\mathbf{A}}_0^{(s/m,m)}$ for an arbitrary integer s. We now claim that we can find functions $\delta_k \in \mathbf{A}_0^{(v_k)}$, v_k integer, $k = 0, 1, \ldots, m-1$, such that

$$a_{n+m} = \sum_{k=0}^{m-1} \delta_k(n)a_{n+k}, \tag{6.6.27}$$

provided a certain matrix is nonsingular. Note that we can always find functions $\delta_k(n)$ not necessarily in $\mathbf{A}_0^{(v_k)}$, v_k integer, for which (6.6.27) holds.

Using $a_{n+1} = c(n)a_n$, we can express (6.6.27) in the form

$$g_m(n)a_n = \sum_{k=0}^{m-1} \delta_k(n)g_k(n)a_n, \tag{6.6.28}$$

where $g_0(n) = 1$ and $g_k(n) = \prod_{j=0}^{k-1} c(n+j)$, $k \geq 1$. We can ensure that (6.6.28) holds if we require

$$g_m(n) = \sum_{k=0}^{m-1} \delta_k(n)g_k(n). \tag{6.6.29}$$

As $g_k \in \tilde{\mathbf{A}}_0^{(ks/m,m)}$, we can decompose it as in

$$g_k(n) = \sum_{i=0}^{m-1} n^{-i/m} g_{ki}(n), \quad g_{ki} \in \mathbf{A}_0^{(\gamma_{ki})} \text{ with } \gamma_{ki} \text{ integer}, \tag{6.6.30}$$

as we showed following Definition 6.6.1. Substituting (6.6.30) in (6.6.29), and equating the coefficients of $n^{-i/m}$ on both sides, we obtain

$$g_{m0}(n) = \delta_0(n) + \sum_{k=1}^{m-1} g_{k0}(n)\delta_k(n)$$

$$g_{mi}(n) = \sum_{k=1}^{m-1} g_{ki}(n)\delta_k(n), \quad i = 1, \dots, m-1. \tag{6.6.31}$$

Therefore, provided $\det[g_{ki}(n)]_{k,i=1}^{m-1} \neq 0$, there is a solution for $\delta_k(n) \in \mathbf{A}_0^{(\nu_k)}$, ν_k integer, $k = 0, 1, \dots, m-1$, for which (6.6.27) is valid, implying that $\{a_n\} \in \mathbf{b}^{(m)}$, with the conditions on $\mathbf{b}^{(m)}$ relaxed as mentioned above.

For $m = 2$, the solution of (6.6.31) is immediate. We have $\delta_1(n) = g_{21}(n)/g_{11}(n)$ and $\delta_0(n) = g_{20}(n) - g_{10}(n)\delta_1(n)$, provided $g_{11}(n) \not\equiv 0$. With this solution, it can be shown at least in some cases that $\{a_n\} \in \tilde{\mathbf{b}}^{(2)}$ and $\sum_{k=1}^{\infty} a_k$ convergent imply that $\{a_n\} \in \mathbf{b}^{(2)}$, with $\mathbf{b}^{(2)}$ as described originally in Definition 6.1.2, that is, $a_n = \sum_{k=1}^{2} p_k(n)\Delta^k a_n$, $p_k \in \mathbf{A}_0^{(k)}$, $k = 1, 2$. Let us now demonstrate this with two examples.

Example 6.6.7 When $c(n) = n^{-1/2}$, we have $p_1(n) = 2\{[n(n+1)]^{-1/2} - 1\}^{-1}$ and $p_2(n) = \frac{1}{2}p_1(n)$ and $p_1, p_2 \in \mathbf{A}^{(0)}$ strictly. Thus, $\{a_n\} \in \mathbf{b}^{(2)}$ as in Definition 6.1.2. Note that $a_n = a_1/\sqrt{(n-1)!}$ in this example.

Example 6.6.8 Assume $c(n) \sim \sum_{i=0}^{\infty} c_i n^{-i/2}$ with $c_0 = 1$ and $c_1 \neq 0$. (Recall that when $\mu = 0$ and $c_0 = 1$, it is necessary that $|c_1| + |c_2| \neq 0$ for $\{a_n\} \in \tilde{\mathbf{b}}^{(2)}$, from Theorem 6.6.5.) After tedious manipulations, it can be shown that

$$p_1(n) = \frac{1}{2c_1^2}(1 - 4c_2) + O(n^{-1}) \quad \text{and} \quad p_2(n) = \frac{1}{c_1^2}n + O(1) \quad \text{as } n \to \infty,$$

so that $p_1 \in \mathbf{A}_0^{(0)}$ and $p_2 \in \mathbf{A}_0^{(1)}$ strictly. Thus, $\{a_n\} \in \mathbf{b}^{(2)}$ as in Definition 6.1.2. Note that $a_n = e^{2c_1\sqrt{n}}u(n)$, where $u \in \tilde{\mathbf{A}}_0^{(\gamma,2)}$ for some γ in this case, as follows from Theorem 6.6.4.

We can now combine the developments above with Heuristics 6.4.1 and 6.4.2 to study sequences $\{a_n\}$ that look more complicated than the ones we encountered in Sections 6.1–6.4 and the new ones we encountered in this section. As an example, let us look at $a_n = g(n)\cos(h(n))$, where $g \in \tilde{\mathbf{A}}_0^{(\gamma,m)}$ and $h \in \tilde{\mathbf{A}}_0^{(1,m)}$. We first notice that $a_n = a_n^+ + a_n^-$, where $a_n^{\pm} = \frac{1}{2}g(n)e^{\pm ih(n)}$. Next, by the fact that $h(n) \sim \sum_{i=0}^{\infty} h_i n^{1-i/m}$ as $n \to \infty$, we can write $h(n) = \tilde{h}(n) + \hat{h}(n)$, where $\tilde{h}(n) = \sum_{i=0}^{m-1} h_i n^{1-i/m}$ and $\hat{h} \in \tilde{\mathbf{A}}_0^{(0,m)}$. As a result, $e^{\pm ih(n)} = e^{\pm i\tilde{h}(n)}e^{\pm i\hat{h}(n)}$. Now since $e^{\pm i\hat{h}(n)} \in \tilde{\mathbf{A}}_0^{(0,m)}$, we have that $a_n^{\pm} = u(n)e^{\pm i\tilde{h}(n)}$ with $u(n) = \frac{1}{2}g(n)e^{i\hat{h}(n)} \in \tilde{\mathbf{A}}_0^{(\gamma,m)}$, satisfies $\{a_n^{\pm}\} \in \tilde{\mathbf{b}}^{(m)}$ by Theorem 6.6.4, and hence $\{a_n^{\pm}\} \in \mathbf{b}^{(m)}$. Finally, $\{a_n = a_n^+ + a_n^-\} \in \mathbf{b}^{(2m)}$ by Heuristic 6.4.1.

We end this section by mentioning that, if $a_n = h(n) \in \tilde{\mathbf{A}}_0^{(\gamma,m)}$, such that $\gamma \neq -1 + i/m$, $i = 0, 1, \dots$, then $\{a_n\} \in \mathbf{b}^{(m)}$ in the strict sense of Definition 6.1.2. This follows from Theorem 6.8.3 of Section 6.8.

6.7 Summary of Properties of Sequences in $\mathbf{b}^{(1)}$

From Heuristic 6.4.1 and Example 6.4.4, it is seen that sequences in $\mathbf{b}^{(1)}$ are important building blocks for sequences in $\mathbf{b}^{(m)}$ with arbitrary m. They also have been a common test ground for different convergence acceleration methods. Therefore, we summarize their properties separately. We start with their characterization theorem that is a consequence of Theorem 6.6.5.

Theorem 6.7.1 *The following statements concerning sequences $\{a_n\}$ are equivalent:*

(i) $\{a_n\} \in \mathbf{b}^{(1)}$, *that is, $a_n = p(n)\Delta a_n$, where $p \in \mathbf{A}_0^{(\sigma)}$ strictly, σ being an integer ≤ 1.*
(ii) $a_{n+1} = c(n)a_n$, *where $c \in \mathbf{A}_0^{(\mu)}$ strictly and μ is an integer, such that $c(n) \neq 1 + O(n^{-2})$ as $n \to \infty$.*
(iii) $a_n = [(n-1)!]^\mu \zeta^n h(n)$ *with μ an integer and $h \in \mathbf{A}_0^{(\gamma)}$ strictly, such that either (a) $\mu = 0$, $\zeta = 1$, and $\gamma \neq 0$, or (b) $\mu = 0$ and $\zeta \neq 1$, or (c) $\mu \neq 0$.*

With σ and μ as in statements (i) and (ii), respectively, we have (a) $\sigma = 1$ when $\mu = 0$, $\zeta = 1$, and $\gamma \neq 0$, (b) $\sigma = 0$ when $\mu = 0$ and $\zeta \neq 1$, and (c) $\sigma = \min\{0, -\mu\}$ when $\mu \neq 0$.

Note that, in statements (iii-a) and (iii-b) of this theorem, we have that $a_n = h(n) \in \mathbf{A}_0^{(\gamma)}$ and $a_n = \zeta^n h(n)$ with $h \in \mathbf{A}_0^{(\gamma)}$, respectively; these are treated in Examples 6.1.10 and 6.1.11, respectively.

The next result on the summation properties of sequences $\{a_n\}$ in $\mathbf{b}^{(1)}$ is a consequence of Theorem 6.6.6, and it combines a few theorems that were originally given by Sidi [270], [273], [294], [295] for the different situations. Here we are using the notation of Theorem 6.7.1.

Theorem 6.7.2 *Let a_n be as in statement (iii) of the previous theorem with $\mu \leq 0$. That is, either (a) $a_n = h(n)$ with $h \in \mathbf{A}_0^{(\gamma)}$, such that $\gamma \neq -1, 0, 1, \ldots$; or (b) $a_n = \zeta^n h(n)$ with $\zeta \neq 1$, $|\zeta| \leq 1$, and $h \in \mathbf{A}_0^{(\gamma)}$ with γ arbitrary; or (c) $a_n = [(n-1)!]^{-r}\zeta^n h(n)$ with $r = 1, 2, \ldots$, and $h \in \mathbf{A}_0^{(\gamma)}$, ζ and γ being arbitrary. Then, there exist a constant $S(\{a_k\})$ and a function $g \in \mathbf{A}_0^{(0)}$ strictly such that*

$$A_{n-1} = S(\{a_k\}) + n^\sigma a_n\, g(n), \tag{6.7.1}$$

where $\sigma \leq 1$ is an integer. With $g(n) \sim \sum_{i=0}^\infty g_i n^{-i}$ as $n \to \infty$, there holds $\sigma = 1$ and $g_0 = \gamma^{-1}$ in case (a); $\sigma = 0$ and $g_0 = (\zeta - 1)^{-1}$ in case (b); while $\sigma = 0$ and $g_0 = -1$, $g_i = 0$, $1 \leq i \leq r - 1$, and $g_r = -\zeta$ in case (c). All this is true whether $\sum_{k=1}^\infty a_k$ converges or not. [In case (c), the series always converges.]

In case (a) of Theorem 6.7.2, we have

$$(A_n - S(\{a_k\}))/(A_{n-1} - S(\{a_k\})) \sim a_{n+1}/a_n \sim 1 \text{ as } n \to \infty.$$

Sequences $\{A_n\}$ with this property are called *logarithmic*, whether they converge or not.

In case (b) of Theorem 6.7.2, we have

$$(A_n - S(\{a_k\}))/(A_{n-1} - S(\{a_k\})) \sim a_{n+1}/a_n \sim \zeta \neq 1 \text{ as } n \to \infty.$$

Sequences $\{A_n\}$ with this property are called *linear*, whether they converge or not. When ζ is real and negative the series $\sum_{k=1}^{\infty} a_k$ is known as an *alternating series*, and when ζ is real and positive it is known as a *linear monotone series*.

In case (c) of Theorem 6.7.2, we have

$$(A_n - S(\{a_k\}))/(A_{n-1} - S(\{a_k\})) \sim a_{n+1}/a_n \sim \zeta n^{-r} \text{ as } n \to \infty.$$

Since $A_n - S(\{a_k\})$ tends to zero practically like $1/(n!)^r$ in this case, sequences $\{A_n\}$ with this property are called *factorial*.

Again, in case (b), the (power) series $\sum_{k=1}^{\infty} a_k$ converges for $|\zeta| < 1$, the limit $S(\{a_k\})$ being an analytic function of ζ inside the circle $|\zeta| < 1$, with a singularity at $\zeta = 1$. Let us denote this function $f(\zeta)$. By restricting the a_n further, we can prove that (6.7.1) is valid also for all $|\zeta| \geq 1$ and $\zeta \notin [1, \infty)$, $S(\{a_k\})$ being the analytic continuation of $f(\zeta)$, whether $\sum_{k=1}^{\infty} a_k$ converges or not. (Recall that convergence takes place when $|\zeta| = 1$ provided $\Re\gamma < 0$, while $\sum_{k=1}^{\infty} a_k$ diverges when $|\zeta| > 1$ with any γ.) This is the subject of the next theorem that was originally given by Sidi [294]. (At this point it may be a good idea to review Example 4.1.8.)

Theorem 6.7.3 *Let* $a_n = \zeta^n h(n)$, *where* $h \in \mathbf{A}_0^{(\gamma)}$ *strictly for some arbitrary* γ *and* $h(n) = n^q w(n)$, *where* $w(n) = \int_0^{\infty} e^{-nt} \psi(t)\,dt$ *with* $\psi(t) = O(e^{ct})$ *as* $t \to \infty$ *for some* $c < 1$ *and* $\psi(t) \sim \sum_{i=0}^{\infty} \alpha_i t^{-\omega-1+i}$ *as* $t \to 0+$, *such that* (i) $q = 0$ *and* $\omega = \gamma$ *if* $\Re\gamma < 0$ *and* (ii) $q = \lfloor \Re\gamma \rfloor + 1$ *and* $\omega = \gamma - q$ *if* $\Re\gamma \geq 0$. *Then, for all complex* ζ *not in the real interval* $[1, +\infty)$, *there holds*

$$A_{n-1} = f(\zeta) + a_n g(n); \quad f(\zeta) = \left(\zeta \frac{d}{d\zeta} \right)^q \int_0^{\infty} \frac{\zeta \psi(t)}{e^t - \zeta}\,dt, \quad g \in \mathbf{A}_0^{(0)} \text{ strictly.} \quad (6.7.2)$$

Clearly, $f(\zeta)$ *is analytic in the complex* ζ-*plane cut along the real interval* $[1, +\infty)$.

Proof. We start with the fact that

$$A_{n-1} = \sum_{k=1}^{n-1} \zeta^k k^q w(k) = \left(\zeta \frac{d}{d\zeta} \right)^q \left[\sum_{k=1}^{n-1} \zeta^k w(k) \right]$$

$$= \left(\zeta \frac{d}{d\zeta} \right)^q \int_0^{\infty} \left[\sum_{k=1}^{n-1} (\zeta e^{-t})^k \right] \psi(t)\,dt.$$

From this and from the identity $\sum_{k=1}^{n-1} z^k = (z - z^n)/(1 - z)$, we obtain

$$A_{n-1} = f(\zeta) - \int_0^{\infty} e^{-nt} \psi(t) \left\{ \left(\zeta \frac{d}{d\zeta} \right)^q \frac{\zeta^n}{1 - \zeta e^{-t}} \right\} dt. \quad (6.7.3)$$

As $\left(\zeta \frac{d}{d\zeta} \right)^q \frac{\zeta^n}{1 - \zeta e^{-t}} = \zeta^n \sum_{i=0}^{q} \beta_i(\zeta, t) n^i$ with $\beta_q(\zeta, t) = (1 - \zeta e^{-t})^{-1} \sim (1 - \zeta)^{-1} \neq 0$ as $t \to 0+$, application of Watson's lemma to the integral term in (6.7.3) produces $A_{n-1} - f(\zeta) = \zeta^n u(n)$ with $u \in \mathbf{A}_0^{(\gamma)}$ strictly. Setting $g(n) = u(n)/h(n)$, the result in (6.7.2) now follows. ∎

Turning things around in Theorem 6.7.2, we next state a theorem concerning logarithmic, linear, and factorial sequences that will be of use in the analysis of other sequence transformations later. Note that the two theorems imply each other.

Theorem 6.7.4

(i) *If* $A_n \sim A + \sum_{i=0}^{\infty} \alpha_i n^{\gamma-i}$ *as* $n \to \infty$, $\alpha_0 \neq 0$, $\gamma \neq 0, 1, \ldots$, *that is*, $\{A_n\}$ *is a logarithmic sequence, then*

$$A_n \sim A + n(\Delta A_n) \sum_{i=0}^{\infty} \beta_i n^{-i} \quad as\ n \to \infty, \quad \beta_0 = \gamma^{-1} \neq 0.$$

(ii) *If* $A_n \sim A + \zeta^n \sum_{i=0}^{\infty} \alpha_i n^{\gamma-i}$ *as* $n \to \infty$, $\zeta \neq 1$, $\alpha_0 \neq 0$, *that is*, $\{A_n\}$ *is a linear sequence, then*

$$A_n \sim A + (\Delta A_n) \sum_{i=0}^{\infty} \beta_i n^{-i} \quad as\ n \to \infty, \quad \beta_0 = (\zeta - 1)^{-1} \neq 0.$$

(iii) *If* $A_n \sim A + (n!)^{-r} \zeta^n \sum_{i=0}^{\infty} \alpha_i n^{\gamma-i}$ *as* $n \to \infty$, $r = 1, 2, \ldots$, $\alpha_0 \neq 0$, *that is*, $\{A_n\}$ *is a factorial sequence, then*

$$A_n \sim A + (\Delta A_n)\left(-1 + \sum_{i=r}^{\infty} \beta_i n^{-i}\right) \quad as\ n \to \infty, \quad \beta_r = -\zeta \neq 0.$$

Remark. In Theorems 6.7.2–6.7.4, we have left out the cases in which $a_n = [(n-1)!]^\mu \zeta^n h(n)$ with $\mu = r = 1, 2, \ldots$, and $h(n) \in \mathbf{A}_0^{(\gamma)}$. Obviously, in such a case, A_n diverges wildly. As shown in Section 19.4, A_n actually satisfies

$$A_n \sim a_n\left(1 + \sum_{i=r}^{\infty} \beta_i n^{-i}\right) \quad as\ n \to \infty, \quad \beta_r = \zeta^{-1}. \tag{6.7.4}$$

In other words, $\{A_n\}$ diverges factorially. Furthermore, when $h(n)$ is independent of ζ and $h(n) = n^\omega \int_0^\infty e^{-nt} \varphi(t)\,dt$ for some integer $\omega \geq 0$ and some $\varphi(t)$ of exponential order, the divergent series $\sum_{k=1}^\infty a_k$ has a (generalized) Borel sum, which, as a function of ζ, is analytic in the ζ-plane cut along the real interval $[0, +\infty)$. This result is a special case of the more general ones proved in Sidi [285]. The fact that A_n satisfies (6.7.4) suggests that the $d^{(1)}$-transformation, and, in particular, the $\mathcal{L}$- and $\mathcal{S}$-transformations, could be effective in summing $\sum_{k=1}^\infty a_k$. Indeed, the latter two turn out to be very effective; they produce approximations to the (generalized) Borel sum of $\sum_{k=1}^\infty a_k$, as suggested by the numerical experiments of Smith and Ford [318], Bhattacharya, Roy, and Bhowmick [22], and Grotendorst [116].

6.8 A General Family of Sequences in $\mathbf{b}^{(m)}$

We now go back to the sequences $\{a_n\}$ considered in Example 6.4.4 and study them more closely. In Theorems 6.8.3 and 6.8.7, we show that two important classes of such sequences are indeed in $\mathbf{b}^{(m)}$ for some m strictly in accordance with Definition 6.1.2. Obviously, this strengthens the conclusions of Example 6.4.4. From Theorems 6.8.4 and 6.8.8, it follows in a rigorous manner that the $d^{(m)}$-transformation can be applied effectively to the infinite series $\sum_{k=1}^\infty a_k$ associated with these sequences. As a pleasant outcome of this study, we also see in Theorems 6.8.5 and 6.8.9 that the d-transformation can be used very effectively to accelerate the convergence of two important classes of sequences even when there is not enough quantitative information about the asymptotic behavior of the sequence elements. We note that all the results of this section are new. We begin with the following useful lemma that was stated and proved as Lemma 1.2 in Sidi [305].

Lemma 6.8.1 *Let* $Q_i(x) = \sum_{j=0}^{i} a_{ij} x^j$, *with* $a_{ii} \neq 0$, $i = 0, 1, \dots, n$, *and let* x_i, $i = 0, 1, \dots, n$, *be arbitrary points. Then*

$$\begin{vmatrix} Q_0(x_0) & Q_0(x_1) & \cdots & Q_0(x_n) \\ Q_1(x_0) & Q_1(x_1) & \cdots & Q_1(x_n) \\ \vdots & \vdots & & \vdots \\ Q_n(x_0) & Q_n(x_1) & \cdots & Q_n(x_n) \end{vmatrix} = \left(\prod_{i=0}^{n} a_{ii} \right) V(x_0, x_1, \dots, x_n), \qquad (6.8.1)$$

where $V(x_0, x_1, \dots, x_n) = \prod_{0 \le i < j \le n} (x_j - x_i)$ *is a Vandermonde determinant.*

Proof. As can easily be seen, it is enough to consider the case $a_{ii} = 1$, $i = 0, 1, \dots, n$. Let us now perform the following elementary row transformations in the determinant on the left-hand side of (6.8.1):

    for $i = 1, 2, \dots, n$ do
        for $j = 0, 1, \dots, i - 1$ do
            multiply $(j + 1)$st row by a_{ij} and subtract from $(i + 1)$st row
        end for
    end for

The result of these transformations is $V(x_0, x_1, \dots, x_n)$.   ∎

The developments we present in the following two subsections parallel each other, even though there are important differences between the sequences that are considered in each. We would also like to note that the results of Theorems 6.8.4 and 6.8.8 below are stronger versions of the result of Theorem 6.1.12 for the sequences $\{a_n\}$ of this section in the sense that they do not assume that $\sum_{k=1}^{\infty} a_k$ converges.

6.8.1 Sums of Logarithmic Sequences

Lemma 6.8.2 *Let* $a_n = \sum_{i=1}^{m} h_i(n)$, *where* $h_i \in \mathbf{A}_0^{(\gamma_i)}$ *strictly for some distinct* $\gamma_i \neq 0, 1, \dots$. *Then, with any integer* $r \ge 0$, *there holds* $h_i(n) = \sum_{k=r}^{m+r-1} v_{ik}(n) n^k \Delta^k a_n$ *for each* i, *where* $v_{ik} \in \mathbf{A}_0^{(0)}$ *for all* i *and* k.

Proof. First, $h_i(n) \sim \alpha_i n^{\gamma_i}$ as $n \to \infty$, for some $\alpha_i \neq 0$. Consequently, $\Delta^k h_i(n) = n^{-k} u_{ik}(n) h_i(n)$ with $u_{ik} \in \mathbf{A}_0^{(0)}$ strictly, $u_{ik}(n) = [\gamma_i]_k + O(n^{-1})$ as $n \to \infty$, where we have defined $[x]_k = x(x-1) \cdots (x-k+1)$, $k = 1, 2, \dots$. (We also define $u_{i0}(n) = 1$ and $[x]_0 = 1$ for every x.) Thus, with any nonnegative integer r, the $h_i(n)$ satisfy the linear system

$$\sum_{i=1}^{m} u_{ik}(n) h_i(n) = n^k \Delta^k a_n, \quad k = r, r+1, \dots, m+r-1,$$

that can be solved for the $h_i(n)$ by Cramer's rule. Obviously, because $u_{ik} \in \mathbf{A}_0^{(0)}$ for all i and k, all the minors of the matrix of this system, namely, of

$$M(n) = \begin{bmatrix} u_{1r}(n) & u_{2r}(n) & \cdots & u_{mr}(n) \\ u_{1,r+1}(n) & u_{2,r+1}(n) & \cdots & u_{m,r+1}(n) \\ \vdots & \vdots & & \vdots \\ u_{1,m+r-1}(n) & u_{2,m+r-1}(n) & \cdots & u_{m,m+r-1}(n) \end{bmatrix},$$

are in $\mathbf{A}_0^{(0)}$. More importantly, its determinant is in $\mathbf{A}_0^{(0)}$ *strictly*. To prove this last point, we replace $u_{ik}(n)$ in $\det M(n)$ by its asymptotic behavior and use the fact that $[x]_{q+r} = [x]_r \cdot [x-r]_q$ to factor out $[\gamma_i]_r$ from the ith column, $i = 1, \ldots, m$. (Note that $[\gamma_i]_r \neq 0$ by our assumptions on the γ_i.) This results in

$$\det M(n) \sim \left(\prod_{i=1}^m [\gamma_i]_r \right) \begin{vmatrix} [\gamma_1 - r]_0 & [\gamma_2 - r]_0 & \cdots & [\gamma_m - r]_0 \\ [\gamma_1 - r]_1 & [\gamma_2 - r]_1 & \cdots & [\gamma_m - r]_1 \\ \vdots & \vdots & & \vdots \\ [\gamma_1 - r]_{m-1} & [\gamma_2 - r]_{m-1} & \cdots & [\gamma_m - r]_{m-1} \end{vmatrix} \quad \text{as } n \to \infty,$$

which, upon invoking Lemma 6.8.1, gives

$$\det M(n) \sim \left(\prod_{i=1}^m [\gamma_i]_r \right) V(\gamma_1, \gamma_2, \ldots, \gamma_m) \neq 0 \text{ as } n \to \infty.$$

Here, we have used the fact that $[x]_k$ is a polynomial in x of degree exactly k and the assumption that the γ_i are distinct. Completing the solution by Cramer's rule, the result follows. ∎

   In the following two theorems, we use the notation of the previous lemma. The first of these theorems is a rigorous version of Example 6.4.4, while the second is concerned with the summation properties of the sequences $\{a_n\}$ that we have considered so far.

Theorem 6.8.3 *Let $a_n = \sum_{i=1}^m h_i(n)$, where $h_i \in \mathbf{A}_0^{(\gamma_i)}$ strictly for some distinct $\gamma_i \neq 0, 1, \ldots$. Then $\{a_n\} \in \mathbf{b}^{(m)}$.*

Proof. Let us first invoke Lemma 6.8.2 with $r = 1$ in $a_n = \sum_{i=1}^m h_i(n)$. We obtain $a_n = \sum_{k=1}^m p_k(n) \Delta^k a_n$ with $p_k(n) = n^k \sum_{i=1}^m v_{ik}(n)$, $k = 1, \ldots, m$. By the fact that v_{ik} are all in $\mathbf{A}_0^{(0)}$, we have that $p_k \in \mathbf{A}_0^{(k)}$ for each k. The result follows by recalling Definition 6.1.2. ∎

Theorem 6.8.4 *Let $a_n = \sum_{i=1}^m h_i(n)$, where $h_i \in \mathbf{A}_0^{(\gamma_i)}$ strictly for some distinct $\gamma_i \neq -1, 0, 1, 2, \ldots$. Then, whether $\sum_{k=1}^\infty a_k$ converges or not, there holds*

$$A_{n-1} = S(\{a_k\}) + \sum_{k=0}^{m-1} n^{k+1}(\Delta^k a_n) g_k(n), \tag{6.8.2}$$

where $S(\{a_k\})$ is the sum of the limits or antilimits of the series $\sum_{k=1}^\infty h_i(k)$, which exist by Theorem 6.7.2, and $g_k \in \mathbf{A}_0^{(0)}$ for each k.

Proof. By Theorem 6.7.2, we first have

$$A_{n-1} = S(\{a_k\}) + \sum_{i=1}^{m} n h_i(n) w_i(n),$$

where $w_i \in \mathbf{A}_0^{(0)}$ for each i. Next, $h_i(n) = \sum_{k=0}^{m-1} v_{ik}(n) n^k \Delta^k a_n$ by Lemma 6.8.2 with $r = 0$. Combining these results, we obtain (6.8.2) with $g_k(n) = \sum_{i=1}^{m} v_{ik}(n) w_i(n)$ for $k = 0, 1, \dots, m-1$.   ∎

A Special Application

The techniques we have developed in this subsection can be used to show that the d-transformation will be effective in computing the limit or antilimit of a sequence $\{S_n\}$, for which

$$S_n = S + \sum_{i=1}^{m} H_i(n), \tag{6.8.3}$$

where $H_i \in \mathbf{A}_0^{(\sigma_i)}$ for some σ_i that are distinct and satisfy $\sigma_i \neq 0, 1, \dots$, and S is the limit or antilimit.

Such sequences arise, for example, when one applies the trapezoidal rule to integrals over a hypercube or a hypersimplex of functions that have algebraic singularities along the edges or on the surfaces of the hypercube or of the hypersimplex. We discuss this subject in some detail in Chapter 25.

If we know the σ_i, then we can use GREP$^{(m)}$ in the standard way described in Chapter 4. If the σ_i are not readily available, then the $d^{(m)}$-transformation for infinite sequences developed in Subsection 6.2.2 serves as a very effective means for computing S. The following theorem provides the rigorous justification of this assertion.

Theorem 6.8.5 *Let the sequence $\{S_n\}$ be as in (6.8.3). Then there holds*

$$S_n = S + \sum_{k=1}^{m} n^k (\Delta^k S_n) g_k(n), \tag{6.8.4}$$

where $g_k \in \mathbf{A}_0^{(0)}$ for each k.

Proof. Applying Lemma 6.8.2 with $r = 1$ to $S_n - S = \sum_{i=1}^{m} H_i(n)$, and realizing that $\Delta^k(S_n - S) = \Delta^k S_n$ for $k \geq 1$, we have $H_i(n) = \sum_{k=1}^{m} v_{ik}(n) n^k \Delta^k S_n$ for each i, where $v_{ik} \in \mathbf{A}_0^{(0)}$ for all i and k. The result follows by substituting this in (6.8.3).   ∎

6.8.2 Sums of Linear Sequences

Lemma 6.8.6 *Let $a_n = \sum_{i=1}^{m} \zeta_i^n h_i(n)$, where $\zeta_i \neq 1$ are distinct and $h_i \in \mathbf{A}_0^{(\gamma_i)}$ for some arbitrary γ_i that are not necessarily distinct. Then, with any integer $r \geq 0$, there holds $\zeta_i^n h_i(n) = \sum_{k=r}^{m+r-1} v_{ik}(n) \Delta^k a_n$ for each i, where $v_{ik} \in \mathbf{A}_0^{(0)}$ for all i and k.*

Proof. Let us write $a_n^{(i)} = \zeta_i^n h_i(n)$ for convenience. By the fact that $\Delta a_n^{(i)} = (\zeta_i - 1)\zeta_i^n h_i(n+1) + \zeta_i^n \Delta h_i(n)$, we have first $\Delta a_n^{(i)} = u_{i1}(n) a_n^{(i)}$, where $u_{i1} \in \mathbf{A}_0^{(0)}$

strictly. In fact, $u_{i1}(n) = (\zeta_i - 1) + O(n^{-1})$ as $n \to \infty$. Consequently, $\Delta^k a_n^{(i)} = u_{ik}(n)a_n^{(i)}$, where $u_{ik} \in \mathbf{A}_0^{(0)}$ strictly and $u_{ik}(n) = (\zeta_i - 1)^k + O(n^{-1})$ as $n \to \infty$. Thus, with any nonnegative integer r, $a_n^{(i)}$, $i = 1, \ldots, m$, satisfy the linear system

$$\sum_{i=1}^{m} u_{ik}(n)a_n^{(i)} = \Delta^k a_n, \quad k = r, r+1, \ldots, m+r-1,$$

that can be solved for the $a_n^{(i)}$ by Cramer's rule. Obviously, since $u_{ik} \in \mathbf{A}_0^{(0)}$ for all i and k, all the minors of the matrix $M(n)$ of this system that is given exactly as in the proof of Lemma 6.8.2 with the present $u_{ik}(n)$ are in $\mathbf{A}_0^{(0)}$. More importantly, its determinant is in $\mathbf{A}_0^{(0)}$ strictly. In fact, substituting the asymptotic behavior of the $u_{ik}(n)$ as $n \to \infty$ in $\det M(n)$, we obtain

$$\det M(n) \sim \left(\prod_{i=1}^{m} (\zeta_i - 1)^r \right) V(\zeta_1, \ldots, \zeta_m) \neq 0 \text{ as } n \to \infty.$$

Completing the solution by Cramer's rule, the result follows.   ∎

We make use of Lemma 6.8.6 in the proofs of the next two theorems that parallel Theorems 6.8.3 and 6.8.4. Because the proofs are similar, we leave them to the reader.

Theorem 6.8.7 *Let $a_n = \sum_{i=1}^{m} \zeta_i^n h_i(n)$, where $\zeta_i \neq 1$ are distinct and $h_i \in \mathbf{A}_0^{(\gamma_i)}$ for some arbitrary γ_i that are not necessarily distinct. Then $\{a_n\} \in \mathbf{b}^{(m)}$.*

Theorem 6.8.8 *Let $a_n = \sum_{i=1}^{m} \zeta_i^n h_i(n)$, where ζ_i are distinct and satisfy $\zeta_i \neq 1$ and $|\zeta| \leq 1$ and $h_i \in \mathbf{A}_0^{(\gamma_i)}$ for some arbitrary γ_i that are not necessarily distinct. Then, whether $\sum_{k=1}^{\infty} a_k$ converges or not, there holds*

$$A_{n-1} = S(\{a_k\}) + \sum_{k=0}^{m-1} (\Delta^k a_n) g_k(n), \tag{6.8.5}$$

where $S(\{a_k\})$ is the sum of the limits or antilimits of the series $\sum_{k=1}^{\infty} \zeta_i^k h_i(k)$, which exist by Theorem 6.7.2, and $g_k \in \mathbf{A}_0^{(0)}$ for each k.

A Special Application

With the help of the techniques developed in this subsection, we can now show that the $d^{(m)}$-transformation can be used for computing the limit or antilimit of a sequence $\{S_n\}$, for which

$$S_n = S + \sum_{i=1}^{m} \zeta_i^n H_i(n), \tag{6.8.6}$$

where $\zeta_i \neq 1$ are distinct, $H_i \in \mathbf{A}_0^{(\sigma_i)}$ for some arbitrary σ_i that are not necessarily distinct, and S is the limit or antilimit.

If we know the ζ_i and σ_i, then we can use GREP$^{(m)}$ in the standard way described in Chapter 4. If the ζ_i and σ_i are not readily available, then the $d^{(m)}$-transformation for infinite sequences developed in Subsection 6.2.2 serves as a very effective means for computing S. The following theorem provides the rigorous justification of this assertion. As its proof is similar to that of Theorem 6.8.5, we skip it.

Theorem 6.8.9 *Let the sequence* $\{S_n\}$ *be as in (6.8.6). Then there holds*

$$S_n = S + \sum_{k=1}^{m} (\Delta^k S_n) g_k(n), \tag{6.8.7}$$

where $g_k \in \mathbf{A}_0^{(0)}$ *for each* k.

6.8.3 Mixed Sequences

We now turn to mixtures of logarithmic and linear sequences that appear to be difficult to handle mathematically. Instead of attempting to extend the theorems proved above, we state the following conjectures.

Conjecture 6.8.10 *Let* $a_n = \sum_{i=1}^{m_1} \zeta_i^n h_i(n) + \sum_{i=1}^{m_2} \tilde{h}_i(n)$, *where* $\zeta_i \neq 1$ *are distinct and* $h_i \in \mathbf{A}_0^{(\gamma_i)}$ *for some arbitrary* γ_i *that are not necessarily distinct and* $\tilde{h}_i \in \mathbf{A}_0^{(\tilde{\gamma}_i)}$ *with* $\tilde{\gamma}_i \neq 0, 1, \dots$, *and distinct. Then* $\{a_n\} \in \mathbf{b}^{(m)}$, *where* $m = m_1 + m_2$.

Conjecture 6.8.11 *If in Conjecture 6.8.10 we have* $|\zeta_i| \leq 1$ *and* $\tilde{\gamma}_i \neq -1$, *in addition to the conditions there, then*

$$A_{n-1} = S(\{a_k\}) + \sum_{k=0}^{m-1} n^{k+1} (\Delta^k a_n) g_k(n), \tag{6.8.8}$$

where $S(\{a_k\})$ *is the sum of the limits or antilimits of the series* $\sum_{k=1}^{\infty} \zeta_i^k h_i(k)$ *and* $\sum_{k=1}^{\infty} \tilde{h}_i(k)$, *which exist by Theorem 6.7.2, and* $g_k \in \mathbf{A}_0^{(0)}$ *for each* k.

Conjecture 6.8.12 *Let the sequence* $\{S_n\}$ *satisfy*

$$S_n = S + \sum_{i=1}^{m_1} \zeta_i^n H_i(n) + \sum_{i=1}^{m_2} \tilde{H}_i(n) \tag{6.8.9}$$

where $\zeta_i \neq 1$ *are distinct and* $H_i \in \mathbf{A}_0^{(\sigma_i)}$ *for some arbitrary* σ_i *that are not necessarily distinct and* $\tilde{H}_i \in \mathbf{A}_0^{(\tilde{\sigma}_i)}$ *with distinct* $\tilde{\sigma}_i \neq 0, 1, \dots$. *Then there holds*

$$S_n = S + \sum_{k=1}^{m} n^k (\Delta^k S_n) g_k(n), \tag{6.8.10}$$

where $g_k \in \mathbf{A}_0^{(0)}$ *for each* k *and* $m = m_1 + m_2$.

7

Recursive Algorithms for GREP

7.1 Introduction

Let us recall the definition of GREP$^{(m)}$ as given in (4.2.1). This definition involves the m form factors (or shape functions) $\phi_k(y)$, $k = 1, \ldots, m$, whose structures may be arbitrary. It also involves the functions $\sum_{i=0}^{n_k-1} \bar{\beta}_i y^{ir_k}$ that behave essentially polynomially in y^{r_k} for $k = 1, \ldots, m$.

These facts enable us to design very efficient recursive algorithms for two cases of GREP:

(i) the *W-algorithm* for GREP$^{(1)}$ with arbitrary y_l in (4.2.1), and

(ii) the *$W^{(m)}$-algorithm* for GREP$^{(m)}$ with $m > 1$ and $r_1 = r_2 = \cdots = r_m$, and with arbitrary y_l in (4.2.1).

In addition, we are able to derive an efficient algorithm for a special case of one of the extensions of GREP considered in Section 4.6:

(iii) the extended W-algorithm (*EW-algorithm*) for the $m = 1$ case of the extended GREP for which the $\beta_k(y)$ are as in (4.6.1), with $y_l = y_0 \omega^l$, $l = 0, 1, \ldots$, and $\omega \in (0, 1)$.

We note that GREP$^{(1)}$ and GREP$^{(m)}$ with $r_1 = \cdots = r_m$ are probably the most commonly occurring forms of GREP. The D-transformation of Chapter 5 and the d-transformation of Chapter 6, for example, are of these forms, with $r_1 = \cdots = r_m = 1$ for both. We also note that the effectiveness of the algorithms of this chapter stems from the fact that they fully exploit the special structure of the underlying extrapolation methods.

As we will see in more detail, GREP can be implemented via the algorithms of Chapter 3. Thus, when $A(y_l)$, $l = 0, 1, \ldots, L$, are given, the computation of those $A_n^{(m,j)}$ that can be derived from them requires $2L^3/3 + O(L^2)$ arithmetic operations when done by the FS-algorithm and about 50% more when done by the E-algorithm. Both algorithms require $O(L^2)$ storage locations. But the algorithms of this chapter accomplish the same task in $O(L^2)$ arithmetic operations and require $O(L^2)$ storage locations. With suitable programming, the storage requirements can be reduced to $O(L)$ locations. In this respect, the algorithms of this chapter are analogous to Algorithm 1.3.1 of Chapter 1.

The W-algorithm was given by Sidi [278], [295], and the $W^{(m)}$-algorithm was developed by Ford and Sidi [87]. The EW-algorithm is unpublished work of the author.

7.2 The W-Algorithm for GREP$^{(1)}$

We start by rewriting (4.2.1) for $m = 1$ in a simpler way. For this, let us first make the substitutions $\phi_1(y) = \phi(y)$, $r_1 = r$, $n_1 = n$, $\bar{\beta}_{1i} = \bar{\beta}_i$, and $A_n^{(1,j)} = A_n^{(j)}$. The equations in (4.2.1) become

$$A(y_l) = A_n^{(j)} + \phi(y_l) \sum_{i=0}^{n-1} \bar{\beta}_i y_l^{ir}, \quad j \leq l \leq j + n. \tag{7.2.1}$$

A further simplification takes place by letting $t = y^r$ and $t_l = y_l^r$, $l = 0, 1, \ldots$, and defining $a(t) \equiv A(y)$ and $\varphi(t) \equiv \phi(y)$. We now have

$$a(t_l) = A_n^{(j)} + \varphi(t_l) \sum_{i=0}^{n-1} \bar{\beta}_i t_l^i, \quad j \leq l \leq j + n. \tag{7.2.2}$$

Let us denote by $D_n^{(j)}\{g(t)\}$ the divided difference of $g(t)$ over the set of points $\{t_j, t_{j+1}, \ldots, t_{j+n}\}$, namely, $g[t_j, t_{j+1}, \ldots, t_{j+n}]$. Then, from the theory of polynomial interpolation it is known that

$$D_n^{(j)}\{g(t)\} = g[t_j, t_{j+1}, \ldots, t_{j+n}] = \sum_{i=0}^{n} c_{ni}^{(j)} g(t_{j+i});$$

$$c_{ni}^{(j)} = \prod_{\substack{k=0 \\ k \neq i}}^{n} \frac{1}{t_{j+i} - t_{j+k}}, \quad 0 \leq i \leq n. \tag{7.2.3}$$

The following theorem forms the basis of the W-algorithm for computing the $A_n^{(j)}$. It also shows how the $\bar{\beta}_i$ can be computed one by one in the order $\bar{\beta}_0, \bar{\beta}_1, \ldots, \bar{\beta}_{n-1}$.

Theorem 7.2.1 *Provided* $\varphi(t_l) \neq 0$, $j \leq l \leq j + n$, *we have*

$$A_n^{(j)} = \frac{D_n^{(j)}\{a(t)/\varphi(t)\}}{D_n^{(j)}\{1/\varphi(t)\}}. \tag{7.2.4}$$

With $A_n^{(j)}$ *and* $\bar{\beta}_0, \ldots, \bar{\beta}_{p-1}$ *already known,* $\bar{\beta}_p$ *can be determined from*

$$\bar{\beta}_p = (-1)^{n-p-1} \left(\prod_{i=j}^{j+n-p-1} t_i \right) D_{n-p-1}^{(j)}\{[a(t) - A_n^{(j)}]t^{-p-1}/\varphi(t) - \sum_{i=0}^{p-1} \bar{\beta}_i t^{i-p-1}\}, \tag{7.2.5}$$

where the summation $\sum_{i=0}^{-1} \mu_i$ *is taken to be zero.*

Proof. We begin by reexpressing the linear equations in (7.2.2) in the form

$$[a(t_l) - A_n^{(j)}]/\varphi(t_l) = \sum_{i=0}^{n-1} \bar{\beta}_i t_l^i, \quad j \leq l \leq j + n. \tag{7.2.6}$$

For $-1 \leq p \leq n - 1$, we multiply the first $n - p$ of the equations in (7.2.6) by $c_{n-p-1,l-j}^{(j)} t_l^{-p-1}$, $l = j, j + 1, \ldots, j + n - p - 1$, respectively. Adding all the resulting equations and invoking (7.2.3), we obtain

$$D^{(j)}_{n-p-1}\{[a(t) - A^{(j)}_n]t^{-p-1}/\varphi(t)\} = D^{(j)}_{n-p-1}\{\sum_{i=0}^{n-1}\bar{\beta}_i t^{i-p-1}\}. \qquad (7.2.7)$$

Let us put $p = -1$ in (7.2.7). Then, we have that $\sum_{i=0}^{n-1}\bar{\beta}_i t^{i-p-1} = \sum_{i=0}^{n-1}\bar{\beta}_i t^i$ is a polynomial of degree at most $n - 1$. But, it is known that

$$D^{(s)}_k\{g(t)\} = 0 \text{ if } g(t) \text{ is a polynomial of degree at most } k - 1. \qquad (7.2.8)$$

By invoking (7.2.8) in (7.2.7) with $p = -1$, we obtain (7.2.4).

Next, let us put $p = 0$ in (7.2.7). Then, we have that $\sum_{i=0}^{n-1}\bar{\beta}_i t^{i-p-1} = \bar{\beta}_0 t^{-1} + \sum_{i=1}^{n-1}\bar{\beta}_i t^{i-1}$, the summation $\sum_{i=1}^{n-1}\bar{\beta}_i t^{i-1}$ being a polynomial of degree at most $n - 2$. Thus, with the help of (7.2.8) again, and by the fact that

$$D^{(j)}_n\{t^{-1}\} = (-1)^n/(t_j t_{j+1} \cdots t_{j+n}), \qquad (7.2.9)$$

we obtain $\bar{\beta}_0$ exactly as given in (7.2.5). [The proof of (7.2.9) can be done by induction with the help of the recursion relation in (7.2.21) below.]

The rest of the proof can now be done similarly. ∎

Remark. From the proof of Theorem 7.2.1, it should become clear that both (7.2.4) and (7.2.5) hold even when the t_l in (7.2.2) are *complex*, although in the extrapolation problems that we usually encounter they are real.

The expression for $A^{(j)}_n$ given in (7.2.4), in addition to forming the basis for the W-algorithm, will prove to be very useful in the convergence and stability studies of GREP$^{(1)}$ in Chapters 8 and 9.

We next treat the problem of assessing the stability of GREP$^{(1)}$ numerically in an efficient manner. Here too divided differences turn out to be very useful. From (7.2.4) and (7.2.3), it is clear that $A^{(j)}_n$ can be expressed in the by now familiar form

$$A^{(j)}_n = \sum_{i=0}^{n}\gamma^{(j)}_{ni} A(y_{j+i}); \quad \sum_{i=0}^{n}\gamma^{(j)}_{ni} = 1, \qquad (7.2.10)$$

with

$$\gamma^{(j)}_{ni} = \frac{1}{D^{(j)}_n\{1/\varphi(t)\}}\frac{c^{(j)}_{ni}}{\varphi(t_{j+i})}, \quad i = 0, 1, \ldots, n. \qquad (7.2.11)$$

Thus, $\Gamma^{(1,j)}_n = \Gamma^{(j)}_n$ is given by

$$\Gamma^{(j)}_n = \sum_{i=0}^{n}|\gamma^{(j)}_{ni}| = \frac{1}{|D^{(j)}_n\{1/\varphi(t)\}|}\sum_{i=0}^{n}\frac{|c^{(j)}_{ni}|}{|\varphi(t_{j+i})|}. \qquad (7.2.12)$$

Similarly, we define

$$\Lambda^{(j)}_n = \sum_{i=0}^{n}|\gamma^{(j)}_{ni}|\,|a(t_{j+i})| = \frac{1}{|D^{(j)}_n\{1/\varphi(t)\}|}\sum_{i=0}^{n}\frac{|c^{(j)}_{ni}|\,|a(t_{j+i})|}{|\varphi(t_{j+i})|}. \qquad (7.2.13)$$

Let us recall briefly the meanings of $\Gamma_n^{(j)}$ and $\Lambda_n^{(j)}$: If the $A(y_i)$ have been computed with *absolute* errors that do not exceed ϵ, and $\bar{A}_n^{(j)}$ is the computed $A_n^{(j)}$, then

$$|\bar{A}_n^{(j)} - A_n^{(j)}| \lesssim \epsilon \Gamma_n^{(j)} \quad \text{and} \quad |\bar{A}_n^{(j)} - A| \lesssim \epsilon \Gamma_n^{(j)} + |A_n^{(j)} - A|.$$

If the $A(y_i)$ have been computed with *relative* errors that do not exceed η, and $\bar{A}_n^{(j)}$ is the computed $A_n^{(j)}$, then

$$|\bar{A}_n^{(j)} - A_n^{(j)}| \lesssim \eta \Lambda_n^{(j)} \quad \text{and} \quad |\bar{A}_n^{(j)} - A| \lesssim \eta \Lambda_n^{(j)} + |A_n^{(j)} - A|.$$

Hence, $\eta \Lambda_n^{(j)}$ is a more refined estimate of the error in the computed value of $A_n^{(j)}$, especially when $A(y)$ is unbounded as $y \to 0+$. See Section 0.5.

Lemma 7.2.2 *Let* u_i, $i = 0, 1, \ldots$, *be scalars, and let the* t_i *in (7.2.3) satisfy* $t_0 > t_1 > t_2 > \cdots$. *Then*

$$\sum_{i=0}^{n} |c_{ni}^{(j)}| \, u_{j+i} = (-1)^j D_n^{(j)}\{v(t)\}, \tag{7.2.14}$$

where $v(t_i) = (-1)^i u_i$, $i = 0, 1, \ldots$, *and* $v(t)$ *is arbitrary otherwise.*

Proof. From (7.2.3), we observe that

$$c_{ni}^{(j)} = (-1)^i |c_{ni}^{(j)}|, \quad i = 0, 1, \ldots, n, \quad \text{for all } j \text{ and } n. \tag{7.2.15}$$

The result now follows by substituting (7.2.15) in (7.2.3). ∎

The following theorem on $\Gamma_n^{(j)}$ and $\Lambda_n^{(j)}$ can be proved by invoking Lemma 7.2.2 in (7.2.12) and (7.2.13).

Theorem 7.2.3 *Define the functions* $P(t)$ *and* $S(t)$ *by*

$$P(t_i) = (-1)^i / |\varphi(t_i)| \quad \text{and} \quad S(t_i) = (-1)^i |a(t_i)/\varphi(t_i)|, \quad i = 0, 1, 2, \ldots, \tag{7.2.16}$$

and arbitrarily for $t \neq t_i$, $i = 0, 1, \ldots$. *Then*

$$\Gamma_n^{(j)} = \frac{|D_n^{(j)}\{P(t)\}|}{|D_n^{(j)}\{1/\varphi(t)\}|} \quad \text{and} \quad \Lambda_n^{(j)} = \frac{|D_n^{(j)}\{S(t)\}|}{|D_n^{(j)}\{1/\varphi(t)\}|}. \tag{7.2.17}$$

We now give the W-algorithm to compute the $A_p^{(j)}$, $\Gamma_p^{(j)}$, and $\Lambda_n^{(j)}$ recursively. This algorithm is directly based on Theorems 7.2.1 and 7.2.3.

Algorithm 7.2.4 (W-algorithm)

1. For $j = 0, 1, \ldots$, set

$$M_0^{(j)} = a(t_j)/\varphi(t_j), \quad N_0^{(j)} = 1/\varphi(t_j), \quad H_0^{(j)} = (-1)^j |N_0^{(j)}|, \quad K_0^{(j)} = (-1)^j |M_0^{(j)}|.$$

2. For $j = 0, 1, \ldots$, and $n = 1, 2, \ldots$, compute $M_n^{(j)}$, $N_n^{(j)}$, $H_n^{(j)}$, and $K_n^{(j)}$ recursively from

$$Q_n^{(j)} = \frac{Q_{n-1}^{(j+1)} - Q_{n-1}^{(j)}}{t_{j+n} - t_j}, \tag{7.2.18}$$

where the $Q_n^{(j)}$ stand for either $M_n^{(j)}$ or $N_n^{(j)}$ or $H_n^{(j)}$ or $K_n^{(j)}$.

3. For all j and n, set

$$A_n^{(j)} = \frac{M_n^{(j)}}{N_n^{(j)}}, \quad \Gamma_n^{(j)} = \left| \frac{H_n^{(j)}}{N_n^{(j)}} \right|, \quad \text{and} \quad \Lambda_n^{(j)} = \left| \frac{K_n^{(j)}}{N_n^{(j)}} \right|. \tag{7.2.19}$$

Note that, when $N_n^{(j)}$ is complex, $|N_n^{(j)}|$ is its *modulus*. [$N_n^{(j)}$ may be complex when $\varphi(t)$ is complex. $H_n^{(j)}$ and $K_n^{(j)}$ are always real.] Also, from the first step of the algorithm it is clear that $\varphi(t_j) \neq 0$ must hold for all j. Obviously, this can be accomplished by choosing the t_j appropriately.

The validity of (7.2.19) is a consequence of the following result.

Theorem 7.2.5 *The $M_n^{(j)}$, $N_n^{(j)}$, $H_n^{(j)}$, and $K_n^{(j)}$ computed by the W-algorithm as in (7.2.18) satisfy*

$$M_n^{(j)} = D_n^{(j)}\{a(t)/\varphi(t)\}, \quad N_n^{(j)} = D_n^{(j)}\{1/\varphi(t)\},$$
$$H_n^{(j)} = D_n^{(j)}\{P(t)\}, \quad K_n^{(j)} = D_n^{(j)}\{S(t)\}. \tag{7.2.20}$$

As a result, (7.2.19) is valid.

Proof. (7.2.20) is a direct consequence of the known recursion relation for divided differences, namely,

$$D_n^{(j)}\{g(t)\} = \frac{D_{n-1}^{(j+1)}\{g(t)\} - D_{n-1}^{(j)}\{g(t)\}}{t_{j+n} - t_j}, \tag{7.2.21}$$

and (7.2.4) and (7.2.17). ∎

In view of the W-algorithm, $M_n^{(j)}$, $N_n^{(j)}$, $H_n^{(j)}$, and $K_n^{(j)}$ can be arranged separately in two-dimensional arrays as in Table 7.2.1, where the $Q_n^{(j)}$ stand for $M_n^{(j)}$ or $N_n^{(j)}$ or $H_n^{(j)}$ or $K_n^{(j)}$, and the computational flow is as described by the arrows.

Obviously, the W-algorithm allows the recursive computation of both the approximations $A_n^{(j)}$ and their accompanying $\Gamma_n^{(j)}$ and $\Lambda_n^{(j)}$ simultaneously and without having to know the $\gamma_{ni}^{(j)}$. This is interesting, as we are not aware of other extrapolation algorithms shown to have this property. Normally, to obtain $\Gamma_n^{(j)}$ and $\Lambda_n^{(j)}$, we would expect to have to determine the $\gamma_{ni}^{(j)}$ separately by using a different algorithm or approach, such as that given by Theorem 3.4.1 in Section 3.4.

Table 7.2.1:

$$Q_0^{(0)}$$
$$Q_0^{(1)} \quad \rightarrow \quad Q_1^{(0)}$$
$$Q_0^{(2)} \quad \rightarrow \quad Q_1^{(1)} \quad \rightarrow \quad Q_2^{(0)}$$
$$Q_0^{(3)} \quad \rightarrow \quad Q_1^{(2)} \quad \rightarrow \quad Q_2^{(1)} \quad \rightarrow \quad Q_3^{(0)}$$
$$\vdots \qquad \vdots \qquad \vdots \qquad \vdots \qquad \ddots$$

7.2.1 A Special Case: $\phi(y) = y^r$

Let us go back to Algorithm 7.2.4. By substituting (7.2.18) in (7.2.19), we obtain a recursion relation for the $A_n^{(j)}$ that reads

$$A_n^{(j)} = \frac{A_{n-1}^{(j+1)} - w_n^{(j)} A_{n-1}^{(j)}}{1 - w_n^{(j)}} = A_{n-1}^{(j)} + \frac{A_{n-1}^{(j+1)} - A_{n-1}^{(j)}}{1 - w_n^{(j)}}, \quad j \geq 0, \quad n \geq 1, \quad (7.2.22)$$

where $w_n^{(j)} = N_{n-1}^{(j)}/N_{n-1}^{(j+1)}$. That is, the W-algorithm now computes tables for the $A_n^{(j)}$ and the $N_n^{(j)}$.

In case $N_n^{(j)}$ are known and do not have to be computed recursively, (7.2.22) provides the $A_n^{(j)}$ by computing only one table. As an example, consider $\phi(y) = y^r$, or, equivalently, $\varphi(t) = t$. Then, from (7.2.20) and (7.2.9), we have

$$N_n^{(j)} = D_n^{(j)}\{t^{-1}\} = (-1)^n/(t_j t_{j+1} \cdots t_{j+n}), \quad (7.2.23)$$

so that $w_n^{(j)}$ becomes

$$w_n^{(j)} = t_{j+n}/t_j, \quad (7.2.24)$$

and (7.2.22) reduces to

$$A_n^{(j)} = \frac{t_j A_{n-1}^{(j+1)} - t_{j+n} A_{n-1}^{(j)}}{t_j - t_{j+n}} = A_{n-1}^{(j)} + \frac{A_{n-1}^{(j+1)} - A_{n-1}^{(j)}}{1 - t_{j+n}/t_j}, \quad j \geq 0, \quad n \geq 1. \quad (7.2.25)$$

GREP$^{(1)}$ in this case is, of course, the polynomial Richardson extrapolation of Chapter 2, and the recursion relation in (7.2.25) is nothing but Algorithm 2.2.1 due to Bulirsch and Stoer [43], which we derived by a different method in Chapter 2.

In this case, we can also give the closed-form expression

$$A_n^{(j)} = \sum_{i=0}^{n} \left(\prod_{\substack{k=0 \\ k \neq i}}^{n} \frac{t_{j+k}}{t_{j+k} - t_{j+i}} \right) a(t_{j+i}). \quad (7.2.26)$$

This expression is obtained by expanding $D_n^{(j)}\{a(t)/t\}$ with the help of (7.2.3) and then dividing by $D_n^{(j)}\{1/t\}$ that is given in (7.2.23). Note that (7.2.26) can also be obtained by recalling that, in this case, $A_n^{(j)} = p_{n,j}(0)$, where $p_{n,j}(t)$ is the polynomial that interpolates $a(t)$ at t_l, $l = j, j+1, \ldots, j+n$, and by setting $t = 0$ in the resulting Lagrange interpolation formula for $p_{n,j}(t)$.

Recall that the computation of the $\Gamma_n^{(j)}$ can be simplified similarly: Set $\Gamma_0^{(j)} = 1$, $j = 0, 1, \dots$, and compute the rest of the $\Gamma_n^{(j)}$ by the recursion relation

$$\Gamma_n^{(j)} = \frac{t_j \Gamma_{n-1}^{(j+1)} + t_{j+n} \Gamma_{n-1}^{(j)}}{t_j - t_{j+n}}, \quad j = 0, 1, \dots, \quad n = 1, 2, \dots . \quad (7.2.27)$$

7.3 The $W^{(m)}$-Algorithm for $GREP^{(m)}$

We now consider the development of an algorithm for $GREP^{(m)}$ when $r_1 = \cdots = r_m = r$. We start by rewriting the equations (4.2.1) that define $GREP^{(m)}$ in a more convenient way as follows: Let $t = y^r$ and $t_l = y_l^r$, $l = 0, 1, \dots$, and define $a(t) \equiv A(y)$ and $\varphi_k(t) \equiv \phi_k(y)$, $k = 1, \dots, m$. Then, (4.2.1) becomes

$$a(t_l) = A_n^{(m,j)} + \sum_{k=1}^{m} \varphi_k(t_l) \sum_{i=0}^{n_k-1} \bar{\beta}_{ki} t_l^i, \quad j \leq l \leq j+N; \quad N = \sum_{k=1}^{m} n_k, \quad (7.3.1)$$

with $n = (n_1, \dots, n_m)$ as usual.

In Section 4.4 on the convergence theory of GREP, we mentioned that those sequences related to Process II in which $n_k \to \infty$, $k = 1, \dots, m$, simultaneously, and, in particular, the sequences $\{A_{q+(\nu,\dots,\nu)}^{(m,j)}\}_{\nu=0}^{\infty}$ with j and $q = (q_1, \dots, q_m)$ fixed, appear to have the best convergence properties. Therefore, we should aim at developing an algorithm for computing such sequences. To keep the treatment simple, we restrict our attention to the sequences $\{A_{(\nu,\dots,\nu)}^{(m,j)}\}_{\nu=0}^{\infty}$ that appear to provide the best accuracy for a given number of the $A(y_i)$. For the treatment of the more general case in which $q_1, \dots, q_m$ are not all 0, we refer the reader to Ford and Sidi [87].

The development of the $W^{(m)}$-algorithm depends heavily on the FS-algorithm discussed in Chapter 3. We freely use the results and notation of Section 3.3 throughout our developments here. Therefore, a review of Section 3.3 is recommended at this point.

One way to compute the sequences $\{A_{(\nu,\dots,\nu)}^{(m,j)}\}_{\nu=0}^{\infty}$ is to "eliminate" first $\varphi_k(t)t^0$, $k = 1, \dots, m$, next $\varphi_k(t)t^1$, $k = 1, \dots, m$, etc., from the expansion of $A(y)$ given in (4.1.1) and (4.1.2). This can be accomplished by ordering the $\varphi_k(t)t^i$ suitably. We begin by considering this issue.

7.3.1 Ordering of the $\varphi_k(t)t^i$

Let us define the sequences $\{g_s(l)\}_{l=0}^{\infty}$, $s = 1, 2, \dots$, as follows:

$$g_{k+im}(l) = \varphi_k(t_l)t_l^i, \quad 1 \leq k \leq m, \quad i \geq 0. \quad (7.3.2)$$

Thus, the sequence g_k is simply $\{\varphi_k(t_l)\}_{l=0}^{\infty}$ for $k = 1, \dots, m$, the sequence g_{m+k} is $\{\varphi_k(t_l)t_l\}_{l=0}^{\infty}$ for $k = 1, \dots, m$, etc., and this takes care of all the sequences g_s, $s = 1, 2, \dots$, and all the $\varphi_k(t)t^i$, $1 \leq k \leq m$, $i \geq 0$. As in Ford and Sidi [87], we call this ordering of the $\varphi_k(t)t^i$ the *normal ordering* throughout this chapter. As a consequence of (7.3.2), we also have

$$g_i(l) = \begin{cases} \varphi_i(t_l), & 1 \leq i \leq m, \\ t_l g_{i-m}(l), & i > m. \end{cases} \quad (7.3.3)$$

Table 7.3.1:

$$
\begin{array}{ccccc}
A_0^{(0)} & & & & \\
A_0^{(1)} & A_1^{(0)} & & & \\
A_0^{(2)} & A_1^{(1)} & A_2^{(0)} & & \\
A_4^{(3)} & A_1^{(2)} & A_2^{(1)} & A_3^{(0)} & \\
\vdots & \vdots & \vdots & \vdots & \ddots
\end{array}
$$

Let us also denote

$$h_k(l) = \frac{\varphi_k(t_l)}{t_l} = \frac{g_k(l)}{t_l}, \quad 1 \le k \le m. \tag{7.3.4}$$

Now every integer p can be expressed as $p = vm + \rho$, where $v = \lfloor p/m \rfloor$ and $\rho \equiv p$ mod m and thus $0 \le \rho < m$. With the normal ordering of the $\varphi_k(t)t^i$ introduced above, $A_p^{(j)}$ that is defined along with the parameters $\bar{\alpha}_1, \ldots, \bar{\alpha}_p$ by the equations

$$a(t_l) = A_p^{(j)} + \sum_{k=1}^{p} \bar{\alpha}_k g_k(l), \quad j \le l \le j + p, \tag{7.3.5}$$

is precisely $A_n^{(m,j)}$ with (i) $n_k = v$, $k = 1, \ldots, m$, if $\rho = 0$, and (ii) $n_k = v + 1$, $k = 1, \ldots, \rho$, and $n_k = v$, $k = \rho + 1, \ldots, m$, if $\rho > 0$. Thus, in the notation of (7.3.1), the equations in (7.3.5) are equivalent to

$$a(t_l) = A_p^{(j)} + \sum_{k=1}^{m} \varphi_k(t_l) \sum_{i=0}^{\lfloor (p-k)/m \rfloor} \bar{\beta}_{ki} t_l^i, \quad j \le l \le j + p. \tag{7.3.6}$$

Consequently, the sequence $\{A_s^{(j)}\}_{s=0}^{\infty}$ contains $\{A_{(v,\ldots,v)}^{(m,j)}\}_{v=0}^{\infty}$ as a subsequence. When $m = 2$, for example, $A_1^{(j)}, A_2^{(j)}, A_3^{(j)}, A_4^{(j)}, \ldots$, are $A_{(1,0)}^{(2,j)}, A_{(1,1)}^{(2,j)}, A_{(2,1)}^{(2,j)}, A_{(2,2)}^{(2,j)}, \ldots$, respectively.

In view of the above, it is convenient to order the $A_n^{(j)}$ as in Table 7.3.1.

As $A_p^{(j)}$ satisfies (7.3.5), it is given by the determinantal formula of (3.2.1), where $a(l)$ stands for $a(t_l) \equiv A(y_l)$, as in Chapter 3. Therefore, $A_p^{(j)}$ could be computed, for all p, by either of the algorithms of Chapter 3. But with the normal ordering defined in (7.3.2), this general approach can profitably be altered. When $p \le m$, the FS-algorithm is used to compute the $A_p^{(j)}$ because the corresponding $g_k(l)$ have no particular structure. The normal ordering has ensured that all these unstructured $g_k(l)$ come first. For $p > m$, however, the $W^{(m)}$-algorithm determines the $A_p^{(j)}$ through a sophisticated set of recursions at a cost much smaller than would be incurred by direct application of the FS-algorithm.

Let us now illuminate the approach taken in the $W^{(m)}$-algorithm. We observe first that the definition of $D_p^{(j)}$ in (3.3.9) involves only the $G_k^{(s)}$. Thus, we look at $G_p^{(j)}$ for $p > m$. For simplicity, we take $m = 2$.

When $m = 2$ and $p \ge 3$, we have by (7.3.2) that

$$G_p^{(j)} = |\varphi_1(t_j) \ \varphi_2(t_j) \ \varphi_1(t_j)t_j \ \varphi_2(t_j)t_j \ \cdots \ g_p(j)|.$$

If we factor out t_{j+i-1} from the ith row, $i = 1, 2, \ldots, p$, we obtain

$$G_p^{(j)} = \left(\prod_{i=1}^{p} t_{j+i-1} \right) |h_1(j)\, h_2(j)\, g_1(j)\, g_2(j) \cdots g_{p-2}(j)|,$$

where $h_i(l)$ are as defined in (7.3.4). The determinant obtained above differs from $G_p^{(j)}$ by two columns, and it is not difficult to see that there are m different columns in the general case. Therefore, we need to develop procedures for evaluating these objects.

7.3.2 Technical Preliminaries

We start this by introducing the following generalizations of $f_p^{(j)}(b)$, $\psi_p^{(j)}(b)$, and $D_p^{(j)}$:

$$F_p^{(j)}(h_1, \ldots, h_q) = |g_1(j) \cdots g_p(j)\, h_1(j) \cdots h_q(j)| \equiv F_p^{(j)}(q), \qquad (7.3.7)$$

$$\Psi_p^{(j)}(h_1, \ldots, h_q) = \frac{F_{p+1-q}^{(j)}(q)}{F_{p+2-q}^{(j)}(q-1)} \equiv \Psi_p^{(j)}(q), \qquad (7.3.8)$$

$$D_p^{(j)}(q) = \frac{F_{p+1-q}^{(j)}(q)\, F_{p-1-q}^{(j+1)}(q)}{F_{p-q}^{(j)}(q)\, F_{p-q}^{(j+1)}(q)}. \qquad (7.3.9)$$

[In these definitions and in Theorems 7.3.1 and 7.3.2 below, the $h_k(l)$ can be arbitrary. They need not be defined by (7.3.4).]

As simple consequences of (7.3.7)–(7.3.9), we obtain

$$F_p^{(j)}(0) = G_p^{(j)}, \quad F_p^{(j)}(1) = f_p^{(j)}(h_1), \qquad (7.3.10)$$

$$\Psi_p^{(j)}(1) = \psi_p^{(j)}(h_1), \qquad (7.3.11)$$

$$D_p^{(j)}(0) = D_p^{(j)}, \qquad (7.3.12)$$

respectively. In addition, we define $F_0^j(0) = 1$. From (7.3.11), it is clear that the algorithm that will be used to compute the different $\psi_p^{(j)}(b)$ can be used to compute the $\Psi_p^{(j)}(1)$ as well.

We also note that, since $F_p^{(j)}(q)$ is defined for $p \geq 0$ and $q \geq 0$, $\Psi_p^{(j)}(q)$ is defined for $1 \leq q \leq p+1$ and $D_p^{(j)}(q)$ is defined for $0 \leq q \leq p-1$, when the $h_k(l)$ are arbitrary,
The following results will be of use in the development of the $W^{(m)}$-algorithm shortly.

Theorem 7.3.1 *The $\Psi_p^{(j)}(q)$ and $D_p^{(j)}(q)$ satisfy*

$$\Psi_p^{(j)}(q) = \frac{\Psi_{p-1}^{(j+1)}(q) - \Psi_{p-1}^{(j)}(q)}{D_p^{(j)}(q-1)}, \quad 1 \leq q \leq p, \qquad (7.3.13)$$

and

$$D_p^{(j)}(q) = \Psi_{p-2}^{(j+1)}(q) \left[\frac{1}{\Psi_{p-1}^{(j)}(q)} - \frac{1}{\Psi_{p-1}^{(j+1)}(q)} \right], \quad 1 \leq q \leq p-1. \qquad (7.3.14)$$

Proof. Consider the $(p + q) \times (p + q)$ determinant $F_p^{(j)}(q)$. Applying Theorem 3.3.1 (the Sylvester determinant identity) to this determinant with $\rho = 1, \rho' = p + q, \sigma = p$, and $\sigma' = p + q$, we obtain

$$F_p^{(j)}(q)F_{p-1}^{(j+1)}(q-1) = F_{p-1}^{(j+1)}(q)F_p^{(j)}(q-1) - F_p^{(j+1)}(q-1)F_{p-1}^{(j)}(q). \quad (7.3.15)$$

Replacing p in (7.3.15) by $p + 1 - q$, and using (7.3.8) and (7.3.9), (7.3.13) follows. To prove (7.3.14), we start with (7.3.9). When we invoke (7.3.8), (7.3.9) becomes

$$D_p^{(j)}(q) = \frac{\Psi_p^{(j)}(q)\Psi_{p-2}^{(j+1)}(q)}{\Psi_{p-1}^{(j)}(q)\Psi_{p-1}^{(j+1)}(q)} D_p^{(j)}(q-1). \quad (7.3.16)$$

When we substitute (7.3.13) in (7.3.16), (7.3.14) follows. ∎

The recursion relations given in (7.3.13) and (7.3.14) will be applied, for a given p, with q increasing from 1. When $q = p - 1$, (7.3.14) requires knowledge of $\Psi_{p-2}^{(j+1)}(p - 1)$, and when $q = p$, (7.3.13) requires knowledge of $\Psi_{p-1}^{(j)}(p)$ and $\Psi_{p-1}^{(j+1)}(p)$. That is to say, we need to have $\Psi_s^{(j)}(s + 1), s \geq 0, j \geq 0$, to be able to complete the recursions in (7.3.14) and (7.3.13). Now, $\Psi_p^{(j)}(p + 1)$ cannot be computed by the recursion relation in (7.3.13), because neither $\Psi_{p-1}^{(j)}(p + 1)$ nor $D_p^{(j)}(p)$ is defined. When the definition of the $h_k(l)$ given in (7.3.4) is invoked, however, $\Psi_p^{(j)}(p + 1)$ can be expressed in simple and familiar terms and computed easily, as we show later.

In addition to the relationships in the previous theorem, we give one more result concerning the $\Psi_p^{(j)}(q)$.

Theorem 7.3.2 *The $\Psi_p^{(j)}(q)$ satisfy the relation*

$$F_{p+1-q}^{(j)}(q) = \Psi_p^{(j)}(1)\Psi_p^{(j)}(2) \cdots \Psi_p^{(j)}(q)G_{p+1}^{(j)}. \quad (7.3.17)$$

Proof. Let us reexpress (7.3.8) in the form

$$F_{p+1-q}^{(j)}(q) = \Psi_p^{(j)}(q)F_{p+1-(q-1)}^{(j)}(q-1). \quad (7.3.18)$$

Invoking (7.3.8), with q replaced by $q - 1$, on the right-hand side of (7.3.18), and continuing, we obtain

$$F_{p+1-q}^{(j)}(q) = \Psi_p^{(j)}(q) \cdots \Psi_p^{(j)}(1)F_{p+1}^{(j)}(0). \quad (7.3.19)$$

The result in (7.3.17) follows from (7.3.19) and (7.3.10). ∎

7.3.3 Putting It All Together

Let us now invoke the normal ordering of the $\varphi_k(t)t^i$. With this ordering, the $g_k(l)$ and $h_k(l)$ are as defined in (7.3.2) and (7.3.4), respectively. This has certain favorable consequences concerning the $D_p^{(j)}$.

Because we have defined $h_k(l)$ only for $1 \leq k \leq m$, we see that $F_p^j(q)$ is defined only for $0 \leq q \leq m$, $p \geq 0$, $\Psi_p^{(j)}(q)$ for $1 \leq q \leq \min\{p + 1, m\}$, and $D_p^{(j)}(q)$ for $0 \leq q \leq$

$\min\{p-1, m\}$. Consequently, the recursion relation in (7.3.13) is valid for $1 \le q \le \min\{p, m\}$, and that in (7.3.14) is valid for $1 \le q \le \min\{p-1, m\}$.

In addition, for $\Psi_p^{(j)}(p+1)$, we have the following results.

Theorem 7.3.3 *With the normal ordering and for $0 \le p \le m-1$, we have*

$$\Psi_p^{(j)}(p+1) = (-1)^p / \psi_p^{(j)}(g_{m+1}). \tag{7.3.20}$$

Proof. First, it is clear that $\Psi_p^{(j)}(p+1)$ is defined only for $0 \le p \le m-1$ now. Next, from (7.3.8), we have

$$\Psi_p^{(j)}(p+1) = \frac{F_0^{(j)}(p+1)}{F_1^{(j)}(p)} = \frac{|h_1(j) \cdots h_{p+1}(j)|}{|g_1(j) h_1(j) \cdots h_p(j)|}. \tag{7.3.21}$$

The result follows by multiplying the ith rows of the $(p+1) \times (p+1)$ numerator and denominator determinants in (7.3.21) by t_{j+i-1}, $i = 1, \ldots, p+1$, and by invoking (7.3.4) and (7.3.3) and the definition of $\psi_p^{(j)}(b)$ from (3.3.5).   ∎

Theorem 7.3.3 implies that $\Psi_p^{(j)}(p+1)$, $0 \le p \le m-1$, and $D_p^{(j)}$, $1 \le p \le m$, can be obtained *simultaneously* by the FS-algorithm. As mentioned previously, we are using the FS-algorithm for computing $D_p^{(j)}$, $1 \le p \le m$, as part of the $W^{(m)}$-algorithm.

Through $g_i(l) = t_l g_{i-m}(l)$, $i > m$, that is given in (7.3.3), when $p \ge m+1$, the normal ordering enables us to relate $F_{p-m}^{(j)}(m)$ to $G_p^{(j)}$ and hence $D_p^{(j)}(m)$ to $D_p^{(j)}(0) = D_p^{(j)}$, the desired quantity, thus reducing the amount of computation for the $W^{(m)}$-algorithm considerably.

Theorem 7.3.4 *With the normal ordering and for $p \ge m+1$, $D_p^{(j)}$ satisfies*

$$D_p^{(j)} = D_p^{(j)}(m). \tag{7.3.22}$$

Proof. First, it is clear that $D_p^{(j)}(m)$ is defined only for $p \ge m+1$. Next, from (7.3.7), we have

$$F_k^{(s)}(m) = (-1)^{mk} \left(\prod_{i=0}^{k+m-1} t_{s+i} \right)^{-1} G_{m+k}^{(s)}. \tag{7.3.23}$$

Finally, (7.3.22) follows by substituting (7.3.23) in the definition of $D_p^{(j)}(m)$ that is given in (7.3.9).

The proof of (7.3.23) can be achieved as follows: Multiplying the ith row of the $(m+k) \times (m+k)$ determinant that represents $F_k^{(s)}(m)$ by t_{s+i-1}, $i = 1, 2, \ldots, m+k$, and using (7.3.3) and (7.3.4), we obtain

$$\left(\prod_{i=0}^{k+m-1} t_{s+i} \right) F_k^{(s)}(m) = |g_{m+1}(s) \, g_{m+2}(s) \, \cdots \, g_{m+k}(s) \, g_1(s) \, g_2(s) \, \cdots \, g_m(s)|.$$

$$\tag{7.3.24}$$

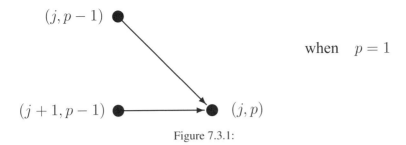

Figure 7.3.1:

By making the necessary column permutations, it can be seen that the right-hand side of (7.3.24) is nothing but $(-1)^{mk} G_{m+k}^{(s)}$.   ∎

From Theorem 7.3.4, it is thus clear that, for $p \geq m + 1$, $D_p^{(j)}$ can be computed by a recursion relation of fixed length m. This implies that the cost of computing $D_p^{(j)}$ is *fixed* and, therefore, independent of j and p. Consequently, the total cost of computing all the $D_p^{(j)}$, $1 \leq j + p \leq L$, is now $O(L^2)$ arithmetic operations, as opposed to $O(L^3)$ for the FS-algorithm.

As we compute the $\psi_p^{(j)}(I)$, we can also apply Theorem 3.4.1 to compute the $\gamma_{pi}^{(j)}$, with the help of which we can determine $\Gamma_p^{(j)} = \sum_{i=0}^{p} |\gamma_{pi}^{(j)}|$ and $\Lambda_p^{(j)} = \sum_{i=0}^{p} |\gamma_{pi}^{(j)}| \, |a(t_{j+i}|$, the quantities of relevance to the numerical stability of $A_p^{(j)}$.

In summary, the results of relevance to the $W^{(m)}$-algorithm are (3.3.6), (3.3.10), (3.3.12), (3.4.1), (3.4.2), (7.3.11), (7.3.13), (7.3.14), (7.3.20), and (7.3.22). With the $A_p^{(j)}$ ordered as in Table 7.3.1, from these results we see that the quantities at the (j, p) location in this table are related to others as in Figure 7.3.1 and Figure 7.3.2.

All this shows that the $W^{(m)}$-algorithm can be programmed so that it computes Table 7.3.1 row-wise, thus allowing the one-by-one introduction of the sets $X_l = \{t_l, a(t_l), \varphi_k(t_l), \ k = 1, \ldots, m\}$, $l = 0, 1, 2, \ldots$, that are necessary for the initial values $\psi_0^{(l)}(a)$, $\psi_0^{(l)}(I)$, and $\psi_0^{(l)}(g_k)$, $k = 1, \ldots, m + 1$. Once a set X_l is introduced, we can compute all the $\psi_p^{(j)}(g_k)$, $\psi_p^{(j)}(a)$, $\psi_p^{(j)}(I)$, $D_p^{(j)}$, $D_p^{(j)}(q)$, and $\Psi_p^{(j)}(q)$ along the row $j + p = l$ in the order $p = 1, 2, \ldots, l$. Also, before introducing the set X_{l+1} for the next row of the table, we can discard some of these quantities and have the remaining ones overwrite the corresponding quantities along the row $j + p = l - 1$. Obviously, this saves a lot of storage. More will be said on this later.

Below by "initialize (l)" we mean

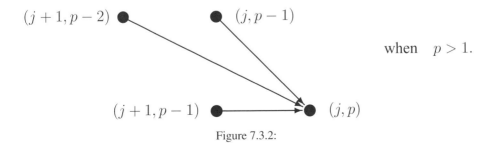

Figure 7.3.2:

$$\{\text{read } t_l, \ a(t_l), \ g_k(l) = \varphi_k(t_l), \ k = 1, \ldots, m, \text{ and set}$$

$$\psi_0^{(l)}(a) = a(l)/g_1(l), \ \psi_0^{(l)}(I) = 1/g_1(l), \ \psi_0^{(l)}(g_k) = g_k(l)/g_1(l), \ k = 2, \ldots, m,$$

$$\psi_0^{(l)}(g_{m+1}) = t_l, \text{ and } \Psi_0^{(l)}(1) = 1/t_l\}.$$

We recall that $a(l)$ stands for $a(t_l) \equiv A(y_l)$ and $g_k(l) = \varphi_k(t_l) \equiv \phi_k(y_l)$, $k = 1, \ldots, m$, and $t_l = y_l^r$.

Algorithm 7.3.5 ($W^{(m)}$-algorithm)
initialize (0)
for $l = 1$ **to** L **do**
   initialize (l)
   for $p = 1$ **to** l **do**
      $j = l - p$
      if $p \leq m$ **then**
         $D_p^{(j)} = \psi_{p-1}^{(j+1)}(g_{p+1}) - \psi_{p-1}^{(j)}(g_{p+1})$
         for $k = p + 2$ **to** $m + 1$ **do**
            $\psi_p^{(j)}(g_k) = [\psi_{p-1}^{(j+1)}(g_k) - \psi_{p-1}^{(j)}(g_k)]/D_p^{(j)}$
         endfor
      endif
      if $p \leq m - 1$ **then**
         $\Psi_p^{(j)}(p + 1) = (-1)^p/\psi_p^{(j)}(g_{m+1})$
      endif
      for $q = 1$ **to** $\min\{p - 1, m - 1\}$ **do**
         $D_p^{(j)}(q) = \Psi_{p-2}^{(j+1)}(q)[1/\Psi_{p-1}^{(j)}(q) - 1/\Psi_{p-1}^{(j+1)}(q)]$
         $q' = q + 1$
         $\Psi_p^{(j)}(q') = [\Psi_{p-1}^{(j+1)}(q') - \Psi_{p-1}^{(j)}(q')]/D_p^{(j)}(q)$
      endfor
      if $p > m$ **then**
         $D_p^{(j)} = D_p^{(j)}(m) = \Psi_{p-2}^{(j+1)}(m)[1/\Psi_{p-1}^{(j)}(m) - 1/\Psi_{p-1}^{(j+1)}(m)]$
      endif
      $\Psi_p^{(j)}(1) = [\Psi_{p-1}^{(j+1)}(1) - \Psi_{p-1}^{(j)}(1)]/D_p^{(j)}$
      $\psi_p^{(j)}(a) = [\psi_{p-1}^{(j+1)}(a) - \psi_{p-1}^{(j)}(a)]/D_p^{(j)}$
      $\psi_p^{(j)}(I) = [\psi_{p-1}^{(j+1)}(I) - \psi_{p-1}^{(j)}(I)]/D_p^{(j)}$
      $A_p^{(j)} = \psi_p^{(j)}(a)/\psi_p^{(j)}(I)$
   endfor
endfor

From the "initialize (l)" statement, it is clear that $\varphi_1(t_l) = g_1(l) \neq 0$ must hold for all l in the algorithm.

As can be seen, not every computed quantity has to be saved throughout the course of computation. Before l is incremented in the statement "**for** $l = 1$ **to** L **do**" the following newly computed quantities are saved: (i) $\psi_p^{(j)}(g_k)$, $p + 2 \leq k \leq m + 1$, for $p \leq m$; (ii) $\Psi_p^{(j)}(q)$, $1 \leq q \leq \min\{p + 1, m\}$; and (iii) $\psi_p^{(j)}(a)$ and $\psi_p^{(j)}(I)$; all with $j \geq 0$,

$p \geq 0$, and $j + p = l$. With suitable programming, these quantities can occupy the storage locations of those computed in the previous stage of this statement. None of the $D_p^{(j)}$ and $D_p^{(j)}(q)$ needs to be saved. Thus, we need to save approximately $m + 2$ vectors of length L when L is large.

As for the operation count, we first observe that there are $L^2/2 + O(L)$ lattice points (j, p) for $0 \leq j + p \leq L$. At each point, we compute $A_p^{(j)}$, $\psi_p^{(j)}(a)$, $\psi_p^{(j)}(I)$, $\Psi_p^{(j)}(q)$ for $1 \leq q \leq \min\{p + 1, m\}$, and $D_p^{(j)}(q)$ for $1 \leq q \leq \min\{p - 1, m\}$. The total number of arithmetic operations then is $(5m + 5)L^2/2 + O(L)$, including the rest of the computation. It can be made even smaller by requiring that only $A_p^{(0)}$ be computed.

We finally mention that SUBROUTINE WMALGM that forms part of the FORTRAN 77 program given in Ford and Sidi [87, Appendix B] implements the $W^{(m)}$-algorithm exactly as in Algorithm 7.3.5. This program (with slight changes, but with the same SUBROUTINE WMALGM) is given in Appendix I of this book.

7.3.4 Simplifications for the Cases $m = 1$ and $m = 2$

The following result is a consequence of Theorem 7.3.2 and (7.3.23).

Theorem 7.3.6 With the normal ordering of the $\varphi_k(t)t^i$, the $\Psi_p^{(j)}(q)$ satisfy

$$(-1)^{mp} \left(\prod_{i=0}^{p} t_{j+i} \right) \prod_{i=1}^{m} \Psi_p^{(j)}(i) = 1, \quad p \geq m - 1. \tag{7.3.25}$$

Application to the Case $m = 1$

If we let $m = 1$ in Theorem 7.3.6, we have $(-1)^p \left(\prod_{i=0}^{p} t_{j+i} \right) \Psi_p^{(j)}(1) = 1$, $p \geq 0$. Thus, solving for $\Psi_p^{(j)}(1)$ and substituting in (7.3.14) with $q = 1$ there, and recalling that $D_p^{(j)} = D_p^{(j)}(1)$ for $p \geq 2$, we obtain $D_p^{(j)} = t_{j+p} - t_j$ for $p \geq 2$. This result is valid also for $p = 1$ as can be shown from (3.3.9). We therefore conclude that the $W^{(1)}$-algorithm is practically identical to the W-algorithm of the preceding section. The difference between the two is that the W-algorithm uses $D_p^{(j)} = t_{j+p} - t_j$ directly, whereas the $W^{(1)}$-algorithm computes $D_p^{(j)}$ with the help of Theorems 7.3.1 and 7.3.4.

Application to the Case $m = 2$

If we let $m = 2$ in Theorem 7.3.6, we have $\left(\prod_{i=0}^{p} t_{j+i} \right) \Psi_p^{(j)}(1)\Psi_p^{(j)}(2) = 1$, $p \geq 1$. Solving for $\Psi_p^{(j)}(2)$ in terms of $\Psi_p^{(j)}(1) = \psi_p^{(j)}(h_1)$, and substituting in (7.3.14) with $q = 2$ there, and recalling that $D_p^{(j)} = D_p^{(j)}(2)$ for $p \geq 3$, we obtain

$$D_p^{(j)} = [t_j \psi_{p-1}^{(j)}(h_1) - t_{j+p} \psi_{p-1}^{(j+1)}(h_1)]/\psi_{p-2}^{(j+1)}(h_1), \tag{7.3.26}$$

which is valid for $p \geq 3$. Using (3.3.9), we can show that this is valid for $p = 2$ as well. As for $p = 1$, we have, again from (3.3.9),

$$D_1^{(j)} = g_2(j+1)/g_1(j+1) - g_2(j)/g_1(j). \tag{7.3.27}$$

This allows us to simplify the $W^{(2)}$-algorithm as follows: Given $\varphi_1(t_l)$ and $\varphi_2(t_l)$, $l = 0, 1, \ldots$, use (7.3.27) and (3.3.10) to compute $\psi_1^{(j)}(h_1)$. Then, for $p = 2, 3, \ldots$, use

(7.3.26) to obtain $D_p^{(j)}$ and (3.3.10) to obtain $\psi_p^{(j)}(a)$, $\psi_p^{(j)}(I)$, and $\psi_p^{(j)}(h_1)$. For the sake of completeness, we give below this simplification of the $W^{(2)}$-algorithm separately.

Algorithm 7.3.7 (Simplified $W^{(2)}$-algorithm)

1. For $j = 0, 1, \ldots$, set $\psi_0^{(j)}(a) = a(t_j)/\varphi_1(t_j)$, $\psi_0^{(j)}(I) = 1/\varphi_1(t_j)$, and $\psi_0^{(j)}(h_1) = 1/t_j$, and

$$D_1^{(j)} = \varphi_2(t_{j+1})/\varphi_1(t_{j+1}) - \varphi_2(t_j)/\varphi_1(t_j).$$

2. For $j = 0, 1, \ldots$, and $p = 1, 2, \ldots$, compute recursively

$$\psi_p^{(j)}(a) = [\psi_{p-1}^{(j+1)}(a) - \psi_{p-1}^{(j)}(a)]/D_p^{(j)},$$

$$\psi_p^{(j)}(I) = [\psi_{p-1}^{(j+1)}(I) - \psi_{p-1}^{(j)}(I)]/D_p^{(j)},$$

$$\psi_p^{(j)}(h_1) = [\psi_{p-1}^{(j+1)}(h_1) - \psi_{p-1}^{(j)}(h_1)]/D_p^{(j)},$$

$$D_{p+1}^{(j)} = [t_j \psi_p^{(j)}(h_1) - t_{j+p+1} \psi_p^{(j+1)}(h_1)]/\psi_{p-1}^{(j+1)}(h_1).$$

3. For all j and p, set $A_p^{(j)} = \psi_p^{(j)}(a)/\psi_p^{(j)}(I)$.

7.4 Implementation of the $d^{(m)}$-Transformation by the $W^{(m)}$-Algorithm

As mentioned in Chapter 6, the $d^{(m)}$-transformation is actually a $GREP^{(m)}$ that can be implemented by the $W^{(m)}$-algorithm. This means that in the "initialize (l)" statement of Algorithm 7.3.5, we need to have the input $t_l = 1/R_l$, $a(t_l) = A_{R_l}$, and $g_k(l) = R_l^k \Delta^{k-1} a_{R_l}$, $k = 1, \ldots, m$. Nothing else is required in the rest of the algorithm.

Similarly, the W-algorithm can be used to implement the $d^{(1)}$-transformation. In this case, we need as input to Algorithm 7.2.4 $t_l = 1/R_l$, $a(t_l) = A_{R_l}$, and $\varphi(t_l) = R_l a_{R_l}$.

As already mentioned, in applying the $W^{(m)}$-algorithm we must make sure that $\varphi_1(t_l) = g_1(l) \neq 0$ for all l. Now, when this algorithm is used to implement the $d^{(m)}$-transformation on an infinite series $\sum_{k=1}^{\infty} a_k$, we can set $\varphi_k(t) = n^k \Delta^{k-1} a_n$, $k = 1, \ldots, m$, where $t = n^{-1}$, as follows from (6.1.18) and (6.2.1). Thus, $\varphi_1(t) = n a_n$. It may happen that $a_n = 0$ for some n, or even for infinitely many values of n. [Consider, for instance, the Fourier series of Example 6.1.7 with $B = 1$ and $C = 0$. When $\theta = \pi/6$, we have $a_{3+6i} = 0$, $i = 0, 1, \ldots$.] Of course, we can avoid $\varphi_1(t_l) = 0$ simply by choosing the R_l such that $a_{R_l} \neq 0$.

Another approach we have found to be very effective that is also automatic is as follows: Set $\varphi_k(t) = n^{m-k+1} \Delta^{m-k} a_n$, $k = 1, \ldots, m$. [Note that $\varphi_1(t), \ldots, \varphi_m(t)$ can be taken as any fixed permutation of $n^k \Delta^{k-1} a_n$, $k = 1, \ldots, m$.] In this case, $\varphi_1(t) = n^m \Delta^{m-1} a_n$, and the chances of $\Delta^{m-1} a_n = 0$ in general are small when $\{a_k\} \in \mathbf{b}^{(m)}$ and $a_n = 0$. This argument applies also to the case $m = 1$, for which we already know that $a_n \neq 0$ for all large n.

The FORTRAN 77 code in Appendix I implements the $d^{(m)}$-transformation exactly as we have just described.

In any case, the zero terms of the series must be kept, because they play the same role as the nonzero terms in the extrapolation process. Without them, the remaining sequence of (nonzero) terms is no longer in $\mathbf{b}^{(m)}$, *and hence the* $d^{(m)}$-transformation cannot be effective.*

7.5 The EW-Algorithm for an Extended GREP[(1)]

Let us consider the class of functions $A(y)$ that have asymptotic expansions of the form

$$A(y) \sim A + \phi(y) \sum_{k=1}^{\infty} \alpha_k y^{\sigma_k} \quad \text{as } y \to 0+, \tag{7.5.1}$$

with

$$\sigma_i \neq 0, \ i = 1, 2, \ldots; \ \Re\sigma_1 < \Re\sigma_2 < \cdots, \ \text{ and } \ \lim_{i \to \infty} \Re\sigma_i = +\infty. \tag{7.5.2}$$

This is the class of functions considered in Section 4.6 and described by (4.1.1) and (4.6.1) with (4.6.2), and with $m = 1$. The extended GREP[(1)] for this class of $A(y)$ is then defined by the linear systems

$$A(y_l) = A_n^{(j)} + \phi(y_l) \sum_{k=1}^{n} \bar{\alpha}_k y_l^{\sigma_k}, \quad j \leq l \leq j + n. \tag{7.5.3}$$

When σ_i are arbitrary, there does not seem to be an efficient algorithm analogous to the W-algorithm. In such a case, we can make use of the FS-algorithm or the E-algorithm to determine the $A_n^{(j)}$. An efficient algorithm becomes possible, however, when y_l are not arbitrary, but $y_l = y_0 \omega^l$, $l = 1, 2, \ldots$, for some $y_0 \in (0, b]$ and $\omega \in (0, 1)$.

Let us rewrite (7.5.3) in the form

$$\frac{A(y_{j+i}) - A_n^{(j)}}{\phi(y_{j+i})} = \sum_{k=1}^{n} \bar{\alpha}_k y_{j+i}^{\sigma_k}, \quad 0 \leq i \leq n. \tag{7.5.4}$$

We now employ the technique used in the proof of Theorem 1.4.5. Set $\omega^{\sigma_k} = c_k$, $k = 1, 2, \ldots$, and let

$$U_n(z) = \prod_{i=1}^{n} \frac{z - c_i}{1 - c_i} \equiv \sum_{i=0}^{n} \rho_{ni} z^i. \tag{7.5.5}$$

Multiplying both sides of (7.5.4) by ρ_{ni}, summing from $i = 0$ to $i = n$, we obtain

$$\sum_{i=0}^{n} \rho_{ni} \frac{A(y_{j+i}) - A_n^{(j)}}{\phi(y_{j+i})} = \sum_{k=1}^{n} \bar{\alpha}_k y_j^{\sigma_k} U_n(c_k) = 0, \tag{7.5.6}$$

from which we have

$$A_n^{(j)} = \frac{\sum_{i=0}^{n} \rho_{ni}[A(y_{j+i})/\phi(y_{j+i})]}{\sum_{i=0}^{n} \rho_{ni}[1/\phi(y_{j+i})]} \equiv \frac{M_n^{(j)}}{N_n^{(j)}}. \tag{7.5.7}$$

Therefore, Algorithm 1.3.1 that was used in the recursive computation of (1.4.5), can be used for computing the $M_n^{(j)}$ and $N_n^{(j)}$.

Algorithm 7.5.1 (EW-algorithm)

1. For $j = 0, 1, \ldots$, set $M_0^{(j)} = A(y_j)/\phi(y_j)$ and $N_0^{(j)} = 1/\phi(y_j)$.
2. For $j = 0, 1, \ldots$, and $n = 1, 2, \ldots$, compute $M_n^{(j)}$ and $N_n^{(j)}$ recursively from

$$M_n^{(j)} = \frac{M_{n-1}^{(j+1)} - c_n M_{n-1}^{(j)}}{1 - c_n} \quad \text{and} \quad N_n^{(j)} = \frac{N_{n-1}^{(j+1)} - c_n N_{n-1}^{(j)}}{1 - c_n}. \tag{7.5.8}$$

3. For all j and n, set $A_n^{(j)} = M_n^{(j)}/N_n^{(j)}$.

Thus the EW-algorithm, just as the W-algorithm, constructs two tables of the form of Table 7.2.1 for the quantities $M_n^{(j)}$ and $N_n^{(j)}$.

We now go on to discuss the computation of $\Gamma_n^{(j)}$ and $\Lambda_n^{(j)}$. By (7.5.7), we have

$$\gamma_{ni}^{(j)} = \frac{\rho_{ni}/\phi(y_{j+i})}{N_n^{(j)}}, \quad i = 0, 1, \ldots, n. \tag{7.5.9}$$

As the ρ_{ni} are independent of j, they can be evaluated inexpensively from (7.5.5). They can then be used to evaluate any of the $\Gamma_n^{(j)}$ and $\Lambda_n^{(j)}$ as part of the EW-algorithm, since $N_n^{(j)}$ and $\phi(y_i)$ are already available. In particular, we may be content with the $\Gamma_n^{(0)}$ and $\Lambda_n^{(0)}, n = 1, 2, \ldots$, associated with the diagonal sequence $\{A_n^{(0)}\}_{n=0}^{\infty}$.

It can be shown, by using (7.5.5) and (7.5.9), that $\Gamma_n^{(j)} = 1$ in the cases (i) c_k are positive and $\phi(y_i)$ alternate in sign, and (ii) c_k are negative and $\phi(y_i)$ have the same sign.

It can also be shown that, when the c_k are either all positive or all negative and the $\phi(y_i)$ are arbitrary, $\Gamma_n^{(j)}$ and $\Lambda_n^{(j)}$ can be computed as part of the EW-algorithm by adding to Algorithm 7.5.1 the following:

(i) Add the following initial conditions to Step 1:

$$H_0^{(j)} = (-1)^j |N_0^{(j)}|, \quad K_0^{(j)} = (-1)^j |M_0^{(j)}|, \quad \text{if } c_k > 0, \ k = 1, 2, \ldots,$$
$$H_0^{(j)} = |N_0^{(j)}|, \quad K_0^{(j)} = |M_0^{(j)}|, \quad \text{if } c_k < 0, \ k = 1, 2, \ldots.$$

(ii) Add $H_n^{(j)} = \dfrac{H_{n-1}^{(j+1)} - c_n H_{n-1}^{(j)}}{1 - c_n}$ and $K_n^{(j)} = \dfrac{K_{n-1}^{(j+1)} - c_n K_{n-1}^{(j)}}{1 - c_n}$ to Step 2.

(iii) Add $\Gamma_n^{(j)} = \left| \dfrac{H_n^{(j)}}{N_n^{(j)}} \right|$ and $\Lambda_n^{(j)} = \left| \dfrac{K_n^{(j)}}{N_n^{(j)}} \right|$ to Step 3.

The proof of this can be accomplished by realizing that, when the c_k are all positive,

$$H_n^{(j)} = \sum_{i=0}^{n} \rho_{ni} \frac{(-1)^{j+i}}{|\phi(y_{j+i})|} \quad \text{and} \quad K_n^{(j)} = \sum_{i=0}^{n} \rho_{ni} \frac{(-1)^{j+i}|A(y_{j+i})|}{|\phi(y_{j+i})|},$$

and, when the c_k are all negative,

$$H_n^{(j)} = \sum_{i=0}^{n} \rho_{ni} \frac{1}{|\phi(y_{j+i})|} \quad \text{and} \quad K_n^{(j)} = \sum_{i=0}^{n} \rho_{ni} \frac{|A(y_{j+i})|}{|\phi(y_{j+i})|}.$$

(cf. Theorem 7.2.3).

Before closing this section, we would like to discuss briefly the application of the extended GREP$^{(1)}$ and the EW-algorithm to a sequence $\{A_m\}$, for which

$$A_m \sim A + g_m \sum_{k=1}^{\infty} \alpha_k c_k^m \quad \text{as } m \to \infty, \tag{7.5.10}$$

where

$$c_k \neq 1, \ k = 1, 2, \ldots, \ |c_1| > |c_2| > \cdots, \ \text{and} \ \lim_{k \to \infty} c_k = 0. \tag{7.5.11}$$

The g_m and the c_k are assumed to be known, and A, the limit or antilimit of $\{A_m\}$, is sought. The extended GREP$^{(1)}$ is now defined through the linear systems

$$A_l = A_n^{(j)} + g_l \sum_{k=1}^{n} \bar{\alpha}_k c_k^l, \ \ j \leq l \leq j + n, \tag{7.5.12}$$

and it can be implemented by Algorithm 7.5.1 by replacing $A(y_m)$ by A_m and $\phi(y_m)$ by g_m in Step 1 of this algorithm.

8

Analytic Study of GREP$^{(1)}$: Slowly Varying $A(y) \in \mathbf{F}^{(1)}$

8.1 Introduction and Error Formula for $A_n^{(j)}$

In Section 4.4, we gave a brief convergence study of GREP$^{(m)}$ for both Process I and Process II. In this study, we treated the cases in which GREP$^{(m)}$ was stable. In addition, we made some practical remarks on stability of GREP$^{(m)}$ in Section 4.5. The aim of the study was to justify the preference given to Process I and Process II as the relevant limiting processes to be used for approximating A, the limit or antilimit of $A(y)$ as $y \rightarrow 0+$. We also mentioned that stability was not necessary for convergence and that convergence could be proved at least in some cases in which the extrapolation process is clearly unstable.

In this chapter as well as the next, we would like to make more refined statements about the convergence and stability properties of GREP$^{(1)}$, the simplest form and prototype of GREP, as it is being applied to functions $A(y) \in \mathbf{F}^{(1)}$.

Before going on, we mention that this chapter is an almost exact reproduction of the recent paper Sidi [306][1].

As we will be using the notation and results of Section 7.2 on the W-algorithm, we believe a review of this material is advisable at this point. We recall that $A(y) \in \mathbf{F}^{(1)}$ if

$$A(y) = A + \phi(y)\beta(y), \quad y \in (0, b] \text{ for some } b > 0, \tag{8.1.1}$$

where y can be a continuous or discrete variable, and $\beta(\xi)$, as a function of the continuous variable ξ, is continuous in $[0, \hat{\xi}]$ for some $\hat{\xi} > 0$ and has a Poincaré-type asymptotic expansion of the form

$$\beta(\xi) \sim \sum_{i=0}^{\infty} \beta_i \xi^{ir} \text{ as } \xi \rightarrow 0+, \text{ for some fixed } r > 0. \tag{8.1.2}$$

We also recall that $A(y) \in \mathbf{F}_\infty^{(1)}$ if the function $B(t) \equiv \beta(t^{1/r})$, as a function of the continuous variable t, is infinitely differentiable in $[0, \hat{\xi}^r]$. [Therefore, in the variable t, $B(t) \in C[0, \hat{t}]$ for some $\hat{t} > 0$ and (8.1.2) reads $B(t) \sim \sum_{i=0}^{\infty} \beta_i t^i$ as $t \rightarrow 0+$.]

Finally, we recall that we have set $t = y^r$, $a(t) = A(y)$, $\varphi(t) = \phi(y)$, and $t_l = y_l^r$, $l = 0, 1, \ldots$, and that $b^r \geq t_0 > t_1 > \cdots > 0$ and $\lim_{l \rightarrow \infty} t_l = 0$. $A_n^{(j)}$ is given by (7.2.4),

[1] First published electronically in *Mathematics of Computation*, November 28, 2001, and later in *Mathematics of Computation*, Volume 71, Number 240, 2002, pp. 1569–1596, published by the American Mathematical Society.

with the divided difference operator $D_n^{(j)}$ as defined by (7.2.3). Similarly, $\Gamma_n^{(j)}$ is given by (7.2.12). Of course, $\varphi(t_l) \neq 0$ for all l.

We aim at presenting various convergence and stability results for $\mathrm{GREP}^{(1)}$ in the presence of different functions $\varphi(t)$ and different collocation sequences $\{t_l\}$. Of course, $\varphi(t)$ are not arbitrary. They are chosen to cover the most relevant cases of the $D^{(1)}$-transformation for infinite integrals and the $d^{(1)}$- and $\tilde{d}^{(m)}$-transformations for infinite series. The sequences $\{t_l\}$ that we consider satisfy (i) $t_{l+1} = \omega t_l$ for all l; or (ii) $\lim_{l\to\infty}(t_{l+1}/t_l) = \omega$; or (iii) $t_{l+1} \leq \omega t_l$ for all l, ω being a fixed constant in $(0, 1)$ in all three cases; or (iv) $\lim_{l\to\infty}(t_{l+1}/t_l) = 1$, in particular, $t_l \sim cl^{-q}$ as $l \to \infty$, for some $c, q > 0$. These are the most commonly used collocation sequences. Thus, by drawing the proper analogies, the results of this chapter and the next apply very naturally to the $D^{(1)}$-, $d^{(1)}$-, and $\tilde{d}^{(m)}$-transformations. We come back to these transformations in Chapter 10, where we show how the conclusions drawn from the study of $\mathrm{GREP}^{(1)}$ can be used to enhance their performance in finite-precision arithmetic.

Surprisingly, $\mathrm{GREP}^{(1)}$ is quite amenable to rigorous and refined analysis, and the conclusions that we draw from the study of $\mathrm{GREP}^{(1)}$ are relevant to $\mathrm{GREP}^{(m)}$ with arbitrary m, in general.

It is important to note that the analytic study of $\mathrm{GREP}^{(1)}$ is made possible by the divided difference representations of $A_n^{(j)}$ and $\Gamma_n^{(j)}$ that are given in Theorems 7.2.1 and 7.2.3. With the help of these representations, we are able to produce results that are optimal or nearly optimal in many cases. We must also add that not all problems associated with $\mathrm{GREP}^{(1)}$ have been solved, however. In particular, various problems concerning Process II are still open.

In this chapter, we concentrate on functions $A(y)$ that vary slowly as $y \to 0+$, while in the next chapter we treat functions $A(y)$ that vary quickly as $y \to 0+$. How $A(y)$ varies depends only on the behavior of $\phi(y)$ as $y \to 0+$ because $\beta(y)$ behaves polynomially in y^r and hence varies slowly as $y \to 0+$. Specifically, $\beta(y) \sim \beta_0$ if $\beta_0 \neq 0$, and $\beta(y) \sim \beta_s y^{sr}$ for some $s > 0$, otherwise. Thus, $A(y)$ and $\phi(y)$ vary essentially in the same manner. If $\phi(y)$ behaves like some power y^γ (γ may be complex), then $A(y)$ varies slowly. If $\phi(y)$ behaves like some exponential function $\exp[w(y)]$, where $w(y) \sim ay^\gamma$, $\gamma < 0$, then $A(y)$ varies quickly.

In the remainder of this section, we present some preliminary results that will be useful in the analysis of $\mathrm{GREP}^{(1)}$ in this chapter and the next.

We start by deriving an error formula for $A_n^{(j)}$.

Lemma 8.1.1 *The error in $A_n^{(j)}$ is given by*

$$A_n^{(j)} - A = \frac{D_n^{(j)}\{B(t)\}}{D_n^{(j)}\{1/\varphi(t)\}}; \quad B(t) \equiv \beta(t^{1/r}). \tag{8.1.3}$$

Proof. The result follows from $a(t) - A = \varphi(t)B(t)$ and from (7.2.4) in Theorem 7.2.1 and from the linearity of $D_n^{(j)}$. ∎

It is clear from Lemma 8.1.1 that the convergence analysis of $\mathrm{GREP}^{(1)}$ on $\mathbf{F}^{(1)}$ is based on the study of $D_n^{(j)}\{B(t)\}$ and $D_n^{(j)}\{1/\varphi(t)\}$. Similarly, the stability analysis is based on the study of $\Gamma_n^{(j)}$ given in (7.2.12), which we reproduce here for the sake of completeness:

$$\Gamma_n^{(j)} = \sum_{i=0}^n |\gamma_{ni}^{(j)}| = \frac{1}{|D_n^{(j)}\{1/\varphi(t)\}|} \sum_{i=0}^n \frac{|c_{ni}^{(j)}|}{|\varphi(t_{j+i})|}; \quad c_{ni}^{(j)} = \prod_{\substack{k=0 \\ k \neq i}}^n \frac{1}{t_{j+i} - t_{j+k}}. \quad (8.1.4)$$

In some of our analyses, we assume the functions $\varphi(t)$ and $B(t)$ to be differentiable; in others, no such requirement is imposed. Obviously, the assumption in the former case is quite strong, and this makes some of the proofs easier.

The following simple result on $A_n^{(j)}$ will become useful shortly.

Lemma 8.1.2 *If $B(t) \in C^\infty[0, t_j]$ and $\psi(t) \equiv 1/\varphi(t) \in C^\infty(0, t_j]$, then for any nonzero complex number c,*

$$A_n^{(j)} - A = \frac{\Re[cB^{(n)}(t'_{jn,1})] + i\Im[cB^{(n)}(t'_{jn,2})]}{\Re[c\psi^{(n)}(t''_{jn,1})] + i\Im[c\psi^{(n)}(t''_{jn,2})]}$$

$$\text{for some } t'_{jn,1}, t'_{jn,2}, t''_{jn,1}, t''_{jn,2} \in (t_{j+n}, t_j). \quad (8.1.5)$$

Proof. It is known that if $f \in C^n[a, b]$ is real and $a \leq x_0 \leq x_1 \leq \cdots \leq x_n \leq b$, then the divided difference $f[x_0, x_1, \ldots, x_n]$ satisfies

$$f[x_0, x_1, \ldots, x_n] = \frac{f^{(n)}(\xi)}{n!} \quad \text{for some } \xi \in (x_0, x_n).$$

Applying this to the real and imaginary parts of the complex-valued function $u(t) \in C^n(0, t_j)$, we have

$$D_n^{(j)}\{u(t)\} = \frac{1}{n!}\left[\Re u^{(n)}(t_{jn,1}) + i\Im u^{(n)}(t_{jn,2})\right], \quad \text{for some } t_{jn,1}, t_{jn,2} \in (t_{j+n}, t_j). \quad (8.1.6)$$

The result now follows. ∎

The constant c in (8.1.5) serves us in the proof of Theorem 8.3.1 in the next section.

Note that, in many of our problems, it is known that $B(t) \in C^\infty[0, \hat{t}]$ for some $\hat{t} > 0$, whereas $\psi(t) \in C^\infty(0, \hat{t}]$ only. That is to say, $B(t)$ has an infinite number of derivatives at $t = 0$, and $\psi(t)$ does not. This is an important observation.

A useful simplification takes place in (8.1.5) for the case $\varphi(t) = t$. In this case, GREP$^{(1)}$ is, of course, nothing but the polynomial Richardson extrapolation process that has been studied most extensively in the literature, which we studied to some extent in Chapter 2.

Lemma 8.1.3 *If $\varphi(t) = t$ in Lemma 8.1.2, then, for some $t'_{jn,1}, t'_{jn,2} \in (t_{j+n}, t_j)$,*

$$A_n^{(j)} - A = (-1)^n \frac{\Re[B^{(n)}(t'_{jn,1})] + i\Im[B^{(n)}(t'_{jn,2})]}{n!}\left(\prod_{i=0}^n t_{j+i}\right). \quad (8.1.7)$$

In this case, it is also true that

$$A_n^{(j)} - A = (-1)^n D_{n+1}^{(j)} \{a(t)\} \left(\prod_{i=0}^{n} t_{j+i} \right), \tag{8.1.8}$$

so that, for some $t_{jn,1}, \ t_{jn,2} \in (t_{j+n}, t_j)$,

$$A_n^{(j)} - A = (-1)^n \frac{\Re[a^{(n+1)}(t_{jn,1})] + i\Im[a^{(n+1)}(t_{jn,2})]}{(n+1)!} \left(\prod_{i=0}^{n} t_{j+i} \right). \tag{8.1.9}$$

Proof. The first part is proved by invoking in (8.1.3) the fact $D_n^{(j)}\{t^{-1}\} = (-1)^n$ $\left(\prod_{i=0}^{n} t_{j+i} \right)^{-1}$ that was used in Chapter 7. We leave the rest to the reader. ∎

Obviously, by imposing suitable growth conditions on $B^{(n)}(t)$, Lemma 8.1.3 can be turned into powerful convergence theorems.

The last result of this section is a slight refinement of Theorem 4.4.2 concerning Process I as it applies to GREP$^{(1)}$.

Theorem 8.1.4 *Let $\sup_j \Gamma_n^{(j)} = \Omega_n < \infty$ and $\phi(y_{j+1}) = O(\phi(y_j))$ as $j \to \infty$. Then, with n fixed,*

$$A_n^{(j)} - A = O(\varphi(t_j) t_j^{n+\mu}) \ \ as \ j \to \infty, \tag{8.1.10}$$

where $\beta_{n+\mu}$ is the first nonzero β_i with $i \geq n$.

Proof. Following the proof of Theorem 4.4.2, we start with

$$A_n^{(j)} - A = \sum_{i=0}^{n} \gamma_{ni}^{(j)} \varphi(t_{j+i})[B(t_{j+i}) - u(t_{j+i})]; \ \ u(t) = \sum_{k=0}^{n-1} \beta_k t^k. \tag{8.1.11}$$

The result now follows by taking moduli on both sides and realizing that $B(t) - u(t) \sim \beta_{n+\mu} t^{n+\mu}$ as $t \to 0+$ and recalling that $t_j > t_{j+1} > t_{j+2} > \cdots$. We leave the details to the reader. ∎

Note that Theorem 8.1.4 does not assume that $B(t)$ and $\varphi(t)$ are differentiable. It does, however, assume that Process I is stable.

As we will see in the sequel, (8.1.10) holds under appropriate conditions on the t_l even when Process I is clearly unstable.

In connection with Process I, we would like to remark that, as in Theorem 3.5.5, under suitable conditions we can obtain a full asymptotic expansion for $A_n^{(j)} - A$ as $j \to \infty$. If we define

$$\Phi_{n,k}^{(j)} = \frac{D_n^{(j)}\{t^k\}}{D_n^{(j)}\{1/\varphi(t)\}}, \ \ k = 0, 1, \ldots, \tag{8.1.12}$$

and recall that $D_n^{(j)}\{t^k\} = 0$, for $k = 0, 1, \ldots, n-1$, then this expansion assumes the simple and elegant form

$$A_n^{(j)} - A \sim \sum_{k=n}^{\infty} \beta_k \Phi_{n,k}^{(j)} \quad \text{as } j \to \infty. \tag{8.1.13}$$

If $\beta_{n+\mu}$ is the first nonzero β_i with $i \geq n$, then $A_n^{(j)} - A$ also satisfies the asymptotic equality

$$A_n^{(j)} - A \sim \beta_{n+\mu} \Phi_{n,n+\mu}^{(j)} \quad \text{as } j \to \infty. \tag{8.1.14}$$

Of course, all this will be true provided that (i) $\{\Phi_{n,k}^{(j)}\}_{k=n}^{\infty}$ is an asymptotic sequence as $j \to \infty$, that is, $\lim_{j\to\infty} \Phi_{n,k+1}^{(j)}/\Phi_{n,k}^{(j)} = 0$ for all $k \geq n$, and (ii) $A_n^{(j)} - A - \sum_{k=n}^{s-1} \beta_k \Phi_{n,k}^{(j)} = O(\Phi_{n,s}^{(j)})$ as $j \to \infty$, for each $s \geq n$. In the next sections, we aim at such results whenever possible. We show that they are possible in most of the cases we treat.

We close this section with the well-known Hermite–Gennochi formula for divided differences, which will be used later.

Lemma 8.1.5 *(Hermite–Gennochi). Let* $f(x)$ *be in* $C^n[a,b]$, *and let* $x_0, x_1, \ldots, x_n$ *be all in* $[a,b]$. *Then*

$$f[x_0, x_1, \ldots, x_n] = \int_{T_n} f^{(n)}\left(\sum_{i=0}^{n} \xi_i x_i\right) d\xi_1 \cdots d\xi_n,$$

where

$$T_n = \left\{ (\xi_1, \ldots, \xi_n) : 0 \leq \xi_i \leq 1, \ i = 1, \ldots, n, \ \sum_{i=1}^{n} \xi_i \leq 1 \right\}; \quad \xi_0 = 1 - \sum_{i=1}^{n} \xi_i.$$

For a proof of this lemma, see, for example, Atkinson [13]. Note that the argument $z = \sum_{i=0}^{n} \xi_i x_i$ of $f^{(n)}$ above is actually a convex combination of $x_0, x_1, \ldots, x_n$ because $0 \leq \xi_i \leq 1$, $i = 0, 1, \ldots, n$, and $\sum_{i=0}^{n} \xi_i = 1$. If we order the x_i such that $x_0 \leq x_1 \leq \cdots \leq x_n$, then $z \in [x_0, x_n] \subseteq [a,b]$.

8.2 Examples of Slowly Varying $a(t)$

Our main concern in this chapter is with functions $a(t)$ that vary slowly as $t \to 0+$. As we mentioned earlier, by this we mean that $\varphi(t) \sim h_0 t^\delta$ as $t \to 0+$ for some $h_0 \neq 0$ and δ that may be complex in general. In other words, $\varphi(t) = t^\delta H(t)$ with $H(t) \sim h_0$ as $t \to 0+$. In most cases, $H(t) \sim \sum_{i=0}^{\infty} h_i t^i$ as $t \to 0+$. When $\Re\delta > 0$, $\lim_{t\to 0+} a(t)$ exists and is equal to A. When $\lim_{t\to 0+} a(t)$ does not exist, we have $\Re\delta \leq 0$ necessarily, and A is the antilimit of $a(t)$ as $t \to 0+$ in this case, with some restriction on δ.

We now present practical examples of functions $a(t)$ that vary slowly.

Example 8.2.1 If $f \in \mathbf{A}^{(\gamma)}$ strictly for some possibly complex $\gamma \neq -1, 0, 1, 2, \ldots$, then we know from Theorem 5.7.3 that

$$F(x) = \int_0^x f(t)\, dt = I[f] + x f(x) g(x); \quad g \in \mathbf{A}^{(0)} \text{ strictly.}$$

This means that $F(x) \leftrightarrow a(t)$, $I[f] \leftrightarrow A$, $x^{-1} \leftrightarrow t$, $xf(x) \leftrightarrow \varphi(t)$ with $\varphi(t)$ as above and with $\delta = -\gamma - 1$. (Recall that $f \in \mathbf{B}^{(1)}$ in this case.)

Example 8.2.2 If $a_n = h(n) \in \mathbf{A}_0^{(\gamma)}$ strictly for some possibly complex $\gamma \neq -1, 0, 1, 2, \dots$, then we know from Theorem 6.7.2 that

$$A_n = \sum_{k=1}^{n} a_k = S(\{a_k\}) + n a_n g(n); \quad g \in \mathbf{A}_0^{(0)} \text{ strictly.}$$

This means that $A_n \leftrightarrow a(t)$, $S(\{a_k\}) \leftrightarrow A$, $n^{-1} \leftrightarrow t$, $n a_n \leftrightarrow \varphi(t)$ with $\varphi(t)$ as above and with $\delta = -\gamma - 1$. (Recall that $\{a_n\} \in \mathbf{b}^{(1)}$ in this case.)

Example 8.2.3 If $a_n = h(n) \in \tilde{\mathbf{A}}_0^{(\gamma, m)}$ strictly for some possibly complex $\gamma \neq -1 + i/m$, $i = 0, 1, \dots$, and a positive integer $m > 1$, then we know from Theorem 6.6.6 that

$$A_n = \sum_{k=1}^{n} a_k = S(\{a_k\}) + n a_n g(n); \quad g \in \tilde{\mathbf{A}}_0^{(0, m)} \text{ strictly.}$$

This means that $A_n \leftrightarrow a(t)$, $S(\{a_k\}) \leftrightarrow A$, $n^{-1/m} \leftrightarrow t$, $n a_n \leftrightarrow \varphi(t)$ with $\varphi(t)$ as above and with $\delta = -\gamma - 1$. (Recall that $\{a_n\} \in \tilde{\mathbf{b}}^{(m)}$ in this case.)

8.3 Slowly Varying $\varphi(t)$ with Arbitrary t_l

We start with the following surprising result that holds for arbitrary $\{t_l\}$. (Recall that so far the t_l satisfy only $t_0 > t_1 > \cdots > 0$ and $\lim_{l \to \infty} t_l = 0$.)

Theorem 8.3.1 *Let $\varphi(t) = t^\delta H(t)$, where δ is in general complex and $\delta \neq 0, -1, -2, \dots$, and $H(t) \in C^\infty[0, \hat{t}]$ for some $\hat{t} > 0$ with $h_0 \equiv H(0) \neq 0$. Let $B(t) \in C^\infty[0, \hat{t}]$, and let $\beta_{n+\mu}$ be the first nonzero β_i with $i \geq n$ in (8.1.2). Then, provided $n \geq -\Re\delta$, we have*

$$A_n^{(j)} - A = O(\varphi(t_j) t_j^{n+\mu}) \text{ as } j \to \infty. \tag{8.3.1}$$

Consequently, if $n > -\Re\delta$, we have $\lim_{j \to \infty} A_n^{(j)} = A$. All this is valid for arbitrary $\{t_l\}$.

Proof. By the assumptions on $B(t)$, we have

$$B^{(n)}(t) \sim \sum_{i=n+\mu}^{\infty} i(i-1) \cdots (i - n + 1) \beta_i t^{i-n} \text{ as } t \to 0+,$$

from which

$$B^{(n)}(t) \sim (\mu + 1)_n \beta_{n+\mu} t^\mu \text{ as } t \to 0+, \tag{8.3.2}$$

and by the assumptions on $\varphi(t)$, we have for $\psi(t) \equiv 1/\varphi(t)$

$$\psi^{(n)}(t) = \sum_{k=0}^{n} \binom{n}{k}(t^{-\delta})^{(k)}[1/H(t)]^{(n-k)} \sim (t^{-\delta})^{(n)}/H(t) \sim (t^{-\delta})^{(n)}/h_0 \ \text{as } t \to 0+,$$

from which

$$\psi^{(n)}(t) \sim (-1)^n h_0^{-1}(\delta)_n t^{-\delta-n} \sim (-1)^n (\delta)_n \psi(t) t^{-n} \ \text{as } t \to 0+. \quad (8.3.3)$$

Also, Lemma 8.1.2 is valid for all sufficiently large j under the present assumptions on $B(t)$ and $\varphi(t)$, because $t_j < \hat{t}$ for all sufficiently large j. Substituting (8.3.2) and (8.3.3) in (8.1.5) with $|c| = 1$ there, we obtain

$$A_n^{(j)} - A = (-1)^n (\mu + 1)_n$$

$$\times \frac{[\Re(c\beta_{n+\mu}) + o(1)](t'_{jn,1})^\mu + i[\Im(c\beta_{n+\mu}) + o(1)](t'_{jn,2})^\mu}{[\Re\alpha_{jn,1} + o(1)](t''_{jn,1})^{-\Re\delta-n} + i[\Im\alpha_{jn,2} + o(1)](t''_{jn,2})^{-\Re\delta-n}} \ \text{as } j \to \infty, \quad (8.3.4)$$

with $\alpha_{jn,s} \equiv ch_0^{-1}(\delta)_n(t''_{jn,s})^{-i\Im\delta}$ and the $o(1)$ terms uniform in c, $|c| = 1$. Here, we have also used the fact that $\lim_{j\to\infty} t'_{jn,s} = \lim_{j\to\infty} t''_{jn,s} = 0$. Next, by $0 < t'_{jn,s} < t_j$ and $\mu \geq 0$, it follows that $(t'_{jn,s})^\mu \leq t_j^\mu$. This implies that the numerator of the quotient in (8.3.4) is $O(t_j^\mu)$ as $j \to \infty$, *uniformly in c, $|c| = 1$.* As for the denominator, we start by observing that $a = h_0^{-1}(\delta)_n \neq 0$. Therefore, either $\Re a \neq 0$ or $\Im a \neq 0$, and we assume without loss of generality that $\Re a \neq 0$. If we now choose $c = (t''_{jn,1})^{i\Im\delta}$, we obtain $\alpha_{jn,1} = a$ and hence $\Re\alpha_{jn,1} = \Re a \neq 0$, as a result of which the modulus of the denominator can be bounded from below by $|\Re a + o(1)|(t''_{jn,1})^{-\Re\delta-n}$, which in turn is bounded below by $|\Re a + o(1)| t_j^{-\Re\delta-n}$, since $0 < t''_{jn,s} < t_j$ and $\Re\delta + n \geq 0$. The result now follows by combining everything in (8.3.4) and by invoking $t^\delta = O(\varphi(t))$ as $t \to 0+$, which follows from $\varphi(t) \sim h_0 t^\delta$ as $t \to 0+$. ∎

Theorem 8.3.1 implies that the column sequence $\{A_n^{(j)}\}_{j=0}^{\infty}$ converges to A if $n > -\Re\delta$; it also gives an upper bound on the rate of convergence through (8.3.1). The fact that convergence takes place for *arbitrary* $\{t_l\}$ and that we are able to prove that it does is quite unexpected.

By restricting $\{t_l\}$ only slightly in Theorem 8.3.1, we can show that $A_n^{(j)} - A$ has the full asymptotic expansion given in (8.1.13) and, as a result, satisfies the asymptotic equality of (8.1.14) as well. We start with the following lemma that turns out to be very useful in the sequel.

Lemma 8.3.2 *Let $g(t) = t^\theta u(t)$, where θ is in general complex and $u(t) \in C^\infty[0, \hat{t}]$ for some $\hat{t} > 0$. Pick the t_l to satisfy, in addition to $t_{l+1} < t_l$, $l = 0, 1, \ldots$, also $t_{l+1} \geq \nu t_l$*

for all sufficiently large l with some $v \in (0, 1)$. Then, the following are true:

(i) *The nonzero members of $\{D_n^{(j)}\{t^{\theta+i}\}\}_{i=0}^\infty$ form an asymptotic sequence as $j \to \infty$.*

(ii) *$D_n^{(j)}\{g(t)\}$ has the bona fide asymptotic expansion*

$$D_n^{(j)}\{g(t)\} \sim \sum_{i=0}^{\infty}{}^* g_i D_n^{(j)}\{t^{\theta+i}\} \text{ as } j \to \infty; \ g_i = u^{(i)}(0)/i!, \ i = 0, 1, \ldots, \quad (8.3.5)$$

where the asterisk on the summation means that only those terms for which $D_n^{(j)}\{t^{\theta+i}\} \neq 0$, that is, for which $\theta + i \neq 0, 1, \ldots, n-1$, are taken into account.

Remark. The extra condition $t_{l+1} \geq v t_l$ for all large l that we have imposed on the t_l is satisfied, for example, when $\lim_{l\to\infty}(t_{l+1}/t_l) = \lambda$ for some $\lambda \in (0, 1]$, and such cases are considered further in the next sections.

Proof. Let α be in general complex and $\alpha \neq 0, 1, \ldots, n-1$. Denote $\binom{\alpha}{n} = M$ for simplicity of notation. Then, by (8.1.6), for any complex number c such that $|c| = 1$, we have

$$c D_n^{(j)}\{t^\alpha\} = \Re\left[cM(t_{jn,1})^{\alpha-n}\right] + i\Im\left[cM(t_{jn,2})^{\alpha-n}\right]$$

$$\text{for some } t_{jn,1}, t_{jn,2} \in (t_{j+n}, t_j), \quad (8.3.6)$$

from which we also have

$$|D_n^{(j)}\{t^\alpha\}| \geq \max\{|\Re\left[cM(t_{jn,1})^{\alpha-n}\right]|, |\Im\left[cM(t_{jn,2})^{\alpha-n}\right]|\}. \quad (8.3.7)$$

Because $M \neq 0$, we have either $\Re M \neq 0$ or $\Im M \neq 0$. Assume without loss of generality that $\Re M \neq 0$ and choose $c = (t_{jn,1})^{-i\Im\alpha}$. Then, $\Re\left[cM(t_{jn,1})^{\alpha-n}\right] = (\Re M)(t_{jn,1})^{\Re\alpha-n}$ and hence

$$|D_n^{(j)}\{t^\alpha\}| \geq |\Re M|(t_{jn,1})^{\Re\alpha-n} \geq |\Re M| \min_{t\in[t_{j+n},t_j]}(t^{\Re\alpha-n}). \quad (8.3.8)$$

Invoking in (8.3.8), if necessary, the fact that $t_{j+n} \geq v^n t_j$, which is implied by the conditions on the t_l, we obtain

$$|D_n^{(j)}\{t^\alpha\}| \geq C_{n1}^{(\alpha)} t_j^{\Re\alpha-n} \text{ for all large } j, \text{ with some constant } C_{n1}^{(\alpha)} > 0. \quad (8.3.9)$$

Similarly, we can show from (8.3.6) that

$$|D_n^{(j)}\{t^\alpha\}| \leq C_{n2}^{(\alpha)} t_j^{\Re\alpha-n} \text{ for all large } j, \text{ with some constant } C_{n2}^{(\alpha)} > 0. \quad (8.3.10)$$

The proof of part (i) can now be achieved by using (8.3.9) and (8.3.10).

To prove part (ii), we need to show [in addition to part (i)] that, for any integer s for which $D_n^{(j)}\{t^{\theta+s}\} \neq 0$, that is, for which $\theta + s \neq 0, 1, \ldots, n-1$, there holds

$$D_n^{(j)}\{g(t)\} - \sum_{i=0}^{s-1} g_i D_n^{(j)}\{t^{\theta+i}\} = O\left(D_n^{(j)}\{t^{\theta+s}\}\right) \text{ as } j \to \infty. \quad (8.3.11)$$

Now $g(t) = \sum_{i=0}^{s-1} g_i t^{\theta+i} + v_s(t) t^{\theta+s}$, where $v_s(t) \in C^{\infty}[0, \hat{t}]$. As a result,

$$D_n^{(j)}\{g(t)\} - \sum_{i=0}^{s-1} g_i D_n^{(j)}\{t^{\theta+i}\} = D_n^{(j)}\{v_s(t) t^{\theta+s}\}. \qquad (8.3.12)$$

Next, by (8.1.6) and by the fact that

$$[v_s(t) t^{\theta+s}]^{(n)} = \sum_{i=0}^{n} \binom{n}{i} [v_s(t)]^{(n-i)} (t^{\theta+s})^{(i)} \sim v_s(t)(t^{\theta+s})^{(n)} \sim g_s(t^{\theta+s})^{(n)} \text{ as } t \to 0+,$$

and by the additional condition on the t_l again, we obtain

$$D_n^{(j)}\{v_s(t) t^{\theta+s}\} = O(t_j^{\Re\theta+s-n}) = O(D_n^{(j)}\{t^{\theta+s}\}) \text{ as } j \to \infty, \qquad (8.3.13)$$

with the last equality being a consequence of (8.3.9). Here we assume that $g_s \neq 0$ without loss of generality. By substituting (8.3.13) in (8.3.12), the result in (8.3.11) follows. This completes the proof. ∎

Theorem 8.3.3 *Let $\varphi(t)$ and $B(t)$ be exactly as in Theorem 8.3.1, and pick the t_l as in Lemma 8.3.2. Then $A_n^{(j)} - A$ has the complete asymptotic expansion given in (8.1.13) and hence satisfies the asymptotic equality in (8.1.14) as well. Furthermore, if $\beta_{n+\mu}$ is the first nonzero β_i with $i \geq n$, then, for all large j, there holds*

$$\Omega_1 |\varphi(t_j)| t_j^{n+\mu} \leq |A_n^{(j)} - A| \leq \Omega_2 |\varphi(t_j)| t_j^{n+\mu}, \quad \text{for some } \Omega_1 > 0 \text{ and } \Omega_2 > 0,$$
$$(8.3.14)$$

whether $n \geq -\Re\delta$ or not.

Proof. The proof of the first part can be achieved by applying Lemma 8.3.2 to $B(t)$ and to $\psi(t) \equiv 1/\varphi(t)$. The proof of the second part can be achieved by using (8.3.9) as well. We leave the details to the reader. ∎

Remark. It is important to make the following observations concerning the behavior of $A_n^{(j)} - A$ as $j \to \infty$ in Theorem 8.3.3. First, any column sequence $\{A_n^{(j)}\}_{j=0}^{\infty}$ converges at least as quickly as (or diverges at most as quickly as) the column sequence $\{A_{n-1}^{(j)}\}_{j=0}^{\infty}$ that precedes it. In other words, each column sequence is at least as good as the one preceding it. In particular, when $\beta_m \neq 0$ but $\beta_{m+1} = \cdots = \beta_{s-1} = 0$ and $\beta_s \neq 0$, we have

$$A_n^{(j)} - A = o(A_m^{(j)} - A) \text{ as } j \to \infty, \quad m+1 \leq n \leq s,$$
$$A_{s+1}^{(j)} - A = o(A_s^{(j)} - A) \text{ as } j \to \infty. \qquad (8.3.15)$$

In addition, for all large j we have

$$\theta_{n1} |A_s^{(j)} - A| \leq |A_n^{(j)} - A| \leq \theta_{n2} |A_s^{(j)} - A|, \quad m+1 \leq n \leq s-1,$$

$$\text{for some } \theta_{n1}, \theta_{n2} > 0, \qquad (8.3.16)$$

which implies that the column sequences $\{A_n^{(j)}\}_{j=0}^{\infty}$, $m + 1 \leq n \leq s$, behave the same way for all large j.

In the next sections, we continue the treatment of Process I by restricting the t_l further, and we treat the issue of stability for Process I as well. In addition, we treat the convergence and stability of Process II.

8.4 Slowly Varying $\varphi(t)$ with $\lim_{l\to\infty}(t_{l+1}/t_l) = 1$

8.4.1 Process I with $\varphi(t) = t^{\delta} H(t)$ and Complex δ

Theorem 8.4.1 *Assume that $\varphi(t)$ and $B(t)$ are exactly as in Theorem 8.3.1. In addition, pick the t_l such that $\lim_{l\to\infty}(t_{l+1}/t_l) = 1$. Then, $A_n^{(j)} - A$ has the complete asymptotic expansion given in (8.1.13) and satisfies (8.1.14) and hence satisfies also the asymptotic equality*

$$A_n^{(j)} - A \sim (-1)^n \frac{(\mu + 1)_n}{(\delta)_n} \beta_{n+\mu} \varphi(t_j) t_j^{n+\mu} \quad as \ j \to \infty, \tag{8.4.1}$$

where, again, $\beta_{n+\mu}$ is the first nonzero β_i with $i \geq n$ in (8.1.2). This result is valid whether $n \geq -\Re\delta$ or not. In addition, Process I is unstable, that is, $\sup_j \Gamma_n^{(j)} = \infty$.

Proof. First, Theorem 8.3.3 applies and thus (8.1.13) and (8.1.14) are valid.

Let us apply the Hermite–Gennochi formula of Lemma 8.1.5 to the function t^{α}, where α may be complex in general. By the assumption that $\lim_{l\to\infty}(t_{l+1}/t_l) = 1$, we have that the argument $z = \sum_{i=0}^{n} \xi_i t_{j+i}$ of the integrand in Lemma 8.1.5 satisfies $z \sim t_j$ as $j \to \infty$. As a result, we obtain

$$D_n^{(j)}\{t^{\alpha}\} \sim \binom{\alpha}{n} t_j^{\alpha-n} \quad as \ j \to \infty, \quad provided \ \alpha \neq 0, 1, \ldots, n - 1. \tag{8.4.2}$$

Next, applying Lemma 8.3.2 to $B(t)$ and to $\psi(t) \equiv 1/\varphi(t)$, and realizing that $\psi(t) \sim \sum_{i=0}^{\infty} \psi_i t^{-\delta+i}$ as $t \to 0+$ for some constants ψ_i with $\psi_0 = h_0^{-1}$, and using (8.4.2) as well, we have

$$D_n^{(j)}\{B(t)\} \sim \sum_{i=n+\mu}^{\infty} \beta_i D_n^{(j)}\{t^i\} \sim \beta_{n+\mu} D_n^{(j)}\{t^{n+\mu}\} \sim \binom{n+\mu}{n} \beta_{n+\mu} t_j^{\mu} \quad as \ j \to \infty,$$

$$\tag{8.4.3}$$

and

$$D_n^{(j)}\{\psi(t)\} \sim \sum_{i=0}^{\infty}{}^{*} \psi_i D_n^{(j)}\{t^{-\delta+i}\} \sim h_0^{-1} D_n^{(j)}\{t^{-\delta}\} \sim \binom{-\delta}{n} \psi(t_j) t_j^{-n} \quad as \ j \to \infty.$$

$$\tag{8.4.4}$$

The result in (8.4.1) is obtained by dividing (8.4.3) by (8.4.4).

For the proof of the second part, we start by observing that when $\lim_{l\to\infty}(t_{l+1}/t_l) = 1$ we also have $\lim_{j\to\infty}(t_{j+k}/t_{j+i}) = 1$ for arbitrary fixed i and k. Therefore, for every

$\epsilon > 0$, there exists a positive integer J, such that

$$|t_{j+i} - t_{j+k}| = \left| 1 - \frac{t_{j+k}}{t_{j+i}} \right| t_{j+i} < \epsilon t_{j+i} \leq \epsilon t_j \text{ for } 0 \leq i, k \leq n \text{ and } j > J. \quad (8.4.5)$$

As a result of this,

$$|c_{ni}^{(j)}| = \prod_{\substack{k=0 \\ k \neq i}}^{n} \frac{1}{|t_{j+i} - t_{j+k}|} > (\epsilon t_j)^{-n} \text{ for } i = 0, 1, \ldots, n, \text{ and } j > J. \quad (8.4.6)$$

Next, by the assumption that $H(t) \sim h_0$ as $t \to 0+$ and by $\lim_{j \to \infty}(t_{j+i}/t_j) = 1$, we have that $\psi(t_{j+i}) \sim \psi(t_j)$ as $j \to \infty$, from which $|\psi(t_{j+i})| \geq K_1 |\psi(t_j)|$ for $0 \leq i \leq n$ and all j, where $K_1 > 0$ is a constant independent of j. Combining this with (8.4.6), we have

$$\sum_{i=0}^{n} |c_{ni}^{(j)}| \, |\psi(t_{j+i})| \geq K_1(n+1)(\epsilon t_j)^{-n}|\psi(t_j)| \text{ for all } j > J. \quad (8.4.7)$$

Similarly, $|D_n^{(j)}\{\psi(t)\}| \leq K_2|\psi(t_j)| \, t_j^{-n}$ for all j, where $K_2 > 0$ is another constant independent of j. (K_2 depends only on n.) Substituting this and (8.4.7) in (8.1.4), we obtain

$$\Gamma_n^{(j)} \geq M_n \epsilon^{-n} \text{ for all } j > J, \text{ with } M_n = (K_1/K_2)(n+1) \text{ independent of } \epsilon \text{ and } j. \quad (8.4.8)$$

Since ϵ can be chosen arbitrarily close to 0, (8.4.8) implies that $\sup_j \Gamma_n^{(j)} = \infty$.   ∎

Obviously, the remarks following Theorem 8.3.3 are valid under the conditions of Theorem 8.4.1 too. In particular, (8.3.15) and (8.3.16) hold. Furthermore, (8.3.16) can now be refined to read

$$A_n^{(j)} - A \sim \theta_n(A_s^{(j)} - A) \text{ as } j \to \infty, \quad m+1 \leq n \leq s-1, \text{ for some } \theta_n \neq 0.$$

Finally, the column sequences $\{A_n^{(j)}\}_{j=0}^{\infty}$ with $n > -\Re\delta$ converge even though they are unstable.

In Theorem 8.4.3, we show that the results of Theorem 8.4.1 remain unchanged if we restrict the t_l somewhat while we still require that $\lim_{l \to \infty}(t_{l+1}/t_l) = 1$ but relax the conditions on $\varphi(t)$ and $B(t)$ considerably. In fact, we do not put any differentiability requirements either on $\varphi(t)$ or on $B(t)$ this time, and we obtain an asymptotic equality for $\Gamma_n^{(j)}$ as well.

The following lemma that is analogous to Lemma 8.3.2 will be useful in the proof of Theorem 8.4.3.

Lemma 8.4.2 *Let $g(t) \sim \sum_{i=0}^{\infty} g_i t^{\theta+i}$ as $t \to 0+$, where $g_0 \neq 0$ and θ is in general complex, and let the t_l satisfy*

$$t_l \sim cl^{-q} \text{ and } t_l - t_{l+1} \sim cpl^{-q-1} \text{ as } l \to \infty, \text{ for some } c > 0, \, p > 0, \text{ and } q > 0. \quad (8.4.9)$$

Then, the following are true:

(i) *The nonzero members of $\{D_n^{(j)}\{t^{\theta+i}\}\}_{i=0}^{\infty}$ form an asymptotic sequence as $j \to \infty$.*
(ii) *$D_n^{(j)}\{g(t)\}$ has the bona fide asymptotic expansion*

$$D_n^{(j)}\{g(t)\} \sim \sum_{i=0}^{\infty}{}^* g_i D_n^{(j)}\{t^{\theta+i}\} \quad \text{as } j \to \infty, \tag{8.4.10}$$

where the asterisk on the summation means that only those terms for which $D_n^{(j)}\{t^{\theta+i}\} \neq 0$, that is, for which $\theta + i \neq 0, 1, \dots, n-1$, are taken into account.

Remark. Note that $\lim_{l\to\infty}(t_{l+1}/t_l) = 1$ under (8.4.9), and that (8.4.9) is satisfied by $t_l = c(l+\eta)^{-q}$, for example. Also, the first part of (8.4.9) does not necessarily imply the second part.

Proof. Part (i) is true by Lemma 8.3.2, because $\lim_{l\to\infty}(t_{l+1}/t_l) = 1$. In particular, (8.4.2) holds. To prove part (ii), we need to show in addition that, for any integer s for which $D_n^{(j)}\{t^{\theta+s}\} \neq 0$, there holds

$$D_n^{(j)}\{g(t)\} - \sum_{i=0}^{s-1} g_i D_n^{(j)}\{t^{\theta+i}\} = O\left(D_n^{(j)}\{t^{\theta+s}\}\right) \quad \text{as } j \to \infty. \tag{8.4.11}$$

Now, $g(t) = \sum_{i=0}^{m-1} g_i t^{\theta+i} + v_m(t)t^{\theta+m}$, where $|v_m(t)| \leq C_m$ for some constant $C_m > 0$ and for all t sufficiently close to 0, and this holds for every m. Let us fix s and take $m > \max\{s + n/q, -\Re\theta\}$. We can write

$$D_n^{(j)}\{g(t)\} = \sum_{i=0}^{s-1} g_i D_n^{(j)}\{t^{\theta+i}\} + \sum_{i=s}^{m-1} g_i D_n^{(j)}\{t^{\theta+i}\} + D_n^{(j)}\{v_m(t)t^{\theta+m}\}. \tag{8.4.12}$$

Let us assume, without loss of generality, that $g_s \neq 0$. Then, by part (i) of the lemma,

$$\sum_{i=s}^{m-1} g_i D_n^{(j)}\{t^{\theta+i}\} \sim g_s D_n^{(j)}\{t^{\theta+s}\} \sim g_s \binom{\theta+s}{n} t_j^{\theta+s-n} \quad \text{as } j \to \infty.$$

Therefore, the proof will be complete if we show that $D_n^{(j)}\{v_m(t)t^{\theta+m}\} = O\left(t_j^{\theta+s-n}\right)$ as $j \to \infty$. Using also the fact that $t_{j+i} \sim t_j$ as $j \to \infty$, we first have that

$$|D_n^{(j)}\{v_m(t)t^{\theta+m}\}| \leq C_m \sum_{i=0}^{n} |c_{ni}^{(j)}| t_{j+i}^{\Re\theta+m} \leq C_m t_j^{\Re\theta+m}\left(\sum_{i=0}^{n} |c_{ni}^{(j)}|\right). \tag{8.4.13}$$

Next, from (8.4.9),

$$t_{j+i} - t_{j+k} \sim cp(k-i)j^{-q-1} \sim p(k-i)j^{-1}t_j \quad \text{as } j \to \infty, \tag{8.4.14}$$

as a result of which,

$$c_{ni}^{(j)} = \prod_{\substack{k=0 \\ k \neq i}}^{n} \frac{1}{t_{j+i} - t_{j+k}} \sim (-1)^i \frac{1}{n!} \binom{n}{i} \left(\frac{j}{pt_j} \right)^n, \quad 0 \leq i \leq n, \quad \text{and}$$

$$\sum_{i=0}^{n} |c_{ni}^{(j)}| \sim \frac{1}{n!} \left(\frac{2j}{pt_j} \right)^n \quad \text{as } j \to \infty. \quad (8.4.15)$$

Substituting (8.4.15) in (8.4.13) and noting that (8.4.9) implies $j \sim (t_j/c)^{-1/q}$ as $j \to \infty$, we obtain

$$D_n^{(j)}\{v_m(t)t^{\theta+m}\} = O(t_j^{\Re\theta+m-n-n/q}) = O(t_j^{\Re\theta+s-n}) \quad \text{as } j \to \infty, \quad (8.4.16)$$

by the fact that $\Re\theta + m - n - n/q > \Re\theta + s - n$. The result now follows. ∎

Theorem 8.4.3 *Assume that $\varphi(t) = t^\delta H(t)$, with δ in general complex and $\delta \neq 0, -1, -2, \ldots$, $H(t) \sim \sum_{i=0}^{\infty} h_i t^i$ as $t \to 0+$ and $B(t) \sim \sum_{i=0}^{\infty} \beta_i t^i$ as $t \to 0+$. Let us pick the t_l to satisfy (8.4.9). Then $A_n^{(j)} - A$ has the complete asymptotic expansion given in (8.1.13) and satisfies (8.1.14) and hence also satisfies the asymptotic equality in (8.4.1). In addition, $\Gamma_n^{(j)}$ satisfies the asymptotic equality*

$$\Gamma_n^{(j)} \sim \frac{1}{|(\delta)_n|} \left(\frac{2j}{p} \right)^n \quad \text{as } j \to \infty. \quad (8.4.17)$$

That is to say, Process I is unstable.

Proof. The assertion concerning $A_n^{(j)}$ can be proved by applying Lemma 8.4.2 to $B(t)$ and to $\psi(t) \equiv 1/\varphi(t)$ and proceeding as in the proof of Theorem 8.4.1.

We now turn to the analysis of $\Gamma_n^{(j)}$. To prove the asymptotic equality in (8.4.17), we need the precise asymptotic behaviors of $\sum_{i=0}^{n} |c_{ni}^{(j)}||\psi(t_{j+i})|$ and $D_n^{(j)}\{\psi(t)\}$ as $j \to \infty$. By (8.4.15) and by the fact that $\psi(t_{j+i}) \sim \psi(t_j)$ as $j \to \infty$ for all fixed i, we obtain

$$\sum_{i=0}^{n} |c_{ni}^{(j)}||\psi(t_{j+i})| \sim \left(\sum_{i=0}^{n} |c_{ni}^{(j)}| \right) |\psi(t_j)| \sim \frac{1}{n!} \left(\frac{2j}{pt_j} \right)^n |\psi(t_j)| \quad \text{as } j \to \infty. \quad (8.4.18)$$

Combining now (8.4.18) and (8.4.4) in (8.1.4), the result in (8.4.17) follows. ∎

So far all our results have been on Process I. What characterizes these results is that they are all obtained by considering only the *local* behavior of $B(t)$ and $\varphi(t)$ as $t \to 0+$. The reason for this is that $A_n^{(j)}$ is determined only by $a(t_l)$, $j \leq l \leq j+n$, and that in Process I we are letting $j \to \infty$ or, equivalently, $t_l \to 0$, $j \leq l \leq j+n$. In Process II, on the other hand, we are holding j fixed and letting $n \to \infty$. This means, of course, that $A_n^{(j)}$ is being influenced by the behavior of $a(t)$ on the *fixed* interval $(0, t_j]$. Therefore, we need to use *global* information on $a(t)$ in order to analyze Process II. It is precisely this point that makes Process II much more difficult to study than Process I.

An additional source of difficulty when analyzing Process II with $\varphi(t) = t^\delta H(t)$ is complex values of δ. Indeed, except for Theorem 8.5.2 in the next section, we do not have any results on Process II under the assumption that δ is complex. Our analysis in the remainder of this section assumes real δ.

8.4.2 Process II with $\varphi(t) = t$

We now would like to present results pertaining to Process II. We start with the case $\varphi(t) = t$. Our first result concerns convergence and follows trivially from Lemma 8.1.3 as follows:

Assuming that $B(t) \in C^\infty[0, t_j]$ and letting

$$\|B^{(n)}\| = \max_{0 \le t \le t_j} |B^{(n)}(t)| \tag{8.4.19}$$

we have

$$\left| \frac{D_n^{(j)}\{B(t)\}}{D_n^{(j)}\{t^{-1}\}} \right| \le \frac{\|B^{(n)}\|}{n!} \left(\prod_{i=0}^{n} t_{j+i} \right), \tag{8.4.20}$$

from which we have that $\lim_{n\to\infty} A_n^{(j)} = A$ when $\varphi(t) = t$ provided that

$$\frac{\|B^{(n)}\|}{n!} = o\left(\prod_{i=0}^{n} t_{j+i}^{-1} \right) \quad \text{as } n \to \infty. \tag{8.4.21}$$

In the special case $t_l = c/(l + \eta)^q$ for some positive c, η, and q, this condition reads

$$\|B^{(n)}\| = o((n!)^{q+1} c^{-n} n^{(j+\eta)q}) \quad \text{as } n \to \infty. \tag{8.4.22}$$

We are thus assured of convergence in this case under a very generous growth condition on $\|B^{(n)}\|$, especially when $q \ge 1$.

Our next result in the theorem below pertains to stability of Process II.

Theorem 8.4.4 *Consider $\varphi(t) = t$ and pick the t_l such that $\lim_{l\to\infty}(t_{l+1}/t_l) = 1$. Then, Process II is unstable, that is, $\sup_n \Gamma_n^{(j)} = \infty$. If the t_l are as in (8.4.9), then $\Gamma_n^{(j)} \to \infty$ as $n \to \infty$ faster than n^σ for every $\sigma > 0$. If, in particular, $t_l = c/(l + \eta)^q$ for some positive c, η, and q, then*

$$\Gamma_n^{(j)} > E_q^{(j)} n^{-1/2} \left(\frac{e}{q} \right)^{qn} \quad \text{for some } E_q^{(j)} > 0, \quad q = 1, 2. \tag{8.4.23}$$

Proof. We already know that, when $\varphi(t) = t$, we can compute the $\Gamma_n^{(j)}$ by the recursion relation in (7.2.27), which can also be written in the form

$$\Gamma_n^{(j)} = \Gamma_{n-1}^{(j+1)} + \frac{w_n^{(j)}}{1 - w_n^{(j)}} \left(\Gamma_{n-1}^{(j)} + \Gamma_{n-1}^{(j+1)} \right); \quad w_n^{(j)} = \frac{t_{j+n}}{t_j} < 1. \tag{8.4.24}$$

Hence,

$$\Gamma_n^{(j)} \ge \Gamma_{n-1}^{(j+1)} \ge \Gamma_{n-2}^{(j+2)} \ge \cdots, \tag{8.4.25}$$

from which $\Gamma_n^{(j)} \geq \Gamma_s^{(j+n-s)}$ for arbitrary fixed s. Applying now Theorem 8.4.1, we have $\lim_{n\to\infty} \Gamma_s^{(j+n-s)} = \infty$ for $s \geq 1$, from which $\lim_{n\to\infty} \Gamma_n^{(j)} = \infty$ follows. When the t_l are as in (8.4.9), we have from (8.4.17) that $\Gamma_s^{(j+n-s)} \sim \frac{1}{s!} \left(\frac{2}{p}\right)^s n^s$ as $n \to \infty$. From this and from the fact that s is arbitrary, we now deduce that $\Gamma_n^{(j)} \to \infty$ as $n \to \infty$ faster than n^σ for every $\sigma > 0$. To prove the last part, we start with

$$\gamma_{ni}^{(j)} = \prod_{\substack{k=0 \\ k \neq i}}^{n} \frac{t_{j+k}}{t_{j+k} - t_{j+i}}, \quad i = 0, 1, \ldots, n, \tag{8.4.26}$$

which follows from (7.2.26). The result in (8.4.23) follows from $\Gamma_n^{(j)} > |\gamma_{nn}^{(j)}|$. ∎

We can expand on the last part of Theorem 8.4.4 by deriving upper bounds on $\Gamma_n^{(j)}$ when $t_l = c/(l + \eta)^q$ for some positive c, η, and q. In this case, we first show that $w_n^{(j+1)} \geq w_n^{(j)}$, from which we can prove by induction, and with the help of (8.4.24), that $\Gamma_n^{(j)} \leq \Gamma_n^{(j+1)}$. Using this in (8.4.24), we obtain the inequality $\Gamma_n^{(j)} \leq \Gamma_{n-1}^{(j+1)}$ $[(1 + w_n^{(j)})/(1 - w_n^{(j)})]$, and, by induction,

$$\Gamma_n^{(j)} \leq \Gamma_0^{(j+n)} \left(\prod_{i=0}^{n-1} \frac{1 + w_{n-i}^{(j+i)}}{1 - w_{n-i}^{(j+i)}}\right). \tag{8.4.27}$$

Finally, we set $\Gamma_0^{(j+n)} = 1$ and bound the product in (8.4.27). It can be shown that, when $q = 2$, $\Gamma_n^{(j)} = O(n^{-1/2}(e^2/3)^n)$ as $n \to \infty$. This result is due to Laurie [159]. On the basis of this result, Laurie concludes that the error propagation in Romberg integration with the harmonic sequence of stepsizes is relatively mild.

8.4.3 Process II with $\varphi(t) = t^\delta H(t)$ and Real δ

We now want to extend Theorem 8.4.4 to the general case in which $\varphi(t) = t^\delta H(t)$, where δ is real and $\delta \neq 0, -1, -2, \ldots$, and $H(t) \in C^\infty[0, t_j]$ with $H(t) \neq 0$ on $[0, t_j]$. To do this, we need additional analytical tools. We make use of these tools in the next sections as well. The results we have obtained for the case $\varphi(t) = t$ will also prove to be very useful in the sequel.

Lemma 8.4.5 *Let δ_1 and δ_2 be two real numbers and $\delta_1 \neq \delta_2$. Define $\Delta_i(t) = t^{-\delta_i}$, $i = 1, 2$. Then, provided $\delta_1 \neq 0, -1, -2, \ldots$,*

$$D_n^{(j)}\{\Delta_2(t)\} = \frac{(\delta_2)_n}{(\delta_1)_n} \tilde{t}_{jn}^{\delta_1 - \delta_2} D_n^{(j)}\{\Delta_1(t)\} \text{ for some } \tilde{t}_{jn} \in (t_{j+n}, t_j).$$

Proof. From Lemma 8.1.5, with $z \equiv \sum_{i=0}^{n} \xi_i t_{j+i}$,

$$D_n^{(j)}\{\Delta_2(t)\} = \int_{T_n} \Delta_2^{(n)}(z) d\xi_1 \cdots d\xi_n = \int_{T_n} \left[\frac{\Delta_2^{(n)}(z)}{\Delta_1^{(n)}(z)}\right] \Delta_1^{(n)}(z) d\xi_1 \cdots d\xi_n.$$

Because $\Delta_1^{(n)}(z)$ is of one sign on T_n, we can apply the mean value theorem to the second integral to obtain

$$
\begin{aligned}
D_n^{(j)}\{\Delta_2(t)\} &= \frac{\Delta_2^{(n)}(\tilde{t}_{jn})}{\Delta_1^{(n)}(\tilde{t}_{jn})}\int_{T_n}\Delta_1^{(n)}(z)d\xi_1\cdots d\xi_n \\
&= \frac{\Delta_2^{(n)}(\tilde{t}_{jn})}{\Delta_1^{(n)}(\tilde{t}_{jn})}D_n^{(j)}\{\Delta_1(t)\}\ \text{ for some }\tilde{t}_{jn}\in(t_{j+n},t_j).
\end{aligned}
$$

This proves the lemma.   ∎

Corollary 8.4.6 *Let $\delta_1 > \delta_2$ in Lemma 8.4.5. Then, for arbitrary $\{t_l\}$,*

$$
\left|\frac{(\delta_2)_n}{(\delta_1)_n}\right|t_{j+n}^{\delta_1-\delta_2} \le \frac{|D_n^{(j)}\{\Delta_2(t)\}|}{|D_n^{(j)}\{\Delta_1(t)\}|} \le \left|\frac{(\delta_2)_n}{(\delta_1)_n}\right|t_j^{\delta_1-\delta_2},
$$

from which we also have

$$
\frac{|D_n^{(j)}\{\Delta_2(t)\}|}{|D_n^{(j)}\{\Delta_1(t)\}|} \le Kn^{\delta_2-\delta_1}t_j^{\delta_1-\delta_2} = o(1)\ \text{ as }j\to\infty\ \text{ and/or as }n\to\infty,
$$

for some constant $K > 0$ independent of j and n. Consequently, for arbitrary real θ and arbitrary $\{t_l\}$, the nonzero members of $\{D_n^{(j)}\{t^{\theta+i}\}\}_{i=0}^{\infty}$ form an asymptotic sequence as $j\to\infty$.

Proof. The first part follows directly from Lemma 8.4.5, whereas the second part is obtained by substituting in the first the known relation

$$
\frac{(a)_n}{(b)_n} = \frac{\Gamma(b)}{\Gamma(a)}\frac{\Gamma(n+a)}{\Gamma(n+b)} \sim \frac{\Gamma(b)}{\Gamma(a)}n^{a-b}\ \text{ as }n\to\infty. \tag{8.4.28}
$$

∎

The next lemma expresses $\Gamma_n^{(j)}$ and $A_n^{(j)} - A$ in factored forms. Analyzing each of the factors makes it easier to obtain good bounds from which powerful results on Process II can be obtained.

Lemma 8.4.7 *Consider $\varphi(t) = t^\delta H(t)$ with δ real and $\delta \ne 0, -1, -2, \dots$. Define*

$$
X_n^{(j)} = \frac{D_n^{(j)}\{t^{-1}\}}{D_n^{(j)}\{t^{-\delta}\}}\ \text{ and }\ Y_n^{(j)} = \frac{D_n^{(j)}\{t^{-\delta}\}}{D_n^{(j)}\{t^{-\delta}/H(t)\}}. \tag{8.4.29}
$$

Define also

$$
\check{\Gamma}_n^{(j)}(\delta) = \frac{1}{|D_n^{(j)}\{t^{-\delta}\}|}\sum_{i=0}^{n}|c_{ni}^{(j)}|\,t_{j+i}^{-\delta}. \tag{8.4.30}
$$

Then

$$\check{\Gamma}_n^{(j)}(\delta) = |X_n^{(j)}|(\check{t}_{jn})^{1-\delta} \check{\Gamma}_n^{(j)}(1) \ \text{for some } \check{t}_{jn} \in (t_{j+n}, t_j), \tag{8.4.31}$$

$$\Gamma_n^{(j)} = |Y_n^{(j)}| \, |H(\hat{t}_{jn})|^{-1} \check{\Gamma}_n^{(j)}(\delta) \ \text{for some } \hat{t}_{jn} \in (t_{j+n}, t_j), \tag{8.4.32}$$

and

$$A_n^{(j)} - A = X_n^{(j)} Y_n^{(j)} \frac{D_n^{(j)}\{B(t)\}}{D_n^{(j)}\{t^{-1}\}}. \tag{8.4.33}$$

In addition,

$$X_n^{(j)} = \frac{n!}{(\delta)_n}(\tilde{t}_{jn})^{\delta-1} \ \text{for some } \tilde{t}_{jn} \in (t_{j+n}, t_j), \tag{8.4.34}$$

These results are valid for all choices of $\{t_l\}$.

Proof. To prove (8.4.31), we start by writing (8.4.30) in the form

$$\check{\Gamma}_n^{(j)}(\delta) = |X_n^{(j)}| \frac{1}{|D_n^{(j)}\{t^{-1}\}|} \sum_{i=0}^{n} \left(|c_{ni}^{(j)}| \, t_{j+i}^{-1} \right) t_{j+i}^{1-\delta}.$$

The result follows by observing that, by continuity of $t^{1-\delta}$ for $t > 0$,

$$\sum_{i=0}^{n} \left(|c_{ni}^{(j)}| \, t_{j+i}^{-1} \right) t_{j+i}^{1-\delta} = \left(\sum_{i=0}^{n} |c_{ni}^{(j)}| \, t_{j+i}^{-1} \right) (\check{t}_{jn})^{1-\delta} \ \text{for some } \check{t}_{jn} \in (t_{j+n}, t_j),$$

and by invoking (8.4.30) with $\delta = 1$. The proof of (8.4.32) proceeds along the same lines, while (8.4.33) is a trivial identity. Finally, (8.4.34) follows from Lemma 8.4.5. ∎

   In the next two theorems, we adopt the notation and definitions of Lemma 8.4.7. The first of these theorems concerns the stability of Process II, and the second concerns its convergence.

Theorem 8.4.8 *Let δ be real and $\delta \neq 0, -1, -2, \ldots$, and let the t_l be as in (8.4.9). Then, the following are true:*

(i) *$\check{\Gamma}_n^{(j)}(\delta) \to \infty$ faster than n^σ for every $\sigma > 0$, that is, Process II for $\varphi(t) = t^\delta$ is unstable.*
(ii) *Let $\varphi(t) = t^\delta H(t)$ with $H(t) \in C^\infty[0, t_j]$ and $H(t) \neq 0$ on $[0, t_j]$. Assume that*

$$|Y_n^{(j)}| \geq C_1 n^{\alpha_1} \ \text{for all } n; \ C_1 > 0 \text{ and } \alpha_1 \text{ constants.} \tag{8.4.35}$$

Then, $\Gamma_n^{(j)} \to \infty$ as $n \to \infty$ faster than n^σ for every $\sigma > 0$, that is, Process II is unstable.

Proof. Substituting (8.4.34) in (8.4.31), we obtain

$$\check{\Gamma}_n^{(j)}(\delta) = \frac{n!}{|(\delta)_n|} \left(\frac{\tilde{t}_{jn}}{\check{t}_{jn}}\right)^{\delta-1} \check{\Gamma}_n^{(j)}(1). \tag{8.4.36}$$

Invoking the asymptotic equality of (8.4.28) in (8.4.36), we have for all large n

$$\check{\Gamma}_n^{(j)}(\delta) \geq K(\delta) n^{1-\delta} \left(\frac{t_{j+n}}{t_j}\right)^{|\delta-1|} \check{\Gamma}_n^{(j)}(1) \text{ for some constant } K(\delta) > 0. \tag{8.4.37}$$

Now, $t_{j+n}^{|\delta-1|} \sim c^{|\delta-1|} n^{-q|\delta-1|}$ as $n \to \infty$ and, by Theorem 8.4.4, $\check{\Gamma}_n^{(j)}(1) \to \infty$ as $n \to \infty$ faster than n^σ for every $\sigma > 0$. Consequently, $\check{\Gamma}_n^{(j)}(\delta) \to \infty$ as $n \to \infty$ faster than n^σ for every $\sigma > 0$ as well. The assertion about $\Gamma_n^{(j)}$ can now be proved by using this result in (8.4.32) along with (8.4.35) and the fact that $|H(\hat{t}_{jn})|^{-1} \geq \left(\max_{t \in [0,t_j]} |H(t)|\right)^{-1} > 0$ independently of n.   ∎

The purpose of the next theorem is to give as good a bound as possible for $|A_n^{(j)} - A|$ in Process II. A convergence result can then be obtained by imposing suitable and liberal growth conditions on $\|B^{(n)}\|$ and $Y_n^{(j)}$ and recalling that $D_n^{(j)}\{t^{-1}\} = (-1)^n / \left(\prod_{i=0}^n t_{j+i}\right)$.

Theorem 8.4.9 *Assume that $B(t) \in C^\infty[0, t_j]$ and define $\|B^{(n)}\|$ as in (8.4.19). Let $\varphi(t)$ be as in Theorem 8.4.8. Then, for some constant $L > 0$,*

$$|A_n^{(j)} - A| \leq L |Y_n^{(j)}| \left(\max_{t \in [t_{j+n}, t_j]} t^{\delta-1}\right) n^{1-\delta} \frac{\|B^{(n)}\|}{n!} \left(\prod_{i=0}^n t_{j+i}\right). \tag{8.4.38}$$

Note that it is quite reasonable to assume that $Y_n^{(j)}$ is bounded as $n \to \infty$. The "justification" for this assumption is that $Y_n^{(j)} = \Delta^{(n)}(t'_{jn})/\psi^{(n)}(t''_{jn})$ for some t'_{jn} and $t''_{jn} \in (t_{j+n}, t_j)$, where $\Delta(t) = t^{-\delta}$ and $\psi(t) = 1/\varphi(t) = t^{-\delta}/H(t)$, and that $\Delta^{(n)}(t)/\psi^{(n)}(t) \sim H(0)$ as $t \to 0+$. Indeed, when $1/H(t)$ is a polynomial in t, we have precisely $D_n^{(j)}\{t^{-\delta}/H(t)\} \sim D_n^{(j)}\{t^{-\delta}\}/H(0)$ as $n \to \infty$, as can be shown with the help of Corollary 8.4.6, from which $Y_n^{(j)} \sim H(0)$ as $n \to \infty$. See also Lemma 8.6.5. Next, with the t_l as in (8.4.9), we also have that $(\max_{t \in [t_{j+n}, t_j]} t^{\delta-1})$ grows at most like $n^{q|\delta-1|}$ as $n \to \infty$. Thus, the product $|Y_n^{(j)}|(\max_{t \in [t_{j+n}, t_j]} t^{\delta-1})$ in (8.4.38) grows at most like a power of n as $n \to \infty$, and, consequently the main behavior of $|A_n^{(j)} - A|$ as $n \to \infty$ is determined by $(\|B^{(n)}\|/n!) \left(\prod_{i=0}^n t_{j+i}\right)$. Also note that the strength of (8.4.38) is primarily due to the factor $\prod_{i=0}^n t_{j+i}$ that tends to zero as $n \to \infty$ essentially like $(n!)^{-q}$ when the t_l satisfy (8.4.9). Recall that what produces this important factor is Lemma 8.4.5.

8.5 Slowly Varying $\varphi(t)$ with $\lim_{l \to \infty}(t_{l+1}/t_l) = \omega \in (0, 1)$

As is clear from our results in Section 8.4, both Process I and Process II are unstable when $\varphi(t)$ is slowly changing and the t_l satisfy (8.4.9) or, at least in some cases, when the t_l satisfy even the weaker condition $\lim_{l \to \infty}(t_{l+1}/t_l) = 1$. These results also show that convergence will take place in Process II nevertheless under rather liberal growth conditions for $B^{(n)}(t)$. The implication of this is that a required level of accuracy in the numerically computed $A_n^{(j)}$ may be achieved by computing the $a(t_l)$ with sufficiently high accuracy. This strategy is quite practical and has been used successfully in numerical calculation of multiple integrals.

In case the accuracy with which $a(t)$ is computed is fixed and the $A_n^{(j)}$ are required to have comparable numerical accuracy, we need to choose the t_l such that the $A_n^{(j)}$ can be computed stably. When $\varphi(t) = t^\delta H(t)$ with $H(0) \neq 0$ and $H(t)$ continuous in a right neighborhood of $t = 0$, best results for $A_n^{(j)}$ and $\Gamma_n^{(j)}$ are obtained by picking $\{t_l\}$ such that $t_l \to 0$ as $l \to \infty$ *exponentially* in l. There are a few ways to achieve this and each of them has been used successfully in various problems.

Our first results with such $\{t_l\}$ given in Theorem 8.5.1 concern Process I, and, like those of Theorems 8.4.1 and 8.4.3, they are best asymptotically.

Theorem 8.5.1 *Let* $\varphi(t) = t^\delta H(t)$ *with* δ *in general complex and* $\delta \neq 0, -1, -2, \ldots$, *and* $H(t) \sim H(0) \neq 0$ *as* $t \to 0+$. *Pick the* t_l *such that* $\lim_{l \to \infty}(t_{l+1}/t_l) = \omega$ *for some fixed* $\omega \in (0, 1)$. *Define*

$$c_k = \omega^{\delta+k-1}, \quad k = 1, 2, \ldots . \tag{8.5.1}$$

Then, for fixed n, (8.1.13) and (8.1.14) hold, and we also have

$$A_n^{(j)} - A \sim \left(\prod_{i=1}^n \frac{c_{n+\mu+1} - c_i}{1 - c_i}\right) \beta_{n+\mu} \varphi(t_j) t_j^{n+\mu} \quad as \quad j \to \infty, \tag{8.5.2}$$

where $\beta_{n+\mu}$ *is the first nonzero* β_i *with* $i \geq n$ *in (8.1.2). This result is valid whether* $n \geq -\Re\delta$ *or not. Also,*

$$\lim_{j \to \infty} \sum_{i=0}^n \gamma_{ni}^{(j)} z^i = \prod_{i=1}^n \frac{z - c_i}{1 - c_i} \equiv \sum_{i=0}^n \rho_{ni} z^i, \tag{8.5.3}$$

so that $\lim_{j \to \infty} \Gamma_n^{(j)}$ *exists and*

$$\lim_{j \to \infty} \Gamma_n^{(j)} = \sum_{i=0}^n |\rho_{ni}| = \prod_{i=1}^n \frac{1 + |c_i|}{|1 - c_i|}, \tag{8.5.4}$$

and hence Process I is stable.

Proof. The proof can be achieved by making suitable substitutions in Theorems 3.5.3, 3.5.5, and 3.5.6 of Chapter 3.   ∎

Note that Theorem 8.5.1 is valid also when $\varphi(t)$ satisfies $\varphi(t) \sim h_0 t^\delta |\log t|^\gamma$ as $t \to 0+$ with arbitrary γ. Obviously, this is a weaker condition than the one imposed on $\varphi(t)$ in the theorem.

Upon comparing Theorem 8.5.1 with Theorem 8.4.1, we realize that the remarks that follow the proof of Theorem 8.4.1 and that concern the convergence of column sequences also are valid without any changes under the conditions of Theorem 8.5.1.

So far we do not have results on Process II with $\varphi(t)$ and $\{t_l\}$ as in Theorem 8.5.1. We are able to provide some analysis for the cases in which δ is real, however. This is the subject of the next section.

We are able to give very strong results on Process II for the case in which $\{t_l\}$ is a truly geometric sequence. The conditions we impose on $\varphi(t)$ in this case are extremely weak in the sense that $\varphi(t) = t^\delta H(t)$ with δ complex in general and $H(t)$ not necessarily differentiable at $t = 0$.

Theorem 8.5.2 *Let $\varphi(t) = t^\delta H(t)$ with δ in general complex and $\delta \neq 0, -1, -2, \ldots$, and $H(t) = H(0) + O(t^\theta)$ as $t \to 0+$, with $H(0) \neq 0$ and $\theta > 0$. Pick the t_l such that $t_l = t_0 \omega^l$, $l = 0, 1, \ldots$, for some $\omega \in (0, 1)$. Define $c_k = \omega^{\delta+k-1}$, $k = 1, 2, \ldots$. Then, for any fixed j, Process II is both stable and convergent whether $\lim_{t\to 0+} a(t)$ exists or not. In particular, we have $\lim_{n\to\infty} A_n^{(j)} = A$ with*

$$A_n^{(j)} - A = O(\omega^{\sigma n}) \text{ as } n \to \infty, \text{ for every } \sigma > 0, \tag{8.5.5}$$

and $\sup_n \Gamma_n^{(j)} < \infty$ with

$$\lim_{n\to\infty} \Gamma_n^{(j)} = \prod_{i=1}^{\infty} \frac{1 + |c_i|}{|1 - c_i|} < \infty. \tag{8.5.6}$$

The convergence result of (8.5.5) can be refined as follows: With $B(t) \in C[0, \hat{t}]$ for some $\hat{t} > 0$, define

$$\hat{\beta}_s = \max_{t\in[0,\hat{t}]} \left(|B(t) - \sum_{i=0}^{s-1} \beta_i t^i|/t^s \right), \; s = 0, 1, \ldots, \tag{8.5.7}$$

or when $B(t) \in C^\infty[0, \hat{t}]$, define

$$\tilde{\beta}_s = \max_{t\in[0,\hat{t}]} (|B^{(s)}(t)|/s!), \; s = 0, 1, \ldots. \tag{8.5.8}$$

If $\hat{\beta}_n$ or $\tilde{\beta}_n$ is $O(e^{\sigma n^\tau})$ as $n \to \infty$ for some $\sigma > 0$ and $\tau < 2$, then, for any $\epsilon > 0$ such that $\omega + \epsilon < 1$,

$$A_n^{(j)} - A = O\left((\omega + \epsilon)^{n^2/2}\right) \text{ as } n \to \infty. \tag{8.5.9}$$

We refer the reader to Sidi [300] for proofs of the results in (8.5.5), (8.5.6), and (8.5.9).

We make the following observations about Theorem 8.5.2. First, note that all the results in this theorem are independent of θ, that is, of the details of $\varphi(t) - H(0)t^\delta$ as $t \to 0+$. Next, (8.5.5) implies that all diagonal sequences $\{A_n^{(j)}\}_{n=0}^\infty$, $j = 0, 1, \ldots$, converge, and the error $A_n^{(j)} - A$ tends to 0 as $n \to \infty$ faster than $e^{-\lambda n}$ for *every* $\lambda > 0$, that is, the convergence is *superlinear*. Under the additional growth condition imposed on $\hat{\beta}_n$ or $\tilde{\beta}_n$, we have that $A_n^{(j)} - A$ tends to 0 as $n \to \infty$ at the rate of $e^{-\kappa n^2}$ for *some* $\kappa > 0$. Note that this condition is very liberal and is satisfied in most practical situations. It holds, for example, when $\hat{\beta}_n$ or $\tilde{\beta}_n$ are $O((pn)!)$ as $n \to \infty$ for *some* $p > 0$. Also, it is quite interesting that $\lim_{n\to\infty} \Gamma_n^{(j)}$ is independent of j, as seen from (8.5.6).

Finally, note that Theorem 8.5.1 pertaining to Process I holds under the conditions of Theorem 8.5.2 without any changes as $\lim_{l\to\infty}(t_{l+1}/t_l) = \omega$ is obviously satisfied because $t_{l+1}/t_l = \omega$ for all l.

8.6 Slowly Varying $\varphi(t)$ with Real δ and $t_{l+1}/t_l \leq \omega \in (0, 1)$

In this section, we consider the convergence and stability properties of Process II when $\{t_l\}$ is not necessarily a geometric sequence as in Theorem 8.5.2 or $\lim_{l\to\infty}(t_{l+1}/t_l)$ does not necessarily exist as in Theorem 8.5.1. We are now concerned with the choice

$$t_{l+1}/t_l \leq \omega, \ l = 0, 1, \ldots, \quad \text{for some fixed } \omega \in (0, 1). \tag{8.6.1}$$

If $\lim_{l\to\infty}(t_{l+1}/t_l) = \lambda$ for some $\lambda \in (0, 1)$, then given $\epsilon > 0$ such that $\omega = \lambda + \epsilon < 1$, there exists an integer $L > 0$ such that

$$\lambda - \epsilon < t_{l+1}/t_l < \lambda + \epsilon \ \text{ for all } \ l \geq L. \tag{8.6.2}$$

Thus, if $t_0, t_1, \ldots, t_{L-1}$ are chosen appropriately, the sequence $\{t_l\}$ automatically satisfies (8.6.1). Consequently, the results of this section apply also to the case in which $\lim_{l\to\infty}(t_{l+1}/t_l) = \lambda \in (0, 1)$.

8.6.1 The Case $\varphi(t) = t$

The case that has been studied most extensively under (8.6.1) is that of $\varphi(t) = t$, and we treat this case first. The outcome of this treatment will prove to be very useful in the study of the general case.

We start with the stability problem. As in Lemma 8.4.7, we denote the $\Gamma_n^{(j)}$ corresponding to $\varphi(t) = t^\delta$ by $\check{\Gamma}_n^{(j)}(\delta)$ and its corresponding $\gamma_{ni}^{(j)}$ by $\check{\gamma}_{ni}^{(j)}(\delta)$.

Theorem 8.6.1 *With $\varphi(t) = t$ and the t_l as in (8.6.1), we have for all j and n*

$$\check{\Gamma}_n^{(j)}(1) = \sum_{i=0}^{n} \left| \check{\gamma}_{ni}^{(j)}(1) \right| \leq \Omega_n \equiv \prod_{i=1}^{n} \frac{1 + \omega^i}{1 - \omega^i} < \prod_{i=1}^{\infty} \frac{1 + \omega^i}{1 - \omega^i} < \infty. \tag{8.6.3}$$

Therefore, both Process I and Process II are stable. Furthermore, for each fixed i, we have $\lim_{n\to\infty} \check{\gamma}_{ni}^{(j)}(1) = 0$, with

$$\check{\gamma}_{ni}^{(j)}(1) = O(\omega^{n^2/2+d_i n}) \ \text{ as } n \to \infty, \ d_i \text{ a constant.} \tag{8.6.4}$$

Proof. Let $\bar{t}_l = t_0 \omega^l, \ l = 0, 1, \ldots$, and denote the $\gamma_{ni}^{(j)}$ and $\Gamma_n^{(j)}$ appropriate for $\varphi(t) = t$ and the $\bar{t}_l$, respectively, by $\bar{\gamma}_{ni}^{(j)}$ and $\bar{\Gamma}_n^{(j)}$. By (8.4.26), we have that

$$
\left| \check{\gamma}_{ni}^{(j)}(1) \right| = \left(\prod_{k=0}^{i-1} \frac{1}{1 - t_{j+i}/t_{j+k}} \right) \left(\prod_{k=i+1}^{n} \frac{1}{t_{j+i}/t_{j+k} - 1} \right)
$$

$$
\leq \left(\prod_{k=0}^{i-1} \frac{1}{1 - \omega^{i-k}} \right) \left(\prod_{k=i+1}^{n} \frac{1}{\omega^{i-k} - 1} \right)
$$

$$
= \left(\prod_{k=0}^{i-1} \frac{1}{1 - \bar{t}_{j+i}/\bar{t}_{j+k}} \right) \left(\prod_{k=i+1}^{n} \frac{1}{\bar{t}_{j+i}/\bar{t}_{j+k} - 1} \right) = \left| \bar{\gamma}_{ni}^{(j)} \right|. \tag{8.6.5}
$$

Therefore, $\check{\Gamma}_n^{(j)}(1) \leq \bar{\Gamma}_n^{(j)}$. But $\bar{\Gamma}_n^{(j)} = \Omega_n$ by Theorem 1.4.3. The relation in (8.6.4) is a consequence of the fact that

$$|\check{\gamma}_{ni}^{(j)}| = \left(\prod_{k=1}^{n}(1 - c_k)\right)^{-1} \sum_{1 \leq k_1 < \cdots < k_{n-i} \leq n} c_{k_1} \cdots c_{k_{n-i}}, \tag{8.6.6}$$

with $c_k = \omega^k$, $k = 1, 2, \ldots$, which in turn follows from $\sum_{i=0}^{n} \check{\gamma}_{ni}^{(j)} z^i = \prod_{k=1}^{n}(z - c_k)/(1 - c_k)$. ∎

The fact that $\check{\Gamma}_n^{(j)}(1)$ is bounded uniformly both in j and in n was originally proved by Laurent [158]. The refined bound in (8.6.3) was mentioned without proof in Sidi [295].

Now that we have proved that Process I is stable, we can apply Theorem 8.1.4 and conclude that $\lim_{j \to \infty} A_n^{(j)} = A$ with

$$A_n^{(j)} - A = O(t_j^{n+\mu+1}) \quad \text{as } j \to \infty, \tag{8.6.7}$$

without assuming that $B(t)$ is differentiable in a right neighborhood of $t = 0$.

Since Process II satisfies the conditions of the Silverman–Toeplitz theorem (Theorem 0.3.3), we also have $\lim_{n \to \infty} A_n^{(j)} = A$. We now turn to the convergence issue for Process II to provide realistic rates of convergence for it. We start with the following important lemma due to Bulirsch and Stoer [43]. The proof of this lemma is very lengthy and difficult and we, therefore, refer the reader to the original paper.

Lemma 8.6.2 *With $\varphi(t) = t$ and the t_l as in (8.6.1), we have for each integer $s \in \{0, 1, \ldots, n\}$*

$$\sum_{i=0}^{n} |\check{\gamma}_{ni}^{(j)}(1)| \, t_{j+i}^{s+1} \leq M\left(\prod_{k=n-s}^{n} t_{j+k}\right) \tag{8.6.8}$$

for some constant $M > 0$ independent of j, n, and s.

This lemma becomes very useful in the proof of the convergence of Process II.

Lemma 8.6.3 *Let us pick the t_l to satisfy (8.6.1). Then, with j fixed,*

$$\frac{|D_n^{(j)}\{B(t)\}|}{|D_n^{(j)}\{t^{-1}\}|} = O(\omega^{\sigma n}) \quad \text{as } n \to \infty, \text{ for every } \sigma > 0. \tag{8.6.9}$$

This result can be refined as follows: Define $\hat{\beta}_s$ exactly as in Theorem 8.5.2. If $\hat{\beta}_n = O(e^{\sigma n^{\tau}})$ as $n \to \infty$ for some $\sigma > 0$ and $\tau < 2$, then, for any $\epsilon > 0$ such that $\omega + \epsilon < 1$,

$$\frac{|D_n^{(j)}\{B(t)\}|}{|D_n^{(j)}\{t^{-1}\}|} = O((\omega + \epsilon)^{n^2/2}) \quad \text{as } n \to \infty. \tag{8.6.10}$$

Proof. From (8.1.11), we have for each $s \leq n$

$$Q_n^{(j)} \equiv \frac{|D_n^{(j)}\{B(t)\}|}{|D_n^{(j)}\{t^{-1}\}|} \leq \sum_{i=0}^{n} |\check{\gamma}_{ni}^{(j)}(1)| \, E_s(t_{j+i}) \, t_{j+i} \, ; \quad E_s(t) \equiv |B(t) - \sum_{k=0}^{s-1} \beta_k t^k|.$$

(8.6.11)

By (8.1.2), there exist constants $\eta_s > 0$ such that $E_s(t) \leq \eta_s t^s$ when $t \in [0, \hat{t}]$ for some $\hat{t} > 0$ and also when $t = t_l > \hat{t}$. (Note that there are at most finitely many $t_l > \hat{t}$.) Therefore, (8.6.11) becomes

$$Q_n^{(j)} \leq \eta_s \sum_{i=0}^{n} |\check{\gamma}_{ni}^{(j)}(1)| \, t_{j+i}^{s+1} \leq M\eta_s \left(\prod_{k=n-s}^{n} t_{j+k} \right),$$

(8.6.12)

the last inequality being a consequence of Lemma 8.6.2. The result in (8.6.9) follows from (8.6.12) once we observe by (8.6.1) that $\prod_{k=n-s}^{n} t_{j+k} = O(\omega^{n(s+1)})$ as $n \to \infty$ with s fixed but arbitrary.

To prove the second part, we use the definition of $\hat{\beta}_s$ to rewrite (8.6.11) (with $s = n$) in the form

$$Q_n^{(j)} \leq \sum_{t_{j+i} > \hat{t}} |\check{\gamma}_{ni}^{(j)}(1)| \, E_n(t_{j+i}) \, t_{j+i} + \hat{\beta}_n \sum_{t_{j+i} \leq \hat{t}} |\check{\gamma}_{ni}^{(j)}(1)| \, t_{j+i}^{n+1}.$$

(8.6.13)

Since $E_n(t) \leq |B(t)| + \sum_{k=0}^{n-1} |\beta_k| \, t^k$, $|\beta_k| \leq \hat{\beta}_k$ for each k, and $\hat{\beta}_n$ grows at most like $e^{\sigma n^\tau}$ for $\tau < 2$, $\check{\gamma}_{ni}^{(j)}(1) = O(\omega^{n^2/2 + d_i n})$ as $n \to \infty$ from (8.6.4), and there are at most finitely many $t_l > \hat{t}$, we have that the first summation on the right-hand side of (8.6.13) is $O((\omega + \epsilon)^{n^2/2})$ as $n \to \infty$, for any $\epsilon > 0$. Using Lemma 8.6.2, we obtain for the second summation

$$\hat{\beta}_n \sum_{t_{j+i} \leq \hat{t}} |\check{\gamma}_{ni}^{(j)}(1)| \, t_{j+i}^{n+1} \leq \hat{\beta}_n \sum_{i=0}^{n} |\check{\gamma}_{ni}^{(j)}(1)| \, t_{j+i}^{n+1} \leq M\hat{\beta}_n \left(\prod_{i=0}^{n} t_{j+i} \right),$$

which, by (8.6.1), is also $O((\omega + \epsilon)^{n^2/2})$ as $n \to \infty$, for any $\epsilon > 0$. Combining the above in (8.6.13), we obtain (8.6.10). ∎

The following theorem is a trivial rewording of Lemma 8.6.3.

Theorem 8.6.4 *Let $\varphi(t) = t$ and pick the t_l to satisfy (8.6.1). Then, with j fixed, $\lim_{n \to \infty} A_n^{(j)} = A$, and*

$$A_n^{(j)} - A = O(\omega^{\sigma n}) \text{ as } n \to \infty, \text{ for every } \sigma > 0.$$

(8.6.14)

This result can be refined as follows: Define $\hat{\beta}_s$ exactly as in Theorem 8.5.2. If $\hat{\beta}_n = O(e^{\sigma n^\tau})$ as $n \to \infty$ for some $\sigma > 0$ and $\tau < 2$, then, for any $\epsilon > 0$ such that $\omega + \epsilon < 1$,

$$A_n^{(j)} - A = O((\omega + \epsilon)^{n^2/2}) \text{ as } n \to \infty.$$

(8.6.15)

Theorem 8.6.4 first implies that all diagonal sequences $\{A_n^{(j)}\}_{n=0}^{\infty}$ converge to A and that $|A_n^{(j)} - A| \to 0$ as $n \to \infty$ faster than $e^{-\lambda n}$ for *every* $\lambda > 0$. It next implies that, with a suitable and liberal growth rate on the $\hat{\beta}_n$, it is possible to achieve $|A_n^{(j)} - A| \to 0$ as $n \to \infty$ practically like $e^{-\kappa n^2}$ for *some* $\kappa > 0$.

8.6.2 *The Case* $\varphi(t) = t^\delta H(t)$ *with Real* δ *and* $t_{l+1}/t_l \leq \omega \in (0, 1)$

We now come back to the general case in which $\varphi(t) = t^\delta H(t)$ with δ real, $\delta \neq 0$, $-1, -2, \ldots$, and $H(t) \sim H(0) \neq 0$ as $t \to 0+$. We assume only that $H(t) \in C[0, \hat{t}]$ and $H(t) \neq 0$ when $t \in [0, \hat{t}]$ for some $\hat{t} > 0$ and that $H(t) \sim \sum_{i=0}^\infty h_i t^i$ as $t \to 0+$, $h_0 \neq 0$. Similarly, $B(t) \in C[0, \hat{t}]$ and $B(t) \sim \sum_{i=0}^\infty \beta_i t^i$ as $t \to 0+$, as before. We do not impose any differentiability conditions on $B(t)$ or $H(t)$. Finally, unless stated otherwise, we require the t_l to satisfy

$$\nu \leq t_{l+1}/t_l \leq \omega, \quad l = 0, 1, \ldots, \quad \text{for some fixed } \nu \text{ and } \omega, \ 0 < \nu < \omega < 1,$$
$$(8.6.16)$$

instead of (8.6.1) only. Recall from the remark following the statement of Lemma 8.3.2 that the additional condition $\nu \leq t_{l+1}/t_l$ is naturally satisfied, for example, when $\lim_{l\to\infty}(t_{l+1}/t_l) = \lambda \in (0, 1)$; cf. also (8.6.2). It also enables us to overcome some problems in the proofs of our main results.

We start with the following lemma that is analogous to Lemma 8.3.2 and Lemma 8.4.2.

Lemma 8.6.5 *Let* $g(t) \sim \sum_{i=0}^\infty g_i t^{\theta+i}$ *as* $t \to 0+$, *where* $g_0 \neq 0$ *and* θ *is real, such that* $g(t)t^{-\theta} \in C[0, \hat{t}]$ *for some* $\hat{t} > 0$, *and pick the* t_l *to satisfy* (8.6.16). *Then, the following are true:*

(i) *The nonzero members of* $\{D_n^{(j)}\{t^{\theta+i}\}\}_{i=0}^\infty$ *form an asymptotic sequence both as* $j \to \infty$ *and as* $n \to \infty$.

(ii) $D_n^{(j)}\{g(t)\}$ *has the bona fide asymptotic expansion*

$$D_n^{(j)}\{g(t)\} \sim \sum_{i=0}^\infty {}^* g_i D_n^{(j)}\{t^{\theta+i}\} \quad \text{as } j \to \infty, \qquad (8.6.17)$$

where the asterisk on the summation means that only those terms for which $D_n^{(j)}\{t^{\theta+i}\} \neq 0$, *that is, for which* $\theta + i \neq 0, 1, \ldots, n-1$, *are taken into account.*

(iii) *When* $\theta \neq 0, 1, \ldots, n-1$, *we also have*

$$D_n^{(j)}\{g(t)\} \sim g_0 D_n^{(j)}\{t^\theta\} \quad \text{as } n \to \infty. \qquad (8.6.18)$$

When $\theta < 0$, *the condition in* (8.6.1) *is sufficient for* (8.6.18) *to hold.*

Proof. Part (i) follows from Corollary 8.4.6. For the proof of part (ii), we follow the steps of the proof of part (ii) of Lemma 8.3.2. For arbitrary m, we have $g(t) = \sum_{i=0}^{m-1} g_i t^{\theta+i} + v_m(t)t^{\theta+m}$, where $|v_m(t)| \leq C_m$ for some constant $C_m > 0$, whenever $t \in [0, \hat{t}]$ and also $t = t_l > \hat{t}$. (Recall again that there are at most finitely many $t_l > \hat{t}$.)

Thus,

$$D_n^{(j)}\{g(t)\} = \sum_{i=0}^{m-1} g_i D_n^{(j)}\{t^{\theta+i}\} + D_n^{(j)}\{v_m(t)t^{\theta+m}\}. \tag{8.6.19}$$

Therefore, we have to show that $D_n^{(j)}\{v_m(t)t^{\theta+m}\} = O(D_n^{(j)}\{t^{\theta+m}\})$ as $j \to \infty$ when $\theta + m \neq 0, 1, \ldots, n-1$.

By the fact that $\check{\gamma}_{ni}^{(j)}(1) = c_{ni}^{(j)} t_{j+i}^{-1}/D_n^{(j)}\{t^{-1}\}$, we have

$$|D_n^{(j)}\{v_m(t)t^{\theta+m}\}| \leq \sum_{i=0}^{n} |c_{ni}^{(j)}| \, |v_m(t_{j+i})| \, t_{j+i}^{\theta+m}$$

$$\leq C_m |D_n^{(j)}\{t^{-1}\}| \sum_{i=0}^{n} |\check{\gamma}_{ni}^{(j)}(1)| \, t_{j+i}^{\theta+m+1}. \tag{8.6.20}$$

Now, taking s to be any integer that satisfies $0 \leq s \leq \min\{\theta+m, n\}$, and applying Lemma 8.6.2, we obtain

$$\sum_{i=0}^{n} |\check{\gamma}_{ni}^{(j)}(1)| \, t_{j+i}^{\theta+m+1} \leq \left(\sum_{i=0}^{n} |\check{\gamma}_{ni}^{(j)}(1)| \, t_{j+i}^{s+1} \right) t_j^{\theta+m-s} \leq M \left(\prod_{k=n-s}^{n} t_{j+k} \right) t_j^{\theta+m-s}. \tag{8.6.21}$$

Consequently, under (8.6.1) only,

$$|D_n^{(j)}\{v_m(t)t^{\theta+m}\}| \leq M C_m |D_n^{(j)}\{t^{-1}\}| \left(\prod_{k=n-s}^{n} t_{j+k} \right) t_j^{\theta+m-s}. \tag{8.6.22}$$

Recalling that $|D_n^{(j)}\{t^{-1}\}| = (\prod_{k=0}^{n} t_{j+k})^{-1}$ and $t_{l+1} \geq \nu t_l$, and invoking (8.3.9) that is valid in the present case, we obtain from (8.6.22)

$$D_n^{(j)}\{v_m(t)t^{\theta+m}\} = O(t_j^{\theta+m-n}) = O(D_n^{(j)}\{t^{\theta+m}\}) \quad \text{as } j \to \infty. \tag{8.6.23}$$

This completes the proof of part (ii).

As for part (iii), we first note that, by Corollary 8.4.6,

$$\lim_{n \to \infty} \sum_{i=0}^{m-1} g_i D_n^{(j)}\{t^{\theta+i}\}/D_n^{(j)}\{t^{\theta}\} = g_0.$$

Therefore, the proof will be complete if we show that $\lim_{n \to \infty} D_n^{(j)}\{v_m(t)t^{\theta+m}\}/D_n^{(j)}\{t^{\theta}\} = 0$. By (8.6.22), we have

$$T_n^{(j)} \equiv \frac{|D_n^{(j)}\{v_m(t)t^{\theta+m}\}|}{|D_n^{(j)}\{t^{\theta}\}|} \leq M C_m \frac{|D_n^{(j)}\{t^{-1}\}|}{|D_n^{(j)}\{t^{\theta}\}|} \left(\prod_{k=n-s}^{n} t_{j+k} \right) t_j^{\theta+m-s}. \tag{8.6.24}$$

By Corollary 8.4.6, again,

$$\frac{|D_n^{(j)}\{t^{-1}\}|}{|D_n^{(j)}\{t^{\theta}\}|} \leq K_1 n^{1+\theta} \left(\max_{t \in [t_{j+n}, t_j]} t^{-1-\theta} \right), \tag{8.6.25}$$

and by (8.6.1),

$$\prod_{k=n-s}^{n} t_{j+k} \leq K_2 t_j^s t_{j+n} \omega^{ns}, \tag{8.6.26}$$

where K_1 and K_2 are some positive constants independent of n. Combining these in (8.6.24), we have

$$T_n^{(j)} \leq L V_n^{(j)} n^{1+\theta} t_{j+n} \omega^{ns}; \quad V_n^{(j)} \equiv \max_{t \in [t_{j+n}, t_j]} t^{-1-\theta}, \tag{8.6.27}$$

for some constant $L > 0$ independent of n. Now, (a) for $\theta \leq -1$, $V_n^{(j)} = t_j^{-1-\theta}$; (b) for $-1 < \theta < 0$, $V_n^{(j)} = t_{j+n}^{-1-\theta}$; while (c) for $\theta > 0$, $V_n^{(j)} = t_{j+n}^{-1-\theta} \leq t_j^{-1-\theta} \nu^{-n(1+\theta)}$ by (8.6.16). Thus, (a) if $\theta \leq -1$, then $T_n^{(j)} = O(n^{1+\theta} t_{j+n} \omega^{ns}) = o(1)$ as $n \to \infty$; (b) if $-1 < \theta < 0$, then $T_n^{(j)} = O(n^{1+\theta} t_{j+n}^{|\theta|} \omega^{ns}) = o(1)$ as $n \to \infty$; and (c) if $\theta > 0$, then $T_n^{(j)} = O(n^{1+\theta} \nu^{-n(1+\theta)} \omega^{ns}) = o(1)$ as $n \to \infty$, provided we take s sufficiently large in this case, which is possible because m is arbitrary and n tends to infinity. This completes the proof.   ∎

Our first major result concerns Process I.

Theorem 8.6.6 *Let $B(t)$, $\varphi(t)$, and $\{t_l\}$ be as in the first paragraph of this subsection. Then $A_n^{(j)} - A$ satisfies (8.1.13) and (8.1.14), and hence $A_n^{(j)} - A = O(\varphi(t_j) t_j^{n+\mu})$ as $j \to \infty$. In addition, $\sup_j \Gamma_n^{(j)} < \infty$, that is, Process I is stable.*

Proof. The assertions about $A_n^{(j)} - A$ follow by applying Lemma 8.6.5 to $B(t)$ and to $\psi(t) \equiv 1/\varphi(t)$. As for $\Gamma_n^{(j)}$, we proceed as follows. By (8.4.31) and (8.4.34) and (8.6.16), we first have that

$$\check{\Gamma}_n^{(j)}(\delta) \leq \frac{n!}{|(\delta)_n|} \left(\frac{t_j}{t_{j+n}} \right)^{|\delta-1|} \check{\Gamma}_n^{(j)}(1) \leq \frac{n!}{|(\delta)_n|} \nu^{-n|\delta-1|} \check{\Gamma}_n^{(j)}(1). \tag{8.6.28}$$

By Theorem 8.6.1, it therefore follows that $\sup_j \check{\Gamma}_n^{(j)}(\delta) < \infty$. Next, by Lemma 8.6.5 again, we have that $Y_n^{(j)} \sim h_0$ as $j \to \infty$, and $|H(t)|^{-1}$ is bounded for all t close to 0. Combining these facts in (8.4.32), it follows that $\sup_j \Gamma_n^{(j)} < \infty$.   ∎

As for Process II, we do not have a stability theorem for it under the conditions of Theorem 8.6.6. [The upper bound on $\check{\Gamma}_n^{(j)}(\delta)$ given in (8.6.28) tends to infinity as $n \to \infty$.] However, we do have a strong convergence theorem for Process II.

Theorem 8.6.7 *Let $B(t)$, $\varphi(t)$, and $\{t_l\}$ be as in the first paragraph of this subsection. Then, for any fixed j, Process II is convergent whether $\lim_{t \to 0+} a(t)$ exists or not. In particular, we have $\lim_{n \to \infty} A_n^{(j)} = A$ with*

$$A_n^{(j)} - A = O(\omega^{\sigma n}) \text{ as } n \to \infty, \text{ for every } \sigma > 0. \tag{8.6.29}$$

This result can be refined as follows: Define $\hat{\beta}_s$ exactly as in Theorem 8.5.2. If $\hat{\beta}_n = O(e^{\sigma n^\tau})$ as $n \to \infty$ for some $\sigma > 0$ and $\tau < 2$, then for any $\epsilon > 0$ such that $\omega + \epsilon < 1$

$$A_n^{(j)} - A = O((\omega + \epsilon)^{n^2/2}) \; as \; n \to \infty. \tag{8.6.30}$$

When $\delta > 0$, these results are valid under (8.6.1).

Proof. First, by (8.4.29) and part (iii) of Lemma 8.6.5, we have that $Y_n^{(j)} \sim H(0)$ as $n \to \infty$. The proof can now be completed by also invoking (8.4.34) and Lemma 8.6.3 in (8.4.33). We leave the details to the reader. ∎

Analytic Study of GREP$^{(1)}$: Quickly Varying $A(y) \in \mathbf{F}^{(1)}$

9.1 Introduction

In this chapter, we continue the analytical study of GREP$^{(1)}$, which we began in the preceding chapter. We treat those functions $A(y) \in \mathbf{F}^{(1)}$ whose associated $\phi(y)$ vary quickly as $y \to 0+$. Switching to the variable t as we did previously, by $\varphi(t)$ varying quickly as $t \to 0+$ we now mean that $\varphi(t)$ is of one of the three forms

$$
\begin{array}{lll}
\text{(a)} & \varphi(t) = e^{u(t)} h(t) \\[2mm]
\text{(b)} & \varphi(t) = [\Gamma(t^{-\hat{s}})]^{-\nu} h(t) \\[2mm]
\text{(c)} & \varphi(t) = [\Gamma(t^{-\hat{s}})]^{-\nu} e^{u(t)} h(t), & (9.1.1)
\end{array}
$$

where

(i) $u(t)$ behaves like

$$
u(t) \sim \sum_{k=0}^{\infty} u_k t^{k-s} \quad \text{as } t \to 0+, \quad u_0 \neq 0, \quad s > 0 \text{ integer}; \qquad (9.1.2)
$$

(ii) $h(t)$ behaves like

$$
h(t) \sim h_0 t^{\delta} \quad \text{as } t \to 0+, \quad \text{for some } h_0 \neq 0 \text{ and } \delta; \qquad (9.1.3)
$$

(iii) $\Gamma(z)$ is the Gamma function, $\nu > 0$, and $\hat{s}$ is a positive integer; and, finally,

(iv) $|\varphi(t)|$ is bounded or grows at worst like a negative power of t as $t \to 0+$. The implications of this are as follows: If $\varphi(t)$ is as in (9.1.1) (a), then $\lim_{t \to 0+} \Re u(t) \neq +\infty$. If $\varphi(t)$ is as in (9.1.1) (c), then $s \leq \hat{s}$. No extra conditions are imposed when $\varphi(t)$ is as in (9.1.1) (b). Thus, in case (a), either $\lim_{t \to 0+} \varphi(t) = 0$, or $\varphi(t)$ is bounded as $t \to 0+$, or it grows like $t^{\Re \delta}$ when $\Re \delta < 0$, whereas in cases (b) and (c), we have $\lim_{t \to 0+} \varphi(t) = 0$ always.

Note also that we have not put any restriction on δ that may now assume any real or complex value.

As for the function $B(t)$, in Section 9.3, we assume only that $B(t) \sim \sum_{k=0}^{\infty} \beta_k t^k$ as $t \to 0+$, without imposing on $B(t)$ any differentiability conditions. In Section 9.4, we assume that $B(t) \in C^{\infty}[0, \hat{t}]$ for some $\hat{t}$.

Throughout this chapter the t_l are chosen to satisfy

$$t_l \sim cl^{-q} \quad \text{and} \quad t_l - t_{l+1} \sim cpl^{-q-1} \quad \text{as } l \to \infty, \quad \text{for some } c > 0, \ p > 0, \ \text{and } q > 0.$$
$$(9.1.4)$$

[This is (8.4.9).]

We also use the notation $\psi(t) \equiv 1/\varphi(t)$ freely as before.

9.2 Examples of Quickly Varying $a(t)$

We now present practical examples of functions $a(t)$ that vary quickly.

Example 9.2.1 Let $f(x) = e^{\theta(x)} w(x)$, where $\theta \in \mathbf{A}^{(m)}$ strictly for some positive integer m and $w \in \mathbf{A}^{(\gamma)}$ strictly for some arbitrary and possibly complex γ. Then, we know from Theorem 5.7.3 that

$$F(x) = \int_a^x f(t)\, dt = I[f] + x^{1-m} f(x) g(x); \quad g \in \mathbf{A}^{(0)} \text{ strictly.}$$

This means that $F(x) \leftrightarrow a(t)$, $I[f] \leftrightarrow A$, $x^{-1} \leftrightarrow t$, $x^{1-m} f(x) \leftrightarrow \varphi(t)$ with $\varphi(t)$ as in (9.1.1)–(9.1.3) and with $s = m$ and $\delta = m - 1 - \gamma$. [Recall that $f \in \mathbf{B}^{(1)}$ in this case and that we can replace $\varphi(t)$ by $xf(x)$. This changes δ only.]

Example 9.2.2 Let a_n be of one of the three forms: (a) $a_n = \zeta^n w(n)$, or (b) $a_n = [\Gamma(n)]^\mu w(n)$, or (c) $a_n = [\Gamma(n)]^\mu \zeta^n w(n)$. Here, $w \in \mathbf{A}_0^{(\gamma)}$ strictly for some arbitrary and possibly complex γ, μ is a negative integer, and ζ is a possibly complex scalar that is arbitrary in case (c) and that satisfies $|\zeta| \leq 1$ and $\zeta \neq 1$ in case (a). Then, we know from Theorem 6.7.2 that

$$A_n = \sum_{k=1}^n a_k = S(\{a_k\}) + a_n g(n); \quad g \in \mathbf{A}_0^{(0)} \text{ strictly.}$$

This means that $A_n \leftrightarrow a(t)$, $S(\{a_k\}) \leftrightarrow A$, $n^{-1} \leftrightarrow t$, $a_n \leftrightarrow \varphi(t)$ with $\varphi(t)$ as in (9.1.1)–(9.1.3) and with $\nu = -\mu$, $\hat{s} = 1$, $s = 1$, and some δ. [Recall that $\{a_n\} \in \mathbf{b}^{(1)}$ in this case and that we can replace $\varphi(t)$ by na_n. This changes δ only.]

Example 9.2.3 Let a_n be of one of the three forms: (a) $a_n = e^{\theta(n)} w(n)$, or (b) $a_n = [\Gamma(n)]^\mu w(n)$, or (c) $a_n = [\Gamma(n)]^\mu e^{\theta(n)} w(n)$. Here, $\theta \in \tilde{\mathbf{A}}_0^{(1-r/m,m)}$ strictly for some integers $m > 0$ and $r \in \{0, 1, \dots, m-1\}$, $w \in \tilde{\mathbf{A}}_0^{(\gamma,m)}$ strictly for some arbitrary and possibly complex γ, and $\mu = \tau/m$ with $\tau < 0$ an integer. Then, we know from Theorem 6.6.6 that

$$A_n = \sum_{k=1}^n a_k = S(\{a_k\}) + n^\sigma a_n g(n); \quad g \in \tilde{\mathbf{A}}_0^{(0,m)} \text{ strictly,}$$

where $\sigma = \rho/m$ for some $\rho \in \{0, 1, \dots, m-1\}$. This means that $A_n \leftrightarrow a(t)$, $S(\{a_k\}) \leftrightarrow A$, $n^{-1/m} \leftrightarrow t$, $n^\sigma a_n \leftrightarrow \varphi(t)$ with $\varphi(t)$ as in (9.1.1)–(9.1.3) and with $\nu = -\mu$, $\hat{s} = m$, $s = m - r$ and some δ. [Recall that $\{a_n\} \in \tilde{\mathbf{b}}^{(m)}$ in this case and that again we can replace $\varphi(t)$ by na_n. Again, this changes δ only.]

9.3 Analysis of Process I

The first two theorems that follow concern $\varphi(t)$ as in (9.1.1) (a).

Theorem 9.3.1 *Let $\varphi(t)$ be as in (9.1.1) (a), and assume that $B(t) \sim \sum_{k=0}^{\infty} \beta_k t^k$ as $t \to 0+$, and pick the constants c, p, and q in (9.1.4) such that $q = 1/s$ and $\xi \equiv \exp(-spu_0/c^s) \neq 1$. Then*

$$A_n^{(j)} - A \sim \frac{p^n(\mu+1)_n}{(1-\xi)^n}\beta_{n+\mu}\varphi(t_j)t_j^{n+\mu}j^{-n} \quad as \ j \to \infty, \tag{9.3.1}$$

where $\beta_{n+\mu}$ is the first nonzero β_i with $i \geq n$. In addition, Process I is stable, and we have

$$\Gamma_n^{(j)} \sim \left(\frac{1+|\xi|}{|1-\xi|}\right)^n \quad as \ j \to \infty. \tag{9.3.2}$$

Proof. First, Lemma 8.4.2 applies to $B(t)$ and, therefore, $D_n^{(j)}\{B(t)\}$ satisfies (8.4.3), namely,

$$D_n^{(j)}\{B(t)\} \sim \frac{(\mu+1)_n}{n!}\beta_{n+\mu}t_j^{\mu} \quad as \ j \to \infty. \tag{9.3.3}$$

We next recall (8.4.14), from which we have

$$t_j^r - t_{j+i}^r = irpj^{-1}(cj^{-q})^r + o(j^{-qr-1}) \quad as \ j \to \infty, \ \text{for every } r. \tag{9.3.4}$$

Finally, we recall (8.4.15), namely,

$$c_{ni}^{(j)} \sim (-1)^i\frac{1}{n!}\binom{n}{i}\left(\frac{j}{pt_j}\right)^n \quad as \ j \to \infty. \tag{9.3.5}$$

All these are valid for any $q > 0$. Now, by (9.3.4), and after a delicate analysis, we have with $q = 1/s$

$$u(t_j) - u(t_{j+i}) \sim u_0(t_j^{-s} - t_{j+i}^{-s}) \sim -ispj^{-1}(cj^{-1/s})^{-s}u_0 = -ispu_0c^{-s} \quad as \ j \to \infty.$$

Consequently,

$$\exp[-u(t_{j+i})] \sim \xi^i \exp[-u(t_j)] \quad as \ j \to \infty, \tag{9.3.6}$$

from which we have

$$D_n^{(j)}\{1/\varphi(t)\} = \sum_{i=0}^{n} c_{ni}^{(j)}e^{-u(t_{j+i})}/h(t_{j+i})$$

$$\sim \frac{1}{n!}\left(\frac{j}{pt_j}\right)^n\left[\sum_{i=0}^{n}(-1)^i\binom{n}{i}\xi^i\right]\frac{e^{-u(t_j)}}{h_0t_j^{\delta}} \quad as \ j \to \infty$$

$$\sim \frac{1}{n!}\left(\frac{j}{pt_j}\right)^n(1-\xi)^n\psi(t_j) \quad as \ j \to \infty. \tag{9.3.7}$$

Similarly,

$$
\sum_{i=0}^{n} |c_{ni}^{(j)}|/|\varphi(t_{j+i})| = \sum_{i=0}^{n} |c_{ni}^{(j)}| \, |e^{-u(t_{j+i})}|/|h(t_{j+i})|
$$

$$
\sim \frac{1}{n!}\left(\frac{j}{pt_j}\right)^n \left[\sum_{i=0}^{n}\binom{n}{i}|\xi|^i\right]\frac{|e^{-u(t_j)}|}{|h_0 t_j^{\delta}|} \quad \text{as } j \to \infty
$$

$$
\sim \frac{1}{n!}\left(\frac{j}{pt_j}\right)^n (1+|\xi|)^n |\psi(t_j)| \quad \text{as } j \to \infty. \tag{9.3.8}
$$

Combining (9.3.3) and (9.3.7) in (8.1.3), (9.3.1) follows, and combining (9.3.7) and (9.3.8) in (8.1.4), (9.3.2) follows.   ∎

A lot can be learned from the analysis of GREP$^{(1)}$ for Process I when GREP$^{(1)}$ is being applied as in Theorem 9.3.1. From (9.3.2), it is clear that Process I will be increasingly stable as ξ (as a complex number) gets farther from 1. Recall that $\xi = \exp(-spu_0/c^s)$, that is, ξ is a function of both $\varphi(t)$ and $\{t_l\}$. Now, the behavior of $\varphi(t)$ is determined by the given $a(t)$ and the user can do nothing about it. The t_l, however, are chosen by the user. Thus, ξ can be controlled effectively by picking the t_l as in (9.1.4) with appropriate c. For example, if u_0 is purely imaginary and $\exp(-spu_0)$ is very close to 1 (note that $|\exp(-spu_0)| = 1$ in this case), then by picking c sufficiently small we can cause $\xi = [\exp(-spu_0)]^{c^{-s}}$ to be sufficiently far from 1, even though $|\xi| = 1$. It is also important to observe from (9.3.1) that the term $(1 - \xi)^{-n}$ also appears as a factor in the dominant behavior of $A_n^{(j)} - A$. Thus, by improving the stability of GREP$^{(1)}$, we are also improving the accuracy of the $A_n^{(j)}$.

In the next theorem, we show that, by fixing the value of q differently, we can cause the behavior of Process I to change completely.

Theorem 9.3.2 *Let $\varphi(t)$ and $B(t)$ be as in Theorem 9.3.1, and let $u(t) = v(t) + iw(t)$, with $v(t)$ and $w(t)$ real and $v(t) \sim \sum_{k=0}^{\infty} v_k t^{k-s}$ as $t \to 0+$, $v_0 \neq 0$. Choose $q > 1/s$ in (9.1.4). Then*

$$
A_n^{(j)} - A \sim (-1)^n p^n (\mu + 1)_n \beta_{n+\mu} \varphi(t_{j+n}) t_j^{n+\mu} j^{-n} \quad \text{as } j \to \infty, \tag{9.3.9}
$$

where $\beta_{n+\mu}$ is the first nonzero β_i with $i \geq n$. In addition, Process I is stable, and we have

$$
\Gamma_n^{(j)} \sim 1 \quad \text{as } j \to \infty. \tag{9.3.10}
$$

Proof. We start by noting that (9.3.3)–(9.3.5) are valid with any $q > 0$ in (9.1.4). Next, we note that by (9.1.2) and the condition on $v(t)$ imposed now, we have

$$
v(t_{j+1}) - v(t_j) \sim spj^{-1}(cj^{-q})^{-s} v_0 = v_0(sp/c^s) j^{qs-1} \quad \text{as } j \to \infty.
$$

Consequently,

$$\left|\frac{\psi(t_j)}{\psi(t_{j+1})}\right| \sim \exp[v(t_{j+1}) - v(t_j)] = \exp[v_0(sp/c^s)j^{qs-1} + o(j^{qs-1})] \text{ as } j \to \infty.$$

(9.3.11)

Since $\lim_{t\to 0+} \Re u(t) = \lim_{t\to 0+} v(t) = -\infty$, we must have $v_0 < 0$, and since $q > 1/s$, we have $qs - 1 > 0$, so that $\lim_{j\to\infty} |\psi(t_j)/\psi(t_{j+1})| = 0$. This and (9.3.5) imply that

$$D_n^{(j)}\{1/\varphi(t)\} = \sum_{i=0}^{n} c_{ni}^{(j)} \psi(t_{j+i}) \sim c_{nn}^{(j)} \psi(t_{j+n}) \sim \frac{(-1)^n}{n!} \left(\frac{j}{pt_j}\right)^n \psi(t_{j+n}) \text{ as } j \to \infty,$$

(9.3.12)

and

$$\sum_{i=0}^{n} |c_{ni}^{(j)}| |\psi(t_{j+i})| \sim |c_{nn}^{(j)}| |\psi(t_{j+n})| \sim \frac{1}{n!} \left(\frac{j}{pt_j}\right)^n |\psi(t_{j+n})| \text{ as } j \to \infty. \quad (9.3.13)$$

The proof can be completed as before. ∎

The technique we have employed in proving Theorem 9.3.2 can be used to treat the cases in which $\varphi(t)$ is as in (9.1.1) (b) and (c).

Theorem 9.3.3 *Let $\varphi(t)$ be as in (9.1.1) (b) or (c). Then, (9.3.9) and (9.3.10) hold if $q > 1/\hat{s}$.*

Proof. First, (9.3.3) is valid as before. Next, after some lengthy manipulation of the Stirling formula for the Gamma function, we can show that, in the cases we are considering, $\lim_{j\to\infty} |\psi(t_j)/\psi(t_{j+1})| = 0$, from which (9.3.12) and (9.3.13) hold. The results in (9.3.9) and (9.3.10) now follow. We leave the details to the reader. ∎

Note the difference between the theorems of this section and those of Chapter 8 pertaining to process I when GREP$^{(1)}$ is applied to slowly varying sequences. Whereas in Chapter 8 $A_n^{(j)} - A = O(\varphi(t_j)t_j^n)$ as $j \to \infty$, in this chapter $A_n^{(j)} - A = O(\varphi(t_j)t_j^n j^{-n})$ and $A_n^{(j)} - A = O(\varphi(t_{j+n})t_j^n j^{-n})$ as $j \to \infty$. This suggests that it is easier to accelerate the convergence of quickly varying $a(t)$.

9.4 Analysis of Process II

As mentioned before, the analysis of Process II turns out to be much more complicated than that of Process I. What complicates things most appears to be the term $D_n^{(j)}\{1/\varphi(t)\}$, which, even for simple $\varphi(t)$ such as $\varphi(t) = t^\delta e^{a/t}$, turns out to be extremely difficult to study. Because of this, we restrict our attention to the single case in which

$$\text{sgn } \varphi(t_l) \sim (-1)^l e^{i\alpha} \text{ as } l \to \infty, \text{ for some real } \alpha, \quad (9.4.1)$$

where $\text{sgn } \xi \equiv e^{i \arg \xi}$.

The results of this section are refined versions of corresponding results in Sidi [288].

The reader may be wondering whether there exist sequences $\{t_l\}$ that satisfy (9.1.4) and ensure the validity of (9.4.1) at the same time. The answer to this question is in the affirmative when $\lim_{t \to 0+} |\Im u(t)| = \infty$. We return to this point in the next section.

We start with the following simple lemma.

Lemma 9.4.1 *Let $a_1, \ldots, a_n$ be real positive and let $\mu_1, \ldots, \mu_n$ be in $[\mu', \mu'']$ where μ' and μ'' are real numbers that satisfy $0 \le \mu'' - \mu' < \pi/2$. Then*

$$\left| \sum_{k=1}^{n} a_k e^{i\mu_k} \right| \ge \sqrt{\cos(\mu'' - \mu')} \left(\sum_{k=1}^{n} a_k \right) > 0.$$

Proof. We have

$$\left| \sum_{k=1}^{n} a_k e^{i\mu_k} \right|^2 = \left(\sum_{k=1}^{n} a_k \cos \mu_k \right)^2 + \left(\sum_{k=1}^{n} a_k \sin \mu_k \right)^2$$

$$= \sum_{k=1}^{n} \sum_{l=1}^{n} a_k a_l \cos(\mu_k - \mu_l)$$

$$\ge \left(\sum_{k=1}^{n} \sum_{l=1}^{n} a_k a_l \right) \cos(\mu'' - \mu') = \left(\sum_{k=1}^{n} a_k \right)^2 \cos(\mu'' - \mu')$$

The result now follows. ∎

We address stability first.

Theorem 9.4.2 *Let $\varphi(t)$ and $\{t_l\}$ be as described in Section 9.1 and assume that (9.4.1) holds. Then, for each fixed i, there holds $\lim_{n \to \infty} \gamma_{ni}^{(j)} = 0$. Specifically,*

$$\gamma_{ni}^{(j)} = O(e^{-\lambda n}) \text{ as } n \to \infty, \text{ for every } \lambda > 0. \tag{9.4.2}$$

Furthermore, Process II is stable and we have

$$\Gamma_n^{(j)} \sim 1 \text{ as } n \to \infty. \tag{9.4.3}$$

If $\operatorname{sgn}[\varphi(t_{j+1})/\varphi(t_j)] = -1$ for $j \ge J$, then $\Gamma_n^{(j)} = 1$ for $j \ge J$ as well.

Proof. Let $\operatorname{sgn} \varphi(t_l) = (-1)^l e^{i\mu_l}$. By (9.4.1), $\lim_{l \to \infty} \mu_l = \alpha$. Therefore, for arbitrary $\eta \in (0, \pi/4)$, there exists a positive integer M such that $\mu_l \in (\alpha - \eta, \alpha + \eta)$ when $l \ge M$. Let us define

$$\Sigma_1 = \sum_{i=0}^{M-j-1} c_{ni}^{(j)} \psi(t_{j+i}) \text{ and } \Sigma_2 = \sum_{i=M-j}^{n} c_{ni}^{(j)} \psi(t_{j+i}) \tag{9.4.4}$$

and

$$\Omega_1 = \sum_{i=0}^{M-j-1} |c_{ni}^{(j)}| |\psi(t_{j+i})| \text{ and } \Omega_2 = \sum_{i=M-j}^{n} |c_{ni}^{(j)}| |\psi(t_{j+i})|. \tag{9.4.5}$$

Obviously, $|\Sigma_2| \le \Omega_2$. But we also have from Lemma 9.4.1 that $0 < K\Omega_2 \le |\Sigma_2|$, $K = \sqrt{\cos 2\eta} > 0$. Next, let us define $\tau_{ik} = |t_{j+i} - t_{j+k}|$. Then

$$e_{ni}^{(j)} \equiv \left| \frac{c_{ni}^{(j)} \psi(t_{j+i})}{c_{nn}^{(j)} \psi(t_{j+n})} \right| = \left(\prod_{\substack{k=0 \\ k \ne i}}^{n-1} \frac{\tau_{nk}}{\tau_{ik}} \right) \left| \frac{\varphi(t_{j+n})}{\varphi(t_{j+i})} \right|. \tag{9.4.6}$$

Now, because $\lim_{l \to \infty} t_l = 0$, given $\epsilon > 0$, there exists a positive integer N for which $\tau_{pq} < \epsilon$ if $p, q \ge N$. Without loss of generality, we pick $\epsilon < \tau_{i,i+1}$ and $N > i$. Also, because $\{t_l\} \to 0$ monotonically, we have $\tau_{ik} \ge \tau_{i,i+1}$ for all $k \ge i+1$. Combining all this in (9.4.6), we obtain with $n > N$

$$e_{ni}^{(j)} < \left(\prod_{\substack{k=0 \\ k \ne i}}^{N-1} \frac{\tau_{nk}}{\tau_{ik}} \right) \left(\frac{\epsilon}{\tau_{i,i+1}} \right)^{n-N} \left| \frac{\varphi(t_{j+n})}{\varphi(t_{j+i})} \right|. \tag{9.4.7}$$

Since i is fixed and $\varphi(t_{j+n}) = O(t_n^\delta) = O(n^{-q\delta})$ as $n \to \infty$, and since ϵ is arbitrary, there holds

$$e_{ni}^{(j)} = O(e^{-\lambda n}) \quad \text{as } n \to \infty, \quad \text{for every } \lambda > 0. \tag{9.4.8}$$

Substituting (9.4.8) in

$$\frac{|\Sigma_1|}{|\Sigma_2|} \le \frac{\Omega_1}{K\Omega_2} \le \frac{\Omega_1}{K|c_{nn}^{(j)}| \, |\psi(t_{j+n})|} = \frac{1}{K} \sum_{i=0}^{M-j-1} e_{ni}^{(j)}, \tag{9.4.9}$$

we have $\lim_{n \to \infty} \Sigma_1 / \Sigma_2 = 0 = \lim_{n \to \infty} \Omega_1 / \Omega_2$. With the help of this, we obtain for all large n that

$$|\gamma_{ni}^{(j)}| = \frac{|c_{ni}^{(j)} / \varphi(t_{j+i})|}{|\Sigma_1 + \Sigma_2|} \le \frac{|c_{ni}^{(j)} / \varphi(t_{j+i})|}{|\Sigma_2|(1 - |\Sigma_1 / \Sigma_2|)} \le \frac{e_{ni}^{(j)}}{K(1 - |\Sigma_1 / \Sigma_2|)}, \tag{9.4.10}$$

and this, along with (9.4.8), results in (9.4.2). To prove (9.4.3), we start with

$$1 \le \Gamma_n^{(j)} = \frac{\Omega_1 + \Omega_2}{|\Sigma_1 + \Sigma_2|} \le \frac{1}{K} \frac{1 + \Omega_1 / \Omega_2}{1 - |\Sigma_1 / \Sigma_2|}, \tag{9.4.11}$$

from which we obtain $1 \le \limsup_{n \to \infty} \Gamma_n^{(j)} \le 1/K$. Since η can be picked arbitrarily close to 0, K is arbitrarily close to 1. As a result, $\limsup_{n \to \infty} \Gamma_n^{(j)} = 1$. In exactly the same way, $\liminf_{n \to \infty} \Gamma_n^{(j)} = 1$. Therefore, $\lim_{n \to \infty} \Gamma_n^{(j)} = 1$, proving (9.4.3). The last assertion follows from the fact that $\gamma_{ni}^{(j)} > 0$ for $0 \le i \le n$ when $j \ge J$. ∎

We close with the following convergence theorem that is a considerably refined version of Theorem 4.4.3.

Theorem 9.4.3 *Let $\varphi(t)$ and $\{t_l\}$ be as in Section 9.1, and let $B(t) \in C^\infty[0, \hat{t}]$ for some $\hat{t} > 0$. Then, $\lim_{n \to \infty} A_n^{(j)} = A$ whether A is the limit or antilimit of $a(t)$ as $t \to 0+$. We actually have*

$$A_n^{(j)} - A = O(n^{-\lambda}) \quad \text{as } n \to \infty, \quad \text{for every } \lambda > 0. \tag{9.4.12}$$

Proof. Following the proof of Theorem 4.4.3, we start with

$$A_n^{(j)} - A = \sum_{i=0}^{n} \gamma_{ni}^{(j)} \varphi(t_{j+i})[B(t_{j+i}) - v_{n-1}(t_{j+i})], \qquad (9.4.13)$$

where $v_m(t) = \sum_{k=0}^{m} f_k T_k(2t/\hat{t} - 1)$ is the mth partial sum of the Chebyshev series of $B(t)$ over $[0, \hat{t}]$. Denoting $V_n(t) = B(t) - v_{n-1}(t)$, we can write (9.4.13) in the form

$$A_n^{(j)} - A = \sum_{t_{j+i} > \hat{t}} \gamma_{ni}^{(j)} \varphi(t_{j+i}) V_n(t_{j+i}) + \sum_{t_{j+i} \leq \hat{t}} \gamma_{ni}^{(j)} \varphi(t_{j+i}) V_n(t_{j+i}) \qquad (9.4.14)$$

(If $t_j \leq \hat{t}$, then the first summation is empty.) By the assumption that $B(t) \in C^{\infty}[0, \hat{t}]$, we have that $\max_{t \in [0, \hat{t}]} |V_n(t)| = O(n^{-\lambda})$ as $n \to \infty$ for every $\lambda > 0$. As a result, in the second summation in (9.4.14) we have $\max_{t_{j+i} \leq \hat{t}} |V_n(t_{j+i})| = O(n^{-\lambda})$ as $n \to \infty$ for every $\lambda > 0$. Next, by (9.1.4) and $\varphi(t) = O(t^{\delta})$ as $t \to 0+$, we have that $\max_{0 \leq i \leq n} |\varphi(t_{j+i})|$ is either bounded independently of n or grows at worst like $n^{-q\Re\delta}$ (when $\Re\delta < 0$). Next, $\sum_{t_{j+i} \leq \hat{t}} |\gamma_{ni}^{(j)}| \leq \Gamma_n^{(j)} \sim 1$ as $n \to \infty$, as we showed in the preceding theorem. Combining all this, we have that

$$\sum_{t_{j+i} \leq \hat{t}} \gamma_{ni}^{(j)} \varphi(t_{j+i}) V_n(t_{j+i}) = O(n^{-\lambda}) \quad \text{as } n \to \infty, \quad \text{for every } \lambda > 0. \quad (9.4.15)$$

As for the first summation (assuming it is not empty), we first note that the number of terms in it is finite, and each of the $\gamma_{ni}^{(j)}$ there satisfies (9.4.2). The $\varphi(t_{j+i})$ there are independent of n. As for $V_n(t_{j+i})$, we have

$$\max_{t_{j+i} > \hat{t}} |V_n(t_{j+i})| \leq \max_{t_{j+i} > \hat{t}} |B(t_{j+i})| + \sum_{k=0}^{n-1} |f_k| |T_k(2t_j/\hat{t} - 1)|. \qquad (9.4.16)$$

Now, by $B(t) \in C^{\infty}[0, \hat{t}]$, $f_n = O(n^{-\sigma})$ as $n \to \infty$ for every $\sigma > 0$, and when $z \notin [-1, 1]$, $T_n(z) = O(e^{\kappa n})$ as $n \to \infty$ for some $\kappa > 0$ that depends on z. Therefore,

$$\sum_{t_{j+i} > \hat{t}} \gamma_{ni}^{(j)} \varphi(t_{j+i}) V_n(t_{j+i}) = O(e^{-\lambda n}) \quad \text{as } n \to \infty \quad \text{for every } \lambda > 0. \quad (9.4.17)$$

Combining (9.4.15) and (9.4.17) in (9.4.14), the result in (9.4.12) follows.   ∎

9.5 Can $\{t_l\}$ Satisfy (9.1.4) and (9.4.1) Simultaneously?

We return to the question we raised at the beginning of the preceding section whether the t_l can satisfy (9.1.4) and (9.4.1) simultaneously. To answer this, we start with the fact that $\Im u(t) = w(t) \sim \sum_{k=0}^{\infty} w_k t^{k-m}$ as $t \to 0+$, for some $w_0 \neq 0$ and integer $m \leq s$, and pick the t_l to be consecutive zeros in $(0, b^r]$ of $\sin \hat{w}(t)$ or of $\cos \hat{w}(t)$ or of a linear combination of them, where $\hat{w}(t) = \sum_{k=0}^{m-1} w_k t^{k-m}$. Without loss of generality, we assume that $w_0 > 0$.

Lemma 9.5.1 *Let t_0 be the largest zero in $(0, b^r)$ of $\sin(\hat{w}(t) - \lambda_0 \pi)$ for some λ_0. This means that t_0 is a solution of the equation $\hat{w}(t) - \lambda_0 \pi = \nu_0 \pi$ for some integer ν_0. Next,* let t_l be the smallest positive solution of the equation $\hat{w}(t) - \lambda_0 \pi = (v_0 + l)\pi$ for each $l \geq 1$. Then, $t_{l+1} < t_l$, $l = 0, 1, \dots$, and t_l has the (convergent) expansion

$$t_l = \sum_{i=0}^{\infty} a_i l^{-(i+1)/m} \text{ for all large } l; \ a_0 = (w_0/\pi)^{1/m} > 0, \qquad (9.5.1)$$

and hence

$$t_l - t_{l+1} = \sum_{i=0}^{\infty} \hat{a}_i l^{-1-(i+1)/m} \text{ for all large } l; \ \hat{a}_0 = \frac{a_0}{m}, \qquad (9.5.2)$$

so that

$$t_l \sim a_0 l^{-1/m} \text{ and } t_l - t_{l+1} \sim \hat{a}_0 l^{-1-1/m} \text{ as } l \to \infty. \qquad (9.5.3)$$

Proof. We start by observing that $\hat{w}(t) \sim w_0 t^{-m}$ and hence $\hat{w}(t) \to +\infty$ as $t \to 0+$, as a result of which the equation $\hat{w}(t) - \lambda_0 \pi = \xi$, for all sufficiently large ξ, has a unique solution $t(\xi)$ that is positive and satisfies $t(\xi) \sim (w_0/\xi)^{1/m}$ as $\xi \to \infty$. Therefore, being a solution of $\hat{w}(t) - \lambda_0 \pi = (v_0 + l)\pi$, t_l is unique for all large l and satisfies $t_l \sim (w_0/\pi)^{1/m} l^{-1/m}$ as $l \to \infty$. Letting $\epsilon = l^{-1/m}$ and substituting $t_l = \epsilon \tau$ in this equation, we see that τ satisfies the polynomial equation $\sum_{k=0}^{m} c_k \tau^k = 0$, where $c_m = 1$ and $c_k = -w_k \epsilon^k / [\pi + \pi(v_0 + \lambda_0)\epsilon^m]$, $k = 0, 1, \dots, m-1$. Note that the coefficients $c_0, \dots, c_{m-1}$ are analytic functions of ϵ about $\epsilon = 0$. Therefore, τ is an analytic function of ϵ about $\epsilon = 0$, and, for all ϵ sufficiently close to zero, there exists a convergent expansion of the form $\tau = \sum_{i=0}^{\infty} a_i \epsilon^i$. The a_i can be obtained by substituting this expansion in $\sum_{k=0}^{m} c_k \tau^k = 0$ and equating the coefficients of the powers ϵ^i to zero for each i. This proves (9.5.1). We can obtain (9.5.2) as a direct consequence of (9.5.1). The rest is immediate. ∎

Clearly, the t_l constructed as in Lemma 9.5.1 satisfy (9.1.4) with $c = (w_0/\pi)^{1/m}$ and $p = q = 1/m$. They also satisfy (9.4.1), because $\exp[u(t_l)] \sim (-1)^{l+v_0} e^{i(w_m + \lambda_0 \pi)} e^{\Re u(t_l)}$ as $l \to \infty$. As a result, when δ is real, $\varphi(t)$ satisfies (9.4.1) with these t_l. We leave the details to the reader.

Efficient Use of GREP$^{(1)}$: Applications to the $D^{(1)}$-, $d^{(1)}$-, and $\tilde{d}^{(m)}$-Transformations

10.1 Introduction

In the preceding two chapters, we presented a detailed analysis of the convergence and stability properties of GREP$^{(1)}$. In this analysis, we considered *all* possible forms of $\varphi(t)$ that may arise from infinite-range integrals of functions in $\mathbf{B}^{(1)}$ and infinite series whose terms form sequences in $\mathbf{b}^{(1)}$ and $\tilde{\mathbf{b}}^{(m)}$. We also considered various forms of $\{t_l\}$ that have been used in applications. In this chapter, we discuss the practical implications of the results of Chapters 8 and 9 and derive operational conclusions about how the $D^{(1)}$-, $d^{(1)}$-, and $\tilde{d}^{(m)}$-transformations should be used to obtain the best possible outcome in different situations involving slowly or quickly varying $a(t)$. It is worth noting again that the conclusions we derive here and that result from our analysis of GREP$^{(1)}$ appear to be valid in many situations involving GREP$^{(m)}$ with $m > 1$ as well.

As is clear from Chapters 8 and 9, GREP$^{(1)}$ behaves in completely different ways depending on whether $\varphi(t)$ varies slowly or quickly as $t \to 0+$. This implies that different strategies are needed for these two classes of $\varphi(t)$. In the next two sections, we dwell on this issue and describe the possible strategies pertinent to the $D^{(1)}$-, $d^{(1)}$-, and $\tilde{d}^{(m)}$-transformations.

Finally, the conclusions that we draw in this chapter concerning the $d^{(1)}$- and $\tilde{d}^{(m)}$-transformations are relevant for other sequence transformations, as will become clear later in this book.

10.2 Slowly Varying $a(t)$

We recall that $a(t)$ is slowly varying when $\varphi(t) \sim \sum_{i=0}^{\infty} h_i t^{\delta+i}$ as $t \to 0+$, $h_0 \neq 0$, and δ may be complex such that $\delta \neq 0, -1, -2, \ldots$. We also advise the reader to review the examples of slowly varying $a(t)$ in Section 8.2.

In Chapter 8, we discussed the application of GREP$^{(1)}$ to slowly varying $a(t)$ with two major choices of $\{t_l\}$: (i) $\lim_{l\to\infty}(t_{l+1}/t_l) = 1$ and (ii) $\nu \leq t_{l+1}/t_l \leq \omega$, $l = 0, 1, \ldots$, for $0 < \nu \leq \omega < 1$. We now consider each of these choices separately.

10.2.1 Treatment of the Choice $\lim_{l\to\infty}(t_{l+1}/t_l) = 1$

The various results concerning the case $\lim_{l\to\infty}(t_{l+1}/t_l) = 1$ show that numerical instabilities occur in the computation of the $A_n^{(j)}$ in finite-precision arithmetic; that is, the precision of the computed $A_n^{(j)}$ is less than the precision with which the $a(t_l)$ are computed, and it deteriorates with increasing j and/or n. The easiest way to remedy this problem is to increase (e.g., double) the accuracy of the finite-precision arithmetic that is used, without changing $\{t_l\}$.

When we are not able to increase the accuracy of our finite-precision arithmetic, we can deal with the problem by changing $\{t_l\}$ suitably. For example, when $\{t_l\}$ is chosen to satisfy (8.4.9), namely,

$$t_l \sim cl^{-q} \quad \text{and} \quad t_l - t_{l+1} \sim cpl^{-q-1} \quad \text{as } l \to \infty, \quad \text{for some } c > 0, \ p > 0, \text{ and } q > 0,$$
(10.2.1)

then, by Theorem 8.4.3, we can increase p to make $\Gamma_n^{(j)}$ smaller because $\Gamma_n^{(j)}$ is proportional to p^{-n} as $j \to \infty$. Now, the easiest way of generating such $\{t_l\}$ is by taking $t_l = c/(l+\eta)^q$ with some $\eta > 0$. In this case $p = q$, as can easily be shown; hence, we increase q to make $\Gamma_n^{(j)}$ smaller for fixed n and increasing j, even though we still have $\lim_{j\to\infty} \Gamma_n^{(j)} = \infty$. Numerical experience suggests that, by increasing q, we make $\Gamma_n^{(j)}$ smaller also for fixed j and increasing n, even though $\lim_{n\to\infty} \Gamma_n^{(j)} = \infty$, at least in the cases described in Theorems 8.4.4 and 8.4.8.

In applying the $D^{(1)}$-transformation to the integral $\int_0^\infty f(t)\,dt$ in Example 8.2.1 with this strategy, we can choose the x_l according to $x_l = (l+\eta)^q/c$ with arbitrary $c > 0$, $q \geq 1$, and $\eta > 0$ without any problem since the variable t is continuous in this case. With this choice, we have $p = q$.

When we apply the $d^{(1)}$-transformation to the series $\sum_{k=1}^\infty a_k$ in Example 8.2.2, however, we have to remember that t now is discrete and takes on the values $1, 1/2, 1/3, \ldots$, only, so that $t_l = 1/R_l$, with $\{R_l\}$ being an increasing sequence of positive integers. Similarly, when we apply the $\tilde{d}^{(m)}$-transformation to the series $\sum_{k=1}^\infty a_k$ in Example 8.2.3, t takes on the discrete values $1, 1/2^{1/m}, 1/3^{1/m}, \ldots$, so that $t_l = 1/R_l^{1/m}$, with $\{R_l\}$ being again an increasing sequence of positive integers. The obvious question then is whether we can pick $\{R_l\}$ such that the corresponding $\{t_l\}$ satisfies (10.2.1). We can, of course, maintain (10.2.1) with $R_l = \kappa(l+1)^r$, where κ and r are both positive integers. But this causes R_l to increase very rapidly when $r = 2, 3, \ldots$, thus increasing the cost of extrapolation considerably. In view of this, we may want to have R_l increase like l^r with smaller (hence noninteger) values of $r \in (1, 2)$, for example, if this is possible at all. To enable this in both applications, we propose to choose the R_l as follows:

pick $\kappa > 0$, $r \geq 1$, and $\eta > 0$, and the integer $R_0 \geq 1$, and set
$$R_l = \begin{cases} R_{l-1} + 1 & \text{if } \lfloor \kappa(l+\eta)^r \rfloor \leq R_{l-1}, \\ \lfloor \kappa(l+\eta)^r \rfloor & \text{otherwise}, \end{cases} \quad l = 1, 2, \ldots .$$
(10.2.2)

Note that κ, η, and r need *not* be integers now.

Concerning these R_l, we have the following results.

Lemma 10.2.1 *Let $\{R_l\}$ be as in (10.2.2), with arbitrary $\kappa > 0$ when $r > 1$ and with $\kappa \geq 1$ when $r = 1$. Then, $R_l = \lfloor \kappa(l+\eta)^r \rfloor$ for all sufficiently large l, from which it follows that $\{R_l\}$ is an increasing sequence and that $R_l \sim \kappa l^r$ as $l \to \infty$. In addition,*

$$R_l = \kappa l^r + \kappa r \eta l^{r-1} + o(l^{r-1}) \quad \text{as } l \to \infty, \quad \text{if } r > 1,$$
(10.2.3)

and

$$R_l = \kappa l + R_0, \quad l = 1, 2, \ldots, \quad \text{if } r = 1 \text{ and } \kappa = 1, 2, \ldots . \quad (10.2.4)$$

Proof. That $\{R_l\}$ is an increasing sequence and $R_l = \lfloor \kappa(l+\eta)^r \rfloor$ for all sufficiently large l is obvious. Next, making use of the fact that $x - 1 < \lfloor x \rfloor \leq x$, we can write

$$\kappa(l+\eta)^r - 1 < R_l \leq \kappa(l+\eta)^r.$$

Dividing these inequalities by κl^r and taking the limit as $l \to \infty$, we obtain $\lim_{l\to\infty} [R_l/(\kappa l^r)] = 1$, from which $R_l \sim \kappa l^r$ as $l \to \infty$ follows. To prove (10.2.3), we proceed as follows: First, we have

$$\kappa(l+\eta)^r = \kappa l^r + \kappa r \eta l^{r-1} + \rho_l; \ \rho_l \equiv \kappa \frac{r(r-1)}{2} \eta^2 (1+\tilde\theta)^{r-2} l^{r-2} \text{ for some } \tilde\theta \in (0, \eta/l).$$

Note that $\kappa r \eta l^{r-1} \to \infty$ as $l \to \infty$, and that $\rho_l > 0$ for $l \geq 1$. Next, from

$$\kappa l^r + \kappa r \eta l^{r-1} + \rho_l - 1 < R_l \leq \kappa l^r + \kappa r \eta l^{r-1} + \rho_l.$$

we obtain $|R_l - (\kappa l^r + \kappa r \eta l^{r-1})| < \rho_l + 1$, which implies (10.2.3). Finally, the result in (10.2.4) is a trivial consequence of the fact that κl is an integer under the conditions imposed on κ and r there. ∎

The following lemma is a direct consequence of Lemma 10.2.1. We leave its proof to the reader.

Lemma 10.2.2 *Let $\{R_l\}$ be as in (10.2.2) with $r > 1$, or with $r = 1$ and $\kappa = 1, 2, \ldots$. Then, the following assertions hold:*

(i) *If $t_l = 1/R_l$, then (10.2.1) is satisfied with $c = 1/\kappa$ and $p = q = r$.*
(ii) *If $t_l = 1/R_l^{1/m}$ where m is a positive integer, then (10.2.1) is satisfied with $c = (1/\kappa)^{1/m}$ and $p = q = r/m$.*

The fact that the t_l in Lemma 10.2.2 satisfy (10.2.1) guarantees that Theorems 8.4.1 and 8.4.3 hold for the $d^{(1)}$- and $\tilde{d}^{(m)}$-transformations as these are applied to Example 8.2.2 and Example 8.2.3, respectively.

Note that the case in which R_l is as in (10.2.2) but with $r = 1$ and $\kappa \geq 1$ not an integer is not covered in Lemmas 10.2.1 and 10.2.2. In this case, we have instead of (10.2.4)

$$R_l = \kappa l + \epsilon_l, \quad l = 0, 1, \ldots; \quad \kappa\eta - 1 \leq \epsilon_l \leq \kappa\eta. \quad (10.2.5)$$

As a result, we also have instead of (10.2.1)

$$t_l \sim cl^{-q} \text{ as } l \to \infty \text{ and } K_1 l^{-q-1} \leq t_l - t_{l+1} \leq K_2 l^{-q-1} \text{ for some } K_1, K_2 > 0, \quad (10.2.6)$$

with $q = 1$ when $t_l = 1/R_l$ and with $q = 1/m$ when $t_l = 1/R_l^{1/m}$. In this case too, Theorem 8.4.1 holds when the $d^{(1)}$- and $\tilde{d}^{(m)}$-transformations are applied to Example 8.2.2 and Example 8.2.3, respectively. As for Theorem 8.4.3, with the exception of (8.4.17), its results remain unchanged, and (8.4.17) now reads

$$G_{n1} j^n \leq \Gamma_n^{(j)} \leq G_{n2} j^n \quad \text{for some } G_{n1}, G_{n2} > 0, \ n \text{ fixed.} \tag{10.2.7}$$

All this can be shown by observing that Lemma 8.4.2 remains unchanged and that $L_{n1}(j/t_j)^n \leq |c_{ni}^{(j)}| \leq L_{n2}(j/t_j)^n$ for some $L_{n1}, L_{n2} > 0$, which follows from (10.2.6). We leave the details to the interested reader.

From the preceding discussion, it follows that both the $d^{(1)}$- and the $\tilde{d}^{(m)}$-transformations can be made effective by picking the R_l as in (10.2.2) with a suitable and moderate value of r. In practice, we can start with $r = 1$ and increase it gradually if needed. This strategy enables us to increase the accuracy of $A_n^{(j)}$ for j or n large and also improve its numerical stability, because it causes the corresponding $\Gamma_n^{(j)}$ to decrease. That is, by increasing r we can achieve more accuracy in the computed $A_n^{(j)}$, even when we are limited to a fixed precision in our arithmetic. Now, the computation of $A_n^{(j)}$ involves the terms a_k, $1 \leq k \leq R_{j+n}$, of $\sum_{k=1}^{\infty} a_k$ as it is defined in terms of the partial sums A_{R_l}, $j \leq l \leq j+n$. Because R_l increases like the power l^r, we see that by increasing r gradually, and not necessarily through integer values, we are able to increase the number of the terms a_k used for computing $A_n^{(j)}$ gradually as well. Obviously, this is an advantage offered by the choice of the R_l as in (10.2.2), with r not necessarily an integer.

Finally, again from Theorem 8.4.3, both $\Gamma_n^{(j)}$ and $|A_n^{(j)} - A|$ are inversely proportional to $|(\delta)_n|$. This suggests that it is easier to extrapolate when $|(\delta)_n|$ is large, as this causes both $\Gamma_n^{(j)}$ and $|A_n^{(j)} - A|$ to become small. One practical situation in which this becomes relevant is that of small $|\Re\delta|$ but large $|\Im\delta|$. Here, the larger $|\Im\delta|$, the better the convergence and stability properties of $A_n^{(j)}$ for $j \to \infty$, despite the fact that $A_n^{(j)} - A = O(\varphi(t_j)t_j^{\Re\delta+n})$ as $j \to \infty$ for *every* value of $\Im\delta$. Thus, extrapolation is easier when $|\Im\delta|$ is large. (Again, numerical experience suggests that this is so both for Process I and for process II.) Consequently, in applying the $D^{(1)}$-transformation to $\int_0^\infty f(t)\,dt$ in Example 8.2.1 with large $|\Im\gamma|$, it becomes sufficient to use $x_l = \kappa(l + \eta)^q$ with a low value of q; e.g., $q = 1$. Similarly, in applying the $d^{(1)}$-transformation in Example 8.2.2 with large $|\Im\gamma|$, it becomes sufficient to choose R_l as in (10.2.2) with a low value of r; e.g., $r = 1$ or slightly larger. The same can be achieved in applying the $\tilde{d}^{(m)}$-transformation in Example 8.2.3, taking $r/m = 1$ or slightly larger.

10.2.2 Treatment of the Choice $0 < \nu \leq t_{l+1}/t_l \leq \omega < 1$

Our results concerning the case $0 < \nu \leq t_{l+1}/t_l \leq \omega < 1$ for all l show that the $A_n^{(j)}$ can be computed stably in finite-precision arithmetic with such a choice of $\{t_l\}$. Of course, $t_0 \nu^l \leq t_l \leq t_0 \omega^l$, $l = 0, 1, \ldots$, and, therefore, $t_l \to 0$ as $l \to \infty$ exponentially in l.

In applying the $D^{(1)}$-transformation to $\int_0^\infty f(t)\,dt$, we can choose the x_l according to $x_l = \xi/\omega^l$ for some $\xi > 0$ and $\omega \in (0, 1)$. With this choice and by $t_l = 1/x_l$, we have $\lim_{l\to\infty}(t_{l+1}/t_l) = \omega$, and Theorems 8.5.1 and 8.5.2 apply. As is clear from (8.5.4) and (8.5.6), $\Gamma_n^{(j)}$, despite its boundedness both as $j \to \infty$ and as $n \to \infty$, may become large if some of the c_i there are very close to $0, -1, -2, \ldots$. Suppose, for instance, that $\delta \approx 0$. Then, $c_1 = \omega^\delta$ is close to 1 if we choose ω close to 1. In this case, we can cause c_1 to separate from 1 by making ω smaller, *whether δ is complex or not*.

In applying the $d^{(1)}$- and $\tilde{d}^{(m)}$-transformations with the present choice of $\{t_l\}$, we should again remember that t is discrete and $t_l = 1/R_l$ for the $d^{(1)}$-transformation and

$t_l = 1/R_l^{1/m}$ for the $\tilde{d}^{(m)}$-transformation, and $\{R_l\}$ is an increasing sequence of positive integers. Thus, the requirement that $t_l \to 0$ exponentially in l forces that $R_l \to \infty$ exponentially in l. This, in turn, implies that the number of terms of the series $\sum_{k=1}^{\infty} a_k$ required for computing $A_n^{(j)}$, namely, the integer R_{j+n}, grows exponentially with $j + n$. To keep this growth to a reasonable and economical level, we should aim at achieving $R_l = O(\sigma^l)$ as $l \to \infty$ for some reasonable $\sigma > 1$ that is not necessarily an integer. The following choice, which is essentially due to Ford and Sidi [87, Appendix B], has proved very useful:

$$\text{pick the scalar } \sigma > 1 \text{ and the integer } R_0 \geq 1, \text{ and set}$$
$$R_l = \begin{cases} R_{l-1} + 1 \text{ if } \lfloor \sigma R_{l-1} \rfloor = R_{l-1}, \\ \lfloor \sigma R_{l-1} \rfloor \text{ otherwise,} \end{cases} \quad l = 1, 2, \ldots . \quad (10.2.8)$$

(The R_l given in [87] are slightly different but have the same asymptotic properties, which is the most important aspect.)

The next two lemmas, whose proofs we leave to the reader, are analogous to Lemmas 10.2.1 and 10.2.2. Again, the fact that $x - 1 < \lfloor x \rfloor \leq x$ becomes useful in part of the proof.

Lemma 10.2.3 *Let $\{R_l\}$ be as in (10.2.8). Then, $R_l = \lfloor \sigma R_{l-1} \rfloor$ for all sufficiently large l, from which it follows that $g\sigma^l \leq R_l \leq \sigma^l$ for some $g \leq 1$. Thus, $\{R_l\}$ is an exponentially increasing sequence of integers that satisfies $\lim_{l\to\infty} (R_{l+1}/R_l) = \sigma$.*

Lemma 10.2.4 *Let $\{R_l\}$ be as in (10.2.8). Then, the following assertions hold:*

(i) *If $t_l = 1/R_l$, then $\lim_{l\to\infty}(t_{l+1}/t_l) = \omega$ with $\omega = \sigma^{-1} \in (0, 1)$.*
(ii) *If $t_l = 1/R_l^{1/m}$, where m is a positive integer, then $\lim_{l\to\infty}(t_{l+1}/t_l) = \omega$ with $\omega = \sigma^{-1/m} \in (0, 1)$.*

The fact that the t_l in Lemma 10.2.4 satisfy $\lim_{l\to\infty}(t_{l+1}/t_l) = \omega$ with $\omega \in (0, 1)$ guarantees that Theorems 8.5.1, 8.5.2, 8.6.1, 8.6.4, 8.6.6, and 8.6.7 hold for the $d^{(1)}$- and $\tilde{d}^{(m)}$-transformations, as these are applied to Examples 8.2.2 and 8.2.3, respectively. Again, in case $\Gamma_n^{(j)}$ is large, we can make it smaller by decreasing ω (equivalently, by increasing σ).

Before closing this section, we recall that the $d^{(m)}$- and $\tilde{d}^{(m)}$-transformations, by their definitions, are applied to subsequences $\{A_{R_l}\}$ of $\{A_l\}$. This amounts to *sampling* the sequence $\{A_l\}$. Let us consider now the choice of the R_l as in (10.2.8). Then the sequence $\{R_l\}$ grows as a geometric progression. On the basis of this, we refer to this choice of the R_l as the *geometric progression sampling* and denote it GPS for short.

10.3 Quickly Varying $a(t)$

We recall that $a(t)$ is quickly varying when $\varphi(t)$ is of one of three forms: (a) $\varphi(t) = e^{u(t)}h(t)$, or (b) $\varphi(t) = [\Gamma(t^{-\hat{s}})]^{-\nu}h(t)$, or (c) $\varphi(t) = [\Gamma(t^{-\hat{s}})]^{-\nu}e^{u(t)}h(t)$, where $h(t) \sim h_0 t^\delta$ as $t \to 0+$ for some arbitrary and possibly complex δ, $u(t) \sim \sum_{i=0}^{\infty} u_i t^{i-s}$ as $t \to 0+$ for some positive integer s, $\Gamma(z)$ is the Gamma function, $\hat{s}$ is a positive integer,

$\nu > 0$, and $\hat{s} \geq s$ in case (c). In addition, $\varphi(t)$ may be oscillatory and/or decaying as $t \to 0+$ and $|\varphi(t)| = O(t^{\Re\delta})$ as $t \to 0+$ at worst. At this point, we advise the reader to review the examples of quickly varying $a(t)$ given in Section 9.2.

In Chapter 9, we discussed the application of GREP$^{(1)}$ to quickly varying $a(t)$ with choices of $\{t_l\}$ that satisfy (10.2.1) only and were able to prove that both convergence and stability prevail with such $\{t_l\}$. Numerical experience shows that other choices of $\{t_l\}$ are not necessarily more effective.

We first look at the application of the $D^{(1)}$-transformation to the integral $\int_0^\infty f(t)\,dt$ in Example 9.2.1. The best strategy appears to be the one that enables the conditions of Theorems 9.4.2 and 9.4.3 to be satisfied. In this strategy, we choose the x_l in accordance with Lemma 9.5.1. Thus, if $\Im\theta(x) \sim \sum_{i=0}^\infty w_i x^{m-i}$ as $x \to \infty$, $w_0 > 0$, in Example 9.2.1, we set $\bar{w}(x) = \sum_{i=0}^{m-1} w_i x^{m-i}$ and choose x_0 to be the smallest zero greater than a of $\sin(\bar{w}(x) - \lambda_0\pi)$ for some λ_0. Thus, x_0 is a solution of the polynomial equation $\bar{w}(x) - \lambda_0\pi = \nu_0\pi$ for some integer ν_0. Once x_0 has been found, x_l for $l = 1, 2, \dots$, is determined as the largest root of the polynomial equation $\bar{w}(x) - \lambda_0\pi = (\nu_0 + l)\pi$. Determination of $\{x_l\}$ is an easy task since $\bar{w}(x)$ is a polynomial; it is easiest when $m = 1$ or $m = 2$. If we do not want to bother with extracting $\bar{w}(x)$ from $\theta(x)$, then we can apply the preceding strategy directly to $\Im\theta(x)$ instead of $\bar{w}(x)$. All this forms the basis of the W-transformation for oscillatory infinite-range integrals that we introduce later in this book.

Let us consider the application of the $d^{(1)}$- and $\tilde{d}^{(m)}$-transformations to the series $\sum_{k=1}^\infty a_k$ in Examples 9.2.2 and 9.2.3, respectively. We again recall that $t_l = 1/R_l$ for the $d^{(1)}$-transformation and $t_l = 1/R_l^{1/m}$ for the $\tilde{d}^{(m)}$-transformation, where $\{R_l\}$ is an increasing sequence of integers. We choose the R_l as in (10.2.2), because we already know from Lemma 10.2.2 that, with this choice of $\{R_l\}$, t_l satisfies (10.2.1) when $r > 1$, or when $r = 1$ but κ an η are integers there. (When $r = 1$ but κ or η is not an integer, we have only $t_l \sim cl^{-q}$ as $l \to \infty$, as mentioned before.) We have only to fix the parameter q as described in Theorems 9.3.1–9.3.3.

Finally, we note that when R_l are as in (10.2.2) with $r = 1$, the sequence $\{R_l\}$ grows as an arithmetic progression. On the basis of this, we refer to this choice of the R_l as the *arithmetic progression sampling* and denote it APS for short. We make use of APS in the treatment of power series, Fourier series, and generalized Fourier series in Chapters 12 and 13.

APS with integer κ was originally suggested in Levin and Sidi [165] and was incorporated in the computer program of Ford and Sidi [87]. It was studied rigorously in Sidi [294], where it was shown when and how to use it most efficiently.

11

Reduction of the D-Transformation for Oscillatory Infinite-Range Integrals: The $\bar{D}$-, $\tilde{D}$-, W-, and mW-Transformations

11.1 Reduction of GREP for Oscillatory $A(y)$

Let us recall that, when $A(y) \in \mathbf{F}^{(m)}$ for some m, we have

$$A(y) \sim A + \sum_{k=1}^{m} \phi_k(y) \sum_{i=0}^{\infty} \beta_{ki} y^{ir_k} \quad \text{as} \quad y \to 0+, \tag{11.1.1}$$

where A is the limit or antilimit of $A(y)$ as $y \to 0+$ and $\phi_k(y)$ are known shape functions that contain the asymptotic behavior of $A(y)$ as $y \to 0+$. Consider the approximations $A_{(v,\ldots,v)}^{(m,0)} \equiv C_v$, $v = 1, 2, \ldots$, that are produced by GREP$^{(m)}$ as the latter is applied to $A(y)$. We consider the sequence $\{C_v\}_{v=1}^{\infty}$, because it has excellent convergence properties. Now C_v is defined via the linear system

$$A(y_l) = C_v + \sum_{k=1}^{m} \phi_k(y_l) \sum_{i=0}^{v-1} \bar{\beta}_{ki} y_l^{ir_k}, \quad l = 0, 1, \ldots, mv, \tag{11.1.2}$$

and is, heuristically, the result of "eliminating" the mv terms $\phi_k(y) y^{ir_k}$, $i = 0, 1, \ldots, v-1$, $k = 1, \ldots, m$, from the asymptotic expansion of $A(y)$ given in (11.1.1). Thus, the number of the $A(y_l)$ needed to achieve this "elimination" process is $mv + 1$.

From this discussion, we conclude that, the smaller the value of m, the cheaper the extrapolation process. [By "cheaper" we mean that the number of function values $A(y_l)$ needed to "eliminate" v terms from each $\beta_k(y)$ is smaller.]

It turns out that, for functions $A(y) \in \mathbf{F}^{(m)}$ that oscillate an infinite number of times as $y \to 0+$, by choosing $\{y_l\}$ judiciously, we are able to *reduce* GREP, which means that we are able to use GREP$^{(q)}$ with suitable $q < m$ to approximate A, the limit or antilimit of $A(y)$ as $y \to 0+$, thus saving a lot in the computation of $A(y)$. This is done as follows:

As $A(y)$ oscillates an infinite number of times as $y \to 0+$, we have that at least one of the form factors $\phi_k(y)$ vanishes at an infinite number of points $\bar{y}_l$, such that $\bar{y}_0 > \bar{y}_1 > \cdots > 0$ and $\lim_{l \to \infty} \bar{y}_l = 0$. Suppose exactly $m - q$ of the $\phi_k(y)$ vanish on the set $\bar{Y} = \{\bar{y}_0, \bar{y}_1, \ldots\}$. Renaming the remaining q shape functions $\phi_k(y)$ if necessary, we have that

$$A(y) = A + \sum_{k=1}^{q} \phi_k(y) \beta_k(y), \quad y \in \bar{Y} = \{\bar{y}_0, \bar{y}_1, \ldots\}. \tag{11.1.3}$$

In other words, when y is a *discrete* variable restricted to the set $\bar{Y}$, $A(y) \in \mathbf{F}^{(q)}$ with

$q < m$. Consequently, choosing $\{y_l\} \subseteq \bar{Y}$, we are able to apply GREP$^{(q)}$ to $A(y)$ and obtain good approximations to A, even though $A(y) \in \mathbf{F}^{(m)}$ to begin with.

Example 11.1.1 The preceding idea can be illustrated very simply via Examples 4.1.5 and 4.1.6 of Chapter 4. We have, with $A(y) \leftrightarrow F(x) = \int_0^x (\sin t/t) \, dt$ and $A \leftrightarrow I = F(\infty)$, that

$$F(x) = I + \frac{\cos x}{x} H_1(x) + \frac{\sin x}{x} H_2(x), \quad H_1, H_2 \in \mathbf{A}^{(0)}.$$

Thus, $A(y) \in \mathbf{F}_\infty^{(2)}$ with $y \leftrightarrow x^{-1}$, $\phi_1(y) \leftrightarrow \cos x/x$, and $\phi_2(y) \leftrightarrow \sin x/x$. Here, y is continuous.

Obviously, both $\phi_1(y) = y \cos(1/y)$ and $\phi_2(y) = y \sin(1/y)$ oscillate an infinite number of times as $y \to 0+$, and $\phi_2(y) = 0$ when $y = \bar{y}_i = 1/[(i+1)\pi]$, $i = 0, 1, \ldots$. Thus, when $y \in \bar{Y} = \{\bar{y}_0, \bar{y}_1, \ldots\}$, $A(y) \in \mathbf{F}_\infty^{(1)}$ with $A(y) = I + y \cos(1/y)\beta_1(y)$.

Example 11.1.2 As another illustration, consider the infinite integral $I = \int_0^\infty J_0(t) \, dt$ that was considered in Example 5.1.13. With $F(x) = \int_0^x J_0(t) \, dt$, we already know that

$$F(x) = I + x^{-1} J_0(x)g_0(x) + J_1(x)g_1(x), \quad g_0, g_1 \in \mathbf{A}^{(0)}.$$

Thus, $A(y) \in \mathbf{F}_\infty^{(2)}$ with $y \leftrightarrow x^{-1}$, $\phi_1(y) \leftrightarrow J_0(x)/x$, $\phi_2(y) \leftrightarrow J_1(x)$, and $A \leftrightarrow I$.

Let $\bar{y}_l = x_l^{-1}$, $l = 0, 1, \ldots$, where x_l are the consecutive zeros of $J_0(x)$ that are greater than 0. Thus, when $y \in \bar{Y} = \{\bar{y}_0, \bar{y}_1, \ldots\}$, $A(y) \in \mathbf{F}_\infty^{(1)}$ with $A(y) = I + J_1(1/y)\beta_2(y)$

The purpose of this chapter is to derive *reductions* of the D-transformation for integrals $\int_0^\infty f(t) \, dt$ whose integrands have an infinite number of oscillations at infinity in the way described. The reduced forms thus obtained have proved to be extremely efficient in computing, among others, integral transforms with oscillatory kernels, such as Fourier, Hankel, and Kontorovich–Lebedev transforms. The importance and usefulness of asymptotic analysis in deriving these economical extrapolation methods are demonstrated several times throughout this chapter.

Recall that we are taking $\alpha = 0$ in the definition of the D-transformation; we do so with its reductions as well. In case the integral to be computed is $\int_a^\infty f(t) \, dt$ with $a \neq 0$, we apply these reductions to the integral $\int_0^\infty \tilde{f}(t) \, dt$, where $\tilde{f}(x) = f(a + x)$ for $x \geq 0$.

11.1.1 Review of the W-Algorithm for Infinite-Range Integrals

As we will see in the next sections, the following linear systems arise from various reductions of the D-transformation for some commonly occurring oscillatory infinite-range integrals.

$$F(x_l) = A_n^{(j)} + \psi(x_l) \sum_{i=0}^{n-1} \frac{\bar{\beta}_i}{x_l^i}, \quad j \leq l \leq j + n. \tag{11.1.4}$$

Here, $F(x) = \int_0^x f(t) \, dt$ and $I[f] = \int_0^\infty f(t) \, dt$, and x_l and the form factor $\psi(x_l)$ depend on the method being used.

The W-algorithm for these equations assumes the following form:

1. For $j = 0, 1, \ldots$, set

$$M_0^{(j)} = \frac{F(x_j)}{\psi(x_j)}, \quad N_0^{(j)} = \frac{1}{\psi(x_j)}, \quad H_0^{(j)} = (-1)^j |N_0^{(j)}|, \quad K_0^{(j)} = (-1)^j |M_0^{(j)}|.$$

2. For $j = 0, 1, \ldots$, and $n = 1, 2, \ldots$, compute

$$Q_n^{(j)} = \frac{Q_{n-1}^{(j+1)} - Q_{n-1}^{(j)}}{x_{j+n}^{-1} - x_j^{-1}},$$

where $Q_n^{(j)}$ stand for $M_n^{(j)}$ or $N_n^{(j)}$ or $H_n^{(j)}$ or $K_n^{(j)}$.

3. For all j and n, set

$$A_n^{(j)} = \frac{M_n^{(j)}}{N_n^{(j)}}, \quad \Gamma_n^{(j)} = \left| \frac{H_n^{(j)}}{N_n^{(j)}} \right|, \quad \text{and} \quad \Lambda_n^{(j)} = \left| \frac{K_n^{(j)}}{N_n^{(j)}} \right|.$$

11.2 The $\bar{D}$-Transformation

11.2.1 Direct Reduction of the D-Transformation

We begin by reducing the D-transformation directly. When $f \in \mathbf{B}^{(m)}$ and is as in Theorem 5.1.12, there holds

$$F(x) = I[f] + \sum_{k=0}^{m-1} x^{\bar{\rho}_k} f^{(k)}(x) g_k(x), \tag{11.2.1}$$

where $F(x) = \int_0^x f(t)\, dt$, $I[f] = \int_0^\infty f(t)\, dt$, and $\bar{\rho}_k \leq k + 1$ are some integers that satisfy (5.1.9), and $g_k \in \mathbf{A}^{(0)}$. [Note that we have replaced the ρ_k in (5.1.8) by $\bar{\rho}_k$, which is legitimate since $\bar{\rho}_k \geq \rho_k$ for each k.]

Suppose that $f(x)$ oscillates an infinite number of times as $x \to \infty$ and that there exist x_l, $0 < x_0 < x_1 < \cdots$, $\lim_{l \to \infty} x_l = \infty$, for which

$$f^{(k_i)}(x_l) = 0, \quad l = 0, 1, \ldots; \quad 0 \leq k_1 < k_2 < \cdots < k_p \leq m - 1. \tag{11.2.2}$$

Letting $E = \{0, 1, \ldots, m - 1\}$ and $E_p = \{k_1, \ldots, k_p\}$ and $X = \{x_0, x_1, \ldots\}$, (11.2.1) then becomes

$$F(x) = I[f] + \sum_{k \in E \setminus E_p} x^{\bar{\rho}_k} f^{(k)}(x) g_k(x), \quad x \in X. \tag{11.2.3}$$

Using (11.2.3), Sidi [274] proposed the $\bar{D}$-transformation as in the following definition.

Definition 11.2.1 Let $f(x)$, E, E_p, and X be as before, and denote $q = m - p$. Then, the $\bar{D}^{(q)}$-transformation for the integral $\int_0^\infty f(t)\, dt$ is defined via the linear system

$$F(x_l) = \bar{D}_{\bar{n}}^{(q,j)} + \sum_{k \in E \setminus E_p} x^{\sigma_k} f^{(k)}(x) \sum_{i=0}^{n_k - 1} \frac{\bar{\beta}_{ki}}{x_l^i}, \quad j \leq l \leq j + \bar{N}; \quad \bar{N} = \sum_{k \in E \setminus E_p} n_k,$$

$$\tag{11.2.4}$$

where $\bar{n} = (n_{s_1}, n_{s_2}, \ldots, n_{s_q})$ with $\{s_1, s_2, \ldots, s_q\} = E \backslash E_p$. When the $\bar{\rho}_k$ in (11.1.1) are known, we can choose $\sigma_k = \bar{\rho}_k$; otherwise, we can take for σ_k any known upper bound of $\bar{\rho}_k$. In case nothing is known about $\bar{\rho}_k$, we can take $\sigma_k = k + 1$, as was done in the original definition of the D-transformation.

It is clear from Definition 11.2.1 that the $\bar{D}$-transformation, just as the D-transformation, can be implemented by the W-algorithm (when $q = 1$) and by the $W^{(q)}$-algorithm (when $q > 1$). Both transformations have been used by Safouhi, Pinchon, and Hoggan [253], and by Safouhi and Hoggan [250], [251], [252] in the accurate evaluation of some very complicated infinite-range oscillatory integrals that arise in molecular structure calculations and have proved to be more effective than others.

We now illustrate the use of the $\bar{D}$-transformation with two examples.

Example 11.2.2 Consider the case in which $m = 2$, $f(x) = u(x)Q(x)$, and $Q(x)$ vanishes and changes sign an infinite number of times as $x \to \infty$. If we choose the x_l such that $Q(x_l) = 0$, $l = 0, 1, \ldots$, then $f(x_l) = 0$, $l = 0, 1, \ldots$, too. Also, $f'(x_l) = u(x_l)Q'(x_l)$ for all l. The resulting $\bar{D}$-transformation is, therefore, defined via the equations

$$F(x_l) = \bar{D}_\nu^{(1,j)} + x_l^{\bar{\rho}_1} u(x_l)Q'(x_l) \sum_{i=0}^{\nu-1} \frac{\bar{\beta}_i}{x_l^i}, \quad j \leq l \leq j + \nu, \qquad (11.2.5)$$

which can be solved by the W-algorithm. Note that only the derivative of $Q(x)$ is needed in (11.2.5), whereas that of $u(x)$ is not. Also, the integer $\bar{\rho}_1$ turns out to be ≤ 0 in such cases and thus can be replaced by 0. In the next sections, we give specific examples of this application.

Example 11.2.3 A practical application of the the $\bar{D}$-transformation is to integrals $I[f] = \int_0^\infty f(t)\,dt$ with $f(x) = u(x)[M(x)]^\mu$, where $u \in \mathbf{A}^{(\gamma)}$ for some γ, $M \in \mathbf{B}^{(r)}$ for some r, and $\mu \geq 1$ is an integer. From Heuristic 5.4.3, we have that $(M)^\mu \in \mathbf{B}^{(m)}$ with $m \leq \binom{r+\mu-1}{\mu}$. Since $u \in \mathbf{B}^{(1)}$ in addition, $f \in \mathbf{B}^{(m)}$ as well. If we choose the x_l such that $f(x_l) = 0$, $l = 0, 1, \ldots$, we also have $f^{(k)}(x_l) = 0$, $l = 0, 1, \ldots$, for $k = 1, \ldots, \mu - 1$. Therefore, $I[f]$ can be computed by the $\bar{D}^{(m-\mu)}$-transformation.

11.2.2 Reduction of the sD-Transformation

Another form of the $\bar{D}$-transformation can be obtained by recalling the developments of Section 5.3 on the simplified $D^{(m)}$-transformation, namely, the $sD^{(m)}$-transformation. In case $f(x) = u(x)Q(x)$, where $u \in \mathbf{A}^{(\gamma)}$ for some γ and $Q(x)$ are simpler to deal with than $f(x)$, in the sense that its derivatives are available more easily than those of $f(x)$, we showed that

$$F(x) = I[f] + \sum_{k=0}^{m-1} x^{\bar{\rho}_k}[u(x)Q^{(k)}(x)]\tilde{h}_k(x), \qquad (11.2.6)$$

with $\bar{\rho}_k$ as before and for some $\tilde{h}_k \in \mathbf{A}^{(0)}$. [Here, for convenience, we replaced the ρ_k' of (5.3.2) by their upper bounds $\bar{\rho}_k$.] Because $u(x)$ is monotonic as $x \to \infty$, the oscillatory behavior of $f(x)$ is contained in $Q(x)$. Therefore, suppose there exist x_l, $0 < x_0 < x_1 < \cdots$, $\lim_{l\to\infty} x_l = \infty$, for which

$$Q^{(k_i)}(x_l) = 0, \quad l = 0, 1, \ldots; \quad 0 \le k_1 < k_2 < \cdots < k_p \le m - 1. \quad (11.2.7)$$

Letting $E = \{0, 1, \ldots, m-1\}$ and $E_p = \{k_1, \ldots, k_p\}$ and $X = \{x_0, x_1, \ldots\}$ as before, (11.2.6) becomes

$$F(x) = I[f] + \sum_{k\in E\backslash E_p} x^{\bar{\rho}_k}[u(x)Q^{(k)}(x)]\tilde{h}_k(x), \quad x \in X. \quad (11.2.8)$$

In view of (11.2.8), we give the following simplified definition of the $\bar{D}$-transformation, which was given essentially in Sidi [299].

Definition 11.2.4 Let $f(x) = u(x)Q(x)$, E, E_p, and X be as before, and denote $q = m - p$. Then, the $s\bar{D}^{(q)}$-transformation, the simplified $\bar{D}^{(q)}$-transformation for the integral $\int_0^\infty f(t)\,dt$ is defined via the linear systems

$$F(x_l) = \bar{D}_{\bar{n}}^{(q,j)} + \sum_{k\in E\backslash E_p} x_l^{\sigma_k}[u(x_l)Q^{(k)}(x_l)] \sum_{i=0}^{n_k - 1} \frac{\bar{\beta}_{ki}}{x_l^i}, \quad j \le l \le j + \bar{N}; \quad \bar{N} = \sum_{k\in E\backslash E_p} n_k,$$

$$(11.2.9)$$

where $\bar{n} = (n_{s_1}, \ldots, n_{s_q})$ with $\{s_1, \ldots, s_q\} = E\backslash E_p$. The σ_k can be chosen exactly as in Definition 11.2.1.

Obviously, the equations in (11.2.9) can be solved via the $W^{(q)}$-algorithm.

11.3 Application of the $\bar{D}$-Transformation to Fourier Transforms

Let $f(x)$ be of the form $f(x) = \mathcal{T}(x)u(x)$, where $\mathcal{T}(x) = A\cos x + B\sin x$, $|A| + |B| \ne 0$, and $u(x) = e^{\phi(x)}h(x)$, where $\phi(x)$ is real and $\phi \in \mathbf{A}^{(k)}$ strictly for some integer $k \ge 0$ and $h \in \mathbf{A}^{(\gamma)}$ for some γ. Assume that, when $k > 0$, $\lim_{x\to\infty}\phi(x) = -\infty$, so that $f(x)$ is integrable of infinity. By the fact that $\mathcal{T} = f/u$ and $\mathcal{T}'' + \mathcal{T} = 0$, it follows that $f = p_1 f' + p_2 f''$ with

$$p_1 = \frac{2(\phi' + h'/h)}{w}, \quad p_2 = -\frac{1}{w}; \quad w = 1 + (\phi' + h'/h)^2 - (\phi' + h'/h)'. \quad (11.3.1)$$

When $k > 0$, we have $\phi' \in \mathbf{A}^{(k-1)}$ strictly. Therefore, $w \in \mathbf{A}^{(2k-2)}$ strictly. Consequently, $p_1 \in \mathbf{A}^{(-k+1)}$ and $p_2 \in \mathbf{A}^{(-2k+2)}$. When $k = 0$, we have $\phi' \equiv 0$, as a result of which, $w \in \mathbf{A}^{(0)}$ strictly, so that $p_1 \in \mathbf{A}^{(-1)}$ and $p_2 \in \mathbf{A}^{(0)}$. In all cases then, $f \in \mathbf{B}^{(2)}$ and Theorem 5.1.12 applies and (11.2.1) holds with

$$\begin{array}{lll} \bar{\rho}_0 = -k + 1, & \bar{\rho}_1 = -2k + 2 & \text{when } k > 0, \\ \bar{\rho}_0 = -1, & \bar{\rho}_1 = 0 & \text{when } k = 0. \end{array} \quad (11.3.2)$$

Note that, in any case, $\bar{\rho}_0, \bar{\rho}_1 \le 0$. If we now pick the x_l to be consecutive zeros of $\mathcal{T}(x)$, $0 < x_0 < x_1 < \cdots$, then the $\bar{D}^{(1)}$-transformation can be used to compute $I[f]$

via the equations

$$F(x_l) = \bar{D}_\nu^{(1,j)} + x_l^{\bar{\rho}_1} u(x_l) \mathcal{T}'(x_l) \sum_{i=0}^{\nu-1} \frac{\bar{\beta}_i}{x_l^i}, \quad j \le l \le j + \nu. \qquad (11.3.3)$$

If we pick the x_l to be consecutive zeros of $\mathcal{T}'(x)$, $0 < x_0 < x_1 < \cdots$, then the $s\bar{D}^{(1)}$-transformation can be used via the equations

$$F(x_l) = \bar{D}_\nu^{(1,j)} + x_l^{\bar{\rho}_0} u(x_l) \mathcal{T}(x_l) \sum_{i=0}^{\nu-1} \frac{\bar{\beta}_i}{x_l^i}, \quad j \le l \le j + \nu. \qquad (11.3.4)$$

These equations can be solved by the W-algorithm.

Note that, in both cases, $x_l = x_0 + l\pi$, $l = 0, 1, \ldots$, with suitable x_0.

11.4 Application of the $\bar{D}$-Transformation to Hankel Transforms

Let $f(x)$ be of the form $f(x) = \mathcal{C}_\nu(x)u(x)$ with $\mathcal{C}_\nu(x) = AJ_\nu(x) + BY_\nu(x)$, where $J_\nu(x)$ and $Y_\nu(x)$ are the Bessel functions of (real) order ν of the first and second kinds, respectively, and $|A| + |B| \ne 0$, and $u(x) = e^{\phi(x)}h(x)$, where $\phi(x)$ is real and $\phi \in \mathbf{A}^{(k)}$ strictly for some integer $k \ge 0$ and $h \in \mathbf{A}^{(\gamma)}$ for some γ. As in the preceding section, $\lim_{x \to \infty} \phi(x) = -\infty$ when $k > 0$, to guarantee that $f(x)$ is integrable at infinity.

By the fact that $\mathcal{C}_\nu(x)$ satisfies the Bessel equation $\mathcal{C}_\nu = \frac{x}{\nu^2 - x^2}\mathcal{C}'_\nu + \frac{x^2}{\nu^2 - x^2}\mathcal{C}''_\nu$, and that $\mathcal{C}_\nu = f/u$, it follows that $f = p_1 f' + p_2 f''$ with

$$p_1 = \frac{2x^2(\phi' + h'/h) - x}{w} \quad \text{and} \quad p_2 = -\frac{x^2}{w},$$

where

$$w = x^2[(\phi' + h'/h)^2 - (\phi' + h'/h)'] - x(\phi' + h'/h) + x^2 - \nu^2.$$

From this, it can easily be shown that $p_1 \in \mathbf{A}^{(i_1)}$ and $p_2 \in \mathbf{A}^{(i_2)}$ with i_1 and i_2 exactly as in the preceding section so that $f \in \mathbf{B}^{(2)}$. Consequently, Theorem 5.1.12 applies and (11.2.1) holds with $m = 2$ and with $\bar{\rho}_0$ and $\bar{\rho}_1$ exactly as in the preceding section, namely,

$$\begin{array}{llll} \bar{\rho}_0 = -k + 1, & \bar{\rho}_1 = -2k + 2 & \text{when } k > 0, \\ \bar{\rho}_0 = -1, & \bar{\rho}_1 = 0 & \text{when } k = 0. \end{array} \qquad (11.4.1)$$

Note that, in any case, $\bar{\rho}_0, \bar{\rho}_1 \le 0$.

(i) If we choose the x_l to be consecutive zeros of $\mathcal{C}_\nu(x)$, $0 < x_0 < x_1 < \cdots$, then the $\bar{D}^{(1)}$-transformation can be used to compute $I[f]$ via the equations

$$F(x_l) = \bar{D}_\nu^{(1,j)} + x_l^{\bar{\rho}_1} u(x_l) \mathcal{C}'_\nu(x_l) \sum_{i=0}^{\nu-1} \frac{\bar{\beta}_i}{x_l^i}, \quad j \le l \le j + \nu. \qquad (11.4.2)$$

Here, we can make use of the known fact that $\mathcal{C}'_\nu(x) = (\nu/x)\mathcal{C}_\nu(x) - \mathcal{C}_{\nu+1}(x)$, so that $\mathcal{C}'_\nu(x_l) = -\mathcal{C}_{\nu+1}(x_l)$, which simplifies things further.

(ii) If we choose the x_l to be consecutive zeros of $\mathcal{C}'_\nu(x)$, $0 < x_0 < x_1 < \cdots$, then the $s\bar{D}^{(1)}$-transformation can be used to compute $I[f]$ via the equations

$$F(x_l) = \bar{D}^{(1,j)}_\nu + x_l^{\bar{\rho}_0} u(x_l)\mathcal{C}_\nu(x_l) \sum_{i=0}^{\nu-1} \frac{\bar{\beta}_i}{x_l^i}, \quad j \le l \le j + \nu. \tag{11.4.3}$$

(iii) If we choose the x_l to be consecutive zeros of $\mathcal{C}_{\nu+1}(x)$, $0 < x_0 < x_1 < \cdots$, then we obtain another form of the $s\bar{D}$-transformation defined again via the equations in (11.4.3). To see this, we rewrite (11.2.6) in the form

$$\begin{aligned} F(x) &= I[f] + x^{\bar{\rho}_0}u(x)\mathcal{C}_\nu(x)\tilde{h}_0(x) + x^{\bar{\rho}_1}u(x)\left[\frac{\nu}{x}\mathcal{C}_\nu(x) - \mathcal{C}_{\nu+1}(x)\right]\tilde{h}_1(x) \\ &= I[f] + x^{\bar{\rho}_0}u(x)\mathcal{C}_\nu(x)\hat{h}_0(x) + x^{\bar{\rho}_1}u(x)\mathcal{C}_{\nu+1}(x)\hat{h}_1(x), \end{aligned} \tag{11.4.4}$$

where $\hat{h}_0(x) = \tilde{h}_0(x) + x^{\bar{\rho}_1 - \bar{\rho}_0 - 1}\tilde{h}_1(x) \in \mathbf{A}^{(0)}$ since $\bar{\rho}_1 - \bar{\rho}_0 - 1 \le 0$, and $\hat{h}_1(x) = -\tilde{h}_1(x) \in \mathbf{A}^{(0)}$ as well. Now, pick the x_l to be consecutive zeros of $\mathcal{C}_{\nu+1}(x)$ in (11.4.4) to obtain the method defined via (11.4.3). [Note that the second expansion in (11.4.4) is in agreement with Remark 5 in Section 4.3.]

These developments are due to Sidi [299]. All three methods above have proved to be very effective for large, as well as small, values of ν. In addition, they can all be implemented via the W-algorithm.

11.5 The $\tilde{D}$-Transformation

The $\bar{D}$-transformation of the previous sections is defined by reducing the D-transformation with the help of the zeros of one or more of the $f^{(k)}(x)$. When these zeros are not readily available or are difficult to compute, the D-transformation can be reduced via the so-called $\tilde{D}$-transformation of Sidi [274] that is a very flexible device. What we have here is really a *general approach* that is based on Remark 5 of Section 4.3, within which a multitude of methods can be defined. What is needed for this approach is some amount of asymptotic analysis of $f(x)$ as $x \to \infty$. Part of this analysis is quantitative, and the remainder is qualitative only.

Our starting point is again (11.2.1). By expressing $f(x)$ and its derivatives as combinations of simple functions when possible, we rewrite (11.2.1) in the form

$$F(x) = I[f] + \sum_{k=0}^{m-1} v_k(x)\tilde{g}_k(x), \tag{11.5.1}$$

where $\tilde{g}_k \in \mathbf{A}^{(0)}$ for all k. Here, the functions $v_k(x)$ have much simpler forms than $f^{(k)}(x)$, and their zeros are readily available. Now, choose the x_l, $0 < x_0 < x_1 < \cdots$, $\lim_{l\to\infty} x_l = \infty$, for which

$$v_k(x_l) = 0, \quad l = 0, 1, \ldots; \quad 0 \le k_1 < k_2 < \cdots < k_p \le m - 1. \tag{11.5.2}$$

Letting $E = \{0, 1, \ldots, m - 1\}$ and $E_p = \{k_1, \ldots, k_p\}$ and $X = \{x_0, x_1, \ldots\}$ again,

(11.5.1) becomes

$$F(x) = I[f] + \sum_{k \in E \setminus E_p} v_k(x) \tilde{g}(x), \quad x \in X. \tag{11.5.3}$$

In view of (11.5.3), we give the following definition analogously to Definitions 11.2.1 and 11.2.4.

Definition 11.5.1 Let $f(x)$, E, E_p, and X be as before, and denote $q = m - p$. Then the $\tilde{D}^{(q)}$-transformation for the integral $\int_0^\infty f(t)\,dt$ is defined via the linear equations

$$F(x_l) = \tilde{D}_{\bar{n}}^{(q,j)} + \sum_{k \in E \setminus E_p} v_k(x_l) \sum_{i=0}^{n_k-1} \frac{\bar{\beta}_{ki}}{x_l^i}, \quad j \le l \le j + \bar{N}; \quad \bar{N} = \sum_{k \in E \setminus E_p} n_k, \tag{11.5.4}$$

where $\bar{n} = (n_{s_1}, \ldots, n_{s_q})$ with $\{s_1, \ldots, s_q\} = E \setminus E_p$.

It is obvious from Definition 11.5.1 that the $\tilde{D}$- transformation, like the $\bar{D}$-transformations, can be implemented by the W-algorithm (when $q = 1$) and by the $W^{(q)}$-algorithm (when $q > 1$).

The best way to clarify what we have done so far is with practical examples, to which we now turn.

11.6 Application of the $\tilde{D}$-Transformation to Hankel Transforms

Consider the integral $\int_0^\infty \mathcal{C}_\nu(t) u(t)\,dt$, where $\mathcal{C}_\nu(x)$ and $u(x)$ are exactly as described in Section 11.4. Now it is known that

$$\mathcal{C}_\nu(x) = \alpha_1(x) \cos x + \alpha_2(x) \sin x, \quad \alpha_1, \alpha_2 \in \mathbf{A}^{(-1/2)}. \tag{11.6.1}$$

The derivatives of $\mathcal{C}_\nu(x)$ are of precisely the same form as $\mathcal{C}_\nu(x)$ but with different $\alpha_1, \alpha_2 \in \mathbf{A}^{(-1/2)}$. Substituting these in (11.2.6) with $Q(x) = \mathcal{C}_\nu(x)$ there, we obtain

$$F(x) = I[f] + v_0(x) \tilde{g}_0(x) + v_1(x) \tilde{g}_1(x), \quad \tilde{g}_0, \tilde{g}_1 \in \mathbf{A}^{(0)}, \tag{11.6.2}$$

where

$$v_0(x) = x^{\rho - 1/2} u(x) \cos x, \quad v_1(x) = x^{\rho - 1/2} u(x) \sin x; \quad \rho = \max\{\bar{\rho}_0, \bar{\rho}_1\}, \tag{11.6.3}$$

with $\bar{\rho}_0$ and $\bar{\rho}_1$ as in Section 11.4.

We now pick the x_l to be consecutive zeros of $\cos x$ or of $\sin x$. The x_l are thus equidistant with $x_l = x_0 + l\pi$, $x_0 > 0$. The equations that define the $\tilde{D}$-transformation then become

$$F(x_l) = \tilde{D}_\nu^{(1,j)} + (-1)^l x_l^{\rho - 1/2} u(x_l) \sum_{i=0}^{\nu-1} \frac{\bar{\beta}_i}{x_l^i}, \quad j \le l \le j + \nu. \tag{11.6.4}$$

Here, we have a method that does not require specific knowledge of the zeros or extrema of $\mathcal{C}_\nu(x)$ and that has proved to be very effective for low to moderate values of ν. The equations in (11.6.4) can be solved via the W-algorithm.

11.7 Application of the $\tilde{D}$-Transformation to Integrals
of Products of Bessel Functions

Consider the integral $\int_0^\infty f(t)\,dt$, where $f(x) = \mathcal{C}_\mu(x)\mathcal{C}_\nu(x)u(x)$ with $\mathcal{C}_\nu(x)$ and $u(x)$ exactly as in Section 11.4 again. Because both $\mathcal{C}_\mu(x)$ and $\mathcal{C}_\nu(x)$ are in $\mathbf{B}^{(2)}$ and $u(x)$ is in $\mathbf{B}^{(1)}$, we conclude, by part (i) of Heuristic 5.4.1, that $f \in \mathbf{B}^{(4)}$. Recalling that $\mathcal{C}_\nu(x) = \alpha_{\nu 1}(x)\cos x + \alpha_{\nu 2}(x)\sin x$, where $\alpha_{\nu i} \in \mathbf{A}^{(-1/2)}$ for $i = 1, 2$ and for all ν, we see that $f(x)$ is, in fact, of the following simple form:

$$f(x) = u(x)[w_0(x) + w_1(x)\cos 2x + w_2(x)\sin 2x], \quad w_i \in \mathbf{A}^{(-1)}, \quad i = 0, 1, 2.$$
(11.7.1)

Now, the sum $w_1(x)\cos 2x + w_2(x)\sin 2x$ can also be expressed in the form $\hat{w}_+(x)e^{i2x} + \hat{w}_-(x)e^{-i2x}$ with $\hat{w}_\pm \in \mathbf{A}^{(-1)}$. Thus, $f(x) = f_0(x) + f_+(x) + f_-(x)$ with $f_0(x) = u(x)w_0(x)$, and $f_\pm(x) = u(x)\hat{w}_\pm(x)e^{\pm i2x}$, $f_0, f_\pm \in \mathbf{B}^{(1)}$. By part (ii) of Heuristic 5.4.1, we therefore conclude that $f \in \mathbf{B}^{(3)}$. Let us now write $F(x) = F_0(x) + F_+(x) + F_-(x)$, where $F_0(x) = \int_0^x f_0(t)\,dt$ and $F_\pm(x) = \int_0^x f_\pm(t)\,dt$. Recalling Theorem 5.7.3, we have

$$F(x) = I[f] + x^{\rho_0}f_0(x)\tilde{h}_0(x) + x^\rho f_+(x)\tilde{h}_+(x) + x^\rho f_-(x)\tilde{h}_-(x), \quad (11.7.2)$$

where

$$\rho_0 = \begin{cases} 1 & \text{if } k = 0 \\ -k+1 & \text{if } k > 0 \end{cases}, \quad \rho = -k+1, \quad \text{and} \quad \tilde{h}_0, \tilde{h}_\pm \in \mathbf{A}^{(0)}, \quad (11.7.3)$$

which can be rewritten in the form

$$F(x) = I[f] + x^{\rho_0-1}u(x)\hat{h}_0(x) + x^{\rho-1}u(x)\hat{h}_1(x)\cos 2x$$
$$+ x^{\rho-1}u(x)\hat{h}_2(x)\sin 2x, \quad \hat{h}_0, \hat{h}_1, \hat{h}_2 \in \mathbf{A}^{(0)}. \quad (11.7.4)$$

Now, let us choose the x_l to be consecutive zeros of $\cos 2x$ or $\sin 2x$. These are equidistant with $x_l = x_0 + l\pi/2$, $x_0 > 0$. The equations that define the $\tilde{D}$-transformation then become

$$F(x_l) = \tilde{D}_{\bar{n}}^{(2,j)} + x_l^{\rho_0-1}u(x_l)\sum_{i=0}^{n_1-1}\frac{\bar{\beta}_{1i}}{x_l^i} + (-1)^l x_l^{\rho-1}u(x_l)\sum_{i=0}^{n_2-1}\frac{\bar{\beta}_{2i}}{x_l^i}, \quad j \leq l \leq j + n_1 + n_2.$$
(11.7.5)

(For simplicity, we can take $\rho_0 = \rho = 1$ throughout.) Obviously, the equations in (11.7.5) can be solved via the $W^{(2)}$-algorithm.

Consider now the integral $\int_0^\infty f(t)\,dt$, where $f(x) = \mathcal{C}_\mu(ax)\mathcal{C}_\nu(bx)u(x)$ with $a \neq b$. Using the preceding technique, we can show that

$$F(x) = I[f] + x^{-1}u(x)\big[\hat{h}_+^c(x)\cos(a+b)x + \hat{h}_-^c(x)\cos(a-b)x$$
$$+ \hat{h}_+^s(x)\sin(a+b)x + \hat{h}_-^s(x)\sin(a-b)x\big], \quad \hat{h}_\pm^c, \hat{h}_\pm^s \in \mathbf{A}^{(0)}. \quad (11.7.6)$$

This means that $f \in \mathbf{B}^{(4)}$ precisely. We can now choose the x_l as consecutive zeros of $\sin(a+b)x$ or of $\sin(a-b)x$ and compute $I[f]$ by the $\tilde{D}^{(3)}$-transformation. If $(a+b)/(a-b)$ is an integer, by choosing the x_l as consecutive zeros of $\sin(a-b)x$

[which make $\sin(a+b)x$ vanish as well], we can even compute $I[f]$ by the $\tilde{D}^{(2)}$-transformation.

11.8 The *W*- and *mW*-Transformations for Very Oscillatory Integrals

11.8.1 Description of the Class $\hat{\mathbf{B}}$

The philosophy behind the $\tilde{D}$-transformation can be extended further to cover a very large family of oscillatory infinite-range integrals. We start with the following definition of the relevant family of integrands.

Definition 11.8.1 We say that a function $f(x)$ belongs to the class $\hat{\mathbf{B}}$ if it can be expressed in the form

$$f(x) = \sum_{j=1}^{r} u_j(\theta_j(x)) \exp(\phi_j(x)) h_j(x), \tag{11.8.1}$$

where u_j, θ_j, ϕ_j and h_j are as follows:

1. $u_j(z)$ is either e^{iz} or e^{-iz} or any linear combination of these (like $\cos z$ or $\sin z$).
2. $\theta_j(x)$ and $\phi_j(x)$ are real and $\theta_j \in \mathbf{A}^{(m)}$ strictly and $\phi_j \in \mathbf{A}^{(k)}$ strictly, where m is a positive integer and k is a nonnegative integer, and

$$\theta_j(x) = \bar{\theta}(x) + \Delta_j(x) \text{ and } \phi_j(x) = \bar{\phi}(x) + \Lambda_j(x) \tag{11.8.2}$$

   with

$$\bar{\theta}(x) = \sum_{i=0}^{m-1} \mu_i x^{m-i} \text{ and } \bar{\phi}(x) = \sum_{i=0}^{k-1} \nu_i x^{k-i} \text{ and } \Delta_j, \Lambda_j \in \mathbf{A}^{(0)}. \tag{11.8.3}$$

   Also, when $k \geq 1$, we have $\lim_{x\to\infty} \bar{\phi}(x) = -\infty$. We assume, without loss of generality, that $\mu_0 > 0$.
3. $h_j \in \mathbf{A}^{(\gamma_j)}$ for some possibly complex γ_j such that $\gamma_j - \gamma_{j'} =$ integer for every j and j'. Denote by γ that γ_j whose real part is largest. (Note that $\Im\gamma_j$ are all the same.)

Now, the class $\hat{\mathbf{B}}$ is the union of two mutually exclusive sets: $\mathbf{B}_c$ and $\mathbf{B}_d$, where $\mathbf{B}_c$ contains the functions in $\hat{\mathbf{B}}$ that are integrable at infinity, while $\mathbf{B}_d$ contains the ones whose integrals diverge but are defined in the sense of Abel summability. (For this point, see the remarks at the end of Section 5.7.)

When $m = 1$, $f(x)$ is a linear combination of functions that have a simple sinusoidal behavior of the form $e^{\pm i\mu_0 x}$ as $x \to \infty$, that is, the period of its oscillations is fixed as $x \to \infty$. When $m \geq 2$, however, $f(x)$ oscillates much more rapidly [like $e^{\pm i\bar{\theta}(x)}$] as $x \to \infty$, with the "period" of these oscillations tending to zero. In the sequel, we call such functions $f(x)$ *very oscillatory*.

In the remainder of this section, we treat only functions in $\hat{\mathbf{B}}$.

Note that, from Definition 11.8.1, it follows that $f(x)$, despite its complicated appearance in (11.8.1), is of the form

$$f(x) = e^{\bar{\phi}(x)}[e^{i\bar{\theta}(x)} h_+(x) + e^{-i\bar{\theta}(x)} h_-(x)], \quad h_\pm \in \mathbf{A}^{(\gamma)}. \tag{11.8.4}$$

This is so because $e^{\alpha(x)} \in \mathbf{A}^{(0)}$ for $\alpha \in \mathbf{A}^{(0)}$. If we denote $f_{\pm}(x) = e^{\bar{\phi}(x) \pm i\bar{\theta}(x)} h_{\pm}(x)$ and $F_{\pm}(x) = \int_0^x f_{\pm}(t)\, dt$, then, by Theorem 5.7.3, we have

$$F_{\pm}(x) = I[f_{\pm}] + x^{\rho} f_{\pm}(x) g_{\pm}(x), \quad \rho = 1 - \max\{m, k\}, \quad g_{\pm} \in \mathbf{A}^{(0)} \text{ strictly}. \quad (11.8.5)$$

From this and from the fact that $f(x) = f_{+}(x) + f_{-}(x)$, it follows that

$$F(x) = I[f] + x^{\rho+\gamma} e^{\bar{\phi}(x)} [\cos(\bar{\theta}(x)) b_1(x) + \sin(\bar{\theta}(x)) b_2(x)], \quad b_1, b_2 \in \mathbf{A}^{(0)}. \quad (11.8.6)$$

Here, $I[f]$ is $\int_0^{\infty} f(t)\, dt$ when this integral converges; it is the Abel sum of $\int_0^{\infty} f(t)\, dt$ when the latter diverges.

Before going on, we would like to note that the class of functions $\hat{\mathbf{B}}$ is very comprehensive in the sense that it includes many integrands of oscillatory integrals that arise in important engineering, physics, and chemistry applications.

11.8.2 The W- and mW-Transformations

We now use (11.8.6) to define the W- and mW-transformations of Sidi [281] and [288], respectively. Throughout, we assume that $f(x)$ behaves in a "regular" manner, starting with $x = 0$. By this, we mean that there do not exist intervals of $(0, \infty)$ in which $f(x)$, relative to its average behavior, has large and sudden changes or remains almost constant. [One simple way to test the "regularity" of $f(x)$ is by plotting its graph.] We also assume that $\bar{\theta}(x)$ and its first few derivatives are strictly increasing on $(0, \infty)$. This implies that there exists a unique sequence $\{x_l\}_{l=0}^{\infty}$, $0 < x_0 < x_1 < \cdots$, such that x_l are consecutive roots of the equation $\sin \bar{\theta}(x) = 0$ [or of $\cos \bar{\theta}(x) = 0$] and behave "regularly" as well. This also means that x_0 is the unique simple root of the polynomial equation $\bar{\theta}(x) = q\pi$ for some integer (or half integer) q, and, for each $l > 0$, x_l is the unique simple root of the polynomial equation $\bar{\theta}(x) = (q + l)\pi$. We also have $\lim_{l \to \infty} x_l = \infty$.

These assumptions make the discussion simpler and the W- and mW-transformations more effective. If $f(x)$ and $\bar{\theta}(x)$ start behaving as described only for $x \gtrsim c$ for some $c > 0$, then a reasonable strategy is to apply the transformations to the integral $\int_0^{\infty} \tilde{f}(t)\, dt$, where $\tilde{f}(x) = f(c + x)$, and add to it the finite-range integral $\int_0^c f(t)\, dt$ that is computed separately.

We begin with the W-transformation.

Definition 11.8.2 Choose the x_l as in the previous paragraph, i.e., as consecutive positive zeros of $\sin \bar{\theta}(x)$ [or of $\cos \bar{\theta}(x)$]. The W-transformation is then defined via the linear equations

$$F(x_l) = W_n^{(j)} + (-1)^l x_l^{\rho+\gamma} e^{\bar{\phi}(x_l)} \sum_{i=0}^{n-1} \frac{\bar{\beta}_i}{x_l^i}, \quad j \leq l \leq j + n, \quad (11.8.7)$$

where $W_n^{(j)}$ is the approximation to $I[f]$. (Obviously, the $W_n^{(j)}$ can be determined via the W-algorithm.)

As is clear, in order to apply the W-transformation to functions in $\hat{\mathbf{B}}$, we need to analyze the $\theta_j(x)$, $\phi_j(x)$, and $h_j(x)$ in (11.8.1) quantitatively as $x \to \infty$ in order to extract $\bar{\theta}(x)$, $\bar{\phi}(x)$, and γ. The modification of the W-transformation, denoted the mW-transformation, requires one to analyze only $\theta_j(x)$; thus, it is very user-friendly. The expansion in (11.8.6) forms the basis of this transformation just as it forms the basis of the W-transformation.

Definition 11.8.3 Choose the x_l exactly as in Definition 11.8.2. The mW-transformation is defined via the linear equations

$$F(x_l) = W_n^{(j)} + \psi(x_l) \sum_{i=0}^{n-1} \frac{\bar{\beta}_i}{x_l^i}, \quad j \le l \le j+n, \tag{11.8.8}$$

where

$$\psi(x_l) = F(x_{l+1}) - F(x_l) = \int_{x_l}^{x_{l+1}} f(t)\,dt. \tag{11.8.9}$$

(These $W_n^{(j)}$ can also be determined via the W-algorithm.)

Remarks.

1. Even though $F(x_{l+1}) - F(x_l)$ is a function of both x_l and x_{l+1}, we have denoted it $\psi(x_l)$. This is not a mistake as x_{l+1} is a function of l, which, in turn, is a function of x_l (there is a 1–1 correspondence between x_l and l), so that x_{l+1} is also a function of x_l.

2. Because $F(x_l) = \sum_{i=0}^{l} \psi(x_{i-1})$ with $x_{-1} = 0$ in (11.8.9), the mW-transformation can be viewed as a convergence acceleration method for the infinite series $\sum_{i=0}^{\infty} \psi(x_{i-1})$. (As we show shortly, when γ is real, this series is ultimately *alternating* too.) In this sense, the mW-transformation is akin to a method of Longman [170], [171], in which one integrates $f(x)$ between its consecutive zeros or extrema $y_1 < y_2 < \cdots$ to obtain the integrals $v_i = \int_{y_i}^{y_i+1} f(t)\,dt$, where $y_0 = 0$, and accelerates the convergence of the alternating infinite series $\sum_{i=0}^{\infty} v_i$ by a suitable sequence transformation. Longman used the Euler transformation for this purpose. Of course, other transformations, such as the iterated Δ^2-process, the Shanks transformation, etc., can also be used. We discuss these methods later.

3. The quantities $\psi(x_l)$ can also be defined by $\psi(x_l) = \int_{x_{l-1}}^{x_l} f(t)\,dt$ without changing the quality of the $W_n^{(j)}$ appreciably.

Even though the W-transformation was obtained directly from the expansion of $F(x)$ given in (11.8.6), the mW-transformation has not relied on an analogous expansion so far. Therefore, we need to provide a rigorous justification of the mW-transformation. For this, it suffices to show that, with $\psi(x_l)$ as in (11.8.9),

$$F(x_l) = I[f] + \psi(x_l)g(x_l), \quad g \in \mathbf{A}^{(0)}. \tag{11.8.10}$$

As we discuss in the next theorem, this is equivalent to showing that this $\psi(x_l)$ is of the form

$$\psi(x_l) = (-1)^l x_l^{\rho+\gamma} e^{\bar{\phi}(x_l)} \tilde{b}(x_l), \quad \tilde{b} \in \mathbf{A}^{(0)}. \tag{11.8.11}$$

An important implication of (11.8.11) is that, when γ is real, the $\psi(x_l)$ alternate in sign for all large l.

Theorem 11.8.4 *With $F(x)$, $\{x_l\}$, and $\psi(x_l)$ as described before, $\psi(x_l)$ satisfies (11.8.11). Consequently, $F(x_l)$ satisfies (11.8.10).*

Proof. Without loss of generality, let us take the x_l as the zeros of $\sin \bar{\theta}(x)$. Let us set $x = x_l$ in (11.8.6). We have $\sin \bar{\theta}(x_l) = 0$ and $\cos \bar{\theta}(x_l) = (-1)^{q+l}$. Consequently, (11.8.6) becomes

$$F(x_l) = I[f] + (-1)^{q+l} x_l^{\rho+\gamma} e^{\bar{\phi}(x_l)} b_1(x_l). \tag{11.8.12}$$

Substituting this in (11.8.9), we obtain

$$\psi(x_l) = (-1)^{q+l+1} \left[x_l^{\rho+\gamma} e^{\bar{\phi}(x_l)} b_1(x_l) + x_{l+1}^{\rho+\gamma} e^{\bar{\phi}(x_{l+1})} b_1(x_{l+1}) \right]$$

$$= (-1)^{q+l+1} x_l^{\rho+\gamma} e^{\bar{\phi}(x_l)} b_1(x_l)[1 + w(x_l)], \tag{11.8.13}$$

where

$$w(x_l) = R(x_l) e^{\Delta \bar{\phi}(x_l)}; \quad R(x_l) = \left(\frac{x_{l+1}}{x_l} \right)^{\rho+\gamma} \frac{b_1(x_{l+1})}{b_1(x_l)}, \quad \Delta \bar{\phi}(x_l) = \bar{\phi}(x_{l+1}) - \bar{\phi}(x_l). \tag{11.8.14}$$

Comparing (11.8.13) with (11.8.11), we identify $\tilde{b}(x) = (-1)^{q+1} b_1(x)[1 + w(x)]$. We have to show only that $\tilde{b} \in \mathbf{A}^{(0)}$.

From Lemma 9.5.1 at the end of Chapter 9, we know that $x_l = \sum_{i=0}^{\infty} a_i l^{(1-i)/m}$ for all large l, $a_0 = (\pi/\mu_0)^{1/m} > 0$. From this, some useful results follow:

$$x_{l+1} = \sum_{i=0}^{\infty} a_i' l^{(1-i)/m}, \ a_i' = a_i, \ i = 0, 1, \ldots, m-1, \ a_m' = a_m + a_0/m,$$

$$x_{l+1} - x_l = \sum_{i=0}^{\infty} \hat{a}_i l^{-1+(1-i)/m}, \ \hat{a}_0 = \frac{a_0}{m}, \tag{11.8.15}$$

$$x_{l+1}^p - x_l^p = x_l^p \left\{ \left(1 + \frac{x_{l+1} - x_l}{x_l} \right)^p - 1 \right\} = \sum_{i=0}^{\infty} \tilde{a}_i^{(p)} l^{-1+(p-i)/m}, \ \tilde{a}_0^{(p)} = \frac{p}{m} a_0^p.$$

Using these, we can show after some manipulation that

$$R(x_l) \sim 1 + l^{-1} \sum_{i=0}^{\infty} d_i l^{-i/m} \text{ as } l \to \infty, \tag{11.8.16}$$

and, for $k \neq 0$,

$$\Delta \bar{\phi}(x_l) = \sum_{i=0}^{k-1} v_i (x_{l+1}^{k-i} - x_l^{k-i}) = l^{-1+k/m} \sum_{i=0}^{\infty} e_i l^{-i/m}, \ e_0 = v_0 \frac{k}{m} a_0^k. \tag{11.8.17}$$

[For $k = 0$, we have $\bar{\phi}(x) \equiv 0$ so that $\Delta \bar{\phi}(x_l) \equiv 0$ too.]

From the fact that $x_l = \sum_{i=0}^{\infty} a_i l^{(1-i)/m}$ for all large l, we can show that any function of l with an asymptotic expansion of the form $\sum_{i=0}^{\infty} \kappa_i l^{(r-i)/m}$ as $l \to \infty$ also has an asymptotic expansion of the form $\sum_{i=0}^{\infty} \hat{\kappa}_i x_l^{r-i}$ as $l \to \infty$, $\hat{\kappa}_0 = \kappa_0/a_0^r$. In addition, if one of these expansions converges, then so does the other. Consequently, (11.8.16) and (11.8.17) can be rewritten, respectively, as

$$R(x_l) \sim 1 + \hat{d}_1 x_l^{-1} + \hat{d}_2 x_l^{-2} + \cdots \quad \text{as } l \to \infty, \tag{11.8.18}$$

and

$$\Delta \bar{\phi}(x_l) = x_l^{k-m} \sum_{i=0}^{\infty} \hat{e}_i x_l^{-i}, \quad \hat{e}_0 = \frac{k}{m} v_0 a_0^m. \tag{11.8.19}$$

Thus, $R \in \mathbf{A}^{(0)}$ strictly and $\Delta\bar{\phi} \in \mathbf{A}^{(k-m)}$ strictly. When $k \leq m$, $e^{\Delta\bar{\phi}(x)} \in \mathbf{A}^{(0)}$ strictly and $e^{\Delta\bar{\phi}(x)} \sim 1$ as $x \to \infty$. Consequently, $1 + w(x) \in \mathbf{A}^{(0)}$ strictly, hence $\tilde{b} \in \mathbf{A}^{(0)}$. For $k > m$, we have that $e^{\Delta\bar{\phi}(x)} \to 0$ as $x \to \infty$ essentially like $\exp(\hat{e}_0 x^{k-m})$ since $\lim_{x \to \infty} \Delta\bar{\phi}(x) = -\infty$ due to the fact that $\hat{e}_0 < 0$ in (11.8.19). (Recall that $a_0 > 0$ and $v_0 < 0$ by assumption.) Consequently, $1 + w(x) \in \mathbf{A}^{(0)}$ strictly, from which $\tilde{b} \in \mathbf{A}^{(0)}$ again. This completes the proof of (11.8.11). The result in (11.8.10) follows from the additional observation that $g(x) = -1/[1 + w(x)] \in \mathbf{A}^{(0)}$ whether $k \leq m$ or $k > m$. ∎

Theorem 11.8.4 shows clearly that $\psi(x_l)$ is a true shape function for $F(x_l)$, and that the mW-transformation is a true GREP$^{(1)}$.

11.8.3 Application of the mW-Transformation to Fourier, Inverse Laplace, and Hankel Transforms

Fourier Transforms

If $f(x) = u(x)\mathcal{T}(x)$ with $u(x)$ and $\mathcal{T}(x)$ exactly as in Section 11.3, then $f \in \hat{\mathbf{B}}$ with $\bar{\theta}(x) = x$. Now, choose $x_l = x_0 + l\pi$, $l = 1, 2, \ldots$, $x_0 > 0$, and apply the mW-transformation.

Hasegawa and Sidi [127] devised an efficient automatic integration method for a large class of oscillatory integrals $\int_0^{\infty} f(t)\,dt$, such as Hankel transforms, in which these integrals are expressed as sums of integrals of the form $\int_0^{\infty} e^{i\omega t} g(t)\,dt$, and the mW-transformation just described is used to compute the latter. An important ingredient of this procedure is a fast method for computing the indefinite integral $\int_0^x e^{i\omega t} g(t)\,dt$, described in Hasegawa and Torii [128].

A variant of the mW-transformation for Fourier transforms $\int_0^{\infty} u(t)\mathcal{T}(t)\,dt$ was proposed by Ehrenmark [73]. In this variant, the x_l are obtained by solving some nonlinear equations involving asymptotic information coming from the function $u(x)$. See [73] for details.

Inverse Laplace Transforms

An interesting application is to the inversion of the Laplace transform by the Bromwich integral. If $\hat{u}(z)$ is the Laplace transform of $u(t)$, that is, $\hat{u}(z) = \int_0^{\infty} e^{-zt} u(t)\,dt$, then

$$\frac{u(t+) + u(t-)}{2} = \frac{1}{2\pi i} \int_{c-i\infty}^{c+i\infty} e^{zt} \hat{u}(z)\,dz, \quad t > 0,$$

where the contour of integration is the straight line $\Re z = c$, and $\hat{u}(z)$ has all its singularities to the left of this line. Making the substitution $z = c + i\xi$ in the Bromwich integral, we obtain

$$\frac{u(t+) + u(t-)}{2} = \frac{e^{ct}}{2\pi} \left[\int_0^\infty e^{i\xi t} \hat{u}(c + i\xi) \, d\xi + \int_0^\infty e^{-i\xi t} \hat{u}(c - i\xi) \, d\xi \right].$$

Assume that $\hat{u}(z) = e^{-zt_0} \hat{u}_0(z)$, with $\hat{u}_0(z) \in \mathbf{A}^{(\gamma)}$ for some γ. Now, we can apply the mW-transformation to the last two integrals with $\xi_l = (l + 1)\pi/(t - t_0)$, $l = 0, 1, \ldots$. In case $u(t)$ is real, only one of these integrals needs to be computed because now

$$\frac{u(t+) + u(t-)}{2} = \frac{e^{ct}}{\pi} \Re \left[\int_0^\infty e^{i\xi t} \hat{u}(c + i\xi) \, d\xi \right].$$

This approach has proved to be very effective in inverting Laplace transforms when $\hat{u}(z)$ is known for complex z.

Hankel Transforms

If $f(x) = u(x)\mathcal{C}_\nu(x)$ with $u(x)$ and $\mathcal{C}_\nu(x)$ exactly as in Section 11.4, then $f \in \hat{\mathbf{B}}$ with $\bar{\theta}(x) = x$ again. Choose $x_l = x_0 + l\pi$, $l = 1, 2, \ldots$, $x_0 > 0$, and apply the mW-transformation.

This approach was found to be one of the most effective means for computing Hankel transforms when the order ν is of moderate size; see Lucas and Stone [189].

Variants for Hankel Transforms

Again, let $f(x) = u(x)\mathcal{C}_\nu(x)$ with $u(x)$ and $\mathcal{C}_\nu(x)$ exactly as in Section 11.4. When ν, the order of the Bessel function $\mathcal{C}_\nu(x)$, is large, it seems more appropriate to choose the x_l in the mW-transformation as the zeros of $\mathcal{C}_\nu(x)$ or of $\mathcal{C}'_\nu(x)$ or of $\mathcal{C}_{\nu+1}(x)$, exactly as was done in Section 11.4, where we developed the relevant $\bar{D}$- and $s\bar{D}$-transformations. This use of the mW-transformation was suggested by Lucas and Stone [189], who chose the x_l as the zeros or extrema of $J_\nu(x)$ for the integrals $\int_0^\infty u(t)J_\nu(t) \, dt$. The choice of the x_l as the zeros of $\mathcal{C}_{\nu+1}(x)$ was proposed by Sidi [299]. The resulting methods produce some of the best results, as concluded in [189] and [299]. The same conclusion is reached in the extensive comparative study of Michalski [211] for the so-called Sommerfeld type integrals that are simply Hankel transforms.

The question, of course, is whether $\psi(x_l)$ is a true shape function for $F(x_l)$. The answer to this question is in the affirmative as $x_l \sim \sum_{i=0}^\infty a_i l^{1-i}$ as $l \to \infty$, $a_0 = \pi$. Even though this asymptotic expansion need not be convergent, the developments of Theorem 11.8.4 remain valid. Let us show this for the case in which x_l are consecutive zeros of $\mathcal{C}_\nu(x)$. For this, we start with

$$F(x) = I[f] + x^{\bar{\rho}_0} u(x)\mathcal{C}_\nu(x)b_0(x) + x^{\bar{\rho}_1} [u(x)\mathcal{C}_\nu(x)]' b_1(x), \qquad (11.8.20)$$

which is simply (11.2.1). Letting $x = x_l$, this reduces to

$$F(x_l) = I[f] + x_l^{\bar{\rho}_1} u(x_l)\mathcal{C}'_\nu(x_l)b_1(x_l). \qquad (11.8.21)$$

Thus,

$$\psi(x_l) = F(x_{l+1}) - F(x_l) = x_l^{\bar{\rho}_1} u(x_l) \mathcal{C}_\nu'(x_l) b_1(x_l) [w(x_l) - 1], \quad (11.8.22)$$

where

$$w(x_l) = \left(\frac{x_{l+1}}{x_l}\right)^{\bar{\rho}_1} \frac{u(x_{l+1})}{u(x_l)} \frac{b_1(x_{l+1})}{b_1(x_l)} \frac{\mathcal{C}_\nu'(x_{l+1})}{\mathcal{C}_\nu'(x_l)}. \quad (11.8.23)$$

Now, differentiating both sides of (11.6.1), we obtain

$$\mathcal{C}_\nu'(x) = \tilde{\alpha}_1(x) \cos x + \tilde{\alpha}_2(x) \sin x, \quad \tilde{\alpha}_1, \tilde{\alpha}_2 \in \mathbf{A}^{(-1/2)}. \quad (11.8.24)$$

Recalling that $x_l \sim \pi l + a_1 + a_2 l^{-1} + \cdots$ as $l \to \infty$, and substituting this in (11.8.24), we get

$$\mathcal{C}_\nu'(x_l) \sim (-1)^l \sum_{i=0}^{\infty} v_i l^{-1/2-i} \quad \text{as } l \to \infty, \quad (11.8.25)$$

from which

$$\frac{\mathcal{C}_\nu'(x_{l+1})}{\mathcal{C}_\nu'(x_l)} \sim -1 + \sum_{i=1}^{\infty} \tau_i l^{-i} \quad \text{as } l \to \infty. \quad (11.8.26)$$

Using the fact that $u(x) = e^{\phi(x)} h(x)$, and proceeding as in the proof of Theorem 11.8.4 [starting with (11.8.15)], and invoking (11.8.25) as well, we can show that

$$w(x) \sim -1 + w_1 x^{-1} + w_2 x^{-2} + \cdots \quad \text{as } x \to \infty \text{ if } k \leq 1,$$

$$w(x) = e^{P(x)}, \quad P(x) \sim \sum_{i=0}^{\infty} p_i x^{k-1-i} \quad \text{as } x \to \infty, \quad p_0 < 0, \text{ if } k > 1. \quad (11.8.27)$$

Combining (11.8.21), (11.8.22), and (11.8.26), we finally obtain

$$F(x_l) = I[f] + \psi(x_l) g(x_l), \quad (11.8.28)$$

where

$$g(x) \sim \sum_{i=0}^{\infty} \hat{b}_i x^{-i} \quad \text{as } x \to \infty \text{ if } k \leq 1,$$

$$g(x) = -1 + e^{P(x)} \text{ if } k > 1. \quad (11.8.29)$$

Note that when $k > 1$, $e^{P(x)} \to 0$ as $x \to \infty$ essentially like $\exp(\hat{e}_0 x^{k-1})$ with $\hat{e}_0 < 0$. Consequently, $g \in \mathbf{A}^{(0)}$ in any case.

We have thus shown that $\psi(x_l)$ is a true shape function for $F(x_l)$ with these variants of the mW-transformation as well.

Variants for General Integral Transforms with Oscillatory Kernels

In view of the variants of the mW-transformation for Hankel transforms we have just discussed, we now propose variants for general integral transforms $\int_0^\infty u(t) K(t) \, dt$, where $u(x)$ does not oscillate as $x \to \infty$ or it oscillates very slowly, for example, like

$\exp(\pm ic(\log x)^s)$, with c and s real, and $K(x)$, the kernel of the transform, can be expressed in the form $K(x) = v(x)\cos\theta(x) + w(x)\sin\theta(x)$, with $\theta(x)$ real and $\theta \in \mathbf{A}^{(m)}$ for some positive integer m, and $v, w \in \mathbf{A}^{(\sigma)}$ for some σ. Let $\bar{\theta}(x)$ be the polynomial part of the asymptotic expansion of $\theta(x)$ as $x \to \infty$. Obviously, $K(x)$ oscillates like $\exp[\pm i\bar{\theta}(x)]$ infinitely many times as $x \to \infty$.

Definition 11.8.5 Choose the x_l to be consecutive zeros of $K(x)$ or of $K'(x)$ on $(0, \infty)$, and define the mW-transformation via the linear equations

$$F(x_l) = W_n^{(j)} + \psi(x_l)\sum_{i=0}^{n-1}\frac{\bar{\beta}_i}{x_l^i}, \quad j \le l \le j+n, \tag{11.8.30}$$

where, with $f(x) = u(x)K(x)$,

$$F(x_l) = \int_0^{x_l} f(t)\,dt, \quad \psi(x_l) = F(x_{l+1}) - F(x_l) = \int_{x_l}^{x_{l+1}} f(t)\,dt. \tag{11.8.31}$$

These variants of the mW-transformation should be effective in case $K(x)$ starts to behave like $\exp[\pm i\bar{\theta}(x)]$ only for sufficiently large x, as is the case, for example, when $K(x) = \mathcal{C}_\nu(x)$ with very large ν.

11.8.4 Further Variants of the mW-Transformation

Let $f \in \hat{\mathbf{B}}$ as in Definition 11.8.1, and let $I[f]$ be the value of the integral $\int_0^\infty f(t)\,dt$ or its Abel sum. In case we do not want to bother with the analysis of $\theta_j(x)$, we can compute $I[f]$ by applying variants of the mW-transformation that are again defined via the equations (11.8.8) and (11.8.9), with x_l now being chosen as consecutive zeros of either $f(x)$ or $f'(x)$. Determination of these x_l may be more expensive than those in Definition 11.8.3. Again, the corresponding $W_n^{(j)}$ can be computed via the W-algorithm.

These new methods can also be justified by proving a theorem analogous to Theorem 11.8.4 for their corresponding $\psi(x_l)$, at least in some cases for which x_l satisfy $x_l \sim \sum_{i=0}^\infty a_i l^{(1-i)/m}$ as $l \to \infty$, $a_0 = (\pi/\mu_0)^{1/m} > 0$.

The performances of these variants of the mW-transformation are similar to those of the $\bar{D}$- and $s\bar{D}$-transformations, which use the same sets of x_l.

Another variant of the mW-transformation can be defined in case $x_l \sim \sum_{i=0}^\infty a_i l^{(1-i)/m}$ as $l \to \infty$, $a_0 > 0$, which is satisfied when the x_l are zeros of $\sin\bar{\theta}(x)$ or of $\cos\bar{\theta}(x)$ or of special functions such as Bessel functions. In such a case, any quantity μ_l that has an asymptotic expansion of the form $\mu_l \sim \sum_{i=0}^\infty e_i x_l^{-i}$ as $l \to \infty$ has an asymptotic expansion also of the form $\mu_l \sim \sum_{i=0}^\infty e_i l^{-i/m}$ as $l \to \infty$, and this applies to the function $g(x)$ in (11.8.10). The new variant of the mW-transformation is then defined via the linear systems

$$F(x_l) = W_n^{(j)} + \psi(x_l)\sum_{i=0}^{n-1}\frac{\bar{\beta}_i}{(l+1)^{i/m}}, \quad j \le l \le j+n. \tag{11.8.32}$$

Again, the $W_n^{(j)}$ can be computed with the help of the W-algorithm. Note that the resulting method is analogous to the $\tilde{d}^{(m)}$-transformation. Numerical results indicate that this new method is also very effective.

11.8.5 Further Applications of the mW-Transformation

Recall that, in applying the mW-transformation, we consider only the dominant (polynomial) part $\bar{\theta}(x)$ of the phase of oscillations of $f(x)$ and need not concern ourselves with the modulating factors $h_j(x)e^{\phi_j(x)}$. This offers a significant advantage, as it suggests (see Sidi [288]) that we could at least attempt to apply the mW-transformation to *all* (very) oscillatory integrals $\int_0^\infty f(t)\,dt$ whose integrands are of the form

$$f(x) = \sum_{j=1}^r u_j(\theta_j(x))H_j(x),$$

where $u_j(z)$ and $\theta_j(x)$ are exactly as described in Definition 11.8.1, and $H_j(x)$ are arbitrary functions that do not oscillate as $x \to \infty$ or that may oscillate slowly, that is, slower than $e^{\theta_j(x)}$. Note that *not all* such functions $f(x)$ are in the class $\hat{\mathbf{B}}$.

Numerical results indicate that the mW-transformation is as efficient on such integrals as it is on those integrals with integrands in $\hat{\mathbf{B}}$. It was used successfully in the numerical inversion of general Kontorovich-Lebedev transforms by Ehrenmark [74], [75], who also provided a rigorous justification for this usage.

If we do not want to bother with the analysis of $\theta_j(x)$, we can apply the variants of the mW-transformation discussed in the preceding subsection to integrals of such $f(x)$ with the same effectiveness. Again, determination of the corresponding x_l may be more expensive than before.

11.9 Convergence and Stability

In this section, we consider briefly the convergence and stability of the reduced transformations we developed so far. The theory of Section 4.4 of Chapter 4 applies in general, because the transformations of this chapter are all GREPs.

For the $\bar{D}^{(1)}$-, $s\bar{D}^{(1)}$-, $\tilde{D}^{(1)}$-, W-, and mW-transformations for (very) oscillatory integrals, the theory of Chapter 9 applies with $t_j = x_j^{-1}$ throughout. For example, (i) the $\bar{D}^{(1)}$-transformations for Fourier transforms defined in Section 11.3, (ii) all three $\bar{D}^{(1)}$-transformations for Hankel transforms in Section 11.4, (iii) the $\tilde{D}$-transformations for Hankel transforms in Section 11.5, and (iv) the W- and mW-transformations for integrals of functions in the class $\hat{\mathbf{B}}$, can all be treated within the framework of Chapter 9. In particular, Theorems 9.3.1–9.3.3, 9.4.2, and 9.4.3 hold with appropriate $\psi(t)$ and $B(t)$ there. These can be identified from the form of $F(x_l) - I[f]$ in each case. We note only that the diagonal sequences $\{\bar{D}_n^{(1,j)}\}_{n=0}^\infty$, $\{\tilde{D}_n^{(1,j)}\}_{n=0}^\infty$, and $\{W_n^{(j)}\}_{n=0}^\infty$, when γ is real everywhere, are stable and converge. Actually, we have

$$\lim_{n\to\infty} \Gamma_n^{(j)} = 1 \quad \text{and} \quad A_n^{(j)} - I[f] = O(n^{-\mu}) \text{ as } n \to \infty \text{ for every } \mu > 0,$$

as follows from Theorems 9.4.2 and 9.4.3. Here $A_n^{(j)}$ stands for $\bar{D}_n^{(1,j)}$ or $\tilde{D}_n^{(1,j)}$ or $W_n^{(j)}$, depending on the method being used.

11.10 Extension to Products of Oscillatory Functions

In this section, we show how the reduced methods can be applied to integrands that are given as products of different oscillating functions in $\hat{\mathbf{B}}$.

As an example, let us consider $f(x) = u(x) \prod_{r=1}^{s} f_r(x)$, where

$$u(x) = e^{\phi(x)} h(x), \quad \phi(x) \text{ real and } \phi \in \mathbf{A}^{(k)}, \quad k \geq 0 \text{ integer}, \quad h \in \mathbf{A}^{(\gamma)}, \quad (11.10.1)$$

and

$$f_r(x) = \alpha_r^{(+)} U_r^{(+)}(x) + \alpha_r^{(-)} U_r^{(-)}(x), \tag{11.10.2}$$

where $\alpha_r^{(\pm)}$ are constants and $U_r^{(\pm)}(x)$ are of the form

$$U_r^{(\pm)}(x) = e^{v_r(x) \pm i w_r(x)} V_r^{(\pm)}(x),$$

$$v_r \in \mathbf{A}^{(k_r)}, \quad w_r \in \mathbf{A}^{(m_r)}, \quad k_r, \ m_r \geq 0 \text{ integers}; \quad v_r(x), \ w_r(x) \text{ real},$$

$$V_r^{(\pm)} \in \mathbf{A}^{(\delta_r)} \text{ for some } \delta_r. \tag{11.10.3}$$

We assume that $u(x)$, $\alpha_r^{(\pm)}$, and $U_r^{(\pm)}(x)$ are all known and that the polynomial parts $\bar{w}_r(x)$ of the $w_r(x)$ are available. In other words, if $w_r(x) \sim \sum_{i=0}^{\infty} w_{ri} x^{m_r - i}$ as $x \to \infty$, then $\bar{w}_r(x) = \sum_{i=0}^{m_r - 1} w_{ri} x^{m_r - i}$ is known explicitly for each r.

(i) We now propose to compute $\int_0^{\infty} f(t) \, dt$ by expanding the product $u \prod_{r=1}^{s} f_r$ in terms of the $U_r^{(\pm)}$ and applying to each term in this expansion an appropriate GREP$^{(1)}$. Note that there are 2^s terms in this expansion and each of them is in $\mathbf{B}^{(1)}$. This approach is very inexpensive and produces very high accuracy.

To illustrate the procedure above, let us look at the case $s = 2$. We have

$$f = u \left[\alpha_1^{(+)} \alpha_2^{(+)} U_1^{(+)} U_2^{(+)} + \alpha_1^{(+)} \alpha_2^{(-)} U_1^{(+)} U_2^{(-)} \right.$$

$$\left. + \alpha_1^{(-)} \alpha_2^{(+)} U_1^{(-)} U_2^{(+)} + \alpha_1^{(-)} \alpha_2^{(-)} U_1^{(-)} U_2^{(-)} \right].$$

Each of the four terms in this summation is in $\mathbf{B}^{(1)}$ since $u, U_1^{(\pm)}, U_2^{(\pm)} \in \mathbf{B}^{(1)}$. It is sufficient to consider the functions $f^{(+,\pm)} \equiv u U_1^{(+)} U_2^{(\pm)}$, the remaining functions $f^{(-,\pm)} = u U_1^{(-)} U_2^{(\pm)}$ being similar. We have

$$f^{(+,\pm)}(x) = e^{\Phi(x) + i\Theta(x)} g(x),$$

with

$$\Phi = \phi + v_1 + v_2, \quad \Theta = w_1 \pm w_2, \quad g = h V_1^{(+)} V_2^{(\pm)}.$$

Obviously, $\Phi \in \mathbf{A}^{(K)}$ and $\Theta \in \mathbf{A}^{(M_\pm)}$ for some integers $K \geq 0$, $M_\pm \geq 0$, and $g \in \mathbf{A}^{(\gamma + \delta_1 + \delta_2)}$.

Two different possibilities can occur:

(a) If $M_\pm > 0$, then $f^{(+,\pm)}(x)$ is oscillatory, and we can apply to $\int_0^{\infty} f^{(+,\pm)}(t) \, dt$ the $D^{(1)}$- or the mW-transformation with x_l as the consecutive zeros of $\sin[\bar{w}_1(x) \pm \bar{w}_2(x)]$ or $\cos[\bar{w}_1(x) \pm \bar{w}_2(x)]$. We can also apply the variants of the mW-transformation by choosing the x_l to be the consecutive zeros of $\Re f^{(+,\pm)}(x)$ or $\Im f^{(+,\pm)}(x)$.

(b) If $M_\pm = 0$, then $f^{(+,\pm)}(x)$ is not oscillatory, and we can apply to $\int_0^\infty f^{(+,\pm)}(t)\,dt$
  the $D^{(1)}$-transformation with $x_l = \xi e^{cl}$ for some $\xi > 0$ and $c > 0$.

(ii) Another method we propose is based on the observation that the expansion of
  $u \prod_{r=1}^s f_r$ can also be expressed as the sum of 2^{s-1} functions, at least some of
  which are in $\hat{\mathbf{B}}$. Thus, the mW-transformation and its variants can be applied to
  each of these functions.

  To illustrate this procedure, we turn to the previous example and write f in the
  form $f = G_+ + G_-$, where

$$G_+ = u\left[\alpha_1^{(+)}\alpha_2^{(+)}U_1^{(+)}U_2^{(+)} + \alpha_1^{(-)}\alpha_2^{(-)}U_1^{(-)}U_2^{(-)}\right]$$

and

$$G_- = u\left[\alpha_1^{(+)}\alpha_2^{(-)}U_1^{(+)}U_2^{(-)} + \alpha_1^{(-)}\alpha_2^{(+)}U_1^{(-)}U_2^{(+)}\right].$$

If $M_\pm > 0$, then $G_\pm \in \hat{\mathbf{B}}$ with phase of oscillation $w_1 \pm w_2$. In this case, we can apply the mW-transformation to $\int_0^\infty G_\pm(t)\,dt$ by choosing the x_l to be the consecutive
zeros of either (a) $\sin[\bar{w}_1(x) \pm \bar{w}_2(x)]$ or $\cos[\bar{w}_1(x) \pm \bar{w}_2(x)]$ or of (b) $\Re G_\pm(x)$ or
$\Im G_\pm(x)$. If $M_\pm \leq 0$, then $G_\pm \in \mathbf{B}^{(1)}$ but is not oscillatory. In this case, we can apply
the $D^{(1)}$-transformation to $\int_0^\infty G_\pm(t)\,dt$.

As an example, we consider integrals of products of two Bessel functions, namely,
$\int_0^\infty u(t)\mathcal{C}_\mu(rt)\mathcal{C}_\nu(st)\,dt$. In this example, $U_1^{(\pm)}(x) = H_\mu^{(\pm)}(rx)$ and $U_2^{(\pm)}(x) = H_\nu^{(\pm)}(sx)$,
where $H_\mu^{(+)}(z)$ and $H_\mu^{(-)}(z)$ stand for the Hankel functions of the first and second kinds,
respectively, namely, $H_\mu^{(\pm)}(z) = J_\mu(z) \pm iY_\mu(z)$. Thus, $v_1(x) = v_2(x) = 0$ and $\bar{w}_1(x) = rx$ and $\bar{w}_2(x) = sx$ and $\delta_1 = \delta_2 = -1/2$. Therefore, we need to be concerned with
computation of the integrals of $u(x)H_\mu^{(+)}(rx)H_\nu^{(+)}(sx)$, etc., by the mW-transformation.

This approach was used by Lucas [188] for computing integrals of the form
$\int_0^\infty u(t)J_\mu(rt)J_\nu(st)\,dt$, which is a special case of those considered here.

12

Acceleration of Convergence of Power Series by the
d-Transformation: Rational d-Approximants

12.1 Introduction

In this chapter, we are concerned with the Sidi–Levin rational d-approximants and efficient summation of (convergent or divergent) power series $\sum_{k=1}^{\infty} c_k z^{k-1}$ by the d-transformation. We assume that $\{c_n\} \in \mathbf{b}^{(m)}$ for some m and analyze the consequences of this.

One of the difficulties in accelerating the convergence of power series has been the lack of stability and acceleration near points of singularity of the functions $f(z)$ represented by the series. We show in this chapter via a rigorous analysis how to tackle this problem and stabilize the acceleration process at will.

As we show in the next chapter, the results of our study of power series are useful for Fourier series, orthogonal polynomial expansions, and other series of special functions.

Most of the treatment of the subject we give in this chapter is based on the work of Sidi and Levin [312] and Sidi [294].

12.2 The d-Transformation on Power Series

Consider the application of the d-transformation to the power series $\sum_{k=1}^{\infty} c_k z^{k-1}$, where c_k satisfy the $(m+1)$-term recursion relation

$$c_{n+m} = \sum_{s=0}^{m-1} q_s(n) c_{n+s}, \quad q_s \in \mathbf{A}_0^{(\mu_s)} \text{ strictly}, \quad \mu_s \text{ integer}, \quad s = 0, 1, \ldots, m-1.$$

$$(12.2.1)$$

(This will be the case when $\{c_n\} \in \mathbf{b}^{(m)}$ in the sense of Definition 6.1.2.) We have the following interesting result on the sequence $\{c_n z^{n-1}\}$:

Theorem 12.2.1 *Let the sequence $\{c_n\}$ be as in the preceding paragraph. Then, the sequence $\{a_n\}$, where $a_n = c_n z^{n-1}$, $n = 1, 2, \ldots$, is in $\mathbf{b}^{(m)}$ exactly in accordance with Definition 6.1.2, except when $z \in Z = \{\tilde{z}_1, \ldots, \tilde{z}_\nu\}$, for some $\tilde{z}_i$ and $0 \le \nu \le m$. Actually, for $z \notin Z$, there holds $a_n = \sum_{k=1}^{m} p_k(n) \Delta^k a_n$ with $p_k \in \mathbf{A}_0^{(0)}$, $k = 1, \ldots, m$.*

Proof. Let us first define $q_m(n) \equiv -1$ and rewrite (12.2.1) in the form $\sum_{s=0}^{m} q_s(n) c_{n+s} = 0$. Multiplying this equation by z^{n+m-1} and invoking $c_k = a_k z^{-k+1}$ and (6.1.7), we obtain

$$\sum_{s=0}^{m} q_s(n) z^{m-s} \sum_{k=0}^{s} \binom{s}{k} \Delta^k a_n = 0. \qquad (12.2.2)$$

Interchanging the order of the summations, we have

$$\sum_{k=0}^{m} \left[\sum_{s=k}^{m} \binom{s}{k} q_s(n) z^{m-s} \right] \Delta^k a_n = 0. \qquad (12.2.3)$$

Rearranging things, we obtain $a_n = \sum_{k=1}^{m} p_k(n) \Delta^k a_n$ with

$$p_k(n) = -\frac{\sum_{s=k}^{m} \binom{s}{k} q_s(n) z^{m-s}}{\sum_{s=0}^{m} q_s(n) z^{m-s}} \equiv \frac{N_k(n)}{D(n)}, \quad k = 1, \dots, m. \qquad (12.2.4)$$

We realize that, when viewed as functions of n (z being held fixed), the numerator $N_k(n)$ and the denominator $D(n)$ of $p_k(n)$ in (12.2.4) have asymptotic expansions of the form

$$N_k(n) \sim \sum_{i=0}^{\infty} r_{ki}(z) n^{\rho_k - i} \quad \text{and} \quad D(n) \sim \sum_{i=0}^{\infty} \hat{r}_i(z) n^{\sigma - i} \quad \text{as } n \to \infty, \qquad (12.2.5)$$

where $r_{ki}(z)$ and $\hat{r}_i(z)$ are polynomials in z, of degree at most $m - k$ and m, respectively, and are independent of n, and

$$\rho_k = \max_{k \leq s \leq m} \{\mu_s\} \quad \text{and} \quad \sigma = \max_{0 \leq s \leq m} \{\mu_s\}; \quad \mu_m = 0. \qquad (12.2.6)$$

Obviously, $r_{k0}(z) \not\equiv 0$ and $\hat{r}_0(z) \not\equiv 0$ and $\rho_k \leq \sigma$ for each k. Therefore, $N_k \in \mathbf{A}_0^{(\rho_k)}$ strictly provided $r_{k0}(z) \neq 0$ and $D \in \mathbf{A}_0^{(\sigma)}$ strictly provided $\hat{r}_0(z) \neq 0$. Thus, as long as $\hat{r}_0(z) \neq 0$ [note that $\hat{r}_0(z) = 0$ for at most m values of z], we have $p_k \in \mathbf{A}_0^{(i_k)}$, $i_k \leq \rho_k - \sigma \leq 0$; hence, $p_k \in \mathbf{A}_0^{(0)}$. When z is such that $\hat{r}_0(z) = 0$, we have that $D \in \mathbf{A}_0^{(\sigma-1)}$, and this may cause an increase in i_k. This completes the proof. ■

Remarks.

1. One immediate consequence of Theorem 12.2.1 is that the $d^{(m)}$-transformation can be applied to the series $\sum_{k=1}^{\infty} c_k z^{k-1}$ with $\rho_k = 0$, $k = 0, 1, \dots, m$.
2. Recall that if $c_n = h(n)$ with $h \in \mathbf{A}_0^{(\gamma)}$ for some γ, and $a_n = c_n z^{n-1}$, then $\{a_n\} \in \mathbf{b}^{(1)}$ with $a_n = p(n) \Delta a_n$, $p \in \mathbf{A}_0^{(0)}$ as long as $z \neq 1$ while $p \in \mathbf{A}_0^{(1)}$ when $z = 1$. As mentioned in Chapter 6, if $|z| < 1$, $\sum_{k=1}^{\infty} c_k z^{k-1}$ converges to a function that is analytic for $|z| < 1$, and this function is singular at $z = 1$. Therefore, we conjecture that the zeros of the polynomial $\hat{r}_0(z)$ in the proof of Theorem 12.2.1 are points of singularity of the function $f(z)$ that is represented by $\sum_{k=1}^{\infty} c_k z^{k-1}$ if this function has singularities in the finite plane. We invite the reader to verify this claim for the case in which $c_n = \alpha^n/n + \beta^n$, $\alpha \neq \beta$, the function represented by $\sum_{k=1}^{\infty} c_k z^{k-1}$ being $-z^{-1} \log(1 - \alpha z) + \beta(1 - \beta z)^{-1}$, with singularities at $z = 1/\alpha$ and $z = 1/\beta$.

12.3 Rational Approximations from the *d*-Transformation

We now apply the $d^{(m)}$-transformation to the infinite series $\sum_{k=1}^{\infty} c_k z^{k-1}$ assuming that $\{c_n\}$ is as in Theorem 12.2.1 and that $\sum_{k=1}^{\infty} c_k z^{k-1}$ has a positive radius of convergence ρ. Now, with $a_n = c_n z^{n-1}$, we have $\{a_n\} \in \mathbf{b}^{(m)}$ and $a_n = \sum_{k=1}^{m} p_k(n) \Delta^k a_n$ with $p_k \in \mathbf{A}_0^{(0)}$ for all k, by Theorem 12.2.1. Therefore, when $|z| < \rho$ and hence $\sum_{k=1}^{\infty} c_k z^{k-1}$ converges, Theorem 6.1.12 holds, and we have the asymptotic expansion

$$A_{n-1} \sim A + \sum_{k=0}^{m-1} (\Delta^k a_n) \sum_{i=0}^{\infty} \frac{g_{ki}}{n^i} \quad \text{as} \quad n \to \infty, \tag{12.3.1}$$

where $A_r = \sum_{k=1}^{r} a_k$ and A is the sum of $\sum_{k=1}^{\infty} a_k$. Invoking (6.1.5), we can rewrite (12.3.1) in the form

$$A_{n-1} \sim A + \sum_{k=0}^{m-1} a_{n+k} \sum_{i=0}^{\infty} \frac{g'_{ki}}{n^i} \quad \text{as} \quad n \to \infty, \tag{12.3.2}$$

and redefine the $d^{(m)}$-transformation by the linear systems

$$A_{R_l-1} = d_n^{(m,j)} + \sum_{k=1}^{m} a_{R_l+k-1} \sum_{i=0}^{n_k-1} \frac{\bar{\beta}_{ki}}{R_l^i}, \quad j \le l \le j+N; \quad N = \sum_{k=1}^{m} n_k, \tag{12.3.3}$$

where $n = (n_1, \dots, n_m)$ as before and $A_0 \equiv 0$. Also, $d_{(0,\dots,0)}^{(m,j)} = A_j$ for all j.

12.3.1 Rational d-Approximants

From this definition of the $d^{(m)}$-transformation, we can derive the following interesting result:

Theorem 12.3.1 *Let $d_n^{(m,j)}$ be as defined in (12.3.3) with $R_l = l+1$, $l = 0, 1, \dots$, that is,*

$$A_l = d_n^{(m,j)} + \sum_{k=1}^{m} a_{l+k} \sum_{i=0}^{n_k-1} \frac{\bar{\beta}_{ki}}{(l+1)^i}, \quad l = j, j+1, \dots, j+N. \tag{12.3.4}$$

Then, $d_n^{(m,j)} \equiv d_n^{(m,j)}(z)$ is of the form

$$d_n^{(m,j)}(z) = \frac{u(z)}{v(z)} = \frac{\sum_{i=0}^{N} \lambda_{ni}^{(j)} z^{N-i} A_{j+i}}{\sum_{i=0}^{N} \lambda_{ni}^{(j)} z^{N-i}} \tag{12.3.5}$$

for some constants $\lambda_{ni}^{(j)}$. Thus, $d_n^{(m,j)}$ is a rational function in z, whose numerator and denominator polynomials $u(z)$ and $v(z)$ are of degree at most $j+N-1$ and N, respectively.

Remark. In view of Theorem 12.3.1, we call the approximations $d_n^{(m,j)}$ to the sum of $\sum_{k=1}^{\infty} a_k z^{k-1}$ (limit or antilimit of $\{A_k\}$) that are defined via the equations in (12.3.4) *rational d-approximants*.

Proof. Let us substitute $a_n = c_n z^{n-1}$ in (12.3.4) and multiply both sides by z^{j+N-l}. This results in the linear system

$$z^{j+N-l} A_l = z^{j+N-l} d_n^{(m,j)} + \sum_{k=1}^{m} c_{l+k} \sum_{i=0}^{n_k-1} \frac{\tilde{\beta}_{ki}}{(l+1)^i}, \quad l = j, j+1, \ldots, j+N,$$

(12.3.6)

where $\tilde{\beta}_{ki} = z^{j+N+k-1} \bar{\beta}_{ki}$ for all k and i are the new auxiliary unknowns. Making $d_n^{(m,j)}$ the last component of the vector of unknowns, the matrix of this system assumes the form

$$M(z) = \left[M_0 \left| \begin{array}{c} z^N \\ z^{N-1} \\ \vdots \\ z^0 \end{array} \right. \right], \quad M_0 : (N+1) \times N \text{ and independent of } z, \quad (12.3.7)$$

and the right-hand side is the vector

$$[z^N A_j, z^{N-1} A_{j+1}, \ldots, z^0 A_{j+N}]^T.$$

The result in (12.3.5) follows if we solve (12.3.6) by Cramer's rule and expand the resulting numerator and denominator determinants with respect to their last columns. For each i, we can take $\lambda_{ni}^{(j)}$ to be the cofactor of z^{N-i} in the last column of $M(z)$ in (12.3.7). ∎

As is clear from (12.3.5), if the $\lambda_{ni}^{(j)}$ are known, then $d_n^{(m,j)}$ is known completely. The $\lambda_{ni}^{(j)}$ can be determined in a systematic way as shown next in Theorem 12.3.2. Note that the $\lambda_{ni}^{(j)}$ are unique up to a common multiplicative factor.

Theorem 12.3.2 *Up to a common multiplicative factor, the coefficients $\lambda_{ni}^{(j)}$ of the denominator polynomial $v(z)$ of $d_n^{(m,j)}$ in (12.3.5) satisfy the linear system*

$$[M_0|w]^T \lambda = e_1, \quad (12.3.8)$$

where M_0 is as in (12.3.7), $\lambda = [\lambda_{n0}^{(j)}, \lambda_{n1}^{(j)}, \ldots, \lambda_{nN}^{(j)}]^T$, and $e_1 = [1, 0, \ldots, 0]^T$, and w is an arbitrary column vector for which $[M_0|w]$ is nonsingular.

Proof. Since $\lambda_{ni}^{(j)}$ can be taken as the cofactor of z^{N-i} in the last column of the matrix $M(z)$, we have that λ satisfies the $N \times (N+1)$ homogeneous system $M_0^T \lambda = 0$, namely,

$$\sum_{i=0}^{N} \lambda_{ni}^{(j)} c_{j+k+i} / (j+i+1)^r = 0, \quad 0 \le r \le n_k - 1, \quad 1 \le k \le m. \quad (12.3.9)$$

Appending to this system the scaling $w^T \lambda = 1$, we obtain (12.3.8). ∎

12.3.2 Closed-Form Expressions for $m = 1$

Using the results of Section 6.3, we can give closed-form expressions for $d_n^{(m,j)}(z)$ when $m = 1$. The $d^{(1)}$-transformation now becomes the Levin $\mathcal{L}$-transformation and $d_n^{(1,j)}(z)$

becomes $\mathcal{L}_n^{(j)}(z)$. By (6.3.2) and (6.3.3), we have

$$\mathcal{L}_n^{(j)}(z) = \frac{\sum_{i=0}^{n} \lambda_{ni}^{(j)} z^{n-i} A_{j+i}}{\sum_{i=0}^{n} \lambda_{ni}^{(j)} z^{n-i}}; \quad \lambda_{ni}^{(j)} = (-1)^i \binom{n}{i} \frac{(j+i+1)^{n-1}}{c_{j+i+1}}. \quad (12.3.10)$$

A similar expression can be given with the $d^{(1)}$-transformation replaced by the Sidi $\mathcal{S}$-transformation. By (6.3.4) and (6.3.5), we have

$$\mathcal{S}_n^{(j)}(z) = \frac{\sum_{i=0}^{n} \lambda_{ni}^{(j)} z^{n-i} A_{j+i}}{\sum_{i=0}^{n} \lambda_{ni}^{(j)} z^{n-i}}; \quad \lambda_{ni}^{(j)} = (-1)^i \binom{n}{i} \frac{(j+i+1)_{n-1}}{c_{j+i+1}}. \quad (12.3.11)$$

12.4 Algebraic Properties of Rational d-Approximants

The rational d-approximants of the preceding section have a few interesting algebraic properties to which we now turn.

12.4.1 Padé-like Properties

Theorem 12.4.1 *Let $d_n^{(m,j)}$ be the rational function defined via (12.3.4). Then $d_{(1,\dots,1)}^{(m,j)}$ is the $[j+m-1/m]$ Padé approximant from the power series $\sum_{k=1}^{\infty} c_k z^{k-1}$.*

Proof. Letting $n_1 = \cdots = n_m = 1$ in (12.3.4), the latter reduces to

$$A_l = d_{(1,\dots,1)}^{(m,j)} + \sum_{k=1}^{m} \bar{\beta}_k a_{l+k}, \quad l = j, j+1, \dots, j+m, \quad (12.4.1)$$

and the result follows from the definition of the Padé approximants that are considered later.   ∎

Our next result concerns not only the rational function $d_n^{(m,j)}(z)$ but all those rational functions that are of the general form given in (12.3.5). Because the proof of this result is straightforward, we leave it to the reader.

Theorem 12.4.2 *Let $A_0 = 0$ and $A_r = \sum_{k=1}^{r} c_k z^{k-1}$, $r = 1, 2, \dots$, and let $R(z)$ be a rational function of the form*

$$R(z) = \frac{U(z)}{V(z)} = \frac{\sum_{i=0}^{s} \mu_i z^{s-i} A_{q+i}}{\sum_{i=0}^{s} \mu_i z^{s-i}}. \quad (12.4.2)$$

Then

$$V(z) \sum_{k=1}^{\infty} c_k z^{k-1} - U(z) = O(z^{q+s}) \ as \ z \to 0. \quad (12.4.3)$$

If $\mu_s \neq 0$ in addition, then

$$\sum_{k=1}^{\infty} c_k z^{k-1} - R(z) = O(z^{q+s}) \ as \ z \to 0. \quad (12.4.4)$$

It is clear from (12.3.5) that Theorem 12.4.2 applies with $q = j$, $s = N$, and $R(z) = d_n^{(m,j)}$. A better result is possible, again with the help of Theorem 12.4.2, and this is achieved in Theorem 12.4.3. The improved result of this theorem exhibits a Padé-like property of $d_n^{(m,j)}$. To help understand this property, we pause to give a one-sentence definition of Padé approximants: the $[M/N]$ Padé approximant $f_{M,N}(z)$ to the formal power series $f(z) := \sum_{k=1}^{\infty} c_k z^{k-1}$, when it exists, is the rational function $U(z)/V(z)$, with $U(z)$ and $V(z)$ being polynomials of degree at most M and N, respectively, with $V(0) = 1$, such that $f(z) - f_{M,N}(z) = O(z^{M+N+1})$ as $z \to 0$. As we show later, $f_{M,N}(z)$ is determined (uniquely) by the terms $c_1, c_2, \dots, c_{M+N+1}$ of $f(z)$. Thus, all Padé approximants $R(z)$ to $f(z)$ that are defined by the terms $c_1, c_2, \dots, c_L$ satisfy $f(z) - R(z) = O(z^L)$ as $z \to 0$. Precisely this is the Padé-like property of $d_n^{(m,j)}$ we are alluding to, which we prove next. Before we do that, we want to point out that Padé approximants, like rational d-approximants, are of the form of $R(z)$ in (12.4.2) with appropriate q and s.

Theorem 12.4.3 *Let $d_n^{(m,j)}$ be as in Theorem 12.3.1. Then*

$$d_n^{(m,j)}(z) = \frac{\sum_{i=0}^{N} \lambda_{ni}^{(j)} z^{N-i} A_{j+m+i}}{\sum_{i=0}^{N} \lambda_{ni}^{(j)} z^{N-i}}, \tag{12.4.5}$$

with $\lambda_{ni}^{(j)}$ exactly as in (12.3.5) and as described in the proof of Theorem 12.3.1. Therefore,

$$v(z) \sum_{k=1}^{\infty} c_k z^{k-1} - u(z) = O(z^{j+N+m}) \quad as \ z \to 0. \tag{12.4.6}$$

If $\lambda_{nN}^{(j)} \neq 0$ in addition, then

$$\sum_{k=1}^{\infty} c_k z^{k-1} - d_n^{(m,j)}(z) = O(z^{j+N+m}) \quad as \ z \to 0. \tag{12.4.7}$$

Proof. We begin with the linear system in (12.3.4). If we add the terms $a_{l+1}, \dots, a_{l+m}$ to both sides of (12.3.4), we obtain the system

$$A_{l+m} = d_n^{(m,j)} + \sum_{k=1}^{m} a_{l+k} \sum_{i=0}^{n_k-1} \frac{\bar{\beta}_{ki}'}{(l+1)^i}, \quad l = j, j+1, \dots, j+N, \tag{12.4.8}$$

where $\bar{\beta}_{ki}' = \bar{\beta}_{ki}$, $i \neq 0$, and $\bar{\beta}_{k0}' = \bar{\beta}_{k0} + 1$, $k = 1, \dots, m$, and $d_n^{(m,j)}$ remains unchanged. As a result, the system (12.3.6) now becomes

$$z^{j+N-l} A_{l+m} = z^{j+N-l} d_n^{(m,j)} + \sum_{k=1}^{m} c_{l+k} \sum_{i=0}^{n_k-1} \frac{\tilde{\beta}_{ki}}{(l+1)^i}, \quad l = j, j+1, \dots, j+N,$$

$$\tag{12.4.9}$$

where $\tilde{\beta}_{ki} = z^{j+N+k-1} \bar{\beta}_{ki}'$ for all k and i are the new auxiliary unknowns, and (12.3.5) becomes (12.4.5) with $\lambda_{ni}^{(j)}$ exactly as in (12.3.5) and as described in the proof of Theorem 12.3.1.

With (12.4.5) available, we now apply Theorem 12.4.2 to $d_n^{(m,j)}$ with $s = N$ and $q = j + m$. ∎

The reader may be led to think that the two expressions for $d_n^{(m,j)}(z)$ in (12.3.5) and (12.4.5) are contradictory. As we showed in the proof of Theorem 12.4.3, they are not. We used (12.3.5) to conclude that the numerator and denominator polynomials $u(z)$ and $v(z)$ of $d_n^{(m,j)}(z)$ have degrees at most $j + N - 1$ and N, respectively. We used (12.4.5) to conclude that $d_n^{(m,j)}$, which is obtained from the terms c_k, $1 \leq k \leq j + N + m$, satisfies (12.4.6) and (12.4.7), which are Padé-like properties of $d_n^{(j,m)}$.

12.4.2 Recursive Computation by the $W^{(m)}$-Algorithm

Because the approximation $d_n^{(m,j)}$ satisfies the linear system in (12.3.6), the $W^{(m)}$-algorithm can be used to compute the coefficients $\lambda_{ni}^{(j)}$ of the denominator polynomials $v(z)$ recursively, as has been shown in Ford and Sidi [87]. [The coefficients of the numerator polynomial $u(z)$ can then be determined by (12.3.5) in Theorem 12.3.1.]

Let us first write the equations in (12.3.6) in the form

$$z^{-l} A_l = z^{-l} d_n^{(m,j)} + \sum_{k=1}^{m} c_{l+k} \sum_{i=0}^{n_k-1} \frac{\hat{\beta}_{ki}}{(l+1)^i}, \quad l = j, j+1, \ldots, j+N. \quad (12.4.10)$$

For each l, let $t_l = 1/(l+1)$, and define the $g_k(l) = \varphi_k(t_l)$ *in the normal ordering* through [cf. (7.3.3)]

$$g_k(l) = \varphi_k(t_l) = c_{l+k}, \quad k = 1, \ldots, m,$$

$$g_k(l) = t_l g_{k-m}(l), \quad k = m+1, m+2, \ldots. \quad (12.4.11)$$

Defining $G_p^{(j)}$, $D_p^{(j)}$, $f_p^{(j)}(b)$, and $\psi_p^{(j)}(b)$ as in Section 3.3 of Chapter 3, with the present $g_k(l)$, we realize that $d_n^{(m,j)}$ is given by

$$d_n^{(m,j)} = \frac{\psi_N^{(j)}(\xi)}{\psi_N^{(j)}(\eta)}, \quad (12.4.12)$$

where

$$n_k = \lfloor (N-k)/m \rfloor, \quad k = 1, \ldots, m, \quad \text{and} \quad N = \sum_{k=1}^{m} n_k, \quad (12.4.13)$$

and

$$\xi(l) = z^{-l} A_l \quad \text{and} \quad \eta(l) = z^{-l}. \quad (12.4.14)$$

Since $\psi_N^{(j)}(\eta) = f_N^{(j)}(\eta)/G_{N+1}^{(j)}$ and $f_N^{(j)}(\eta)$ is a polynomial in z^{-1} of degree N and $G_{N+1}^{(j)}$ is independent of z, we have that

$$\psi_N^{(j)}(\eta) = \sum_{i=0}^{N} \kappa_{Ni}^{(j)} z^{-j-i}, \quad (12.4.15)$$

where $\kappa_{Ni}^{(j)} = \lambda_{ni}^{(j)}$, $i = 0, 1, \ldots, N$, up to scaling. By the fact that

$$\psi_p^{(j)}(\eta) = \frac{\psi_{p-1}^{(j+1)}(\eta) - \psi_{p-1}^{(j)}(\eta)}{D_p^{(j)}}, \quad (12.4.16)$$

with $D_p^{(j)}$ independent of z, we see that the $\kappa_{pi}^{(j)}$ can be computed recursively from

$$\kappa_{pi}^{(j)} = \frac{\kappa_{p-1,i-1}^{(j+1)} - \kappa_{p-1,i}^{(j)}}{D_p^{(j)}}, \quad i = 0, 1, \dots, p, \tag{12.4.17}$$

with $\kappa_{pi}^{(j)} = 0$ when $i < 0$ and $i > p$. It is clear from (12.4.17) that the $W^{(m)}$-algorithm is used *only* to obtain the $D_p^{(j)}$.

12.5 Prediction Properties of Rational *d*-Approximants

So far we have used extrapolation methods to approximate (or *predict*) the limit or antilimit S of a given sequence $\{S_n\}$ from a finite number of the S_n. We now show how they can be used to predict the terms $S_{r+1}, S_{r+2}, \dots$, when $S_1, S_2, \dots, S_r$ are given. An approach to the solution of this problem was first formulated in Gilewicz [98], where the Maclaurin expansion of a Padé approximant $f_{M,N}(z)$ from $\sum_{k=0}^{\infty} c_k z^{k-1}$ is used to approximate the coefficients c_k, $k \geq M + N + 2$. [Recall that $f_{M,N}(z)$ is determined only by $c_k, k = 1, \dots, M + N + 1$, and that $f_{M,N}(z) = \sum_{k=1}^{M+N+1} c_k z^{k-1} + \sum_{k=M+N+2}^{\infty} \hat{c}_k z^{k-1}$.] The approach we describe here with the help of the rational *d*-approximants is valid for *all* sequence transformations and was first given by Sidi and Levin [312], [313]. It was applied recently by Weniger [355] to some known sequence transformations that naturally produce rational approximations when applied to partial sums of power series. (For different approaches to the issue of prediction, see Brezinski and Redivo Zaglia [41, pp. 390] and Vekemans [345].)

Let $c_1 = S_1$, $c_k = S_k - S_{k-1}$, $k = 2, 3, \dots$, and assume that $\{c_n\}$ is as in Section 12.2. By Theorem 12.2.1, we have that $\{c_n z^{n-1}\} \in \mathbf{b}^{(m)}$, so we can apply the $d^{(m)}$-transformation to the sequence $\{A_r\}$, where $A_r = \sum_{k=1}^{r} c_k z^{k-1}$, to obtain the rational *d*-approximant $R_N^{(j)}(z) \equiv d_n^{(m,j)}$, with $n = (n_1, \dots, n_m)$ and $N = \sum_{k=1}^{m} n_k$ as before.

Because $R_N^{(j)}(z)$ is a good approximation to the sum of $\sum_{k=1}^{\infty} c_k z^{k-1}$, its Maclaurin expansion $\sum_{k=1}^{\infty} \hat{c}_k z^{k-1}$ is likely to be close to $\sum_{k=1}^{\infty} c_k z^{k-1}$ for small z. Indeed, from Theorem 12.4.3, we already know that $\hat{c}_k = c_k$, $k = 1, 2, \dots, \tilde{k}$, where $\tilde{k} = j + N + m$. Thus,

$$\sum_{k=1}^{\infty} c_k z^{k-1} - R_N^{(j)}(z) = \sum_{k=\tilde{k}+1}^{\infty} \delta_k z^{k-1}; \quad \delta_k = c_k - \hat{c}_k, \ k \geq \tilde{k} + 1, \tag{12.5.1}$$

and we expect at least the first few of the δ_k, $k \geq \tilde{k} + 1$, to be very small. In case this happens, we will have that for the first few values of $k \geq \tilde{k} + 1$, $\hat{c}_k$ will be very close to the corresponding c_k, and hence $S_k \approx S_{\tilde{k}} + \sum_{i=\tilde{k}+1}^{k} \hat{c}_i$. Precisely this is the *prediction property* we alluded to above.

For a theoretical justification of the preceding procedure, see Sidi and Levin [312], where the prediction properties of the rational $d^{(m)}$- and Padé approximants are compared by a nontrivial example. For a thorough and rigorous treatment of the case in which $m = 1$, see Sidi and Levin [313], who provide precise rates of convergence of the $\hat{c}_k$ to the corresponding c_k, along with convincing theoretical and numerical examples. The results of [313] clearly show that $\hat{c}_k \to c_k$ very quickly under both Process I and Process II.

Finally, it is clear that the rational function $R_N^{(j)}(z)$ can be replaced by any other approximation (not necessarily rational) that has Padé-like properties.

12.6 Approximation of Singular Points

Assume that the series $\sum_{k=1}^{\infty} c_k z^{k-1}$ has a positive radius of convergence ρ so that it is the Maclaurin expansion of a function $f(z)$ that is analytic for $|z| < \rho$. In case ρ is finite, $f(z)$ has singularities on $|z| = \rho$. It may have additional singularities for $|z| > \rho$. It has been observed numerically that some or all of the poles of the rational functions obtained by applying sequence transformations to the power series of $f(z)$ serve as approximations to the points of singularity of $f(z)$. This is so for rational d-approximants and Padé approximants as well. Sidi and Levin [312] use the poles of the rational $d^{(m)}$- and Padé approximants from the power series of a function $f(z)$ that has both polar and branch singularities to approximate these singular points.

In the case of Padé approximants, the generalized Koenig theorem, which is discussed in Chapter 17, provides a theoretical justification of this use.

To clarify this point, we consider two examples of $\{c_n\} \in \mathbf{b}^{(1)}$ under Process I. In particular, we consider the approximations $d_1^{(1,j)}$ (which are also Padé approximants) from $\sum_{k=1}^{\infty} c_k z^{k-1}$ that are given by

$$d_1^{(1,j)}(z) = \frac{c_{j+1} z A_j - c_j A_{j+1}}{c_{j+1} z - c_j}.$$

Thus, each $d_1^{(1,j)}(z)$ has one simple pole at $\hat{z}^{(j)} = c_j / c_{j+1}$.

Example 12.6.1 When $c_n = h(n) \in \mathbf{A}_0^{(\gamma)}$ strictly, $\gamma \neq 0$, we have $\{c_n\} \in \mathbf{b}^{(1)}$. As mentioned earlier, the series $\sum_{k=1}^{\infty} c_k z^{k-1}$ converges for $|z| < 1$ and represents a function analytic for $|z| < 1$ with a singularity at $z = 1$. We have

$$\frac{c_j}{c_{j+1}} = 1 - \gamma j^{-1} + O(j^{-2}) \text{ as } j \to \infty.$$

That is, $\hat{z}^{(j)}$, the pole of $d_1^{(1,j)}(z)$, satisfies $\lim_{j \to \infty} \hat{z}^{(j)} = 1$, thus verifying the claim we have made.

Example 12.6.2 When $c_n = (n!)^{-\mu} h(n)$, where μ is a positive integer and $h \in \mathbf{A}_0^{(\gamma)}$ strictly for some γ, we have $\{c_n\} \in \mathbf{b}^{(1)}$. In this case, the series $\sum_{k=1}^{\infty} c_k z^{k-1}$ converges for all z and represents a function that is analytic everywhere with singularities only at infinity. We have

$$\frac{c_j}{c_{j+1}} \sim j^{\mu} \text{ as } j \to \infty.$$

That is, $\hat{z}^{(j)}$, the pole of $d_1^{(1,j)}(z)$, satisfies $\lim_{j \to \infty} \hat{z}^{(j)} = \infty$, verifying the claim we have made.

12.7 Efficient Application of the d-Transformation to Power Series with APS

Let the infinite series $\sum_{k=1}^{\infty} c_k z^{k-1}$ have a positive but finite radius of convergence. As mentioned earlier, this series is the Maclaurin expansion of a function $f(z)$ that is analytic for $|z| < \rho$ and has singularities for some z with $|z| = \rho$ and possibly with $|z| > \rho$ as well.

When a suitable convergence acceleration method is applied directly to the sequence of the partial sums $A_r = \sum_{k=1}^{r} c_k z^{k-1}$, accurate approximants to $f(z)$ can be obtained as long as z is not close to a singularity of $f(z)$. If z is close to a singularity, however, we are likely to face severe stability and accuracy problems in applying convergence acceleration. In case the $d^{(m)}$-transformation is applicable, the following strategy involving APS (discussed in Section 10.3) has been observed to be very effective in coping with the problems of stability and accuracy: (i) For z not close to a singularity, choose $R_l = l + 1$, $l = 0, 1, \ldots$. (ii) For z close to a singularity, choose $R_l = \kappa(l + 1)$, $l = 0, 1, \ldots$, where κ is a positive integer ≥ 2; the closer z is to a singularity, the larger κ should be. The $d^{(1)}$-transformation with APS was used successfully by Hasegawa [126] for accelerating the convergence of some slowly converging power series that arise in connection with a numerical quadrature problem.

This strategy is not ad hoc by any means, and its theoretical justification has been given in [294, Theorems 4.3 and 4.4] for the case in which $\{c_n\} \in \mathbf{b}^{(1)}$ and $\sum_{k=1}^{\infty} c_k z^{k-1}$ has a positive but finite radius of convergence so that $f(z)$, the limit or antilimit of $\sum_{k=1}^{\infty} c_k z^{k-1}$, has a singularity as in Example 12.6.1. Recall that the sequence of partial sums of such a series is linear. We now present the mathematical treatment of APS that was given by Sidi [294]. The next theorem combines Theorems 4.2–4.4 of [294].

Theorem 12.7.1 *Let $c_n = \lambda^n h(n)$, where λ is some scalar and $h \in \mathbf{A}_0^{(\gamma)}$ for some γ, and λ and γ may be complex in general. Let z be such that $|\lambda z| \leq 1$ and $\lambda z \neq 1$. Denote by $f(z)$ the limit or antilimit of $\sum_{k=1}^{\infty} c_k z^{k-1}$. Then,*

$$A_n = f(z) + c_n z^n g(n), \ g \in \mathbf{A}_0^{(0)} \ strictly. \tag{12.7.1}$$

If we apply the $d^{(1)}$-transformation to the power series $\sum_{k=1}^{\infty} c_k z^{k-1}$ with $R_l = \kappa(l + 1)$, $l = 0, 1, \ldots$, where κ is some positive integer, then

$$d_n^{(1,j)} - f(z) = O(c_{\kappa j} z^{\kappa j} j^{-2n}) \ as \ j \to \infty, \tag{12.7.2}$$

an asymptotically optimal version of which reads

$$d_n^{(1,j)} - f(z) \sim \frac{\kappa^{-n-\mu}(\mu + 1)_n}{(1 - \xi)^n} g_{n+\mu} c_{\kappa(j+1)} z^{\kappa(j+1)} j^{-2n-\mu} \ as \ j \to \infty, \tag{12.7.3}$$

where $g_{n+\mu}$ is the first nonzero g_i with $i \geq n$ in the asymptotic expansion $g(n) \sim \sum_{i=0}^{\infty} g_i n^{-i}$ as $n \to \infty$, and

$$\xi = (\lambda z)^{-\kappa}, \tag{12.7.4}$$

and

$$\Gamma_n^{(j)} \sim \left(\frac{1 + |\xi|}{|1 - \xi|} \right)^n \ as \ j \to \infty, \tag{12.7.5}$$

provided, of course, that $\xi \neq 1$. [Here $d_n^{(1,j)}$ is the solution to the equations $A_{R_l} = d_n^{(1,j)} + c_{R_l} z^{R_l} \sum_{i=0}^{n-1} \bar{\beta}_i / R_l^i$, $l = j, j+1, \ldots, j+N$.]

Proof. The validity of (12.7.1) is already a familiar fact that follows from Theorem 6.7.2.

Because $c_n z^n = (\lambda z)^n h(n) = e^{n \log(\lambda z)} h(n)$, (12.7.1) can be rewritten in the form $A(t) = A + \varphi(t) B(t)$, where $t = n^{-1}$, $A(t) \leftrightarrow A_n$, $\varphi(t) = e^{u(t)} t^\delta H(t)$ with $u(t) = \log(\lambda z)/t$, $\delta = -\gamma$, $H(t) \leftrightarrow n^{-\gamma} h(n)$, and $B(t) \leftrightarrow g(n)$.

We also know that applying the d-transformation is the same as applying GREP with $t_l = 1/R_l$. With $R_l = \kappa(l+1)$, these t_l satisfy (9.1.4) with $c = 1/\kappa$ and $p = q = 1$ there.

From these observations, it is clear that Theorem 9.3.1 applies with $\xi = \exp(-\kappa \log(\lambda z)) = (\lambda z)^{-\kappa}$, and this results in (12.7.2)–(12.7.5). ∎

From Theorem 12.7.1, it is clear that the $d^{(1)}$-transformation accelerates the convergence of the series $\sum_{k=1}^{\infty} c_k z^{k-1}$:

$$\max_{0 \leq i \leq n} \left| \frac{d_n^{(1,j)} - f(z)}{A_{\kappa(j+i)} - f(z)} \right| = O(j^{-2n}) \text{ as } j \to \infty.$$

We now turn to the explanation of why APS with the choice $R_l = \kappa(l+1)$ guarantees more stability and accuracy in the application of the $d^{(1)}$-transformation to the preceding power series.

As we already know, the numerical stability of $d_n^{(1,j)}$ is determined by $\Gamma_n^{(j)}$. As $\Gamma_n^{(j)}$ becomes large, $d_n^{(1,j)}$ becomes less stable. Therefore, we should aim at keeping $\Gamma_n^{(j)}$ close to 1, its lowest possible value. Now, by (12.7.5), $\lim_{j \to \infty} \Gamma_n^{(j)}$ is proportional to $|1 - \xi|^{-n}$, which, from $\xi = (\lambda z)^{-\kappa}$, is unbounded as $z \to \lambda^{-1}$, the point of singularity of $f(z)$. Thus, if we keep κ fixed ($\kappa = 1$ say) and let z get close to λ^{-1}, then ξ gets close to 1, and this causes numerical instabilities in acceleration. On the other hand, if we increase κ, we cause $\xi = (\lambda z)^{-\kappa}$ to separate from 1 in modulus and/or in phase so that $|1 - \xi|$ increases, thus causing $\Gamma_n^{(j)}$ to stay bounded. This provides the theoretical justification for the introduction of the integer κ in the choice of the R_l. We can even see that, as z approaches the point of singularity λ^{-1}, if we keep $\xi = (\lambda z)^{-\kappa}$ approximately fixed by increasing κ gradually, we can maintain an almost fixed and small value for $\Gamma_n^{(j)}$.

We now look at the error in $d_n^{(1,j)}$ which, by (12.7.3), can be written in the form $d_n^{(1,j)} - f(z) \sim K_n c_{\kappa(j+1)} z^{\kappa(j+1)} j^{-2n}$ as $j \to \infty$. The size of K_n gives a good indication of whether the acceleration is effective. Surprisingly, K_n, just as $\lim_{j \to \infty} \Gamma_n^{(j)}$, is proportional to $|1 - \xi|^{-n}$, as is seen from (12.7.3). Again, when z is close to λ^{-1}, the point of singularity of $f(z)$, we can cause K_n to stay bounded by increasing κ. It is thus interesting that, by forcing the acceleration process to become stable numerically, we are also preserving the quality of the theoretical error $d_n^{(1,j)} - f(z)$.

If λz is real and negative, then it is enough to take $\kappa = 1$. This produces excellent results and $\Gamma_n^{(j)} \sim 1$ as $j \to \infty$. If λz is real or complex and very close to 1, then we need to take larger values of κ. In case $\{c_n\}$ is as in Theorem 12.7.1, we can use c_{n+1}/c_n as an estimate of λ, because $\lim_{n \to \infty} c_{n+1}/c_n = \lambda$. On this basis, we can take $(c_{n+1}/c_n)z$ to be an estimate of λz and decide on an appropriate value for κ.

Table 12.7.1: *Effect of APS with the $d^{(1)}$-transformation on the series $\sum_{k=1}^{\infty} z^{k-1}/k$.*
The relevant R_l are chosen as $R_l = \kappa(l+1)$. Here, $\bar{E}_n(z) = |\bar{d}_n^{(1,0)}(z) - f(z)|$,
and the $\bar{d}_n^{(1,j)}(z)$ are the computed $d_n^{(1,j)}(z)$. Computations have been
done in quadruple-precision arithmetic

s	z	$\bar{E}_n(z)$ "smallest" error	$\bar{E}_{28}(z)$ with $\kappa = 1$	$\bar{E}_{28}(z)$ with $\kappa = s$
1	0.5	$4.97D - 30 \ (n = 28)$	$4.97D - 30$	$4.97D - 30$
2	$0.5^{1/2}$	$1.39D - 25 \ (n = 35)$	$5.11D - 21$	$3.26D - 31$
3	$0.5^{1/3}$	$5.42D - 23 \ (n = 38)$	$4.70D - 17$	$4.79D - 31$
4	$0.5^{1/4}$	$1.41D - 21 \ (n = 41)$	$1.02D - 14$	$5.74D - 30$
5	$0.5^{1/5}$	$1.31D - 19 \ (n = 43)$	$3.91D - 13$	$1.59D - 30$
6	$0.5^{1/6}$	$1.69D - 18 \ (n = 43)$	$5.72D - 12$	$2.60D - 30$
7	$0.5^{1/7}$	$1.94D - 17 \ (n = 44)$	$4.58D - 11$	$4.72D - 30$

We end this section by mentioning that we can replace the integer κ by an arbitrary real number ≥ 1, and change $R_l = \kappa(l+1), l = 0, 1, \ldots,$ to

$$R_l = \lfloor \kappa(l+1) \rfloor, \quad l = 0, 1, \ldots, \tag{12.7.6}$$

[cf. (10.2.2)]. It is easy to show that, with $\kappa \geq 1$, there holds $R_0 < R_1 < R_2 < \cdots$, and $R_l \sim \kappa l$ as $l \to \infty$. Even though Theorem 12.7.1 does not apply to this case, the numerical results obtained by the d-transformation with such values of κ appear to be just as good as those obtained with integer values.

Example 12.7.2 We illustrate the advantage of APS by applying it to the power series $\sum_{k=1}^{\infty} z^{k-1}/k$ whose sum is $f(z) = -z^{-1} \log(1 - z)$ when $|z| \leq 1, z \neq 1$. The $d^{(1)}$-transformation can be used in summing this series because $\{1/n\} \in \mathbf{b}^{(1)}$. Now, $f(z)$ has a singularity at $z = 1$. Thus, as z approaches 1 it becomes difficult to preserve numerical stability and good convergence if the $d^{(1)}$-transformation is used with $R_l = l + 1$. Use of APS, however, improves things substantially, precisely as explained before. We denote by $d_n^{(1,j)}(z)$ also the $d_n^{(1,j)}$ with APS.

In Table 12.7.1, we present the results obtained for $z = 0.5^{1/s}, s = 1, 2, \ldots$. The computations have been done in quadruple-precision arithmetic. Let us denote the computed $d_n^{(1,j)}(z)$ by $\bar{d}_n^{(1,j)}(z)$ and set $\bar{E}_n(z) = |\bar{d}_n^{(1,0)}(z) - f(z)|$. Then, the $\bar{d}_n^{(1,0)}(z)$ associated with the $\bar{E}_n(z)$ in the third column seem to have the highest accuracy when $\kappa = 1$ in APS. The results of the third and fifth columns together clearly show that, as z approaches 1, the quality of both $d_n^{(1,0)}(z)$ and $\bar{d}_n^{(1,0)}(z)$, with n fixed or otherwise, declines. The results of the last column, on the other hand, verify the claim that, by applying APS with κ chosen such that z^κ remains almost fixed, the quality of $d_n^{(1,0)}(z)$, with n fixed, is preserved. In fact, the $\bar{d}_{28}^{(1,0)}(0.5^{1/s})$ with $\kappa = s$ turn out to be almost the best approximations in quadruple-precision arithmetic and are of the same size for all s.

12.8 The *d*-Transformation on Factorial Series

Our next result concerns the application of the d-transformation to infinite power series whose partial sums form factorial sequences. This result essentially forms part (iii) of Theorem 19.2.3, and Theorem 6.7.2 plays an important role in its proof.

Table 12.9.1: *"Smallest" errors in* $d_n^{(m,0)}(z)$, $n = (\nu, \ldots, \nu)$, *and* $\varepsilon_{2s}^{(0)}(z) = f_{s,s}(z)$ *for the functions* $f^r(z)$ *with z on the circles of convergence. The first $m(\nu+1)$ of the coefficients c_k are used in constructing* $d_{(\nu,\ldots,\nu)}^{(m,0)}(z)$, *and $2s+1$ coefficients are used in constructing* $\varepsilon_{2s}^{(0)}(z)$. *Computations have been done in quadruple-precision arithmetic*

$z = e^{i\pi/6}$	$\|d_n^{(1,0)}(z) - f^1(z)\| = 1.15D - 24$ $(\nu = 42)$	$\|\varepsilon_{160}^{(0)}(z) - f^1(z)\| = 5.60D - 34$
$z = \frac{1}{2}e^{i\pi/6}$	$\|d_n^{(2,0)}(z) - f^2(z)\| = 1.96D - 33$ $(\nu = 23)$	$\|\varepsilon_{26}^{(0)}(z) - f^2(z)\| = 4.82D - 14$
$z = \frac{1}{3}e^{i\pi/6}$	$\|d_n^{(4,0)}(z) - f^4(z)\| = 5.54D - 33$ $(\nu = 20)$	$\|\varepsilon_{26}^{(0)}(z) - f^4(z)\| = 1.11D - 09$

Theorem 12.8.1 *Let $c_n = [(n-1)!]^{-r}\lambda^n h(n)$, where r is a positive integer, λ is some scalar, and $h \in \mathbf{A}_0^{(\gamma)}$ for some γ, and λ and γ may be complex in general. Denote by $f(z)$ the limit of $\sum_{k=1}^{\infty} c_k z^{k-1}$. Then*

$$A_n = f(z) + n^{-r}c_n z^n g(n), \quad g \in \mathbf{A}_0^{(0)} \quad strictly. \tag{12.8.1}$$

If we apply the $d^{(1)}$-transformation to the power series $\sum_{k=1}^{\infty} c_k z^{k-1}$ with $R_l = l+1$, then

$$d_n^{(1,j)} - f(z) = \begin{cases} O(a_{j+n+1}j^{-2n}) \text{ as } j \to \infty, \quad n \ge r+1, \\ O(a_{j+n+1}j^{-r-1}) \text{ as } j \to \infty, \quad n < r+1, \end{cases} \tag{12.8.2}$$

and

$$\Gamma_n^{(j)} \sim 1 \quad as \ j \to \infty. \tag{12.8.3}$$

The $d^{(1)}$-transformation of this theorem is nothing but the $\mathcal{L}$-transformation, as mentioned earlier.

12.9 Numerical Examples

In this section, we illustrate the effectiveness of the d-transformation in accelerating the convergence of power series $f(z) := \sum_{k=1}^{\infty} c_k z^{k-1}$ such that $\{c_k\} \in \mathbf{b}^{(m)}$ for various values of m. We compare the results obtained from the d-transformation with those obtained from the Padé table. In particular, we compare the sequences $\{d_{(\nu,\ldots,\nu)}^{(m,0)}(z)\}_{\nu=0}^{\infty}$ with the sequences $\{f_{N,N}(z)\}_{N=0}^{\infty}$ of Padé approximants that have the best convergence properties for all practical purposes. We have computed the $d_{(\nu,\ldots,\nu)}^{(m,0)}(z)$ with the help of the computer code given in Appendix I after modifying the latter to accommodate complex arithmetic. Thus, the equations that define the $d_{(\nu,\ldots,\nu)}^{(m,0)}(z)$ are

$$A_l = d_{(\nu,\ldots,\nu)}^{(m,0)}(z) + \sum_{k=1}^{m} l^k \Delta^{k-1} a_l \sum_{i=0}^{\nu-1} \frac{\bar{\beta}_{ki}}{l^i}, \quad l = 1, 2, \ldots, m\nu+1.$$

Here, $a_k = c_k z^{k-1}$ for all k, as before. The $f_{N,N}(z)$ have been computed via the ε-algorithm of Wynn to which we come in Chapters 16 and 17. [We mention only that the quantities $\varepsilon_k^{(j)}$ generated by the ε-algorithm on the sequence of the partial sums of $\sum_{k=1}^{\infty} c_k z^{k-1}$ are related to the Padé table via $\varepsilon_{2s}^{(j)} = f_{j+s,s}(z)$.] In other words, we

Table 12.9.2: *"Smallest" errors in* $d_n^{(m,0)}(z)$, $n = (\nu, \dots, \nu)$, *and* $\varepsilon_{2s}^{(0)}(z) = f_{s,s}(z)$ *for the functions* $f^r(z)$ *with z outside the circles of convergence. The first $m(\nu + 1)$ of the coefficients c_k are used in constructing* $d_{(\nu,\dots,\nu)}^{(m,0)}(z)$, *and $2s + 1$ coefficients are used in constructing* $\varepsilon_{2s}^{(0)}(z)$. *Computations have been done in quadruple-precision arithmetic*

$z = \frac{3}{2}e^{i\pi/6}$	$\lvert d_n^{(1,0)}(z) - f^1(z) \rvert = 6.15D - 18$ $(\nu = 44)$	$\lvert \varepsilon_{200}^{(0)}(z) - f^1(z) \rvert = 4.25D - 20$
$z = \frac{3}{4}e^{i\pi/6}$	$\lvert d_n^{(2,0)}(z) - f^2(z) \rvert = 7.49D - 27$ $(\nu = 26)$	$\lvert \varepsilon_{26}^{(0)}(z) - f^2(z) \rvert = 1.58D - 09$
$z = \frac{1}{2}e^{i\pi/6}$	$\lvert d_n^{(4,0)}(z) - f^4(z) \rvert = 1.62D - 22$ $(\nu = 20)$	$\lvert \varepsilon_{26}^{(0)}(z) - f^4(z) \rvert = 1.48D - 06$

have not computed our approximations explicitly as rational functions. To have a good comparative picture, all our computations were done in quadruple-precision arithmetic.

As our test cases, we chose the following three series:

$$f^1(z) := \sum_{k=1}^{\infty} z^{k-1}/k$$

$$f^2(z) := \sum_{k=1}^{\infty} [1 - (-2)^k] z^{k-1}/k,$$

$$f^4(z) := \sum_{k=1}^{\infty} [1 - (-2)^k + (3i)^k - (-3i)^k] z^{k-1}/k.$$

The series $f^1(z)$, $f^2(z)$, and $f^4(z)$ converge for $\lvert z \rvert < \rho$, where $\rho = 1$, $\rho = 1/2$, and $\rho = 1/3$, respectively, and represent functions that are analytic for $\lvert z \rvert < \rho$. Denoting these functions by $f^1(z)$, $f^2(z)$, and $f^4(z)$ as well, we have specifically

$f^1(z) = -z^{-1} \log(1 - z)$, $\lvert z \rvert < 1$,

$f^2(z) = z^{-1}[-\log(1 - z) + \log(1 + 2z)]$, $\lvert z \rvert < 1/2$,

$f^4(z) = z^{-1}[-\log(1 - z) + \log(1 + 2z) - \log(1 - 3iz) + \log(1 + 3iz)]$, $\lvert z \rvert < 1/3$.

In addition, each series converges also when $\lvert z \rvert = \rho$, except when z is a branch point of the corresponding limit function. The branch points are at $z = 1$ for $f^1(z)$, at $z = 1, -1/2$ for $f^2(z)$, and at $z = 1, -1/2, \pm i/3$ for $f^4(z)$.

As $\{c_k\}$ is in $\mathbf{b}^{(1)}$ for $f^1(z)$, in $\mathbf{b}^{(2)}$ for $f^2(z)$, and in $\mathbf{b}^{(4)}$ for $f^4(z)$, we can apply the $d^{(1)}$-, $d^{(2)}$-, and $d^{(4)}$-transformations to approximate the sums of these series when $\lvert z \rvert \leq \rho$ but z is not a branch point.

Numerical experiments suggest that both the d-approximants and the Padé approximants continue the functions $f^r(z)$ analytically outside the circles of convergence of their corresponding power series $f^r(z)$. The functions $f^r(z)$ are defined via the principal values of the logarithms $\log(1 - \alpha z)$ involved. [The principal value of the logarithm of ζ is given as $\log \zeta = \log \lvert \zeta \rvert + i \arg \zeta$, and $-\pi < \arg \zeta \leq \pi$. Therefore, $\log(1 - \alpha z)$ has its branch cut along the ray $(\alpha^{-1}, \infty e^{-i \arg \alpha})$.] For the series $f^1(z)$, this observation is consistent with the theory of Padé approximants from Stieltjes series, a topic we discuss in Chapter 17. It is consistent also with the findings of Example 4.1.8 and Theorem 6.8.8.

In Table 12.9.1, we present the "best" numerical results obtained from the d-transformation and the ε-algorithm with z on the circles of convergence. The partial

Table 12.9.3: *"Smallest" errors in* $d_n^{(m,0)}(z)$, $n = (\nu, \ldots, \nu)$, *for the functions* $f^r(z)$ *with* z *on the circles of convergence and close to a branch point, with and without APS. Here* $R_l = \kappa(l+1)$. *The first* $\kappa(m\nu + 1) + m - 1$ *of the coefficients* c_k *are used in constructing* $d_{(\nu,\ldots,\nu)}^{(m,0)}(z)$. *Computations have been done in quadruple-precision arithmetic*

$z = e^{i0.05\pi}$	$\|d_n^{(1,0)}(z) - f^1(z)\| = 1.67D - 16$ $(\kappa = 1, \ \nu = 48)$
	$\|d_n^{(1,0)}(z) - f^1(z)\| = 2.88D - 27$ $(\kappa = 5, \ \nu = 37)$
$z = \frac{1}{2}e^{i0.95\pi}$	$\|d_n^{(2,0)}(z) - f^2(z)\| = 9.85D - 19$ $(\kappa = 1, \ \nu = 37)$
	$\|d_n^{(2,0)}(z) - f^2(z)\| = 6.99D - 28$ $(\kappa = 9, \ \nu = 100)$
$z = \frac{1}{3}e^{i0.45\pi}$	$\|d_n^{(4,0)}(z) - f^4(z)\| = 8.23D - 19$ $(\kappa = 1, \ \nu = 33)$
	$\|d_n^{(4,0)}(z) - f^4(z)\| = 8.86D - 29$ $(\kappa = 3, \ \nu = 39)$

sums converge extremely slowly for such z. The ε-algorithm, although very effective for $f^1(z)$, appears to suffer from stability problems in the case of $f^2(z)$ and $f^4(z)$.

As Table 12.9.1 shows, the best result that can be obtained from the $d^{(1)}$-transformation for $f^1(z)$ with $z = e^{i\pi/6}$ has 24 correct digits. By using APS with $R_l = 3(l+1)$, we can achieve almost machine accuracy; in fact, $|d_{33}^{(1,0)}(z) - f^1(z)| = 3.15 \times 10^{-32}$ for this z, and the number of terms used for this is 102.

Table 12.9.2 shows results obtained for z outside the circles of convergence. As we increase ν, the results deteriorate. The reason is that the partial sums diverge quickly, and hence the floating-point errors in their computation diverge as well.

From both Table 12.9.1 and Table 12.9.2, the d-transformations appear to be more effective than the Padé approximants for these examples.

Finally, in Table 12.9.3, we compare results obtained from the d-transformation with and without APS when z is on the circles of convergence and very close to a branch point. Again, the partial sums converge extremely slowly. We denote the approximations $d_n^{(m,j)}$ with APS also $d_n^{(m,j)}(z)$.

For further examples, we refer the reader to Sidi and Levin [312].

13

Acceleration of Convergence of Fourier and Generalized Fourier Series by the d-Transformation: The Complex Series Approach with APS

13.1 Introduction

In this chapter, we extend the treatment we gave to power series in the preceding chapter to Fourier series and their generalizations, whether convergent or divergent. In particular, we are concerned with Fourier cosine and sine series, orthogonal polynomial expansions, series that arise from Sturm–Liouville problems, such as Fourier–Bessel series, and other general special function series.

Several convergence acceleration methods have been used on such series, with limited success. An immediate problem many of these methods face is that they do not produce any acceleration when applied to Fourier and generalized Fourier series. The transformations of Euler and of Shanks discussed in the following chapters and the d-transformation are exceptions. See the review paper by Smith and Ford [318] and the paper by Levin and Sidi [165]. With those methods that do produce acceleration, another problem one faces in working with such series is the lack of stability and acceleration near points of singularity of the functions that serve as limits or antilimits of these series. Recall that the same problem occurs in dealing with power series.

In this chapter, we show how the d-transformation can be used effectively to accelerate the convergence of these series. The approach we are about to propose has two main ingredients that can be applied also with some of the other sequence transformations. (This subject is considered in detail later.) The first ingredient involves the introduction of what we call *functions of the second kind*. This decreases the cost of acceleration to half of what it would be if extrapolation were applied to the original series. The second ingredient involves the use of APS with $R_l = \kappa(l + 1)$, where $\kappa > 1$ is an integer, near points of singularity, as was done in the case of power series in the preceding chapter. [We can also use APS by letting $R_l = \lfloor \kappa(l + 1) \rfloor$, where $\kappa > 1$ and is not an integer necessarily, as we suggested at the end Section 12.7.]

The contents of this chapter are based on the paper by Sidi [294]. As the approach we describe here has been illustrated with several numerical examples [294], we do not include any examples here, and refer the reader to Sidi [294]. See also the numerical examples given by Levin and Sidi [165, Section 7] that are precisely of the type we treat here.

253

13.2 Introducing Functions of Second Kind

13.2.1 General Background

Assume we want to accelerate the convergence of an infinite series $F(x)$ given in the form

$$F(x) := \sum_{k=1}^{\infty} [b_k \phi_k(x) + c_k \psi_k(x)], \qquad (13.2.1)$$

where the functions $\phi_n(x)$ and $\psi_n(x)$ satisfy

$$\rho_n^{\pm}(x) \equiv \phi_n(x) \pm i\psi_n(x) = e^{\pm in\omega x} g_n^{\pm}(x), \qquad (13.2.2)$$

ω being some fixed real positive constant, and

$$g_n^{\pm}(x) \sim n^{\epsilon} \sum_{j=0}^{\infty} \delta_j^{\pm}(x) n^{-j} \quad \text{as } n \to \infty, \qquad (13.2.3)$$

for some fixed ϵ that can be complex in general. (Here $i = \sqrt{-1}$ is the imaginary unit, as usual.) In other words, $g_n^+(x)$ and $g_n^-(x)$, as functions of n, are in $\mathbf{A}_0^{(\epsilon)}$.

From (13.2.2) and (13.2.3) and Example 6.1.11, it is clear that $\{\rho_n^{\pm}(x)\} \in \mathbf{b}^{(1)}$. By the fact that

$$\phi_n(x) = \frac{1}{2} \left[\rho_n^+(x) + \rho_n^-(x) \right] \quad \text{and} \quad \psi_n(x) = \frac{1}{2i} \left[\rho_n^+(x) - \rho_n^-(x) \right], \qquad (13.2.4)$$

and by Theorem 6.8.7, we also have that $\{\phi_n(x)\} \in \mathbf{b}^{(2)}$ and $\{\psi_n(x)\} \in \mathbf{b}^{(2)}$ exactly in accordance with Definition 6.1.2, as long as $e^{i\omega x} \neq \pm 1$.

The simplest and most widely treated members of the series above are the *classical Fourier series*

$$F(x) := \sum_{k=0}^{\infty} (b_k \cos k\omega x + c_k \sin k\omega x),$$

for which $\phi_n(x) = \cos n\omega x$ and $\psi_n(x) = \sin n\omega x$, so that $\rho_n^{\pm}(x) = e^{\pm in\omega x}$ and hence $g_n^{\pm}(x) \equiv 1$, $n = 0, 1, \ldots$. More examples are provided in the next section.

In general, $\phi_n(x)$ and $\psi_n(x)$ may be (some linear combinations of) the nth eigenfunction of a Sturm–Liouville problem and the corresponding second linearly independent solution of the relevant O.D.E., that is, the corresponding *function of the second kind*. In most cases of interest, $c_n = 0$, $n = 1, 2, \ldots$, so that $F(x) := \sum_{k=1}^{\infty} b_k \phi_k(x)$.

13.2.2 Complex Series Approach

The approach we propose for accelerating the convergence of the series $F(x)$, *assuming that the coefficients b_n and c_n are known*, is as follows:

1. Define the series $B^{\pm}(x)$ and $C^{\pm}(x)$ by

$$B^{\pm}(x) := \sum_{k=1}^{\infty} b_k \rho_k^{\pm}(x) \quad \text{and} \quad C^{\pm}(x) := \sum_{k=1}^{\infty} c_k \rho_k^{\pm}(x), \qquad (13.2.5)$$

and observe that

$$F_\phi(x) := \sum_{k=1}^\infty b_k \phi_k(x) = \frac{1}{2}\left[B^+(x) + B^-(x)\right] \text{ and}$$

$$F_\psi(x) := \sum_{k=1}^\infty c_k \psi_k(x) = \frac{1}{2i}\left[C^+(x) - C^-(x)\right], \tag{13.2.6}$$

and that

$$F(x) = F_\phi(x) + F_\psi(x). \tag{13.2.7}$$

2. Apply the d-transformation to the series $B^\pm(x)$ and $C^\pm(x)$ and then invoke (13.2.6) and (13.2.7). Near the points of singularity of $B^\pm(x)$ and $C^\pm(x)$ use APS with $R_l = \kappa(l+1)$.

When $\phi_n(x)$ and $\psi_n(x)$ are real, $\rho_n^\pm(x)$ are necessarily complex. For this reason, we call this the *complex series approach*.

In connection with this approach, we note that when the functions $\phi_n(x)$ and $\psi_n(x)$ and the coefficients b_n and c_n are all *real*, it is enough to treat the two complex series $B^+(x)$ and $C^+(x)$, as $B^-(x) = \overline{B^+(x)}$ and $C^-(x) = \overline{C^+(x)}$, so that $F_\phi(x) = \Re B^+(x)$ and $F_\psi(x) = \Im C^+(x)$ in such cases.

We also note that the series $B^\pm(x)$ and $C^\pm(x)$ can be viewed as power series because

$$B^\pm(x) := \sum_{k=1}^\infty \left[b_k g_k^\pm(x)\right] z^k \text{ and } C^\pm(x) := \sum_{k=1}^\infty \left[c_k g_k^\pm(x)\right] z^k; \quad z = e^{\pm i\omega x}.$$

The developments of Section 12.7 of the preceding chapter, including APS, thus become very useful in dealing with the series of this chapter.

Note that the complex series approach, in connection with the summation of classical Fourier series by the Shanks transformation, was first suggested by Wynn [374]. Wynn proposed that a real cosine series $\sum_{k=0}^\infty b_k \cos k\omega x$ be written as $\Re(\sum_{k=0}^\infty b_k z^k)$ with $z = e^{i\omega x}$, and then the ε-algorithm be used to accelerate the convergence of the complex power series $\sum_{k=0}^\infty b_k z^k$. Later, Sidi [273, Section 3] proposed that the Levin transformations be used to accelerate the convergence of Fourier series in their complex power series form $\sum_{k=0}^\infty b_k z^k$ when $\{b_k\} \in \mathbf{b}^{(1)}$, providing a convergence analysis for Process I at the same time.

13.2.3 Justification of the Complex Series Approach and APS

We may wonder why the complex series approach is needed. After all, we can apply a suitable extrapolation method directly to $F(x)$. As we discuss next, this approach is economical in terms of the number of the b_k and c_k used in acceleration.

For simplicity, we consider the case in which $F(x) = F_\phi(x)$, that is, $c_n = 0$ for all n. We further assume that the b_n satisfy [cf. (12.2.1)]

$$b_{n+m} = \sum_{s=0}^{m-1} q_s(n) b_{n+s}, \quad q_s \in \mathbf{A}_0^{(\mu_s)}, \quad \mu_s \text{ integer}, \quad s = 0, 1, \dots, m-1. \tag{13.2.8}$$

Thus, $b_n \rho_n^{\pm}(x) = \hat{b}_n z^n$ with $z = e^{\pm i\omega x}$ and $\hat{b}_n = b_n g_n^{\pm}(x)$. As a result, (13.2.8) becomes

$$\hat{b}_{n+m} = \sum_{s=0}^{m-1} \hat{q}_s(n)\hat{b}_{n+s}, \quad \hat{q}_s \in \mathbf{A}_0^{(\mu_s)}, \quad s = 0, 1, \ldots, m-1, \qquad (13.2.9)$$

with the same integers μ_s as in (13.2.8), because $\hat{q}_s(n) = q_s(n)g_{n+m}^{\pm}(x)/g_{n+s}^{\pm}(x)$, $s = 0, 1, \ldots, m-1$. Therefore, Theorem 12.2.1 applies and we have $\{b_n\rho_n^{\pm}(x)\} = \{\hat{b}_n z^n\} \in \mathbf{b}^{(m)}$, except for at most m values of x for which the limit or antilimit of (the power series) $B^{\pm}(x) := \sum_{k=1}^{\infty} \hat{b}_k z^k$ is singular, and the $d^{(m)}$-transformation can be applied to accelerate the convergence of $B^{\pm}(x)$. Finally, we use (13.2.6) to compute $F_\phi(x)$. Because $\{b_n\rho_n^{\pm}(x)\} \in \mathbf{b}^{(m)}$ and because $\phi_n(x) = \frac{1}{2}[\rho_n^+(x) + \rho_n^-(x)]$, we have that $\{b_n\phi_n(x)\} \in \mathbf{b}^{(2m)}$ by part (ii) of Heuristic 6.4.1, and the $d^{(2m)}$-transformation can be applied to accelerate the convergence of $F_\phi(x)$ directly.

From our discussion in Subsection 4.4.5, we now conclude heuristically that, with $R_l = \kappa(l+1)$ and fixed ν, the approximations $d_{(\nu,\ldots,\nu)}^{(m,0)}$ obtained by applying the $d^{(m)}$-transformation to the series $B^{\pm}(x)$ and the approximation $d_{(\nu,\ldots,\nu)}^{(2m,0)}$ obtained by applying the $d^{(2m)}$-transformation directly to the series $F_\phi(x)$ have comparable accuracy. Now $d_{(\nu,\ldots,\nu)}^{(m,0)}$ is obtained by using the coefficients b_k, $1 \leq k \leq m\kappa\nu + \kappa + m - 1$, while $d_{(\nu,\ldots,\nu)}^{(2m,0)}$ is obtained by using b_k, $1 \leq k \leq 2m\kappa\nu + \kappa + 2m - 1$. That is, if the first M coefficients b_k are needed to achieve a certain accuracy when applying the d-transformation with he complex series approach, about $2M$ coefficients b_k are needed to achieve the same accuracy from the application of the d-transformation directly to $F_\phi(x)$. This suggests that the cost of the complex series approach is about half that of the direct approach, when costs are measured in terms of the number of coefficients used. This is the proper measure of cost as the series are defined solely via their coefficients, and the functions $\phi_n(x)$ and $\psi_n(x)$ are readily available in most cases of interest.

13.3 Examples of Generalized Fourier Series

In the preceding section, we mentioned the classical Fourier series as the simplest example of the class of series that is characterized through (13.2.1)–(13.2.3). In this section, we give further examples involving "nonclassical" Fourier series, orthogonal polynomial expansions, and Fourier–Bessel series. Additional examples involving other special functions can also be given.

13.3.1 Chebyshev Series

$$F_T(x) := \sum_{k=0}^{\infty} d_k T_k(x) \quad \text{and} \quad F_U(x) := \sum_{k=0}^{\infty} e_k U_k(x), \quad -1 \leq x \leq 1, \qquad (13.3.1)$$

where $T_k(x)$ and $U_k(x)$ are the Chebyshev polynomials of degree k, of the first and second kinds respectively.

Defining $x = \cos\theta$, $0 \leq \theta \leq \pi$, we have $T_n(x) = \cos n\theta$ and $U_n(x) = \sin(n+1)\theta/\sin\theta$. Therefore, $\phi_n(x) = T_n(x) = \cos n\theta$ and $\psi_n(x) = \sqrt{1-x^2}U_{n-1}(x) = \sin n\theta$,

$n = 0, 1, \ldots$, with $U_{-1}(x) \equiv 0$. Both series, $F_T(x)$ and $\sqrt{1 - x^2} F_U(x)$, can thus be treated as ordinary Fourier series.

13.3.2 "Nonclassical" Fourier Series

$$F(x) := \sum_{k=1}^{\infty} (b_k \cos \lambda_k x + c_k \sin \lambda_k x), \qquad (13.3.2)$$

where

$$\lambda_n \sim n \sum_{j=0}^{\infty} \alpha_j n^{-j} \text{ as } n \to \infty, \ \alpha_0 > 0, \qquad (13.3.3)$$

so that $\lambda_n \sim \alpha_0 n$ as $n \to \infty$.

Functions $\phi_n(x)$ and $\psi_n(x)$ such as the ones here arise, for example, when one solves an eigenvalue problem associated with a boundary value problem involving the O.D.E. $u'' + \lambda^2 u = 0$ on the interval $[0, l]$, which, in turn, may arise when one uses separation of variables in the solution of some appropriate heat equation, wave equation, or Laplace's equation. For instance, the eigenfunctions of the problem

$$u'' + \lambda^2 u = 0, \ 0 < x < l; \ u(0) = 0, \ u'(l) = -hu(l), \quad h > 0,$$

are $\sin \lambda_n x$, $n = 1, 2, \ldots$, where λ_n is the nth positive solution of the nonlinear equation $\lambda \cos \lambda l = -h \sin \lambda l$. By straightforward asymptotic techniques, it can be shown that

$$\lambda_n \sim \left(n - \frac{1}{2}\right)\frac{\pi}{l} + e_1 n^{-1} + e_2 n^{-2} + \cdots \text{ as } n \to \infty.$$

Consequently, we also have that $\omega = \alpha_0 = \pi/l$ and $\epsilon = 0$ in (13.2.2) and (13.2.3).

13.3.3 Fourier–Legendre Series

$$F_P(x) := \sum_{k=0}^{\infty} d_k P_k(x) \text{ or } F_Q(x) := \sum_{k=0}^{\infty} e_k Q_k(x), \ -1 < x < 1, \quad (13.3.4)$$

where $P_n(x)$ is the Legendre polynomial of degree n and $Q_n(x)$ is the associated Legendre function of the second kind of order 0 of degree n. They are both generated by the recursion relation

$$M_{n+1}(x) = \frac{2n + 1}{n + 1} x M_n(x) - \frac{n}{n + 1} M_{n-1}(x), \ n = 1, 2, \ldots, \quad (13.3.5)$$

where $M_n(x)$ is either $P_n(x)$ or $Q_n(x)$, with the initial conditions

$$P_0(x) = 1, \ P_1(x) = x \text{ and}$$

$$Q_0(x) = \frac{1}{2} \log \frac{1 + x}{1 - x}, \ Q_1(x) = x Q_0(x) - 1, \text{ when } |x| < 1. \quad (13.3.6)$$

We now show that, with $\theta = \cos^{-1} x$, $\phi_n(\theta) = P_n(x)$ and $\psi_n(\theta) = -\frac{2}{\pi} Q_n(x)$ so that $\rho_n^{\pm}(\theta) = P_n(x) \mp \mathrm{i}\frac{2}{\pi} Q_n(x)$, and $\omega = 1$ in (13.2.2) and $\epsilon = -1/2$ in (13.2.3).

First, we recall that $P_n(x) = P_n^0(x)$ and $Q_n(x) = Q_n^0(x)$, where $P_\nu^\mu(x)$ and $Q_\nu^\mu(x)$ are the associated Legendre functions of degree ν and of order μ. Next, letting $x = \cos\theta$, $0 < \theta < \pi$, and $u = n + 1/2$ in [223, p. 473, Ex. 13.3], it follows that, for any fixed real m, there exist two asymptotic expansions $A_m(\theta; u)$ and $B_m(\theta; u)$,

$$A_m(\theta; u) := \sum_{s=0}^{\infty} \frac{A_s^{-m}(\theta^2)}{u^{2s}} \quad \text{as } n \to \infty,$$

$$B_m(\theta; u) := \sum_{s=0}^{\infty} \frac{B_s^{-m}(\theta^2)}{u^{2s}} \quad \text{as } n \to \infty, \tag{13.3.7}$$

such that

$$P_n^{-m}(\cos\theta) \mp i\frac{2}{\pi} Q_n^{-m}(\cos\theta) \sim \frac{1}{u^m} \left(\frac{\theta}{\sin\theta}\right)^{1/2}$$

$$\times \left\{ H_m^{(\pm)}(u\theta) A_m(\theta; u) + \frac{\theta}{u} H_{m-1}^{(\pm)}(u\theta) B_m(u; \theta) \right\}. \tag{13.3.8}$$

Here $H_m^{(+)}(z)$ and $H_m^{(-)}(z)$ stand for the Hankel functions $H_m^{(1)}(z)$ and $H_m^{(2)}(z)$, respectively. Finally, we also have

$$H_m^{(\pm)}(z) \sim e^{\pm iz} \sum_{j=0}^{\infty} \frac{C_{mj}^{\pm}}{z^{j+1/2}} \quad \text{as } z \to +\infty. \tag{13.3.9}$$

Substituting (13.3.7) and (13.3.9) in (13.3.8), letting $m = 0$ there, and invoking $u = n + 1/2$, the result now follows. We leave the details to the interested reader.

It is now also clear that, in the series $F_P(x)$ and $F_Q(x)$, $P_n(x)$ and $Q_n(x)$ can be replaced by $P_n^\mu(x)$ and $Q_n^\mu(x)$, where μ is an arbitrary real number.

13.3.4 Fourier–Bessel Series

$$F(x) := \sum_{k=1}^{\infty} [d_k J_\nu(\lambda_k x) + e_k Y_\nu(\lambda_k x)], \quad 0 < x \le r, \quad \text{some } r, \tag{13.3.10}$$

where $J_\nu(z)$ and $Y_\nu(z)$ are Bessel functions of order $\nu \ge 0$ of the first and second kinds, respectively, and λ_n are scalars satisfying (13.3.3). Normally, such λ_n result from boundary value problems involving the Bessel equation

$$\frac{d}{dx}\left(x\frac{du}{dx}\right) + \left(\lambda^2 x - \frac{\nu^2}{x}\right)u = 0,$$

and r and α_0 are related through $\alpha_0 r = \pi$. For example, λ_n can be the nth positive zero of $J_\nu(z)$ or of $J_\nu'(z)$ or of some linear combination of them. In these cases, for all ν, $\alpha_0 = \pi$ in (13.3.3).

We now show that $\phi_n(x) = J_\nu(\lambda_n x)$ and $\psi_n(x) = Y_\nu(\lambda_n x)$ so that $\rho_n^{\pm}(x) = J_\nu(\lambda_n x) \pm iY_\nu(\lambda_n x)$, and $\omega = \alpha_0$ in (13.2.2) and $\epsilon = -1/2$ in (13.2.3). From (13.3.9) and the fact that $\lambda_n \to +\infty$ as $n \to \infty$, it follows that

$$\rho_n^\pm(x) = H_\nu^{(\pm)}(\lambda_n x) \sim e^{\pm i\lambda_n x} \sum_{j=0}^\infty \frac{c_{\nu j}^\pm}{(\lambda_n x)^{j+1/2}} \quad \text{as } n \to \infty, \qquad (13.3.11)$$

which, when combined with the asymptotic expansion in (13.3.3), leads to (13.2.3).

13.4 Convergence and Stability when $\{b_n\} \in \mathbf{b}^{(1)}$

Let $\{b_n\} \in \mathbf{b}^{(1)}$ with $b_n = \lambda^n h(n)$ for some λ and some $h \in \mathbf{A}_0^{(\gamma)}$, where λ and γ may be complex in general. In this section, we consider the stability and convergence of the complex series approach to the summation of the series $F_\phi(x) := \sum_{k=1}^\infty b_k \phi_k(x)$. We recall that in this approach we apply the d-transformation to the series $B^\pm(x) := \sum_{k=1}^\infty b_k \rho_k^\pm(x)$ and then use the fact that $F_\phi(x) = \frac{1}{2}[B^+(x) + B^-(x)]$.

We have seen that $\{b_n \rho_n^\pm(x)\} \in \mathbf{b}^{(1)}$ so that the $d^{(1)}$-transformation can be used in accelerating the convergence of $B^\pm(x)$. As we saw earlier, $B^\pm(x)$ can also be viewed as power series $B^\pm(x) := \sum_{k=1}^\infty \hat{b}_k(\lambda z)^k$ where $\hat{b}_n$, as a function of n, is in $\mathbf{A}_0^{(\gamma+\epsilon)}$ and $z = e^{\pm i\omega x}$. Thus, for $|\lambda| < 1$, $B^\pm(x)$ converge to analytic functions of λ, say $G^\pm(x; \lambda)$. When $|\lambda| = 1$ but $\lambda z \neq 1$, $B^\pm(x)$ converge also when $\Re(\gamma + \epsilon) < 0$. When $|\lambda| = 1$ but $\Re(\gamma + \epsilon) \geq 0$, $B^\pm(x)$ diverge, but they have antilimits by Theorem 6.7.2, and these antilimits are Abelian means of $B^\pm(x)$, namely, $\lim_{\tau \to 1-} \sum_{k=1}^\infty b_k \rho_k^\pm(x) \tau^k$. They are also generalized functions in x. When $|\lambda| = 1$, the functions $G^\pm(x; \lambda)$, as functions of x, are singular when $\lambda z = 1$, that is, at $x = x^\pm = \mp(\arg \lambda)/\omega$. For $|\lambda| > 1$ the series $B^\pm(x)$ diverge, but we assume that $G^\pm(x; \lambda)$ can be continued analytically to $|\lambda| \geq 1$, with singularities removed.

From this information, we conclude that when $|\lambda| = 1$ and x is close to $x^\pm$ or when $|\lambda| \approx 1$ and $\lambda e^{\pm i\omega x} \approx 1$, we can apply the $d^{(1)}$-transformation with APS via $R_l = \kappa(l + 1)$ to the series $B^\pm(x)$, where κ is a positive integer whose size depends on how close $\lambda e^{\pm i\omega x}$ is to 1. This results in approximations that enjoy excellent stability and convergence properties by Theorem 12.7.1. They also have a low cost.

13.5 Direct Approach

In the preceding sections, we introduced the complex series approach to the summation of Fourier and generalized Fourier series of the form (13.2.1)–(13.2.3). We would like to emphasize that this approach can be used when the coefficients b_n and c_n are available. In fact, the complex series approach coupled with APS is the most efficient approach in this case.

When b_n and c_n are not available by themselves, but we are given $a_n = b_n \phi_n(x) + c_n \psi_n(x)$ *in one piece* and there is no way of determining the b_n and c_n separately, the complex series approach is not applicable. In such a case, we are forced to apply extrapolation methods directly to $\sum_{k=1}^\infty a_k$. Again, the d-transformation can be applied directly to the series $\sum_{k=1}^\infty a_k$ with APS through $R_l = \kappa(l + 1)$; the size of κ depends on how close x is to the singularities of the limit or antilimit of $\sum_{k=1}^\infty a_k$.

13.6 Extension of the Complex Series Approach

The complex series approach of Section 13.2, after appropriate preparatory work, can be applied to series of products of functions of the form

$$H(x, y) := \sum_{k=1}^{\infty} [b_k \phi_{1k}(x)\phi_{2k}(y) + c_k \phi_{1k}(x)\psi_{2k}(y)$$

$$+ d_k \psi_{1k}(x)\phi_{2k}(y) + e_k \psi_{1k}(x)\psi_{2k}(y)] , \tag{13.6.1}$$

where

$$\rho_{pn}^{\pm}(x) \equiv \phi_{pn}(x) \pm i\psi_{pn}(x) = e^{\pm in\omega_p x} g_{pn}^{\pm}(x), \tag{13.6.2}$$

with ω_p real positive constants and

$$g_{pn}^{\pm}(x) \sim n^{\epsilon_p} \sum_{j=0}^{\infty} \delta_{pj}^{\pm}(x) n^{-j} \quad \text{as } n \to \infty, \tag{13.6.3}$$

for some fixed ϵ_p that can be complex in general. Thus, as a function of n, $g_{pn}^{\pm}(x)$ is a function in $\mathbf{A}_0^{(\epsilon_p)}$. We assume the coefficients b_n, c_n, d_n, and e_n are available.

We have already seen that $\{\phi_{pn}(x)\}_{n=1}^{\infty} \in \mathbf{b}^{(2)}$. From Heuristic 6.4.1, we have that $\{\phi_{pn}(x)\phi_{qn}(y)\}_{n=1}^{\infty} \in \mathbf{b}^{(r)}$, where $r = 3$ or $r = 4$. Thus, if the d-transformation can be applied to $H(x, y)$, this is possible with $m \geq 4$ in general, and we need a large number of the terms of $H(x, y)$ to accelerate its convergence by the d-transformation.

Employing in (13.6.1) the fact that

$$\phi_{pn}(x) = \frac{1}{2}\left[\rho_{pn}^{+}(x) + \rho_{pn}^{-}(x)\right] \text{ and } \psi_{pn}(x) = \frac{1}{2i}\left[\rho_{pn}^{+}(x) - \rho_{pn}^{-}(x)\right], \tag{13.6.4}$$

we can express $H(x, y)$ in the form

$$H(x, y) = H^{+,+}(x, y) + H^{+,-}(x, y) + H^{-,+}(x, y) + H^{-,-}(x, y), \tag{13.6.5}$$

where

$$H^{\pm,\pm}(x, y) := \sum_{k=1}^{\infty} w_k^{\pm,\pm} \rho_{1k}^{\pm}(x)\rho_{2k}^{\pm}(y),$$

$$H^{\pm,\mp}(x, y) := \sum_{k=1}^{\infty} w_k^{\pm,\mp} \rho_{1k}^{\pm}(x)\rho_{2k}^{\mp}(y), \tag{13.6.6}$$

with appropriate coefficients $w_k^{\pm,\pm}$ and $w_k^{\pm,\mp}$. For example, when $H(x, y) := \sum_{k=1}^{\infty} b_k \phi_{1k}(x)\phi_{2k}(y)$, we have $w_k^{\pm,\pm} = w_k^{\pm,\mp} = \frac{1}{4} b_k$ for all k.

By (13.6.2), we have that

$$\rho_{1n}^{\pm}(x)\rho_{2n}^{\pm}(y) = e^{\pm in(\omega_1 x + \omega_2 y)} u^{\pm,\pm}(n), \quad u^{\pm,\pm} \in \mathbf{A}_0^{(\epsilon_1 + \epsilon_2)},$$

$$\rho_{1n}^{\pm}(x)\rho_{2n}^{\mp}(y) = e^{\pm in(\omega_1 x - \omega_2 y)} u^{\pm,\mp}(n), \quad u^{\pm,\mp} \in \mathbf{A}_0^{(\epsilon_1 + \epsilon_2)}. \tag{13.6.7}$$

Thus, $\{\rho_{1n}^{\pm}(x)\rho_{2n}^{\pm}(y)\}$ and $\{\rho_{1n}^{\pm}(x)\rho_{2n}^{\mp}(y)\}$ are in $\mathbf{b}^{(1)}$. This means that we can apply the $d^{(m)}$-transformation with a low value of m to each of the four series $H^{\pm,\pm}(x, y)$ and $H^{\pm,\mp}(x, y)$. In particular, if $b_n = \lambda^n h(n)$ with $h(n) \in \mathbf{A}_0^{(\gamma)}$, where λ and γ are in general complex, all four series can be treated by the $d^{(1)}$-transformation. As long as $\lambda e^{\pm i(\omega_1 x + \omega_2 y)} \neq 1$ $[\lambda e^{\pm i(\omega_1 x - \omega_2 y)} \neq 1]$, the series $B^{\pm,\pm}(x, y)$ $[B^{\pm,\mp}(x, y)]$ are oscilla-tory and hence can be treated by applying the $d^{(1)}$-transformation with APS. When $\lambda e^{\pm i(\omega_1 x + \omega_2 y)} = 1$ $[\lambda e^{\pm i(\omega_1 x - \omega_2 y)} = 1]$, however, $B^{\pm,\pm}(x, y)$ $[B^{\pm,\mp}(x, y)]$ are slowly varying and can be treated by applying the $d^{(1)}$-transformation, with GPS if necessary.

This use of the d-transformation is much more economical than its application directly to $H(x, y)$.

Finally, the approach of this section can be extended further to products of three or more functions $\phi_{pn}(x)$ and/or $\psi_{pn}(x)$ without any conceptual changes.

13.7 The $\mathcal{H}$-Transformation

A method, called the $\mathcal{H}$-*transformation*, to accelerate the convergence of Fourier sine and cosine series was proposed by Homeier [135]. We include this transformation in this chapter, as it is simply a GREP$^{(2)}$ and a variant of the $d^{(2)}$-transformation.

Let

$$F(x) := \sum_{k=0}^{\infty} (b_k \cos kx + c_k \sin kx),$$

be the given Fourier series and let its partial sums be

$$S_n = \sum_{k=0}^{n} (b_k \cos kx + c_k \sin kx), \quad n = 0, 1, \ldots,$$

Then, the approximation $\mathcal{H}_n^{(j)}$ to the sum of this series is defined via the linear system

$$S_l = \mathcal{H}_n^{(j)} + r_l \left[\cos lx \sum_{i=0}^{n-1} \frac{\bar{\beta}_i}{(l+\delta)^i} + \sin lx \sum_{i=0}^{n-1} \frac{\bar{\gamma}_i}{(l+\delta)^i} \right], \quad j \leq l \leq j + 2n, \quad (13.7.1)$$

where

$$r_n = (n+1)M(b_n, c_n), \quad M(p, q) = \begin{cases} p \text{ if } |p| > |q|, \\ q \text{ otherwise,} \end{cases} \quad (13.7.2)$$

and δ is some fixed constant. As before, $\bar{\beta}_i$ and $\bar{\gamma}_i$ are additional auxiliary unknowns. Homeier has given an elegant recursive algorithm for implementing the $\mathcal{H}$-transformation that is very economical.

Unfortunately, this transformation has two drawbacks:

1. The class of Fourier series to which it applies successfully is quite limited. This can be seen as follows: The equations in (13.7.1) should be compared with those that define the $d_{(n,n)}^{(2,j)}$, namely,

$$S_{R_l} = d_{(n,n)}^{(2,j)} + a_{R_l} \sum_{i=0}^{n-1} \frac{\bar{\beta}_i}{R_l^i} + \Delta a_{R_l} \sum_{i=0}^{n-1} \frac{\bar{\gamma}_i}{R_l^i}, \quad j \leq l \leq j + 2n, \quad (13.7.3)$$

where $a_n = b_n \cos nx + c_n \sin nx$, with the special choice of the R_l, namely, $R_l = l + 1, \; l = 0, 1, \ldots$. Thus, $d_{(n,n)}^{(2,j)}$ and $\mathcal{H}_n^{(j)}$ use almost the same number of terms of $F(x)$.

The equations in (13.7.1) immediately suggest that the $\mathcal{H}$-transformation can be effective when

$$S_n \sim S + r_n \left[\cos nx \sum_{i=0}^{\infty} \frac{\beta_i}{n^i} + \sin nx \sum_{i=0}^{\infty} \frac{\gamma_i}{n^i} \right] \text{ as } n \to \infty,$$

that is, when S_n is associated with a function $A(y) \in \mathbf{F}^{(2)}$. This situation is possible only when $\{b_n\}$ and $\{c_n\}$ are both in $\mathbf{b}^{(1)}$. In view of this, it is clear that, when either $\{b_n\}$ or $\{c_n\}$ or both are in $\mathbf{b}^{(s)}$ with $s > 1$, the $\mathcal{H}$-transformation ceases to be effective. In contrast, the $d^{(m)}$-transformation for some appropriate value of $m > 2$ is effective, as we mentioned earlier.

As an example, let us consider the cosine series $F(x) := \sum_{k=0}^{\infty} b_k \cos kx$, where $b_n = P_n(t)$ are the Legendre polynomials. Because $\{b_n\} \in \mathbf{b}^{(2)}$, we have that $\{b_n \cos nx\} \in \mathbf{b}^{(4)}$. The $d^{(4)}$-transformation can be applied directly to $F(x)$. The $d^{(2)}$-transformation with the complex series approach can also be applied at about half the cost of the direct approach. The $\mathcal{H}$-transformation is ineffective.

2. By the way r_n is defined, it is clear that the b_n and c_n are assumed to be available. In this case, as explained before, the $d^{(1)}$-transformation with $R_l = l + 1$ (that is nothing but the Levin transformation) coupled with the complex series approach achieves the required accuracy at about half the cost of the $\mathcal{H}$-transformation, when the latter is applicable. (As mentioned in Section 13.2, the application of the Levin transformation with the complex series approach was suggested and analyzed earlier in Sidi [273, Section 3].) Of course, better stability and accuracy is achieved by the $d^{(1)}$-transformation with APS near points of singularity.

14

Special Topics in Richardson Extrapolation

14.1 Confluence in Richardson Extrapolation

14.1.1 The Extrapolation Process and the SGRom-Algorithm

In Chapters 1 and 2, we considered functions $A(y)$ that satisfy (1.1.2). In this section, we consider functions that we now denote $B(y)$ and that have asymptotic expansions of the form

$$B(y) \sim B + \sum_{k=1}^{\infty} Q_k(\log y) y^{\sigma_k} \quad \text{as } y \to 0+, \qquad (14.1.1)$$

where y is a discrete or continuous variable, σ_k are distinct and in general complex scalars satisfying

$$\sigma_k \neq 0, \quad k = 1, 2, \ldots; \quad \Re \sigma_1 \leq \Re \sigma_2 \leq \cdots; \quad \lim_{k \to \infty} \Re \sigma_k = +\infty, \qquad (14.1.2)$$

and $Q_k(x)$ are some polynomials given as

$$Q_k(x) = \sum_{i=0}^{q_k} \alpha_{ki} x^i \quad \text{for some integer } q_k \geq 0. \qquad (14.1.3)$$

From (14.1.2), it is clear that there can be only a finite number of σ_k with equal real parts. When $\Re \sigma_1 > 0$, $\lim_{y \to 0+} B(y)$ exists and is equal to B. When $\Re \sigma_1 \leq 0$ and $Q_1(x) \not\equiv 0$, however, $\lim_{y \to 0+} B(y)$ does not exist, and B in this case is the antilimit of $B(y)$ as $y \to 0+$.

We assume that $B(y)$ is known (and hence is computable) for all possible $y > 0$ and that the σ_k and q_k are also known. Note that q_k is an upper bound for ∂Q_k, the degree of $Q_k(x)$, and that ∂Q_k need not be known exactly. We assume that B and the α_{ki} are not necessarily known and that B is being sought.

Clearly, the problem we have here generalizes that of Chapter 1 (i) by allowing the σ_k to have equal real parts and (ii) by replacing the *constants* α_k in (1.1.2) by the *polynomials* Q_k in $\log y$.

Note that we can also think of the expansion $\sum_{k=1}^{\infty} Q_k(\log y) y^{\sigma_k}$ as being obtained by letting $\tau_{ki} \to \sigma_k$, $i = 0, 1, \ldots, q_k$, in the expansion $\sum_{k=1}^{\infty} \sum_{i=0}^{q_k} \beta_{ki} y^{\tau_{ki}}$. Thus, we view the extrapolation process we are about to propose for determining B as a Richardson extrapolation process "with confluence."

Let us order the functions $y^{\sigma_k}(\log y)^i$ as follows:

$$\phi_i(y) = y^{\sigma_1}(\log y)^{i-1}, \qquad 1 \le i \le \nu_1 \equiv q_1 + 1,$$

$$\phi_{\nu_1+i}(y) = y^{\sigma_2}(\log y)^{i-1}, \qquad 1 \le i \le \nu_2 \equiv q_2 + 1,$$

$$\phi_{\nu_1+\nu_2+i}(y) = y^{\sigma_3}(\log y)^{i-1}, \qquad 1 \le i \le \nu_3 \equiv q_3 + 1, \qquad (14.1.4)$$

and so on.

Let us choose $y_0 > y_1 > \cdots > 0$ such that $\lim_{l\to\infty} y_l = 0$. Then we define the generalization of the Richardson extrapolation process for the present problem through the linear systems

$$B(y_l) = B_n^{(j)} + \sum_{k=1}^{n} \bar{\alpha}_k \phi_k(y_l), \quad j \le l \le j + n. \qquad (14.1.5)$$

As always, we have $B_n^{(j)} = \sum_{i=0}^{n} \theta_{ni}^{(j)} B(y_{j+i})$ with $\sum_{i=0}^{n} \theta_{ni}^{(j)} = 1$, and we define $\Theta_n^{(j)} = \sum_{i=0}^{n} |\theta_{ni}^{(j)}|$. (Note that we have changed our usual notation slightly and written $\theta_{ni}^{(j)}$ instead of $\gamma_{ni}^{(j)}$ and $\Theta_n^{(j)}$ instead of $\Gamma_n^{(j)}$.)

In this section, we summarize the treatment given to this problem by Sidi [298] with $y_l = y_0\omega^l$, $l = 0, 1, \ldots$, for some $\omega \in (0, 1)$. For details and numerical examples, we refer the reader to [298].

We start with the following recursive algorithm for computing the $B_n^{(j)}$. This algorithm is denoted the *SGRom-algorithm*.

Algorithm 14.1.1 (SGRom-algorithm)

1. Let $c_k = \omega^{\sigma_k}$, $k = 1, 2, \ldots$, and set

$$\lambda_i = c_1, \qquad 1 \le i \le \nu_1,$$

$$\lambda_{\nu_1+i} = c_2, \qquad 1 \le i \le \nu_2,$$

$$\lambda_{\nu_1+\nu_2+i} = c_3, \qquad 1 \le i \le \nu_3, \qquad (14.1.6)$$

and so on.
2. Set

$$B_0^{(j)} = B(y_j), \quad j = 0, 1, \ldots.$$

3. Compute $B_n^{(j)}$ by the recursion relation

$$B_n^{(j)} = \frac{B_{n-1}^{(j+1)} - \lambda_n B_{n-1}^{(j)}}{1 - \lambda_n}, \quad j = 0, 1, \ldots, \ n = 1, 2, \ldots.$$

Our first result concerns the $\theta_{ni}^{(j)}$.

Theorem 14.1.2 *The $\theta_{ni}^{(j)}$ are independent of j and satisfy*

$$\sum_{i=0}^{n} \theta_{ni}^{(j)} z^i = \prod_{i=1}^{n} \frac{z - \lambda_i}{1 - \lambda_i} \equiv U_n(z). \qquad (14.1.7)$$

Hence $\Theta_n^{(j)}$ *is independent of j. In addition,*

$$\Theta_n^{(j)} \leq \prod_{i=1}^{n} \frac{1 + |\lambda_i|}{|1 - \lambda_i|} \quad \text{for all } j, n, \tag{14.1.8}$$

with equality in (14.1.8) when the c_k all have the same phase or, equivalently, when the σ_k all have the same imaginary part.

14.1.2 Treatment of Column Sequences

Let us first note that, for any integer $n \geq 0$, there exist two unique nonnegative integers t and s, such that

$$n = \sum_{k=1}^{t} \nu_k + s, \quad 0 \leq s \leq \nu_{t+1} - 1, \tag{14.1.9}$$

where $\nu_k = q_k + 1$ as before. We also define $\sum_{k=1}^{t} \nu_k$ to be zero when $t = 0$. With n, t, and s as in (14.1.9), we next define the two sets of integers S_n and T_n as in

$$S_n = \{(k, r) : 0 \leq r \leq q_k, \ 1 \leq k \leq t, \text{ and } 0 \leq r \leq s - 1, \ k = t + 1\},$$
$$T_n = \{(k, r) : 0 \leq r \leq q_k, \ k \geq 1\} \setminus S_n. \tag{14.1.10}$$

Theorem 14.1.3 concerns the convergence and stability of the column sequences $\{B_n^{(j)}\}_{j=0}^{\infty}$. Note only that $|c_k| = e^{\Re \sigma_k}$ for all k and that the ordering of the σ_k in (14.1.2) implies

$$c_k \neq 1, \ k = 1, 2, \ldots \ ; \ |c_1| \geq |c_2| \geq \cdots ; \ \lim_{k \to \infty} c_k = 0, \tag{14.1.11}$$

and that $|c_i| = |c_j|$ if and only if $\Re \sigma_i = \Re \sigma_j$. Also, with $y_l = y_0 \omega^l$, the asymptotic expansion in (14.1.1) assumes the form

$$B(y_m) \sim B + \sum_{k=1}^{\infty} \left(\sum_{i=0}^{q_k} \beta_{ki} m^i \right) c_k^m \quad \text{as } m \to \infty, \tag{14.1.12}$$

where β_{ki} depend linearly on the α_{kr}, $i \leq r \leq q_k$, and $\beta_{k,q_k} = \alpha_{k,q_k} y_0^{\sigma_k} (\log \omega)^{q_k}$.

It is clear that the extrapolation method of this section can be applied via the SGRom-algorithm with no changes to sequences $\{X_m\}$ when the asymptotic expansion of X_m is exactly as in the right-hand side of (14.1.12).

Theorem 14.1.3

(i) *With $U_n(z)$ as in (14.1.7), and S_n and T_n as in (14.1.10), for fixed n, we have the complete asymptotic expansion*

$$B_n^{(j)} - B \sim \sum_{(k,r) \in T_n}^{\infty} \beta_{kr} \left\{ \left(z \frac{d}{dz} \right)^r \left[z^j U_n(z) \right] \Big|_{z=c_k} \right\} \quad \text{as } j \to \infty. \tag{14.1.13}$$

If we let μ be that integer for which $|c_{t+1}| = \cdots = |c_{t+\mu}| > |c_{t+\mu+1}|$ and let $\bar{q} = \max\{q_{t+1} - s, q_{t+2}, \dots, q_{t+\mu}\}$, then (14.1.13) gives

$$B_n^{(j)} - B = O(|c_{t+1}|^j j^{\bar{q}}) \text{ as } j \to \infty. \tag{14.1.14}$$

(ii) *The process is stable in the sense that $\sup_j \Theta_n^{(j)} < \infty$. [Recall that $\Theta_n^{(j)}$ is independent of j and satisfies (14.1.8).]*

When $\mu = 1$ and $\beta_{t+1,q_{t+1}} \neq 0$, (14.1.13) gives the asymptotic equality

$$B_n^{(j)} - B \sim \beta_{t+1,q_{t+1}} \binom{q_{t+1}}{s} U_n^{(s)}(c_{t+1}) c_{t+1}^{j+s} j^{q_{t+1}-s} \text{ as } j \to \infty. \tag{14.1.15}$$

This is the case, in particular, when $|c_1| > |c_2| > \cdots$ and $\beta_{k,q_k} \neq 0$ for all k.

In any case, Theorem 14.1.3 implies that each column in the extrapolation table is at least as good as the one preceding it.

14.1.3 Treatment of Diagonal Sequences

The treatment of the diagonal sequences turns out to be much more involved, as usual. To proceed, we need to introduce some new notation. First, let $1 = k_1 < k_2 < k_3 < \cdots$ be the (smallest) integers for which

$$\Re\sigma_{k_i} < \Re\sigma_{k_{i+1}} \text{ and } \Re\sigma_m = \Re\sigma_{k_i}, \quad k_i \leq m \leq k_{i+1} - 1, \quad i = 1, 2, \dots, \tag{14.1.16}$$

which implies that

$$|c_{k_i}| > |c_{k_{i+1}}| \text{ and } |c_m| = |c_{k_i}|, \quad k_i \leq m \leq k_{i+1} - 1, \quad i = 1, 2, \dots. \tag{14.1.17}$$

Then, let

$$N_i = \sum_{r=k_i}^{k_{i+1}-1} v_r, \quad i = 1, 2, \dots. \tag{14.1.18}$$

In other words, N_i is the sum of the "multiplicities" v_r of all the c_r that have modulus equal to $|c_{k_i}|$.

Part (i) of the following theorem is new and its proof can be achieved as that of part (ii) of Theorem 1.5.4. Part (ii) is given in [298].

Theorem 14.1.4 *Assume that the σ_k satisfy*

$$\Re\sigma_{k_{i+1}} - \Re\sigma_{k_i} \geq d > 0 \text{ for all } i. \tag{14.1.19}$$

Assume also that the σ_k and q_k are such that there exist constants $E > 0$ and $b \geq 0$ for which

$$\limsup_{i \to \infty} N_i / i^b = E. \tag{14.1.20}$$

(i) *Under these conditions, the sequence $\{B_n^{(j)}\}_{n=0}^{\infty}$ converges to B as indicated in*

$$B_n^{(j)} - B = O(\omega^\mu) \text{ as } n \to \infty \text{ for every } \mu > 0. \tag{14.1.21}$$

(ii) *The process is stable in the sense that $\sup_n \Theta_n^{(j)} < \infty$.*

It is worth mentioning that the proof of this theorem relies on the fact that $\sum_{i=1}^{\infty} |\lambda_i|$ converges under the conditions in (14.1.19) and (14.1.20).

By imposing suitable growth conditions on the α_{ki} and $\Re\sigma_k$ and assuming, in addition to (14.1.20), that

$$\liminf_{i\to\infty} N_i/i^a = D > 0, \quad 0 \le a \le b, \quad a+2 > b, \tag{14.1.22}$$

Sidi [298] shows that

$$|B_n^{(j)} - B| \le (K+\epsilon)^{r^{(a+2)}} \le (L+\epsilon)^{n^{(a+2)/(b+1)}} \quad \text{for all large } n, \tag{14.1.23}$$

where $\epsilon > 0$ is arbitrary, r is that integer for which $k_{r+1} \le t+1 < k_{r+2}$, and

$$K = \omega^{\eta}, \quad \eta = \frac{Dd}{a+2}; \quad L = \omega^{\tau}, \quad \tau = \frac{Dd}{a+2}\left(\frac{b+1}{E}\right)^{(a+2)/(b+1)}. \tag{14.1.24}$$

We note that in many cases of interest $a = b$ in general and $a = b = 0$ in particular. For example, when $N_i \sim Fi^{\nu}$ as $i \to \infty$, we have $a = b = \nu$ and $D = E = F$. In such cases, (14.1.23) shows that $B_n^{(j)} - B = O(e^{-\kappa n^u})$ as $n \to \infty$, for some $\kappa > 0$, with $u = (a+2)/(a+1) > 1$. This is a refinement of (14.1.21).

We also note that, for all practical purposes, $B_n^{(j)} - B$ is $O(\prod_{i=1}^{n} |\lambda_i|)$ as $n \to \infty$. Very realistic information about the error can be obtained by analyzing the behavior of the product $\prod_{i=1}^{n} |\lambda_i|$ for large n. See Sidi [301] for examples.

It is useful at this point to mention that the trapezoidal rule approximation $T(h)$ for the singular integral in Example 3.1.2 has all the characteristics of the functions $B(y)$ we have discussed so far. From the asymptotic expansion in (3.1.6) and from (3.1.7), it is clear that, in this example, $q_k = 0$ or $q_k = 1$, $a = b = 0$, and $D = 1$ and $E = 2$. This example has been generalized in Sidi [298].

14.1.4 A Further Problem

Before we end this section, we would like to mention that a problem similar to the one considered in Subsections 14.1.1–14.1.3, but of a more general nature, has been treated by Sidi [297]. In this problem, we consider a function $B(y)$ that has the asymptotic expansion

$$B(y) \sim B + \sum_{k=1}^{\infty} \psi_k(y) Q_k(\log y) \quad \text{as } y \to 0+, \tag{14.1.25}$$

where B is the limit or antilimit of $B(y)$ as $y \to 0+$, $Q_k(x)$ is a polynomial in x of degree at most q_k, $k = 1, 2, \ldots$, as before, and

$$\psi_{k+1}(y) = O(\psi_k(y)) \quad \text{as } y \to 0+, \quad k = 1, 2, \ldots. \tag{14.1.26}$$

The y_l are chosen to satisfy $\lim_{l\to\infty}(y_{l+1}/y_l) = \omega \in (0,1)$, and it is assumed that

$$\lim_{l\to\infty} \frac{\psi_k(y_{l+1})}{\psi_k(y_l)} = c_k \ne 1, \quad k = 1, 2, \ldots, \quad \text{and} \quad c_i \ne c_j \text{ if } i \ne j. \tag{14.1.27}$$

Thus, $|c_1| \geq |c_2| \geq |c_3| \geq \cdots$. [It also follows that if $|c_{k+1}| < |c_k|$, then (14.1.26) is actually $\psi_{k+1}(y) = o(\psi_k(y))$ as $y \to 0+$.] In addition, we assume that there are finitely many c_k with the same modulus.

The Richardson extrapolation is generalized to this problem in a suitable fashion in [297] and the column sequences are analyzed with respect to convergence and stability. In particular, the approximation $B_n^{(j)}$ to B, where $n = \sum_{k=1}^{t}(q_k + 1)$, are defined through

$$B(y_l) = B_n^{(j)} + \sum_{k=1}^{t} \psi_k(y_l) \sum_{i=0}^{q_k} \bar{\alpha}_{ki}(\log y_l)^i, \quad j \leq l \leq j + n. \quad (14.1.28)$$

Again, $B_n^{(j)} = \sum_{i=0}^{n} \theta_{ni}^{(j)} A(y_{j+i})$ with $\sum_{i=0}^{n} \theta_{ni}^{(j)} = 1$, and we define $\Theta_n^{(j)} = \sum_{i=0}^{n} |\theta_{ni}^{(j)}|$. We then have the following convergence result for the column sequence $\{B_n^{(j)}\}_{j=0}^{\infty}$:

$$B_n^{(j)} - B = O(R_t(y_j)) \text{ as } j \to \infty, \quad (14.1.29)$$

where

$$R_t(y) = B(y) - B - \sum_{k=1}^{t} \psi_k(y) \sum_{i=0}^{q_k} Q_k(\log y). \quad (14.1.30)$$

Note that

$$R_t(y) = O(\psi_{t+1}(y)(\log y)^{\hat{q}}) \text{ as } y \to 0+; \quad \hat{q} = \max\{q_k : |c_k| = |c_{t+1}|, \ k \geq t + 1\}. \quad (14.1.31)$$

The process is also stable because

$$\lim_{j \to \infty} \sum_{i=0}^{n} \theta_{ni}^{(j)} z^i = \prod_{k=1}^{t} \left(\frac{z - c_k}{1 - c_k}\right)^{q_k+1} \equiv \sum_{i=0}^{n} \tilde{\theta}_{ni} z^i, \quad (14.1.32)$$

from which we have

$$\lim_{j \to \infty} \Theta_n^{(j)} = \sum_{i=0}^{n} |\tilde{\theta}_{ni}| \leq \prod_{k=1}^{t} \left(\frac{1 + |c_k|}{|1 - c_k|}\right)^{q_k+1} < \infty. \quad (14.1.33)$$

These results are special cases of those proved in Sidi [297], where $B_n^{(j)}$ are defined and analyzed for *all* $n = 1, 2, \ldots$. For details see [297].

It is clear that the function $B(y)$ in (14.1.1) and (14.1.2) considered in Subsections 14.1.1–14.1.3 is a special case of the one we consider in this subsection with $\psi_k(y) = y^{\sigma_k}$. Because the y_l are chosen to satisfy $\lim_{l \to \infty}(y_{l+1}/y_l) = \omega$ in the extrapolation process, we also have $c_k = \omega^{\sigma_k}$ in (14.1.27). Thus, the theory of [297] in general and (14.1.29)–(14.1.33) in particular provide additional results for the function $B(y)$ in (14.1.1)–(14.1.3).

14.2 Computation of Derivatives of Limits and Antilimits

In some common applications, functions $B(y)$ (and their limits or antilimits B) of the form we treated in the preceding section arise as derivatives with respect to some parameter ξ of functions $A(y)$ (and their limits or antilimits A) that we treated in Chapter 1. The trapezoidal rule approximation $T(h)$ of Example 3.1.2, which we come back to shortly, is one such function.

Also, numerical examples and the rigorous theoretical explanation given in Sidi [301] show that, when $A(y)$ and $B(y)$ have asymptotic expansions that are essentially different, it is much cheaper to use extrapolation on $A(y)$ to approximate its limit or antilimit A than on $B(y)$ to approximate its limit or antilimit B. By this we mean that, for a given level of required accuracy, more function values $B(y)$ than $A(y)$ need to be computed. The question that arises then is whether it is possible to reduce the large computational cost of extrapolating $B(y)$ to approximate B when we know that $B(y) = \frac{d}{d\xi} A(y)$ and $B = \frac{d}{d\xi} A$.

This question was raised recently by Sidi [301], who also proposed an effective approach to its solution. In this section, we summarize the approach of [301] and give the accompanying theoretical results that provide its justification. For the details and numerical examples, we refer the reader to this paper.

Let us denote by E_0 the extrapolation process used on $A(y)$ in approximating A. When the asymptotic expansion of $B(y) = \frac{d}{d\xi} A(y)$ is essentially different from that of $A(y)$, it is proposed *to differentiate with respect to ξ the approximations to A produced by E_0 on $A(y)$ and take these derivatives as approximations to B.* As a way of implementing this procedure numerically, it is also proposed *to differentiate with respect to ξ the recursion relations used in implementing E_0.* (In doing that we should also differentiate the initial conditions.)

Of course, the process proposed here can be applied to computation of higher-order derivatives of limits and antilimits as well. It can also be applied to computation of partial derivatives with respect to several variables.

These ideas can best be demonstrated via the Richardson extrapolation process of Chapter 1.

14.2.1 Derivative of the Richardson Extrapolation Process

Let $A(y)$ be exactly as in Chapter 1, that is,

$$A(y) \sim A + \sum_{k=1}^{\infty} \alpha_k y^{\sigma_k} \text{ as } y \to 0+, \qquad (14.2.1)$$

where

$$\sigma_k \neq 0, \quad k = 1, 2, \ldots; \quad \Re\sigma_1 < \Re\sigma_2 < \cdots; \quad \lim_{k \to \infty} \Re\sigma_k = \infty, \qquad (14.2.2)$$

and assume that $A(y)$, A, the α_k and the σ_k depend on a parameter ξ in addition. Let us also assume that $\dot{A}(y) \equiv \frac{d}{d\xi} A(y)$ has an asymptotic expansion as $y \to 0+$ that is obtained by differentiating that of $A(y)$ termwise, that is,

$$\dot{A}(y) \sim \dot{A} + \sum_{k=1}^{\infty} (\dot{\alpha}_k + \alpha_k \dot{\sigma}_k \log y) y^{\sigma_k} \text{ as } y \to 0+. \qquad (14.2.3)$$

Here we have also denoted $\frac{d}{d\xi} A = \dot{A}$, $\frac{d}{d\xi} \alpha_k = \dot{\alpha}_k$ and $\frac{d}{d\xi} \sigma_k = \dot{\sigma}_k$. Note that when the $\dot{\sigma}_k$ do not all vanish, the asymptotic expansion in (14.2.3) is essentially different from that in (14.2.1). Also, the asymptotic expansion of $\dot{A}(y)$ in (14.2.3) is exactly of the form given in (14.1.1) for $B(y)$ with $q_k = 1$ for all k in the latter.

Let us apply the Richardson extrapolation process, with $y_l = y_0\omega^l$, $l = 0, 1, \ldots$, to the function $A(y)$ to obtain the approximations $A_n^{(j)}$ to A. Next, let us differentiate the $A_n^{(j)}$ to obtain the approximations $\dot{A}_n^{(j)}$ to $\dot{A}$. The necessary algorithm for this is obtained by differentiating the recursion relation satisfied by the $A_n^{(j)}$ that is given in Algorithm 1.3.1. Thus, letting $c_n = \omega^{\sigma_n}$ and $\dot{c}_n = \frac{d}{d\xi}c_n = (\log \omega)\dot{\sigma}_n c_n$, we obtain the following recursion relation for the $\dot{A}_n^{(j)}$:

$$A_0^{(j)} = A(y_j) \quad \text{and} \quad \dot{A}_0^{(j)} = \dot{A}(y_j), \quad j = 0, 1, \ldots,$$

$$A_n^{(j)} = \frac{A_{n-1}^{(j+1)} - c_n A_{n-1}^{(j)}}{1 - c_n} \quad \text{and}$$

$$\dot{A}_n^{(j)} = \frac{\dot{A}_{n-1}^{(j+1)} - c_n \dot{A}_{n-1}^{(j)}}{1 - c_n} + \frac{\dot{c}_n}{1 - c_n}(A_n^{(j)} - A_{n-1}^{(j)}), \quad j = 0, 1, \ldots, \quad n = 1, 2, \ldots.$$

Remark. It is clear that we need both $A(y_l)$ and $\dot{A}(y_l)$, $j \leq l \leq j + n$, for determining $\dot{A}_n^{(j)}$, and this may seem expensive at first. However, in most problems of interest, the computation of $\dot{A}(y_l)$ can be done simultaneously with that of $A(y_l)$, and at almost no additional cost. Thus, in such problems, the computation of $\dot{A}_n^{(j)}$ entails practically the same cost as that of $A_n^{(j)}$, in general. This makes the proposed approach desirable.

Because $A_n^{(j)} = \sum_{i=0}^{n} \gamma_{ni}^{(j)} A(y_{j+i})$, we have that

$$\dot{A}_n^{(j)} = \sum_{i=0}^{n} \gamma_{ni}^{(j)} \dot{A}(y_{j+i}) + \sum_{i=0}^{n} \dot{\gamma}_{ni}^{(j)} A(y_{j+i}),$$

as a result of which we conclude that the propagation of errors (roundoff and other) in the $A(y_l)$ and $\dot{A}(y_l)$, $j \leq l \leq j + n$, into $\dot{A}_n^{(j)}$ is controlled by the quantity $\Omega_n^{(j)}$ that is defined as in

$$\Omega_n^{(j)} = \sum_{i=0}^{n} |\gamma_{ni}^{(j)}| + \sum_{i=0}^{n} |\dot{\gamma}_{ni}^{(j)}|. \tag{14.2.4}$$

The following theorem summarizes the convergence and stability of the column and diagonal sequences in the extrapolation table of the $\dot{A}_n^{(j)}$. In this theorem, we make use of the fact that the $\gamma_{ni}^{(j)}$ and hence the $\dot{\gamma}_{ni}^{(j)}$ are independent of j in the case being considered.

Theorem 14.2.1

(i) *For fixed n, the error $\dot{A}_n^{(j)} - \dot{A}$ has the complete asymptotic expansion*

$$\dot{A}_n^{(j)} - \dot{A} \sim \sum_{k=n+1}^{\infty} \left\{ \frac{d}{d\xi}[U_n(c_k)\alpha_k] + U_n(c_k)\alpha_k\dot{\sigma}_k \log y_j \right\} y_j^{\sigma_k} \quad \text{as} \quad j \to \infty, \tag{14.2.5}$$

from which we conclude that $\dot{A}_n^{(j)} - \dot{A} = O(j|c_{n+1}|^j)$ as $j \to \infty$. Here $U_n(z) = \prod_{i=1}^{n}(z - c_i)/(1 - c_i)$. In addition, $\sup_j \Omega_n^{(j)} < \infty$ since $\Omega_n^{(j)}$ are all independent of j. That is, column sequences are stable.

(ii) *Let us assume, in addition to (14.2.2), that the σ_k also satisfy*

$$\Re\sigma_{k+1} - \Re\sigma_k \geq d > 0, \quad k = 1, 2, \ldots, \quad \text{for some fixed } d, \quad (14.2.6)$$

and that $\sum_{i=1}^{\infty} |\dot{c}_i| < \infty$. Then, for fixed j, the error $\dot{A}_n^{(j)} - \dot{A}$ satisfies

$$\dot{A}_n^{(j)} - \dot{A} = O(\omega^{\mu}) \text{ as } n \to \infty, \quad \text{for every } \mu > 0. \quad (14.2.7)$$

Also, $\sup_n \Omega_n^{(j)} < \infty$, which implies that diagonal sequences are stable.

Part (ii) of Theorem 14.2.1 says that, as $n \to \infty$, $\dot{A}_n^{(j)} - \dot{A} = O(e^{-\lambda n})$ for *every* $\lambda > 0$. By imposing some mild growth conditions on the α_k and the $\Re\sigma_k$, and assuming further that $|\dot{c}_i| \leq K_i |c_i|$ for all i, such that $K_i = O(i^a)$ as $i \to \infty$, for some $a \geq -1$, the result in (14.2.7) can be improved to read

$$|\dot{A}_n^{(j)} - \dot{A}| \leq (\omega + \epsilon)^{dn^2/2} \text{ as } n \to \infty, \quad \text{with arbitrary } \epsilon > 0. \quad (14.2.8)$$

This means that $\dot{A}_n^{(j)} - \dot{A} = O(e^{-\kappa n^2})$ as $n \to \infty$ for *some* $\kappa > 0$.

We also note that, for all practical purposes, $\dot{A}_n^{(j)} - \dot{A}$, just like $A_n^{(j)} - A$, is $O(\prod_{i=1}^{n} |c_i|)$ as $n \to \infty$. Very realistic information on both can be obtained by analyzing the behavior of the product $\prod_{i=1}^{n} |c_i|$ as $n \to \infty$.

The important conclusion we draw from both parts of Theorem 14.2.1 is that the accuracy of $\dot{A}_n^{(j)}$ as an approximation to $\dot{A}$ is almost the same as that of $A_n^{(j)}$ as an approximation to A. This clearly shows that the approach we suggested for determining $\dot{A}$ is very efficient. In fact, we can now show rigorously that it is more efficient than the approach of the preceding section involving the SGRom-algorithm. (However, we should keep in mind that the problem we are treating here is a special one and that the approach with the SGRom-algorithm can be applied to a larger class of problems.)

Let us denote $\dot{A}(y) = B(y)$ and $\dot{A} = B$ in (14.2.3) and apply the SGRom-algorithm to $B(y)$ to obtain the approximations $B_n^{(j)}$ to B. Let us assume that the conditions of Theorem 14.2.1 are satisfied. Then, the conditions of Theorem 14.1.4 are satisfied as well, and $N_i \leq 2$ for all i there. Assuming that $N_i = 2$ for all i, we have $D = E = 2$ and $a = b = 0$, so that $L = \omega^{d/4}$ in (14.1.24). As a result, by (14.1.23), $B_n^{(j)} - B$ is of order $\omega^{dn^2/4}$ as $n \to \infty$, for all practical purposes. For the approximations $\dot{A}_n^{(j)}$ to $\dot{A}$, on the other hand, we have that $\dot{A}_n^{(j)} - \dot{A}$ is of order $\omega^{dn^2/2}$ as $n \to \infty$, for all practical purposes. It is clear that $B_{\lfloor \sqrt{2}n \rfloor}^{(j)}$ will have an accuracy comparable to that of $\dot{A}_n^{(j)}$; its computational cost will, of course, be higher.

An Application to Numerical Quadrature

Let us consider the numerical approximation of the integral $B = \int_0^1 (\log x) x^{\xi} g(x) dx$, $\Re\xi > -1$, with $g \in C^{\infty}[0, 1]$. Clearly, $B = \frac{d}{d\xi} A \equiv \dot{A}$, where $A = \int_0^1 x^{\xi} g(x) dx$. Let us set $h = 1/n$, where n is a positive integer, and define the trapezoidal rule approximations with stepsize h to A and B, respectively, by

$$A(h) = h \left[\sum_{j=1}^{n-1} G(jh) + \frac{1}{2} G(1) \right]; \quad G(x) \equiv x^{\xi} g(x), \quad (14.2.9)$$

and

$$B(h) = h \left[\sum_{j=1}^{n-1} H(jh) + \frac{1}{2} H(1) \right]; \ \ H(x) \equiv (\log x) x^{\xi} g(x). \quad (14.2.10)$$

Note that $B(h) = \dot{A}(h)$ because $H(x) = \dot{G}(x)$. Then we have the following extensions of the classical Euler–Maclaurin expansion for $A(h)$ and $B(h)$:

$$A(h) \sim A + \sum_{i=1}^{\infty} a_i h^{2i} + \sum_{i=0}^{\infty} b_i h^{\xi+i+1} \ \text{ as } \ h \to 0, \quad (14.2.11)$$

and

$$B(h) \sim B + \sum_{i=1}^{\infty} \dot{a}_i h^{2i} + \sum_{i=0}^{\infty} (\dot{b}_i + b_i \log h) h^{\xi+i+1} \ \text{ as } \ h \to 0, \quad (14.2.12)$$

with

$$a_i = \frac{B_{2i}}{(2i)!} G^{(2i-1)}(1), \ i = 1, 2, \dots; \ \ b_i = \frac{\zeta(-\xi - i)}{i!} g^{(i)}(0), \ i = 0, 1, \dots,$$

$$\quad (14.2.13)$$

and $\dot{a}_i = \frac{d}{d\xi} a_i$ and $\dot{b}_i = \frac{d}{d\xi} b_i$. As before, B_k are the Bernoulli numbers and should not be confused with $B(h)$ or with $B_n^{(j)}$ below. The expansion in (14.2.12) is obtained by differentiating that in (14.2.11). [Note that $G(x)$ depends on ξ but $g(x)$ does not.] For these expansions, see Appendix D.

Let us now consider the case $-1 < \Re\xi < 0$. Then $A(h)$ is of the form described in (14.2.1) and treated throughout, with $\sigma_1, \sigma_2, \dots$, as in

$$\sigma_{3i-2} = \xi + 2i - 1, \ \sigma_{3i-1} = \xi + 2i, \ \sigma_{3i} = 2i, \ i = 1, 2, \dots, \quad (14.2.14)$$

so that (14.2.6) is satisfied with $d = \min(-\Re\xi, 1 + \Re\xi) > 0$.

Let us also apply the Richardson extrapolation process to the sequence $\{A(h_l)\}$ with $h_l = \omega^l$, $l = 0, 1, \dots$, for some $\omega \in \{1/2, 1/3, \dots\}$. (Recall that other more economical choices of $\{h_l\}$ are possible; but we stick with $h_l = \omega^l$ here as it enables us to use the theory of [301] to make rigorous statements about convergence, convergence rates, and stability of diagonal sequences.)

Recall now that $|A_n^{(j)} - A|$ is practically $O(\prod_{i=1}^{n} |c_i|)$ as $n \to \infty$. This implies that $|A_{3m}^{(j)} - A| \to 0$ as $m \to \infty$ practically like ω^{Σ_m}, where $\Sigma_m = \sum_{i=1}^{3m} \Re\sigma_i = 3m^2 + O(m)$ as $m \to \infty$. Thus, $|A_{3m}^{(j)} - A| \to 0$ as $m \to \infty$ practically like ω^{3m^2}.

Since $\dot{\sigma}_{3i-2} = \dot{\sigma}_{3i-1} = 1$, $\dot{\sigma}_{3i} = 0$, $i = 1, 2, \dots$, we have $\dot{c}_{3i-2} = (\log \omega)c_{3i-2}$, $\dot{c}_{3i-1} = (\log \omega)c_{3i-1}$, $\dot{c}_{3i} = 0$, $i = 1, 2, \dots$, and thus $\sum_{i=1}^{\infty} |\dot{c}_i| < \infty$ and $K_i \leq |\log \omega|$ for all i. Consequently, Theorem 14.2.1 applies, and we have that $|\dot{A}_{3m}^{(j)} - \dot{A}| \to 0$ as $m \to \infty$ like ω^{3m^2} practically, just like $|A_{3m}^{(j)} - A|$.

Similarly, $B(h) = \dot{A}(h)$ is of the form given in (14.1.1) with σ_i as in (14.2.14) and $q_{3i-2} = q_{3i-1} = 1$, $q_{3i} = 0$, $i = 1, 2, \dots$. Let us also apply the generalized Richardson extrapolation process of the previous section (via the SGRom-algorithm) to the sequence $\{B(h_l)\}$, also with $h_l = \omega^l$, $l = 0, 1, \dots$. By Theorem 14.1.4, the sequence $\{B_n^{(j)}\}_{n=0}^{\infty}$ converges to B, and $|B_{5m}^{(j)} - B|$ is practically $O(\prod_{i=1}^{5m} |\lambda_i|)$, hence $O(\prod_{i=1}^{3m} |c_i|^{q_i+1})$, as

$m \to \infty$. This implies that $|B_{5m}^{(j)} - B| \to 0$ as $m \to \infty$ practically like ω^{Λ_m}, where $\Lambda_m = \sum_{i=1}^{3m} (q_i + 1)(\Re\sigma_i) = 5m^2 + O(m)$ as $m \to \infty$. Therefore, $|B_{5m}^{(j)} - B|$ is $O(\omega^{5m^2})$ as $m \to \infty$ practically speaking.

Thus, $|\dot{A}_n^{(j)} - \dot{A}|$ and $|B_n^{(j)} - B|$ tend to 0 as $n \to \infty$ like $\omega^{n^2/3}$ and $\omega^{n^2/5}$, respectively, for all practical purposes. That is to say, of the two diagonal sequences $\{\dot{A}_n^{(j)}\}_{n=0}^{\infty}$ and $\{B_n^{(j)}\}_{n=0}^{\infty}$, the former has superior convergence properties. Also, $B_{\lfloor\sqrt{5/3}\,n\rfloor}^{(j)}$ has an accuracy comparable to that of $\dot{A}_n^{(j)}$. Of course, $B_{\lfloor\sqrt{5/3}\,n\rfloor}^{(j)}$ is much more expensive to compute than $\dot{A}_n^{(j)}$. [Recall that the computation of $A(y_l)$ and/or $B(y_l)$ involves ω^{-l} integrand evaluations.]

The preceding comparative study suggests, therefore, that computation of integrals of the form $\int_0^1 (\log x) x^\xi g(x) dx$ by first applying the Richardson extrapolation process to the integral $\int_0^1 x^\xi g(x) dx$ and then differentiating the resulting approximations with respect to ξ may be a preferred method if we intend to use extrapolation methods in the first place. This approach may be used in multidimensional integration of integrands that have logarithmic corner, or surface, or line singularities, for which appropriate extensions of the Euler–Maclaurin expansion can be found in the works of Lyness [196], Lyness and Monegato [201], Lyness and de Doncker [199], and Sidi [283]. All these expansions are obtained by term-by-term differentiation of other simpler expansions, and this is what makes the approach of this section appropriate. Since the computation of the trapezoidal rule approximations for multidimensional integrals becomes very expensive as the dimension increases, the economy that can be achieved with this approach should make it especially attractive.

The new approach can also be used to compute the singular integrals $I_r = \int_0^1 (\log x)^r x^\xi g(x) dx$, where $r = 2, 3, \ldots$, by realizing that $I_r = \frac{d^r}{d\xi^r} A$. (Note that $B = I_1$.) The approximations produced by the generalized Richardson extrapolation process have convergence properties that deteriorate in quality as r becomes large, whereas the $\frac{d^r}{d\xi^r} A_n^{(j)}$ maintain the high-quality convergence properties of the $A_n^{(j)}$. For application of the generalized Richardson extrapolation process to such integrals, see Sidi [298]. Again, the extension to multidimensional singular integrals is immediate.

Before closing, we make the following interesting observation that is analogous to an observation of Bauer, Rutishauser, and Stiefel [20] about Romberg integration (see also Davis and Rabinowitz [63]): The approximation $\dot{A}_n^{(j)}$ to B can be expressed as a sort of "numerical quadrature formula" with stepsize h_{j+n} of the form

$$\dot{A}_n^{(j)} = \sum_{k=0}^{\nu^{j+n}} w_{jnk}^{(0)} H(k h_{j+n}) + \sum_{k=0}^{\nu^{j+n}} w_{jnk}^{(1)} G(k h_{j+n}); \quad \nu = 1/\omega, \quad (14.2.15)$$

in which the "weights" $w_{jnk}^{(0)}$ and $w_{jnk}^{(1)}$ depend on j and n, and satisfy

$$\sum_{k=0}^{\nu^{j+n}} w_{jnk}^{(0)} = 1 \quad \text{and} \quad \sum_{k=0}^{\nu^{j+n}} w_{jnk}^{(1)} = 0. \quad (14.2.16)$$

These follow from the facts that $\sum_{i=0}^{n} \rho_{ni} = 1$ and $\sum_{i=0}^{n} \dot{\rho}_{ni} = 0$. [We have obtained (14.2.15) and (14.2.16) by adding the terms $\frac{1}{2} G(0) h$ and $\frac{1}{2} H(0) h$ to the right-hand sides of (14.2.9) and (14.2.10), respectively, with the understanding that $G(0) \equiv 0$ and

$H(0) \equiv 0.$] Also, these formulas are stable numerically in the sense that

$$\sum_{k=0}^{v^{j+n}} \left(|w_{jnk}^{(0)}| + |w_{jnk}^{(1)}| \right) \le \Omega_n^{(j)} < \infty \quad \text{for all } j \text{ and } n. \tag{14.2.17}$$

14.2.2 $\frac{d}{d\xi}$GREP$^{(1)}$: *Derivative of GREP$^{(1)}$*

We end by showing how the approach of Sidi [301] can be applied to GREP$^{(1)}$. The resulting method that produces the $\dot{A}_n^{(j)}$ has been denoted $\frac{d}{d\xi}$GREP$^{(1)}$. The following material is from Sidi [302].

If $a(t)$ has an asymptotic expansion of the form

$$a(t) \sim A + \varphi(t) \sum_{i=0}^{\infty} \beta_i t^i \quad \text{as } t \to 0+, \tag{14.2.18}$$

and $a(t)$, A, $\varphi(t)$, and the β_i depend on ξ, and $\frac{d}{d\xi} a(t) \equiv \dot{a}(t)$ has an asymptotic expansion that can be obtained by termwise differentiation of that in (14.2.18), then

$$\dot{a}(t) \sim \dot{A} + \varphi(t) \sum_{i=0}^{\infty} \dot{\beta}_i t^i + \dot{\varphi}(t) \sum_{i=0}^{\infty} \beta_i t^i \quad \text{as } t \to 0+. \tag{14.2.19}$$

Recalling the developments of Section 7.2 concerning the W-algorithm, we observe that $\frac{d}{d\xi} D_n^{(j)}\{g(t)\} = D_n^{(j)}\{\frac{d}{d\xi} g(t)\}$. This fact is now used to differentiate the W-algorithm and to derive what we call the $\frac{d}{d\xi}$W-algorithm for computing the $\dot{A}_n^{(j)}$:

Algorithm 14.2.2 ($\frac{d}{d\xi}$W-algorithm)

1. For $j = 0, 1, \dots$, set

$$M_0^{(j)} = \frac{a(t_j)}{\varphi(t_j)}, \quad N_0^{(j)} = \frac{1}{\varphi(t_j)}, \quad H_0^{(j)} = (-1)^j |N_0^{(j)}|, \quad \text{and}$$

$$\dot{M}_0^{(j)} = \frac{\dot{a}(t_j)}{\varphi(t_j)} - \frac{a(t_j)\dot{\varphi}(t_j)}{[\varphi(t_j)]^2}, \quad \dot{N}_0^{(j)} = -\frac{\dot{\varphi}(t_j)}{[\varphi(t_j)]^2}, \quad \tilde{H}_0^{(j)} = (-1)^j |\dot{N}_0^{(j)}|.$$

2. For $j = 0, 1, \dots$, and $n = 1, 2, \dots$, compute $M_n^{(j)}, N_n^{(j)}, H_n^{(j)}, \dot{M}_n^{(j)}, \dot{N}_n^{(j)}$, and $\tilde{H}_n^{(j)}$ recursively from

$$Q_n^{(j)} = \frac{Q_{n-1}^{(j+1)} - Q_{n-1}^{(j)}}{t_{j+n} - t_j}.$$

3. For all j and n, set

$$A_n^{(j)} = \frac{M_n^{(j)}}{N_n^{(j)}}, \quad \Gamma_n^{(j)} = \frac{|H_n^{(j)}|}{|N_n^{(j)}|}, \quad \text{and}$$

$$\dot{A}_n^{(j)} = \frac{\dot{M}_n^{(j)}}{N_n^{(j)}} - A_n^{(j)} \frac{\dot{N}_n^{(j)}}{N_n^{(j)}}, \quad \tilde{\Omega}_n^{(j)} = \frac{|\tilde{H}_n^{(j)}|}{|N_n^{(j)}|} + \left(1 + \frac{|\dot{N}_n^{(j)}|}{|N_n^{(j)}|} \right) \Gamma_n^{(j)}.$$

Here $\tilde{\Omega}_n^{(j)}$ is an upper bound on $\Omega_n^{(j)}$ defined as in (14.2.4), and it turns out to be quite tight.

The convergence and stability of column sequences for the case in which the t_l are chosen such that

$$\lim_{l \to \infty} (t_{l+1}/t_l) = \omega \text{ for some } \omega \in (0, 1) \tag{14.2.20}$$

and $\varphi(t)$ satisfies

$$\lim_{l \to \infty} \varphi(t_{l+1})/\varphi(t_l) = \omega^\delta \text{ for some (complex) } \delta \neq 0, -1, -2, \ldots, \tag{14.2.21}$$

and

$$\dot{\varphi}(t) = \varphi(t)[K \log t + L + o(1)] \text{ as } t \to 0+, \text{ for some } K \neq 0 \text{ and } L, \tag{14.2.22}$$

have been investigated in a thorough manner in [302], where an application to the $d^{(1)}$-transformation is also provided. Recall that we have already encountered the condition in (14.2.21) in Theorem 8.5.1. The condition in (14.2.22) was formulated in [302] and is crucial to the analysis of $\frac{d}{d\xi}\text{GREP}^{(1)}$. Theorem 14.2.3 summarizes the main results of Sidi [302, Section 3].

Theorem 14.2.3 *Under the conditions given in (14.2.20)–(14.2.22), the following hold:*

(i) *The $\gamma_{ni}^{(j)}$ and $\dot{\gamma}_{ni}^{(j)}$ have well-defined and finite limits as $j \to \infty$. Specifically,*

$$\lim_{j \to \infty} \sum_{i=0}^{n} \gamma_{ni}^{(j)} z^i = U_n(z),$$

$$\lim_{j \to \infty} \sum_{i=0}^{n} \dot{\gamma}_{ni}^{(j)} z^i = K(\log \omega)[U_n(z)U_n'(1) - zU_n'(z)],$$

where $U_n(z) = \prod_{i=1}^{n}(z - c_i)/(1 - c_i)$ with $c_i = \omega^{\delta+i-1}$, $i = 1, 2, \ldots$, and $U_n'(z) = \frac{d}{dz}U_n(z)$. Consequently, $\sup_j \Omega_n^{(j)} < \infty$, implying that the column sequences $\{\dot{A}_n^{(j)}\}_{j=0}^{\infty}$ are stable.

(ii) *For $\dot{A}_n^{(j)}$ we have that*

$$\dot{A}_n^{(j)} - \dot{A} = O(\varphi(t_j)t_j^n \log t_j) \text{ as } j \to \infty.$$

For more refined statements of the results of Theorem 14.2.3 and for additional developments, we refer the reader to Sidi [302].

An Application to the $d^{(1)}$-Transformation: The $\frac{d}{d\xi}d^{(1)}$-Transformation

The preceding approach can be applied to the problem of determining the derivative with respect to a parameter ξ of sums of infinite series, whether convergent or divergent. Consider the infinite series $\sum_{k=1}^{\infty} v_k$, where

$$v_n \sim \sum_{i=0}^{\infty} \theta_i n^{\rho-i} \text{ as } n \to \infty; \quad \theta_0 \neq 0, \quad \rho + 1 \neq 0, 1, 2, \ldots, \tag{14.2.23}$$

and let us define $S_n = \sum_{k=1}^{n} v_k$, $n = 1, 2, \ldots$. Then we already know that

$$S_n \sim S + n v_n \sum_{i=0}^{\infty} \beta_i n^{-i} \text{ as } n \to \infty. \tag{14.2.24}$$

Here $S = \lim_{n\to\infty} S_n$ when the series converges; S is the antilimit of $\{S_n\}$ otherwise. Because $\{v_n\} \in \mathbf{b}^{(1)}$ as well, excellent approximations to S can be obtained by applying the $d^{(1)}$-transformation to $\sum_{k=1}^{\infty} v_k$, with GPS if necessary. Let us denote the resulting approximations to S by $S_n^{(j)}$.

If the asymptotic expansion in (14.2.24) can be differentiated with respect to ξ term by term, then we have

$$\dot{S}_n \sim \dot{S} + n\dot{v}_n \sum_{i=0}^{\infty} \beta_i n^{-i} + n v_n \sum_{i=0}^{\infty} \dot{\beta}_i n^{-i} \text{ as } n \to \infty. \qquad (14.2.25)$$

Here $\dot{S}_n = \frac{d}{d\xi} S_n$, $\dot{S} = \frac{d}{d\xi} S$, etc. Let us now differentiate the $S_n^{(j)}$ with respect to ξ and take $\dot{S}_n^{(j)}$ as the approximations to $\dot{S}$. In other words, let us make in the $\frac{d}{d\xi}$ W-algorithm the substitutions $t_j = 1/R_j$, $a(t_j) = S_{R_j}$, $\dot{a}(t_j) = \dot{S}_{R_j}$, $\varphi(t_j) = R_j v_{R_j}$, and $\dot{\varphi}(t_j) = R_j \dot{v}_{R_j}$ for the input, and the substitutions $A_n^{(j)} = S_n^{(j)}$ and $\dot{A}_n^{(j)} = \dot{S}_n^{(j)}$ for the output. We call the extrapolation method thus obtained the $\frac{d}{d\xi} d^{(1)}$-transformation.

The $\dot{S}_n^{(j)}$ converge to $\dot{S}$ as quickly as $S_n^{(j)}$ converge to S when the v_n satisfy

$$\dot{v}_n = v_n[K' \log n + L' + o(1)] \text{ as } n \to \infty, \text{ for some } K' \neq 0 \text{ and } L'. \quad (14.2.26)$$

In addition, good stability and accuracy is achieved by using GPS, that is, by choosing the R_l as in

$$R_0 \geq 1, \quad R_l = \begin{cases} R_{l-1} + 1 \text{ if } \lfloor \sigma R_{l-1} \rfloor = R_{l-1}, \\ \lfloor \sigma R_{l-1} \rfloor \text{ otherwise,} \end{cases} \quad l = 1, 2, \dots; \quad \sigma > 1.$$

(For $\sigma = 1$, we have $R_l = l + 1$, and we recall that, in this case, the $d^{(1)}$-transformation reduces to the Levin u-transformation.)

As we have seen earlier, $t_l = 1/R_l$ satisfy $\lim_{l\to\infty}(t_{l+1}/t_l) = \sigma^{-1}$. Consequently, from Theorem 8.5.1 and Theorem 14.2.3, there holds

$$S_n^{(j)} - S = O(v_{R_j} R_j^{-n+1}) \text{ as } j \to \infty,$$

$$\dot{S}_n^{(j)} - \dot{S} = O(v_{R_j} R_j^{-n+1} \log R_j) \text{ as } j \to \infty.$$

An immediate application of the $\frac{d}{d\xi} d^{(1)}$-transformation is to the summation of series of the form $\sum_{k=1}^{\infty} v_k \log k$, where v_n are as in (14.2.23). This is possible because $v_k \log k = \frac{d}{d\xi}[v_k k^{\xi}]|_{\xi=0}$. Thus, we should make the following substitutions in the $\frac{d}{d\xi}$ W-algorithm: $t_j = 1/R_j$, $a(t_j) = S_{R_j}$, $\dot{a}(t_j) = \dot{S}_{R_j}$, $\varphi(t_j) = R_j v_{R_j}$, and $\dot{\varphi}(t_j) = R_j v_{R_j} \log R_j$, where $S_n = \sum_{k=1}^{n} v_k$, $\dot{S}_n = \sum_{k=1}^{n} v_k \log k$. The series $\sum_{k=1}^{\infty} v_k \log k$ can also be summed by the $d^{(2)}$-transformation, but this is more costly.

For further examples, see Sidi [302].

One important assumption we made here is that the asymptotic expansion of S_n can be differentiated with respect to ξ term by term. This assumption seems to hold in general. In Appendix E, we prove rigorously that it does for the partial sums $S_n = \sum_{k=1}^{n}(k + \theta - 1)^{-\xi}$ of the generalized Zeta function $\zeta(\xi, \theta)$.

Part II

Sequence Transformations

15

The Euler Transformation, Aitken Δ^2-Process, and Lubkin W-Transformation

15.1 Introduction

In this chapter, we begin the treatment of sequence transformations. As mentioned in the Introduction, a sequence transformation operates on a given sequence $\{A_n\}$ and produces another sequence $\{\hat{A}_n\}$ that hopefully converges more quickly than the former. We also mentioned there that a sequence transformation is useful only when $\hat{A}_n$ is constructed from a *finite* number of the A_k.

Our purpose in this chapter is to review briefly a few transformations that have been in existence longer than others and that have been applied successfully in various situations. These are the Euler transformation, which is linear, the Aitken Δ^2-process and Lubkin W-transformation, which are nonlinear, and a few of the more recent generalizations of the latter two. As stated in the Introduction, linear transformations are usually less effective than nonlinear ones, and they have been considered extensively in other places. For these reasons, we do not treat them in this book. The Euler transformation is an exception to this in that it is one of the most effective of the linear methods and also one of the oldest acceleration methods. What we present here is a general version of the Euler transformation known as the Euler–Knopp transformation. A good source for this transformation on which we have relied is Hardy [123].

15.2 The Euler–Knopp (E, q) Method

15.2.1 Derivation of the Method

Given the infinite sequence $\{A_n\}_{n=0}^\infty$, let us set $a_0 = A_0$ and $a_n = A_n - A_{n-1}, n = 1, 2, \ldots$. Thus, $A_n = \sum_{k=0}^n a_k, n = 0, 1, \ldots$. We give two different derivations of the *Euler–Knopp transformation*:

1. Let us define the operator E by $Ea_k = a_{k+1}$ and $E^i a_k = a_{k+i}$ for all $i = 0, 1, \ldots$. Thus, formally,

$$\sum_{k=0}^\infty a_k = \left(\sum_{k=0}^\infty E^k\right) a_0 = (1 - E)^{-1} a_0. \tag{15.2.1}$$

279

Next, let us expand $(1 - E)^{-1}$ formally as in

$$(1 - E)^{-1} = [(1 + q) - (q + E)]^{-1}$$

$$= (1 + q)^{-1} \left(1 - \frac{q + E}{1 + q} \right)^{-1} = \sum_{k=0}^{\infty} \frac{(q + E)^k}{(1 + q)^{k+1}}. \qquad (15.2.2)$$

Since

$$(q + E)^k a_0 = \left[\sum_{i=0}^{k} \binom{k}{i} q^{k-i} E^i \right] a_0 = \sum_{i=0}^{k} \binom{k}{i} q^{k-i} a_i, \qquad (15.2.3)$$

we finally have formally

$$\sum_{k=0}^{\infty} a_k = \sum_{k=0}^{\infty} \frac{1}{(1 + q)^{k+1}} \sum_{i=0}^{k} \binom{k}{i} q^{k-i} a_i. \qquad (15.2.4)$$

The double sum on the right-hand side of (15.2.4) is known as the (E, q) *sum* of $\sum_{k=0}^{\infty} a_k$, whether the latter converges or not. If the (E, q) sum of $\sum_{k=0}^{\infty} a_k$ is finite, we say that $\sum_{k=0}^{\infty} a_k$ is *summable* (E, q).

2. Consider the power series $\sum_{i=0}^{\infty} a_i x^{i+1}$ with small x. Consider the bilinear transformation $x = y/(1 - qy)$. Thus, $y = x/(1 + qx)$, so that $y \to 0$ as $x \to 0$ and $y \to (1 + q)^{-1}$ as $x \to 1$. Substituting this in $\sum_{i=0}^{\infty} a_i x^{i+1}$ and expanding in powers of y, we obtain

$$\sum_{i=0}^{\infty} a_i x^{i+1} = \sum_{k=0}^{\infty} \left(\frac{x}{1 + qx} \right)^{k+1} \sum_{i=0}^{k} \binom{k}{i} q^{k-i} a_i, \qquad (15.2.5)$$

which, upon setting $x = 1$, becomes (15.2.4).

By setting $q = 1$ in (15.2.4), we obtain

$$\sum_{k=0}^{\infty} a_k = \sum_{k=0}^{\infty} \frac{(-1)^k}{2^{k+1}} (\Delta^k b_0); \quad b_i = (-1)^i a_i, \quad i = 0, 1, \ldots. \qquad (15.2.6)$$

The right-hand side of (15.2.6), the (E, 1) sum of $\sum_{k=0}^{\infty} a_k$, is known as the *Euler transformation* of the latter.

From (15.2.4), it is clear that the (E, q) method produces a sequence of approximations $\hat{A}_n$ to the sum of $\sum_{k=0}^{\infty} a_k$, where

$$\hat{A}_n = \sum_{k=0}^{n} \frac{1}{(1 + q)^{k+1}} \sum_{i=0}^{k} \binom{k}{i} q^{k-i} a_i, \quad n = 0, 1, \ldots. \qquad (15.2.7)$$

Obviously, $\hat{A}_n$ depends solely on $a_0, a_1, \ldots, a_n$ (hence on $A_0, A_1, \ldots, A_n$), which makes the (E, q) method an acceptable sequence transformation.

15.2.2 Analytic Properties

It follows from (15.2.7) that the (E, q) method is a linear summability method, namely,

$$\hat{A}_n = \sum_{i=0}^{n} \mu_{ni} A_i, \quad \mu_{ni} = \frac{1}{(1+q)^{n+1}} \binom{n+1}{i+1} q^{n-i}, \quad 0 \le i \le n. \quad (15.2.8)$$

Theorem 15.2.1 *Provided $q > 0$, the (E, q) method is a regular summability method, i.e., if $\{A_n\}$ has limit A, then so does $\{\hat{A}_n\}$.*

Proof. The proof can be achieved by showing that the Silverman–Toeplitz theorem (Theorem 0.3.3) applies. ∎

Actually, when $q > 0$, we have $\sum_{i=0}^{n} |\mu_{ni}| = \sum_{i=0}^{n} \mu_{ni} = 1 - (1+q)^{-n-1} < 1$ for all n, which implies that the (E, q) method is also a stable summability method. In the sequel, we assume that $q > 0$.

The following result shows that the class of Euler–Knopp transformations is closed under composition.

Theorem 15.2.2 *Denote by $\{\hat{A}_n\}$ the sequence obtained by the (E, q) method on $\{A_n\}$, and denote by $\{\tilde{A}_n\}$ the sequence obtained by the (E, r) method on $\{\hat{A}_n\}$. Then $\{\tilde{A}_n\}$ is also the sequence obtained by the $(E, q + r + qr)$ method on $\{A_n\}$.*

The next result shows the beneficial effect of increasing q.

Theorem 15.2.3 *If $\{A_n\}$ is summable (E, q), then it is also summable (E, q') with $q' > q$.*

The proof of Theorem 15.2.2 follows from (15.2.8), and the proof of Theorem 15.2.3 follows from Theorem 15.2.2. We leave the details to the reader.

We next would like to comment on the convergence and acceleration properties of the Euler–Knopp transformation. The geometric series turns out to be very instructive for this purpose. Letting $a_k = z^k$ in (15.2.7), we obtain $\hat{A}_n = (q + 1)^{-1} \sum_{k=0}^{n} \left(\frac{q+z}{q+1}\right)^k$. The sequence $\{\hat{A}_n\}$ converges to $(1 - z)^{-1}$ provided $z \in D_q = B(-q; q + 1)$, where $B(c; \rho) = \{z : |z - c| < \rho\}$. Since $\sum_{k=0}^{\infty} z^k$ converges for $D_0 = B(0; 1)$ and since $D_0 \subset D_q$, we see that the (E, q) method has enlarged the domain of convergence of $\sum_{k=0}^{\infty} z^k$. Also, D_q expands as q increases, since $D_q \subset D_{q'}$ for $q < q'$. This does not mean that the (E, q) method always accelerates the convergence of $\sum_{k=0}^{\infty} z^k$, however. Acceleration takes place only when $|q + z| < (q + 1)|z|$, that is, only when z is in the exterior of $\hat{D}$, where $\hat{D} = B(c; \rho)$ with $c = 1/(q + 2)$ and $\rho = (q + 1)/(q + 2)$. Note that $D_0 \supset \hat{D}$, which means that not for every $z \in D_0$ the (E, q) method accelerates convergence. For $z \in \partial \hat{D}$, $\{A_n\}$ and $\{\hat{A}_n\}$ converge at the same rate, whereas for $z \in \hat{D}$, $\{\hat{A}_n\}$ converges less rapidly than $\{A_n\}$. Let us consider the case $q = 1$. In this case, $\hat{D} = B(1/3; 2/3)$, so that for $z = -1/4$, $z = -1/3$, and $z = -1/2$, the series $\sum_{k=0}^{\infty} z^k$ and its $(E, 1)$ sum are, respectively, $\sum_{k=0}^{\infty} (-1/4)^k$ and $\frac{1}{2} \sum_{k=0}^{\infty} (3/8)^k$, $\sum_{k=0}^{\infty} (-1/3)^k$ and $\frac{1}{2} \sum_{k=0}^{\infty} (1/3)^k$, and

$\sum_{k=0}^{\infty} (-1/2)^k$ and $\frac{1}{2} \sum_{k=0}^{\infty} (1/4)^k$. Note that $\sum_{k=0}^{\infty} z^k$ is an alternating series for these values of z.

Another interesting example for which we can give $\hat{A}_n$ in closed form is the logarithmic series $\sum_{k=0}^{\infty} z^{k+1}/(k+1)$, which converges to $\log(1-z)^{-1}$ for $|z| \le 1$, $z \ne 1$. Using the fact that $a_i = z^{i+1}/(i+1) = z^{i+1} \int_0^1 t^i dt$, we have

$$\sum_{i=0}^{k} \binom{k}{i} q^{k-i} a_i = z \int_0^1 (q + zt)^k dt = \frac{(q+z)^{k+1}}{k+1} - \frac{q^{k+1}}{k+1}, \quad k = 0, 1, \dots,$$

so that

$$\hat{A}_n = \sum_{k=0}^{n} \frac{1}{k+1} \left[\left(\frac{q+z}{q+1} \right)^{k+1} - \left(\frac{q}{q+1} \right)^{k+1} \right], \quad n = 0, 1, \dots.$$

The (E, q) method thus enlarges the domain of convergence of $\{A_n\}$ from $B(0; 1)\backslash\{1\}$ to $B(-q; q+1)\backslash\{1\}$, as in the previous example. We leave the issue of the domain of acceleration to the reader.

The following result, whose proof is given in Knopp [152, p. 263], gives a sufficient condition for the (E, 1) method to accelerate the convergence of alternating series.

Theorem 15.2.4 *Let $a_k = (-1)^k b_k$ and let the b_k be positive and satisfy $\lim_{k\to\infty} b_k = 0$ and $(-1)^i \Delta^i b_k \ge 0$ for all i and k. In addition, assume that $b_{k+1}/b_k \ge \alpha > 1/2$, $k = 0, 1, \dots$. Then the sequence $\{\hat{A}_n\}$ generated by the (E, 1) method converges more rapidly than $\{A_n\}$. If $A = \lim_{n\to\infty} A_n$, then*

$$|A_n - A| \ge \frac{1}{2} b_{n+1} \ge \frac{1}{2} b_0 \alpha^{n+1} \quad \text{and} \quad |\hat{A}_n - A| \le b_0 2^{-n-1},$$

so that

$$\frac{|\hat{A}_n - A|}{|A_n - A|} \le \frac{1}{\alpha} \left(\frac{1}{2\alpha} \right)^n.$$

Remark. Sequences $\{b_k\}$ as in Theorem 15.2.4 are called *totally monotonic* and we treat them in more detail in the next chapter. Here, we state only the fact that if $b_k = f(k)$, where $f(x) \in C^{\infty}[0, \infty)$ and $(-1)^i f^{(i)}(x) \ge 0$ on $[0, \infty)$ for all i, then $\{b_k\}$ is totally monotonic.

Theorem 15.2.4 suggests that the Euler transformation is especially effective on alternating series $\sum_{k=0}^{\infty} (-1)^k b_k$ when $b_{k+1}/b_k \to 1$ as $k \to \infty$. As an illustration, let us apply the (E, 1) method to the series $\sum_{k=0}^{\infty} (-1)^k/(k+1)$ whose sum is $\log 2$. This results in the series $\sum_{k=0}^{\infty} 2^{-k-1}/(k+1)$ that converges much more quickly. In this case, $|\hat{A}_n - A|/|A_n - A| = O(2^{-n})$ as $n \to \infty$, in agreement with Theorem 15.2.4.

Being linear, the Euler transformation is also effective on infinite series of the form $\sum_{k=0}^{\infty} \left(\sum_{i=1}^{p} \alpha_i c_k^{(i)} \right)$, where each of the sequences $\{c_k^{(i)}\}_{k=0}^{\infty}$, $i = 1, \dots, p$, is as in Theorem 15.2.4.

Further results on the Euler–Knopp (E, q) method as this is applied to power series can be found in Scraton [261], Niethammer [220], and Gabutti [90].

15.2.3 A Recursive Algorithm

The sequence $\{\hat{A}_n\}$ generated by the Euler–Knopp (E, q) method according to (15.2.7) can also be computed by using the following recursive algorithm due to Wynn [375], which resembles very much Algorithm 1.3.1 for the Richardson extrapolation process:

$$A_{-1}^{(j)} = A_{j-1}, \quad j = 0, 1, \dots ; \quad (A_{-1} = 0)$$

$$A_n^{(j)} = \frac{A_{n-1}^{(j+1)} + q A_{n-1}^{(j)}}{1 + q}, \quad j, n = 0, 1, \dots . \tag{15.2.9}$$

It can be shown by induction that

$$A_n^{(j)} = A_{j-1} + \frac{(-q)^j}{1+q} \sum_{k=0}^{n} \left(\frac{-q}{1+q}\right)^k \Delta^k \left((-1)^j \frac{a_j}{q^j}\right), \tag{15.2.10}$$

with a_n as before. It is easy to verify that, for each $j \geq 0$, $\{A_n^{(j)}\}_{n=0}^{\infty}$ is the sequence obtained by applying the (E, q) method to the sequence $\{A_{j+n}\}_{n=0}^{\infty}$. In particular, $A_n^{(0)} = \hat{A}_n$, $n = 0, 1, \dots$, with $\hat{A}_n$ as in (15.2.7). It is also clear from (15.2.9) that the $\hat{A}_n$ generated by the $(E, 1)$ method (the Euler transformation) are obtained from the A_n by the repeated application of a simple averaging process.

15.3 The Aitken Δ^2-Process

15.3.1 General Discussion of the Δ^2-Process

We began our discussion of the *Aitken Δ^2-process* already in Example 0.1.1 of the Introduction. Let us recall that the sequence $\{\hat{A}_m\}$ generated by applying the Δ^2-process to $\{A_m\}$ is defined via

$$\hat{A}_m = \phi_m(\{A_s\}) = \frac{A_m A_{m+2} - A_{m+1}^2}{A_m - 2A_{m+1} + A_{m+2}} = \frac{\begin{vmatrix} A_m & A_{m+1} \\ \Delta A_m & \Delta A_{m+1} \end{vmatrix}}{\begin{vmatrix} 1 & 1 \\ \Delta A_m & \Delta A_{m+1} \end{vmatrix}}. \tag{15.3.1}$$

Drummond [68] has shown that $\hat{A}_m$ can also be expressed as in

$$\hat{A}_m = \phi_m(\{A_s\}) = \frac{\Delta(A_m/\Delta A_m)}{\Delta(1/\Delta A_m)}. \tag{15.3.2}$$

Computationally stable forms of $\hat{A}_m$ are

$$\hat{A}_m = A_m - \frac{(\Delta A_m)^2}{\Delta^2 A_m} = A_{m+1} - \frac{(\Delta A_m)(\Delta A_{m+1})}{\Delta^2 A_m}. \tag{15.3.3}$$

It is easy to verify that $\hat{A}_m$ is also the solution of the equations $A_{r+1} - \hat{A}_m = \lambda(A_r - \hat{A}_m)$, $r = m, m + 1$, where λ is an additional auxiliary unknown. Therefore, when $\{A_m\}$ is such that $A_{m+1} - A = \lambda(A_m - A)$, $m = 0, 1, 2, \ldots$, (equivalently, $A_m = A + C\lambda^m$, $m = 0, 1, 2, \ldots$), for some $\lambda \neq 0, 1$, we have $\hat{A}_m = A$ for *all m*.

The Δ^2-process has been analyzed extensively for general sequences $\{A_m\}$ by Lubkin [187] and Tucker [339], [340]. The following results are due to Tucker.

1. If $\{A_m\}$ converges to A, then there exists a subsequence of $\{\hat{A}_m\}$ that converges to A too.
2. If $\{A_m\}$ and $\{\hat{A}_m\}$ converge, then their limits are the same.
3. If $\{A_m\}$ converges to A and if $|(\Delta A_m/\Delta A_{m-1}) - 1| \geq \delta > 0$ for all m, then $\{\hat{A}_m\}$ converges to A too.

Of these results the third is quite easy to prove, and the second follows from the first. For the first, we refer the reader to Tucker [339]. Note that none of these results concerns convergence acceleration. Our next theorem shows convergence acceleration for general linear sequences. The result in part (i) of this theorem appears in Henrici [130, p. 73, Theorem 4.5], and that in part (ii) (already mentioned in the Introduction) is a special case of a more general result of Wynn [371] on the Shanks transformation that we discuss in the next chapter.

Theorem 15.3.1

(i) If $\lim_{m\to\infty}(A_{m+1} - A)/(A_m - A) = \lambda$ for some $\lambda \neq 0, 1$, then $\lim_{m\to\infty}(\hat{A}_m - A)/(A_m - A) = 0$, whether $\{A_m\}$ converges or not.

(ii) If $A_m = A + a\lambda^m + b\mu^m + O(\nu^m)$ as $m \to \infty$, where $a, b \neq 0$, $|\lambda| > |\mu|$, and $\lambda, \mu \neq 0, 1$, and $|\nu| < \min\{1, |\mu|\}$, then $\hat{A}_m - A \sim b(\frac{\lambda-\mu}{\lambda-1})^2\mu^m$ as $m \to \infty$, whether $\{A_m\}$ converges or not.

Proof. We start with the error formula

$$\hat{A}_m - A = \frac{1}{\Delta^2 A_m} \begin{vmatrix} A_m - A & A_{m+1} - A \\ \Delta A_m & \Delta A_{m+1} \end{vmatrix} \tag{15.3.4}$$

that follows from (15.3.1). By elementary row transformations, we next have

$$\frac{\hat{A}_m - A}{A_m - A} = \frac{1}{r_m - 1} \begin{vmatrix} 1 & R_m \\ 1 & r_m \end{vmatrix}; \quad R_m \equiv \frac{A_{m+1} - A}{A_m - A}, \quad r_m \equiv \frac{\Delta A_{m+1}}{\Delta A_m}. \tag{15.3.5}$$

Now $\lim_{m\to\infty} R_m = \lambda \neq 0, 1$ implies $\lim_{m\to\infty} r_m = \lambda$ as well, since $r_m = R_m(R_{m+1} - 1)/(R_m - 1)$. This, in turn, implies that the right-hand side of the equality in (15.3.5) tends to 0 as $m \to \infty$. This proves part (i). The proof of part (ii) can be done by making suitable substitutions in (15.3.4). ∎

Let us now consider the case in which the condition $|\lambda| > |\mu|$ of part (ii) of Theorem 15.3.1 is not satisfied. In this case, we have $|\lambda| = |\mu|$ only, and we cannot make a definitive statement on the convergence of $\{\hat{A}_m\}$. We can show instead that at least a subsequence of $\{\hat{A}_m\}$ satisfies $\hat{A}_m - A = O(\lambda^m)$ as $m \to \infty$; hence, no convergence acceleration takes place. This follows from the fact that $\Delta^2 A_m = a(\lambda - 1)^2\lambda^m \left[1 + \frac{b}{a}\left(\frac{\mu-1}{\lambda-1}\right)^2 e^{im\theta} + O\left(|\nu/\mu|^m\right) \right]$ as $m \to \infty$, where θ is real and defined by $e^{i\theta} = \mu/\lambda$.

Before proceeding further, we give the following definition, which will be useful in the remainder of the book. At this point, it is worth reviewing Theorem 6.7.4.

Definition 15.3.2

(i) We say that $\{A_m\} \in \mathbf{b}^{(1)}/\text{LOG}$ if

$$A_m \sim A + \sum_{i=0}^{\infty} \alpha_i m^{\gamma-i} \quad \text{as } m \to \infty, \quad \gamma \neq 0, 1, \dots, \quad \alpha_0 \neq 0. \qquad (15.3.6)$$

(ii) We say that $\{A_m\} \in \mathbf{b}^{(1)}/\text{LIN}$ if

$$A_m \sim A + \zeta^m \sum_{i=0}^{\infty} \alpha_i m^{\gamma-i} \quad \text{as } m \to \infty, \quad \zeta \neq 1, \quad \alpha_0 \neq 0. \qquad (15.3.7)$$

(iii) We say that $\{A_m\} \in \mathbf{b}^{(1)}/\text{FAC}$ if

$$A_m \sim A + (m!)^{-r}\zeta^m \sum_{i=0}^{\infty} \alpha_i m^{\gamma-i} \quad \text{as } m \to \infty, \quad r = 1, 2, \dots, \quad \alpha_0 \neq 0. \quad (15.3.8)$$

In cases (i) and (ii), $\{A_m\}$ may be convergent or divergent, and A is either the limit or antilimit of $\{A_m\}$. In case (iii), $\{A_m\}$ is always convergent and $A = \lim_{m\to\infty} A_m$.

Note that, if $\{A_m\}$ is as in Definition 15.3.2, then $\{\Delta A_m\} \in \mathbf{b}^{(1)}$. Also recall that the sequences in Definition 15.3.2 satisfy Theorem 6.7.4. This fact is used in the analysis of the different sequence transformations throughout the rest of this work.

The next theorem too concerns convergence acceleration when the Δ^2-process is applied to sequences $\{A_m\}$ described in Definition 15.3.2. Note that the first of the results in part (ii) of this theorem was already mentioned in the Introduction.

Theorem 15.3.3

(i) *If* $\{A_m\} \in \mathbf{b}^{(1)}/\text{LOG}$, *then* $\hat{A}_m - A \sim \sum_{i=0}^{\infty} w_i m^{\gamma-i}$ *as* $m \to \infty$, $w_0 = \frac{\alpha_0}{1-\gamma} \neq 0$.

(ii) *If* $\{A_m\} \in \mathbf{b}^{(1)}/\text{LIN}$, *then* (a) $\hat{A}_m - A \sim \zeta^m \sum_{i=0}^{\infty} w_i m^{\gamma-2-i}$ *as* $m \to \infty$, $w_0 = -\alpha_0\gamma(\frac{\zeta}{\zeta-1})^2 \neq 0$, *if* $\gamma \neq 0$ *and* (b) $\hat{A}_m - A \sim \zeta^m \sum_{i=0}^{\infty} w_i m^{-3-i}$ *as* $m \to \infty$, $w_0 = 2\alpha_1(\frac{\zeta}{\zeta-1})^2 \neq 0$, *if* $\gamma = 0$ *and* $\alpha_1 \neq 0$.

(iii) *If* $\{A_m\} \in \mathbf{b}^{(1)}/\text{FAC}$, *then* $\hat{A}_m - A \sim (m!)^{-r}\zeta^m \sum_{i=0}^{\infty} w_i m^{\gamma-2r-1-i}$ *as* $m \to \infty$, $w_0 = -\alpha_0\zeta^2 r \neq 0$.

Here we have adopted the notation of Definition 15.3.2.

Proof. All three parts can be proved by using the error formula in (15.3.4). We leave the details to the reader. ∎

It is easy to show that $\lim_{m\to\infty}(\hat{A}_m - A)/(A_{m+2} - A) = 0$ in parts (ii) and (iii) of this theorem, which implies convergence acceleration for linear and factorial sequences. Part (i) shows that there is no convergence acceleration for logarithmic sequences.

Examples in which the Δ^2-process behaves in peculiar ways can also be constructed. Let $A_m = \sum_{k=0}^{m}(-1)^{\lfloor k/2\rfloor}/(k+1), m = 0, 1, \ldots$. Thus, A_m is the mth partial sum of the convergent infinite series $1 + 1/2 - 1/3 - 1/4 + 1/5 + 1/6 - \cdots$ whose sum is $A = \frac{\pi}{4} + \frac{1}{2}\log 2$. The sequence $\{\hat{A}_m\}$ generated by the Δ^2-process does not converge in this case. We actually have

$$\hat{A}_{2m} = A_{2m} + (-1)^m \frac{2m+3}{(2m+2)(4m+5)} \quad \text{and} \quad \hat{A}_{2m+1} = A_{2m+1} + (-1)^{m+1}\frac{2m+4}{2m+3},$$

from which it is clear that the sequence $\{\hat{A}_m\}$ has A, $A + 1$, and $A - 1$ as its limit points. (However, the Shanks transformation that we discuss in the next chapter and the $d^{(2)}$-transformation are effective on this sequence.) This example is due to Lubkin [187, Example 7].

15.3.2 Iterated Δ^2-Process

The Δ^2-process on $\{A_m\}$ can be iterated as many times as desired. This results in the following method, which we denote the *iterated Δ^2-process*:

$$B_0^{(j)} = A_j, \quad j = 0, 1, \ldots,$$

$$B_{n+1}^{(j)} = \phi_j(\{B_n^{(s)}\}), \quad n, j = 0, 1, \ldots . \tag{15.3.9}$$

[Thus, $B_1^{(j)} = \hat{A}_j$ with $\hat{A}_j$ as in (15.3.1).] Note that $B_n^{(j)}$ is determined by A_k, $j \le k \le j + 2n$.

The use of this method can be justified with the help of Theorems 15.3.1 and 15.3.3 in the following cases:

1. If $A_m \sim A + \sum_{k=1}^{\infty}\alpha_k\lambda_k^m$ as $m \to \infty$, where $\lambda_k \neq 1$ for all k, $|\lambda_1| > |\lambda_2| > \cdots$, $\lim_{k\to\infty}\lambda_k = 0$ and $\alpha_k \neq 0$ for all k, then it can be shown by expanding $\hat{A}_m$ properly that $\hat{A}_m \sim A + \sum_{k=1}^{\infty}\beta_k\xi_k^m$ as $m \to \infty$, where ξ_k are related to the λ_k and satisfy $|\xi_1| > |\xi_2| > \ldots$, $\lim_{k\to\infty}\xi_k = 0$, and $\xi_1 = \lambda_2$, in addition. Because $\{\hat{A}_m\}$ converges more rapidly (or diverges less rapidly) than $\{A_m\}$ and because $\hat{A}_m$ has an asymptotic expansion of the same form as that of A_m, the Δ^2-process can be applied to $\{\hat{A}_m\}$ very effectively, provided $\xi_k \neq 1$ for all k.
2. If $\{A_m\} \in \mathbf{b}^{(1)}/\text{LIN}$ or $\{A_m\} \in \mathbf{b}^{(1)}/\text{FAC}$, then, by parts (ii) and (iii) of Theorem 15.3.3, $\hat{A}_m$ has an asymptotic expansion of precisely the same form as that of A_m and converges more quickly than A_m. This implies that the Δ^2-process can be applied to $\{\hat{A}_m\}$ very effectively. This is the subject of the next theorem.

Theorem 15.3.4

(i) *If* $\{A_m\} \in \mathbf{b}^{(1)}/\text{LOG}$, *then as* $j \to \infty$

$$B_n^{(j)} - A \sim \sum_{i=0}^{\infty} w_{ni}\, j^{\gamma - i}, \quad w_{n0} = \frac{\alpha_0}{(1-\gamma)^n} \neq 0.$$

(ii) *If* $\{A_m\} \in \mathbf{b}^{(1)}/\text{LIN}$ *with* $\gamma \neq 0, 2, \ldots, 2n-2$, *then as* $j \to \infty$

$$B_n^{(j)} - A \sim \zeta^j \sum_{i=0}^{\infty} w_{ni} j^{\gamma-2n-i}, \quad w_{n0} = (-1)^n \alpha_0 \left[\prod_{i=0}^{n-1}(\gamma-2i)\right]\left(\frac{\zeta}{\zeta-1}\right)^{2n} \neq 0.$$

For the remaining values of γ *we have* $B_n^{(j)} - A = O(\zeta^j j^{\gamma-2n-1})$.

(iii) *If* $\{A_m\} \in \mathbf{b}^{(1)}/\text{FAC}$, *then as* $j \to \infty$

$$B_n^{(j)} - A \sim (j!)^{-r} \zeta^j \sum_{i=0}^{\infty} w_{ni} j^{\gamma-(2r+1)n-i}, \quad w_{n0} = (-1)^n \alpha_0 (\zeta^2 r)^n \neq 0.$$

Here we have adopted the notation of Definition 15.3.2.

It can be seen from this theorem that the sequence $\{B_{n+1}^{(j)}\}_{j=0}^{\infty}$ converges faster than $\{B_n^{(j)}\}_{j=0}^{\infty}$ in parts (ii) and (iii), but no acceleration takes place in part (i).

We end with Theorem IV of Shanks [264], which shows that the iterated Δ^2-process may also behave peculiarly.

Theorem 15.3.5 *Let* $\{A_m\}$ *be the sequence of partial sums of the Maclaurin series for* $f(z) = 1/[(z-z_0)(z-z_1)], 0 < |z_0| < |z_1|$, *and define* $z_r = z_0(z_1/z_0)^r, r = 2, 3, \ldots$. *Apply the iterated* Δ^2-*process to* $\{A_m\}$ *as in (15.3.9). Then (i) provided* $|z| < |z_n|$ *and* $z \neq z_r, r = 0, 1, \ldots, B_n^{(j)} \to f(z)$ *as* $j \to \infty$, *(ii) for* $z = z_0$ *and* $z = z_1, B_n^{(j)} \to \infty$ *as* $j \to \infty$, *and (iii) for* $z = z_2 = z_1^2/z_0$ *and* $n \geq 2, B_n^{(j)} \to \left[1 - \frac{z_0 z_1}{(z_0+z_1)^2}\right] f(z_2) \neq f(z_2)$ *as* $j \to \infty$.

Note that in this theorem, $A_m = A + a\lambda^m + b\mu^m$ with $A = f(z), \lambda = z/z_0$, and $\mu = z/z_1, |\lambda| > |\mu|$, and suitable $a \neq 0$ and $b \neq 0$. Then $\hat{A}_j = B_1^{(j)}$ is of the form $B_1^{(j)} = A + \sum_{k=1}^{\infty} \beta_k \xi_k^j$, where $\xi_k = \mu(\mu/\lambda)^{k-1}, k = 1, 2, \ldots$, and $\beta_k \neq 0$ for all k. If $z = z_2$, then $\xi_2 = \mu^2/\lambda = 1$, and this implies that $B_1^{(j)} = B + \sum_{k=1}^{\infty} \hat{\beta}_k \hat{\xi}_k^j$, where $B = A + \beta_2 \neq A$ and $\hat{\xi}_1 = \xi_1$ and $\hat{\xi}_k = \xi_{k+1}, k = 2, 3, \ldots$, and $\hat{\xi}_k \neq 1$ for all k. Furthermore, $|\hat{\xi}_k| < 1$ for $k = 2, 3, \ldots$, so that $B_2^{(j)} \to B$ as $j \to \infty$ for $z = z_2$. This explains why $\{B_n^{(j)}\}_{j=0}^{\infty}$ converges, and to the wrong answer, when $z = z_2$, for $n \geq 2$.

15.3.3 Two Applications of the Iterated Δ^2-Process

Two common applications of the Δ^2-process are to the power method for the matrix eigenvalue problem and to the iterative solution of nonlinear equations.

In the *power method* for an $N \times N$ matrix Q, we start with an arbitrary vector x_0 and generate $x_1, x_2, \ldots$, via $x_{m+1} = Qx_m$. For simplicity, assume that Q is diagonalizable. Then, the vector x_m is of the form $x_m = \sum_{k=1}^{p} v_k \mu_k^m$, where $Qv_k = \mu_k v_k$ for each k and μ_k are distinct and nonzero and $p \leq N$. If $|\mu_1| > |\mu_2| \geq |\mu_3| \geq \cdots$, and if the vector y is such that $y^* v_1 \neq 0$, then

$$\rho_m = \frac{y^* x_{m+1}}{y^* x_m} = \frac{\sum_{k=1}^{p} \gamma_k \mu_k^{m+1}}{\sum_{k=1}^{p} \gamma_k \mu_k^m} = \mu_1 + \sum_{k=1}^{\infty} \alpha_k \lambda_k^m; \quad \gamma_k = y^* v_k, \quad k = 1, 2, \ldots.$$

where λ_k are related to the μ_k and $1 > |\lambda_1| \geq |\lambda_2| \geq \cdots$ and $\lambda_1 = \mu_2/\mu_1$. Thus, the iterated Δ^2-process can be applied to $\{\rho_m\}$ effectively to produce good approximations to μ_1.

If Q is a normal matrix, then better results can be obtained by choosing $y = x_m$. In this case, ρ_m is called a *Rayleigh quotient*, and

$$\rho_m = \frac{x_m^* x_{m+1}}{x_m^* x_m} = \frac{\sum_{k=1}^{p} \delta_k |\mu_k|^{2m} \mu_k}{\sum_{k=1}^{p} \delta_k |\mu_k|^{2m}} = \mu_1 + \sum_{k=1}^{\infty} \sigma_k |\lambda_k|^{2m}; \quad \delta_k = v_k^* v_k, \quad k = 1, 2, \ldots,$$

with the λ_k exactly as before. Again, the iterated Δ^2-process can be applied to $\{\rho_m\}$, but it is much more effective than before.

In the *fixed-point iterative solution* of a nonlinear equation $x = g(x)$, we begin with an arbitrary approximation x_0 to the solution s and generate the sequence of approximations $\{x_m\}$ via $x_{m+1} = g(x_m)$. It is known that, provided $|g'(s)| < 1$ and x_0 is sufficiently close to s, the sequence $\{x_m\}$ converges to s linearly in the sense that $\lim_{m \to \infty}(x_{m+1} - s)/(x_m - s) = g'(s)$. There is, however, a very elegant result concerning the asymptotic expansion of the x_m when $g(x)$ is infinitely differentiable in a neighborhood of s, and this result reads

$$x_m \sim s + \sum_{k=1}^{\infty} \alpha_k \lambda^{km} \text{ as } m \to \infty, \quad \alpha_1 \neq 0, \quad \lambda = g'(s),$$

for some α_k that depend only on $g(x)$ and x_0. (See de Bruijn [42, pp. 151–153] and also Meinardus [210].) If $\{\hat{x}_m\}$ is the sequence obtained by applying the Δ^2-process to $\{x_m\}$, then by a careful analysis of $\hat{x}_m$ it follows that

$$\hat{x}_m \sim s + \sum_{k=1}^{\infty} \beta_k \lambda^{(k+1)m} \text{ as } m \to \infty.$$

It is clear in this problem as well that the iterated Δ^2-process is very effective, and we have

$$B_n^{(j)} - s = O(\lambda^{(n+1)j}) \text{ as } j \to \infty.$$

In conjunction with the iterative method for the nonlinear equation $x = g(x)$, we would like to mention a different usage of the Δ^2-process that reads as follows:

> Pick u_0 and set $x_0 = u_0$.
> **for** $m = 1, 2, \ldots,$ **do**
>   Compute $x_1 = g(x_0)$ and $x_2 = g(x_1)$.
>   Compute $u_m = (x_0 x_2 - x_1^2)/(x_0 - 2x_1 + x_2)$.
>   Set $x_0 = u_m$.
> **end do**

This is known as *Steffensen's method*. When $g(x)$ is twice differentiable in a neighborhood of s, provided x_0 is sufficiently close to s and $g'(s) \neq 1$, the sequence $\{u_m\}$ converges to s quadratically, i.e., $\lim_{n \to \infty}(u_{m+1} - s)/(u_m - s)^2 = C \neq \pm\infty$.

15.3.4 A Modified Δ^2-Process for Logarithmic Sequences

For sequences $\{A_m\} \in \mathbf{b}^{(1)}/\mathrm{LOG}$, the Δ^2-process is not effective, as we have already seen in Theorem 15.3.3. To remedy this, Drummond [69] has proposed (in the notation of Definition 15.3.2) the following modification of the Δ^2-process, which we denote the $\Delta^2(\gamma)$-*process*:

$$\hat{A}_m = \psi_m(\{A_s\}; \gamma) = A_{m+1} - \frac{\gamma - 1}{\gamma} \frac{(\Delta A_m)(\Delta A_{m+1})}{\Delta^2 A_m}, \quad m = 0, 1, \ldots . \quad (15.3.10)$$

[Note that we have introduced the factor $(\gamma - 1)/\gamma$ in the Aitken formula (15.3.3).] We now give a new derivation of (15.3.10), through which we also obtain a convergence result for $\{\hat{A}_m\}$.

Let us set $a_0 = A_0$ and $a_m = A_m - A_{m-1}$, $m \geq 1$, and define $p(m) = a_m/\Delta a_m$, $m \geq 0$. Using summation by parts, we obtain [see (6.6.16) and (6.6.17)],

$$A_m = \sum_{k=0}^{m} p(k)\Delta a_k = p(m)a_{m+1} - \sum_{k=0}^{m} [\Delta p(k-1)]a_k, \quad p(k) = 0 \text{ if } k < 0. \quad (15.3.11)$$

Now $a_m = h(m) \in \mathbf{A}_0^{(\gamma-1)}$ strictly, so that $p(m) \in \mathbf{A}_0^{(1)}$ strictly with $p(m) \sim \frac{1}{\gamma-1}m + \sum_{i=0}^{\infty} c_i m^{-i}$ as $m \to \infty$. As a result, $\Delta p(m) = \frac{1}{\gamma-1} + q(m)$, where $q(m) \in \mathbf{A}_0^{(-2)}$. Thus, (15.3.11) becomes

$$A_m = p(m)a_{m+1} - \frac{1}{\gamma - 1}A_m - \sum_{k=0}^{m} q(k-1)a_k,$$

from which

$$A_m = \frac{\gamma - 1}{\gamma}p(m)a_{m+1} + B_m, \quad B_m = \frac{1-\gamma}{\gamma}\sum_{k=0}^{m} q(k-1)a_k. \quad (15.3.12)$$

Noting that $q(m-1)a_m = q(m-1)h(m) \in \mathbf{A}_0^{(\gamma-3)}$ and applying Theorem 6.7.2, we first obtain $B_m = B + H(m)$, where B is the limit or antilimit of $\{B_m\}$ and $H(m) \in \mathbf{A}_0^{(\gamma-2)}$. Because $p(m)a_{m+1} \in \mathbf{A}_0^{(\gamma)}$ strictly, we have that $\frac{\gamma-1}{\gamma}p(m)a_{m+1} + H(m) \in \mathbf{A}_0^{(\gamma)}$ strictly. Invoking now that $\gamma \neq 0, 1, \ldots$, we therefore have that B is nothing but A, namely, the limit or antilimit of $\{A_m\}$. Thus, we have

$$A_m - \frac{\gamma - 1}{\gamma}p(m)a_{m+1} = A + H(m) = A + O(m^{\gamma-2}) \text{ as } m \to \infty. \quad (15.3.13)$$

Now the left-hand side of (15.3.13) is nothing but $\hat{A}_{m-1}$ in (15.3.10), and we have also proved the following result:

Theorem 15.3.6 *Let $\{A_m\} \in \mathbf{b}^{(1)}/\mathrm{LOG}$ in the notation of Definition 15.3.2 and $\hat{A}_m$ be as in (15.3.10). Then*

$$\hat{A}_m \sim A + \sum_{i=0}^{\infty} w_i m^{\gamma-2-i} \quad \text{as } m \to \infty. \quad (15.3.14)$$

In view of Theorem 15.3.6, we can iterate the $\Delta^2(\gamma)$-process as in the next theorem, whose proof is left to the reader.

Theorem 15.3.7 *With $\psi_m(\{A_s\}; \gamma)$ as in (15.3.10), define*

$$B_0^{(j)} = A_j, \quad j = 0, 1, \ldots ,$$

$$B_{n+1}^{(j)} = \psi_j(\{B_n^{(s)}\}; \gamma - 2n), \quad n, j = 0, 1, \ldots , \tag{15.3.15}$$

When $\{A_m\} \in \mathbf{b}^{(1)}/\mathrm{LOG}$, we have that

$$B_n^{(j)} - A \sim \sum_{i=0}^{\infty} w_{ni} j^{\gamma - 2n - i} \quad \text{as } j \to \infty, \quad n \text{ fixed.} \tag{15.3.16}$$

Here $w_{ni} = 0, 0 \le i \le I$, for some $I \ge 0$ is possible.

The use of the $\Delta^2(\gamma)$-process of (15.3.10) in iterated form was suggested by Drummond [69], who also alluded to the result in (15.3.16). The results in (15.3.14) and (15.3.16) were also given by Bjørstad, Dahlquist, and Grosse [26].

We call the method defined through (15.3.15) the *iterated $\Delta^2(\gamma)$-process*.

It is clear from (15.3.10) and (15.3.15) that to be able to use the $\Delta^2(\gamma)$- and iterated $\Delta^2(\gamma)$-processes we need to have precise knowledge of γ. In this sense, this method is less user-friendly than other methods (for example, the Lubkin W-transformation, which we treat in the next section) that do not need any such input. To remedy this deficiency, Drummond [69] proposed to approximate γ in (15.3.10) by γ_m defined by

$$\gamma_m = 1 - \frac{(\Delta a_m)(\Delta a_{m+1})}{a_m a_{m+2} - a_{m+1}^2} = 1 + \frac{1}{\Delta(a_m/\Delta a_m)}. \tag{15.3.17}$$

It is easy to show (see Bjørstad, Dahlquist, and Grosse [26]) that

$$\gamma_m \sim \gamma + \sum_{i=0}^{\infty} e_i m^{-2-i} \quad \text{as } m \to \infty. \tag{15.3.18}$$

Using an approach proposed by Osada [225] in connection with some modifications of the Wynn ρ-algorithm, we can use the iterated $\Delta^2(\gamma)$-process as follows: Taking (15.3.18) into account, we first apply the iterated $\Delta^2(-2)$-process to the sequence $\{\gamma_m\}$ to obtain the best possible estimate $\hat{\gamma}$ to γ. Following that, we apply the iterated $\Delta^2(\hat{\gamma})$-process to the sequence $\{A_m\}$ to obtain approximations to A. Osada's work is reviewed in Chapter 20.

15.4 The Lubkin W-Transformation

The *W-transformation* of Lubkin [187], when applied to a sequence $\{A_m\}$, produces the sequence $\{\hat{A}_m\}$, whose members are given by

$$\hat{A}_m = W_m(\{A_s\}) = A_{m+1} + \frac{(\Delta A_{m+1})(1 - r_{m+1})}{1 - 2r_{m+1} + r_m r_{m+1}}, \quad r_m \equiv \frac{\Delta A_{m+1}}{\Delta A_m}. \tag{15.4.1}$$

It can easily be verified (see Drummond [68] and Van Tuyl [343]) that $\hat{A}_m$ can also be expressed as in

$$\hat{A}_m = W_m(\{A_s\}) = \frac{\Delta^2(A_m/\Delta A_m)}{\Delta^2(1/\Delta A_m)} = \frac{\Delta(A_{m+1} \times \Delta(1/\Delta A_m))}{\Delta^2(1/\Delta A_m)}. \quad (15.4.2)$$

Note also that the column $\{\theta_2^{(j)}\}_{j=0}^{\infty}$ of the θ-algorithm and the column $\{u_2^{(j)}\}$ of the Levin u-transformation are identical to the Lubkin W-transformation.

Finally, with the help of the first formula in (15.4.2), it can be shown that $W_m \equiv W_m(\{A_s\})$ can also be expressed in terms of the $\phi_k \equiv \phi_k(\{A_s\})$ of the Δ^2-process as follows:

$$W_m = \frac{\phi_{m+1} - q_m \phi_m}{1 - q_m}, \quad q_m = r_{m+1} \frac{1 - r_m}{1 - r_{m+1}}. \quad (15.4.3)$$

The following are known:

1. $\hat{A}_m = A$ for all $m = 0, 1, \ldots$, if and only if A_m is of the form

$$A_m = A + C \prod_{k=1}^{m} \frac{ak + b + 1}{ak + b}, \quad C \neq 0, \ a \neq 1, \ ak + b \neq 0, -1, k = 1, 2, \ldots .$$

   Obviously, $A_m = A + C\lambda^m$, $\lambda \neq 0, 1$, is a special case. This result on the kernel of the Lubkin transformation is due to Cordellier [56]. We obtain it as a special case of a more general one on the Levin transformations later.
2. Wimp [366] has shown that, if $|r_m - \lambda| \leq \delta < \delta^*$ for all m, where $0 < |\lambda| < 1$ and $\delta^* = [|\lambda - 1|^2 + (|\lambda| + 1)^2]^{1/2} - (|\lambda| + 1)$, and if $\{A_m\} \to A$, then $\{\hat{A}_m\} \to A$ too.
3. Lubkin [187] has shown that, if $\lim_{m\to\infty}(A_{m+1} - A)/(A_m - A) = \lambda$ for some $\lambda \neq 0, 1$, then $\lim_{m\to\infty}(\hat{A}_m - A)/(A_m - A) = 0$. [This can be proved by combining (15.4.3), part (i) of Theorem 15.3.3, and the fact that $\lim_{m\to\infty} r_m = \lambda$.]

For additional results of a general nature on convergence acceleration, see Lubkin [187] and Tucker [339], [340]. The next theorem shows that the W-transformation accelerates the convergence of all sequences discussed in Definition 15.3.2 and is thus analogous to Theorem 15.3.3. We leave its proof to the reader. Note that parts (i) and (ii) of this theorem are special cases of results corresponding to the Levin transformation that were given by Sidi [273]. Part (i) of this theorem was also given by Sablonnière [248] and Van Tuyl [343]. Part (iii) was given recently by Sidi [307]. For details, see [307].

Theorem 15.4.1

(i) *If $\{A_m\} \in \mathbf{b}^{(1)}/\text{LOG}$, then $\hat{A}_m - A \sim \sum_{i=0}^{\infty} w_i m^{\hat{\gamma}-i}$ as $m \to \infty$, such that $\hat{\gamma} - \gamma$ is an integer ≤ -2. Thus $\hat{A}_m - A = O(m^{\gamma-2})$ as $m \to \infty$.*

(ii) *If $\{A_m\} \in \mathbf{b}^{(1)}/\text{LIN}$, then $\hat{A}_m - A \sim \zeta^m \sum_{i=0}^{\infty} w_i m^{\hat{\gamma}-i}$ as $m \to \infty$, such that $\hat{\gamma} - \gamma$ is an integer ≤ -3. Thus $\hat{A}_m - A = O(\zeta^m m^{\gamma-3})$ as $m \to \infty$.*

(iii) *If $\{A_m\} \in \mathbf{b}^{(1)}/\text{FAC}$, then $\hat{A}_m - A \sim (m!)^{-r} \zeta^m \sum_{i=0}^{\infty} w_i m^{\gamma-3r-2-i}$ as $m \to \infty$, $w_0 = \alpha_0 \zeta^3 r(r+1) \neq 0$. Thus, $\hat{A}_m - A = O((m!)^{-r} \zeta^m m^{\gamma-3r-2})$ as $m \to \infty$.*

Here we have adopted the notation of Definition 15.3.2.

It is easy to show that $\lim_{m\to\infty}(\hat{A}_m - A)/(A_{m+3} - A) = 0$ in all three parts of this theorem, which implies convergence acceleration for linear, factorial, and logarithmic sequences.

Before we proceed further, we note that, when applied to the sequence $\{A_m\}$, $A_m = \sum_{k=0}^{m}(-1)^{\lfloor k/2\rfloor}/(k+1), m = 0, 1, \ldots$, the W-transformation converges, but is not better than the sequence itself, as mentioned in [187].

It follows from Theorem 15.4.1 that the Lubkin W-transformation can also be iterated. This results in the following method, which we denote the *iterated Lubkin transformation*:

$$B_0^{(j)} = A_j, \quad j = 0, 1, \ldots,$$
$$B_{n+1}^{(j)} = W_j(\{B_n^{(s)}\}), \quad n, j = 0, 1, \ldots . \tag{15.4.4}$$

[Thus, $B_1^{(j)} = \hat{A}_j$ with $\hat{A}_j$ as in (15.3.1).] Note that $B_n^{(j)}$ is determined by A_k, $j \leq k \leq j + 3n$.

The results of iterating the W-transformation are given in the next theorem. Again, part (i) of this theorem was given in [248] and [343], and parts (ii) and (iii) were given recently in [307].

Theorem 15.4.2

(i) *If $\{A_m\} \in \mathbf{b}^{(1)}/\text{LOG}$, then there exist constants γ_k such that $\gamma_0 = \gamma$ and $\gamma_k - \gamma_{k-1}$ are integers ≤ -2, for which, as $j \to \infty$,*

$$B_n^{(j)} - A \sim \sum_{i=0}^{\infty} w_{ni} j^{\gamma_n - i}, \quad w_{n0} \neq 0.$$

(ii) *If $\{A_m\} \in \mathbf{b}^{(1)}/\text{LIN}$, then there exist constants γ_k such that $\gamma_0 = \gamma$ and $\gamma_k - \gamma_{k-1}$ are integers ≤ -3, for which, as $j \to \infty$,*

$$B_n^{(j)} - A \sim \zeta^j \sum_{i=0}^{\infty} w_{ni} j^{\gamma_n - i}, \quad w_{n0} \neq 0.$$

(iii) *If $\{A_m\} \in \mathbf{b}^{(1)}/\text{FAC}$, then as $j \to \infty$*

$$B_n^{(j)} - A \sim (j!)^{-r} \zeta^j \sum_{i=0}^{\infty} w_{ni} j^{\gamma - (3r+2)n - i}, \quad w_{n0} = \alpha_0[\zeta^3 r(r+1)]^n \neq 0.$$

Here we have adopted the notation of Definition 15.3.2.

It can be seen from this theorem that the sequence $\{B_{n+1}^{(j)}\}_{j=0}^{\infty}$ converges faster than $\{B_n^{(j)}\}_{j=0}^{\infty}$ for all sequences considered.

15.5 Stability of the Iterated Δ^2-Process and Lubkin Transformation

15.5.1 Stability of the Iterated Δ^2-Process

From (15.3.1), we can write $\hat{A}_m = \theta_0 A_m + \theta_1 A_{m+1}$, where $\theta_0 = \Delta A_{m+1}/\Delta^2 A_m$ and $\theta_1 = -\Delta A_m/\Delta^2 A_m$. Consequently, we can express $B_n^{(j)}$ as in

$$B_{n+1}^{(j)} = \lambda_n^{(j)} B_n^{(j)} + \mu_n^{(j)} B_n^{(j+1)}, \tag{15.5.1}$$

where

$$\lambda_n^{(j)} = \Delta B_n^{(j+1)}/\Delta^2 B_n^{(j)} \quad \text{and} \quad \mu_n^{(j)} = -\Delta B_n^{(j)}/\Delta^2 B_n^{(j)}, \tag{15.5.2}$$

and $\Delta B_n^{(j)} = B_n^{(j+1)} - B_n^{(j)}$ for all j and n. Thus, we can write

$$B_n^{(j)} = \sum_{i=0}^{n} \gamma_{ni}^{(j)} A_{j+i}, \tag{15.5.3}$$

where the $\gamma_{ni}^{(j)}$ satisfy the recursion relation

$$\gamma_{n+1,i}^{(j)} = \lambda_n^{(j)} \gamma_{ni}^{(j)} + \mu_n^{(j)} \gamma_{n,i-1}^{(j+1)}, \quad i = 0, 1, \ldots, n+1. \tag{15.5.4}$$

Here we define $\gamma_{ni}^{(j)} = 0$ for $i < 0$ and $i > n$. In addition, from the fact that $\lambda_n^{(j)} + \mu_n^{(j)} = 1$, it follows that $\sum_{i=0}^{n} \gamma_{ni}^{(j)} = 1$.

Let us now define

$$P_n^{(j)}(z) = \sum_{i=0}^{n} \gamma_{ni}^{(j)} z^i \quad \text{and} \quad \Gamma_n^{(j)} = \sum_{i=0}^{n} |\gamma_{ni}^{(j)}|. \tag{15.5.5}$$

As we have done until now, we take $\Gamma_n^{(j)}$ as a measure of propagation of the errors in the A_m into $B_n^{(j)}$. Our next theorem, which seems to be new, shows how $P_n^{(j)}(z)$ and $\Gamma_n^{(j)}$ behave as $j \to \infty$. As before, we adopt the notation of Definition 15.3.2.

Theorem 15.5.1

(i) *If* $\{A_m\} \in \mathbf{b}^{(1)}/\text{LIN}$, *with* $\gamma \neq 0, 2, \ldots, 2n-2$, *then*

$$\lim_{j\to\infty} P_n^{(j)}(z) = \left(\frac{z-\zeta}{1-\zeta}\right)^n \quad \text{and} \quad \lim_{j\to\infty} \Gamma_n^{(j)} = \left(\frac{1+|\zeta|}{|1-\zeta|}\right)^n.$$

(ii) *If* $\{A_m\} \in \mathbf{b}^{(1)}/\text{FAC}$, *then*

$$\lim_{j\to\infty} P_n^{(j)}(z) = z^n \quad \text{and} \quad \lim_{j\to\infty} \Gamma_n^{(j)} = 1.$$

Proof. Combining Theorem 15.3.4 and (15.5.2), we first obtain that (i) $\lambda_n^{(j)} \sim \zeta/(\zeta-1)$ and $\mu_n^{(j)} \sim -1/(\zeta-1)$ as $j \to \infty$ for all n in part (i), and (ii) $\lim_{j\to\infty} \lambda_n^{(j)} = 0$ and $\lim_{j\to\infty} \mu_n^{(j)} = 1$ for all n in part (ii). We leave the rest of the proof to the reader. ∎

Note that we have not included the case in which $\{A_m\} \in \mathbf{b}^{(1)}/\text{LOG}$ in this theorem. The reason for this is that we would not use the iterated Δ^2-process on such sequences, and hence discussion of stability for such sequences is irrelevant. For linear and factorial sequences Theorem 15.5.1 shows stability.

15.5.2 Stability of the Iterated Lubkin Transformation

From the second formula in (15.4.2), we can write $\hat{A}_m = \theta_1 A_{m+1} + \theta_2 A_{m+2}$, where $\theta_1 = -\Delta(1/\Delta A_m)/\Delta^2(1/\Delta A_m)$ and $\theta_2 = \Delta(1/\Delta A_{m+1})/\Delta^2(1/\Delta A_m)$. Consequently, we can express $B_n^{(j)}$ as in

$$B_{n+1}^{(j)} = \lambda_n^{(j)} B_n^{(j+1)} + \mu_n^{(j)} B_n^{(j+2)}, \tag{15.5.6}$$

where

$$\lambda_n^{(j)} = -\frac{\Delta(1/\Delta B_n^{(j)})}{\Delta^2(1/\Delta B_n^{(j)})} \quad \text{and} \quad \mu_n^{(j)} = \frac{\Delta(1/\Delta B_n^{(j+1)})}{\Delta^2(1/\Delta B_n^{(j)})}, \tag{15.5.7}$$

and $\Delta B_n^{(j)} = B_n^{(j+1)} - B_n^{(j)}$ for all j and n. Thus, we can write

$$B_n^{(j)} = \sum_{i=0}^{n} \gamma_{ni}^{(j)} A_{j+n+i}, \tag{15.5.8}$$

where the $\gamma_{ni}^{(j)}$ satisfy the recursion relation

$$\gamma_{n+1,i}^{(j)} = \lambda_n^{(j)} \gamma_{ni}^{(j+1)} + \mu_n^{(j)} \gamma_{n,i-1}^{(j+2)}, \quad i = 0, 1, \ldots, n+1. \tag{15.5.9}$$

Here we define $\gamma_{ni}^{(j)} = 0$ for $i < 0$ and $i > n$. Again, from the fact that $\lambda_n^{(j)} + \mu_n^{(j)} = 1$, there holds $\sum_{i=0}^{n} \gamma_{ni}^{(j)} = 1$.

Let us now define again

$$P_n^{(j)}(z) = \sum_{i=0}^{n} \gamma_{ni}^{(j)} z^i \quad \text{and} \quad \Gamma_n^{(j)} = \sum_{i=0}^{n} |\gamma_{ni}^{(j)}|. \tag{15.5.10}$$

Again, we take $\Gamma_n^{(j)}$ as a measure of propagation of the errors in the A_m into $B_n^{(j)}$. The next theorem, due to Sidi [307], shows how $P_n^{(j)}(z)$ and $\Gamma_n^{(j)}$ behave as $j \to \infty$. The proof of this theorem is almost identical to that of Theorem 15.5.1. Only this time we invoke Theorem 15.4.2.

Theorem 15.5.2

(i) *If* $\{A_m\} \in \mathbf{b}^{(1)}/\text{LOG}$, *then*

$$P_n^{(j)}(z) \sim \left(\prod_{k=0}^{n-1} \gamma_k\right)^{-1} (1-z)^n j^n \quad \text{and} \quad \Gamma_n^{(j)} \sim \left|\prod_{k=0}^{n-1} \gamma_k\right|^{-1} (2j)^n \quad \text{as } j \to \infty,$$

where γ_k *are as in part (i) of Theorem 15.4.2.*

(ii) *If* $\{A_m\} \in \mathbf{b}^{(1)}/\text{LIN}$, *then*

$$\lim_{j \to \infty} P_n^{(j)}(z) = \left(\frac{z - \zeta}{1 - \zeta}\right)^n \quad \text{and} \quad \lim_{j \to \infty} \Gamma_n^{(j)} = \left(\frac{1 + |\zeta|}{|1 - \zeta|}\right)^n.$$

(iii) *If* $\{A_m\} \in \mathbf{b}^{(1)}/\text{FAC}$, *then*

$$\lim_{j \to \infty} P_n^{(j)}(z) = z^n \quad \text{and} \quad \lim_{j \to \infty} \Gamma_n^{(j)} = 1.$$

Thus, for logarithmic sequences the iterated Lubkin transformation is not stable, whereas for linear and factorial sequences it is.

15.6 Practical Remarks

As we see from Theorems 15.5.1 and 15.5.2, for both of the iterated transformations, $\Gamma_n^{(j)}$ is proportional to $(1 - \zeta)^{-n}$ as $j \to \infty$ when $\{A_m\} \in \mathbf{b}^{(1)}/\text{LIN}$. This implies that, as ζ approaches 1, the singular point of A, the methods suffer from diminishing stability. In addition, the theoretical error $B_n^{(j)} - A$ for such sequences, as $j \to \infty$, is proportional to $(1 - \zeta)^{-2n}$ in the case of the iterated Δ^2-process and to $(1 - \zeta)^{-n}$ in the case of the iterated Lubkin transformation. (This is explicit in Theorem 15.3.4 for the former, and it can be shown for the latter by refining the proof of Theorem 15.4.2 as in [307].) That is, as $\zeta \to 1$, the theoretical error $B_n^{(j)} - A$ and $\Gamma_n^{(j)}$ grow simultaneously. We can remedy both problems by using arithmetic progression sampling (APS), that is, by applying the methods to a subsequence $\{A_{\kappa m + \eta}\}$, where κ and η are fixed integers with $\kappa \geq 2$. The justification for this follows from the fact that the factor $(1 - \zeta)^{-n}$ is now replaced everywhere by $(1 - \zeta^\kappa)^{-n}$, which can be kept at a moderate size by a judicious choice of κ that removes ζ from 1 in the complex plane sufficiently. Numerical experience shows that this strategy works very well.

From Theorems 15.3.4 and 15.4.2, only the iterated Lubkin transformation is useful for sequences $\{A_m\} \in \mathbf{b}^{(1)}/\text{LOG}$. From Theorem 15.5.2, this method is unstable even though it is convergent mathematically. The most direct way of obtaining more accuracy from the $B_n^{(j)}$ in finite-precision arithmetic is by increasing the precision being used (doubling it, for example). Also, note that, when $|\Im\gamma|$ is sufficiently large, the product $\prod_{k=0}^{n-1} |\gamma_k|$ increases in size, which makes $\Gamma_n^{(j)}$ small (even though $\Gamma_n^{(j)} \to \infty$ as $j \to \infty$). This implies that good numerical accuracy can be obtained for the $B_n^{(j)}$ without having to increase the precision of the arithmetic being used in such a case.

As for sequences $\{A_m\} \in \mathbf{b}^{(1)}/\text{FAC}$, both methods are effective and stable numerically, as the relevant theorems show.

Finally, the diagonal sequences $\{B_n^{(j)}\}_{n=0}^\infty$ with fixed j (e.g., $j = 0$) have better convergence properties than the column sequences $\{B_n^{(j)}\}_{j=0}^\infty$ with fixed n.

15.7 Further Convergence Results

In this section, we state two more convergence results that concern the application of the iterated modified Δ^2-process (in a suitably modified form) and the iterated Lubkin transformation to sequences $\{A_m\}$ for which

$$A_m \sim A + \sum_{i=0}^{\infty} \alpha_i m^{\gamma - i/p} \text{ as } m \to \infty, \quad p \geq 2 \text{ integer},$$

$$\gamma \neq \frac{i}{p}, \ i = 0, 1, \ldots, \quad \alpha_0 \neq 0. \quad (15.7.1)$$

Obviously, $A_m - A = h(m) \in \tilde{\mathbf{A}}_0^{(\gamma, p)}$ strictly. (Recall that the case with $p = 1$ is already treated in Theorems 15.3.7 and 15.4.2.)

The following convergence acceleration results were proved by Sablonnière [248] with $p = 2$ and by Van Tuyl [343] with general p.

Theorem 15.7.1 *Consider the sequence* $\{A_m\}$ *described in (15.7.1) and, with* $\psi_m(\{A_s\}; \gamma)$ *as defined in (15.3.10), set*

$$B_0^{(j)} = A_j, \quad j = 0, 1, \ldots ,$$

$$B_{n+1}^{(j)} = \psi_j(\{B_n^{(s)}\}; \gamma - n/p), \quad n, j = 0, 1, \ldots . \tag{15.7.2}$$

Then

$$B_n^{(j)} - A \sim \sum_{i=0}^{\infty} w_{ni} j^{\gamma - (n+i)/p)} \ \text{as } j \to \infty. \tag{15.7.3}$$

Here $w_{ni} = 0, 0 \le i \le I,$ *for some* $I \ge 0$ *is possible.*

We denote the method described by (15.7.2) the *iterated* $\Delta^2(\gamma, p)$-*process*. It is important to realize that this method does not reduce to the iterated $\Delta^2(\gamma)$-process upon setting $p = 1$.

Theorem 15.7.2 *Consider the sequence* $\{A_m\}$ *described in (15.7.1) and let* $B_n^{(j)}$ *be as defined via the iterated Lubkin transformation in (15.4.4). Then there exist scalars* γ_k *such that* $\gamma_0 = \gamma$ *and* $(\gamma_k - \gamma_{k-1})p$ *are integers* ≤ -1, $k = 1, 2, \ldots$, *for which*

$$B_n^{(j)} - A \sim \sum_{i=0}^{\infty} w_{ni} j^{\gamma_n - i/p} \ \text{as } j \to \infty, \quad w_{n0} \ne 0. \tag{15.7.4}$$

16

The Shanks Transformation

16.1 Derivation of the Shanks Transformation

In our discussion of algorithms for extrapolation methods in the Introduction, we presented a brief treatment of the Shanks transformation and the ε-algorithm as part of this discussion. This transformation was originally derived by Schmidt [258] for solving linear systems by iteration. After being neglected for a long time, it was resurrected by Shanks [264], who also gave a detailed study of its remarkable properties. Shanks' paper was followed by that of Wynn [368], in which the ε-algorithm, the most efficient implementation of the Shanks transformation, was presented. The papers of Shanks and Wynn made an enormous impact and paved the way for more research in sequence transformations.

In this chapter, we go into the details of the Shanks transformation, one of the most useful sequence transformations to date. We start with its derivation.

Let $\{A_m\}$ be a given sequence with limit or antilimit A. Assume that $\{A_m\}$ satisfies

$$A_m \sim A + \sum_{k=1}^{\infty} \alpha_k \lambda_k^m \quad \text{as } m \to \infty, \tag{16.1.1}$$

for some nonzero constants α_k and λ_k independent of m, with λ_k distinct and $\lambda_k \neq 1$ for all k, and $|\lambda_1| \geq |\lambda_2| \geq \cdots$, such that $\lim_{k\to\infty} \lambda_k = 0$. Assume, furthermore, that the λ_k are not necessarily known. Note that the condition that $\lim_{k\to\infty} \lambda_k = 0$ implies that there can be only a *finite* number of λ_k that have the same modulus.

Obviously, when $|\lambda_1| < 1$, the sequence $\{A_m\}$ converges and $\lim_{m\to\infty} A_m = A$. When $|\lambda_1| = 1$, $\{A_m\}$ diverges but is bounded. When $|\lambda_1| > 1$, $\{A_m\}$ diverges and is unbounded.

In case the summation in (16.1.1) is finite and, therefore,

$$A_r = A + \sum_{k=1}^{n} \alpha_k \lambda_k^r, \quad r = 0, 1, \ldots, \tag{16.1.2}$$

we can determine A and the $2n$ parameters α_k and λ_k by solving the following system of nonlinear equations:

$$A_r = A + \sum_{k=1}^{n} \alpha_k \lambda_k^r, \quad r = j, j+1, \ldots, j+2n, \quad \text{for arbitrary fixed } j. \tag{16.1.3}$$

297

Now, $\lambda_1, \dots, \lambda_n$ are the zeros of some polynomial $P(\lambda) = \sum_{i=0}^{n} w_i \lambda^i$, $w_0 \neq 0$ and $w_n \neq 0$. Here the w_i are unique up to a multiplicative constant. Then, from (16.1.3), we have that

$$\sum_{i=0}^{n} w_i (A_{r+i} - A) = 0, \quad r = j, j+1, \dots, j+n. \tag{16.1.4}$$

Note that we have eliminated the parameters $\lambda_1, \dots, \lambda_n$ from equations (16.1.3). It follows from (16.1.4) that

$$\sum_{i=0}^{n} w_i A_{r+i} = \left(\sum_{i=0}^{n} w_i \right) A, \quad r = j, j+1, \dots, j+n. \tag{16.1.5}$$

Now, because $\lambda_k \neq 1$ for all k, we have $\sum_{i=0}^{n} w_i = P(1) \neq 0$. This enables us to scale the w_i such that $\sum_{i=0}^{n} w_i = 1$, as a result of which (16.1.5) can be written in the form

$$\sum_{i=0}^{n} w_i A_{r+i} = A, \quad r = j, j+1, \dots, j+n; \quad \sum_{i=0}^{n} w_i = 1. \tag{16.1.6}$$

This is a linear system in the unknowns A and $w_0, w_1, \dots, w_n$. Using the fact that $A_{r+1} = \Delta A_r + A_r$, we have that

$$\sum_{i=0}^{n} w_i A_{r+i} = \left(\sum_{i=0}^{n} w_i \right) A_r + \sum_{i=0}^{n-1} \hat{w}_i \Delta A_{r+i}; \quad \hat{w}_i = \sum_{p=i+1}^{n} w_p, \quad i = 0, 1, \dots, n-1,$$

$$\tag{16.1.7}$$

which, when substituted in (16.1.6), results in

$$A = A_r + \sum_{i=0}^{n-1} \hat{w}_i \Delta A_{r+i}, \quad r = j, j+1, \dots, j+n. \tag{16.1.8}$$

Finally, by making the substitution $\beta_i = -\hat{w}_{i-1}$ in (16.1.8), we obtain the linear system

$$A_r = A + \sum_{i=1}^{n} \beta_i \Delta A_{r+i-1}, \quad r = j, j+1, \dots, j+n. \tag{16.1.9}$$

On the basis of (16.1.9), we now give the definition of the Shanks transformation, hoping that it accelerates the convergence of sequences $\{A_m\}$ that satisfy (16.1.1).

Definition 16.1.1 Let $\{A_m\}$ be an arbitrary sequence. Then the *Shanks transformation* on this sequence is defined via the linear systems

$$A_r = e_n(A_j) + \sum_{i=1}^{n} \beta_i \Delta A_{r+i-1}, \quad r = j, j+1, \dots, j+n, \tag{16.1.10}$$

where $e_n(A_j)$ are the approximations to the limit or antilimit of $\{A_m\}$ and β_i are additional auxiliary unknowns.

Solving (16.1.10) for $e_n(A_j)$ by Cramer's rule, we obtain the following determinant representation for $A_n^{(j)}$:

$$e_n(A_j) = \frac{\begin{vmatrix} A_j & A_{j+1} & \cdots & A_{j+n} \\ \Delta A_j & \Delta A_{j+1} & \cdots & \Delta A_{j+n} \\ \vdots & \vdots & & \vdots \\ \Delta A_{j+n-1} & \Delta A_{j+n} & \cdots & \Delta A_{j+2n-1} \end{vmatrix}}{\begin{vmatrix} 1 & 1 & \cdots & 1 \\ \Delta A_j & \Delta A_{j+1} & \cdots & \Delta A_{j+n} \\ \vdots & \vdots & & \vdots \\ \Delta A_{j+n-1} & \Delta A_{j+n} & \cdots & \Delta A_{j+2n-1} \end{vmatrix}}. \qquad (16.1.11)$$

[Obviously, $e_1(A_j) = \phi_j(\{A_s\})$, defined in (15.3.1), that is, the e_1-transformation is the Aitken Δ^2-process.] By performing elementary row transformations on the numerator and denominator determinants in (16.1.11), we can express $e_n(A_j)$ also in the form

$$e_n(A_j) = \frac{H_{n+1}^{(j)}(\{A_s\})}{H_n^{(j)}(\{\Delta^2 A_s\})}, \qquad (16.1.12)$$

where $H_p^{(m)}(\{u_s\})$ is a Hankel determinant defined by

$$H_p^{(m)}(\{u_s\}) = \begin{vmatrix} u_m & u_{m+1} & \cdots & u_{m+p-1} \\ u_{m+1} & u_{m+2} & \cdots & u_{m+p} \\ \vdots & \vdots & & \vdots \\ u_{m+p-1} & u_{m+p} & \cdots & u_{m+2p-2} \end{vmatrix}; \quad H_0^{(m)}(\{u_s\}) = 1. \quad (16.1.13)$$

Before proceeding further, we mention that the Shanks transformation as given in (16.1.11) was derived by Tucker [341] through an interesting geometric approach. Tucker's approach generalizes that of Todd [336, p. 260], which concerns the Δ^2-process.

The following result on the kernel of the Shanks transformation, which is completely related to the preceding developments, is due to Brezinski and Crouzeix [40]. We note that this result also follows from Theorem 17.2.5 concerning the Padé table for rational functions.

Theorem 16.1.2 *A necessary and sufficient condition for $e_n(A_j) = A$, $j = J, J + 1, \ldots$, to hold with minimal n is that there exist constants $w_0, w_1, \ldots, w_n$ such that $w_n \neq 0$ and $\sum_{i=0}^n w_i \neq 0$, and that*

$$\sum_{i=0}^n w_i(A_{r+i} - A) = 0, \quad r \geq J, \qquad (16.1.14)$$

which means that

$$A_r = A + \sum_{k=1}^t P_k(r)\lambda_k^r, \quad r = J, J + 1, \ldots, \qquad (16.1.15)$$

where $P_k(r)$ is a polynomial in r of degree exactly p_k, $\sum_{i=1}^t (p_i + 1) = n$, and λ_k are distinct and $\lambda_k \neq 1$ for all k.

Proof. Let us first assume that $e_n(A_j) = A$ for every $j \geq J$. Then the equations in (16.1.10) become $A_r - A = \sum_{i=1}^{n} \beta_i \Delta A_{r+i-1}, j \leq r \leq j+n$. The β_i are thus the solution of n of these $n+1$ equations. For $j = J$, we take these n equations to be those for which $j+1 \leq r \leq j+n$, and for $j = J+1$, we take them such that $j \leq r \leq j+n-1$. But these two systems are identical. Therefore, the β_i are the same for $j = J$ and for $j = J+1$. By induction, the β_i are the same for all $j \geq J$. Also, since n is smallest, $\beta_n \neq 0$ necessarily. Thus, (16.1.10) is the same as (16.1.9) for any $j \geq J$, with the β_i independent of j and $\beta_n \neq 0$.

Working backward from this, we reach (16.1.4), in which the w_i satisfy $w_n \neq 0$ and $\sum_{i=0}^{n} w_i \neq 0$ and are independent of j when $j \geq J$; this is the same as (16.1.14). Let us next assume, conversely, that (16.1.14) holds with $w_n \neq 0$ and $\sum_{i=0}^{n} w_i \neq 0$. This implies (16.1.4) for all $j \geq J$. Working forward from this, we reach (16.1.9) for all $j \geq J$, and hence (16.1.10) for all $j \geq J$, with $e_n(A_j) = A$. Finally, because (16.1.14) is an $(n+1)$-term recursion relation for $\{A_r - A\}_{r \geq J}$ with constant coefficients, its solutions are all of the form (16.1.15).   ∎

Two types of sequences are of interest in the application of the Shanks transformation:

1. $\{e_n(A_j)\}_{j=0}^{\infty}$ with n fixed. In analogy with the first generalization of the Richardson extrapolation process, we call them *column sequences*.
2. $\{e_n(A_j)\}_{n=0}^{\infty}$ with j fixed. In analogy with the first generalization of the Richardson extrapolation process, we call them *diagonal sequences*.

In general, diagonal sequences appear to have much better convergence properties than column sequences.

Before closing, we recall that in case the λ_k in (16.1.1) are known, we can also use the Richardson extrapolation process for infinite sequences of Section 1.9 to approximate A. This is very effective and turns out to be much less expensive than the Shanks transformation. [Formally, it takes $2n+1$ sequence elements to "eliminate" the terms $\alpha_k \lambda_k^m, k = 1, \ldots, n$, from (16.1.1) by the Shanks transformation, whereas the same task is achieved by the Richardson extrapolation process with only $n+1$ sequence elements.] When the λ_k are *not* known and can be *arbitrary*, the Shanks transformation appears to be the only extrapolation method that can be used.

We close this section with the following result due to Shanks, which can be proved by applying appropriate elementary row and column transformations to the numerator and denominator determinants in (16.1.11).

Theorem 16.1.3 *Let $A_m = \sum_{k=0}^{m} c_k z^k$, $m = 0, 1, \ldots$. If the Shanks transformation is applied to $\{A_m\}$, then the resulting $e_n(A_j)$ turns out to be $f_{j+n,n}(z)$, the $[j+n/n]$ Padé approximant from the infinite series $f(z) := \sum_{k=0}^{\infty} c_k z^k$.*

Note. Recall that $f_{m,n}(z)$ is a rational function whose numerator and denominator polynomials have degrees at most m and n, respectively, that satisfies, and is uniquely determined by, the requirement $f_{m,n}(z) - f(z) = O(z^{m+n+1})$ as $z \to 0$. The subject of Padé approximants is considered in some detail in the next chapter.

16.2 Algorithms for the Shanks Transformation

Comparing (16.1.10), the equations defining $e_n(A_j)$, with (3.1.4), we realize that $e_n(A_j)$ can be expressed in the form

$$e_n(A_j) = \frac{f_n^{(j)}(a)}{f_n^{(j)}(I)} = \frac{|g_1(j) \cdots g_n(j) \, a(j)|}{|g_1(j) \cdots g_n(j) \, I(j)|}, \tag{16.2.1}$$

where

$$a(l) = A_l, \quad \text{and} \quad g_k(l) = \Delta A_{k+l-1} \quad \text{for all } l \geq 0 \text{ and } k \geq 1. \tag{16.2.2}$$

It is possible to use the FS-algorithm or the E-algorithm to compute the $e_n(A_j)$ recursively. Direct application of these algorithms without taking the special nature of the $g_k(l)$ into account results in expensive computational procedures, however. When $A_0, A_1, \ldots, A_L$ are given, the costs of these algorithms are $O(L^3)$, as explained in Chapter 3. When the special structure of the $g_k(l)$ in (16.2.2) is taken into account, these costs can be reduced to $O(L^2)$ by the rs- and FS/qd-algorithms that are discussed in Chapter 21 on the G-transformation.

The most elegant and efficient algorithm for implementing the Shanks transformation is the famous ε-*algorithm* of Wynn [368]. As the derivation of this algorithm is very complicated, we do not give it here. We refer the reader to Wynn [368] or to Wimp [366, pp. 244–247] instead. Here are the steps of the ε-algorithm.

Algorithm 16.2.1 (ε-algorithm)

1. Set $\varepsilon_{-1}^{(j)} = 0$ and $\varepsilon_0^{(j)} = A_j, j = 0, 1, \ldots$.
2. Compute the $\varepsilon_k^{(j)}$ by the recursion

$$\varepsilon_{k+1}^{(j)} = \varepsilon_{k-1}^{(j+1)} + \frac{1}{\varepsilon_k^{(j+1)} - \varepsilon_k^{(j)}}, \quad j, k = 0, 1, \ldots \, .$$

Wynn has shown that

$$\varepsilon_{2n}^{(j)} = e_n(A_j) \quad \text{and} \quad \varepsilon_{2n+1}^{(j)} = \frac{1}{e_n(\Delta A_j)} \quad \text{for all } j \text{ and } n. \tag{16.2.3}$$

[Thus, if $\{A_m\}$ has a limit and the $\varepsilon_{2n}^{(j)}$ converge, then we should be able to observe that $|\varepsilon_{2n+1}^{(j)}| \to \infty$ both as $j \to \infty$ and as $n \to \infty$.]

Commonly, the $\varepsilon_k^{(j)}$ are arranged in a two-dimensional array as in Table 16.2.1. Note that the sequences $\{\varepsilon_{2n}^{(j)}\}_{j=0}^{\infty}$ form the columns of the epsilon table, and the sequences $\{\varepsilon_{2n}^{(j)}\}_{n=0}^{\infty}$ form its diagonals.

It is easy to see from Table 16.2.1 that, given $A_0, A_1, \ldots, A_K$, the ε-algorithm computes $\varepsilon_k^{(j)}$ for $0 \leq j + k \leq K$. As the number of these $\varepsilon_k^{(j)}$ is $K^2/2 + O(K)$, the cost of this computation is $K^2 + O(K)$ additions, $K^2/2 + O(K)$ divisions, and no multiplications. (We show in Chapter 21 that the FS/qd-algorithm has about the same cost as the ε-algorithm.)

Table 16.2.1: *The ε-table*

$$
\begin{array}{ccccc}
\varepsilon_{-1}^{(0)} & & & & \\
 & \varepsilon_{0}^{(0)} & & & \\
\varepsilon_{-1}^{(1)} & & \varepsilon_{1}^{(0)} & & \\
 & \varepsilon_{0}^{(1)} & & \varepsilon_{2}^{(0)} & \\
\varepsilon_{-1}^{(2)} & & \varepsilon_{1}^{(1)} & & \varepsilon_{3}^{(0)} \\
 & \varepsilon_{0}^{(2)} & & \varepsilon_{2}^{(1)} & \ddots \\
\varepsilon_{-1}^{(3)} & & \varepsilon_{1}^{(2)} & & \varepsilon_{3}^{(1)} \\
\vdots & \varepsilon_{0}^{(3)} & & \varepsilon_{2}^{(2)} & \ddots \\
\vdots & \vdots & \varepsilon_{1}^{(3)} & & \varepsilon_{3}^{(2)} \\
\vdots & \vdots & \vdots & \varepsilon_{2}^{(3)} & \ddots \\
\vdots & \vdots & \vdots & \vdots & \varepsilon_{3}^{(3)} \\
\vdots & \vdots & \vdots & \vdots & \vdots & \ddots
\end{array}
$$

Since we are interested only in the $\varepsilon_{2n}^{(j)}$ by (16.2.3), and since the $\varepsilon_{2n+1}^{(j)}$ are auxiliary quantities, we may ask whether it is possible to obtain a recursion relation among the $\varepsilon_{2n}^{(j)}$ only. The answer to this question, which is in the affirmative, was given again by Wynn [372], the result being the so-called *cross rule*:

$$
\frac{1}{\varepsilon_{2n+2}^{(j-1)} - \varepsilon_{2n}^{(j)}} + \frac{1}{\varepsilon_{2n-2}^{(j+1)} - \varepsilon_{2n}^{(j)}} = \frac{1}{\varepsilon_{2n}^{(j-1)} - \varepsilon_{2n}^{(j)}} + \frac{1}{\varepsilon_{2n}^{(j+1)} - \varepsilon_{2n}^{(j)}}
$$

with the initial conditions

$$
\varepsilon_{-2}^{(j)} = \infty \quad \text{and} \quad \varepsilon_{0}^{(j)} = A_j, \quad j = 0, 1, \ldots .
$$

Another implementation of the Shanks transformation proceeds through the qd-algorithm that is related to continued fractions and that is discussed at length in the next chapter. The connection between the ε- and qd-algorithms was discovered and analyzed by Bauer [18], [19], who also developed another algorithm, denoted the η-algorithm, that is closely related to the ε-algorithm. We do not go into the η-algorithm here, but refer the reader to [18] and [19]. See also the description given in Wimp [366, pp. 160–165].

16.3 Error Formulas

Before we embark on the error analysis of the Shanks transformation, we need appropriate error formulas for $e_n(A_j) = \varepsilon_{2n}^{(j)}$.

Lemma 16.3.1 *Let* $C_m = A_m - A$, $m = 0, 1, \ldots$. *Then*

$$
\varepsilon_{2n}^{(j)} - A = \frac{H_{n+1}^{(j)}(\{C_s\})}{H_n^{(j)}(\{\Delta^2 C_s\})}. \tag{16.3.1}
$$

If we let $C_m = \zeta^m D_m, m = 0, 1, \ldots$, *then*

$$\varepsilon_{2n}^{(j)} - A = \zeta^{j+2n} \frac{H_{n+1}^{(j)}(\{D_s\})}{H_n^{(j)}(\{E_s\})}; \quad E_m = D_m - 2\zeta D_{m+1} + \zeta^2 D_{m+2} \text{ for all } m. \quad (16.3.2)$$

Proof. Let us first subtract A from both sides of (16.1.11). This changes the first row of the numerator determinant to $[C_j, C_{j+1}, \ldots, C_{j+n}]$. Realizing now that $\Delta A_s = \Delta C_s$ in the other rows, and adding the first row to the second, the second to the third, etc., in this order, we arrive at $H_{n+1}^{(j)}(\{C_s\})$. We already know that the denominator is $H_n^{(j)}(\{\Delta^2 A_s\})$. But we also have that $\Delta^2 A_m = \Delta^2 C_m$. This completes the proof of (16.3.1). To prove (16.3.2), substitute $C_m = \zeta^m D_m$ everywhere and factor out the powers of ζ from the rows and columns in the determinants of both the numerator and the denominator. This completes the proof of (16.3.2). ∎

Both (16.3.1) and (16.3.2) are used in the analysis of the column sequences $\{\varepsilon_{2n}^{(j)}\}_{j=0}^{\infty}$ in the next two sections.

16.4 Analysis of Column Sequences When $A_m \sim A + \sum_{k=1}^{\infty} \alpha_k \lambda_k^m$

Let us assume that $\{A_m\}$ satisfies

$$A_m \sim A + \sum_{k=1}^{\infty} \alpha_k \lambda_k^m \text{ as } m \to \infty, \quad (16.4.1)$$

with $\alpha_k \neq 0$ and $\lambda_k \neq 0$ constants independent of m and

$$\lambda_k \text{ distinct, } \lambda_k \neq 1 \text{ for all } k; \; |\lambda_1| \geq |\lambda_2| \geq \cdots, \quad \lim_{k \to \infty} \lambda_k = 0. \quad (16.4.2)$$

What we mean by (16.4.1) is that, for any integer $N \geq 1$, there holds

$$A_m - A - \sum_{k=1}^{N-1} \alpha_k \lambda_k^m = O(\lambda_N^m) \text{ as } m \to \infty.$$

Also, $\lim_{k \to \infty} \lambda_k = 0$ implies that there can be only a finite number of λ_k with the same modulus. We say that such sequences are *exponential*.

Because the Shanks transformation was derived with the specific intention of accelerating the convergence of sequences $\{A_m\}$ that behave as in (16.4.1), it is natural to first analyze its behavior when applied to such sequences.

We first recall from Theorem 16.1.2 that, when $A_m = A + \sum_{k=1}^{n} \alpha_k \lambda_k^m$ for all m, we have $A = \varepsilon_{2n}^{(j)}$ for all j. From this, we can expect the Shanks transformation to perform well when applied to the general case in (16.4.1). This indeed turns out to be the case, but the theory behind it is not simple. In addition, the existing theory pertains only to column sequences; nothing is known about diagonal sequences so far.

We carry out the analysis of the column sequences $\{\varepsilon_{2n}^{(j)}\}_{j=0}^{\infty}$ with the help of the following lemma of Sidi, Ford, and Smith [309, Lemma A.1].

Lemma 16.4.1 *Let* $i_1, \ldots, i_k$ *be positive integers, and assume that the scalars* $v_{i_1,\ldots,i_k}$ *are odd under an interchange of any two of the indices* $i_1, \ldots, i_k$. *Let* $t_{i,j}, i \geq 1, 1 \leq j \leq k$, be scalars. Define $I_{k,N}$ and $J_{k,N}$ by

$$I_{k,N} = \sum_{i_1=1}^{N} \cdots \sum_{i_k=1}^{N} \left(\prod_{p=1}^{k} t_{i_p,p} \right) v_{i_1,\ldots,i_k} \qquad (16.4.3)$$

and

$$J_{k,N} = \sum_{1 \leq i_1 < i_2 < \cdots < i_k \leq N} \begin{vmatrix} t_{i_1,1} & t_{i_2,1} & \cdots & t_{i_k,1} \\ t_{i_1,2} & t_{i_2,2} & \cdots & t_{i_k,2} \\ \vdots & \vdots & & \vdots \\ t_{i_1,k} & t_{i_2,k} & \cdots & t_{i_k,k} \end{vmatrix} v_{i_1,\ldots,i_k}. \qquad (16.4.4)$$

Then

$$I_{k,N} = J_{k,N}. \qquad (16.4.5)$$

Proof. Let Σ_k be the set of all permutations π of the index set $\{1, 2, \ldots, k\}$. Then, by the definition of determinants,

$$J_{k,N} = \sum_{1 \leq i_1 < i_2 < \cdots < i_k \leq N} \sum_{\pi \in \Sigma_k} (\operatorname{sgn} \pi) \left(\prod_{p=1}^{k} t_{i_{\pi(p)},p} \right) v_{i_1,\ldots,i_k}. \qquad (16.4.6)$$

Here $\operatorname{sgn} \pi$, the signature of π, is $+1$ or -1 depending on whether π is even or odd, respectively. The notation $\pi(p)$, where $p \in \{1, 2, \ldots, k\}$, designates the image of π as a function operating on the index set. Now $\pi^{-1}\pi(p) = p$ when $1 \leq p \leq k$, and $\operatorname{sgn} \pi^{-1} = \operatorname{sgn} \pi$ for any permutation $\pi \in \Sigma_k$. Hence

$$J_{k,N} = \sum_{1 \leq i_1 < i_2 < \cdots < i_k \leq N} \sum_{\pi \in \Sigma_k} (\operatorname{sgn} \pi^{-1}) \left(\prod_{p=1}^{k} t_{i_{\pi(p)},p} \right) v_{i_{\pi^{-1}\pi(1)},\ldots,i_{\pi^{-1}\pi(k)}}. \qquad (16.4.7)$$

By the oddness of $v_{i_1,\ldots,i_k}$, we have

$$(\operatorname{sgn} \pi^{-1}) v_{i_{\pi^{-1}\pi(1)},\ldots,i_{\pi^{-1}\pi(k)}} = v_{i_{\pi(1)},\ldots,i_{\pi(k)}}. \qquad (16.4.8)$$

Substituting (16.4.8) in (16.4.7), we obtain

$$J_{k,N} = \sum_{1 \leq i_1 < i_2 < \cdots < i_k \leq N} \sum_{\pi \in \Sigma_k} \left(\prod_{p=1}^{k} t_{i_{\pi(p)},p} \right) v_{i_{\pi(1)},\ldots,i_{\pi(k)}}. \qquad (16.4.9)$$

Because $v_{i_1,\ldots,i_k}$ is odd under an interchange of the indices $i_1, \ldots, i_k$, it vanishes when any two of the indices are equal. Using this fact in (16.4.3), we see that $I_{k,N}$ is just the sum over all permutations of the *distinct* indices $i_1, \ldots, i_k$. The result now follows by comparison with (16.4.9). ∎

We now prove the following central result:

Lemma 16.4.2 *Let $\{f_m\}$ be such that*

$$f_m \sim \sum_{k=1}^{\infty} e_k \lambda_k^m \quad \text{as } m \to \infty, \qquad (16.4.10)$$

with the λ_k exactly as in (16.4.2). Then, $H_n^{(j)}(\{f_s\})$ satisfies

$$H_n^{(j)}(\{f_s\}) \sim \sum_{1 \le k_1 < k_2 < \cdots < k_n} \left(\prod_{p=1}^{n} e_{k_p} \lambda_{k_p}^{j}\right) [V(\lambda_{k_1}, \ldots, \lambda_{k_n})]^2 \ \ as \ j \to \infty. \quad (16.4.11)$$

Proof. Let us substitute (16.4.10) in $H_n^{(j)}(\{f_s\})$. We obtain

$$H_n^{(j)}(\{f_s\}) \sim \begin{vmatrix} \sum_{k_1} e_{k_1} \lambda_{k_1}^{j} & \sum_{k_1} e_{k_1} \lambda_{k_1}^{j+1} & \cdots & \sum_{k_1} e_{k_1} \lambda_{k_1}^{j+n-1} \\ \sum_{k_2} e_{k_2} \lambda_{k_2}^{j+1} & \sum_{k_2} e_{k_2} \lambda_{k_2}^{j+2} & \cdots & \sum_{k_2} e_{k_2} \lambda_{k_2}^{j+n} \\ \vdots & \vdots & & \vdots \\ \sum_{k_n} e_{k_n} \lambda_{k_n}^{j+n-1} & \sum_{k_n} e_{k_n} \lambda_{k_n}^{j+n} & \cdots & \sum_{k_n} e_{k_n} \lambda_{k_n}^{j+2n-2} \end{vmatrix}, \quad (16.4.12)$$

where we have used $\sum_k$ to mean $\sum_{k=1}^{\infty}$ and we have also used $P(j) \sim Q(j)$ to mean $P(j) \sim Q(j)$ as $j \to \infty$. We continue to do so below.

By the multilinearity property of determinants with respect to their rows, we can move the summations outside the determinant. Factoring out $e_{k_i} \lambda_{k_i}^{j+i-1}$ from the ith row of the remaining determinant, and making use of the definition of the Vandermonde determinant given in (3.5.4), we obtain

$$H_n^{(j)}(\{f_s\}) \sim \sum_{k_1} \cdots \sum_{k_n} \left(\prod_{p=1}^{n} \lambda_{k_p}^{p-1}\right)\left[\left(\prod_{p=1}^{k} e_{k_p} \lambda_{k_p}^{j}\right) V(\lambda_{k_1}, \ldots, \lambda_{k_n})\right]. \quad (16.4.13)$$

Because the term inside the square brackets is odd under an interchange of any two of the indices $k_1, \ldots, k_n$, Lemma 16.4.1 applies, and by invoking the definition of the Vandermonde determinant again, the result follows. ∎

The following is the first convergence result of this section. In this result, we make use of the fact that there can be only a finite number of λ_k with the same modulus.

Theorem 16.4.3 *Assume that $\{A_m\}$ satisfies (16.4.1) and (16.4.2). Let n and r be positive integers for which*

$$|\lambda_n| > |\lambda_{n+1}| = \cdots = |\lambda_{n+r}| > |\lambda_{n+r+1}|. \quad (16.4.14)$$

Then

$$\varepsilon_{2n}^{(j)} - A = \sum_{p=n+1}^{n+r} \alpha_p \left(\prod_{i=1}^{n} \frac{\lambda_p - \lambda_i}{1 - \lambda_i}\right)^2 \lambda_p^{j} + o(\lambda_{n+1}^{j}) \ \ as \ j \to \infty. \quad (16.4.15)$$

Consequently,

$$\varepsilon_{2n}^{(j)} - A = O(\lambda_{n+1}^{j}) \ \ as \ j \to \infty. \quad (16.4.16)$$

All this is valid whether $\{A_m\}$ converges or not.

Proof. Applying Lemma 16.4.2 to $H_{n+1}^{(j)}(\{C_s\})$, we obtain

$$H_{n+1}^{(j)}(\{C_s\}) \sim \sum_{1 \le k_1 < k_2 < \cdots < k_{n+1}} \left(\prod_{p=1}^{n+1} \alpha_{k_p} \lambda_{k_p}^j \right) \left[V(\lambda_{k_1}, \lambda_{k_2}, \ldots, \lambda_{k_{n+1}}) \right]^2. \quad (16.4.17)$$

The most dominant terms in (16.4.17) are those with $k_1 = 1, k_2 = 2, \ldots, k_n = n$, and $k_{n+1} = \lambda_{n+i}, i = 1, \ldots, r$, their number being r. Thus,

$$H_{n+1}^{(j)}(\{C_s\}) = \left(\prod_{p=1}^{n} \alpha_p \lambda_p^j \right) \left\{ \sum_{p=n+1}^{n+r} \alpha_p \lambda_p^j \left[V(\lambda_1, \ldots, \lambda_n, \lambda_p) \right]^2 + o(\lambda_{n+1}^j) \right\} \quad \text{as } j \to \infty.$$

$$(16.4.18)$$

Observing that

$$\Delta^2 C_m \sim \sum_{k=1}^{\infty} \beta_k \lambda_k^m \quad \text{as } m \to \infty; \quad \beta_k = \alpha_k (\lambda_k - 1)^2 \quad \text{for all } k, \quad (16.4.19)$$

and applying Lemma 16.4.2 again, we obtain

$$H_n^{(j)}(\{\Delta^2 C_s\}) \sim \sum_{1 \le k_1 < k_2 < \cdots < k_n} \left(\prod_{p=1}^{n} \beta_{k_p} \lambda_{k_p}^j \right) \left[V(\lambda_{k_1}, \lambda_{k_2}, \ldots, \lambda_{k_n}) \right]^2. \quad (16.4.20)$$

The most dominant term in (16.4.20) is that with $k_1 = 1, k_2 = 2, \ldots, k_n = n$. Thus,

$$H_n^{(j)}(\{\Delta^2 C_s\}) \sim \left(\prod_{p=1}^{n} \beta_p \lambda_p^j \right) [V(\lambda_1, \ldots, \lambda_n)]^2. \quad (16.4.21)$$

Dividing (16.4.18) by (16.4.21), which is allowed because the latter is an asymptotic equality, and recalling (16.3.1), we obtain (16.4.15), from which (16.4.16) follows immediately. ∎

This theorem was stated, subject to the condition that either $\lambda_1 > \lambda_2 > \cdots > 0$ or $\lambda_1 < \lambda_2 < \cdots < 0$, in an important paper by Wynn [371]. The general result in (16.4.15) was given by Sidi [296].

Corollary 16.4.4 *If $r = 1$ in Theorem 16.4.3, i.e., if $|\lambda_n| > |\lambda_{n+1}| > |\lambda_{n+2}|$, then (16.4.15) can be replaced by the asymptotic equality*

$$\varepsilon_{2n}^{(j)} - A \sim \alpha_{n+1} \left(\prod_{i=1}^{n} \frac{\lambda_{n+1} - \lambda_i}{1 - \lambda_i} \right)^2 \lambda_{n+1}^j \quad \text{as } j \to \infty. \quad (16.4.22)$$

It is clear from Theorem 16.4.3 that the column sequence $\{\varepsilon_{2n}^{(j)}\}_{j=0}^{\infty}$ converges when $|\lambda_{n+1}| < 1$ whether $\{A_m\}$ converges or not. When $|\lambda_{n+1}| \ge 1$, $\{\varepsilon_{2n}^{(j)}\}_{j=0}^{\infty}$ diverges, but more slowly than $\{A_m\}$.

Note. Theorem 16.4.3 and Corollary 16.4.4 concern the convergence of $\{\varepsilon_{2n}^{(j)}\}_{j=0}^{\infty}$ subject to the condition $|\lambda_n| > |\lambda_{n+1}|$, but they do not apply when $|\lambda_n| = |\lambda_{n+1}|$, which is the only remaining case. In this case, we can show at best that a *subsequence* of $\{\varepsilon_{2n}^{(j)}\}_{j=0}^{\infty}$

converges under certain conditions and there holds $\varepsilon_{2n}^{(j)} - A = O(\lambda_{n+1}^{j})$ as $j \to \infty$ for this subsequence. It can be shown that such a subsequence exists when $|\lambda_{n-1}| > |\lambda_n| = |\lambda_{n+1}|$.

We see from Theorem 16.4.3 that the Shanks transformation accelerates the convergence of $\{A_m\}$ under the prescribed conditions. Indeed, assuming that $|\lambda_1| > |\lambda_2|$, and using the asymptotic equality $A_m - A \sim \alpha_1 \lambda_1^m$ as $m \to \infty$ that follows from (16.4.1) and (16.4.2), and observing that $|\lambda_{n+1}/\lambda_1| < 1$ by (16.4.14), we have, for any fixed i,

$$\frac{\varepsilon_{2n}^{(j)} - A}{A_{j+i} - A} = O(|\lambda_{n+1}/\lambda_1|^j) = o(1) \text{ as } j \to \infty,$$

The next result, which appears to be new, concerns the stability of the Shanks transformation under the conditions of Theorem 16.4.3. Before stating this result, we recall Theorems 3.2.1 and 3.2.2 of Chapter 3, according to which

$$\varepsilon_{2n}^{(j)} = \sum_{i=0}^{n} \gamma_{ni}^{(j)} A_{j+i}, \tag{16.4.23}$$

and

$$\sum_{i=0}^{n} \gamma_{ni}^{(j)} z^i = \frac{\begin{vmatrix} 1 & z & \cdots & z^n \\ \Delta A_j & \Delta A_{j+1} & \cdots & \Delta A_{j+n} \\ \vdots & \vdots & & \vdots \\ \Delta A_{j+n-1} & \Delta A_{j+n} & \cdots & \Delta A_{j+2n-1} \end{vmatrix}}{\begin{vmatrix} 1 & 1 & \cdots & 1 \\ \Delta A_j & \Delta A_{j+1} & \cdots & \Delta A_{j+n} \\ \vdots & \vdots & & \vdots \\ \Delta A_{j+n-1} & \Delta A_{j+n} & \cdots & \Delta A_{j+2n-1} \end{vmatrix}} \equiv \frac{R_n^{(j)}(z)}{R_n^{(j)}(1)}. \tag{16.4.24}$$

Theorem 16.4.5 *Under the conditions of Theorem 16.4.3, there holds*

$$\lim_{j \to \infty} \sum_{i=0}^{n} \gamma_{ni}^{(j)} z^i = \prod_{i=}^{n} \frac{z - \lambda_i}{1 - \lambda_i} \equiv \sum_{i=0}^{n} \rho_{ni} z^i. \tag{16.4.25}$$

That is, $\lim_{j \to \infty} \gamma_{ni}^{(j)} = \rho_{ni}, i = 0, 1, \ldots, n$. Consequently, the Shanks transformation is stable with respect to its column sequences, and we have

$$\lim_{j \to \infty} \Gamma_n^{(j)} = \sum_{i=0}^{n} |\rho_{ni}| \leq \prod_{i=1}^{n} \frac{1 + |\lambda_i|}{|1 - \lambda_i|}. \tag{16.4.26}$$

Equality holds in (16.4.26) when $\lambda_1, \ldots, \lambda_n$ have the same phase. When λ_k are all real negative, $\lim_{j \to \infty} \Gamma_n^{(j)} = 1$.

Proof. Applying to the determinant $R_n^{(j)}(z)$ the technique that was used in the proof of Lemma 16.4.2 in the analysis of $H_n^{(j)}(\{f_s\})$, we can show that $R_n^{(j)}(z)$ satisfies the asymptotic equality

$$R_n^{(j)}(z) \sim \left[\prod_{p=1}^{n} \alpha_p(\lambda_p - 1)\lambda_p^j\right] V(\lambda_1, \ldots, \lambda_n) V(z, \lambda_1, \ldots, \lambda_n) \text{ as } j \to \infty. \quad (16.4.27)$$

Using this asymptotic equality in (16.4.24), the result in (16.4.25) follows. The rest can be completed as the proof of Theorem 3.5.6. ∎

As the results of Theorems 16.4.3 and 16.4.5 are best asymptotically, they can be used to draw some important conclusions about efficient use of the Shanks transformation. We see from (16.4.15) and (16.4.26) that both $\varepsilon_{2n}^{(j)} - A$ and $\Gamma_n^{(j)}$ are large if some of the λ_k are too close to 1. That is, poor stability and poor accuracy in $\varepsilon_{2n}^{(j)}$ occur simultaneously. Thus, making the transformation more stable results in more accuracy as well. [We reached the same conclusion when we treated the application of the d-transformation to power series and (generalized) Fourier series close to points of singularity of their limit functions.]

In most cases of interest, λ_1 and possibly a few of the succeeding λ_k are close to 1. For some positive integer q, the numbers λ_k^q separate from 1. In view of this, we propose to apply the Shanks transformation to the subsequence $\{A_{qm+s}\}$, where $q > 1$ and $s \geq 0$ are some fixed integers. This achieves the desired result of increased stability as

$$A_{qm+s} \sim A + \sum_{k=1}^{\infty} \beta_k \mu_k^m \text{ as } m \to \infty; \quad \mu_k = \lambda_k^q, \quad \beta_k = \alpha_k \lambda_k^s \text{ for all } k, \quad (16.4.28)$$

by which the λ_k in (16.4.15), (16.4.16), (16.4.22), (16.4.25), and (16.4.26) are replaced by the corresponding μ_k, which are further away from 1 than the λ_k. This strategy is nothing but arithmetic progression sampling (APS).

Before leaving this topic, we revisit two problems we discussed in detail in Section 15.3 of the preceding chapter in connection with the iterated Δ^2-process, namely, the power method and fixed-point iterative solution of nonlinear equations.

Consider the sequence $\{\rho_m\}$ generated by the power method for a matrix Q. Recall that, when Q is diagonalizable, ρ_m has an expansion of the form $\rho_m = \mu + \sum_{k=1}^{\infty} \alpha_k \lambda_k^m$, where μ is the largest eigenvalue of Q and $1 > |\lambda_1| \geq |\lambda_2| \geq \cdots$. Therefore, the Shanks transformation can be applied to the sequence $\{\rho_m\}$, and Theorem 16.4.3 holds.

Consider next the equation $x = g(x)$, whose solution we denote s. Starting with x_0, we generate $\{x_m\}$ via $x_{m+1} = g(x_m)$. Recall that, provided $0 < |g'(s)| < 1$ and x_0 is sufficiently close to s, there holds

$$x_m \sim s + \sum_{k=1}^{\infty} \alpha_k \lambda^{km} \text{ as } m \to \infty, \quad \alpha_1 \neq 0; \quad \lambda = g'(s).$$

Thus, the Shanks transformation can be applied to $\{x_m\}$ successfully. By Theorem 16.4.3, this results in

$$\varepsilon_{2n}^{(j)} - s = O(\lambda^{(n+1)j}) \text{ as } j \to \infty,$$

whether some of the α_k vanish or not.

16.4.1 Extensions

So far, we have been concerned with analyzing the performance of the Shanks transformation on sequences that satisfy (16.4.1). We now wish to extend this analysis to sequences $\{A_m\}$ that satisfy

$$A_m \sim A + \sum_{k=1}^{\infty} P_k(m)\lambda_k^m \quad \text{as } m \to \infty, \qquad (16.4.29)$$

where

$$\lambda_k \text{ distinct}; \quad \lambda_k \neq 1 \text{ for all } k; \quad |\lambda_1| \geq |\lambda_2| \geq \cdots; \quad \lim_{k\to\infty} \lambda_k = 0, \quad (16.4.30)$$

and, for each k, $P_k(m)$ is a polynomial in m of degree exactly $p_k \geq 0$ and with leading coefficient $e_k \neq 0$. We recall that the assumption that $\lim_{k\to\infty} \lambda_k = 0$ implies that there can be only a finite number of λ_k with the same modulus. We say that such sequences are *exponential with confluence*.

Again, we recall from Theorem 16.1.2 that, when $A_m = A + \sum_{k=1}^{t} P_k(m)\lambda_k^m$ for all m, and n is chosen such that $n = \sum_{k=1}^{t}(p_k + 1)$, then $\varepsilon_{2n}^{(j)} = A$ for all j. From this, we expect the Shanks transformation also to perform well when applied to the general case in (16.4.29). This turns out to be the case, but its theory is much more complicated than we have seen in this section so far.

We state in the following without proof theorems on the convergence and stability of column sequences generated by the Shanks transformation. These are taken from Sidi [296]. Of these, the first three generalize the results of Theorem 16.4.3, Corollary 16.4.4, and Theorem 16.4.5.

Theorem 16.4.6 *Let $\{A_m\}$ be exactly as in the first paragraph of this subsection and let t and r be positive integers for which*

$$|\lambda_1| \geq \cdots \geq |\lambda_t| > |\lambda_{t+1}| = \cdots = |\lambda_{t+r}| > |\lambda_{t+r+1}|, \qquad (16.4.31)$$

and order the λ_k such that

$$\bar{p} \equiv p_{t+1} = \cdots = p_{t+\mu} > p_{t+\mu+1} \geq \cdots \geq p_{t+r}. \qquad (16.4.32)$$

Set

$$\omega_k = p_k + 1, \quad k = 1, 2, \ldots, \qquad (16.4.33)$$

and let

$$n = \sum_{k=1}^{t} \omega_k. \qquad (16.4.34)$$

Then

$$\varepsilon_{2n}^{(j)} - A = j^{\bar{p}} \sum_{s=t+1}^{t+\mu} e_s \left[\prod_{i=1}^{t} \left(\frac{\lambda_s - \lambda_i}{1 - \lambda_i} \right)^{2\omega_i} \right] \lambda_s^j + o(j^{\bar{p}}\lambda_{t+1}^j) \quad \text{as } j \to \infty,$$

$$= O(j^{\bar{p}}\lambda_{t+1}^j) \quad \text{as } j \to \infty. \qquad (16.4.35)$$

Corollary 16.4.7 *When $\mu = 1$ in Theorem 16.4.6, the result in (16.4.35) becomes*

$$\varepsilon_{2n}^{(j)} - A \sim e_{t+1}\left[\prod_{i=1}^{t}\left(\frac{\lambda_{t+1} - \lambda_i}{1 - \lambda_i}\right)^{2\omega_i}\right]j^{p_{t+1}}\lambda_{t+1}^{j} \quad as \; j \to \infty. \quad (16.4.36)$$

Theorem 16.4.8 *Under the conditions of Theorem 16.4.6, there holds*

$$\lim_{j \to \infty}\sum_{i=0}^{n}\gamma_{ni}^{(j)}z^i = \prod_{i=1}^{n}\left(\frac{z - \lambda_i}{1 - \lambda_i}\right)^{\omega_i} \equiv \sum_{i=0}^{n}\rho_{ni}z^i. \quad (16.4.37)$$

That is, $\lim_{j \to \infty}\gamma_{ni}^{(j)} = \rho_{ni}$, $i = 0, 1, \ldots, n$. Consequently, the Shanks transformation is stable with respect to its column sequences, and we have

$$\lim_{j \to \infty}\Gamma_n^{(j)} = \sum_{i=0}^{n}|\rho_{ni}| \leq \prod_{i=1}^{n}\left(\frac{1 + |\lambda_i|}{|1 - \lambda_i|}\right)^{\omega_i}. \quad (16.4.38)$$

Equality holds in (16.4.38) when $\lambda_1, \ldots, \lambda_t$ have the same phase. When λ_k are all negative, $\lim_{j \to \infty}\Gamma_n^{(j)} = 1$.

Note that the second of the results in (16.4.35), namely, $\varepsilon_{2n}^{(j)} - A = O(j^{\bar{p}}\lambda_{t+1}^{j})$ as $j \to \infty$, was first given by Sidi and Bridger [308, Theorem 3.1 and Note on p. 42].

Now, Theorem 16.4.6 covers only the cases in which n is as in (16.4.34). It does not cover the remaining cases, namely, those for which $\sum_{k=1}^{t}\omega_k < n < \sum_{k=1}^{t+r}\omega_k$, which turn out to be more involved but very interesting. The next theorem concerns the convergence of column sequences in these cases.

Theorem 16.4.9 *Assume that the λ_k are as in Theorem 16.4.6 with the same notation. Let n be such that*

$$\sum_{k=1}^{t}\omega_k < n < \sum_{k=1}^{t+r}\omega_k \quad (16.4.39)$$

and let

$$\tau = n - \sum_{k=1}^{t}\omega_k \quad (16.4.40)$$

This time, however, also allow $t = 0$ and define $\sum_{k=1}^{0}\omega_k = 0$. Denote by $\mathrm{IP}(\tau)$ the nonlinear integer programming problem

$$maximize \; g(\vec{\sigma}); \quad g(\vec{\sigma}) = \sum_{k=t+1}^{t+r}(\omega_k\sigma_k - \sigma_k^2)$$

$$subject \; to \; \sum_{k=t+1}^{t+r}\sigma_k = \tau \; and \; 0 \leq \sigma_k \leq \omega_k, \; t+1 \leq k \leq t+r, \quad (16.4.41)$$

and denote by $G(\tau)$ the (optimal) value of $g(\vec{\sigma})$ at the solution to $\mathrm{IP}(\tau)$.

Provided $\mathrm{IP}(\tau)$ has a unique solution for σ_k, $k = t+1, \ldots, t+r$, $\varepsilon_{2n}^{(j)}$ satisfies

$$\varepsilon_{2n}^{(j)} - A = O(j^{G(\tau+1)-G(\tau)}\lambda_{t+1}^{j}) \quad as \; j \to \infty, \quad (16.4.42)$$

whether $\{A_m\}$ converges or not. [Here $\mathrm{IP}(\tau + 1)$ is not required to have a unique solution.]

Corollary 16.4.10 *When $r = 1$, i.e., $|\lambda_t| > |\lambda_{t+1}| > |\lambda_{t+2}|$, $p_{t+1} > 1$, and $\sum_{k=1}^{t} \omega_k < n < \sum_{k=1}^{t+1} \omega_k$, a unique solution to $\mathrm{IP}(\tau)$ exists, and we have*

$$\varepsilon_{2n}^{(j)} - A \sim Cj^{p_{t+1} - 2\tau} \lambda_{t+1}^{j} \quad as \ j \to \infty, \tag{16.4.43}$$

where $\tau = n - \sum_{k=1}^{t} \omega_k$, hence $0 < \tau < \omega_{t+1}$, and C is a constant given by

$$C = (-1)^{\tau} \frac{p_{t+1}! \, \tau!}{(p_{t+1} - \tau)!} e_{t+1} \left(\frac{\lambda_{t+1}}{1 - \lambda_{t+1}} \right)^{2\tau} \left[\prod_{i=1}^{t} \left(\frac{\lambda_{t+1} - \lambda_i}{1 - \lambda_i} \right)^{2\omega_i} \right]. \tag{16.4.44}$$

Therefore, the sequence $\{\varepsilon_{2n}^{(j)}\}_{j=0}^{\infty}$ is better then $\{\varepsilon_{2n-2}^{(j)}\}_{j=0}^{\infty}$. In particular, if $|\lambda_1| > |\lambda_2| > |\lambda_3| > \cdots$, then this is true for all $n = 1, 2, \ldots$.

All the above hold whether $\{A_m\}$ converges or not.

It follows from Corollary 16.4.10 that, when $1 > |\lambda_1| > |\lambda_2| > \cdots$, *all* column sequences $\{\varepsilon_{2n}^{(j)}\}_{j=0}^{\infty}$ converge, and $\{\varepsilon_{2n}^{(j)}\}_{j=0}^{\infty}$ converges faster than $\{\varepsilon_{2n-2}^{(j)}\}_{j=0}^{\infty}$ for each n.

Note. In case the problem $\mathrm{IP}(\tau)$ does not have a unique solution, the best we can say is that, under certain conditions, there may exist a subsequence of $\{\varepsilon_{2n}^{(j)}\}_{j=0}^{\infty}$ that satisfies (16.4.42).

In connection with $\mathrm{IP}(\tau)$, we would like to mention that algorithms for its solution have been given by Parlett [228] and by Kaminski and Sidi [148]. These algorithms also enable one to decide in a simple manner whether the solution is unique. A direct solution of $\mathrm{IP}(\tau)$ has been given by Liu and Saff [169]. Some properties of the solutions to $\mathrm{IP}(\tau)$ have been given by Sidi [292], and we mention them here for completeness.

Denote $J = \{t + 1, \ldots, t + r\}$, and let $\sigma_k, k \in J$, be a solution of $\mathrm{IP}(\tau)$:

1. $\sigma_k' = \omega_k - \sigma_k$, $k \in J$, is a solution of $\mathrm{IP}(\tau')$ with $\tau' = \sum_{k=t+1}^{t+r} \omega_k - \tau$.
2. If $\omega_{k'} = \omega_{k''}$ for some $k', k'' \in J$, and if $\sigma_{k'} = \delta_1$ and $\sigma_{k''} = \delta_2$ in a solution to $\mathrm{IP}(\tau)$, $\delta_1 \neq \delta_2$, then there is another solution to $\mathrm{IP}(\tau)$ with $\sigma_{k'} = \delta_2$ and $\sigma_{k''} = \delta_1$. Consequently, a solution to $\mathrm{IP}(\tau)$ cannot be unique unless $\sigma_{k'} = \sigma_{k''}$. One implication of this is that, for $\omega_{t+1} = \cdots = \omega_{t+r} = \bar{\omega} > 1$, $\mathrm{IP}(\tau)$ has a unique solution only for $\tau = qr, q = 1, \ldots, \bar{\omega} - 1$, and in this solution $\sigma_k = q, k \in J$. For $\omega_{t+1} = \cdots = \omega_{t+r} = 1$, no unique solution to $\mathrm{IP}(\tau)$ exists with $1 \leq \tau \leq r - 1$. Another implication is that, for $\omega_{t+1} = \cdots = \omega_{t+\mu} > \omega_{t+\mu+1} \geq \cdots \geq \omega_{t+r}$, $\mu < r$, no unique solution to $\mathrm{IP}(\tau)$ exists for $\tau = 1, \ldots, \mu - 1$, and a unique solution exists for $\tau = \mu$, this solution being $\sigma_{t+1} = \cdots = \sigma_{t+\mu} = 1$, $\sigma_k = 0$, $t + \mu + 1 \leq k \leq t + r$.
3. A unique solution to $\mathrm{IP}(\tau)$ exists when $\omega_k, k \in J$, are all even or all odd, and $\tau = qr + \frac{1}{2} \sum_{k=t+1}^{t+r} (\omega_k - \omega_{t+r})$, $0 \leq q \leq \omega_{t+r}$. This solution is given by $\sigma_k = q + \frac{1}{2}(\omega_k - \omega_{t+r})$, $k \in J$.
4. Obviously, when $r = 1$, a unique solution to $\mathrm{IP}(\tau)$ exists for all possible τ and is given as $\sigma_{t+1} = \tau$. When $r = 2$ and $\omega_1 + \omega_2$ is odd, a unique solution to $\mathrm{IP}(\tau)$ exists for *all* possible τ, as shown by Kaminski and Sidi [148].

16.4.2 Application to Numerical Quadrature

Sequences of the type treated in this section arise naturally when one approximates finite-range integrals with endpoint singularities via the trapezoidal rule or the midpoint rule. Consequently, the Shanks transformation has been applied to accelerate the convergence of sequences of these approximations.

Let us go back to Example 4.1.4 concerning the integral $I[G] = \int_0^1 G(x)\,dx$, where $G(x) = x^s g(x)$, $\Re s > -1$, s not an integer, and $g \in C^\infty[0, 1]$. Setting $h = 1/n$, n a positive integer, this integral is approximated by $Q(h)$, where $Q(h)$ stands for either the trapezoidal rule or the midpoint rule that are defined in (4.1.3), and $Q(h)$ has the asymptotic expansion given by (4.1.4). Letting $h_m = 1/2^m$, $m = 0, 1, \ldots$, we realize that $A_m \equiv Q(h_m)$ has the asymptotic expansion

$$A_m \sim A + \sum_{k=1}^{\infty} \alpha_k \lambda_k^m \text{ as } m \to \infty,$$

where $A = I[G]$, and the λ_k are obtained by ordering 4^{-i} and 2^{-s-i}, $i = 1, 2, \ldots$, in decreasing order according to their moduli. Thus, the λ_k satisfy

$$1 > |\lambda_1| > |\lambda_2| > |\lambda_3| > \cdots.$$

If we now apply the Shanks transformation to the sequence $\{A_m\}$, Corollary 16.4.4 applies, and we have that each column of the epsilon table converges to $I[G]$ faster than the one preceding it.

Let us next consider the approximation of the integral $I[G] = \int_0^1 G(x)\,dx$, where $G(x) = x^s (\log x)^p g(x)$, with $\Re s > -1$, p a positive integer, and $g \in C^\infty[0, 1]$. [Recall that the case $p = 1$ has already been considered in Example 3.1.2.] Let $h = 1/n$, and define the trapezoidal rule to $I[G]$ by $T(h) = h \sum_{i=1}^{n-1} G(ih)$. Then, $T(h)$ has an asymptotic expansion of the form

$$T(h) \sim I[G] + \sum_{i=1}^{\infty} a_i h^{2i} + \sum_{i=0}^{\infty} \left[\sum_{j=0}^{p} b_{ij} (\log h)^j \right] h^{s+i+1} \text{ as } h \to 0,$$

where a_i and b_{ij} are constants independent of h. [For $p = 1$, this asymptotic expansion is nothing but that of (3.1.6) in Example 3.1.2.] Again, letting $h_m = 1/2^m$, $m = 0, 1, \ldots$, we realize that $A_m \equiv Q(h_m)$ has the asymptotic expansion

$$A_m \sim A + \sum_{k=1}^{\infty} P_k(m) \lambda_k^m \text{ as } m \to \infty,$$

where $A = I[G]$ and the λ_k are precisely as before, while $P_k(m)$ are polynomials in m. The degree of $P_k(m)$ is 0 if λ_k is 4^{-i} for some i, and it is at most p if λ_k is 2^{-s-i} for some i. If we now apply the Shanks transformation to the sequence $\{A_m\}$, Corollaries 16.4.7 and 16.4.10 apply, and we have that each column of the epsilon table converges to $I[G]$ faster than the one preceding it. For the complete details of this example with $p = 1$ and $s = 0$, and for the treatment of the general case in which p is an arbitrary positive integer, we refer the reader to Sidi [296, Example 5.2].

This use of the Shanks transformation for the integrals $\int_0^1 x^s (\log x)^p g(x)\,dx$ with $p = 0$ and $p = 1$ was originally suggested by Chisholm, Genz, and Rowlands [49] and by Kahaner [147]. The only convergence result known at that time was Corollary 16.4.4, which is valid only for $p = 0$, and this was mentioned by Genz [94]. The treatment of the general case with $p = 1, 2, \ldots$, was given later by Sidi [296].

16.5 Analysis of Column Sequences When $\{\Delta A_m\} \in \mathbf{b}^{(1)}$

In the preceding section, we analyzed the behavior of the Shanks transformation on sequences that behave as in (16.4.1) and showed that it is very effective on such sequences. This effectiveness was "expected" in view of the fact that the derivation of the transformation was actually based on (16.4.1). It is interesting that the effectiveness of the Shanks transformation is not limited to sequences that are as in (16.4.1). In this section, we present some results pertaining to those sequences $\{A_m\}$ that were discussed in Definition 15.3.2 and for which $\{\Delta A_m\} \in \mathbf{b}^{(1)}$. These results show that the Shanks transformation is effective on linear and factorial sequences, but it is ineffective on logarithmic sequences. Actually, they are completely analogous to the results of Chapter 15 on the iterated Δ^2-process. Throughout this section, we assume the notation of Definition 15.3.2.

16.5.1 Linear Sequences

The following theorem is due to Garibotti and Grinstein [92].

Theorem 16.5.1 *Let* $\{A_m\} \in \mathbf{b}^{(1)}/\text{LIN}$. *Provided* $\gamma \neq 0, 1, \ldots, n - 1$, *there holds*

$$\varepsilon_{2n}^{(j)} - A \sim (-1)^n \alpha_0 \frac{n! \, [\gamma]_n}{(\zeta - 1)^{2n}} \zeta^{j+2n} j^{\gamma-2n} \quad as \; j \to \infty, \tag{16.5.1}$$

where $[\gamma]_n = \gamma(\gamma - 1) \cdots (\gamma - n + 1)$ *when* $n > 0$ *and* $[\gamma]_0 = 1$, *as before. This result is valid whether* $\{A_m\}$ *converges or not.*

Proof. To prove (16.5.1), we make use of the error formula in (16.3.2) with $D_m = (A_m - A)\zeta^{-m}$ precisely as in Lemma 16.3.1. We first observe that

$$H_p^{(m)}(\{u_s\}) = H_p^{(m)}(\{\Delta^{s-m} u_m\}) = \begin{vmatrix} u_m & \Delta u_m & \ldots & \Delta^{p-1} u_m \\ \Delta u_m & \Delta^2 u_m & \ldots & \Delta^p u_m \\ \vdots & \vdots & & \vdots \\ \Delta^{p-1} u_m & \Delta^p u_m & \ldots & \Delta^{2p-2} u_m \end{vmatrix}, \tag{16.5.2}$$

which can be proved by performing a series of elementary row and column transformations on $H_p^{(m)}(\{u_s\})$ in (16.1.13). As a result, (16.3.2) becomes

$$\varepsilon_{2n}^{(j)} - A = \zeta^{j+2n} \frac{H_{n+1}^{(j)}(\{\Delta^{s-j} D_j\})}{H_n^{(j)}(\{\Delta^{s-j} E_j\})}; \quad E_m = D_m - 2\zeta D_{m+1} + \zeta^2 D_{m+2} \;\; \text{for all } m. \tag{16.5.3}$$

Next, we have $D_m \sim \sum_{i=0}^{\infty} \alpha_i m^{\gamma-i}$ as $m \to \infty$, so that

$$\Delta^r D_m \sim \alpha_0 [\gamma]_r m^{\gamma-r} \text{ as } m \to \infty,$$

$$\Delta^r E_m \sim (\zeta - 1)^2 \alpha_0 [\gamma]_r m^{\gamma-r} \sim (\zeta - 1)^2 D_m \text{ as } m \to \infty. \qquad (16.5.4)$$

Substituting (16.5.4) in the determinant $H_{n+1}^{(j)}(\{\Delta^{s-j} D_j\})$, and factoring out the powers of j, we obtain

$$H_{n+1}^{(j)}(\{\Delta^{s-j} D_j\}) \sim \alpha_0^{n+1} j^{\sigma_n} K_n \text{ as } j \to \infty, \qquad (16.5.5)$$

where $\sigma_n = \sum_{i=0}^n (\gamma - 2i)$ and

$$K_n = \begin{vmatrix} [\gamma]_0 & [\gamma]_1 & \cdots & [\gamma]_n \\ [\gamma]_1 & [\gamma]_2 & \cdots & [\gamma]_{n+1} \\ \vdots & \vdots & & \vdots \\ [\gamma]_n & [\gamma]_{n+1} & \cdots & [\gamma]_{2n} \end{vmatrix}, \qquad (16.5.6)$$

provided, of course, that $K_n \neq 0$. Using the fact that $[x]_{q+r} = [x]_r \cdot [x - r]_q$, we factor out $[\gamma]_0$ from the first column, $[\gamma]_1$ from the second, $[\gamma]_2$ from the third, etc. (Note that all these factors are nonzero by our assumption on γ.) Applying Lemma 6.8.1 as we did in the proof of Lemma 6.8.2, we obtain

$$K_n = \left(\prod_{i=0}^n [\gamma]_i \right) V(\gamma, \gamma - 1, \ldots, \gamma - n) = \left(\prod_{i=0}^n [\gamma]_i \right) V(0, -1, -2, \ldots, -n).$$

$$(16.5.7)$$

Consequently,

$$H_{n+1}^{(j)}(\{\Delta^{s-j} D_j\}) \sim \alpha_0^{n+1} \left(\prod_{i=0}^n [\gamma]_i \right) V(0, -1, \ldots, -n) j^{\sigma_n} \text{ as } j \to \infty. \quad (16.5.8)$$

Similarly, by (16.5.4) again, we also have

$$H_n^{(j)}(\{\Delta^{s-j} E_j\}) \sim (\zeta - 1)^{2n} H_n^{(j)}(\{\Delta^{s-j} D_j\}) \text{ as } j \to \infty. \qquad (16.5.9)$$

Combining (16.5.8) and (16.5.9) in (16.5.3), we obtain (16.5.1). ∎

Note that $(\varepsilon_{2n}^{(j)} - A)/(A_{j+2n} - A) \sim K j^{-2n}$ as $j \to \infty$ for some constant K. That is, the columns of the epsilon table converge faster than $\{A_m\}$, and each column converges faster than the one preceding it.

The next result that appears to be new concerns the stability of the Shanks transformation under the conditions of Theorem 16.5.1.

Theorem 16.5.2 *Under the conditions of Theorem 16.5.1, there holds*

$$\lim_{j \to \infty} \sum_{i=0}^n \gamma_{ni}^{(j)} z^i = \left(\frac{z - \zeta}{1 - \zeta} \right)^n \equiv \sum_{i=0}^n \rho_{ni} z^i. \qquad (16.5.10)$$

That is, $\lim_{j\to\infty} \gamma_{ni}^{(j)} = \rho_{ni}$, $i = 0, 1, \dots, n$. *Consequently, the Shanks transformation is stable with respect to its column sequences and we have*

$$\lim_{j\to\infty} \Gamma_n^{(j)} = \left(\frac{1 + |\zeta|}{|1 - \zeta|}\right)^n. \tag{16.5.11}$$

When $\zeta = -1$, *we have* $\lim_{j\to\infty} \Gamma_n^{(j)} = 1$.

Proof. Substituting $\Delta A_m = \alpha_0(\zeta - 1)\zeta^m B_m$ in the determinant $R_n^{(j)}(z)$, and performing elementary row and column transformation, we obtain

$$R_n^{(j)}(z) = [\alpha_0(\zeta - 1)]^n \zeta^\tau \begin{vmatrix} 1 & w & \dots & w^n \\ B_j & \Delta B_j & \dots & \Delta^n B_j \\ \Delta B_j & \Delta^2 B_j & \dots & \Delta^{n+1} B_j \\ \vdots & \vdots & & \vdots \\ \Delta^{n-1} B_j & \Delta^n B_j & \dots & \Delta^{2n-1} B_j \end{vmatrix}, \tag{16.5.12}$$

where $\tau = n + \sum_{i=0}^{n-1}(j + 2i)$ and $w = z/\zeta - 1$. Now, since $B_m \sim m^\gamma$ as $m \to \infty$, we also have $\Delta^k B_m \sim [\gamma]_k m^{\gamma - k}$ as $m \to \infty$. Substituting this in (16.5.12), and proceeding as in the proof of Theorem 16.5.1, we obtain

$$R_n^{(j)}(z) \sim L_n^{(j)}(z/\zeta - 1)^n \text{ as } j \to \infty, \tag{16.5.13}$$

for some $L_n^{(j)}$ that is nonzero for all large j and independent of z. The proof of (16.5.10) can now be completed easily. The rest is also easy and we leave it to the reader. ∎

As Theorem 16.5.1 covers the convergence of the Shanks transformation for the cases in which $\gamma \neq 0, 1, \dots, n - 1$, we need a separate result for the cases in which γ is an integer in $\{0, 1, \dots, n - 1\}$. The following result, which covers these remaining cases, is stated in Garibotti and Grinstein [92].

Theorem 16.5.3 *If* γ *is an integer,* $0 \leq \gamma \leq n - 1$, *and* $\alpha_{\gamma+1} \neq 0$, *then*

$$\varepsilon_{2n}^{(j)} - A \sim \alpha_{\gamma+1} \frac{(n - \gamma - 1)!(n + \gamma + 1)!}{(\zeta - 1)^{2n}} \zeta^{j+2n} j^{-2n-1} \text{ as } j \to \infty. \tag{16.5.14}$$

Important conclusions can be drawn from these results concerning the application of the Shanks transformation to sequences $\{A_m\} \in \mathbf{b}^{(1)}/\text{LIN}$.

As is obvious from Theorems 16.5.1 and 16.5.3, all the column sequences $\{\varepsilon_{2n}^{(j)}\}_{j=0}^\infty$ converge to A when $|\zeta| < 1$, with each column converging faster than the one preceding it. When $|\zeta| = 1$ but $\zeta \neq 1$, $\{\varepsilon_{2n}^{(j)}\}_{j=0}^\infty$ converges to A (even when $\{A_m\}$ diverges) provided (i) $n > \Re\gamma/2$ when $\gamma \neq 0, 1, \dots, n - 1$, or (ii) $n \geq \gamma + 1$ when γ is a nonnegative integer. In all other cases, $\{\varepsilon_{2n}^{(j)}\}_{j=0}^\infty$ diverges. From (16.5.1) and (16.5.11) and (16.5.14), we see that both the theoretical accuracy of $\varepsilon_{2n}^{(j)}$ and its stability properties deteriorate when ζ approaches 1, because $(\zeta - 1)^{-2n} \to \infty$ as $\zeta \to 1$. Recalling that ζ^q for a positive integer q is farther away from 1 when ζ is close to 1, we propose to apply the Shanks transformation to the subsequence $\{A_{qm+s}\}$, where $q > 1$ and $s \geq 0$ are fixed integers. This improves the quality of the $\varepsilon_{2n}^{(j)}$ as approximations to A. We already mentioned that this strategy is APS.

16.5.2 Logarithmic Sequences

The Shanks transformation turns out to be completely ineffective on logarithmic sequences $\{A_m\}$ for which $\{\Delta A_m\} \in \mathbf{b}^{(1)}$.

Theorem 16.5.4 *Let* $\{A_m\} \in \mathbf{b}^{(1)}/LOG$. *Then, there holds*

$$\varepsilon_{2n}^{(j)} - A \sim (-1)^n \alpha_0 \frac{n!}{[\gamma - 1]_n} j^\gamma \quad as \ j \to \infty. \tag{16.5.15}$$

In other words, no acceleration of convergence takes place for logarithmic sequences.

The result in (16.5.15) can be proved by using the technique of the proof of Theorem 16.5.1. We leave it to the interested reader.

16.5.3 Factorial Sequences

We end with the following results concerning the convergence and stability of the Shanks transformation on factorial sequences $\{A_m\}$. These results appear to be new.

Theorem 16.5.5 *Let* $\{A_m\} \in \mathbf{b}^{(1)}/FAC$. *Then*

$$\varepsilon_{2n}^{(j)} - A \sim (-1)^n \alpha_0 r^n \frac{\zeta^{j+2n}}{(j!)^r} j^{\gamma - 2nr - n} \quad as \ j \to \infty, \tag{16.5.16}$$

and

$$\lim_{j \to \infty} \sum_{i=0}^{n} \gamma_{ni}^{(j)} z^i = z^n \quad and \quad \lim_{j \to \infty} \Gamma_n^{(j)} = 1. \tag{16.5.17}$$

Note that $(\varepsilon_{2n}^{(j)} - A)/(A_{j+2n} - A) \sim K j^{-n}$ as $j \to \infty$ for some constant K. Thus, in this case too the columns of the epsilon table converge faster than the sequence $\{A_m\}$, and each column converges faster than the one preceding it.

The technique we used in the proofs of Theorems 16.5.1 and 16.5.4 appears to be too difficult to apply to Theorem 16.5.5, so we developed a totally different approach. We start with the recursion relations of the ε-algorithm. We first have

$$e_1(A_j) = \varepsilon_2^{(j)} = \frac{A_{j+1}\Delta(1/\Delta A_j) + 1}{\Delta(1/\Delta A_j)}. \tag{16.5.18}$$

Substituting the expression for $\varepsilon_{2n+1}^{(j)}$ in that for $\varepsilon_{2n+2}^{(j)}$ and invoking (16.5.18), we obtain

$$\varepsilon_{2n+2}^{(j)} = \frac{e_1(\varepsilon_{2n}^{(j)})\Delta(1/\Delta\varepsilon_{2n}^{(j)}) + \varepsilon_{2n}^{(j+1)}\Delta\varepsilon_{2n-1}^{(j+1)}}{\Delta\varepsilon_{2n+1}^{(j)}}. \tag{16.5.19}$$

From this, we obtain the error formula

$$\varepsilon_{2n+2}^{(j)} - A = \frac{[e_1(\varepsilon_{2n}^{(j)}) - A]\Delta(1/\Delta\varepsilon_{2n}^{(j)}) + [\varepsilon_{2n}^{(j+1)} - A]\Delta\varepsilon_{2n-1}^{(j+1)}}{\Delta\varepsilon_{2n+1}^{(j)}}. \quad (16.5.20)$$

Now use induction on n along with part (iii) of Theorem 15.3.3 to prove convergence. To prove stability, we begin by paying attention to the fact that (16.5.19) can also be expressed as

$$\varepsilon_{2n+2}^{(j)} = \lambda_n^{(j)}\varepsilon_{2n}^{(j)} + \mu_n^{(j)}\varepsilon_{2n}^{(j+1)}, \quad (16.5.21)$$

with appropriate $\lambda_n^{(j)}$ and $\mu_n^{(j)}$. Now proceed as in Theorems 15.5.1 and 15.5.2.

This technique can also be used to prove Theorem 16.5.1. We leave the details to the reader.

16.6 The Shanks Transformation on Totally Monotonic and Totally Oscillating Sequences

We now turn to application of the Shanks transformation to two important classes of real sequences, namely, *totally monotonic sequences* and *totally oscillating sequences*. The importance of these sequences stems from the fact that they arise in many applications. This treatment was started by Wynn [371] and additional contributions to it were made by Brezinski [34], [35].

16.6.1 Totally Monotonic Sequences

Definition 16.6.1 A sequence $\{\mu_m\}_{m=0}^{\infty}$ is said to be *totally monotonic* if it is real and if $(-1)^k\Delta^k\mu_m \geq 0$, $k, m = 0, 1, \ldots$, and we write $\{\mu_m\} \in TM$.

The sequences $\{(m + \beta)^{-1}\}_{m=0}^{\infty}$, $\beta > 0$, and $\{\lambda^m\}_{m=0}^{\infty}$, $\lambda \in (0, 1)$, are totally monotonic.

The following lemmas follow from this definition in a simple way.

Lemma 16.6.2 *If $\{\mu_m\} \in TM$, then*

(i) $\{\mu_m\}_{m=0}^{\infty}$ *is a nonnegative and nonincreasing sequence and hence has a nonnegative limit,*

(ii) $\{(-1)^k\Delta^k\mu_m\}_{m=0}^{\infty} \in TM$, $k = 1, 2, \ldots$.

Lemma 16.6.3

(i) *If $\{\mu_m\} \in TM$ and $\{\nu_m\} \in TM$, and $\alpha, \beta > 0$, then $\{\alpha\mu_m + \beta\nu_m\} \in TM$ as well. Also, $\{\mu_m\nu_m\} \in TM$. These can be extended to an arbitrary number of sequences.*

(ii) *Suppose that $\{\mu_m^{(i)}\}_{m=0}^{\infty} \in TM$ and $c_i \geq 0$, $i = 0, 1, \ldots$, and that $\sum_{i=0}^{\infty} c_i\mu_0^{(i)}$ converges, and define $\tau_m = \sum_{i=0}^{\infty} c_i\mu_m^{(i)}$, $m = 0, 1, \ldots$. Then $\{\tau_m\} \in TM$.*

(iii) *Suppose that $\{\mu_m\} \in TM$ and $c_i \geq 0$, $i = 0, 1, \ldots$, and $\sum_{i=0}^{\infty} c_i\mu_0^i$ converges, and define $\tau_m = \sum_{i=0}^{\infty} c_i\mu_m^i$, $m = 0, 1, \ldots$. Then $\{\tau_m\} \in TM$.*

Obviously, part (iii) of Lemma 16.6.3 is a corollary of parts (i) and (ii).

By Lemma 16.6.3, the sequence $\{\lambda^m/(m + \beta)\}_{m=0}^{\infty}$ with $\lambda \in (0, 1)$ and $\beta > 0$ is totally monotonic. Also, all functions $f(z)$ that are analytic at $z = 0$ and have Maclaurin series $\sum_{i=0}^{\infty} c_i z^i$ with $c_i \geq 0$ for all i render $\{f(\mu_m)\} \in TM$ when $\{\mu_m\} \in TM$, provided $\sum_{i=0}^{\infty} c_i \mu_0^i$ converges.

The next theorem is one of the fundamental results in the Hausdorff moment problem.

Theorem 16.6.4 *The sequence* $\{\mu_m\}_{m=0}^{\infty}$ *is totally monotonic if and only if there exists a function* $\alpha(t)$ *that is bounded and nondecreasing in* $[0, 1]$ *such that* $\mu_m = \int_0^1 t^m d\alpha(t)$, $m = 0, 1, \ldots$, *these integrals being defined in the sense of Stieltjes.*

The sequence $\{(m + \beta)^{-\nu}\}_{m=0}^{\infty}$ with $\beta > 0$ and $\nu > 0$ is totally monotonic, because $(m + \beta)^{-\nu} = \int_0^1 t^{m+\beta-1}(\log t^{-1})^{\nu-1} dt / \Gamma(\nu)$.

As an immediate corollary of Theorem 16.6.4, we have the following result on Hankel determinants that will be of use later.

Theorem 16.6.5 *Let* $\{\mu_m\} \in TM$. *Then* $H_p^{(m)}(\{\mu_s\}) \geq 0$ *for all* m, $p = 0, 1, 2, \ldots$. *If the function* $\alpha(t)$ *in* $\mu_m = \int_0^1 t^m d\alpha(t)$, $m = 0, 1, \ldots$, *has an infinite number of points of increase on* $[0, 1]$, *then* $H_p^{(m)}(\{\mu_s\}) > 0$, $m, p = 0, 1, \ldots$.

Proof. Let $P(t) = \sum_{i=0}^{k} c_i t^i$ be an arbitrary polynomial. Then

$$\sum_{i=0}^{k} \sum_{j=0}^{k} \mu_{m+i+j} c_i c_j = \int_0^1 t^m [P(t)]^2 d\alpha(t) \geq 0,$$

with strict inequality when $\alpha(t)$ has an infinite number of points of increase. The result now follows by a well-known theorem on quadratic forms. ∎

An immediate consequence of Theorem 16.6.5 is that $\mu_m \mu_{m+2} - \mu_{m+1}^2 \geq 0$ for all m if $\{\mu_m\} \in TM$. If $\hat{\mu} = \lim_{m \to \infty} \mu_m$ in this case, then $\{\mu_m - \hat{\mu}\} \in TM$ too. Therefore, $(\mu_m - \hat{\mu})(\mu_{m+2} - \hat{\mu}) - (\mu_{m+1} - \hat{\mu})^2 \geq 0$ for all m, and hence

$$0 < \frac{\mu_1 - \hat{\mu}}{\mu_0 - \hat{\mu}} \leq \frac{\mu_2 - \hat{\mu}}{\mu_1 - \hat{\mu}} \leq \cdots \leq 1,$$

provided $\mu_m \neq \hat{\mu}$ for all m. As a result of this, we also have that

$$\lim_{m \to \infty} \frac{\mu_{m+1} - \hat{\mu}}{\mu_m - \hat{\mu}} = \zeta \in (0, 1].$$

16.6.2 The Shanks Transformation on Totally Monotonic Sequences

The next theorem can be proved by invoking Theorem 16.1.3 on the relation between the Shanks transformation and the Padé table and by using the so-called *two-term identities* in the Padé table. We come back to these identities later.

Theorem 16.6.6 *When the Shanks transformation is applied to the sequence* $\{A_m\}$*, the following identities hold among the resulting approximants* $e_k(A_m) = \varepsilon_{2k}^{(m)}$*, provided the relevant* $\varepsilon_{2k}^{(m)}$ *exist:*

$$\varepsilon_{2n+2}^{(j)} - \varepsilon_{2n}^{(j)} = -\frac{[H_{n+1}^{(j)}(\{\Delta A_s\})]^2}{H_n^{(j)}(\{\Delta^2 A_s\})H_{n+1}^{(j)}(\{\Delta^2 A_s\})}, \tag{16.6.1}$$

$$\varepsilon_{2n}^{(j+1)} - \varepsilon_{2n}^{(j)} = \frac{H_{n+1}^{(j)}(\{\Delta A_s\})H_n^{(j+1)}(\{\Delta A_s\})}{H_n^{(j)}(\{\Delta^2 A_s\})H_n^{(j+1)}(\{\Delta^2 A_s\})}, \tag{16.6.2}$$

$$\varepsilon_{2n+2}^{(j)} - \varepsilon_{2n}^{(j+1)} = -\frac{H_{n+1}^{(j)}(\{\Delta A_s\})H_{n+1}^{(j+1)}(\{\Delta A_s\})}{H_{n+1}^{(j)}(\{\Delta^2 A_s\})H_n^{(j+1)}(\{\Delta^2 A_s\})}, \tag{16.6.3}$$

$$\varepsilon_{2n+2}^{(j)} - \varepsilon_{2n}^{(j+2)} = -\frac{[H_{n+1}^{(j+1)}(\{\Delta A_s\})]^2}{H_{n+1}^{(j)}(\{\Delta^2 A_s\})H_n^{(j+2)}(\{\Delta^2 A_s\})}. \tag{16.6.4}$$

An immediate corollary of this theorem that concerns $\{A_m\} \in TM$ is the following:

Theorem 16.6.7 *Let* $\{A_m\} \in TM$ *in the previous theorem. Then* $\varepsilon_{2n}^{(j)} \geq 0$ *for each* j *and* n*. In addition, both the column sequences* $\{\varepsilon_{2n}^{(j)}\}_{j=0}^{\infty}$ *and the diagonal sequence* $\{\varepsilon_{2n}^{(j)}\}_{n=0}^{\infty}$ *are nonincreasing and converge to* $A = \lim_{m\to\infty} A_m$*. Finally,* $(\varepsilon_{2n}^{(j)} - A)/(A_{j+2n} - A) \to 0$ *both as* $j \to \infty$*, and as* $n \to \infty$*, when* $\lim_{m\to\infty}(A_{m+1} - A)/(A_m - A) = \zeta \neq 1$*.*

Proof. That $\varepsilon_{2n}^{(j)} \geq 0$ follows from (16.1.12) and from the assumption that $\{A_m\} \in TM$. To prove the rest, it is sufficient to consider the case in which $A = 0$ since $\{A_m - A\} \in TM$ and $e_n(A_j) = A + e_n(A_j - A)$. From (16.6.3), it follows that $0 \leq \varepsilon_{2n}^{(j)} \leq \varepsilon_{2n-2}^{(j+1)} \leq \cdots \leq \varepsilon_0^{(j+n)} = A_{j+n}$. Thus, $\lim_{j\to\infty} \varepsilon_{2n}^{(j)} = 0$ and $\lim_{n\to\infty} \varepsilon_{2n}^{(j)} = 0$. That $\{\varepsilon_{2n}^{(j)}\}_{n=0}^{\infty}$ and $\{\varepsilon_{2n}^{(j)}\}_{j=0}^{\infty}$ are nonincreasing follows from (16.6.1) and (16.6.2), respectively. The last part follows from part (i) of Theorem 15.3.1 on the Aitken Δ^2-process, which says that $\varepsilon_2^{(j)} - A = o(A_j - A)$ as $j \to \infty$ under the prescribed condition. ∎

It follows from Theorem 16.6.7 that, if $A_m = cB_m + d$, $m = 0, 1, \ldots$, for some constants $c \neq 0$ and d, and if $\{B_m\} \in TM$, then the Shanks transformation on $\{A_m\}$ produces approximations that satisfy $\lim_{j\to\infty} \varepsilon_{2n}^{(j)} = A$ and $\lim_{n\to\infty} \varepsilon_{2n}^{(j)} = A$, where $A = \lim_{m\to\infty} A_m$. Also, $(\varepsilon_{2n}^{(j)} - A)/(A_{j+2n} - A) \to 0$ both as $j \to \infty$ and as $n \to \infty$, when $\lim_{m\to\infty}(A_{m+1} - A)/(A_m - A) = \zeta \neq 1$.

It must be noted that Theorem 16.6.7 gives almost no information about rates of convergence or convergence acceleration. To be able to say more, extra conditions need to be imposed on $\{A_m\}$. As an example, let us consider the sequence $\{A_m\} \in TM$ with $A_m = 1/(m + \beta)$, $m = 0, 1, \ldots$, $\beta > 0$, whose limit is 0. We already know from Theorem 16.5.4 that the column sequences are only as good as the sequence $\{A_m\}$ itself. Wynn [371] has given the following closed-form expression for $\varepsilon_{2n}^{(j)}$:

$$\varepsilon_{2n}^{(j)} = \frac{1}{(n+1)(j+n+\beta)}.$$

It is clear that the diagonal sequences are only slightly better than $\{A_m\}$ itself, even though they converge faster; $\varepsilon_{2n}^{(j)} \sim n^{-2}$ as $n \to \infty$. No one diagonal sequence converges more quickly than the one preceding it, however. Thus, neither the columns nor the diagonals of the epsilon table are effective for $A_m = 1/(m + \beta)$ even though $\{A_m\}$ is totally monotonic. (We note that the ε-algorithm is very unstable when applied to this sequence.)

Let us now consider the sequence $\{A_m\}$, where $A_m = \zeta^{m+1}/(1 - \zeta^{m+1})$, $m = 0, 1, \ldots$, where $0 < \zeta < 1$. It is easy to see that $A_m = \sum_{k=1}^{\infty} \zeta^{k(m+1)}$, which by Lemma 16.6.3 is totally monotonic. Therefore, Theorem 16.6.7 applies. We know from Corollary 16.4.4 that $\varepsilon_{2n}^{(j)} = O(\zeta^{(n+1)j})$ as $j \to \infty$; that is, each column of the epsilon table converges (to 0) faster than the one preceding it. Even though we have no convergence theory for diagonal sequences that is analogous to Theorem 16.4.3, in this example we have a closed-form expression for $\varepsilon_{2n}^{(j)}$, namely,

$$\varepsilon_{2n}^{(j)} = A_{j(n+1)+n(n+2)}.$$

This expression is the same as that given by Brezinski and Redivo Zaglia [41, p. 288] for the epsilon table of the sequence $\{S_m\}$, where $S_0 = 1$ and $S_{m+1} = 1 + a/S_m$ for some $a \notin (-\infty - 1/4]$. It follows that $\varepsilon_{2n}^{(j)} \sim \zeta^{j(n+1)+n(n+2)+1}$ as $n \to \infty$, so that $\lim_{n\to\infty} \varepsilon_{2n}^{(j)} = 0$. In other words, the diagonal sequences $\{\varepsilon_{2n}^{(j)}\}_{n=0}^{\infty}$ converge superlinearly, while $\{A_m\}_{m=0}^{\infty}$ converges only linearly. Also, each diagonal converges more quickly than the one preceding it.

We end this section by mentioning that the remark following the proof of Theorem 16.6.7 applies to sequences of partial sums of the convergent series $\sum_{i=0}^{\infty} c_i x^i$ for which $\{c_m\} \in TM$ with $\lim_{m\to\infty} c_m = 0$ and $x > 0$. This is so because $A_m = \sum_{i=0}^{m} c_i x^i = A - R_m$, where $R_m = \sum_{i=m+1}^{\infty} c_i x^i$, and $\{R_m\} \in TM$. Here, A is the sum of $\sum_{i=0}^{\infty} c_i x^i$. An example of this is the Maclaurin series of $-\log(1 - x) = \sum_{i=1}^{\infty} x^i/i$.

16.6.3 Totally Oscillating Sequences

Definition 16.6.8 A sequence $\{\mu_m\}_{m=0}^{\infty}$ is said to be *totally oscillating* if $\{(-1)^m \mu_m\}_{m=0}^{\infty}$ is totally monotonic, and we write $\{\mu_m\} \in TO$.

Lemma 16.6.9 *If $\{\mu_m\} \in TO$, then $\{(-1)^k \Delta^k \mu_m\} \in TO$ as well. If $\{\mu_m\}_{m=0}^{\infty}$ is convergent, then its limit is* 0.

The following result is analogous to Theorem 16.6.5 and follows from it.

Theorem 16.6.10 *Let $\{\mu_m\} \in TO$. Then $(-1)^{mp} H_p^{(m)}(\{\mu_s\}) \geq 0$ for all $m, p = 0, 1, \ldots$. If the function $\alpha(t)$ in $(-1)^m \mu_m = \int_0^1 t^m d\alpha(t)$, $m = 0, 1, \ldots$, has an infinite number of points of increase on $[0, 1]$, then $(-1)^{mp} H_p^{(m)}(\{\mu_s\}) > 0$ for all $m, p = 0, 1, \ldots$.*

An immediate consequence of Theorem 16.6.10 is that $\mu_m \mu_{m+2} - (\mu_{m+1})^2 \geq 0$ for all m if $\{\mu_m\} \in TO$; hence,

$$0 > \frac{\mu_1}{\mu_0} \geq \frac{\mu_2}{\mu_1} \geq \cdots \geq -1.$$

As a result, we also have that

$$\lim_{m \to \infty} \frac{\mu_{m+1}}{\mu_m} = \zeta \in [-1, 0).$$

16.6.4 The Shanks Transformation on Totally Oscillating Sequences

The following results follow from (16.1.12) and Theorem 16.6.10. We leave their proof to the interested reader.

Theorem 16.6.11 *Let $\{A_m\} \in TO$ in Theorem 16.6.6. Then $(-1)^j \varepsilon_{2n}^{(j)} \geq 0$ for each j and n. Also, both the column sequences $\{\varepsilon_{2n}^{(j)}\}_{j=0}^{\infty}$ and the diagonal sequences $\{\varepsilon_{2n}^{(j)}\}_{n=0}^{\infty}$ converge (to 0) when $\{A_m\}$ converges, the sequences $\{\varepsilon_{2n}^{(j)}\}_{n=0}^{\infty}$ being monotonic. In addition, when $\{A_m\}$ converges, $(\varepsilon_{2n}^{(j)} - A)/(A_{j+2n} - A) \to 0$ both as $j \to \infty$ and as $n \to \infty$.*

From this theorem, it is obvious that, if $A_m = c B_m + d$, $m = 0, 1 \ldots$, for some constants $c \neq 0$ and d, and if $\{B_m\} \in TO$ converges, then the Shanks transformation on $\{A_m\}$ produces approximations that satisfy $\lim_{j \to \infty} \varepsilon_{2n}^{(j)} = d$ and $\lim_{n \to \infty} \varepsilon_{2n}^{(j)} = d$. For each n, the sequences $\{\varepsilon_{2n}^{(j)}\}_{n=0}^{\infty}$ tend to d monotonically, while the sequences $\{\varepsilon_{2n}^{(j)}\}_{j=0}^{\infty}$ oscillate about d. (Note that $\lim_{m \to \infty} A_m = d$ because $\lim_{m \to \infty} B_m = 0$.)

It must be noted that Theorem 16.6.11, just as Theorem 16.6.7, gives almost no information about rates of convergence or convergence acceleration. Again, to obtain results in this direction, more conditions need to be imposed on $\{A_m\}$.

16.7 Modifications of the ε-Algorithm

As we saw in Theorem 16.5.4, the Shanks transformation is not effective on logarithmic sequences in $\mathbf{b}^{(1)}/\text{LOG}$. For such sequences, Vanden Broeck and Schwartz [344] suggest modifying the ε-algorithm by introducing a parameter η that can be complex in general. This modification reads as follows:

$$\varepsilon_{-1}^{(j)} = 0 \text{ and } \varepsilon_0^{(j)} = A_j, \ j = 0, 1, \ldots,$$

$$\varepsilon_{2n+1}^{(j)} = \eta \varepsilon_{2n-1}^{(j+1)} + \frac{1}{\varepsilon_{2n}^{(j+1)} - \varepsilon_{2n}^{(j)}}, \text{ and}$$

$$\varepsilon_{2n+2}^{(j)} = \varepsilon_{2n}^{(j+1)} + \frac{1}{\varepsilon_{2n+1}^{(j+1)} - \varepsilon_{2n+1}^{(j)}}, \ n, j = 0, 1, \ldots . \tag{16.7.1}$$

When $\eta = 1$ the ε-algorithm of Wynn is recovered, whereas when $\eta = 0$ the iterated Δ^2-process is obtained. Even though the resulting method is defined only via the recursion relations of (16.7.1) when η is arbitrary, it is easy to show that it is quasi-linear in the sense of the Introduction.

As with the ε-algorithm, with the present algorithm too, the $\varepsilon_k^{(j)}$ with odd k can be eliminated, and this results in the identity

$$\frac{1}{\varepsilon_{2n+2}^{(j-1)} - \varepsilon_{2n}^{(j)}} + \frac{\eta}{\varepsilon_{2n-2}^{(j+1)} - \varepsilon_{2n}^{(j)}} = \frac{1}{\varepsilon_{2n}^{(j-1)} - \varepsilon_{2n}^{(j)}} + \frac{1}{\varepsilon_{2n}^{(j+1)} - \varepsilon_{2n}^{(j)}}. \tag{16.7.2}$$

Note that the new $\varepsilon_2^{(j)}$ and Wynn's $\varepsilon_2^{(j)}$ are identical, but the other $\varepsilon_{2n}^{(j)}$ are not. The following results are due to Barber and Hamer [17].

Theorem 16.7.1 *Let $\{A_m\}$ be in $\mathbf{b}^{(1)}$/LOG in the notation of Definition 15.3.2. With $\eta = -1$ in (16.7.1), we have*

$$\varepsilon_4^{(j)} - A = O(j^{\gamma-2}) \ as \ j \to \infty.$$

Theorem 16.7.2 *Let $\{A_m\}$ be such that*

$$A_m = A + C(-1)^m \binom{\delta}{m}, \quad m = 0, 1, \dots ; \ \delta \neq 0, \pm 1, \pm 2, \dots .$$

Then, with $\eta = -1$ in (16.7.1), $\varepsilon_4^{(j)} = A$ for all j. (Because $A_m - A \sim [C/\Gamma(-\delta)]m^{-\delta-1}$ as $m \to \infty$, $A = \lim_{m\to\infty} A_m$ when $\Re\delta > -1$. Otherwise, A is the antilimit of $\{A_m\}$.)

We leave the proof of these to the reader.

Vanden Broeck and Schwartz [344] demonstrated by numerical examples that their modification of the ε-algorithm is effective on some strongly divergent sequences when η is tuned properly. We come back to this in Chapter 18 on generalizations of Padé approximants.

Finally, we can replace η in (16.7.1) by η_k; that is, a different value of η can be used to generate each column in the epsilon table.

Further modifications of the ε-algorithm suitable for some special sequences $\{A_m\}$ that behave logarithmically have been devised in the works of Sedogbo [262] and Sablonnière [247]. These modifications are of the form

$$\varepsilon_{-1}^{(j)} = 0 \ \text{and} \ \varepsilon_0^{(j)} = A_j, \quad j = 0, 1, \dots ,$$

$$\varepsilon_{k+1}^{(j)} = \varepsilon_{k-1}^{(j+1)} + \frac{g_k}{\varepsilon_{2n+1}^{(j+1)} - \varepsilon_{2n+1}^{(j)}}, \quad k, j = 0, 1, \dots . \tag{16.7.3}$$

Here g_k are scalars that depend on the asymptotic expansion of the A_m. For the details we refer the reader to [262] and [247].

17

The Padé Table

17.1 Introduction

In the preceding chapter, we saw that the approximations $e_n(A_j)$ obtained by applying the Shanks transformation to $\{A_m\}$, the sequence of the partial sums of the formal power series $\sum_{k=0}^{\infty} c_k z^k$, are Padé approximants corresponding to this power series. Thus, Padé approximants are a very important tool that can be used in effective summation of power series, whether convergent or divergent. They have been applied as a convergence acceleration tool in diverse engineering and scientific disciplines and they are related to various topics in classical analysis as well as to different methods of numerical analysis, such as continued fractions, the moment problem, orthogonal polynomials, Gaussian integration, the qd-algorithm, and some algorithms of numerical linear algebra. As such, they have been the subject of a large number of papers and books. For these reasons, we present a brief survey of Padé approximants in this book. For more information and extensive bibliographies, we refer the reader to the books by Baker [15], Baker and Graves-Morris [16], and Gilewicz [99], and to the survey paper by Gragg [106]. The bibliography by Brezinski [39] contains over 6000 items. For the subject of continued fractions, see the books by Perron [229], Wall [348], Jones and Thron [144], and Lorentzen and Waadeland [179]. See also Henrici [132, Chapter 12]. For a historical survey of continued fractions, see the book by Brezinski [38].

We start with the modern definition of Padé approximants due to Baker [15].

Definition 17.1.1 Let $f(z) := \sum_{k=0}^{\infty} c_k z^k$, whether this series converges or diverges. The $[m/n]$ *Padé approximant* corresponding to $f(z)$, if it exists, is the rational function $f_{m,n}(z) = P_{m,n}(z)/Q_{m,n}(z)$, where $P_{m,n}(z)$ and $Q_{m,n}(z)$ are polynomials in z of degree at most m and n, respectively, such that $Q_{m,n}(0) = 1$ and

$$f_{m,n}(z) - f(z) = O(z^{m+n+1}) \text{ as } z \to 0. \tag{17.1.1}$$

It is customary to arrange the $f_{m,n}(z)$ in a two-dimensional array, which has been called the *Padé table* and which looks as shown in Table 17.1.1.

The following uniqueness theorem is quite easy to prove.

Theorem 17.1.2 *If $f_{m,n}(z)$ exists, it is unique.*

Table 17.1.1: *The Padé table*

[0/0]	[1/0]	[2/0]	[3/0]	$\cdots$
[0/1]	[1/1]	[2/1]	[3/1]	$\cdots$
[0/2]	[1/2]	[2/2]	[3/2]	$\cdots$
[0/3]	[1/3]	[2/3]	[3/3]	$\cdots$
$\vdots$	$\vdots$	$\vdots$	$\vdots$	$\ddots$

As follows from (17.1.1), the polynomials $P_{m,n}(z)$ and $Q_{m,n}(z)$ also satisfy

$$Q_{m,n}(z)f(z) - P_{m,n}(z) = O(z^{m+n+1}) \text{ as } z \to 0, \quad Q_{m,n}(0) = 1. \quad (17.1.2)$$

If we now substitute $P_{m,n}(z) = \sum_{i=0}^{m} a_i z^i$ and $Q_{m,n}(z) = \sum_{i=0}^{n} b_i z^i$ in (17.1.2), we obtain the linear system

$$\sum_{j=0}^{\min(i,n)} c_{i-j} b_j = a_i, \quad i = 0, 1, \dots, m, \quad (17.1.3)$$

$$\sum_{j=0}^{\min(i,n)} c_{i-j} b_j = 0, \quad i = m+1, \dots, m+n; \quad b_0 = 1. \quad (17.1.4)$$

Obviously, the b_i can be obtained by solving the equations in (17.1.4). With the b_i available, the a_i can be obtained from (17.1.3). Using Cramer's rule to express the b_i, it can be shown that $f_{m,n}(z)$ has the determinant representation

$$f_{m,n}(z) = \frac{u_{m,n}(z)}{v_{m,n}(z)} = \frac{\begin{vmatrix} z^n S_{m-n}(z) & z^{n-1} S_{m-n+1}(z) & \cdots & z^0 S_m(z) \\ c_{m-n+1} & c_{m-n+2} & \cdots & c_{m+1} \\ c_{m-n+2} & c_{m-n+3} & \cdots & c_{m+2} \\ \vdots & \vdots & & \vdots \\ c_m & c_{m+1} & \cdots & c_{m+n} \end{vmatrix}}{\begin{vmatrix} z^n & z^{n-1} & \cdots & z^0 \\ c_{m-n+1} & c_{m-n+2} & \cdots & c_{m+1} \\ c_{m-n+2} & c_{m-n+3} & \cdots & c_{m+2} \\ \vdots & \vdots & & \vdots \\ c_m & c_{m+1} & \cdots & c_{m+n} \end{vmatrix}}, \quad (17.1.5)$$

where $S_p(z) = \sum_{k=0}^{p} c_k z^k$, $p = 0, 1, \dots$, and $c_k = 0$ for $k < 0$.

From $Q_{m,n}(z) = \sum_{i=0}^{n} b_i z^i$ and from (17.1.5), we see that

$$f_{m,n}(z) = \frac{\sum_{i=0}^{n} b_i z^i S_{m-i}(z)}{\sum_{i=0}^{n} b_i z^i}, \quad (17.1.6)$$

that is, once the b_i have been determined, $f_{m,n}(z)$ can be determined completely without actually having to determine the a_i.

17.2 Algebraic Structure of the Padé Table

Definition 17.2.1 We say that the Padé approximant $f_{m,n}(z)$ is *normal* if it occurs exactly once in the Padé table. We say that the Padé table is *normal* if all its entries are normal; that is, no two entries are equal.

Let us define

$$C_{m,0} = 1; \ C_{m,n} = \begin{vmatrix} c_{m-n+1} & c_{m-n+2} & \cdots & c_m \\ c_{m-n+2} & c_{m-n+3} & \cdots & c_{m+1} \\ \vdots & \vdots & & \vdots \\ c_m & c_{m+1} & \cdots & c_{m+n-1} \end{vmatrix}, \ m \geq 0, \ n \geq 1, \quad (17.2.1)$$

where $c_k = 0$ for $k < 0$, as before. Thus, we also have that

$$C_{0,n} = (-1)^{n(n-1)/2} c_0^n, \ n \geq 0. \quad (17.2.2)$$

Let us arrange the determinants $C_{m,n}$ in a two-dimensional table, called the *C-table*, in complete analogy to the Padé table itself. The first row of this table is occupied by $C_{m,0} = 1$, $m = 0, 1, \ldots$, and the second row is occupied by $C_{m,1} = c_m$, $m = 0, 1, \ldots$. The first column is occupied by $C_{0,n} = (-1)^{n(n-1)/2} c_0^n$, $n = 0, 1, \ldots$. When $C_{m,n} \neq 0$ for all m and n, the rest of the C-table can be computed recursively by using the Frobenius identity

$$C_{m,n+1} C_{m,n-1} = C_{m+1,n} C_{m-1,n} - (C_{m,n})^2. \quad (17.2.3)$$

This identity can be proved by applying the Sylvester determinant identity to $C_{m,n+1}$.

The normality of the Padé approximants is closely related to the zero structure of the C-table as the following theorems show.

Theorem 17.2.2 *The following statements are equivalent:*

(i) $f_{m,n}(z)$ *is normal.*
(ii) *The numerator $P_{m,n}(z)$ and the denominator $Q_{m,n}(z)$ of $f_{m,n}(z)$ have degrees exactly m and n, respectively, and the expansion of $Q_{m,n}(z) f(z) - P_{m,n}(z)$ begins exactly with the power z^{m+n+1}.*
(iii) *The determinants $C_{m,n}$, $C_{m,n+1}$, $C_{m+1,n}$, and $C_{m+1,n+1}$ do not vanish.*

Theorem 17.2.3 *A necessary and sufficient condition for the Padé table to be normal is that $C_{m,n} \neq 0$ for all m and n. In particular, $C_{m,1} = c_m \neq 0$, $m = 0, 1, \ldots$, must hold.*

Theorem 17.2.4 *In case the Padé table is not normal, the vanishing $C_{m,n}$ in the C-table appear in square blocks, which are entirely surrounded by nonzero entries (except at infinity). For the $r \times r$ block of zeros in the C-table, say, $C_{\mu,\nu} = 0$, $m + 1 \leq \mu \leq m + r$, $n + 1 \leq \nu \leq n + r$, we have that $f_{\mu,\nu}(z) = f_{m,n}(z)$ for $\mu \geq m$, $\nu \geq n$, and $\mu + \nu \leq m + n + r$, while the rest of the $f_{\mu,\nu}(z)$ in the $(r + 1) \times (r + 1)$ block, $m \leq \mu \leq m + r$, $n \leq \nu \leq n + r$, do not exist.*

Theorem 17.2.5 *Let $\sum_{k=0}^{\infty} c_k z^k$ be the Maclaurin series of the rational function $R(z) = p(z)/q(z)$, where $p(z)$ and $q(z)$ are polynomials of degree exactly m and n, respectively, and have no common factor. Then, $f_{\mu,\nu}(z) = R(z)$ for all $\mu \geq m$ and $\nu \geq n$. In this case, the C-table has an infinite zero block.*

The next theorem shows the existence of an infinite number of Padé approximants (i) along any row, (ii) along any column, and (iii) along any diagonal, and it is proved by considering the zero structure of the C-table.

Theorem 17.2.6 *Given a formal power series $\sum_{k=0}^{\infty} c_k z^k$ with $c_0 \neq 0$, in its corresponding Padé table, there exist infinite sequences (i) $\{f_{\mu_i,n}(z)\}_{i=0}^{\infty}$ for arbitrary fixed n, (ii) $\{f_{m,\nu_i}(z)\}_{i=0}^{\infty}$ for arbitrary fixed m, and (iii) $\{f_{\lambda_i+k,\lambda_i}(z)\}_{i=0}^{\infty}$ for arbitrary fixed k.*

The Padé table enjoys some very interesting duality and invariance properties that are very easy to prove.

Theorem 17.2.7 *Let $f(z) := \sum_{k=0}^{\infty} c_k z^k$ be a formal power series with $c_0 \neq 0$. Denote $g(z) = 1/f(z) := \sum_{k=0}^{\infty} d_k z^k$. [That is, d_k are the solution of the triangular system of equations $c_0 d_0 = 1$ and $\sum_{j=0}^{k} c_{k-j} d_j = 0$, $k = 1, 2, \ldots$.] If $f_{\mu,\nu}(z)$ and $g_{\mu,\nu}(z)$ denote the $[\mu/\nu]$ Padé approximants corresponding to $f(z)$ and $g(z)$, respectively, and if $f_{m,n}(z)$ exists, then $g_{n,m}(z)$ exists and $g_{n,m}(z) = 1/f_{m,n}(z)$.*

Corollary 17.2.8 *If $1/f(z) = f(-z)$, then $f_{m,n}(-z) = 1/f_{n,m}(z)$. Consequently, if $f_{m,n}(z) = P_{m,n}(z)/Q_{m,n}(z)$ with $Q_{m,n}(0) = 1$, then $P_{m,n}(z) = c_0 Q_{n,m}(-z)$ as well.*

A most obvious application of this corollary is to $f(z) = e^z$. Also, $1/f(z) = f(-z)$ holds when $f(z) = s(z)/s(-z)$ with any $s(z)$.

Theorem 17.2.9 *Let $f(z) := \sum_{k=0}^{\infty} c_k z^k$ be a formal power series and let $g(z) = [A + Bf(z)]/[C + Df(z)] := \sum_{k=0}^{\infty} d_k z^k$. Then, provided $C + Dc_0 \neq 0$ and provided $f_{m,m}(z)$ exists, $g_{m,m}(z)$ exists and there holds $g_{m,m}(z) = [A + Bf_{m,m}(z)]/[C + Df_{m,m}(z)]$.*

The result of the following theorem is known as *homographic invariance under an argument transformation.*

Theorem 17.2.10 *Let $f(z) := \sum_{k=0}^{\infty} c_k z^k$ be a formal power series. Define the origin-preserving transformation $w = Az/(1 + Bz)$, $A \neq 0$, and let $g(w) \equiv f(z) = f(w/(A - Bw)) := \sum_{k=0}^{\infty} d_k w^k$. If $f_{\mu,\nu}(z)$ and $g_{\mu,\nu}(w)$ are Padé approximants corresponding to $f(z)$ and $g(w)$, respectively, then $g_{m,m}(w) = f_{m,m}(z)$, provided $f_{m,m}(z)$ exists.*

17.3 Padé Approximants for Some Hypergeometric Functions

As can be concluded from the developments of Section 17.1, in general, it is impossible to obtain closed-form expressions for Padé approximants. Such expressions are possible for some special cases involving hypergeometric series, however. In this section, we treat a few such cases by using a technique of the author that avoids the pain of going through the determinant representation given in (17.1.5). Lemmas 17.3.1 and 17.3.2 that follow are due to Sidi [277] and they form the main tools of this technique. We provide the proof of the first lemma only; that of the second is similar. We make use of Lemma 17.3.1 again in Chapter 19 in the derivation of a simple closed-form expression for the $\mathcal{S}$-transformation of Sidi.

Lemma 17.3.1 *Let T and γ_i, $i = 0, 1, \ldots, n-1$, be defined by the linear equations*

$$A_r = b_r T + c_r \left(\gamma_0 + \sum_{i=1}^{n-1} \frac{\gamma_i}{\alpha + r + i - 1} \right), \quad r = k, k+1, \ldots, k+n, \quad (17.3.1)$$

or

$$A_r = b_r T + c_r \sum_{i=0}^{n-1} \frac{\gamma_i}{(\alpha + r)_i}, \quad r = k, k+1, \ldots, k+n, \quad (17.3.2)$$

where $(x)_0 = 1$ and $(x)_i = x(x+1) \cdots (x+i-1)$, $i \geq 1$. Provided $\alpha + r + i - 1 \neq 0$ for $1 \leq i \leq n-1$ and $k \leq r \leq k+n$, T is given by

$$T = \frac{\Delta^n ((\alpha+k)_{n-1} A_k / c_k)}{\Delta^n ((\alpha+k)_{n-1} b_k / c_k)} = \frac{\sum_{j=0}^n (-1)^j \binom{n}{j} (\alpha+k+j)_{n-1} A_{k+j} / c_{k+j}}{\sum_{j=0}^n (-1)^j \binom{n}{j} (\alpha+k+j)_{n-1} b_{k+j} / c_{k+j}}. \quad (17.3.3)$$

Proof. Let us multiply both sides of (17.3.1) and (17.3.2) by $(\alpha+r)_{n-1}/c_r$. We obtain

$$(\alpha+r)_{n-1} (A_r / c_r) = (\alpha+r)_{n-1} (b_r / c_r) T + P(r), \quad r = k, k+1, \ldots, k+n,$$

where $P(x)$ is a polynomial in x of degree at most $n-1$. Next, let us write $r = k + j$ in this equation and multiply both sides by $(-1)^{n-j} \binom{n}{j}$ and sum from $j = 0$ to $j = n$ (the number of equations is exactly $n+1$). By the fact that $\Delta^n f_k = \sum_{j=0}^n (-1)^{n-j} \binom{n}{j} f_{k+j}$, this results in

$$\Delta^n ((\alpha+k)_{n-1} A_k / c_k) = T \Delta^n ((\alpha+k)_{n-1} b_k / c_k) + \Delta^n P(k).$$

Now, $\Delta^n P(k) = 0$ because the polynomial $P(x)$ is of degree at most $n-1$. The result now follows. ∎

Lemma 17.3.2 *Let T and γ_i, $i = 0, 1, \ldots, n-1$, be defined by the linear equations*

$$A_r = b_r T + c_r \sum_{i=0}^{n-1} \gamma_i p_i(r), \quad r = k, k+1, \ldots, k+n, \quad (17.3.4)$$

where $p_i(x)$ are polynomials in x, each of degree $\leq n - 1$. Then T is given by

$$T = \frac{\Delta^n(A_k/c_k)}{\Delta^n(b_k/c_k)} = \frac{\sum_{j=0}^n (-1)^j \binom{n}{j} A_{k+j}/c_{k+j}}{\sum_{j=0}^n (-1)^j \binom{n}{j} b_{k+j}/c_{k+j}}. \tag{17.3.5}$$

Let us now turn to Padé approximants corresponding to $f(z) := \sum_{k=0}^\infty c_k z^k$. From the determinantal expression for $f_{m,n}(z)$ given in (17.1.5), it follows that $f_{m,n}(z)$, along with the additional parameters $\delta_1, \dots, \delta_n$, satisfies

$$z^{m-r} S_r(z) = z^{m-r} f_{m,n}(z) + \sum_{i=1}^n \delta_i c_{r+i}, \quad m - n \leq r \leq m, \tag{17.3.6}$$

where $S_p(z) = \sum_{k=0}^p c_k z^k$, $p = 0, 1, \dots$.

When c_k have the appropriate form, we can solve (17.3.6) for $f_{m,n}(z)$ in closed form with the help of Lemmas 17.3.1 and 17.3.2.

Example 17.3.3 $f(z) = {}_1F_1(1; \beta; z) = \sum_{j=0}^\infty z^j/(\beta)_j$, $\beta \neq 0, -1, -2, \dots$. Using the fact that $(\beta)_{p+q} = (\beta)_p(\beta + p)_q$, the equations in (17.3.6) can be written in the form

$$z^{m-r} S_r(z) = z^{m-r} f_{m,n}(z) + \frac{1}{(\beta)_{r+1}} \sum_{i=0}^{n-1} \frac{\gamma_i}{(\beta + r + 1)_i}, \quad m - n \leq r \leq m.$$

Thus, Lemma 17.3.1 applies, and after some manipulation it can be shown that

$$f_{m,n}(z) = \frac{\sum_{j=0}^n (-1)^j \binom{n}{j} (\beta)_{m+j} z^{n-j} S_{m-n+j}(z)}{\sum_{j=0}^n (-1)^j \binom{n}{j} (\beta)_{m+j} z^{n-j}}.$$

For $\beta = 1$, we have $f(z) = e^z$, and Corollary 17.2.8 also applies. Thus, we have

$$f_{m,n}(z) = \frac{\sum_{j=0}^m \binom{m}{j} (m + n - j)! z^j}{\sum_{j=0}^n \binom{n}{j} (m + n - j)! (-z)^j}$$

as the $[m/n]$ Padé approximant for $e^z = \sum_{k=0}^\infty z^k/k!$ with very little effort.

Example 17.3.4 $f(z) = {}_2F_1(1, \mu; \beta; z) = \sum_{j=0}^\infty [(\mu)_j/(\beta)_j] z^j$, $\mu, \beta \neq 0, -1, -2, \dots$. Again, using the fact that $(x)_{p+q} = (x)_p(x + p)_q$, we have in (17.3.6) that

$$\sum_{i=1}^n \delta_i c_{r+i} = \frac{(\mu)_{r+1}}{(\beta)_{r+1}} \sum_{i=1}^n \delta_i \frac{(\mu + r + 1)_{i-1}}{(\beta + r + 1)_{i-1}}$$

$$= \frac{(\mu)_{r+1}}{(\beta)_{r+1}} \left(\gamma_0 + \sum_{i=1}^{n-1} \frac{\gamma_i}{\beta + r + i} \right) \quad \text{for some } \gamma_i.$$

The last equality results from the partial fraction decomposition of the rational function $\sum_{i=1}^n \delta_i(\mu + r + 1)_{i-1}/(\beta + r + 1)_{i-1}$. Thus, $f_{m,n}(z)$ satisfies

$$z^{m-r} S_r(z) = z^{m-r} f_{m,n}(z) + \frac{(\mu)_{r+1}}{(\beta)_{r+1}} \left(\gamma_0 + \sum_{i=1}^{n-1} \frac{\gamma_i}{\beta + r + i} \right), \quad m - n \leq r \leq m.$$

Therefore, Lemma 17.3.1 applies, and we obtain

$$f_{m,n}(z) = \frac{\sum_{j=0}^{n} (-1)^j \binom{n}{j} \frac{(\beta)_{m+j}}{(\mu)_{m-n+j+1}} z^{n-j} S_{m-n+j}(z)}{\sum_{j=0}^{n} (-1)^j \binom{n}{j} \frac{(\beta)_{m+j}}{(\mu)_{m-n+j+1}} z^{n-j}}$$

Special cases of this are obtained by assigning to μ and β specific values:

(i) For $\beta = 1$, $f(z) = {}_1F_0(\mu; z) = (1 - z)^{-\mu}$.
(ii) For $\beta = \mu + 1$, $f(z) = \mu \sum_{k=0}^{\infty} z^k/(\mu + k)$. With $\mu = 1$, this reduces to $f(z) = -z^{-1} \log(1 - z)$. With $\mu = 1/2$, it reduces to $f(z) = z^{-1/2} \tanh^{-1} z^{1/2}$.

Example 17.3.5 $f(z) = {}_2F_0(1, \mu; z) = \sum_{j=0}^{\infty} (\mu)_j z^j$, $\mu \neq 0, -1, -2, \ldots$. Again, using the fact that $(x)_{p+q} = (x)_p (x + p)_q$, we have in (17.3.6) that

$$\sum_{i=1}^{n} \delta_i c_{r+i} = (\mu)_{r+1} \sum_{i=1}^{n} \delta_i (\mu + r + 1)_{i-1},$$

the summation on the right being a polynomial in r of degree at most $n - 1$. Thus, $f_{m,n}(z)$ satisfies

$$z^{m-r} S_r(z) = z^{m-r} f_{m,n}(z) + (\mu)_{r+1} \sum_{i=0}^{n-1} \gamma_i r^i, \quad m - n \leq r \leq m.$$

Therefore, Lemma 17.3.2 applies, and we obtain

$$f_{m,n}(z) = \frac{\sum_{j=0}^{n} (-1)^j \binom{n}{j} \frac{1}{(\mu)_{m-n+j+1}} z^{n-j} S_{m-n+j}(z)}{\sum_{j=0}^{n} (-1)^j \binom{n}{j} \frac{1}{(\mu)_{m-n+j+1}} z^{n-j}}.$$

Setting $\mu = 1$, we obtain the Padé approximant for the Euler series, which is related to the *exponential integral* $E_1(-z^{-1})$, where $E_1(\zeta) = \int_{\zeta}^{\infty} (e^{-t}/t) dt$ and $E_1(\zeta) \sim (e^{-\zeta}/\zeta) \sum_{k=0}^{\infty} (-1)^k (k!) \zeta^{-k}$ as $\Re \zeta \to +\infty$.

Some of the preceding examples have also been treated by different techniques by Luke [193] and by Iserles [139]. Padé approximants from so-called q-series have been obtained in closed form by Wynn [373]. For additional treatment of q-elementary functions, see Borwein [28].

17.4 Identities in the Padé Table

Various identities relate neighboring entries in the Padé table. We do not intend to go into these in any detail here. We are content to derive one of the so-called *two-term identities* to which we alluded in our discussion of the Shanks transformation in the previous chapter. There, we mentioned, in particular, that the four identities (16.6.1)–(16.6.4) among the different entries of the epsilon table could be derived from these two-term identities.

Consider $f_{\mu,\nu}(z) = u_{\mu,\nu}(z)/v_{\mu,\nu}(z)$, with $u_{\mu,\nu}(z)$ and $v_{\mu,\nu}(z)$ being, respectively, the numerator and denominator determinants in (17.1.5). Now, by (17.1.1) we have $f_{m+1,n}(z) - f_{m,n+1}(z) = O(z^{m+n+2})$ as $z \to 0$. Next, we have $f_{m+1,n}(z) - f_{m,n+1}(z) = N(z)/D(z)$, where $N(z) = u_{m+1,n}(z)v_{m,n+1}(z) - u_{m,n+1}(z)v_{m+1,n}(z)$ and $D(z) = v_{m+1,n}(z)v_{m,n+1}(z)$. Because $N(z)$ is a polynomial of degree at most $m + n + 2$ and $D(0) \neq 0$, we thus have that $N(z) = \alpha z^{m+n+2}$ for some constant α. Obviously, α is the product of the leading coefficients of $u_{m+1,n}(z)$ and $v_{m,n+1}(z)$, which are both $C_{m+1,n+1}$. Summarizing, we have the two-term identity

$$ f_{m+1,n}(z) - f_{m,n+1}(z) = \frac{(C_{m+1,n+1})^2 z^{m+n+2}}{v_{m+1,n}(z)v_{m,n+1}(z)}. \tag{17.4.1} $$

Note that the identity in (16.6.4) pertaining to the epsilon table is obtained directly from the two-term identity in (17.4.1) by recalling Theorem 16.1.3, which relates the Shanks transformation to the Padé table.

Using the approach that led to (17.4.1), we can obtain additional two-term identities involving $f_{m+1,n+1}(z) - f_{m,n}(z)$, $f_{m+1,n}(z) - f_{m,n}(z)$, and $f_{m,n+1}(z) - f_{m,n}(z)$, from which we can obtain the identities given in (16.6.1)–(16.6.3). We leave the details to the interested reader.

For additional identities involving more entries in the Padé table, see Baker [15].

We close this section by stating the *five-term identity* that follows directly from Wynn's cross rule, which we encountered in the previous chapter. This identity reads

$$ \frac{1}{f_{m+1,n}(z) - f_{m,n}(z)} + \frac{1}{f_{m-1,n}(z) - f_{m,n}(z)} = $$

$$ \frac{1}{f_{m,n+1}(z) - f_{m,n}(z)} + \frac{1}{f_{m,n-1}(z) - f_{m,n}(z)}, \tag{17.4.2} $$

and is known as *Wynn's identity*. Of course, it is valid when $f_{m,n}(z)$ is normal.

17.5 Computation of the Padé Table

Various methods have been developed for computing the Padé table corresponding to a given power series $f(z) := \sum_{k=0}^{\infty} c_k z^k$ when this table is normal. The most straightforward of these methods is provided by Wynn's five-term identity given in (17.4.2). Obviously, this is a recursive method that enables us to compute the Padé table row-wise or columnwise. The initial conditions for this method are $f_{m,0}(z) = S_m(z)$, $f_{m,1}(z) = S_{m-1}(z) + c_m z^m/(1 - (c_{m+1}/c_m)z)$, $m = 0, 1, \ldots$, and $f_{0,n}(z) = 1/T_n(z)$, $n = 0, 1, \ldots$, where $S_p(z)$ and $T_p(z)$ are the pth partial sums of $f(z) := \sum_{k=0}^{\infty} c_k z^k$ and $1/f(z) := \sum_{k=0}^{\infty} d_k z^k$, respectively. Obviously, the d_k can be obtained from $c_0 d_0 = 1$ and $\sum_{j=0}^{k} c_{k-j} d_j = 0$, $k = 1, 2, \ldots$, recursively, in the order $d_0, d_1, \ldots$. Without the $f_{0,n}(z)$, it is possible to compute half the Padé table, namely, those $f_{m,n}(z)$ with $m \geq n$. When the Padé table is not normal, we can use an extension of Wynn's identity due to Cordellier [57]. For a detailed treatment, see Baker and Graves-Morris [16].

We next present a method due to Longman [174] for recursive determination of the coefficients of the numerator polynomial $P_{m,n}(z) = \sum_{k=0}^{m} p_k^{(m,n)} z^k$ and the denominator polynomial $Q_{m,n}(z) = \sum_{k=0}^{n} q_k^{(m,n)} z^k$ of $f_{m,n}(z)$ with the normalization condition $q_0^{(m,n)} = 1$, which implies $p_0^{(m,n)} = c_0$. This method too is valid for normal tables. We start with

$$Q_{m+1,n}(z)f(z) - P_{m+1,n}(z) = O(z^{m+n+2})$$

$$Q_{m,n+1}(z)f(z) - P_{m,n+1}(z) = O(z^{m+n+2}) \qquad (17.5.1)$$

that follow from (17.1.1). Upon subtraction, we obtain from (17.5.1) that

$$\left[Q_{m,n+1}(z) - Q_{m+1,n}(z)\right] f(z) - \left[P_{m,n+1}(z) - P_{m+1,n}(z)\right] = O(z^{m+n+2}).$$

Now, both terms inside the square brackets are polynomials that vanish at $z = 0$. Dividing both sides by z, and invoking (17.1.1) again, we realize that

$$\frac{P_{m,n+1}(z) - P_{m+1,n}(z)}{\left(q_1^{(m,n+1)} - q_1^{(m+1,n)}\right)z} = P_{m,n}(z), \qquad (17.5.2)$$

$$\frac{Q_{m,n+1}(z) - Q_{m+1,n}(z)}{\left(q_1^{(m,n+1)} - q_1^{(m+1,n)}\right)z} = Q_{m,n}(z). \qquad (17.5.3)$$

Let us denote by $\{p_k, q_k\}$, $\{p_k', q_k'\}$, and $\{p_k'', q_k''\}$ the $\{p_k^{(\mu,\nu)}, q_k^{(\mu,\nu)}\}$ corresponding to $f_{m,n}(z)$, $f_{m,n+1}(z)$, and $f_{m+1,n}(z)$, respectively. Then, it follows from (17.5.2) and (17.5.3) that

$$\frac{p_k' - p_k''}{q_1' - q_1''} = p_{k-1}, \quad k = 1, \ldots, m; \quad -\frac{p_{m+1}''}{q_1' - q_1''} = p_m, \qquad (17.5.4)$$

and

$$\frac{q_k' - q_k''}{q_1' - q_1''} = q_{k-1}, \quad k = 1, \ldots, n; \quad \frac{q_{n+1}'}{q_1' - q_1''} = q_n. \qquad (17.5.5)$$

Eliminating $q_1' - q_1''$, we can rewrite (17.5.4) and (17.5.5) in the form

$$p_0' = c_0 \text{ and } p_k' = p_k'' - \frac{p_{m+1}''}{p_m} p_{k-1}, \quad k = 1, \ldots, m, \qquad (17.5.6)$$

and

$$q_0'' = 1 \text{ and } q_k'' = q_k' - \frac{q_{n+1}'}{q_n} q_{k-1}, \quad k = 1, \ldots, n. \qquad (17.5.7)$$

The $p_k^{(\mu,\nu)}$ can be computed with the help of (17.5.6) with the initial conditions $p_k^{(\mu,0)} = c_k$, $k = 0, 1, \ldots, \mu$, and the $q_k^{(\mu,\nu)}$ can be computed with the help of (17.5.7) with the initial conditions $q_k^{(0,\nu)} = \hat{d}_k$, $k = 0, 1 \ldots, \nu$, where the $\hat{d}_k$ are the Maclaurin series coefficients for $c_0/f(z)$ and thus can be computed recursively from $\hat{d}_0 = 1$ and $\hat{d}_k = -\left(\sum_{i=0}^{k-1} c_{k-i}\hat{d}_i\right)/c_0$, $k = 1, 2, \ldots$. Thus, given the terms $c_0, c_1, \ldots, c_M$, this algorithm enables us to obtain the coefficients of $f_{\mu,\nu}(z)$ for $0 \leq \mu + \nu \leq M$.

For additional methods that also treat the issue of numerical stability, see Baker and Graves-Morris [16].

Finally, Padé approximants can also be computed via their connection to continued fractions with the help of the *quotient-difference (qd) algorithm*. We turn to this in the next section.

17.6 Connection with Continued Fractions
17.6.1 Definition and Algebraic Properties of Continued Fractions

By a *continued fraction*, we mean an expression of the form

$$b_0 + K\left(\frac{a_n}{b_n}\right) \equiv b_0 + \cfrac{a_1}{b_1 + \cfrac{a_2}{b_2 + \cfrac{a_3}{b_3 + \ddots}}}, \tag{17.6.1}$$

which we write in the typographically more convenient form

$$b_0 + K\left(\frac{a_n}{b_n}\right) = b_0 + \frac{a_1}{b_1 +}\frac{a_2}{b_2 +}\frac{a_3}{b_3 +}\cdots. \tag{17.6.2}$$

Let us set

$$\frac{A_0}{B_0} = \frac{b_0}{1}, \quad \frac{A_1}{B_1} = b_0 + \frac{a_1}{b_1}, \quad \text{and}$$

$$\frac{A_n}{B_n} = b_0 + \frac{a_1}{b_1 +}\cdots\frac{a_n}{+b_n}, \quad n = 2, 3, \ldots. \tag{17.6.3}$$

We call a_n and b_n, respectively, the *nth partial numerator* and the *nth partial denominator* of $b_0 + K(a_n/b_n)$, and A_n/B_n is its *nth convergent*.

In case $a_n \neq 0$, $n = 1, \ldots, N$, $a_{N+1} = 0$, the continued fraction *terminates* and its value is A_N/B_N. If $a_n \neq 0$ and $b_n \neq 0$ for all n, then it is *infinite*. In this case, if $\lim_{n\to\infty}(A_n/B_n) = G$ exists, then we say that $b_0 + K(a_n/b_n)$ converges to G.

It is easy to show by induction that the A_n and B_n satisfy the recursion relations

$$A_{n+1} = b_{n+1}A_n + a_{n+1}A_{n-1}, \quad n = 1, 2, \ldots, \quad A_{-1} = 1, \ A_0 = b_0,$$

$$B_{n+1} = b_{n+1}B_n + a_{n+1}B_{n-1}, \quad n = 1, 2, \ldots, \quad B_{-1} = 0, \ B_0 = 1. \tag{17.6.4}$$

We call A_n and B_n, respectively, the *nth numerator* and the *nth denominator* of the continued fraction $b_0 + K(a_n/b_n)$.

By (17.6.4), we have

$$A_{n+1}B_n - A_nB_{n+1} = -a_{n+1}(A_nB_{n-1} - A_{n-1}B_n),$$

repeated application of which gives

$$A_{n+1}B_n - A_nB_{n+1} = (-1)^n a_1 a_2 \cdots a_{n+1}. \tag{17.6.5}$$

Therefore,

$$A_{n+1}/B_{n+1} - A_n/B_n = (-1)^n a_1 a_2 \cdots a_{n+1}/(B_n B_{n+1}), \tag{17.6.6}$$

and hence

$$\frac{A_n}{B_n} = b_0 + \frac{a_1}{B_0 B_1} - \frac{a_1 a_2}{B_1 B_2} + \cdots + (-1)^{n-1} \frac{a_1 a_2 \cdots a_n}{B_{n-1} B_n}. \qquad (17.6.7)$$

This last result can be used to prove some interesting convergence theorems for continued fractions. One of these theorems says that $K(1/\beta_n)$, with $\beta_n > 0$, $n = 1, 2, \ldots$, converges if and only if $\sum_{n=1}^{\infty} \beta_n$ diverges. See Henrici [132, Theorem 12.1c].

We say two continued fractions are *equivalent* if they have the same sequence of convergents. It is easy to show that, for any sequence of nonzero complex numbers $\{c_n\}_{n=0}^{\infty}$, $c_0 = 1$, the continued fractions $b_0 + K(a_n/b_n)$ and $b_0 + K(c_{n-1} c_n a_n / c_n b_n)$ are equivalent. In particular, by choosing $c_n = 1/b_n$, $n = 1, 2, \ldots$, $b_0 + K(a_n/b_n)$ becomes equivalent to $b_0 + K(a_n^*/1)$, where $a_1^* = a_1/b_1$, $a_n^* = a_n/(b_{n-1} b_n)$, $n = 2, 3, \ldots$, provided that $b_n \neq 0$, $n = 1, 2, \ldots$. Next, by choosing $c_1 = 1/a_1$ and $c_n = 1/(a_n c_{n-1})$, $n = 2, 3, \ldots$, $b_0 + K(a_n/b_n)$ becomes equivalent to $b_0 + K(1/b_n^*)$, where $b_n^* = b_n c_n$, $n = 1, 2, \ldots$.

By the *even (odd) part* of $b_0 + K(a_n/b_n)$, we mean a continued fraction whose sequence of convergents is $\{A_{2n}/B_{2n}\}_{n=0}^{\infty}$ ($\{A_{2n+1}/B_{2n+1}\}_{n=0}^{\infty}$). For the continued fraction $b_0 + K(a_n/1)$, we have by a few applications of (17.6.4)

$$C_{n+1} = (1 + a_n + a_{n+1}) C_{n-1} - a_n a_{n-1} C_{n-3}, \qquad (17.6.8)$$

where C_n stands for either A_n or B_n. Thus, the even part of $b_0 + K(a_n/1)$ is

$$b_0 + \frac{a_1}{1 + a_2} - \frac{a_2 a_3}{1 + a_3 + a_4} - \frac{a_4 a_5}{1 + a_5 + a_6} - \frac{a_6 a_7}{1 + a_7 + a_8} - \cdots \qquad (17.6.9)$$

and the odd part is

$$(b_0 + a_1) - \frac{a_1 a_2}{1 + a_2 + a_3} - \frac{a_3 a_4}{1 + a_4 + a_5} - \frac{a_5 a_6}{1 + a_6 + a_7} - \cdots. \qquad (17.6.10)$$

17.6.2 Regular C-Fractions and the Padé Table

By a *regular C-fraction*, we mean an infinite continued fraction $z^{-1} K(a_n z/1)$, namely,

$$\frac{a_1}{1} + \frac{a_2 z}{1} + \frac{a_3 z}{1} + \cdots. \qquad (17.6.11)$$

Here, z is a complex variable. A regular C-fraction for which $a_n > 0$, $n = 1, 2, \ldots$, is called an *S-fraction*. To emphasize the dependence of the nth numerator and denominator A_n and B_n on z, we denote them by $A_n(z)$ and $B_n(z)$, respectively.

By (17.6.4), it is clear that $A_n(z)$ and $B_n(z)$ are polynomials in z, and hence the convergents $A_n(z)/B_n(z)$ are rational functions. It is easy to see that $A_{2n}(z)$, $B_{2n}(z)$, $A_{2n+1}(z)$, and $B_{2n+1}(z)$ have degrees at most $n - 1$, n, n, and n, respectively. By (17.6.5), we also have

$$A_{n+1}(z) B_n(z) - A_n(z) B_{n+1}(z) = (-1)^n a_1 a_2 \cdots a_{n+1} z^n. \qquad (17.6.12)$$

From this and from the fact that $A_n(0) = a_1 \neq 0$ and $B_n(0) = 1$ for all n, it can be shown

Table 17.6.1: *The staircase in the Padé*
table from the C-fraction of (17.6.11)

[0/0]		
[0/1]	[1/1]	
	[1/2]	[2/2]
		[2/3] ...
		... ...

that $A_n(z)$ and $B_n(z)$ have no common zeros. It is also clear that

$$\frac{A_{2n+1}(z)}{B_{2n+1}(z)} - \frac{A_{2n}(z)}{B_{2n}(z)} = O(z^{2n}) \text{ as } z \to 0, \tag{17.6.13}$$

and

$$\frac{A_{2n+2}(z)}{B_{2n+2}(z)} - \frac{A_{2n+1}(z)}{B_{2n+1}(z)} = O(z^{2n+1}) \text{ as } z \to 0. \tag{17.6.14}$$

From these, it follows that the Maclaurin expansion of the convergent $A_{2n}(z)/B_{2n}(z)$ has the form $\sum_{i=0}^{2n-1} c_i z^i + O(z^{2n})$, whereas that of $A_{2n+1}(z)/B_{2n+1}(z)$ has the form $\sum_{i=0}^{2n} c_i z^i + O(z^{2n+1})$, where the coefficients c_i are independent of n for all n that satisfies $2n \geq i$. Thus, we conclude that $A_{2n+1}(z)/B_{2n+1}(z)$ is the $[n/n]$ Padé approximant corresponding to the formal power series $\sum_{i=0}^{\infty} c_i z^i$, and $A_{2n}(z)/B_{2n}(z)$ is the $[n-1/n]$ Padé approximant. In other words, the convergents $A_r(z)/B_r(z)$, $r = 1, 2, \dots$, of the regular C-fraction $z^{-1} K(a_n z/1)$ form the *staircase* in the Padé table of $\sum_{i=0}^{\infty} c_i z^i$, which is shown in Table 17.6.1.

We now consider the converse problem: Given a formal power series $f(z) := \sum_{k=0}^{\infty} c_k z^k$, does there exists a regular C-fraction whose convergents are Padé approximants? The following theorem, whose proof can be found in Henrici [132, Theorem 12.4c], summarizes everything.

Theorem 17.6.1 *Given a formal power series $f(z) := \sum_{k=0}^{\infty} c_k z^k$, there exists at most one regular C-fraction corresponding to $f(z)$. There exists precisely one such fraction if and only if $H_n^{(m)}(\{c_s\}) \neq 0$ for $m = 0, 1$ and $n = 1, 2, \dots$. [Here $H_n^{(m)}(\{c_s\})$ are Hankel determinants defined in (16.1.13).]*

Theorem 17.6.1 and the developments preceding it form the basis of the qd-algorithm to which we alluded in the previous section.

Assuming that $H_n^{(m)}(\{c_s\}) \neq 0$ for $m = 0, 1$ and $n = 1, 2, \dots$, we now know that, given $f(z) := \sum_{k=0}^{\infty} c_k z^k$, there exists a *corresponding regular C-fraction*, which we choose to write as

$$F_0(z) := \frac{c_0}{1} - \frac{q_1^{(0)} z}{1} - \frac{e_1^{(0)} z}{1} - \frac{q_2^{(0)} z}{1} - \frac{e_2^{(0)} z}{1} - \dots . \tag{17.6.15}$$

Similarly, if also $H_n^{(m)}(\{c_s\}) \neq 0$ for all $m = 2, 3, \dots$, and $n = 1, 2, \dots$, there exist

Table 17.6.2: *The staircase in the Padé table*
from the C-fraction of (17.6.16)

$[k-1/0]$	$[k/0]$		
	$[k/1]$	$[k+1/1]$	
		$[k+1/2]$	$[k+2/2]$
		$\cdots$	$\cdots$

corresponding regular C-fractions of the form

$$F_k(z) := \sum_{i=0}^{k-1} c_i z^i + \frac{c_k z^k}{1} - \frac{q_1^{(k)} z}{1} - \frac{e_1^{(k)} z}{1} - \frac{q_2^{(k)} z}{1} - \frac{e_2^{(k)} z}{1} - \cdots, \quad (17.6.16)$$

for every $k = 1, 2, \ldots$, and the convergents $A_r(z)/B_r(z)$, $r = 0, 1, \ldots$, of $F_k(z)$ form the staircase in the Padé table of $f(z)$, which is shown in Table 17.6.2.

It is seen that the even convergents of $F_{k+1}(z)$ and the odd convergents of $F_k(z)$ are identical. Now, the even part of $F_{k+1}(z)$ is the continued fraction

$$F_{k+1}^e(z) := \sum_{i=0}^{k} c_i z^i$$

$$+ \frac{c_{k+1} z^{k+1}}{1 - q_1^{(k+1)} z} - \frac{q_1^{(k+1)} e_1^{(k+1)} z^2}{1 - (e_1^{(k+1)} + q_2^{(k+1)}) z} - \frac{q_2^{(k+1)} e_2^{(k+1)} z^2}{1 - (e_2^{(k+1)} + q_3^{(k+1)}) z} - \cdots,$$

$$(17.6.17)$$

and the odd part of $F_k(z)$ is the continued fraction

$$F_k^o(z) := \sum_{i=0}^{k} c_i z^i$$

$$+ \frac{c_k q_1^{(k)} z^{k+1}}{1 - (q_1^{(k)} + e_1^{(k)}) z} - \frac{e_1^{(k)} q_2^{(k)} z^2}{1 - (q_2^{(k)} + e_2^{(k)}) z} - \frac{e_2^{(k)} q_3^{(k)} z^2}{1 - (q_3^{(k)} + e_3^{(k)}) z} - \cdots. \quad (17.6.18)$$

By equating $F_k^o(z)$ and $F_{k+1}^e(z)$, we obtain the relations

$$c_{k+1} = c_k q_1^{(k)}, \quad q_1^{(k+1)} = q_1^{(k)} + e_1^{(k)}, \quad \text{and}$$

$$q_n^{(k+1)} e_n^{(k+1)} = q_{n+1}^{(k)} e_n^{(k)}, \quad e_n^{(k+1)} + q_{n+1}^{(k+1)} = q_{n+1}^{(k)} + e_{n+1}^{(k)}, \quad n = 1, 2, \ldots. \quad (17.6.19)$$

It is easy to see that the $q_n^{(k)}$ and $e_n^{(k)}$ can be computed recursively from these relations, which form the basis of the *qd-algorithm* given next.

Algorithm 17.6.2 (qd-algorithm)

1. Given $c_0, c_1, c_2, \ldots$, set $e_0^{(k)} = 0$ and $q_1^{(k)} = c_{k+1}/c_k$, $k = 0, 1, \ldots$.
2. For $n = 1, 2, \ldots$, and $k = 0, 1, \ldots$, compute $e_n^{(k)}$ and $q_n^{(k)}$ recursively by

$$e_n^{(k)} = e_{n-1}^{(k+1)} + q_n^{(k+1)} - q_n^{(k)}, \qquad q_{n+1}^{(k)} = q_n^{(k+1)} e_n^{(k+1)} / e_n^{(k)}.$$

The quantities $q_n^{(k)}$ and $e_n^{(k)}$ can be arranged in a two-dimensional array as in Table 17.6.3. This table is called the *qd-table*.

Table 17.6.3: *The qd-table*

$$
\begin{array}{cccccc}
 & & q_1^{(0)} & & & \\
e_0^{(1)} & & & e_1^{(0)} & & \\
 & q_1^{(1)} & & & q_2^{(0)} & \\
e_0^{(2)} & & e_1^{(1)} & & & e_2^{(0)} \\
 & q_1^{(2)} & & q_2^{(1)} & & \ddots \\
e_0^{(3)} & & e_1^{(2)} & & e_2^{(1)} & \\
 & q_1^{(3)} & & q_2^{(2)} & & \ddots \\
e_0^{(4)} \quad \vdots & & e_1^{(3)} & & e_2^{(2)} & \\
\vdots & & \vdots & q_2^{(3)} & \vdots & \ddots \\
 & & & \vdots & &
\end{array}
$$

With the $q_n^{(k)}$ and $e_n^{(k)}$ available, we can now compute the convergents of the continued fractions $F_k(z)$.

It can be shown that (see Henrici [131, Theorem 7.6a])

$$
q_n^{(k)} = \frac{H_{n-1}^{(k)} H_n^{(k+1)}}{H_n^{(k)} H_{n-1}^{(k+1)}} \quad \text{and} \quad e_n^{(k)} = \frac{H_{n+1}^{(k)} H_{n-1}^{(k+1)}}{H_n^{(k)} H_n^{(k+1)}}, \tag{17.6.20}
$$

where we have used $H_n^{(k)}$ to denote the Hankel determinant $H_n^{(k)}(\{c_s\})$ for short.

The qd-algorithm was developed by Rutishauser in [243] and discussed further in [242] and [244]. For a detailed treatment of it, we refer the reader to Henrici [131], [132].

17.7 Padé Approximants and Exponential Interpolation

Padé approximants have a very close connection with the problem of interpolation by a sum of exponential functions. The problem, in its simple form, is to find a function $u(x; h) = \sum_{j=1}^{n} \alpha_j e^{\sigma_j x}$ that satisfies the $2n$ interpolation conditions $u(x_0 + ih; h) = c_i$, $i = 0, 1, \dots, 2n - 1$. Here, α_j and σ_j are parameters to be determined. In case a solution exists, it can be constructed with the help of a method due to Prony [232]. It has been shown by Weiss and McDonough [352] that the method of Prony is closely related to the Padé table. As shown in [352], one first constructs the $[n - 1/n]$ Padé approximant $F_{n-1,n}(z)$ from $F(z) = \sum_{k=0}^{2n-1} c_k z^k$. If $F_{n-1,n}(z)$ has the partial fraction expansion $\sum_{j=1}^{n} A_j/(z - z_j)$, assuming simple poles only, then $u(x; h) = \sum_{j=1}^{n} \hat{A}_j \zeta_j^{(x-x_0)/h}$, where $\hat{A}_j = -A_j/z_j$ and $\zeta_j = 1/z_j$, $j = 1, \dots, n$.

In case $F_{n-1,n}(z)$ has multiple poles, the solution of Prony is no longer valid. This case has been treated in detail by Sidi [280]. In such a case, the function $u(x; h)$ needs to be chosen from the set

$$
U_n^h = \left\{ \sum_{j=1}^{r} \sum_{k=1}^{\lambda_j} B_{j,k} x^{k-1} \zeta_j^{x/h} : \zeta_j \text{ distinct}, \ -\pi < \arg \zeta_j \leq \pi, \ \sum_{j=1}^{r} \lambda_j \leq n \right\}.
$$

The following theorem, which concerns this problem, has been proved in [280]:

Theorem 17.7.1

(i) *There exists a unique function $u(x;h)$ in U_n^h that solves the interpolation problem $u(x_0 + ih; h) = c_i$, $i = 0, 1, \ldots, 2n - 1$, if and only if the $[n-1/n]$ Padé approximant $F_{n-1,n}(z)$ from $F(z) = \sum_{k=0}^{2n-1} c_k z^k$ exists and satisfies $\lim_{z\to\infty} F_{n-1,n}(z) = 0$.*

(ii) *In case $F_{n-1,n}(z)$ exists and is as in part (i), it has the partial fraction expansion*

$$F_{n-1,n}(z) = \sum_{j=1}^{s} \sum_{k=1}^{\mu_j} \frac{A_{j,k}}{(z - z_j)^k}, \quad \sum_{j=1}^{s} \mu_j \le n.$$

Then, $u(x;h)$ is given by

$$u(x;h) = \sum_{j=1}^{s} \sum_{k=1}^{\mu_j} E_{j,k} \binom{k + (x - x_0)/h - 1}{k - 1} \zeta_j^{(x-x_0)/h},$$

where $\binom{p}{i} = [p(p - 1)\cdots(p - i + 1)]/i!$ is the binomial coefficient,

$$\zeta_j = z_j^{-1}, \quad E_{j,k} = (-1)^k A_{j,k} z_j^{-k}, \quad 1 \le k \le \mu_j, \ 1 \le j \le s.$$

Another interpolation problem treated by Sidi [280] concerns the case in which $h \to 0$. The result relevant to this problem also involves Padé approximants. To treat this case, we define the function set U_n via

$$U_n = \left\{ \sum_{j=1}^{r} \sum_{k=1}^{\lambda_j} B_{j,k} x^{k-1} e^{\sigma_j x} : \ \sigma_j \text{ distinct}, \ \sum_{j=1}^{r} \lambda_j \le n \right\}.$$

Theorem 17.7.2

(i) *There exists a unique function $v(x)$ in U_n that solves the interpolation problem $v^{(i)}(x_0) = \gamma_i$, $i = 0, 1, \ldots, 2n - 1$, if and only if the $[n-1/n]$ Padé approximant $V_{n-1,n}(\tau)$ from $V(\tau) = \sum_{k=0}^{2n-1} \gamma_k \tau^k$ exists.*

(ii) *In case $V_{n-1,n}(\tau)$ exists, $\bar{V}_{n-1,n}(\sigma) = \sigma^{-1} V_{n-1,n}(\sigma^{-1})$ satisfies $\lim_{\sigma\to\infty} \bar{V}_{n-1,n}(\sigma) = 0$ and has the partial fraction expansion*

$$\bar{V}_{n-1,n}(\sigma) = \sum_{j=1}^{s} \sum_{k=1}^{\nu_j} \frac{B_{j,k}}{(\sigma - \sigma_j)^k}, \quad \sum_{j=1}^{s} \nu_j \le n.$$

Then, $v(x)$ is given by

$$v(x) = \sum_{j=1}^{s} \sum_{k=1}^{\nu_j} \frac{B_{j,k}}{(k - 1)!} (x - x_0)^{k-1} e^{\sigma_j(x-x_0)}.$$

Integral representations for both $u(x;h)$ and $v(x)$ are given in [280], and they involve $F_{n-1,n}(z)$ and $V_{n-1,n}(\tau)$ directly. These representations are used in [280] to prove the following theorem.

Theorem 17.7.3 *Let $f(x)$ be $2n - 1$ times differentiable in a neighborhood of x_0, and set $c_i = f(x_0 + ih)$ and $\gamma_i = f^{(i)}(x_0)$, $i = 0, 1, \ldots, 2n - 1$. Finally, assume that the $[n-1/n]$ Padé approximant $V_{n-1,n}(\tau)$ from $\sum_{k=0}^{2n-1} \gamma_k \tau^k$ exists, its denominator* *polynomial has degree exactly n, and $V_{n-1,n}(\tau)$ is not reducible. Then the following hold:*

(i) *The interpolant $v(x)$ of Theorem 17.7.2 exists.*
(ii) *The interpolant $u(x; h)$ of Theorem 17.7.1 exists for all small h, and satisfies*

$$\lim_{h \to 0} u(x; h) = v(x).$$

In a separate paper by Sidi [284], the interpolation problems above are extended to the cases in which some of the exponents are preassigned. The solutions to these problems are achieved by the so-called *Padé-type approximants*, which we discuss in the next chapter. See [284] for details.

The method of Prony can be generalized by replacing the Padé approximant $F_{n-1,n}(z)$ by the rational function $\phi_n(z) = \sum_{i=0}^{n-1} a'_i z^i / \sum_{j=0}^{n} b'_j z^j$, $b'_0 = 1$, with the b'_j determined via the least-squares solution of the overdetermined linear system

$$\sum_{j=0}^{n} c_{i-j} b'_j = 0, \quad i = n, n+1, \ldots, N,$$

where N is significantly larger than $2n$ [cf. (17.1.4)], and the a'_i computed from

$$\sum_{j=0}^{i} c_{i-j} b'_j = a'_i, \quad i = 0, 1, \ldots, n-1,$$

once the b'_j are determined [cf. (17.1.3)]. If $\phi_n(z) = \sum_{j=1}^{n} A_j/(z - z_j)$ (simple poles assumed for simplicity), then $u(x; h) = \sum_{j=1}^{n} \hat{A}_j \zeta_j^{(x-x_0)/h}$, where $\hat{A}_j = -A_j/z_j$ and $\zeta_j = 1/z_j$, $j = 1, \ldots, n$, is the desired exponential approximation that satisfies $u(x_0 + ih; h) \approx c_i$, $i = 0, 1, \ldots, N$. Essentially this approach and some further extensions of it have been used in problems of signal processing in electrical engineering. (Note that, in such problems, the c_i are given with some errors, and this causes the original method of Prony to perform poorly.)

17.8 Convergence of Padé Approximants from Meromorphic Functions

In this and the next sections, we summarize some of the convergence theory pertaining to the Padé table. In this summary, we do not include the topics of *convergence in measure* and *convergence in capacity*. Some of the theorems we state clearly show convergence acceleration, whereas others show only convergence. For proofs and further results, we refer the reader to the vast literature on the subject.

In this section, we are concerned with the convergence of rows of the Padé table from the Maclaurin series of meromorphic functions.

17.8.1 de Montessus's Theorem and Extensions

The classic result for row sequences $\{f_{m,n}(z)\}_{m=0}^{\infty}$ (with n fixed) of Padé approximants of meromorphic functions is *de Montessus's theorem*, which is a true convergence acceleration result and which reads as follows:

Theorem 17.8.1 *Let $f(z)$ be analytic at $z = 0$ and meromorphic in the disk $K = \{z : |z| < R\}$ and let it have q poles in K counting multiplicities. Then, the row sequence $\{f_{m,q}(z)\}_{m=0}^{\infty}$ converges to $f(z)$ uniformly in any compact subset of K excluding the poles of $f(z)$, such that*

$$\limsup_{m \to \infty} |f(z) - f_{m,q}(z)|^{1/m} \leq |z/R|. \tag{17.8.1}$$

This theorem was originally proved by de Montessus de Ballore [64] and follows from the work of Hadamard [120]. Different proofs of it have been given in Baker [15], Karlsson and Wallin [150], Saff [249], and Sidi [292].

In case the singularities of $f(z)$ on the boundary of the disk K, namely, on $\partial K = \{z : |z| = R\}$, are all poles, Theorem 17.8.1 can be improved substantially. This improvement, presented originally in Sidi [292, Theorem 3.3], is quantitative in nature, and we state it next.

Theorem 17.8.2 *Let $f(z)$ be analytic at $z = 0$ and meromorphic in the disk K and on its boundary ∂K. Let $z_1, \ldots, z_t$ be the poles of $f(z)$ in K and let $\omega_1, \ldots, \omega_t$ be their respective multiplicities. Define $Q(z) = \prod_{j=1}^{t} (1 - z/z_j)^{\omega_j}$ and $q = \sum_{j=1}^{t} \omega_j$. Similarly, let $\hat{z}_1, \ldots, \hat{z}_r$ be the poles of $f(z)$ on ∂K, and let $\hat{\omega}_1, \ldots, \hat{\omega}_r$ be their respective multiplicities. Thus, for each $j \in \{1, \ldots, r\}$, the Laurent expansion of $f(z)$ about $z = \hat{z}_j$ is given by*

$$f(z) = \sum_{i=1}^{\hat{\omega}_j} \frac{\hat{a}_{ji}}{(1 - z/\hat{z}_j)^i} + \Theta_j(z); \quad \hat{a}_{j\hat{\omega}_j} \neq 0, \quad \Theta_j(z) \text{ analytic at } \hat{z}_j. \tag{17.8.2}$$

Let us now order the $\hat{z}_j$ on ∂K such that

$$\hat{\omega}_1 = \hat{\omega}_2 = \cdots = \hat{\omega}_\mu > \hat{\omega}_{\mu+1} \geq \cdots \geq \hat{\omega}_r, \tag{17.8.3}$$

and set $\bar{p} = \hat{\omega}_1 - 1$. Then there holds

$$f(z) - f_{m,q}(z) = \frac{m^{\bar{p}}}{\bar{p}!} \sum_{j=1}^{\mu} \frac{\hat{a}_{j\hat{\omega}_j}}{1 - z/\hat{z}_j} \left[\frac{Q(\hat{z}_j)}{Q(z)} \right]^2 \left(\frac{z}{\hat{z}_j} \right)^{m+q+1} + o(m^{\bar{p}} |z/R|^m)$$

$$= O(m^{\bar{p}} |z/R|^m) \text{ as } m \to \infty, \tag{17.8.4}$$

uniformly in any compact subset of $K \setminus \{z_1, \ldots, z_t\}$. This result is best possible as $m \to \infty$.

Theorem 17.8.2 also shows that, as $m \to \infty$, $f_{m,q}(z)$ diverges everywhere on ∂K. We note that the result in (17.8.4) is best possible in the sense that its right-hand side gives the first term in the asymptotic expansion of $f(z) - f_{m,q}(z)$ as $m \to \infty$ explicitly.

Now, Theorem 17.8.1 concerns the row sequence $\{f_{m,n}(z)\}_{m=0}^{\infty}$, where $n = q$, q being the number of poles of $f(z)$ in K, counted according to their multiplicities. It does not apply to the sequences $\{f_{m,n}(z)\}_{m=0}^{\infty}$ with $n > q$, however. Instead, we have the following weaker result due to Karlsson and Wallin [150] on convergence of *subsequences* of $\{f_{m,n}(z)\}_{m=0}^{\infty}$.

Theorem 17.8.3 *Let $f(z)$ be as in Theorem 17.8.1 with poles $\xi_1, \ldots, \xi_q$ in K that are not necessarily distinct, and let $n > q$. Then, there exist $n - q$ points $\xi_{q+1}, \ldots, \xi_n$ and a subsequence $\{f_{m_k,n}(z)\}_{k=0}^{\infty}$ that converges to $f(z)$ uniformly in any compact subset of $K \backslash \{\xi_1, \ldots, \xi_n\}$, such that*

$$\limsup_{k\to\infty} |f(z) - f_{m_k,n}(z)|^{1/m_k} \leq |z/R|. \tag{17.8.5}$$

It must be clear that a priori we do not have any knowledge of $\xi_{q+1}, \ldots, \xi_n$ and the integers m_k in this theorem.

It is easy to construct examples for which one can show definitely that the sequence $\{f_{m,n}(z)\}_{m=0}^{\infty}$ does not converge under the conditions of Theorem 17.8.3, but a subsequence does precisely as described in Theorem 17.8.3.

Treatment of Intermediate Rows

Similarly, Theorem 17.8.2 is not valid when $q < n < q + \sum_{j=1}^{r} \hat{\omega}_j$. Rows of the Padé table for which n takes on such values are called *intermediate rows*. Note that intermediate rows may appear not only when $n > q$; when $f(z)$ has multiple poles and/or a number of poles with equal modulus in K, they appear with $n < q$ as well. Thus, intermediate rows are at least as common as those treated by de Montessus's theorem.

The convergence problem of intermediate rows was treated partially (for some special cases) in a series of papers by Wilson [358], [359], [360]. Preliminary work on the treatment of the general case was presented by Lin [168]. The complete solution for the general case was given only recently by Sidi [292].

The following convergence result pertaining to the convergence of intermediate rows in the most general case is part of Sidi [292, Theorem 6.1], and it gives a surprisingly simple condition sufficient for the convergence of the *whole* sequence $\{f_{m,n}(z)\}_{m=0}^{\infty}$. This condition involves the nonlinear integer programming problem IP(τ) we discussed in detail in the preceding chapter.

Theorem 17.8.4 *Let $f(z)$ be precisely as in Theorem 17.8.2 and let $n = q + \tau$ with $0 < \tau < \sum_{j=1}^{r} \hat{\omega}_j$. Denote by IP$(\tau)$ the nonlinear integer programming problem*

$$\text{maximize } g(\vec{\sigma}); \quad g(\vec{\sigma}) = \sum_{k=1}^{r} (\hat{\omega}_k \sigma_k - \sigma_k^2)$$

$$\text{subject to } \sum_{k=1}^{r} \sigma_k = \tau \text{ and } 0 \leq \sigma_k \leq \hat{\omega}_k, \ 1 \leq k \leq r. \tag{17.8.6}$$

Then, $\{f_{m,n}(z)\}_{m=0}^{\infty}$ converges uniformly to $f(z)$ in any compact subset of $K \backslash \{z_1, \ldots, z_t\}$, provided IP$(\tau)$ has a unique solution for $\sigma_1, \ldots, \sigma_r$. If we denote by $G(\tau)$ and $G(\tau + 1)$ the (optimal) value of $g(\vec{\sigma})$ at the solutions to IP(τ) and IP$(\tau + 1)$, respectively, then there holds

$$f(z) - f_{m,n}(z) = O(m^{G(\tau+1)-G(\tau)}|z/R|^m) \text{ as } m \to \infty. \tag{17.8.7}$$

[Note that IP$(\tau + 1)$ need not have a unique solution.]

Obviously, when IP(τ) does not have a unique solution, Theorem 17.8.3 applies, and we conclude that a subsequence $\{f_{m_k,n}(z)\}_{k=0}^{\infty}$ converges to $f(z)$ uniformly, as explained in Theorem 17.8.3. Theorems 17.8.2 and 17.8.4, together with Theorem 17.8.3, present a complete treatment of the convergence of row sequences in the Padé table of meromorphic functions with polar singularities on their circles of meromorphy.

Interestingly, in the paper by Liu and Saff [169], the existence of a unique solution to IP(τ) features prominently as a sufficient condition for convergence of the intermediate rows of Walsh arrays of best rational approximations as well.

17.8.2 Generalized Koenig's Theorem and Extensions

The following result concerning the poles of Padé approximants is known as the *generalized Koenig's theorem*.

Theorem 17.8.5 *Let $f(z)$ be as in Theorem 17.8.1 and denote its poles by $\xi_1, \ldots, \xi_q$. Here, the ξ_i are not necessarily distinct and are ordered such that $|\xi_1| \leq \cdots \leq |\xi_q|$. Define $Q(z) = \prod_{j=1}^{q}(1 - z/\xi_j)$. Let $Q_{m,n}(z)$ be the denominator of $f_{m,n}(z)$, normalized such that $Q_{m,n}(0) = 1$. Then*

$$\limsup_{m \to \infty} |Q_{m,q}(z) - Q(z)|^{1/m} \leq |\xi_q/R| \ \text{as } m \to \infty. \tag{17.8.8}$$

The special case in which $q = 1$ was proved originally by Koenig [153]. The general case follows from a closely related theorem of Hadamard [120], and was proved by Golomb [100], and more recently, by Gragg and Householder [108].

If the poles of $f(z)$ in K are as in Theorem 17.8.2, then the result in (17.8.8) can be refined, as shown in Sidi [292], and reads

$$Q_{m,q}(z) - Q(z) = O(m^{\alpha} |\xi_q/R|^m) \ \text{as } m \to \infty, \ \ \alpha \geq 0 \ \text{some integer.} \tag{17.8.9}$$

Of course, what Theorem 17.8.5 implies is that, for all large m, $f_{m,q}(z)$ has precisely q poles that tend to the poles $\xi_1, \ldots, \xi_q$ of $f(z)$ in K. If we let the poles of $f(z)$ and their multiplicities and $\bar{p}$ be as in Theorem 17.8.2, then, for each $j \in \{1, \ldots, t\}$, $f_{m,q}(z)$ has precisely ω_j poles $z_{jl}(m)$, $l = 1, \ldots, \omega_j$, that tend to z_j. Also, the p_jth derivative of $Q_{m,q}(z)$, the denominator of $f_{m,q}(z)$, has a zero $z'_j(m)$ that tends to z_j. More specifically, we have the following quantitative results, whose proofs are given in Sidi [292, Theorem 3.1].

Theorem 17.8.6

(i) *In Theorem 17.8.1,*

$$\limsup_{m \to \infty} \left| z_{jl}(m) - z_j \right|^{1/m} = \left| z_j/R \right|^{1/\omega_j},$$

$$\limsup_{m \to \infty} \left| \frac{1}{\omega_j} \sum_{l=1}^{\omega_j} z_{jl}(m) - z_j \right|^{1/m} = \left| z_j/R \right|,$$

$$\limsup_{m \to \infty} \left| z'_j(m) - z_j \right|^{1/m} = \left| z_j/R \right|. \tag{17.8.10}$$

(ii) *In Theorem 17.8.2, we obtain the following more refined results:*

$$z_{jl}(m) - z_j = O([m^{\bar{p}}|z_j/R|^m]^{1/\omega_j}) \ as \ m \to \infty,$$

$$\frac{1}{\omega_j} \sum_{l=1}^{\omega_j} z_{jl}(m) - z_j = O(m^{\bar{p}}|z_j/R|^m) \ as \ m \to \infty,$$

$$z'_j(m) - z_j = O(m^{\bar{p}}|z_j/R|^m) \ as \ m \to \infty. \quad (17.8.11)$$

The first of the results in (17.8.10) and (17.8.11) were given earlier by Gončar [101]. The version of (17.8.11) that is given in [292] is actually more refined in that it provides the first term of the asymptotic expansion of $z_{jl}(m) - z_j$:

$$z_{jl}(m) \sim z_j + E_{jl}(m)(m^{\bar{p}}|z_j/R|^m)^{1/\omega_j} \ as \ m \to \infty, \quad (17.8.12)$$

where $\{E_{jl}(m)\}_{m=0}^{\infty}$ is some bounded sequence with a subsequence that has a nonzero limit.

A similar result related to intermediate rows of Padé approximants in Theorem 17.8.4 exists and is given as part of [292, Theorem 6.1]. For additional references concerning the theorem of Koenig and its generalizations, we refer the reader to Sidi [292].

Recently, results that form a sort of inverse to the generalized Koenig's theorem have been of interest. The essential question now is the following: Suppose that the function $f(z)$ has a formal power series $\sum_{k=0}^{\infty} c_k z^k$ and that the poles of some sequence of Padé approximants from this series converge to a set X. Does it follow that $f(z)$ (or some continuation of it) is singular on X? Is $f(z)$ analytic off X? The first theorem along these lines that concerns the poles of the sequence $\{f_{m,1}(z)\}_{m=0}^{\infty}$ was given by Fabry [81]. Fabry's result was generalized to the sequences $\{f_{m,n}(z)\}_{m=0}^{\infty}$ with arbitrary fixed n in the works of Gončar [101] and of Suetin [329], [330]. It is shown in [329] and [330], in particular, that if the poles of the sequence $\{f_{m,n}(z)\}_{m=0}^{\infty}$ converge to some complex numbers $\xi_1, \ldots, \xi_n$, not necessarily distinct, then $\sum_{k=0}^{\infty} c_k z^k$ represents a function $f(z)$ analytic at 0 and meromorphic in the disk $K_n = \{z : |z| < R_n\}$, where $R_n = \max\{|\xi_1|, \ldots, |\xi_n|\}$. In addition, it is shown that if $K = \{z : |z| < R\}$ is the actual disk of meromorphy of $f(z)$, then those ξ_i in the interior of K are poles of $f(z)$, while those on the boundary of K are points of singularity of $f(z)$. For additional results and references on this interesting topic, see also Karlsson and Saff [149]. This paper treats both rows and columns of Padé tables of nonmeromorphic as well as meromorphic functions.

17.9 Convergence of Padé Approximants from Moment Series

Note that all our results about Padé approximants from meromorphic functions have been on row sequences. No definitive results are currently known about diagonal sequences. However, there is a detailed convergence theory of diagonal sequences of Padé approximants from moment series associated with Stieltjes and Hamburger functions (also called Markov functions). This theory is also closely related to orthogonal polynomials and Gaussian integration. We present a summary of this subject in this section. For more details and further results, we refer the reader to Baker and Graves-Morris [16] and Stahl and Totik [321]. For orthogonal polynomials and related matters, see the books by Szegő [332] and Freud [88]. For the moment problem, see also the book by Widder [357].

Definition 17.9.1 Let (a, b) be a real interval and let $\psi(t)$ be a real function that is non-decreasing on (a, b) with infinitely many points of increase there.

(a) Define the function $f(z)$ via

$$f(z) = \int_a^b \frac{d\psi(t)}{1 + tz}, \tag{17.9.1}$$

where the integral is defined in the sense of Stieltjes. If $0 \leq a < b \leq \infty$, $f(z)$ is called a *Stieltjes function*, and if $-\infty \leq a < 0 < b \leq \infty$, it is called a *Hamburger function*.

(b) Let $\psi(t)$ be such that its moments f_k defined by

$$f_k = \int_a^b t^k d\psi(t), \quad k = 0, 1, \ldots, \tag{17.9.2}$$

all exist. Then the formal (convergent or divergent) power series $\sum_{k=0}^\infty f_k(-z)^k$ is said to be the *moment series* associated with $\psi(t)$. It is called a *Stieltjes series* if $0 \leq a < b \leq \infty$ and a *Hamburger series* if $-\infty \leq a < 0 < b \leq \infty$.

It is easy to see that a Stieltjes (or Hamburger) function is real analytic in the complex z-plane cut along the real interval $[-a^{-1}, -b^{-1}]$ (or along the real intervals $(-\infty, -b^{-1}]$ and $[-a^{-1}, +\infty)$).

It is also easy to see that the moment series $\sum_{k=0}^\infty f_k(-z)^k$ represents $f(z)$ asymptotically as $z \to 0$, that is,

$$f(z) \sim \sum_{k=0}^\infty f_k(-z)^k \quad \text{as } z \to 0. \tag{17.9.3}$$

Clearly, if (a, b) is finite, $\sum_{k=0}^\infty f_k(-z)^k$ has a positive and finite radius of convergence and is the Maclaurin expansion of $f(z)$. Otherwise, $\sum_{k=0}^\infty f_k(-z)^k$ diverges everywhere. If $\{f_k\}$ is a moment sequence as in Definition 17.9.1, then

$$H_n^{(m)}(\{f_s\}) > 0, \quad m = 0, 1 \text{ and } n = 0, 1, \ldots, \text{ for } 0 \leq a < b \leq \infty, \tag{17.9.4}$$

and

$$H_n^{(m)}(\{f_s\}) > 0, \ m = 0 \text{ and } n = 0, 1, \ldots, \text{ for } -\infty \leq a < 0 \leq b \leq \infty. \tag{17.9.5}$$

Conversely, if $\{f_k\}$ is such that (17.9.4) [or (17.9.5)] holds, then there exists a function $\psi(t)$ as described in Definition 17.9.1 for which f_k are as given in (17.9.2) with $0 \leq a < b \leq \infty$ (or $-\infty \leq a < 0 < b \leq \infty$). The function $\psi(t)$ is unique if

$$\sum_{k=0}^\infty f_k^{-1/(2k)} = \infty \quad \text{when } 0 \leq a < b \leq \infty, \tag{17.9.6}$$

and

$$\sum_{k=0}^{\infty} f_{2k}^{-1/(2k)} = \infty \quad \text{when} \ -\infty \le a < 0 < b \le \infty. \qquad (17.9.7)$$

The conditions in (17.9.6) and (17.9.7) are known as *Carleman's conditions*. It is easy to verify that they are satisfied automatically if $\sum_{k=0}^{\infty} f_k(-z)^k$ has a nonzero radius of convergence. What is implied by Carleman's condition in (17.9.6) [or (17.9.7)] is that there exists a *unique* Stieltjes (or Hamburger) function that admits $\sum_{k=0}^{\infty} f_k(-z)^k$ as its asymptotic expansion as $z \to 0$.

Given a Stieltjes series $\sum_{k=0}^{\infty} f_k(-z)^k$, there exists an S-fraction, namely, $z^{-1} K(\alpha_n z/1)$ with $\alpha_n > 0$, $n = 1, 2, \dots$, whose convergents are the $[n/n]$, $[n/n+1]$, $n = 0, 1, \dots$, entries in the Padé table of $\sum_{k=0}^{\infty} f_k(-z)^k$. In addition, the Padé table in question is normal. The poles of Padé approximants $f_{m,n}(z)$ from a Stieltjes series are all simple and lie on the negative real axis, and the corresponding residues are all positive. Thus,

$$f_{n+j,n}(z) = \sum_{k=0}^{j} f_k(-z)^k + (-z)^{j+1} \sum_{i=1}^{n} \frac{H_i}{1 + t_i z}, \quad t_i > 0 \text{ distinct and } H_i > 0. \quad (17.9.8)$$

Let us also denote by $Q_{m,n}(z)$ the denominator polynomial of $f_{m,n}(z)$. Then, with $j \ge -1$, $\varphi_{j,n}(t) \equiv t^n Q_{n+j,n}(-t^{-1})$ is the nth orthogonal polynomial with respect to the inner product

$$(F, G) \equiv \int_a^b F(t)G(t)t^{j+1}d\psi(t), \qquad (17.9.9)$$

where $\psi(t)$ is the function that gives rise to the moment sequence $\{f_k\}$ as in (17.9.2). The poles and residues of $f_{n+j,n}(z)$ are also related to numerical integration. Specifically, with the t_i and H_i as in (17.9.8), the sum $\sum_{i=1}^{n} H_i g(t_i)$ is the n-point Gaussian quadrature formula for the integral $\int_a^b g(t)t^{j+1}\, d\psi(t)$.

Concerning the Padé approximants from Hamburger series, results analogous to those of the preceding paragraph can be stated. For example, all $f_{n+2j-1,n}(z)$ exist for $j \ge 0$ and there holds

$$f_{n+2j-1,n}(z) = \sum_{k=0}^{2j-1} f_k(-z)^k + z^{2j} \sum_{i=1}^{n} \frac{H_i}{1 + t_i z}, \quad t_i \text{ real distinct and } H_i > 0. \quad (17.9.10)$$

We now state two convergence theorems for the diagonal sequences of Padé approximants from Stieltjes and Hamburger series $\sum_{k=0}^{\infty} f_k(-z)^k$. We assume that, in case $\sum_{k=0}^{\infty} f_k(-z)^k$ diverges everywhere, the f_k satisfy the suitable Carleman condition in (17.9.6) or (17.9.7). We recall that, with this condition, the associated Stieltjes function or Hamburger function $f(z)$ in (17.9.1) is unique. Theorem 17.9.2 concerns the case in which $\sum_{k=0}^{\infty} f_k(-z)^k$ has a positive radius of convergence, while Theorem 17.9.3 concerns the case of zero radius of convergence.

Theorem 17.9.2

(i) *If* $-\infty < a < 0 < b < \infty$ *and* $j \ge 0$, *the sequence* $\{f_{n+2j-1,n}(z)\}_{n=0}^{\infty}$ *converges to* $f(z)$ *in the open set* $\mathcal{D}_0$ *formed from the complex z-plane cut along* $(-\infty, -b^{-1}]$

and $[-a^{-1}, +\infty)$, *the pointwise rate of convergence being given by*

$$\limsup_{n \to \infty} |f(z) - f_{n+2j-1,n}(z)|^{1/n} \le \left| \frac{\sqrt{z^{-1}+b} - \sqrt{z^{-1}+a}}{\sqrt{z^{-1}+b} + \sqrt{z^{-1}+a}} \right| < 1 \quad (17.9.11)$$

with the phase convention that $\sqrt{z^{-1}+b}$ *and* $\sqrt{z^{-1}+a}$ *are positive for* $z^{-1} > -a$. *The convergence is uniform in any compact subset of* $\mathcal{D}_0$.

(ii) *If* $0 \le a < b < \infty$ *and* $j \ge -1$, *the sequence* $\{f_{n+j,n}(z)\}_{n=0}^{\infty}$ *converges to* $f(z)$ *in the open set* $\mathcal{D}_+$ *formed from the* z-*plane cut along* $(-\infty, -b^{-1}]$, *the pointwise rate of convergence being given by*

$$\limsup_{n \to \infty} |f(z) - f_{n+j,n}(z)|^{1/n} \le \left| \frac{\sqrt{1+bz} - 1}{\sqrt{1+bz} + 1} \right| < 1 \quad (17.9.12)$$

with the convention that $\sqrt{1+bz} > 0$ *for* $z > -b^{-1}$. *The convergence is uniform in any compact subset of* $\mathcal{D}_+$.

Theorem 17.9.3

(i) *When* $(a, b) = (-\infty, \infty)$ *and* $j \ge 0$, *the sequence* $\{f_{n+2j-1,n}(z)\}_{n=0}^{\infty}$ *converges to* $f(z)$ *uniformly in* $\{z : |z| \le R$ *and* $|\Im z| \ge \delta > 0\}$ *for arbitrary* $R > 0$ *and small* δ.

(ii) *When* $(a, b) = (0, \infty)$ *and* $j \ge -1$, *the sequence* $\{f_{n+j,n}(z)\}_{n=0}^{\infty}$ *converges to* $f(z)$ *uniformly in* $\{z : |z| \le R$ *and* $(a) |\Im z| \ge \delta$ *if* $\Re z \le 0$ *and* $(b) |z| \ge \delta$ *if* $\Re z \ge 0\}$ *for arbitrary* $R > 0$ *and small* $\delta > 0$.

17.10 Convergence of Padé Approximants from Pólya Frequency Series

Definition 17.10.1 A formal power series $f(z) := \sum_{k=0}^{\infty} c_k z^k$ is said to be a *Pólya frequency series* if

$$(-1)^{m(m-1)/2} C_{m,n} > 0, \quad m, n = 0, 1, \dots . \quad (17.10.1)$$

Note that the sign of $C_{m,n}$ is independent of n, and, for $m = 0, 1, \dots$, the sign pattern $++--++\cdots$ prevails. It is known (see Schönberg [260] and Edrei [72]) that all Pólya frequency series are Maclaurin series of functions of the form

$$f(z) = a_0 e^{\gamma z} \prod_{i=1}^{\infty} \frac{1 + \alpha_i z}{1 - \beta_i z}, \quad (17.10.2)$$

where

$$a_0 > 0, \ \gamma \ge 0, \ \alpha_i \ge 0, \ \beta_i \ge 0, \ i = 1, 2, \dots, \ \sum_{i=1}^{\infty} (\alpha_i + \beta_i) < \infty. \quad (17.10.3)$$

Obviously, the Padé table from a Pólya frequency series is normal. The following convergence theorem of Arms and Edrei [11] concerns rays of entries in the Padé table.

Theorem 17.10.2 *Let $f(z)$ be as in (17.10.2) and (17.10.3). Choose sequences of integers $\{m_k\}$ and $\{n_k\}$ such that $\lim_{k\to\infty}(m_k/n_k) = \omega$. Then, the ray sequence $\{f_{m_k,n_k}(z)\}_{k=0}^{\infty}$ converges to $f(z)$ uniformly in any compact set of the z-plane excluding the poles of $f(z)$ if there are such. Specifically, with $f_{m,n}(z) = P_{m,n}(z)/Q_{m,n}(z)$, $Q_{m,n}(0) = 1$, we have*

$$\lim_{k\to\infty} P_{m_k,n_k}(z) = a_0 \exp[\gamma z/(1+\omega)] \prod_{i=1}^{\infty}(1+\alpha_i z),$$

$$\lim_{k\to\infty} Q_{m_k,n_k}(z) = \exp[-\gamma\omega z/(1+\omega)] \prod_{i=1}^{\infty}(1-\beta_i z),$$

uniformly in any compact set of the z-plane.

Note that Theorem 17.10.2 applies to the function $f(z) = e^z$.

17.11 Convergence of Padé Approximants from Entire Functions

As has been shown by Perron [229, Chapter 4], convergence of Padé approximants from arbitrary entire functions is not guaranteed. Indeed, we can have rows of the Padé table not converging in any given open set of the complex plane.

In this section, we give convergence theorems for Padé approximants from entire functions of very slow and smooth growth. For these and additional theorems on this subject, we refer the reader to the papers by Lubinsky [180], [181], [182], [183], [184]. Our first theorem is from [180].

Theorem 17.11.1 *Let $f(z) = \sum_{k=0}^{\infty} c_k z^k$ be entire with infinitely many $c_k \neq 0$ and*

$$\limsup_{k\to\infty} |c_k|^{1/k^2} = \rho < \frac{1}{3}. \tag{17.11.1}$$

Then, there exists an increasing sequence of positive integers $\{m_0, m_1, \ldots\}$ such that $\{f_{m_k,n_k}(z)\}_{k=0}^{\infty}$ converges to $f(z)$ uniformly in every compact set of the z-plane, where $\{n_0, n_1, \ldots\}$ is arbitrary. This applies, in particular, with $n_k = m_k + j$ and arbitrary j (diagonal sequences). It also applies with $n_0 = n_1 = \cdots$ (row sequences).

Theorem 17.11.1 has been improved in [184] by weakening the condition in (17.11.1) as in Theorem 17.11.2 that follows.

Theorem 17.11.2 *Let $f(z) = \sum_{k=0}^{\infty} c_k z^k$ be entire with infinitely many $c_k \neq 0$ and*

$$\limsup_{k\to\infty} |c_k|^{1/k^2} = \rho < 1. \tag{17.11.2}$$

Then, the diagonal sequence $\{f_{n,n}(z)\}$ converges to $f(z)$ uniformly in every compact set of the z-plane.

The next theorem gives another improvement of Theorem 17.11.1 and was proved in [181].

Theorem 17.11.3 *Let $f(z) = \sum_{k=0}^{\infty} c_k z^k$ be entire with $c_k \neq 0$, $k = 0, 1, \ldots$, and assume*

$$\left| c_{k-1} c_{k+1} / c_k^2 \right| \leq \rho_0^2, \quad k = 1, 2, \ldots, \tag{17.11.3}$$

where $\rho_0 = 0.4559 \cdots$ is the positive root of the equation

$$2 \sum_{k=1}^{\infty} \rho^{k^2} = 1. \tag{17.11.4}$$

Then, the Padé table of $f(z)$ is normal and, for any nonnegative sequence of integers $\{n_0, n_1, \ldots\}$, the sequence $\{f_{m,n_m}(z)\}_{m=0}^{\infty}$ converges to $f(z)$ uniformly in any compact set of the z-plane. In addition, the constant ρ_0 in (17.11.3) is best possible.

Note that the c_k in Theorem 17.11.3 satisfy $\limsup_{k\to\infty} |c_k|^{1/k^2} \leq \rho_0$. It is also in this sense that Theorem 17.11.3 is an improvement over Theorem 17.11.1.

Lubinsky [183] showed that both rows and diagonals of the Padé table converge when the condition in (17.11.3) is replaced by

$$\lim_{k\to\infty} c_{k-1} c_{k+1} / c_k^2 = q, \quad |q| < 1, \tag{17.11.5}$$

for some possibly complex scalar q.

Generalizations of Padé Approximants

18.1 Introduction

In this chapter, we consider some of the many generalizations of Padé approximants. We describe the general ideas and show how the relevant approximations are constructed. We do not go into their algebraic properties and the theory of their convergence, however. For these subjects and an extensive bibliography, we refer the reader to Baker and Graves-Morris [16].

The generalizations we mention are the so-called Padé-type approximants, multi-point Padé approximants, algebraic and differential Hermite–Padé approximants, Padé approximants from orthogonal polynomial expansions, Baker–Gammel approximants, and Padé–Borel approximants.

What we present here in no way exhausts the existing arsenal of approaches and methods. For example, we leave out the vector and matrix Padé approximants. For these topics, again, we refer the reader to [16].

18.2 Padé-Type Approximants

We begin with the formal power series $f(z) := \sum_{k=0}^{\infty} c_k z^k$ and denote its $[m/n]$ Padé approximant by $f_{m,n}(z)$ as before. We recall that $f_{m,n}(z)$ is completely determined by the equations in (17.1.3) and (17.1.4). Clearly, all the zeros and poles of $f_{m,n}(z)$ are determined by these equations as well. In some situations, in addition to the c_k, we may have information about some or all of the zeros and/or poles of the function represented by the series $f(z)$, and we would like to incorporate this information in our rational approximants. This can be achieved via the so-called Padé-type approximants defined next.

Definition 18.2.1 Let the formal power series $f(z) := \sum_{k=0}^{\infty} c_k z^k$ be given. The $[m/n]$ *Padé-type approximant* corresponding to $f(z)$ and having known zeros $z_1, \ldots, z_\mu$ and known poles $\xi_1, \ldots, \xi_\nu$, with respective multiplicities $\sigma_1, \ldots, \sigma_\mu$ and $\omega_1, \ldots, \omega_\nu$, such that $z_i \neq 0$ and $\xi_i \neq 0$, if it exists, is the rational function $\hat{f}_{m,n}(z) = u(z)[p(z)/q(z)]$, where $u(z) = \prod_{i=1}^{\mu}(1 - z/z_i)^{\sigma_i} / \prod_{i=1}^{\nu}(1 - z/\xi_i)^{\omega_i}$ and $p(z)$ and $q(z)$ are polynomials of degree at most m and n, respectively, such that $q(0) = 1$ and and

$$\hat{f}_{m,n}(z) - f(z) = O(z^{m+n+1}) \text{ as } z \to 0. \tag{18.2.1}$$

The following elementary result that generalizes another due to Sidi [284] shows that, when the known zeros and poles are fixed, no new theory or justification is needed for Padé-type approximants and that the known theory of Padé approximants applies directly. It also shows that no new algorithms are needed to compute the Padé-type approximants since the algorithms for ordinary Padé approximants discussed in Chapter 17 can be used for this purpose.

Theorem 18.2.2 *Let $\hat{f}_{m,n}(z)$ be precisely as in Definition 18.2.1. The rational function $p(z)/q(z)$ is the $[m/n]$ Padé approximant $g_{m,n}(z)$ from the power series $\sum_{k=0}^{\infty} d_k z^k$ of the quotient $g(z) = f(z)/u(z)$. Thus, $\hat{f}_{m,n}(z)$ is unique as well.*

Proof. Dividing both sides of (18.2.1) by $u(z)$, we observe that $p(z)/q(z)$ must satisfy

$$p(z)/q(z) - g(z) = O(z^{m+n+1}) \text{ as } z \to 0.$$

The result now follows from the definition of Padé approximants, namely, from Definition 17.1.1.   ■

Now, to compute $p(z)$ and $q(z)$, we need to know the d_k. Obviously, the d_k can be computed by multiplying the power series $f(z)$ by the Maclaurin series of $1/u(z)$.

As a result of Theorem 18.2.2, we have

$$\hat{f}_{m,n}(z) - f(z) = [g_{m,n}(z) - g(z)]u(z), \tag{18.2.2}$$

where, as before, $f(z)$ and $g(z)$ are the functions represented by $\sum_{k=0}^{\infty} c_k z^k$ and $\sum_{k=0}^{\infty} d_k z^k$ respectively, and $g(z) = f(z)/u(z)$, of course. Because $u(z)$ is completely known, we can make statements on $\hat{f}_{m,n}(z) - f(z)$ by applying the known convergence theory of Padé approximants to $g_{m,n}(z) - g(z)$.

Finally, note that this idea can be extended with no changes whatsoever to the case in which some or all of the z_i and ξ_i are algebraic branch points, that is, the corresponding σ_i and ω_i are not integers. We can thus define $\hat{f}_{m,n}(z) = u(z)[p(z)/q(z)]$, with $u(z)$, $p(z)$, and $q(z)$ precisely as in Definition 18.2.1. Theorem 18.2.2 applies, in addition. Of course, this time $\hat{f}_{m,n}(z)$ is not a rational function, but has also algebraic branch points.

The approximations mentioned above (and additional ones involving multipoint Padé approximants that we discuss later in this chapter) have been employed with success in the treatment of several problems of interest in the literature of fluid mechanics by Frost and Harper [89].

A different and interesting use of Padé-type approximants has been suggested and analyzed in a series of papers by Ambroladze and Wallin [6], [7], [8], [9]. This use differs from the one just described in that the preassigned poles and their number are not necessarily fixed. They are chosen to enhance the quality of the Padé-type approximants that can be obtained from a number of the coefficients c_k.

For example, Ambroladze and Wallin [6] consider the Padé-type approximation of a function

$$f(\zeta) = \int_a^b \frac{w(t)}{\zeta - t} d\psi(t),$$

where $\psi(t)$ is a real function that is nondecreasing on the finite interval (a, b) with infinitely many points of increase there, and $w(t)$ is an entire function. The moment series, $\sum_{k=0}^{\infty} c_k/\zeta^{k+1}$, where $c_k = \int_a^b t^k w(t) d\psi(t)$, $k = 0, 1, \ldots$, converges to the function $f(\zeta)$ for all large ζ. The Padé-type approximants are constructed from the moment series of $f(\zeta)$, and they are of the form $r_n(\zeta) = p_n(\zeta)/q_n(\zeta)$, where, for each n, $p_n(\zeta)$ is a polynomial of degree at most $n - 1$ and $q_n(\zeta)$ is the nth orthogonal polynomial with respect to the inner product $(F, G) = \int_a^b \overline{F(t)} G(t) d\psi(t)$, and $p_n(\zeta)$ is determined by requiring that

$$f(\zeta) - r_n(\zeta) = O(\zeta^{-n-1}) \text{ as } \zeta \to \infty.$$

That is, the preassigned poles of $r_n(\zeta)$ are the n zeros of $q_n(\zeta)$ [which are distinct and lie in (a, b)] and thus their number tends to infinity as $n \to \infty$. It is shown in [6] that the sequence $\{r_n(\zeta)\}_{n=1}^{\infty}$ converges to $f(\zeta)$ uniformly in any compact subset of the extended complex plane cut along the interval $[a, b]$. Furthermore, when $w(t) \geq 0$ for $t \in (a, b)$, the upper bound on the error in $r_n(\zeta)$ is the same as that in the diagonal ordinary Padé approximant $R_n(\zeta)$ from $\sum_{k=0}^{\infty} c_k/\zeta^{k+1}$, whose numerator and denominator degrees are at most $n - 1$ and n, respectively, and satisfies

$$f(\zeta) - R_n(\zeta) = O(\zeta^{-2n-1}) \text{ as } \zeta \to \infty.$$

[Recall that the convergence of the sequence $\{R_n(\zeta)\}_{n=1}^{\infty}$ is covered completely by Theorem 17.9.2, where we set $z = 1/\zeta$, when $w(t) \geq 0$ for $t \in (a, b)$.] Note that it takes $2n$ coefficients of the series $\sum_{k=0}^{\infty} c_k/\zeta^{k+1}$ to determine $R_n(\zeta)$ as opposed to only n for $r_n(\zeta)$. This implies that, if $c_0, c_1, \ldots, c_{2n}$ are available, we can construct both $R_n(\zeta)$ and $r_{2n}(\zeta)$, but $r_{2n}(\zeta)$ is a better approximation to $f(\zeta)$ than $R_n(\zeta)$ is. In fact, the upper bounds on the errors in $R_n(\zeta)$ and in $r_{2n}(\zeta)$ suggest that the latter converges to $f(\zeta)$ as $n \to \infty$ twice as fast as the former.

Note that no theory of uniform convergence of $\{R_n(\zeta)\}_{n=1}^{\infty}$ exists in case $w(t)$ changes sign on (a, b). In fact, examples can be constructed for which $\{R_n(\zeta)\}_{n=1}^{\infty}$ does not converge locally uniformly anywhere. See Stahl [320].

18.3 Vanden Broeck–Schwartz Approximations

Let $f(z) = \sum_{k=0}^{\infty} c_k z^k$ as before. In Section 16.7, we introduced a generalization of the ε-algorithm due to Vanden Broeck and Schwartz [344] that contains a parameter η. Let us set in (16.7.1) $\varepsilon_0^{(j)} = \sum_{k=0}^{j} c_k z^k$, $j = 0, 1, \ldots$, and generate the $\varepsilon_k^{(j)}$. The entries $\varepsilon_{2n}^{(j)}$, which we now denote $\hat{f}_{m,n}(z)$, are rational approximations to the sum of the series $f(z)$. By (16.7.2), the $\hat{f}_{m,n}(z)$, for $m \geq n$, can be obtained recursively from

$$\frac{1}{\hat{f}_{m+1,n}(z) - \hat{f}_{m,n}(z)} + \frac{1}{\hat{f}_{m-1,n}(z) - \hat{f}_{m,n}(z)} =$$

$$\frac{1}{\hat{f}_{m,n+1}(z) - \hat{f}_{m,n}(z)} + \frac{\eta}{\hat{f}_{m,n-1}(z) - \hat{f}_{m,n}(z)}.$$

Concerning $\hat{f}_{m,n}(z)$, we observe the following:

1. Because $\varepsilon_k^{(j)}$, $k = 0, 1, 2$, are independent of η, the approximations $\hat{f}_{m,0}(z)$ and $\hat{f}_{m,1}(z)$ are simply the ordinary Padé approximants $f_{m,0}(z)$ and $f_{m,1}(z)$ respectively. The rest of the $\hat{f}_{m,n}(z)$ are different from the $f_{m,n}(z)$ when $\eta \neq 1$.

2. Just like $f_{m,n}(z)$, $\hat{f}_{m,n}(z)$ too is determined from the $m + n + 1$ coefficients $c_0, c_1, \ldots, c_{m+n}$ and satisfies

$$\hat{f}_{m,n}(z) - f(z) = O(z^{m+n+1}) \text{ as } z \to 0, \tag{18.3.1}$$

provided

$$c_{m+n}^{(m-1,n)} \neq c_{m+n} \text{ and } c_{m+n}^{(m,n-1)} \neq c_{m+n}, \tag{18.3.2}$$

where $c_k^{(m,n)}$ are the coefficients of the Maclaurin expansion of $\hat{f}_{m,n}(z)$, that is, $\hat{f}_{m,n}(z) = \sum_{k=0}^{\infty} c_k^{(m,n)} z^k$. This can be shown by induction on n. The conditions in (18.3.2) can be violated only in exceptional circumstances because neither $\hat{f}_{m,n-1}(z)$ nor $\hat{f}_{m-1,n}(z)$ depends on c_{m+n}.

Vanden Broeck and Schwartz [344] show with a numerical example that the diagonal approximations $\hat{f}_{n,n}(z)$ with appropriate complex η are quite effective in summing everywhere-divergent asymptotic expansions also on the branch cuts of functions they represent. They report some results for the function $(1/z)e^{1/z}E_1(1/z)$, which we also denote by $f(z)$, which has the (everywhere-divergent) Euler series $\sum_{k=0}^{\infty}(-1)^k k! z^k$ as its asymptotic expansion when $z \to \infty$, $\Re z > 0$. Here $E_1(\zeta) = \int_{\zeta}^{\infty}(e^{-t}/t)\,dt$ is the exponential integral. [Note also that this series is a Stieltjes series that satisfies the Carleman condition, and $f(z)$ is the corresponding Stieltjes function.] The function $f(z)$ is analytic in the z-plane cut along the negative real axis. Now, when the c_k are real and η is chosen to be real, the $\hat{f}_{m,n}(z)$ are real analytic, that is, they are real for z real. Thus, the $\hat{f}_{m,n}(z)$ from the Euler series are real when both $z > 0$ and $z < 0$. Although $f(z)$ is real for $z > 0$, it is not real for $z < 0$, because $\Im E_1(-x \pm i0) = \mp\pi$ for $x > 0$. We can obtain a complex approximation in this case if we choose η to be complex. We refer the reader to [344] for more details. To date we are not aware of any research on the convergence properties of this interesting method of approximation, which certainly deserves serious consideration.

18.4 Multipoint Padé Approximants

We defined a Padé approximant to be a rational function whose Maclaurin expansion agrees with a given power series $\sum_{k=0}^{\infty} c_k z^k$ as far as possible. This idea can be generalized as follows:

Definition 18.4.1 Let the formal power series $F_r(z) := \sum_{k=0}^{\infty} c_{rk}(z - z_r)^k$, $r = 1, \ldots, q$, with z_r distinct, be given. Define $\vec{\mu} = (\mu_1, \ldots, \mu_q)$. Then the rational function $\hat{f}_{\vec{\mu},n}(z)$ with degrees of numerator and denominator at most m and n respectively, such that $m + n + 1 = \sum_{r=1}^{q}(\mu_r + 1)$, is the q-point Padé approximant of type $(\mu_1, \ldots, \mu_q)$ from $\{F_r(z)\}_{r=1}^{q}$ if it satisfies

$$\hat{f}_{\vec{\mu},n}(z) - F_r(z) = O((z - z_r)^{\mu_r+1}) \text{ as } z \to z_r, r = 1, \ldots, q. \tag{18.4.1}$$

Let us write $\hat{f}_{\vec{\mu},n}(z) = \sum_{i=0}^{m} \alpha_i z^i / \sum_{i=0}^{n} \beta_i z^i$. From (18.4.1), it is clear that, if we set $\beta_0 = 1$, then the α_i and β_i can be determined from the linear system of $m + n + 1$ equations that result from (18.4.1).

Multipoint Padé approximants may be very useful when, for each r, the power series $F_r(z)$ represents asymptotically as $z \to z_r$ a single function $f(z)$. In such a case, $\hat{f}_{\vec{\mu},n}(z)$ may be a very good approximation to $f(z)$ in a large domain of the complex plane, assuming that $f(z)$ has suitable analyticity properties.

When $\mu_r = 0$ for all r, $\hat{f}_{\vec{\mu},n}(z)$ becomes simply a rational interpolant that assumes given values at $z_1, z_2, \ldots, z_q$. This rational interpolation problem is known as the Cauchy–Jacobi problem and it can be treated numerically by, for example, Thiele's *reciprocal difference algorithm*. When $q = 1$, $\hat{f}_{\vec{\mu},n}(z)$ reduces to the $[\mu_1 - n/n]$ Padé approximant from $F_1(z)$.

For arbitrary q and μ_r, a determinant expression for $\hat{f}_{\vec{\mu},n}(z)$ is known. See Baker [15].

The convergence of the approximants $\hat{f}_{\vec{\mu},n}(z)$ as $\mu_r \to \infty$ with n fixed can be treated by applying the following important theorem of Saff [249] from which de Montessus's theorem can be derived as a special case.

Theorem 18.4.2 *Let E be a closed bounded set in the z-plane whose complement K, including the point at infinity, is connected and possesses a Green's function $G(z)$ with pole at infinity. For each $\sigma > 1$, let Γ_σ denote the locus $G(z) = \log \sigma$, and let E_σ denote the interior of Γ_σ. Let the sequence of (not necessarily distinct) interpolation points $\{\zeta_k^{(s)}, k = 0, \ldots, s\}_{s=0}^{\infty}$ be given such that it has no limit points in K and*

$$\lim_{s \to \infty} \left| \prod_{k=0}^{s} (z - \zeta_k^{(s)}) \right|^{1/s} = \mathrm{cap}(E) \exp\{G(z)\},$$

uniformly in z on each compact subset of K, where $\mathrm{cap}(E)$ is the capacity of E. Let $f(z)$ be analytic in E and meromorphic with precisely n poles, counting multiplicities, in E_ρ for some $\rho > 1$. Denote by D_ρ the region obtained from E_ρ by deleting the n poles of $f(z)$. Then, for all m sufficiently large, there exists a unique rational function $R_{m,n}(z)$ with numerator of degree at most m and denominator of degree n that interpolates $f(z)$ at $\zeta_k^{(m+n)}$, $k = 0, \ldots, m + n$. Each $R_{m,n}(z)$ has n finite poles that converge to the poles of $f(z)$ in E_ρ. Furthermore, $\lim_{m \to \infty} R_{m,n}(z) = f(z)$ throughout D_ρ and uniformly on every compact subset of D_ρ. Specifically, if S is a compact subset of D_ρ such that $S \subset E_\lambda, 1 < \lambda < \rho$, then

$$\limsup_{m \to \infty} \left(\max_{z \in S} |f(z) - R_{m,n}(z)| \right)^{1/m} \leq \frac{\lambda}{\rho} < 1.$$

In particular

$$\limsup_{m \to \infty} \left(\max_{z \in E} |f(z) - R_{m,n}(z)| \right)^{1/m} \leq \frac{1}{\rho} < 1.$$

The capacity of E is defined via

$$\mathrm{cap}(E) = \lim_{n \to \infty} \left(\min_{p \in \mathcal{P}_n} \max_{z \in E} |p(z)| \right)^{1/n},$$

where $\mathcal{P}_n$ is the set of all polynomials of degree exactly n with leading coefficient unity.

When $\zeta_k^{(s)} = 0$ for all k and s in Theorem 18.4.2, $R_{m,n}(z)$ is nothing but the Padé approximant $f_{m,n}(z)$ from the Maclaurin series of $f(z)$. In this case, $E = \{z : |z| \leq r\}$ for some $r > 0$, and hence $\operatorname{cap}(E) = r$, $G(z) = \log(|z|/r)$, and $E_\rho = \{z : |z| < \rho r\}$, and Saff's theorem reduces to de Montessus's theorem.

18.4.1 Two-Point Padé Approximants

The case that has received more attention than others is that of $q = 2$. This case can be standardized by sending the points z_1 and z_2 to 0 and ∞ by a Moebius transformation. In its "symmetric" form, this case can thus be formulated as follows:

Definition 18.4.3 Let the function $f(z)$ satisfy

$$f(z) \sim \frac{c_0}{2} + c_1 z + c_2 z^2 + \cdots \quad \text{as } z \to 0,$$

$$f(z) \sim -\left(\frac{c_0}{2} + \frac{c_{-1}}{z} + \frac{c_{-2}}{z^2} + \cdots\right) \quad \text{as } z \to \infty; \quad c_0 \neq 0. \tag{18.4.2}$$

For any two integers i and j such that $i + j$ is even, we define the *two-point Padé approximant* $\hat{f}_{i,j}(z)$ to be the rational function

$$\hat{f}_{i,j}(z) = \frac{P(z)}{Q(z)} = \frac{\sum_{k=0}^m \alpha_k z^k}{\sum_{k=0}^m \beta_k z^k}, \quad \beta_0 = 1, \quad m = (i+j)/2, \tag{18.4.3}$$

that satisfies

$$\hat{f}_{i,j}(z) - f(z) = O(z^i, z^{-j-1}) \equiv \begin{cases} O(z^i) & \text{as } z \to 0, \\ O(z^{-j-1}) & \text{as } z \to \infty, \end{cases} \tag{18.4.4}$$

provided $\hat{f}_{i,j}(z)$ exists.

Note that, in case $f(0) + f(\infty) \neq 0$, the asymptotic expansions of $f(z)$ do not have the symmetric form of (18.4.2). The symmetric form can be achieved simply by adding a constant to $f(z)$. For example, if $\phi(z) \sim \sum_{i=0}^\infty \gamma_i z^i$ as $z \to 0$ and $\phi(z) \sim \sum_{i=0}^\infty \delta_i/z^i$ as $z \to \infty$, and $\gamma_0 + \delta_0 \neq 0$, then $f(z) = \phi(z) - (\gamma_0 + \delta_0)/2$ has asymptotic expansions as in (18.4.2), with $c_0 = \gamma_0 - \delta_0$ and $c_i = \gamma_i$ and $c_{-i} = -\delta_i$, $i = 1, 2, \ldots$.

The following results and their proofs can be found in Sidi [276].

Theorem 18.4.4 *When $\hat{f}_{i,j}(z)$ exists, it is unique.*

Theorem 18.4.5 *Let $g(z) = 1/f(z)$ in the sense that $g(z)$ has asymptotic expansions as $z \to 0$ and $z \to \infty$ obtained by inverting those of $f(z)$ in (18.4.2) appropriately. If $\hat{g}_{i,j}(z)$ is the two-point Padé approximant from $g(z)$ precisely as in Definition 18.4.3, then $\hat{g}_{i,j}(z) = 1/\hat{f}_{i,j}(z)$.*

Obviously, α_k and β_k can be obtained by solving the linear equations that result from (18.4.4). In particular, the β_k satisfy the system

$$\sum_{s=0}^m c_{r-s} \beta_s = 0, \quad r = i - m, i - m + 1, \ldots, i - 1. \tag{18.4.5}$$

The approximant $\hat{f}_{i,j}(z)$ has the determinant representation

$$\hat{f}_{i,j}(z) = \frac{\begin{vmatrix} S_{i-m}(z) & zS_{i-m-1}(z) & \cdots & z^m S_{-j}(z) \\ c_{i-1} & c_{i-2} & \cdots & c_{i-m-1} \\ c_{i-2} & c_{i-3} & \cdots & c_{i-m-2} \\ \vdots & \vdots & & \vdots \\ c_{i-m} & c_{i-m-1} & \cdots & c_{-j} \end{vmatrix}}{\begin{vmatrix} 1 & z & \cdots & z^m \\ c_{i-1} & c_{i-2} & \cdots & c_{i-m-1} \\ c_{i-2} & c_{i-3} & \cdots & c_{i-m-2} \\ \vdots & \vdots & & \vdots \\ c_{i-m} & c_{i-m-1} & \cdots & c_{-j} \end{vmatrix}}, \tag{18.4.6}$$

where

$$S_0(z) = \frac{c_0}{2}, \quad S_k(z) = S_{k-1}(z) + c_k z^k, \quad k = \pm 1, \pm 2, \dots. \tag{18.4.7}$$

[Note that $S_k(z) = c_0/2 + \sum_{i=1}^{k} c_i z^i$ and $S_{-k}(z) = -c_0/2 - \sum_{i=1}^{k-1} c_{-i} z^{-i}$ for $k = 1, 2, \dots$.] This representation is obtained by unifying the ones for the cases $i \geq j$ and $i \leq j$ given in Sidi [276, Theorem 3]. As a consequence of (18.4.6), we also have that

$$\hat{f}_{i,j}(z) = \frac{\sum_{k=0}^{m} \beta_k z^k S_{i-m-k}(z)}{\sum_{k=0}^{m} \beta_k z^k}. \tag{18.4.8}$$

This implies that, once the β_k have been determined, the approximant $\hat{f}_{i,j}(z)$ is known for all practical purposes.

Two-point Padé approximants are convergents of certain continued fractions that are known as M-fractions and T-fractions. (See Sidi [276] and McCabe and Murphy [208].) For example, the approximants $\hat{f}_{r,r}(z)$, $\hat{f}_{r+1,r-1}(z)$, $r = 1, 2, \dots$, are consecutive convergents of a continued fraction of the form

$$c + \frac{dz}{1+ez} + \frac{\lambda_1}{\mu_1+} \frac{w_1 z}{1} + \frac{\lambda_2}{\mu_2+} \frac{w_2 z}{1} + \dots \; ; \; \lambda_i + \mu_i = 1 \text{ for all } i,$$

with $\hat{f}_{1,1}(z) = c + dz/(1+ez)$, etc.

Being convergents of M-fractions, the $\hat{f}_{i,j}(z)$ can be computed by the F-G algorithm of McCabe and Murphy [208] that is of the quotient-difference type. They can also be computed by a different method proposed in Sidi [276] that is based on the recursive computation of the β_s.

18.5 Hermite–Padé Approximants

We can view the Padé approximant $f_{m,n}(z)$ as being obtained in the following manner: Determine the polynomials $P_{m,n}(z)$ and $Q_{m,n}(z)$ by solving (17.1.2), namely,

$$Q_{m,n}(z)f(z) - P_{m,n}(z) = O(z^{m+n+1}) \text{ as } z \to 0, \quad Q_{m,n}(0) = 1,$$

and then obtain $f_{m,n}(z)$ as the solution to the equation $Q_{m,n} f_{m,n} - P_{m,n} = 0$. This has been generalized with Hermite–Padé polynomials in two different ways, which we now present.

Definition 18.5.1 Let $h_1(z), \ldots, h_r(z)$ be given formal power series. The *Hermite–Padé polynomials* $[Q_1(z), \ldots, Q_r(z)]$ *of type* $(\mu_1, \ldots, \mu_r)$, where $Q_k(z)$ is a polynomial of degree at most μ_k, $k = 1, \ldots, r$, are the solution of

$$\sum_{k=1}^{r} Q_k(z) h_k(z) = O(z^{\hat{\mu}-1}) \text{ as } z \to 0, \quad \hat{\mu} = \sum_{k=1}^{r} (\mu_k + 1).$$

18.5.1 Algebraic Hermite–Padé Approximants

Let $f(z) := \sum_{k=0}^{\infty} c_k z^k$ be a given formal series. Define the Hermite–Padé polynomials $[Q_0(z), Q_1(z), \ldots, Q_s(z)]$ of type $(m_0, m_1, \ldots, m_s)$ via

$$\sum_{k=0}^{s} Q_k(z)[f(z)]^k = O(z^{\hat{m}-1}) \text{ as } z \to 0, \quad \hat{m} = \sum_{k=0}^{s} (m_k + 1). \quad (18.5.1)$$

Thus, $Q_k(z) = \sum_{i=0}^{m_k} q_{ki} z^i$ for each k and we set $q_{s0} = 1$. (Obviously, the q_{ki} satisfy a linear system of $\hat{m} - 1$ equations.) With the $Q_k(z)$ available, solve the equation

$$\sum_{k=0}^{s} Q_k(z) \xi^k = 0 \quad (18.5.2)$$

for ξ. Obviously, this is a polynomial equation in ξ that has s solutions that are functions of z. The solution whose Maclaurin expansion agrees with $\sum_{k=0}^{\infty} c_k z^k$ as far as possible is the *algebraic Hermite–Padé approximant of type* $\vec{m} = (m_0, m_1, \ldots, m_s)$, and we denote it $\hat{f}_{\vec{m}}(z)$. Obviously, we recover the ordinary Padé approximant $f_{m_0, m_1}(z)$ when $s = 1$.

When $s = 2$, after determining $Q_0(z)$, $Q_1(z)$, and $Q_2(z)$, we obtain

$$\xi = \psi(z) = \left(-Q_1(z) + \sqrt{[Q_1(z)]^2 - 4Q_0(z)Q_2(z)} \right) / [2Q_2(z)],$$

whose singularities may be poles and branch points of square-root type. Provided we pick the right branch for the square-root, we obtain $\hat{f}_{\vec{m}}(z) = \psi(z)$. The $\hat{f}_{\vec{m}}(z)$ obtained this way are known as the *quadratic approximants* of Shafer [263].

18.5.2 Differential Hermite–Padé Approximants

Let $f(z) := \sum_{k=0}^{\infty} c_k z^k$ be a given formal series. Define the Hermite–Padé polynomials $[Q_{-1}(z), Q_0(z), \ldots, Q_s(z)]$ of type $(m_{-1}, m_0, \ldots, m_s)$ via

$$\sum_{k=0}^{s} Q_k(z) f^{(k)}(z) - Q_{-1}(z) = O(z^{\hat{m}-1}) \text{ as } z \to 0, \quad \hat{m} = \sum_{k=-1}^{s} (m_k + 1). \quad (18.5.3)$$

Thus, $Q_k(z) = \sum_{i=0}^{m_k} q_{ki} z^i$ for each k and we set $q_{s0} = 1$. What is meant by $f^{(j)}(z)$ is the formal power series $\sum_{k=j}^{\infty} k(k-1) \cdots (k-j+1) c_k z^{k-j}$ that is obtained by differentiating $f(z)$ formally termwise. (The q_{ki} satisfy a linear system of $\hat{m} - 1$ equations in this case too.) With the $Q_k(z)$ determined, solve the linear ordinary differential equation

$$\sum_{k=0}^{s} Q_k(z) y^{(k)} = Q_{-1}(z), \quad y^{(j)}(0) = c_j(j!), \quad j = 0, 1, \dots, s-1, \quad (18.5.4)$$

for $y(z)$, provided the $Q_k(z)$ allow such a solution to exist. When it exists, $y(z)$ is the *differential Hermite–Padé approximant of type* $\vec{m} = (m_{-1}, m_0, \dots, m_s)$, and we denote it $\hat{f}_{\vec{m}}(z)$. It must be noted that a priori it is not clear what one should take for the integers m_k to guarantee that (18.5.4) has a solution. Clearly, $\hat{f}_{\vec{m}}(z)$ is not necessarily a rational function. Obviously, we recover the ordinary Padé approximant $f_{m_{-1},m_0}(z)$ when $s = 0$.

Differential Hermite–Padé approximants seem to have originated in so-called "series analysis" in statistical mechanics, for the case $s = 2$; the resulting approximants are known as Gammel–Guttmann–Gaunt–Joyce (G^3J) approximants. These were given in the works of Guttmann and Joyce [119], Joyce and Guttmann [146], and Gammel [91]. For a detailed review, we refer the reader also to Guttmann [118].

18.6 Padé Approximants from Orthogonal Polynomial Expansions

Let $\theta(x)$ be a real function that is nondecreasing in $[a, b]$ with infinitely many points of increase there. Let $\phi_k(x)$ be the kth orthogonal polynomial with respect to the inner product $(\cdot, \cdot)$ defined by

$$(F, G) = \int_a^b F(x)G(x)\, d\theta(x). \quad (18.6.1)$$

When a formal expansion $f(x) := \sum_{k=0}^{\infty} c_k \phi_k(x)$ is given, we would like to find a rational function $R_{m,n}(x) = P_{m,n}(x)/Q_{m,n}(x)$, with degrees of $P_{m,n}(x)$ and $Q_{m,n}(x)$ at most m and n, respectively, such that $R_{m,n}(x)$ is a good approximation to the sum of the given expansion when the latter converges. Two different types of rational approximation procedures have been discussed in the literature.

18.6.1 Linear (Cross-Multiplied) Approximations

Define $R_{m,n}(x)$ by demanding that

$$Q_{m,n}(x)f(x) - P_{m,n}(x) := \sum_{k=m+n+1}^{\infty} \tilde{c}_k \phi_k(x). \quad (18.6.2)$$

In other words, we are requiring $Q_{m,n}f - P_{m,n}$ to be orthogonal to ϕ_k, $k = 0, 1, \dots,$ $m + n$, with respect to $(\cdot, \cdot)$ in (18.6.1). Thus, (18.6.2) is the same as

$$(Q_{m,n}f - P_{m,n}, \phi_k) = 0, \quad k = 0, 1, \dots, m,$$
$$(Q_{m,n}f, \phi_k) = 0, \quad k = m+1, \dots, m+n, \quad (18.6.3)$$

supplemented by the normalization condition

$$Q_{m,n}(\xi_0) = 1 \text{ for some } \xi_0 \in [a, b]. \quad (18.6.4)$$

[Obviously, (18.6.4) makes sense as the existence of $R_{m,n}(x)$ depends also on its being free of singularities on $[a, b]$.]

The $R_{m,n}(x)$ defined this way are called *linear* or *cross-multiplied* approximants. They were originally developed by Maehly [205] for Chebyshev polynomial expansions. Cheney [47] generalized them to series of arbitrary orthogonal polynomials, and Holdeman [134] considered more general expansions. Fleischer [84] applied these approximations to Legendre polynomial expansions. For their convergence theory see Lubinsky and Sidi [185] and Gončar, Rakhmanov, and Suetin [102]. [Note that Lubinsky and Sidi [185] allow $d\theta(x)$ to have (a finite number of) sign changes on (a, b).]

Expressing $P_{m,n}(x)$ and $Q_{m,n}(x)$ in the form

$$P_{m,n}(x) = \sum_{i=0}^{m} p_i \phi_i(x) \text{ and } Q_{m,n}(x) = \sum_{i=0}^{n} q_i \phi_i(x), \tag{18.6.5}$$

and using the fact that

$$\phi_i(x)\phi_j(x) = \sum_{k=|i-j|}^{i+j} \alpha_{ijk} \phi_k(x), \quad \alpha_{ijk} \text{ constants}, \tag{18.6.6}$$

it can be shown that, unlike the $[m/n]$ Padé approximant from a power series, $R_{m,n}(x)$ is determined by the first $m + 2n + 1$ terms of the series $f(x)$, namely, by $c_0, c_1, \ldots, c_{m+2n}$. To see this, we start by noting that

$$f(x)\phi_k(x) := \sum_{l=0}^{\infty} A_{kl}\phi_l(x); \quad A_{kl} = \sum_{j=|l-k|}^{l+k} c_j \alpha_{jkl}, \tag{18.6.7}$$

which follows by invoking (18.6.6). Next, observing that $(Q_{m,n}f, \phi_k) = (Q_{m,n}, f\phi_k)$, and assuming for simplicity of notation that the $\phi_k(x)$ are normalized such that $(\phi_k, \phi_k) = 1$ for all k, we realize that (18.6.3) and (18.6.4) can be expressed as in

$$p_k = \sum_{l=0}^{n} A_{kl}q_l, \quad k = 0, 1, \ldots, m,$$

$$\sum_{l=0}^{n} A_{kl}q_l = 0, \quad k = m+1, \ldots, m+n,$$

$$\sum_{l=0}^{n} q_l \phi_l(\xi_0) = 1. \tag{18.6.8}$$

It is clear that the A_{kl} in these equations are constructed from c_k, $0 \le k \le m + 2n$. It is also clear that the q_l are computed first, and the p_k are computed with the help of the q_l. Making use of the fact that

$$P_{m,n}(x) = \sum_{l=0}^{n} q_l \left(\sum_{k=0}^{m} A_{kl}\phi_k(x) \right) \text{ and } Q_{m,n}(x) = \sum_{l=0}^{n} q_l \phi_l(x), \tag{18.6.9}$$

and of the conditions $\sum_{l=0}^{n} A_{kl}q_l = 0$, $k = m+1, \ldots, m+n$, we can conclude that

$R_{m,n}(x)$ has the interesting determinant representation

$$R_{m,n}(x) = \frac{\begin{vmatrix} S_{m,0}(x) & S_{m,1}(x) & \dots & S_{m,n}(x) \\ A_{m+1,0} & A_{m+1,1} & \cdots & A_{m+1,n} \\ A_{m+2,0} & A_{m+2,1} & \cdots & A_{m+2,n} \\ \vdots & \vdots & & \vdots \\ A_{m+n,0} & A_{m+n,1} & \cdots & A_{m+n,n} \end{vmatrix}}{\begin{vmatrix} \phi_0(x) & \phi_1(x) & \dots & \phi_n(x) \\ A_{m+1,0} & A_{m+1,1} & \cdots & A_{m+1,n} \\ A_{m+2,0} & A_{m+2,1} & \cdots & A_{m+2,n} \\ \vdots & \vdots & & \vdots \\ A_{m+n,0} & A_{m+n,1} & \cdots & A_{m+n,n} \end{vmatrix}}, \qquad (18.6.10)$$

where

$$S_{m,l}(x) = \sum_{k=0}^{m} A_{kl}\phi_k(x). \qquad (18.6.11)$$

18.6.2 Nonlinear (Properly Expanded) Approximations

The linear approximations were derived by working with the expansion of $Q_{m,n}f - P_{m,n}$. The nonlinear ones are derived by considering the expansion of $f - R_{m,n}$ directly. We now require, instead of (18.6.2), that

$$f(x) - \frac{P_{m,n}(x)}{Q_{m,n}(x)} := \sum_{k=m+n+1}^{\infty} \tilde{c}_k \phi_k(x). \qquad (18.6.12)$$

This is equivalent to

$$c_k(\phi_k, \phi_k) = (\phi_k, f) = (\phi_k, P_{m,n}/Q_{m,n}), \quad k = 0, 1, \dots, m+n, \quad (18.6.13)$$

and these equations are nonlinear in the coefficients of the polynomials $P_{m,n}$ and $Q_{m,n}$. Furthermore, only the first $m + n + 1$ coefficients of $f(x)$, namely, $c_0, c_1, \dots, c_{m+n}$, are needed now.

The approximations $R_{m,n}(x)$ defined this way are called *nonlinear* or *properly expanded* approximants and were developed by Fleischer [86]. They are also called by the names of the orthogonal polynomials involved; for example, Legendre–Padé, Chebyshev–Padé, etc. They turn out to be much more effective than the linear approximations. For their convergence theory, see Suetin [328], Lubinsky and Sidi [185], and Gončar, Rakhmanov, and Suetin [102]. Some results on the convergence of the so-called Chebyshev–Padé approximations are also given in Gragg [107].

One way of determining $R_{m,n}$ is by expressing it as a sum of partial fractions and solving for the poles and residues. Let us assume, for simplicity, that $m \geq n - 1$ and that all the poles of $R_{m,n}(x)$ are simple. Then

$$R_{m,n}(x) = r(x) + \sum_{j=1}^{n} \frac{H_j}{x - \xi_j}, \quad r(x) = \sum_{k=0}^{m-n} r_k \phi_k(x). \qquad (18.6.14)$$

Therefore, the equations in (18.6.13) become

$$r_k(\phi_k, \phi_k) + \sum_{j=1}^{n} H_j \psi_k(\xi_j) = c_k(\phi_k, \phi_k), \quad k = 0, 1, \ldots, m - n,$$

$$\sum_{j=1}^{n} H_j \psi_k(\xi_j) = c_k(\phi_k, \phi_k), \quad k = m - n + 1, \ldots, m + n, \qquad (18.6.15)$$

where $\psi_k(\xi)$ are the functions of the second kind that are defined by

$$\psi_k(\xi) = \int_a^b \frac{\phi_k(x)}{x - \xi} \, d\theta(x), \quad \xi \notin [a, b]. \qquad (18.6.16)$$

[Note again that $\xi_j \notin [a, b]$ because otherwise $R_{m,n}(x)$ does not exist.] Clearly, the H_j and ξ_j can be determined from the last $2n$ equations in (18.6.15) and, following that, the r_k can be computed from the first $m - n + 1$ of these equations.

Because the equations for the H_j and ξ_j are nonlinear, more than one solution for them may be obtained. Numerical computations with Legendre polynomial expansions have shown that only one solution with $\xi_j \notin [a, b]$ is obtained. This point can be understood with the help of the following uniqueness theorem due to Sidi [269].

Theorem 18.6.1 *Let $f(x)$ be a real function in $C[a, b]$ such that $f(x) := \sum_{k=0}^{\infty} c_k \phi_k(x)$, $c_k = (\phi_k, f)/(\phi_k, \phi_k)$ for all k. Let $R_{m,n}(x)$ be the $[m/n]$ nonlinear Padé approximant to $f(x)$ such that $R_{m,n}(x)$ has no poles along $[a, b]$. Then $R_{m,n}(x)$ is unique.*

[The more basic version of this theorem is as follows: *There exists at most one real rational function $R_{m,n}(x)$ with degree of numerator at most m and degree of denominator at most n and with no poles in $[a, b]$, such that $(\phi_k, R_{m,n}) = c_k$, $k = 0, 1, \ldots, m + n$, where the c_k are given real numbers.*]

Chebyshev–Padé Table

A very elegant solution for $R_{m,n}(x)$ has been given by Clenshaw and Lord [53] for Chebyshev polynomial expansions. This solution circumvents the nonlinear equations in a clever fashion. If $f(x) := \sum_{k=0}^{\infty}{}' c_k T_k(x)$ and $R_{m,n}(x) = \sum_{i=0}^{m}{}' p_i T_i(x) / \sum_{i=0}^{n}{}' q_i T_i(x)$, $q_0 = 2$, where $T_k(x)$ is the kth Chebyshev polynomial of the first kind and $\sum_{i=0}^{s}{}' \alpha_i = \frac{1}{2}\alpha_0 + \sum_{i=1}^{s} \alpha_i$, then the coefficients p_i and q_i can be determined as follows:

1. Solve $\sum_{s=0}^{n} \gamma_s c_{|r-s|} = 0$, $r = m + 1, \ldots, m + n$; $\gamma_0 = 1$.
2. Compute $q_s = \mu \sum_{i=0}^{n-s} \gamma_i \gamma_{s+i}$, $s = 1, \ldots, n$; $\mu^{-1} = \frac{1}{2} \sum_{i=0}^{n} \gamma_i^2$.
3. Compute $p_r = \frac{1}{2} \sum_{s=0}^{n}{}' q_s (c_{r+s} + c_{|r-s|})$, $r = 0, 1, \ldots, m$.

The expensive part of this procedure for increasing n is determining the $\gamma_s \equiv \gamma_s^{(m,n)}$. Clenshaw and Lord [53] have given a recursive algorithm for the $\gamma_s^{(m,n)}$ that enables their computation for $m \geq n$. A different algorithm has been given by Sidi [267] and it enables $\gamma_s^{(m,n)}$ to be computed for all m and n. Here are the details of this algorithm.

1. Set

$$\gamma_0^{(m,n)} = 1 \text{ and } \gamma_{-1}^{(m,n)} = 0 = \gamma_{n+1}^{(m,n)} \text{ for all } m, \ n.$$

2. Compute

$$\omega^{(m,n+1)} = \frac{\sum_{s=0}^{n} \gamma_s^{(m+1,n)} c_{|m+1-s|}}{\sum_{s=0}^{n} \gamma_s^{(m,n)} c_{|m-s|}}.$$

3. Compute $\gamma_s^{(m,n+1)}$ recursively from

$$\gamma_s^{(m,n+1)} = \gamma_s^{(m+1,n)} + \omega^{(m,n+1)} \gamma_{s-1}^{(m,n)}, \quad s = 0, 1, \ldots, n+1.$$

Of course, this algorithm is successful as long as $\sum_{s=0}^{n} \gamma_s^{(m,n)} c_{|m-s|} \neq 0$. A similar condition can be formulated for the algorithm of Clenshaw and Lord [53].

Another way of obtaining the $R_{m,n}(x)$ from $f(x) := \sum'^{\infty}_{k=0} c_k T_k(x)$ with $m \geq n$ makes use of the fact that

$$T_k(x) = \frac{1}{2}(t^k + t^{-k}), \quad k = 0, 1, \ldots; \quad x = \cos\theta, \quad t = e^{i\theta}.$$

Let us rewrite $f(x)$ in the form $f(x) = \frac{1}{2}[g(t) + g(t^{-1})]$, where $g(t) := \sum'^{\infty}_{k=0} c_k t^k$, and compute the $[m/n]$ Padé approximant $g_{m,n}(t)$ to $g(t)$. Then $R_{m,n}(x) = \frac{1}{2}[g_{m,n}(t) + g_{m,n}(t^{-1})]$. According to Fleischer [85], this approach to the Chebyshev–Padé table was first proposed by Gragg. It is also related to the Laurent–Padé table of Gragg [107]. We do not go into the details of the latter here.

The block structure of the Chebyshev–Padé table has been analyzed by Geddes [93] and by Trefethen and Gutknecht [337], [338].

18.7 Baker–Gammel Approximants

Another approach to accelerating the convergence of orthogonal polynomial expansions makes use of the generating functions of these polynomials and the resulting approximations are called *Baker–Gammel approximants*. This approach can be applied to any formal expansion $f(x) := \sum_{k=0}^{\infty} c_k \phi_k(x)$, when some generating function $G(z, x)$ for $\{\phi_k(x)\}_{k=0}^{\infty}$, namely,

$$G(z, x) = \sum_{k=0}^{\infty} \epsilon_k z^k \phi_k(x), \quad \epsilon_k \text{ constants}, \tag{18.7.1}$$

is known and it is analytic in a neighborhood of $z = 0$, with x restricted appropriately. Therefore,

$$\epsilon_k \phi_k(x) = \frac{1}{2\pi i} \oint_{|z|=\rho} z^{-k-1} G(z, x) \, dz, \quad k = 0, 1, \ldots, \tag{18.7.2}$$

such that $G(z, x)$ is analytic for $|z| \leq \rho$. Substituting (18.7.2) in $f(x)$ and interchanging formally the summation and the integration, we obtain

$$f(x) := \frac{1}{2\pi i} \oint_{|z|=\rho} \left(\sum_{k=0}^{\infty} \frac{c_k}{\epsilon_k} z^{-k-1} \right) G(z, x) \, dz. \tag{18.7.3}$$

Let $h_{m,n}(t)$ be the $[m/n]$ Padé approximant for the series $h(t) := \sum_{k=0}^{\infty} (c_k/\epsilon_k) t^k$. Assume that the contour $|z| = \rho$ in (18.7.3) can be deformed if necessary to a different contour

C_1 whose interior contains all the poles of $z^{-1}h_{m,n}(z^{-1})$. Then the $[m/n]$ Baker–Gammel approximant from $f(x)$ is defined via

$$R_{m,n}(x) = \frac{1}{2\pi i} \oint_{C_1} z^{-1}h_{m,n}(z^{-1})G(z,x)\,dz. \qquad (18.7.4)$$

If $h_{m,n}(t)$ has the partial fraction expansion (only simple poles assumed for simplicity)

$$h_{m,n}(t) = r(t) + \sum_{j=1}^{n} \frac{H_j}{1 - tz_j}, \quad r(t) = \sum_{k=0}^{m-n} r_k t^k, \qquad (18.7.5)$$

then, by the residue theorem, $R_{m,n}(x)$ becomes

$$R_{m,n}(x) = \sum_{k=0}^{m-n} \epsilon_k r_k \phi_k(x) + \sum_{j=1}^{n} H_j G(z_j, x). \qquad (18.7.6)$$

From the way $R_{m,n}(x)$ is constructed, it is clear that we need to determine the partial fraction expansion of $h_{m,n}(t)$ in (18.7.5) numerically.

Obviously, if $G(z,x)$ is not a rational function in x, $R_{m,n}(x)$ is not either. Interestingly, if we expand $R_{m,n}(x)$ formally in terms of the $\phi_k(x)$, invoking (18.7.1) and using the fact that $h_{m,n}(t) - h(t) = O(t^{m+n+1})$ as $t \to 0$, we obtain

$$R_{m,n}(x) - f(x) := \sum_{k=m+n+1}^{\infty} \tilde{c}_k \phi_k(x), \qquad (18.7.7)$$

analogously to (18.6.12) for nonlinear Padé approximants.

Let us now illustrate this approach with a few examples:

1. $G(z,x) = 1/(1 - zx)$, $\epsilon_k = 1$, $\phi_k(x) = x^k$. The resulting $R_{m,n}(x)$ is nothing but the $[m/n]$ Padé approximant from $f(x) := \sum_{k=0}^{\infty} c_k x^k$.
2. $G(z,x) = (1 - xz)/(1 - 2xz + z^2)$, $\epsilon_k = 1$, $\phi_k(x) = T_k(x)$ (Chebyshev polynomials of the first kind). The resulting $R_{m,n}(x)$ is a rational function and, more importantly, it is the $[m/n]$ Chebyshev–Padé approximant from $\sum_{k=0}^{\infty} c_k T_k(x)$.

   A different generating function for the $T_k(x)$ is $G(z,x) = 1 - \frac{1}{2}\log(1 - 2xz + z^2)$ for which $\epsilon_0 = 1$ and $\epsilon_k = 1/k$ for $k \geq 1$. Of course, the $R_{m,n}(x)$ produced via this $G(z,x)$ has logarithmic branch singularities in the x-plane.
3. $G(z,x) = 1/(1 - 2xz + z^2)$, $\epsilon_k = 1$, $\phi_k(x) = U_k(x)$ (Chebyshev polynomials of the second kind). The resulting $R_{m,n}(x)$ is a rational function and, more importantly, it is the $[m/n]$ Chebyshev–Padé approximant from $\sum_{k=0}^{\infty} c_k U_k(x)$.
4. $G(z,x) = (1 - 2xz + z^2)^{-1/2}$, $\epsilon_k = 1$, $\phi_k(x) = P_k(x)$ (Legendre polynomials). The resulting $R_{m,n}(x)$ is of the form $R_{m,n}(x) = \sum_{k=0}^{m-n} r_k P_k(x) + \sum_{j=1}^{n} H_j(1 - 2xz_j + z_j^2)^{-1/2}$ and thus has branch singularities of the square-root type in the x-plane.
5. $G(z,x) = e^{xz}$, $\epsilon_k = 1/k!$, $\phi_k(x) = x^k$. The resulting $R_{m,n}(x)$ is of the form $R_{m,n}(x) = \sum_{k=0}^{m-n} (r_k/k!)x^k + \sum_{j=1}^{n} H_j e^{xz_j}$. In this case, (18.7.7) is equivalent to $R_{m,n}^{(i)}(0) = f^{(i)}(0), i = 0, 1, \ldots, m + n$. This implies that the approximation $R_{n-1,n}(x)$ is nothing but the solution to the exponential interpolation problem treated in Theorem 17.7.2.

Note that the assumption that the poles of $h_{m,n}(t)$ be simple is not necessary. We made this assumption only for the sake of simplicity. The treatment of multiple poles poses no difficulty in determining $R_{m,n}(x)$ defined via (18.7.4).

Finally, we propose to modify the Baker–Gammel approximants $R_{m,n}(x)$ by replacing the Padé approximants $h_{m,n}(t)$ from $h(t) := \sum_{k=0}^{\infty}(c_k/\epsilon_k)t^k$, whenever possible, by other suitable rational approximations, such as the Sidi–Levin rational d-approximants we discussed earlier. This may produce excellent results when, for example, $\{c_n/\epsilon_n\} \in \mathbf{b}^{(m)}$ for some m.

18.8 Padé–Borel Approximants

Closely related to Baker–Gammel approximants are the *Padé–Borel approximants*, which have proved to be useful in summing divergent series with zero radius of convergence. These approximations were introduced in a paper by Graffi, Grecchi, and Simon [103].

Let $f(z) := \sum_{k=0}^{\infty} c_k z^k$ and assume it is known that $c_k = d_k[(pk)!]$ and $d_k = O(\rho^k)$ as $k \to \infty$ for some $p > 0$ and $\rho > 0$. Invoking in $f(z)$ the fact that $\alpha! := \int_0^{\infty} e^{-t}t^{\alpha}\,dt$, $\Re\alpha > -1$, and interchanging the integration and summation formally, we obtain

$$f(z) := \int_0^{\infty} e^{-t}g(zt^p)\,dt, \quad g(u) := \sum_{k=0}^{\infty} d_k u^k. \tag{18.8.1}$$

Note that the series $g(u)$ has a nonzero radius of convergence r and represents a function that is analytic for $u < r$. Let us denote this function by $g(u)$ as well and assume that it can be continued analytically to all $u > 0$. If, in addition, the integral in (18.8.1) converges, its value is called the *Borel sum* of $f(z)$. The Padé–Borel approximant $\hat{f}_{m,n}(z)$ of $f(z)$ is defined by

$$\hat{f}_{m,n}(z) = \int_0^{\infty} e^{-t}g_{m,n}(zt^p)\,dt, \tag{18.8.2}$$

where $g_{m,n}(u)$ is the $[m/n]$ Padé approximant from $g(u)$. The integral on the right-hand side needs to be computed numerically.

Applying Watson's lemma, it can be shown that

$$\hat{f}_{m,n}(z) - f(z) = O(z^{m+n+1}) \text{ as } z \to 0, \tag{18.8.3}$$

analogously to ordinary Padé approximants.

Of course, the Padé approximant $g_{m,n}(u)$ can be replaced by any other suitable rational approximation in this case too.

19

The Levin $\mathcal{L}$- and Sidi $\mathcal{S}$-Transformations

19.1 Introduction

In this and the next few chapters, we discuss some nonlinear sequence transformations that have proved to be effective on some or all types of logarithmic, linear, and factorial sequences $\{A_m\}$ for which $\{\Delta A_m\} \in \mathbf{b}^{(1)}$. We show how these transformations are derived, and we provide a thorough analysis of their convergence and stability with respect to columns in their corresponding tables, as we did for the iterated Δ^2-process, the iterated Lubkin transformation, and the Shanks transformation. (Analysis of the diagonal sequences turns out to be very difficult, and the number of meaningful results concerning this has remained very small.)

We recall that the sequences mentioned here are in either $\mathbf{b}^{(1)}$/LOG or $\mathbf{b}^{(1)}$/LIN or $\mathbf{b}^{(1)}$/FAC described in Definition 15.3.2. In the remainder of this work, we use the notation of this definition with no changes, as we did in previous chapters.

Before proceeding further, let us define

$$a_1 = A_1 \text{ and } a_m = \Delta A_{m-1} = A_m - A_{m-1}, \quad m = 2, 3, \dots . \quad (19.1.1)$$

Consequently, we also have

$$A_m = \sum_{k=1}^{m} a_k, \quad m = 1, 2, \dots . \quad (19.1.2)$$

19.2 The Levin $\mathcal{L}$-Transformation

19.2.1 Derivation of the $\mathcal{L}$-Transformation

We mentioned in Section 6.3 that the Levin–Sidi $d^{(1)}$-transformation reduces to the Levin u-transformation when the R_l in Definition 6.2.2 are chosen to be $R_l = l + 1$. We now treat the Levin transformations in more detail.

In his original derivation, Levin [161] considered sequences $\{A_m\}$ that behave like

$$A_m = A + \omega_m f(m), \quad f(m) \sim \sum_{i=0}^{\infty} \frac{\beta_i}{m^i} \text{ as } m \to \infty, \quad (19.2.1)$$

for some known $\{\omega_m\}$, and defined the appropriate extrapolation method, which we now call the $\mathcal{L}$-transformation, as was done in Chapter 6, via the linear equations

$$A_r = \mathcal{L}_n^{(j)} + \omega_r \sum_{i=0}^{n-1} \frac{\bar{\beta}_i}{r^i}, \quad J \le r \le J+n; \quad J = j+1. \tag{19.2.2}$$

On multiplying both sides by r^{n-1} and dividing by ω_r, these equations become

$$r^{n-1}A_r/\omega_r = \mathcal{L}_n^{(j)} r^{n-1}/\omega_r + \sum_{i=0}^{n-1} \bar{\beta}_i r^{n-1-r}, \quad J \le r \le J+n; \quad J = j+1. \tag{19.2.3}$$

Because $\sum_{i=0}^{n-1} \bar{\beta}_i r^{n-1-r}$ is a polynomial in r of degree at most $n-1$, we can now proceed as in the proof of Lemma 17.3.1 to obtain the representation

$$\mathcal{L}_n^{(j)} = \frac{\Delta^n \left(J^{n-1} A_J/\omega_J \right)}{\Delta^n \left(J^{n-1}/\omega_J \right)} = \frac{\sum_{i=0}^n (-1)^i \binom{n}{i}(J+i)^{n-1} A_{J+i}/\omega_{J+i}}{\sum_{i=0}^n (-1)^i \binom{n}{i}(J+i)^{n-1}/\omega_{J+i}}; \quad J = j+1.$$

$$\tag{19.2.4}$$

Levin considered three different choices for the ω_m and defined three different sequence transformations:

1. $\omega_m = a_m$ (*t-transformation*)
2. $\omega_m = ma_m$ (*u-transformation*)
3. $\omega_m = a_m a_{m+1}/(a_{m+1} - a_m)$ (*v-transformation*)

Of these, the u-transformation appeared much earlier in work by Bickley and Miller [24, p. 764].

Levin in his paper [161] and Smith and Ford in [317] and [318] (in which they presented an exhaustive comparative study of acceleration methods) concluded that the u- and v-transformations are effective on all three types of sequences, whereas the t-transformation is effective on linear and factorial sequences only. [Actually, all three transformations are the best convergence acceleration methods on alternating series $\sum_{k=1}^\infty (-1)|a_k|$ with $\{a_n\} \in \mathbf{b}^{(1)}$.] The theoretical justification of these conclusions can be supplied with the help of Theorem 6.7.2. That the t-transformation will be effective for linear and factorial sequences is immediate by the fact that $\sigma = 0$ in Theorem 6.7.2 for such sequences. That the u-transformation will be effective on logarithmic sequences is obvious by the fact that $\sigma = 1$ in Theorem 6.7.2 for such sequences. These also explain why the t-transformation is not effective on logarithmic sequences and why the u-transformation is effective for linear and factorial sequences as well.

The justification of the conclusion about the v-transformation is a little involved. (a) In the case of logarithmic sequences, $\{A_m\} \in \mathbf{b}^{(1)}/\text{LOG}$, we have $A_m - A = G(m)$, $G \in \mathbf{A}_0^{(\gamma)}$, $\gamma \ne 0, 1, 2, \ldots$, and we can show that $\omega_m = ma_m q(m)$ for some $q \in \mathbf{A}_0^{(0)}$ strictly. We actually have $q(m) \sim (\gamma - 1)^{-1}$ as $m \to \infty$. (b) In the case of linear sequences, $\{A_m\} \in \mathbf{b}^{(1)}/\text{LIN}$, we have $A_m - A = \zeta^m G(m)$, $\zeta \ne 1$, $G \in \mathbf{A}_0^{(\gamma)}$, with arbitrary γ, and we obtain $\omega_m = a_m q(m)$ for some $q \in \mathbf{A}_0^{(0)}$ strictly. We actually have $q(m) \sim \zeta(\zeta - 1)^{-1}$ as $m \to \infty$. (c) Finally, in the case of factorial sequences, $\{A_m\} \in \mathbf{b}^{(1)}/\text{FAC}$, we have $A_m - A = [(m)!]^{-r}\zeta^m G(m)$, $G \in \mathbf{A}_0^{(\gamma)}$, $r = 1, 2, \ldots$, and ζ and γ are arbitrary, and we can show that $\omega_m = a_{m+1} q(m)$, where $q \in \mathbf{A}_0^{(0)}$ strictly. Actually, $q(m) \sim -1$ as $m \to \infty$. Now, from Theorem 6.7.2, we have that $A_m - A = a_{m+1}g(m+1)$, where

$g \in \mathbf{A}_0^{(0)}$ strictly. Substituting here $a_{m+1} = \omega_m/q(m)$ and noting that $g(m+1)/q(m) \in \mathbf{A}_0^{(0)}$ strictly, we obtain $A_m - A = \omega_m f(m)$ with $f \in \mathbf{A}_0^{(0)}$ as well.

19.2.2 Algebraic Properties

Letting $\omega_m = m a_m$ in (19.2.4) (u-transformation), we realize that

$$\mathcal{L}_n^{(j)} = \frac{\Delta^n \left(J^{n-2} A_J / a_J \right)}{\Delta^n \left(J^{n-2} / a_J \right)} = \frac{\Delta^n \left(J^{n-2} A_j / \Delta A_j \right)}{\Delta^n \left(J^{n-2} / \Delta A_j \right)}, \quad J = j+1, \quad (19.2.5)$$

where the second equality holds for $n \geq 2$. From (19.2.5), we see that $\mathcal{L}_2^{(j)} = W_j(\{A_s\})$, where $\{W_j(\{A_s\})\}$ is the sequence produced by the Lubkin transformation. This observation is due to Bhowmick, Bhattacharya, and Roy [23].

The next theorem concerns the kernel of the u-transformation; it also provides the kernel of the Lubkin transformation discussed in Chapter 15 as a special case.

Theorem 19.2.1 *Let $\mathcal{L}_n^{(j)}$ be produced by the u-transformation on $\{A_m\}$. We have $\mathcal{L}_n^{(j)} = A$ for all $j = 0, 1, \ldots$, and fixed n, if and only if A_m is of the form*

$$A_m = A + C \prod_{k=2}^m \frac{P(k)+1}{P(k)}, \quad P(k) = \sum_{i=0}^{n-1} \beta_i k^{1-i},$$

$$C \neq 0, \ \beta_0 \neq 1, \ P(k) \neq 0, -1, \ k = 2, 3, \ldots. \quad (19.2.6)$$

Proof. Let us denote the $\bar{\beta}_i$ in the equations in (19.2.2) that define $\mathcal{L}_n^{(j)}$ by $\beta_{ni}^{(j)}$. We first note that, by the fact that $\sum_{i=0}^{n-1} \beta_{ni}^{(j)} t^i$ is a polynomial of degree at most $n-1$, the $\beta_{ni}^{(j)}$ are uniquely determined by those equations with the index $r = J+1, \ldots, J+n$, (n in number) in (19.2.2). The same equations determine the $\beta_{ni}^{(j+1)}$ when $\mathcal{L}_n^{(j+1)} = A$. This forces $\beta_{ni}^{(j)} = \beta_{ni}^{(j+1)}, 0 \leq i \leq n-1$. Therefore, $\beta_{ni}^{(j)} = \beta_i', 0 \leq i \leq n-1$, for all j. As a result,

$$A_m = A + (\Delta A_{m-1}) Q(m), \quad Q(m) = \sum_{i=0}^{n-1} \beta_i' m^{1-i}, \quad (19.2.7)$$

must hold for all m. Writing (19.2.7) in the form

$$A_m - A = [\Delta(A_{m-1} - A)] Q(m), \quad (19.2.8)$$

and solving for $A_m - A$, we obtain

$$\frac{A_m - A}{A_{m-1} - A} = \frac{P(m)+1}{P(m)}, \quad P(m) = Q(m) - 1, \quad (19.2.9)$$

from which (19.2.6) follows. The rest is left to the reader. ∎

Smith and Ford [317] have shown that the family of sequences of the partial sums of Euler series is contained in the kernel of the u-transformation. As this is difficult to conclude from (19.2.6) in Theorem 19.2.1, we provide a separate proof of it in Theorem 19.2.2 below.

Theorem 19.2.2 *Let* $A_m = \sum_{k=1}^{m} k^\mu z^k$, $m = 1, 2, \ldots$, *where* μ *is a nonnegative integer and* $z \neq 1$. *Let* $\mathcal{L}_n^{(j)}$ *be produced by the u-transformation on* $\{A_m\}$. *Provided* $n \geq \mu + 2$, *there holds* $\mathcal{L}_n^{(j)} = A$ *for all* j, *where* $A = \left(z\frac{d}{dz}\right)^\mu \frac{1}{1-z}$.

Proof. We start by observing that

$$A_m = \left(z\frac{d}{dz}\right)^\mu \sum_{k=1}^{m} z^k = \left(z\frac{d}{dz}\right)^\mu \frac{z - z^{m+1}}{1 - z}, \quad m = 1, 2, \ldots.$$

Next, it can be shown by induction on μ that

$$\left(z\frac{d}{dz}\right)^\mu \frac{z^{m+1}}{1-z} = z^{m+1}\frac{R(m,z)}{(1-z)^{\mu+1}},$$

where $R(m, z)$ is a polynomial in m and in z of degree μ. From this, we conclude that A_m satisfies (19.2.7). Substituting this in (19.2.5), the result follows. ■

We now turn to algorithms for computing the $\mathcal{L}_n^{(j)}$. First, we can use (19.2.4) as is for this purpose. Being a GREP$^{(1)}$, the $\mathcal{L}$-transformation can also be implemented very conveniently by the W-algorithm of Sidi [278] discussed in Chapter 7. For this, we need only to let $t_l = (l + 1)^{-1}$, $a(t_l) = A_{l+1}$, and $\varphi(t_l) = \omega_{l+1}$, $l = 0, 1, \ldots$, in our input for the W-algorithm. Another recursive algorithm was given independently by Longman [178] and by Fessler, Ford, and Smith [83]. This algorithm was called HURRY in [83], where a computer program that also estimates error propagation in a thorough manner is supplied. It reads as follows:

1. With $J = j + 1$ throughout, and for $j = 0, 1, \ldots$, set

$$P_0^{(j)} = A_J/\omega_J \quad \text{and} \quad Q_0^{(j)} = 1/\omega_J.$$

2. For $j = 0, 1, \ldots$, and $n = 1, 2, \ldots$, compute $P_n^{(j)}$ and $Q_n^{(j)}$ recursively from

$$U_n^{(j)} = U_{n-1}^{(j+1)} - \frac{J}{J+n}\left(\frac{J+n-1}{J+n}\right)^{n-2} U_{n-1}^{(j)},$$

where $U_n^{(j)}$ stands either for $P_n^{(j)}$ or for $Q_n^{(j)}$.

3. For all j and n, set

$$\mathcal{L}_n^{(j)} = P_n^{(j)}/Q_n^{(j)}.$$

Note that here

$$P_n^{(j)} = \frac{\Delta^n(J^{n-1}A_J/\omega_J)}{(J+n)^{n-1}} \quad \text{and} \quad Q_n^{(j)} = \frac{\Delta^n(J^{n-1}/\omega_J)}{(J+n)^{n-1}}.$$

This prevents the $P_n^{(j)}$ and $Q_n^{(j)}$ from becoming too large as n increases.

The $\mathcal{L}$-transformation was extended slightly by Weniger [353] by replacing r^i in (19.2.2) by $(r + \alpha)^i$ for some fixed α. It is easy to see that the solution for $\mathcal{L}_n^{(j)}$ is now obtained by replacing the factors J^{n-1} and $(J + i)^{n-1}$ in the numerator and denominator of (19.2.4) by $(J + \alpha)^{n-1}$ and $(J + \alpha + i)^{n-1}$, respectively. The effect of α on the quality of the resulting approximations is not clear at this time. (See Remark 1 following Definition 6.2.1 in Chapter 6.)

19.2.3 Convergence and Stability

We next summarize the convergence and stability properties of the u-transformation for Process I (concerning $\{\mathcal{L}_n^{(j)}\}_{j=0}^{\infty}$) as it is applied to sequences in $\mathbf{b}^{(1)}/\text{LOG}$ and $\mathbf{b}^{(1)}/\text{LIN}$ and $\mathbf{b}^{(1)}/\text{FAC}$.

Theorem 19.2.3 *Let A_m be as in (19.2.1), and let $\beta_{n+\mu}$ be the first nonzero β_i with $i \geq n$, and let $\mathcal{L}_n^{(j)}$ be as defined above.*

(i) *If $\omega_m \sim \sum_{i=0}^{\infty} h_i m^{-\delta-i}$ as $m \to \infty$, $h_0 \neq 0$, $\delta \neq 0, -1, -2, \dots$, then*

$$\mathcal{L}_n^{(j)} - A \sim (-1)^n \frac{(\mu+1)_n}{(\delta)_n} \beta_{n+\mu} \omega_j j^{-n-\mu} \quad and \quad \Gamma_n^{(j)} \sim \frac{(2j)^n}{|(\delta)_n|} \quad as \ j \to \infty.$$

(ii) *If $\omega_m \sim \zeta^m \sum_{i=0}^{\infty} h_i m^{-\delta-i}$ as $m \to \infty$, $h_0 \neq 0$, $\zeta \neq 1$, then*

$$\mathcal{L}_n^{(j)} - A \sim (\mu+1)_n \left(\frac{\zeta}{\zeta-1}\right)^n \beta_{n+\mu} \omega_{j+1} j^{-2n-\mu} \quad and$$

$$\Gamma_n^{(j)} \sim \left(\frac{1+|\zeta|}{|1-\zeta|}\right)^n \quad as \ j \to \infty.$$

(iii) *If $\omega_m \sim (m!)^{-r} \zeta^m \sum_{i=0}^{\infty} h_i m^{-\delta-i}$ as $m \to \infty$, $h_0 \neq 0$, $r = 1, 2, \dots$, then*

$$\mathcal{L}_n^{(j)} - A \sim (-1)^n (\mu+1)_n \beta_{n+\mu} \omega_{j+n+1} j^{-2n-\mu} \quad and \quad \Gamma_n^{(j)} \sim 1 \quad as \ j \to \infty.$$

Part (i) of this theorem was given in Sidi [270], part (ii) was given in Sidi [273], and part (iii) is new. These results can also be obtained by specializing those given in Theorems 8.4.1, 8.4.3, 9.3.1, and 9.3.2. Finally, they can also be proved by analyzing directly

$$\mathcal{L}_n^{(j)} - A = \frac{\Delta^n \left(J^{n-1} f(J)\right)}{\Delta^n \left(J^{n-1}/\omega_J\right)}; \quad J = j+1,$$

and

$$\Gamma_n^{(j)} = \frac{\sum_{i=0}^n \binom{n}{i}(J+i)^{n-1}/|\omega_{J+i}|}{|\Delta^n \left(J^{n-1}/\omega_J\right)|}; \quad J = j+1.$$

The analysis of Process II (concerning $\{\mathcal{L}_n^{(j)}\}_{n=0}^{\infty}$) turns out to be quite difficult. We can make some statements by making the proper substitutions in the theorems of Sections 8.5 and 9.4. The easiest case concerns the sequences $\{A_m\}$ in $\mathbf{b}^{(1)}/\text{LIN}$ and $\mathbf{b}^{(1)}/\text{FAC}$ and was treated in [270].

Theorem 19.2.4 *Let A_m be as in (19.2.1), with ω_m as in part (ii) or part (iii) of Theorem 19.2.3 and $\omega_m \omega_{m+1} < 0$ for $m > j$. Assume that $B(t) \equiv f(t^{-1}) \in C^{\infty}[0, \hat{t}]$ for some $\hat{t} > 0$. Then $\mathcal{L}_n^{(j)} - A = O(n^{-\lambda})$ as $n \to \infty$ for every $\lambda > 0$, and $\Gamma_n^{(j)} = 1$.*

A different approach to the general case was proposed by Sidi [273]. In this approach, one assumes that $f(m) = m \int_0^{\infty} e^{-mt} \chi(t) \, dt$, where $\chi(t)$ is analytic in a strip about the positive real axis in the t-plane. Subsequently, one analyzes

$$\mathcal{L}_n^{(j)} - A = \frac{\int_0^\infty e^{-Jt}(e^{-t}-1)^n \chi^{(n)}(t)\,dt}{\Delta^n\left(J^{n-1}/\omega_J\right)}; \quad J = j+1.$$

This can be achieved by using some of the properties of Laplace transforms. (See Appendix B.)

This approach was used successfully in the analysis of some special cases by Sidi [273] and [291]. Note that the difficult part of the analysis turns out to be that of $\Delta^n\left(J^{n-1}/\omega_J\right)$. For example, the analysis of the special case in which $\omega_m = \zeta^m/m$ is the subject of a long paper by Lubinsky and Sidi [186]. Writing

$$\Delta^n\left(J^{n-1}/\omega_J\right) = z^{j+1}\psi_n^{(j)}(z); \quad J = j+1, \ z = \zeta^{-1},$$

with

$$\psi_n^{(j)}(z) = \sum_{i=0}^n (-1)^{n-i}\binom{n}{i}(j+i+1)^n z^i,$$

it is shown in [186] that

$$\psi_n^{(0)}(z) \sim \frac{n!\,e^{\phi(z)}}{\sqrt{2\pi n\phi(z)}}\left(ze^{\phi(z)}\right)^n \quad \text{as } n \to \infty,$$

where $\phi(z)$ is the unique solution of the equation $ze^v(1-v) = 1$ in the region $\mathcal{A}$ defined by

$$\mathcal{A} = \{v = a + ib : a \ge 0, \ b \in (-\pi, \pi), \ 0 < |v-1|^2 < (b/\sin b)^2\}.$$

It is interesting (see Sidi [275] and Sidi and Lubinsky [314]) that (i) all the zeros of the polynomials $\psi_n^{(j)}(z)$ are simple and lie in $(0, 1)$, and (ii) $\psi_n^{(j)}(z)$ are orthogonal to $(\log z)^k$, $0 \le k \le n-1$, in the sense that

$$\int_0^1 z^j \psi_n^{(j)}(z)(\log z)^k \, dz = 0, \quad 0 \le k \le n-1.$$

Thus, if we define the scalars $\beta_{ni}^{(j)}$ via the relation

$$\prod_{i=1}^n \left(\xi - \frac{1}{j+i}\right) = \sum_{i=0}^n \beta_{ni}^{(j)}\xi^i,$$

and the polynomials $\Lambda_n^{(j)}(y)$ via

$$\Lambda_n^{(j)}(y) = \sum_{i=0}^n (-1)^i \frac{\beta_{ni}^{(j)}}{i!}\, y^i,$$

then the $\psi_n^{(j)}(z)$ and $\Lambda_n^{(j)}(\log z)$ form a biorthogonal set of functions, in the sense that

$$\int_0^1 z^j \psi_n^{(j)}(z)\Lambda_{n'}^{(j)}(\log z)\,dz = 0, \quad n \ne n'.$$

19.3 The Sidi $\mathcal{S}$-Transformation

We again consider sequences $\{A_m\}$ that behave as in (19.2.1). Now observe that if $K_m \sim \sum_{i=0}^{\infty} \gamma_i/m^i$ as $m \to \infty$, then $K_m \sim \sum_{i=0}^{\infty} \gamma_i'/(m)_i$ as $m \to \infty$, with $(m)_0 = 1$ and $(m)_i = \prod_{k=0}^{i-1}(m+k)$ for $i \geq 1$, as before. It is clear that $\gamma_0' = \gamma_0$, and, for each $i > 0$, γ_i' is uniquely determined by $\gamma_1, \ldots, \gamma_i$. Thus, (19.2.1) can be rewritten in the form

$$A_m = A + \omega_m f(m), \quad f(m) \sim \sum_{i=0}^{\infty} \beta_i'/(m)_i \quad m \to \infty. \tag{19.3.1}$$

We can now define the $\mathcal{S}$-transformation by truncating the asymptotic expansion in (19.3.1) via the linear system

$$A_r = \mathcal{S}_n^{(j)} + \omega_r \sum_{i=0}^{n-1} \frac{\bar{\beta}_i}{(r)_i}, \quad J \leq r \leq J+n; \quad J = j+1. \tag{19.3.2}$$

By Lemma 17.3.1, the solution of this system for $\mathcal{S}_n^{(j)}$ is given by

$$\mathcal{S}_n^{(j)} = \frac{\Delta^n \left((J)_{n-1} A_J/\omega_J \right)}{\Delta^n \left((J)_{n-1}/\omega_J \right)} = \frac{\sum_{i=0}^{n}(-1)^i \binom{n}{i}(J+i)_{n-1} A_{J+i}/\omega_{J+i}}{\sum_{i=0}^{n}(-1)^i \binom{n}{i}(J+i)_{n-1}/\omega_{J+i}}; \quad J = j+1. \tag{19.3.3}$$

The ω_m can be chosen exactly as in the $\mathcal{L}$-transformation. The resulting sequence transformations have numerical properties similar to those of the t-, u-, and v-transformations, except that they are quite inferior on sequences in $\mathbf{b}^{(1)}/\mathrm{LOG}$. The $\mathcal{S}$-transformation is very effective on sequences $\{A_m\}$ in $\mathbf{b}^{(1)}/\mathrm{LIN}/\mathrm{FAC}$; its performance on such sequences is quite similar to that of the $\mathcal{L}$-transformation. Nevertheless, the $\mathcal{L}$-transformation appears to be the best method for handling alternating series $\sum_{k=1}^{\infty}(-1)^k c_k$, $c_k > 0$ for all k.

As acknowledged also by Weniger [354, Section 2], the $\mathcal{S}$-transformation was first used as a bona fide convergence acceleration method in the M.Sc. thesis of Shelef [265], which was done under the supervision of the author. The method was applied in [265] to certain power series for the purpose of deriving some new numerical quadrature formulas for the Bromwich integral. (We consider this topic in some detail in Section 25.7.) Later, Weniger [353] observed that this transformation is a powerful accelerator for some asymptotic power series with zero radius of convergence, such as the Euler series $\sum_{k=0}^{\infty}(-1)^k k! z^{-k}$. We discuss this in the next section. The name $\mathcal{S}$-transformation was introduced by Weniger [353] as well.

Computation of the $\mathcal{S}_n^{(j)}$ can be done by direct use of the formula in (19.3.3). It can also be carried out via the following recursive algorithm that was developed by Weniger [353]:

1. With $J = j+1$ throughout, and for $j = 0, 1, \ldots$, set

$$P_0^{(j)} = A_J/\omega_J \quad \text{and} \quad Q_0^{(j)} = 1/\omega_J.$$

2. For $j = 0, 1, \ldots$, and $n = 1, 2, \ldots$, compute $P_n^{(j)}$ and $Q_n^{(j)}$ recursively from

$$U_n^{(j)} = U_{n-1}^{(j+1)} - \frac{(j+n-1)(j+n)}{(j+2n-2)(j+2n-1)} U_{n-1}^{(j)},$$

where $U_n^{(j)}$ stands for $P_n^{(j)}$ or for $Q_n^{(j)}$.

3. For all j and n, set

$$\mathcal{S}_n^{(j)} = P_n^{(j)}/Q_n^{(j)}.$$

Note that here

$$P_n^{(j)} = \frac{\Delta^n((J)_{n-1}A_J/\omega_J)}{(J+n)_{n-1}} \quad \text{and} \quad Q_n^{(j)} = \frac{\Delta^n((J)_{n-1}/\omega_J)}{(J+n)_{n-1}}.$$

This prevents the $P_n^{(j)}$ and $Q_n^{(j)}$ from becoming too large as n increases.

The treatment of the kernel of the $\mathcal{S}$-transformation is identical to that of the $\mathcal{L}$-transformation given in Theorem 19.2.1. A necessary and sufficient condition under which $\mathcal{S}_n^{(j)} = A$ for all $j = 0, 1, \ldots$, when $\omega_m = ma_m$, is that

$$A_m = A + C \prod_{k=2}^{m} \frac{P(k)+1}{P(k)}, \quad P(k) = \sum_{i=0}^{n-1} \beta_i k/(k)_i,$$

$$C \neq 0, \quad \beta_0 \neq 1, \quad P(k) \neq 0, -1, \quad k = 2, 3, \ldots.$$

We leave the details of its proof to the reader.

We next state a theorem on the convergence and stability of the $\mathcal{S}$-transformation under Process I. The proof can be accomplished in exactly the same way as those of Theorem 19.2.3 and Theorem 19.2.4.

Theorem 19.3.1 *Let A_m be as in (19.3.1), and let $\beta'_{n+\mu}$ be the first nonzero β'_i with $i \geq n$, and let $\mathcal{S}_n^{(j)}$ be as defined above. Then, all the results of Theorem 19.2.3 and that of Theorem 19.2.4 hold with $\mathcal{L}_n^{(j)}$ and $\beta_{n+\mu}$ there replaced by $\mathcal{S}_n^{(j)}$ and $\beta'_{n+\mu}$, respectively.*

The results of this theorem pertaining to the convergence of the method on sequences $\{A_m\}$ in $\mathbf{b}^{(1)}/\text{LOG}$ and $\mathbf{b}^{(1)}/\text{LIN}$ were mentioned by Weniger [353], while that pertaining to sequences in $\mathbf{b}^{(1)}/\text{FAC}$ is new.

For the proof of Theorem 19.3.1, we start with

$$\mathcal{S}_n^{(j)} - A = \frac{\Delta^n\left((J)_{n-1}f(J)\right)}{\Delta^n\left((J)_{n-1}/\omega_J\right)}; \quad J = j+1,$$

and

$$\Gamma_n^{(j)} = \frac{\sum_{i=0}^{n}\binom{n}{i}(J+i)_{n-1}/|\omega_{J+i}|}{|\Delta^n(J_{n-1}/\omega_J)|}; \quad J = j+1,$$

and proceed as before.

The $\mathcal{S}$-transformation was extended slightly by Weniger [353] by replacing $(r)_i$ in (19.3.2) by $(r+\alpha)_i$ for some fixed α. This is also covered by Lemma 17.3.1 with (17.3.2) and (17.3.3). Thus, the solution for $\mathcal{S}_n^{(j)}$ is now obtained by replacing the factors $(J)_{n-1}$ and $(J+i)_{n-1}$ in the numerator and denominator of (19.3.3) by $(J+\alpha)_{n-1}$ and $(J+\alpha+i)_{n-1}$, respectively. The effect of α on the quality of the resulting approximations is not clear at this time. (See Remark 1 following Definition 6.2.1 in Chapter 6 again.)

A more substantial extension was also given by Weniger [354]. For this, one starts by rewriting (19.3.1) in the form

$$A_m = A + \omega_m f(m), \quad f(m) \sim \sum_{i=0}^{\infty} \beta_i'' / (c[m+\alpha])_i \text{ as } m \to \infty, \quad (19.3.4)$$

which is legitimate. (As before, $\beta_0'' = \beta_0$ and, for each $i > 0$, β_i'' is uniquely determined by $\beta_1, \ldots, \beta_i$.) Here, c and α are some fixed constants. We next truncate this expansion to define the desired extension:

$$A_r = C_n^{(j)} + \omega_r \sum_{i=0}^{n-1} \frac{\bar{\beta}_i}{(c[r+\alpha])_i}, \quad J \le r \le J+n; \quad J = j+1. \quad (19.3.5)$$

These equations can also be solved by applying Lemma 17.3.1 to yield

$$C_n^{(j)} = \frac{\Delta^n \left((c[J+\alpha])_{n-1} A_J / \omega_J \right)}{\Delta^n \left((c[J+\alpha])_{n-1} / \omega_J \right)} = \frac{\sum_{i=0}^{n} (-1)^i \binom{n}{i} (c[J+\alpha+i])_{n-1} A_{J+i} / \omega_{J+i}}{\sum_{i=0}^{n} (-1)^i \binom{n}{i} (c[J+\alpha+i])_{n-1} / \omega_{J+i}}.$$
$$(19.3.6)$$

Weniger reports that the size of c may influence the performance of this transformation. For details, we refer the reader to [354].

19.4 A Note on Factorially Divergent Sequences

All our results concerning the $\mathcal{L}$- and $\mathcal{S}$-transformations so far have been on sequences $\{A_m\}$ in $\mathbf{b}^{(1)}$/LOG, $\mathbf{b}^{(1)}$/LIN, and $\mathbf{b}^{(1)}$/FAC. We have not treated those sequences that *diverge factorially*, in the sense that

$$A_m \sim (m!)^r \zeta^m \sum_{i=0}^{\infty} \alpha_i m^{\gamma-i} \text{ as } m \to \infty; \quad r > 0 \text{ integer.} \quad (19.4.1)$$

In keeping with Definition 15.3.2, we denote the class of such sequences $\mathbf{b}^{(1)}$/FACD. Such sequences arise also from partial sums of infinite series $\sum_{k=1}^{\infty} a_k$, where

$$a_m = c_m \zeta^m, \quad c_m = (m!)^r h(m), \quad h(m) \sim \sum_{i=0}^{\infty} \delta_i m^{\gamma-i} \text{ as } m \to \infty; \quad r > 0 \text{ integer.}$$
$$(19.4.2)$$

We show the truth of this statement in the following lemma.

Lemma 19.4.1 *Let a_m be as in (19.4.2) and $A_m = \sum_{k=1}^{m} a_k$. Then*

$$A_m \sim a_m \left(1 + \sum_{i=0}^{\infty} \tau_i m^{-r-i} \right) \text{ as } m \to \infty, \quad \tau_0 = 1/\zeta. \quad (19.4.3)$$

Proof. We start with

$$A_m = \sum_{k=1}^{m} a_k = a_m \left(1 + \frac{a_{m-1}}{a_m} + \cdots + \frac{a_1}{a_m} \right). \quad (19.4.4)$$

Now, for each fixed s,

$$\frac{a_{m-s}}{a_m} \sim \sum_{i=0}^{\infty} \tau_{s,i} m^{-sr-i} \quad \text{as } m \to \infty, \quad \tau_{s,0} = 1/\zeta^s.$$

Therefore, for fixed μ, there holds

$$\frac{a_{m-1}}{a_m} + \cdots + \frac{a_{m-(\mu-1)}}{a_m} = \sum_{i=0}^{\mu r-1} \tau_i m^{-r-i} + O(m^{-\mu r}) \quad \text{as } m \to \infty, \quad (19.4.5)$$

and $\tau_i, i = 0, 1, \ldots \mu r - 1$, remain unchanged if we increase μ. Also,

$$\frac{a_{m-\mu}}{a_m} + \frac{a_{m-(\mu+1)}}{a_m} + \cdots + \frac{a_1}{a_m} = O(ma_{m-\mu}/a_m) = O(m^{1-\mu r}) \quad \text{as } m \to \infty. \quad (19.4.6)$$

Combining (19.4.5) and (19.4.6) in (19.4.4), and realizing that μ is arbitrary, we obtain the result in (19.4.3). ∎

The following theorem on $\mathbf{b}^{(1)}$/FACD class sequences states that the two transformations diverge under Process I but definitely less strongly than the original sequences.

Theorem 19.4.2 *Let A_m be either as in (19.4.1) or $A_m = \sum_{k=1}^{n} a_k$ with a_m as in (19.4.2). With $\omega_m = A_m - A_{m-1}$ or $\omega_m = a_m$ in the $\mathcal{L}$- and $\mathcal{S}$-transformations, and provided $n \geq r + 1$, there holds*

$$T_n^{(j)} \sim K_n j^{-2n-s} A_{j+1} \quad \text{as } j \to \infty, \quad \text{for some } K_n > 0 \text{ and integer } s \geq 0, \quad (19.4.7)$$

where $T_n^{(j)}$ stands for either $\mathcal{L}_n^{(j)}$ or $\mathcal{S}_n^{(j)}$.

Proof. First, if A_m is as in (19.4.1), then

$$\omega_m = A_m - A_{m-1} \sim A_m \left(1 + \sum_{i=0}^{\infty} \eta_i m^{-r-i} \right) \quad \text{as } m \to \infty, \quad \eta_0 = -1/\zeta.$$

If $A_m = \sum_{k=1}^{n} a_k$ with a_m as in (19.4.2), we have from Lemma 19.4.1 similarly

$$A_m \sim a_m \left(1 + \sum_{i=0}^{\infty} \tau_i m^{-r-i} \right) \quad \text{as } m \to \infty, \quad \tau_0 = 1/\zeta.$$

In either case, $\omega_m \sim A_m$ as $m \to \infty$ and

$$\frac{A_m}{\omega_m} \sim \left(1 + \sum_{i=0}^{\infty} \epsilon_i m^{-r-i} \right) \quad \text{as } m \to \infty, \quad \epsilon_0 = 1/\zeta \neq 0.$$

In the case of the $\mathcal{L}$-transformation, we then have

$$\Delta^n (J^{n-1} A_J/\omega_J) \sim K J^{-n-1-s} \quad \text{and} \quad \Delta^n (J^{n-1}/\omega_J) \sim (-1)^n J^{n-1}/\omega_J \quad \text{as } j \to \infty,$$

for some integer $s \geq 0$ and positive constant K. We have the same situation in the case of the $\mathcal{S}$-transformation, namely,

$$\Delta^n ((J)_{n-1} A_J/\omega_J) \sim K J^{-n-1-s} \quad \text{and} \quad \Delta^n ((J)_{n-1}/\omega_J) \sim (-1)^n J^{n-1}/\omega_J \quad \text{as } j \to \infty.$$

The result now follows by invoking (19.2.4) and (19.3.3). ∎

Table 19.4.1: *Absolute errors in $S_n^{(0)}(z)$ and $\mathcal{L}_n^{(0)}(z)$ on the Euler series* $\sum_{k=0}^{\infty}(-1)^k k! z^{-k}$ *for* $z = 1/2, 1, 3$. *Here* $E_n(z;S) = |S_n^{(0)}(z) - U(z)|$ *and* $E_n(z;\mathcal{L}) = |\mathcal{L}_n^{(0)}(z) - U(z)|$, *where* $U(z) = \int_0^{\infty} \frac{e^{-x}}{1+x/z} dx$ *is the Borel sum of the series*

n	$E_n(1/2;S)$	$E_n(1/2;\mathcal{L})$	$E_n(1;S)$	$E_n(1;\mathcal{L})$	$E_n(3;S)$	$E_n(3;\mathcal{L})$
2	$4.97D-02$	$4.97D-02$	$2.49D-02$	$2.49D-02$	$3.64D-03$	$3.64D-03$
4	$9.67D-04$	$4.81D-05$	$3.38D-04$	$9.85D-04$	$2.92D-05$	$1.74D-04$
6	$4.81D-05$	$1.12D-04$	$8.66D-06$	$5.18D-05$	$1.96D-07$	$4.47D-06$
8	$4.62D-06$	$1.76D-05$	$1.30D-07$	$1.26D-06$	$1.73D-09$	$1.52D-08$
10	$5.32D-07$	$2.38D-06$	$2.50D-08$	$7.92D-08$	$6.44D-12$	$3.20D-09$
12	$1.95D-08$	$3.25D-07$	$2.49D-10$	$1.77D-08$	$5.06D-13$	$7.67D-11$
14	$9.74D-09$	$4.64D-08$	$1.31D-10$	$2.23D-09$	$2.13D-14$	$2.42D-12$
16	$7.74D-10$	$7.00D-09$	$4.37D-12$	$2.45D-10$	$2.73D-16$	$2.66D-13$
18	$1.07D-10$	$1.12D-09$	$9.61D-13$	$2.58D-11$	$1.56D-17$	$9.82D-15$
20	$3.74D-11$	$1.86D-10$	$9.23D-14$	$2.70D-12$	$8.42D-19$	$8.76D-17$
22	$3.80D-12$	$3.19D-11$	$5.58D-15$	$2.84D-13$	$6.31D-21$	$4.16D-17$
24	$4.44D-13$	$5.55D-12$	$1.70D-15$	$3.05D-14$	$1.97D-21$	$3.58D-18$
26	$2.37D-13$	$9.61D-13$	$6.84D-17$	$3.39D-15$	$2.04D-23$	$1.87D-19$
28	$3.98D-14$	$1.61D-13$	$2.22D-17$	$3.96D-16$	$5.57D-24$	$3.41D-21$
30	$2.07D-16$	$2.48D-14$	$3.90D-18$	$4.99D-17$	$1.28D-25$	$6.22D-22$
32	$1.77D-15$	$2.91D-15$	$3.87D-20$	$6.89D-18$	$2.08D-26$	$1.04D-22$
34	$5.25D-16$	$2.87D-14$	$8.36D-20$	$4.57D-18$	$1.83D-27$	$9.71D-24$
36	$7.22D-17$	$8.35D-13$	$1.51D-20$	$2.06D-16$	$2.75D-28$	$8.99D-24$
38	$2.84D-17$	$4.19D-11$	$3.74D-20$	$1.05D-15$	$6.64D-27$	$5.68D-23$
40	$2.38D-16$	$2.35D-10$	$1.66D-19$	$1.09D-13$	$2.43D-26$	$3.52D-22$

For a large class of $\{a_m\}$ that satisfy (19.4.2), it is shown by Sidi [285] that the series $\sum_{k=1}^{\infty} a_k = \sum_{k=1}^{\infty} c_k \zeta^k$ have (generalized) Borel sums $A(\zeta)$ that are analytic in the ζ-plane cut along the real interval $[0, +\infty)$. This is the case, for example, when $h(m) = m^{\omega} \int_0^{\infty} e^{-mt} \varphi(t) \, dt$ for some integer $\omega \geq 0$ and some $\varphi(t)$ of exponential order. As mentioned in [285], the numerical results of Smith and Ford [318] suggest that the $\mathcal{L}$-transformation (under Process II) produces approximations to these Borel sums. The same applies to the S-transformation, as indicated by the numerical experiments of Weniger [353].

The numerical testing done by Grotendorst [116] suggests that, on the factorially divergent series considered here, the S-transformation is more effective than the $\mathcal{L}$-transformation, which, in turn, is more effective than the Shanks transformation (equivalently, the Padé table). According to Weniger [354] and Weniger, Čížek, and Vinette [356], computations done in very high precision (up to 1000 digits) suggest that, when applied to certain very wildly divergent asymptotic power series that arise in the Rayleigh-Schrödinger perturbation theory, the $\mathcal{L}$-transformation eventually diverges, while the S-transformation converges.

Unfortunately, currently there is no mathematical theory that explains the observed numerical behavior of the $\mathcal{L}$- and S-transformations on factorially divergent sequences under Process II; this is also true of other methods that can be applied to such sequences. The one exception to this concerns the diagonals of the Padé table from $\sum_{k=0}^{\infty} c_k \zeta^k$ (equivalently, the Shanks transformation on the sequence of the partial sums) when $c_k = \int_0^{\infty} t^k d\psi(t)$, where the function $\psi(t)$ is nondecreasing on $(0, \infty)$ and has an infinite number of points of increase there. By Theorem 17.9.3, provided $\{c_k\}$ satisfies the

Carleman condition, the diagonals (analogous to Process II) converge to the corresponding Stieltjes function, which can be shown to be the Borel sum of the series in many instances.

We close with a numerical example in which we demonstrate the performances of the $\mathcal{L}$- and $\mathcal{S}$-transformations on such series under Process II.

Example 19.4.3 Let us consider the Euler series $0! - 1!z^{-1} + 2!z^{-2} - 3!z^{-3} + \cdots$. Obviously, the elements of this series satisfy (19.4.2). Its Borel sum is $U(z) = ze^z E_1(z)$, where $E_1(z) = \int_z^\infty t^{-1} e^{-t}\, dt$ is the exponential integral, and $z \notin (-\infty, 0]$. In Table 19.4.1, we give the errors in the $\mathcal{L}_n^{(0)}$ and $\mathcal{S}_n^{(0)}$ for $z = 1/2, 1, 3$. The value of $U(z)$ is obtained with machine precision by computing $U(z) = \int_0^\infty \frac{e^{-x}}{1 + x/z}\, dx$ numerically. We have $U(1/2) = 0.4614553162 \cdots$, $U(1) = 0.5963473623 \cdots$, $U(3) = 0.7862512207 \cdots$. Here we have taken $\omega_m = m a_m$ so the $\mathcal{L}$-transformation becomes the u-transformation.

We would like to note that, when applied to the Euler series, the iterated Δ^2-process, the iterated Lubkin transformation, and the Shanks transformation discussed earlier, and the θ-algorithm we discuss in Chapter 20 are able to produce fewer correct significant digits than the $\mathcal{S}$- and $\mathcal{L}$-transformations. Numerical experiments suggest that this seems to be the case generally when these methods are applied to other factorially divergent series.

In connection with the summation of the Euler series, we would like to refer the reader to the classical paper by Rosser [241], where the Euler transformation is applied in a very interesting manner.

20

The Wynn ρ- and Brezinski θ-Algorithms

20.1 The Wynn ρ-Algorithm and Generalizations

20.1.1 The Wynn ρ-Algorithm

As we saw in Chapter 16, the Shanks transformation does not accelerate the convergence of logarithmic sequences in $\mathbf{b}^{(1)}/\mathrm{LOG}$. Simultaneously with his paper [368] on the ε-algorithm for the Shanks transformation, Wynn published another paper [369] in which he developed a different method and its accompanying algorithm. This method is very effective on sequences $\{A_m\}$ for which

$$A_m \sim A + \sum_{i=1}^{\infty} \delta_i m^{-i} \quad \text{as } m \to \infty. \tag{20.1.1}$$

Of course, the set of all such sequences is a subset of $\mathbf{b}^{(1)}/\mathrm{LOG}$.

The idea and motivation behind this method is the following: Since $A_m = h(m) \in \mathbf{A}_0^{(0)}$, $h(m)$ has a smooth behavior as $m \to \infty$. Therefore, near $m = \infty$ $A_m = h(m)$ can be approximated very efficiently by a rational function in m, $R(m)$ say, with degree of numerator equal to degree of denominator, and $\lim_{m \to \infty} R(m)$ can serve as a good approximation for $A = \lim_{m \to \infty} h(m) = \lim_{m \to \infty} A_m$. In particular, $R(m)$ can be chosen to interpolate $h(m)$ at $2n + 1$ points.

Denoting the rational function with degree of numerator and degree of denominator equal to n that interpolates $f(x)$ at the points $x_j, x_{j+1}, \ldots, x_{j+2n}$ by $R_{2n}^{(j)}(x)$, we can use Thiele's continued fraction to compute $\lim_{x \to \infty} R_{2n}^{(j)}(x) \equiv \rho_{2n}^{(j)}$ recursively via

$$\rho_{-1}^{(j)} = 0, \quad \rho_0^{(j)} = f(x_j), \quad j = 0, 1, \ldots,$$

$$\rho_{k+1}^{(j)} = \rho_{k-1}^{(j+1)} + \frac{x_{j+k+1} - x_j}{\rho_k^{(j+1)} - \rho_k^{(j)}}, \quad j, k = 0, 1, \ldots. \tag{20.1.2}$$

In Wynn's method, the function we interpolate is $h(x)$, and $x_i = i, i = 0, 1, \ldots$. In other words, $\rho_{2n}^{(j)}$ is defined by the equations

$$A_m = \frac{\sum_{i=0}^{n-1} e_i m^i + \rho_{2n}^{(j)} m^n}{\sum_{i=0}^{n-1} f_i m^i + m^n}, \quad j \le m \le j + 2n.$$

Of course, the e_i and f_i are the additional (auxiliary) unknowns. Thus, we have the following computational scheme that is known as Wynn's ρ-algorithm:

Algorithm 20.1.1 (ρ-algorithm)

1. Set

$$\rho_{-1}^{(j)} = 0, \quad \rho_0^{(j)} = A_j, \quad j = 0, 1, \dots .$$

2. Compute the $\rho_k^{(j)}$ recursively from

$$\rho_{k+1}^{(j)} = \rho_{k-1}^{(j+1)} + \frac{k+1}{\rho_k^{(j+1)} - \rho_k^{(j)}}, \quad j, k = 0, 1, \dots .$$

The $\rho_k^{(j)}$ can be arranged in a two-dimensional table that is the same as that corresponding to the ε-algorithm. We mention again that only the $\rho_{2n}^{(j)}$ are the desired approximations to A. Wynn has given a determinantal expression for $\rho_{2n}^{(j)}$ in [369]. We also note that this method is quasi-linear, as can be seen easily. Finally, the kernel of the ρ-algorithm is the set of convergent sequences $\{A_m\}$ for which A_m is a rational function of m. Specifically, if $A_m = P(m)/Q(m)$, where $P(m)$ and $Q(m)$ have no common factors, $P(m)$ is of degree at most n, and $Q(m)$ is of degree exactly n, then $\rho_{2n}^{(j)} = \lim_{m\to\infty} A_m$ for all $j = 0, 1, \dots .$

It must be emphasized that the ρ-algorithm is effective only on sequences $\{A_m\}$ that are as in (20.1.1).

20.1.2 Modifications of the ρ-Algorithm

From the way the ρ-algorithm is derived, we can conclude that it will not be effective on sequences $\{A_m\} \in \mathbf{b}^{(1)}/\text{LOG}$ for which

$$A_m \sim A + \sum_{i=0}^{\infty} \alpha_i m^{\gamma-i} \quad \text{as } m \to \infty; \quad \alpha_0 \neq 0, \tag{20.1.3}$$

where γ is not a negative integer. (Recall that $\gamma \neq 0, 1, \dots$, must always be true.) The reason for this is that now $A_m = h(m)$ is not a smooth function of m at $m = \infty$ (it has a branch singularity of the form m^γ there), and we are trying to approximate it by a rational function in m that is smooth at $m = \infty$. Actually, the following theorem about the behavior of the ρ-algorithm can be proved by induction.

Theorem 20.1.2 *Let A_m be as in (20.1.3), where γ is either not an integer or $\gamma = -s$, s being a positive integer.*

(i) *When γ is not an integer, there holds for every n*

$$\rho_{2n}^{(j)} - A \sim K_{2n} j^\gamma, \quad \rho_{2n+1}^{(j)} \sim K_{2n+1} j^{-\gamma+1} \quad \text{as } j \to \infty, \tag{20.1.4}$$

where

$$K_{2n} = \alpha_0 \prod_{i=1}^{n} \frac{i+\gamma}{i-\gamma}, \quad K_{2n+1} = \frac{n+1}{\alpha_0 \gamma} \prod_{i=1}^{n} \frac{i+1-\gamma}{i+\gamma}. \tag{20.1.5}$$

(ii) *When $\gamma = -s$, (20.1.4) holds for $n = 1, 2, \ldots, s-1$, while for $n = s$, there holds*

$$\rho_{2s}^{(j)} - A = O(j^{-s-1}) \ \text{as } j \to \infty. \tag{20.1.6}$$

In other words, when γ is not an integer the ρ-algorithm induces no convergence acceleration, and when $\gamma = -s$, where s is a positive integer, convergence acceleration begins to take place with the column $\rho_{2s}^{(j)}$.

To fix this problem, Osada [225] took an approach that ignores the fact that the ρ-algorithm was obtained from rational interpolation. In this approach, Osada generalized Algorithm 20.1.1 to make it effective for the cases in which $A_m - A = g(m) \in \mathbf{A}_0^{(\gamma)}$ with arbitrary $\gamma \neq 0, 1, \ldots$. The generalized ρ-algorithm on the sequence $\{A_m\}$, which we denote the $\rho(\gamma)$-*algorithm*, reads as follows:

Algorithm 20.1.3 [$\rho(\gamma)$-algorithm]

1. Set

$$\bar{\rho}_{-1}^{(j)} = 0, \ \ \bar{\rho}_0^{(j)} = A_j, \ \ j = 0, 1, \ldots .$$

2. Compute the $\bar{\rho}_k^{(j)}$ recursively from

$$\bar{\rho}_{k+1}^{(j)} = \bar{\rho}_{k-1}^{(j+1)} + \frac{k - \gamma}{\bar{\rho}_k^{(j+1)} - \bar{\rho}_k^{(j)}}, \ \ j, k = 0, 1, \ldots .$$

Obviously, the $\rho(-1)$-algorithm is exactly the Wynn ρ-algorithm.

The following theorem by Osada [225] shows that the $\rho(\gamma)$-algorithm accelerates convergence when $A_m - A = g(m) \in \mathbf{A}_0^{(\gamma)}$ strictly.

Theorem 20.1.4 *Let $A_m - A = g(m) \in \mathbf{A}_0^{(\gamma)}$ strictly, with $\gamma \neq 0, 1, \ldots$, and apply the $\rho(\gamma)$-algorithm to $\{A_m\}$. Then*

$$\bar{\rho}_{2n}^{(j)} - A \sim \sum_{i=0}^{\infty} w_{ni} \, j^{\gamma - 2n - i} \ \text{as } j \to \infty. \tag{20.1.7}$$

Of course, we need to have precise knowledge of γ to be able to use this algorithm. In case we know that $A_m - A = g(m) \in \mathbf{A}_0^{(\gamma)}$, but we do not know γ, Osada proposes to apply the $\rho(\alpha)$-algorithm to $\{A_m\}$ with α chosen as a good estimate of γ, obtained by applying the $\rho(-2)$-algorithm to the sequence $\{\gamma_m\}$, where γ_m is given by

$$\gamma_m = 1 - \frac{(\Delta a_m)(\Delta a_{m+1})}{a_m a_{m+2} - a_{m+1}^2} = 1 + \frac{1}{\Delta (a_m / \Delta a_m)}.$$

Here $a_0 = A_0$ and $a_{m+1} = \Delta A_m$, $m = 0, 1, \ldots$. [Recall that, at the end of Section 15.3, we mentioned that $\gamma_m - \gamma = u(m) \in \mathbf{A}_0^{(-2)}$.]

Osada proposes the following algorithm, which he calls the automatic generalized ρ-algorithm. We call it the *automatic $\rho(\gamma)$-algorithm*.

Algorithm 20.1.5 [Automatic $\rho(\gamma)$-algorithm]

(i) Given the terms $A_0, A_1, \dots, A_r$, apply the $\rho(-2)$-algorithm to $\{\gamma_m\}_{m=0}^{r-2}$. If $r = 2s + 2$, set $\gamma' = \bar{\rho}_{2s}^{(0)}$, whereas if $r = 2s + 3$, set $\gamma' = \bar{\rho}_{2s}^{(1)}$.

(ii) Apply the $\rho(\gamma')$-algorithm to $\{A_m\}$ to obtain the approximation $\bar{\rho}_{2n}^{(j)}$ for $0 \leq j + 2n \leq r$.

We end this discussion by noting that the philosophy of the automatic $\rho(\gamma)$-algorithm can be used in conjunction with the repeated generalized Δ^2-process of Section 15.3, as we have already discussed.

A further modification of the ρ-algorithm that is actually a generalization of the $\rho(\gamma)$-algorithm for sequences $\{A_m\}$ such that

$$A_m \sim A + \sum_{i=0}^{\infty} \alpha_i m^{\gamma - i/p} \quad \text{as } m \to \infty; \quad p \geq 2 \text{ integer},$$

$$\alpha_0 \neq 0, \ \gamma \neq \frac{i}{p}, \ i = 0, 1, \dots . \quad (20.1.8)$$

was given by Van Tuyl [343]. We denote this algorithm the $\rho(\gamma, p)$-algorithm. It reads as follows:

Algorithm 20.1.6 [$\rho(\gamma, p)$-algorithm]

1. Set

$$\hat{\rho}_{-1}^{(j)} = 0, \ \hat{\rho}_0^{(j)} = A_j, \ j = 0, 1, \dots .$$

2. Compute the $\hat{\rho}_k^{(j)}$ recursively from

$$\hat{\rho}_{k+1}^{(j)} = \hat{\rho}_{k-1}^{(j+1)} + \frac{C_k^{(\gamma, p)}}{\hat{\rho}_k^{(j+1)} - \hat{\rho}_k^{(j)}}, \ j, k = 0, 1, \dots ,$$

   where

$$C_{2n}^{(\gamma, p)} = -\gamma + \frac{n}{p}, \ C_{2n+1}^{(\gamma, p)} = -\gamma + \frac{n}{p} + 1.$$

The next theorem by Van Tuyl [343] shows that the $\rho(\gamma, p)$-algorithm accelerates convergence when $A_m - A = h(m) \in \tilde{\mathbf{A}}_0^{(\gamma, p)}$ strictly as in (20.1.8).

Theorem 20.1.7 *With A_m as in (20.1.8), apply the $\rho(\gamma, p)$-algorithm to $\{A_m\}$. Then*

$$\hat{\rho}_{2n}^{(j)} - A \sim \sum_{i=0}^{\infty} w_{ni} j^{\gamma - (n+i)/p} \quad \text{as } j \to \infty.$$

20.2 The Brezinski θ-Algorithm

A different approach to accelerating the convergence of sequences in $\mathbf{b}^{(1)}$/LOG was given by Brezinski [32]. In this approach, Brezinski modifies the ε-algorithm in a clever way we discuss next.

Changing our notation from $\varepsilon_k^{(j)}$ to $\theta_k^{(j)}$, we would like to modify the recursion relation for $\theta_k^{(j)}$, by introducing some parameters ω_n, to read

$$\theta_{-1}^{(j)} = 0, \quad \theta_0^{(j)} = A_j, \quad j = 0, 1, \dots,$$

$$\theta_{2n+1}^{(j)} = \theta_{2n-1}^{(j+1)} + D_{2n}^{(j)}; \quad D_k^{(j)} = 1/\Delta\theta_k^{(j)}, \quad \text{for all } j, k \geq 0,$$

$$\theta_{2n+2}^{(j)} = \theta_{2n}^{(j+1)} + \omega_n D_{2n+1}^{(j)}, \quad n, j = 0, 1, \dots.$$

Here $\Delta F_k^{(j)} = F_k^{(j+1)} - F_k^{(j)}$ for all j and n. Now we would like to choose the ω_n to induce acceleration of convergence. (Of course, with $\omega_n = 1$ for all n we have nothing but the ε-algorithm.) Let us apply the difference operator Δ to both sides of the recursion for $\theta_{2n+2}^{(j)}$ and divide both sides of the resulting relation by $\Delta\theta_{2n}^{(j+1)}$. We obtain

$$\Delta\theta_{2n+2}^{(j)}/\Delta\theta_{2n}^{(j+1)} = 1 + \omega_n \Delta D_{2n+1}^{(j)}/\Delta\theta_{2n}^{(j+1)}.$$

We now require that $\{\theta_{2n+2}^{(j)}\}_{j=0}^{\infty}$ converge more quickly than $\{\theta_{2n}^{(j)}\}_{j=0}^{\infty}$. Assuming for the moment that both sequences are in $\mathbf{b}^{(1)}$/LOG, this is the same as requiring that $\lim_{j\to\infty}(\theta_{2n+2}^{(j)} - A)/(\theta_{2n}^{(j+1)} - A) = 0$. But the latter implies that $\lim_{j\to\infty} \Delta\theta_{2n+2}^{(j)}/\Delta\theta_{2n}^{(j+1)} = 0$ and, therefore,

$$\omega_n = -\lim_{j\to\infty} \Delta\theta_{2n}^{(j+1)}/\Delta D_{2n+1}^{(j)}.$$

Now, we do not know this limit to be able to determine ω_n. Therefore, we replace ω_n by

$$\omega_n^{(j)} = -\Delta\theta_{2n}^{(j+1)}/\Delta D_{2n+1}^{(j)}.$$

As a result, we obtain the θ-algorithm of Brezinski [32].

Algorithm 20.2.1 (θ-algorithm)

1. For $j = 0, 1, \dots$, set

$$\theta_{-1}^{(j)} = 0, \quad \theta_0^{(j)} = A_j.$$

2. For $j, n = 0, 1, \dots$, compute the $\theta_k^{(j)}$ recursively from

$$\theta_{2n+1}^{(j)} = \theta_{2n-1}^{(j+1)} + D_{2n}^{(j)}; \quad D_k^{(j)} = 1/\Delta\theta_k^{(j)}, \quad \text{for all } j, k \geq 0,$$

$$\theta_{2n+2}^{(j)} = \theta_{2n}^{(j+1)} - \frac{\Delta\theta_{2n}^{(j+1)}}{\Delta D_{2n+1}^{(j)}} D_{2n+1}^{(j)}. \tag{20.2.1}$$

Note that only the $\theta_{2n}^{(j)}$ are the relevant approximations to the limit or antilimit of $\{A_m\}$. For given j and n, $\theta_{2n}^{(j)}$ is determined from A_k, $j \leq k \leq j + 3n$, just like $B_n^{(j)}$ in the iterated Lubkin transformation.

It is easy to see that the θ-algorithm is quasi-linear. As mentioned before, $\{\theta_2^{(j)}\}$ is nothing but the sequence produced by the Lubkin transformation, that is, $\theta_2^{(j)} = W_j(\{A_s\})$ in the notation of Section 15.4. It is also the sequence $\{\mathcal{L}_2^{(j)}\}$ produced by the Levin u-transformation. This last observation is due to Bhowmick, Bhattacharya, and Roy [23].

The derivation of the θ-algorithm we gave here is only heuristic, as is the case with all sequence transformations. We need to show that the θ-algorithm is a bona fide convergence acceleration method.

20.2.1 Convergence and Stability of the θ-Algorithm

Because of the complexity of the recursion relation in (20.2.1), the analysis of the θ-algorithm turns out to be quite involved. Specifically, we have two different types of sequences to worry about, namely, $\{\theta_{2n}^{(j)}\}_{j=0}^{\infty}$ and $\{\theta_{2n+1}^{(j)}\}_{j=0}^{\infty}$, and the two are coupled nonlinearly. Luckily, a rigorous analysis of both types of sequences can be given, and we turn to it now.

The first result pertaining to this method was given by Van Tuyl [343] and concerns the convergence of the θ-algorithm on sequences in $\mathbf{b}^{(1)}/\mathrm{LOG}$. Theorems pertaining to convergence of the method on sequences in $\mathbf{b}^{(1)}/\mathrm{LIN}$ and $\mathbf{b}^{(1)}/\mathrm{FAC}$ have been obtained recently by Sidi in [307], where the issue of stability is also given a detailed treatment. Our treatment of the θ-algorithm here follows [307].

We begin by expressing $\theta_{2n}^{(j)}$ in forms that are different from that given in Algorithm 20.2.1, as this enables us to present its analysis in a convenient way.

Lemma 20.2.2 $\theta_{2n+2}^{(j)}$ can be expressed as in

$$\theta_{2n+2}^{(j)} = \frac{1}{\Delta^2\theta_{2n+1}^{(j)}} \left\{ W_j(\{\theta_{2n}^{(s)}\}) \times \Delta^2(1/\Delta\theta_{2n}^{(j)}) + \Delta(\theta_{2n}^{(j+1)} \times \Delta\theta_{2n-1}^{(j+1)}) \right\} \quad (20.2.2)$$

and

$$\theta_{2n+2}^{(j)} = \frac{\Delta(\theta_{2n}^{(j+1)} \times \Delta\theta_{2n+1}^{(j)})}{\Delta^2\theta_{2n+1}^{(j)}}. \quad (20.2.3)$$

Proof. From Algorithm 20.2.1, we first have

$$\theta_{2n+2}^{(j)} = \frac{\theta_{2n}^{(j+1)} \times \Delta^2\theta_{2n+1}^{(j)} + \Delta\theta_{2n}^{(j+1)} \times \Delta\theta_{2n+1}^{(j+1)}}{\Delta^2\theta_{2n+1}^{(j)}}. \quad (20.2.4)$$

Substituting $\Delta\theta_{2n}^{(j+1)} = \theta_{2n}^{(j+2)} - \theta_{2n}^{(j+1)}$ and $\Delta^2\theta_{2n+1}^{(j)} = \Delta\theta_{2n+1}^{(j+1)} - \Delta\theta_{2n+1}^{(j)}$ in the numerator of (20.2.4), we obtain (20.2.3). Next, substituting $\theta_{2n+1}^{(j)} = \theta_{2n-1}^{(j+1)} + 1/\Delta\theta_{2n}^{(j)}$ in the numerator of (20.2.3), we have

$$\theta_{2n+2}^{(j)} = \frac{\Delta(\theta_{2n}^{(j+1)} \times \Delta(1/\Delta\theta_{2n}^{(j)})) + \Delta(\theta_{2n}^{(j+1)} \times \Delta\theta_{2n-1}^{(j+1)})}{\Delta^2\theta_{2n+1}^{(j)}}. \quad (20.2.5)$$

Letting now $n = 0$ and recalling that $\theta_{-1}^{(j)} = 0$, $\theta_0^{(j)} = A_j$, and hence $\theta_1^{(j)} = 1/\Delta A_j$, we have from (20.2.5) and (15.4.2) that

$$\theta_2^{(j)} = \frac{\Delta(A_{j+1} \times \Delta(1/\Delta A_j))}{\Delta^2(1/\Delta A_j)} = W_j(\{A_s\}). \qquad (20.2.6)$$

Using this in (20.2.5), the result in (20.2.2) follows. ∎

The next theorem explains the convergence of the sequences $\{\theta_{2n}^{(j)}\}_{j=0}^\infty$.

Theorem 20.2.3 *Let us assume that the sequence $\{A_m\}$ is as in Definition 15.3.2 with exactly the same notation.*

(i) *If $\{A_m\} \in \mathbf{b}^{(1)}/\text{LOG}$, then there exist constants γ_k such that $\gamma_0 = \gamma$ and $\gamma_k - \gamma_{k-1}$ are integers ≤ -2, for which, as $j \to \infty$,*

$$\theta_{2n}^{(j)} - A \sim \sum_{i=0}^\infty w_{ni} j^{\gamma_n - i}, \quad w_{n0} \neq 0,$$

$$\theta_{2n+1}^{(j)} \sim \sum_{i=0}^\infty h_{ni} j^{-\gamma_n - i + 1}, \quad h_{n0} = 1/(\gamma_n w_{n0}) \neq 0.$$

(ii) *If $\{A_m\} \in \mathbf{b}^{(1)}/\text{LIN}$, then there exist constants γ_k such that $\gamma_0 = \gamma$ and $\gamma_k - \gamma_{k-1}$ are integers ≤ -3, for which, as $j \to \infty$,*

$$\theta_{2n}^{(j)} - A \sim \zeta^j \sum_{i=0}^\infty w_{ni} j^{\gamma_n - i}, \quad w_{n0} \neq 0,$$

$$\theta_{2n+1}^{(j)} \sim \zeta^{-j} \sum_{i=0}^\infty h_{ni} j^{-\gamma_n - i}, \quad h_{n0} = 1/[(\zeta - 1)w_{n0}] \neq 0.$$

(iii) *If $\{A_m\} \in \mathbf{b}^{(1)}/\text{FAC}$, then, as $j \to \infty$,*

$$\theta_{2n}^{(j)} - A \sim \frac{\zeta^j}{(j!)^r} \sum_{i=0}^\infty w_{ni} j^{\gamma_n - i}, \quad w_{n0} = \alpha_0[\zeta^3 r(r+1)]^n \neq 0,$$

$$\theta_{2n+1}^{(j)} \sim \zeta^{-j}(j!)^r \sum_{i=0}^\infty h_{ni} j^{-\gamma_n - i}, \quad h_{n0} = -1/w_{n0} \neq 0,$$

where $\gamma_k = \gamma - k(3r + 2)$, $k = 0, 1, \ldots$.

The proof of this theorem can be carried out by induction on n. For this, we use the error expression

$$\theta_{2n+2}^{(j)} - A = \frac{1}{\Delta^2 \theta_{2n+1}^{(j)}} \left\{ [W_j(\{\theta_{2n}^{(s)}\}) - A]\Delta^2(1/\Delta\theta_{2n}^{(j)}) + \Delta[(\theta_{2n}^{(j+1)} - A)(\Delta\theta_{2n-1}^{(j+1)})] \right\},$$

which follows from (20.2.2), and the definition of $\theta_{2n+1}^{(j)}$ given in Algorithm 20.2.1, and Theorem 15.4.1 on the convergence of the Lubkin transformation. We leave the details to the reader.

Note the similarity of Theorem 20.2.3 on the θ-algorithm to Theorem 15.4.2 on the iterated Lubkin transformation. It must be noted that the γ_k in parts (i) and (ii) of these theorems are not necessarily the same. However, in parts (iii) they are the same, and we have $\lim_{j\to\infty}(\theta_{2n}^{(j)} - A)/(B_n^{(j)} - A) = 1$. Finally, we see that the θ-algorithm accelerates the convergence of all three types of sequences in Theorem 20.2.3.

We now turn to the stability of the θ-algorithm. We first note that, by (20.2.3), $\theta_{2n+2}^{(j)}$ can be expressed as in

$$\theta_{2n+2}^{(j)} = \lambda_n^{(j)}\theta_{2n}^{(j+1)} + \mu_n^{(j)}\theta_{2n}^{(j+2)},$$

$$\lambda_n^{(j)} = -\frac{\Delta\theta_{2n+1}^{(j)}}{\Delta^2\theta_{2n+1}^{(j)}} \quad \text{and} \quad \mu_n^{(j)} = \frac{\Delta\theta_{2n+1}^{(j+1)}}{\Delta^2\theta_{2n+1}^{(j)}}. \tag{20.2.7}$$

Starting from (20.2.7), by $\theta_0^{(j)} = A_j$, $j = 0, 1, \dots$, and by the fact that $\lambda_n^{(j)} + \mu_n^{(j)} = 1$, we can easily see by induction on n that

$$\theta_{2n}^{(j)} = \sum_{i=0}^{n} \gamma_{ni}^{(j)} A_{j+n+i}; \quad \sum_{i=0}^{n} \gamma_{ni}^{(j)} = 1, \tag{20.2.8}$$

and that the $\gamma_{ni}^{(j)}$ satisfy the recursion relation

$$\gamma_{n+1,i}^{(j)} = \lambda_n^{(j)}\gamma_{ni}^{(j+1)} + \mu_n^{(j)}\gamma_{n,i-1}^{(j+2)}, \quad i = 0, 1, \dots, n+1, \tag{20.2.9}$$

where we have defined $\gamma_{n,-1}^{(j)} = \gamma_{n,n+1}^{(j)} = 0$. Next, let us define the polynomials $P_n^{(j)}(z) = \sum_{i=0}^{n} \gamma_{ni}^{(j)} z^i$. It is easy to see from (20.2.9) that the $P_n^{(j)}(z)$ satisfy the recursion

$$P_{n+1}^{(j)}(z) = \lambda_n^{(j)} P_n^{(j+1)}(z) + \mu_n^{(j)} z P_n^{(j+2)}(z). \tag{20.2.10}$$

Then we have the following theorem on the stability of the θ-algorithm.

Theorem 20.2.4 *Let us assume that the sequence $\{A_m\}$ is exactly as in Theorem 20.2.3.*

(i) *If $\{A_m\} \in \mathbf{b}^{(1)}/\text{LOG}$, then*

$$P_n^{(j)}(z) \sim \left(\prod_{k=0}^{n-1} \gamma_k\right)^{-1} (1-z)^n j^n \quad \text{and} \quad \Gamma_n^{(j)} \sim \left|\prod_{k=0}^{n-1} \gamma_k\right|^{-1} (2j)^n \quad \text{as } j \to \infty.$$

$$\tag{20.2.11}$$

(ii) *If $\{A_m\} \in \mathbf{b}^{(1)}/\text{LIN}$, then*

$$\lim_{j\to\infty} P_n^{(j)}(z) = \left(\frac{z-\zeta}{1-\zeta}\right)^n \quad \text{and} \quad \lim_{j\to\infty} \Gamma_n^{(j)} = \left(\frac{1+|\zeta|}{|1-\zeta|}\right)^n. \tag{20.2.12}$$

   (Of course, when $\zeta = -1$, $\Gamma_n^{(j)} \sim 1$ as $j \to \infty$.)

(iii) *If $\{A_m\} \in \mathbf{b}^{(1)}/\text{FAC}$, then*

$$\lim_{j\to\infty} P_n^{(j)}(z) = z^n \quad \text{and} \quad \lim_{j\to\infty} \Gamma_n^{(j)} = 1. \tag{20.2.13}$$

The proof of this theorem can be achieved by first noting that (i) $\lambda_n^{(j)} \sim j/\gamma_n$ and $\mu_n^{(j)} \sim -j/\gamma_n$ as $j \to \infty, n \geq 0$, in part (i), (ii) $\lambda_n^{(j)} \sim \zeta/(\zeta-1)$ and $\mu_n^{(j)} \sim -1/(\zeta-1)$ as $j \to \infty$, $n \geq 0$, in part (ii), and (iii) $\lambda_n^{(j)} = o(1)$ and $\mu_n^{(j)} \sim 1$ as $j \to \infty$, $n \geq 0$, in part (iii). All these follow from Theorem 20.2.3. Next, we combine these with (20.2.9) and (20.2.10) and use induction on n. We leave the details to the reader.

Again, note the similarity of Theorem 20.2.4 on the stability of the θ-algorithm with Theorem 15.5.2 on the stability of the iterated Lubkin transformation.

In view of the results of Theorem 20.2.4, we conclude that the θ-algorithm is stable on linear and factorial sequences but not on logarithmic sequences. The remarks we made in Section 15.6 about the effective use of the iterated Lubkin transformation in finite-precision arithmetic are valid for the θ-algorithm without any modifications. In particular, good accuracy can be achieved on logarithmic sequences by increasing the precision of the arithmetic used. Good accuracy is possible on logarithmic sequences when $|\Im\gamma|$ is sufficiently large, even though $\sup_j \Gamma_n^{(j)} = \infty$. As for linear sequences, in case ζ is very close to 1, $\Gamma_n^{(j)}$ is large, even though $\sup_j \Gamma_n^{(j)} < \infty$. In this case, it is best to use arithmetic progression sampling (APS), that is, we should apply the θ-algorithm to a subsequence $\{A_{\kappa m+\eta}\}$, where κ and η are fixed integers with $\kappa \geq 2$. For more details, we refer the reader to Section 15.6.

20.2.2 A Further Convergence Result

We have shown that the θ-algorithm accelerates the convergence of sequences $\{A_m\}$ for which $\{\Delta A_m\} \in \mathbf{b}^{(1)}$. In case $\{\Delta A_m\} \in \mathbf{b}^{(m)}$ with $m > 1$, however, the θ-algorithm is ineffective in general. It is effective in case A_m satisfies (20.1.8). For this application, Van Tuyl [343] provides the following convergence acceleration result.

Theorem 20.2.5 *With A_m as in (20.1.8), apply the θ-algorithm to $\{A_m\}$. Then there exist scalars γ_k such that $\gamma_0 = \gamma$ and $(\gamma_k - \gamma_{k-1})p$ is an integer ≤ -1, $k = 1, 2, \ldots$, for which*

$$\theta_{2n}^{(j)} - A \sim \sum_{i=0}^{\infty} w_{ni} j^{\gamma_n - i/p} \quad as \ j \to \infty, \quad w_{n0} \neq 0,$$

$$\theta_{2n+1}^{(j)} \sim \sum_{i=0}^{\infty} h_{ni} j^{-\gamma_n - i/p + 1} \quad as \ j \to \infty, \quad h_{n0} = 1/(\gamma_n w_{n0}) \neq 0.$$

The proof of this theorem can be carried out as that of Theorem 20.2.3, and it is left to the reader.

The G-Transformation and Its Generalizations

21.1 The G-Transformation

The G-transformation was designed by Gray and Atchison [110] for the purpose of evaluating infinite integrals of the form $\int_a^\infty f(t)\,dt$. It was later generalized in different ways in [12] and [111], the ultimate generalization being given by Gray, Atchison, and McWilliams [112]. The way it is defined by Gray and Atchison [110], the G-transformation produces an approximation to $I[f]$ that is of the form

$$G(x;h) = \frac{F(x+h) - R(x,h)F(x)}{1 - R(x,h)}, \qquad (21.1.1)$$

where

$$F(x) = \int_a^x f(t)\,dt \quad \text{and} \quad R(x,h) = f(x+h)/f(x). \qquad (21.1.2)$$

It is easy to see that $G(x;h)$ can also be expressed in the form

$$G(x;h) = \frac{\begin{vmatrix} F(x) & F(x+h) \\ f(x) & f(x+h) \end{vmatrix}}{\begin{vmatrix} 1 & 1 \\ f(x) & f(x+h) \end{vmatrix}}, \qquad (21.1.3)$$

and hence is the solution of the linear system

$$F(x) = G(x;h) + \alpha f(x)$$

$$F(x+h) = G(x;h) + \alpha f(x+h) \qquad (21.1.4)$$

Thus, the G-transformation is simply the $D^{(1)}$-transformation with $\rho_0 = 0$ in Definition 5.2.1. By Theorem 5.7.3, we see that it will be effective on functions $f(x)$ that vary exponentially as $x \to \infty$, and that it will be ineffective on those $f(x)$ that vary like some power of x at infinity. This is the subject of the following theorem.

Theorem 21.1.1

(i) *If the function $f(x)$ is of the form $f(x) = e^{cx} H(x)$ with $c \neq 0$ and $\Re c \leq 0$ and $H \in \mathbf{A}^{(\gamma)}$ for arbitrary γ, then, with h fixed such that $e^{ch} \neq 1$, there holds*

$$\frac{G(x;h) - I[f]}{F(x) - I[f]} = O(x^{-2}) \text{ as } x \to \infty.$$

(ii) *If the function* $f(x)$ *is of the form* $f(x) = H(x)$ *with* $H \in \mathbf{A}^{(\gamma)}$ *for some* $\gamma \neq -1, 0, 1, 2, \ldots$, *then*

$$\frac{G(x; h) - I[f]}{F(x) - I[f]} = O(1) \ \text{as} \ x \to \infty.$$

Proof. We start with the error formula

$$G(x; h) - I[f] = \frac{\begin{vmatrix} F(x) - I[f] & F(x + h) - I[f] \\ f(x) & f(x + h) \end{vmatrix}}{\begin{vmatrix} 1 & 1 \\ f(x) & f(x + h) \end{vmatrix}},$$

From Theorem 5.7.3, we then have

$$G(x; h) - I[f] = \frac{\begin{vmatrix} x^\rho f(x)g(x) & (x + h)^\rho f(x + h)g(x + h) \\ f(x) & f(x + h) \end{vmatrix}}{\begin{vmatrix} 1 & 1 \\ f(x) & f(x + h) \end{vmatrix}},$$

with $\rho = 0$ in part (i) and $\rho = 1$ in part (ii) and $g(x) \in \mathbf{A}^{(0)}$. The results follow from a simple analysis of the right-hand side of this equality as $x \to \infty$. ∎

Let us define the kernel of the G-transformation to be the collection of all functions $f(x) \in C^\infty[a, \infty)$ such that $G(x; h) = I[f]$ for all x and h. It is easy to show that $f(x)$ is in this kernel if $f(x) = e^{cx}$ with $c \neq 0$ and $\int_a^\infty f(t) \, dt$ is defined in the sense of Abel summability. (This implies that $\Re c \leq 0$ must hold.)

21.2 The Higher-Order G-Transformation

The higher-order G-transformation of Gray, Atchison, and McWilliams [112] is obtained by generalizing the determinantal representation of the G-transformation as follows:

$$G_n(x; h) = \frac{\begin{vmatrix} F(x) & F(x + h) & \cdots & F(x + nh) \\ f(x) & f(x + h) & \cdots & f(x + nh) \\ \vdots & \vdots & & \vdots \\ f(x + (n - 1)h) & f(x + nh) & \cdots & f(x + (2n - 1)h) \end{vmatrix}}{\begin{vmatrix} 1 & 1 & \cdots & 1 \\ f(x) & f(x + h) & \cdots & f(x + nh) \\ \vdots & \vdots & & \vdots \\ f(x + (n - 1)h) & f(x + nh) & \cdots & f(x + (2n - 1)h) \end{vmatrix}}. \tag{21.2.1}$$

As such, $G_n(x; h)$ is also the solution of the linear system

$$F(x + ih) = G_n(x; h) + \sum_{k=1}^{n} \bar{\alpha}_k f(x + (i + k - 1)h), \quad i = 0, 1, \ldots, n, \quad (21.2.2)$$

where $\bar{\alpha}_k$ are additional unknowns. This fact was noted and exploited in the computation of $G_n(x; h)$ by Levin and Sidi [165].

Let us define the kernel of the higher-order G-transformation to be the collection of all functions $f(x) \in C^{\infty}[a, \infty)$ such that $G_n(x; h) = I[f]$ for all x and h and for some appropriate n. It has been shown in [112] that $f(x)$ is in this kernel if it is integrable at infinity in the sense of Abel and satisfies a linear homogeneous ordinary differential equation with constant coefficients. Thus, $f(x)$ is in this kernel if it is of the form $f(x) = \sum_{k=1}^{r} P_k(x)e^{c_k x}$, where $c_k \neq 0$ are distinct and $\Re c_k \leq 0$, and $P_k(x)$ are polynomials. If p_k is the degree of $P_k(x)$ for each k, and if $\sum_{k=1}^{r}(p_k + 1) = n$, then $G_n(x; h) = I[f]$ for all x and h. On the basis of this result, Levin and Sidi [165] concluded that the higher-order G-transformation is effective on functions of the form $\sum_{k=1}^{s} e^{c_k x} h_k(x)$, where $h_k(x) \in \mathbf{A}^{(\gamma_k)}$ with arbitrary γ_k.

We end this section by mentioning that the papers of Gray, Atchison, and McWilliams [112] and Levin [161] have been an important source of inspiration for the D- and d-transformations.

21.3 Algorithms for the Higher-Order G-Transformation

The first effective algorithm for implementing the higher-order G-transformation was given by Pye and Atchison [233]. Actually, these authors consider the more general problem in which one would like to compute the quantities $A_n^{(j)}$ defined via the linear equations

$$A_l = A_n^{(j)} + \sum_{k=1}^{n} \bar{\alpha}_k u_{k+l-1}, \quad l = j, j + 1, \ldots, j + n, \quad (21.3.1)$$

where the A_i and u_i are known scalars, and the $\bar{\alpha}_k$ are not necessarily known. Before proceeding further, we note that these equations are the same as those in (3.7.1) with $g_k(l) = u_{k+l-1}, l = 0, 1, \ldots$. This suggests that the E- and FS-algorithms of Chapter 3 can be used for computing the $A_n^{(j)}$. Of course, direct application of these algorithms without taking into account the special nature of the $g_k(l)$ is very uneconomical. By taking the nature of the $g_k(l)$ into consideration, fast algorithms for the $A_n^{(j)}$ can be derived. Here we consider two such algorithms: (i) that of Pye and Atchison [233] that has been denoted the *rs-algorithm* and (ii) the *FS/qd-algorithm* of the author that is new.

21.3.1 The rs-Algorithm

The rs-algorithm computes the $A_n^{(j)}$ with the help of two sets of auxiliary quantities, $r_n^{(j)}$ and $s_n^{(j)}$. These quantities are defined by

$$r_n^{(j)} = \frac{H_n^{(j)}}{K_n^{(j)}}, \quad s_n^{(j)} = \frac{K_{n+1}^{(j)}}{H_n^{(j)}}, \quad (21.3.2)$$

where $H_n^{(j)}$, the Hankel determinant $H_n^{(j)}(\{u_s\})$ associated with $\{u_s\}$, and $K_n^{(j)}$ are given as in

$$
H_n^{(j)} = \begin{vmatrix} u_j & u_{j+1} & \cdots & u_{j+n-1} \\ u_{j+1} & u_{j+2} & \cdots & u_{j+n} \\ \vdots & \vdots & & \vdots \\ u_{j+n-1} & u_{j+n} & \cdots & u_{j+2n-2} \end{vmatrix}, \tag{21.3.3}
$$

and

$$
K_n^{(j)} = \begin{vmatrix} 1 & 1 & \cdots & 1 \\ u_j & u_{j+1} & \cdots & u_{j+n-1} \\ u_{j+1} & u_{j+2} & \cdots & u_{j+n} \\ \vdots & \vdots & & \vdots \\ u_{j+n-2} & u_{j+n-1} & \cdots & u_{j+2n-3} \end{vmatrix}. \tag{21.3.4}
$$

The rs-algorithm computes the $A_n^{(j)}$, $r_n^{(j)}$, and $s_n^{(j)}$ simultaneously by efficient recursions as follows:

Algorithm 21.3.1 (rs-algorithm)

1. For $j = 0, 1, \ldots$, set

$$
s_0^{(j)} = 1, \quad r_1^{(j)} = u_j, \quad A_0^{(j)} = A_j, \quad j = 0, 1, \ldots .
$$

2. For $j = 0, 1, \ldots$, and $n = 1, 2, \ldots$, compute recursively

$$
s_n^{(j)} = s_{n-1}^{(j+1)} \left(\frac{r_n^{(j+1)}}{r_n^{(j)}} - 1 \right), \quad r_{n+1}^{(j)} = r_n^{(j+1)} \left(\frac{s_n^{(j+1)}}{s_n^{(j)}} - 1 \right).
$$

3. For $j = 0, 1, \ldots$, and $n = 1, 2, \ldots$, set

$$
A_n^{(j)} = \frac{r_n^{(j)} A_{n-1}^{(j+1)} - r_n^{(j+1)} A_{n-1}^{(j)}}{r_n^{(j)} - r_n^{(j+1)}}.
$$

We now realize that $r_n^{(j)} = (-1)^{n-1} \chi_{n-1}^{(j)}(g_n)$ in the notation of the E-algorithm of Chapter 3. Thus, the rs-algorithm is simply the E-algorithm in which the $\chi_{n-1}^{(j)}(g_n)$, whose determination forms the most expensive part of the E-algorithm, are computed by a fast recursion. For this point and others, see Brezinski and Redivo Zaglia [41, Section 2.4].

21.3.2 The FS/qd-Algorithm

In view of the close connection between the rs- and E-algorithms, it is natural to investigate the possibility of designing another algorithm that is related to the FS-algorithm. This is worth the effort, as the FS-algorithm is more economical than the E-algorithm to begin with. We recall that the most expensive part of the FS-algorithm is the determination of the $D_n^{(j)}$, and we would like to reduce its cost. Fortunately, this can be achieved once we realize that, with $G_n^{(j)}$ defined as in (3.3.4), we have $G_n^{(j)} = H_n^{(j)}$ in the present case, where $H_n^{(j)}$ is as defined in (21.3.3). From this, (3.3.9), and (17.6.20), we obtain the surprising result that

$$D_n^{(j)} = \frac{H_{n+1}^{(j)} H_{n-1}^{(j+1)}}{H_n^{(j)} H_n^{(j+1)}} = e_n^{(j)}. \tag{21.3.5}$$

Here, $e_n^{(j)}$ is a quantity computed by the qd-algorithm (Algorithm 17.6.2) of Chapter 17, along with the quantities $q_n^{(j)}$ given as in

$$q_n^{(j)} = \frac{H_{n-1}^{(j)} H_n^{(j+1)}}{H_n^{(j)} H_{n-1}^{(j+1)}}. \tag{21.3.6}$$

This observation enables us to combine the FS- and qd-algorithms to obtain the following economical implementation, the FS/qd-algorithm, for the higher-order G-transformation. This has been done recently by Sidi [303]. For simplicity of notation, we let $\psi_n^{(j)}(a) = M_n^{(j)}$ and $\psi_n^{(j)}(I) = N_n^{(j)}$ in the FS-algorithm, as we did with the W-algorithm.

Algorithm 21.3.2 (FS/qd-algorithm)

1. For $j = 0, 1, \ldots$, set

$$e_0^{(j)} = 0, \quad q_1^{(j)} = \frac{u_{j+1}}{u_j}, \quad M_0^{(j)} = \frac{A_j}{u_j}, \quad N_0^{(j)} = \frac{1}{u_j}.$$

2. For $j = 0, 1, \ldots$, and $n = 1, 2, \ldots$, compute recursively

$$e_n^{(j)} = q_n^{(j+1)} - q_n^{(j)} + e_{n-1}^{(j+1)}, \quad q_{n+1}^{(j)} = \frac{e_n^{(j+1)}}{e_n^{(j)}} q_n^{(j+1)},$$

$$M_n^{(j)} = \frac{M_{n-1}^{(j+1)} - M_{n-1}^{(j)}}{e_n^{(j)}}, \quad N_n^{(j)} = \frac{N_{n-1}^{(j+1)} - N_{n-1}^{(j)}}{e_n^{(j)}}.$$

3. For $j, n = 0, 1, \ldots$, set

$$A_n^{(j)} = \frac{M_n^{(j)}}{N_n^{(j)}}.$$

We recall that the $e_n^{(j)}$ and $q_n^{(j)}$ are ordered as in Table 17.6.3 (the qd-table) and this table can be computed columnwise in the order $\{e_1^{(j)}\}, \{q_2^{(j)}\}, \{e_2^{(j)}\}, \{q_3^{(j)}\}, \ldots$.

When $A_0, A_1, \ldots, A_L$, and $u_0, u_1, \ldots, u_{2L-1}$ are given, we can determine $A_n^{(j)}$ for $0 \le j + n \le L$. Algorithm 21.3.2 will compute all these except $A_L^{(0)}$, because $A_L^{(0)} = M_L^{(0)}/N_L^{(0)}$, and $M_L^{(0)}$ and $N_L^{(0)}$ require $e_L^{(0)}$, which in turn requires u_{2L}. To avoid this, we act exactly as in (3.3.13). That is, we do not compute $e_L^{(0)}$, but we compute all the $A_n^{(0)}$ (including $A_L^{(0)}$) by

$$A_n^{(0)} = \frac{M_{n-1}^{(1)} - M_{n-1}^{(0)}}{N_{n-1}^{(1)} - N_{n-1}^{(0)}}. \tag{21.3.7}$$

Table 21.3.1: *Operation counts of the rs- and FS/qd-algorithms*

Algorithm	No. of Multiplications	No. of Additions	No. of Divisions
FS/qd	$L^2 + O(L)$	$3L^2 + O(L)$	$5L^2/2 + O(L)$
rs	$3L^2 + O(L)$	$3L^2 + O(L)$	$5L^2/2 + O(L)$

21.3.3 Operation Counts of the rs- and FS/qd-Algorithms

Let us now compare the operation counts of the two algorithms. First, we note that the $r_n^{(j)}$ and $s_n^{(j)}$ can be arranged in a table similar to the qd-table of the $e_n^{(j)}$ and $q_n^{(j)}$. Thus, given $A_0, A_1, \dots, A_L$, and $u_0, u_1, \dots, u_{2L-1}$, we can compute $A_n^{(j)}$ for $0 \le j + n \le L$. Now, the number of the $e_n^{(j)}$ in the relevant qd-table is $L^2 + O(L)$ and so is that of the $q_n^{(j)}$. A similar statement can be made about the $r_n^{(j)}$ and $s_n^{(j)}$. The number of the $A_n^{(j)}$ is $L^2/2 + O(L)$, and so are the numbers of the $M_n^{(j)}$ and the $N_n^{(j)}$. Consequently, we have the operation counts given in Table 21.3.1.

In case only the $A_n^{(0)}$ are needed (as they have the best convergence properties), the number of divisions in the FS/qd-algorithm can be reduced from $5L^2/2 + O(L)$ to $2L^2 + O(L)$. In any case, we see that the operation count of the rs-algorithm is about 30% more than that of the FS/qd-algorithm.

Finally, we observe that, with the substitution $u_k = \Delta A_k$, the higher-order G-transformation reduces to the Shanks transformation, which was discussed in Chapter 16. Actually, $A_n^{(j)} = e_n(A_j)$ by Definition 16.1.1 in this case. Thus, the rs- and FS/qd-algorithms can be used for computing the $e_n(A_j)$. Of course, the best-known algorithm for implementing the Shanks transformation is the ε-algorithm, which was given again in Chapter 16 as Algorithm 16.2.1. Given $A_0, A_1, \dots, A_{2L}$, the ε-algorithm computes $\varepsilon_{2n}^{(j)}$ for $0 \le j + 2n \le 2L$, and the diagonal approximants $\varepsilon_{2n}^{(0)}$ for $0 \le n \le L$, at a cost of $4L^2 + O(L)$ additions, $2L^2 + O(L)$ divisions, and no multiplications. Comparing this with the counts in Table 21.3.1, we see that the FS/qd-algorithm, when applied with $u_k = \Delta A_k$, computes the $\varepsilon_{2n}^{(0)}$ for $0 \le n \le L$, at a cost of $3L^2 + O(L)$ additions, $L^2 + O(L)$ multiplications, and $2L^2 + O(L)$ divisions; thus it compares very favorably with the ε-algorithm.

22

The Transformations of Overholt and Wimp

22.1 The Transformation of Overholt

In Chapter 15, we gave a detailed discussion of the Aitken Δ^2-process. There we saw that one of the uses of this method is in accelerating the convergence of fixed-point iterative procedures for the solution of nonlinear equations. In this chapter, we come back to this problem again and discuss two extensions of the Δ^2-process within the same context.

Recall that, in the iterative solution of a nonlinear equation $x = \varphi(x)$, we begin with an arbitrary approximation x_0 to the solution s and generate the sequence of approximations $\{x_m\}$ via $x_{m+1} = \varphi(x_m)$. It is known that, provided $0 < |\varphi'(s)| < 1$ and x_0 is sufficiently close to s, the sequence $\{x_m\}$ converges to s linearly in the sense that $\lim_{m\to\infty}(x_{m+1} - s)/(x_m - s) = \varphi'(s)$, and that, provided $\varphi(x)$ is infinitely differentiable in a neighborhood of s, x_m has an asymptotic expansion of the form

$$x_m \sim s + \sum_{k=1}^{\infty} \alpha_k \mu^{km} \quad \text{as } m \to \infty, \quad \alpha_1 \neq 0, \quad \mu = \varphi'(s), \qquad (22.1.1)$$

for some α_k that depend only on φ and x_0. Now, if μ were known, we could use the Richardson extrapolation process for infinite sequences (see Section 1.9) and approximate s via the $z_n^{(j)}$ that are given by

$$z_0^{(j)} = x_j, \quad j = 0, 1, \ldots,$$

$$z_n^{(j)} = \frac{z_{n-1}^{(j+1)} - \mu^n z_{n-1}^{(j)}}{1 - \mu^n}, \quad j = 0, 1, \ldots, \quad n = 1, 2, \ldots. \qquad (22.1.2)$$

Since we do not know μ, let us replace it by some suitable approximation. Now μ can be approximated in terms of the Δx_m in different ways. By (22.1.1), we have that

$$\Delta x_m \sim \sum_{k=1}^{\infty} \alpha_k (\mu^k - 1)\mu^{km} \quad \text{as } m \to \infty. \qquad (22.1.3)$$

From this and from the fact that $\alpha_1 \neq 0$, we see that $\lim_{m\to\infty} \Delta x_{m+1}/\Delta x_m = \mu$. On the basis of this, we choose to approximate μ^n in (22.1.2) by $(\Delta x_{j+n}/\Delta x_{j+n-1})^n$. This results in the method of Overholt [227], the first extension of the Δ^2-process:

$$x_0^{(j)} = x_j, \quad j = 0, 1, \ldots,$$

$$x_n^{(j)} = \frac{(\Delta x_{j+n-1})^n x_{n-1}^{(j+1)} - (\Delta x_{j+n})^n x_{n-1}^{(j)}}{(\Delta x_{j+n-1})^n - (\Delta x_{j+n})^n}, \quad j = 0, 1, \ldots, \quad n = 1, 2, \ldots. \quad (22.1.4)$$

A variant of Overholt's method was given by Meinardus [210]. This time, μ^n in (22.1.2) is approximated by $(\Delta x_{j+1}/\Delta x_j)^n$. This results in the following method:

$$x_0^{(j)} = x_j, \quad j = 0, 1, \ldots,$$

$$x_n^{(j)} = \frac{(\Delta x_j)^n x_{n-1}^{(j+1)} - (\Delta x_{j+1})^n x_{n-1}^{(j)}}{(\Delta x_j)^n - (\Delta x_{j+1})^n}, \quad j = 0, 1, \ldots, \quad n = 1, 2, \ldots. \quad (22.1.5)$$

Note that both methods are quasi-linear. Obviously, they can be applied to arbitrary sequences $\{A_m\}$, provided we replace the x_m by the corresponding A_m and the $x_n^{(j)}$ by $A_n^{(j)}$.

Finally, if we let $n = 1$ in (22.1.4) and (22.1.5), we obtain $x_1^{(j)} = \phi_j(\{x_s\})$ with $\phi_j(\{x_s\})$ as defined in (15.3.1). That is, $\{x_1^{(j)}\}$ is the sequence generated by the Δ^2-process on $\{x_m\}$.

22.2 The Transformation of Wimp

Let $\varphi(x)$, s, $\{x_m\}$, and μ be as in the preceding section. Starting with (22.1.3), let us note that, by the fact that $|\mu| < 1$, there holds $(\Delta x_m)^k \sim \sum_{i=0}^{\infty} c_{ki} \mu^{(k+i)m}$ as $m \to \infty$, $c_{k0} = [\alpha_1(\mu - 1)]^k \neq 0$, so that $(\Delta x_m)^k \sim c_{k0} \mu^{km}$ as $m \to \infty$. From this, we conclude that (i) $\{(\Delta x_m)^k\}_{k=1}^{\infty}$ is an asymptotic scale as $m \to \infty$, and (ii) we can reexpress (22.1.1) in the form

$$x_m \sim s + \sum_{k=1}^{\infty} \beta_k (\Delta x_m)^k \quad \text{as } m \to \infty, \quad \beta_1 = 1/(\mu - 1). \quad (22.2.1)$$

Here β_k is uniquely determined by $\alpha_1, \ldots, \alpha_k$ for each k. On the basis of (22.2.1), it is easy to see that we can apply the polynomial Richardson extrapolation to the sequence $\{x_m\}$. This results in the following scheme that was first presented by Wimp in [361] and rediscovered later by Germain-Bonne [96]:

$$x_0^{(j)} = x_j, \quad j = 0, 1, \ldots,$$

$$x_n^{(j)} = \frac{(\Delta x_j) x_{n-1}^{(j+1)} - (\Delta x_{j+n}) x_{n-1}^{(j)}}{\Delta x_j - \Delta x_{j+n}}, \quad j = 0, 1, \ldots, \quad n = 1, 2, \ldots. \quad (22.2.2)$$

Note that Wimp's method can also be obtained by approximating μ^n in (22.1.2) by $\Delta x_{j+n}/\Delta x_j$.

In addition to these developments, Wimp's method also has its basis in the solution of the nonlinear equation $f(x) = \varphi(x) - x = 0$ by *inverse interpolation*. This is also discussed in the book by Wimp [366, pp. 73–75, pp. 105–112]. For the subject of inverse interpolation, see, for example, Ralston and Rabinowitz [235]. Assuming that the inverse function $x = h(y) \equiv f^{-1}(y)$ exists in a neighborhood of $y = 0$, we have that $h(0) = s$. Let us denote by $q_{n,j}(y)$ the polynomial of interpolation of degree at most n to $h(y)$ at the points $y_j, y_{j+1}, \ldots , y_{j+n}$ in this neighborhood. From the Neville–Aitken interpolation formula, we thus have

$$q_{n,j}(y) = \frac{(y - y_j)q_{n-1,j+1}(y) - (y - y_{j+n})q_{n-1,j}(y)}{y_{j+n} - y_j}.$$

Letting $y = 0$ in this formula and defining $x_n^{(j)} = q_{n,j}(0)$, we obtain

$$x_n^{(j)} = \frac{y_j x_{n-1}^{(j+1)} - y_{j+n} x_{n-1}^{(j)}}{y_j - y_{j+n}}. \tag{22.2.3}$$

Now, if the y_i are sufficiently close to 0, we will have $q_{n,j}(0) \approx h(0) = s$. In other words, the $x_n^{(j)}$ will be approximations to s. What remains is to choose the y_i appropriately. With the sequence $\{x_m\}$ generated by $x_{i+1} = \varphi(x_i), i = 0, 1, \ldots ,$ let us take $y_i = f(x_i)$, $i = 0, 1, \ldots .$ Thus, $y_i = x_{i+1} - x_i = \Delta x_i$ for each i. [Furthermore, because $\{x_m\}$ converges to s when x_0 is sufficiently close to s, we have that $\lim_{i \to \infty} y_i = 0$.] Combining this with (22.2.3), we obtain the method given in (22.2.2).

Note that, just like the methods of Overholt and Meinardus, the method of Wimp is also quasi-linear. It can also be applied to arbitrary sequences $\{A_m\}$ with proper substitutions in (22.2.2).

In addition, the kernel of the method of Wimp is the set of all sequences $\{A_m\}$ for which

$$A_m = A + \sum_{k=1}^{n} \beta_k (\Delta A_m)^k, \quad m = 0, 1, \ldots .$$

Finally, if we let $n = 1$ in (22.2.2), we obtain $x_1^{(j)} = \phi_j(\{x_s\})$ with $\phi_j(\{x_s\})$ as defined in (15.3.1). That is, $\{x_1^{(j)}\}$ is the sequence generated by the Δ^2-process on $\{x_m\}$.

22.3 Convergence and Stability

Because of the way they are derived, we expect the methods we discussed in the preceding two sections to be effective on sequences $\{A_m\}$ that satisfy

$$A_m \sim A + \sum_{k=1}^{\infty} \alpha_k \mu^{km} \text{ as } m \to \infty; \quad \alpha_1 \neq 0, \ |\mu| < 1. \tag{22.3.1}$$

Therefore, we are going to investigate the convergence and stability of these methods on such sequences.

22.3.1 Analysis of Overholt's Method

We start with the following lemma.

Lemma 22.3.1 *Let r be a positive integer, μ a complex scalar, $|\mu| < 1$, and let $\{B_m\}$ and $\{C_m\}$ be such that*

$$B_m \sim \sum_{i=0}^{\infty} b_i \mu^{(r+i)m} \text{ as } m \to \infty,$$

$$C_m \sim \sum_{i=0}^{\infty} c_i \mu^{(r+i)m} \text{ as } m \to \infty; \quad c_0 \neq 0.$$

Then

$$\hat{B}_m \equiv \frac{B_m C_{m+1} - B_{m+1} C_m}{C_{m+1} - C_m} \sim \sum_{i=0}^{\infty} \hat{b}_i \mu^{(r+i+1)m} \text{ as } m \to \infty, \qquad (22.3.2)$$

where

$$\hat{b}_0 = \frac{b_0 c_1 - b_1 c_0}{c_0} \frac{\mu^r(\mu - 1)}{\mu^r - 1}.$$

In case $b_0 c_1 - b_1 c_0 \neq 0$, we have $\hat{b}_0 \neq 0$ and hence $\hat{B}_m \sim \hat{b}_0 \mu^{(r+1)m}$ as $m \to \infty$. In case $b_i = 0, 0 \leq i \leq p - 1, b_p \neq 0$, for some integer $p > 0$, we have $\hat{b}_i = 0, 0 \leq i \leq p - 1$, $\hat{b}_p \neq 0$ and hence $\hat{B}_m \sim \hat{b}_p \mu^{(r+p)m}$ as $m \to \infty$. In any case, we have $\hat{B}_m = O(\mu^{(r+1)m})$ as $m \to \infty$.

Proof. We first observe that

$$\hat{B}_m = \frac{\begin{vmatrix} B_m & B_{m+1} \\ C_m & C_{m+1} \end{vmatrix}}{\Delta C_m} \equiv \frac{X_m}{Y_m}.$$

Now, $Y_m \sim \sum_{i=0}^{\infty} c_i (\mu^{r+i} - 1) \mu^{(r+i)m}$ as $m \to \infty$, and because $c_0 \neq 0$ and $|\mu| < 1$, we also have that $Y_m \sim c_0(\mu^r - 1)\mu^{rm}$ as $m \to \infty$. Next, applying Lemma 16.4.1 to the determinant X_m, we obtain

$$X_m \sim \sum_{0 \leq i < j}^{\infty} \begin{vmatrix} b_i & b_j \\ c_i & c_j \end{vmatrix} (\mu^j - \mu^i)\mu^r \mu^{(2r+i+j)m} \text{ as } m \to \infty.$$

The result in (22.3.2) follows by combining the two asymptotic expansions. We leave the rest of the proof to the reader. ∎

Using Lemma 22.3.1 and induction on n, we can prove the following theorem on the convergence of the column sequences $\{A_n^{(j)}\}_{j=0}^{\infty}$ in the methods of Overholt and of Meinardus.

Theorem 22.3.2 *Let the sequence $\{A_m\}$ satisfy (22.3.1). Let $A_n^{(j)}$ be as in (22.1.4) or (22.1.5). Then*

$$A_n^{(j)} - A \sim \sum_{i=0}^{\infty} a_{ni} \mu^{(n+i+1)j} \text{ as } j \to \infty. \qquad (22.3.3)$$

Thus, $A_n^{(j)} - A = O(\mu^{(n+1)j})$ as $j \to \infty$.

As for stability of column sequences, we proceed exactly as in Section 15.5 on the iterated Δ^2-process. We first write

$$A_{n+1}^{(j)} = \lambda_n^{(j)} A_n^{(j)} + \mu_n^{(j)} A_n^{(j+1)}, \tag{22.3.4}$$

where

$$\lambda_n^{(j)} = -\frac{\sigma_n^{(j)}}{1 - \sigma_n^{(j)}} \quad \text{and} \quad \mu_n^{(j)} = \frac{1}{1 - \sigma_n^{(j)}}, \tag{22.3.5}$$

where $\sigma_n^{(j)} = (\Delta A_{j+n+1}/\Delta A_{j+n})^{n+1}$ for Overholt's method and $\sigma_n^{(j)} = (\Delta A_{j+1}/\Delta A_j)^{n+1}$ for Meinardus's method. Consequently, we can write

$$A_n^{(j)} = \sum_{i=0}^{n} \gamma_{ni}^{(j)} A_{j+i}; \quad \sum_{i=0}^{n} \gamma_{ni}^{(j)} = 1, \tag{22.3.6}$$

where the $\gamma_{ni}^{(j)}$ satisfy the recursion relation

$$\gamma_{n+1,i}^{(j)} = \lambda_n^{(j)} \gamma_{ni}^{(j)} + \mu_n^{(j)} \gamma_{n,i-1}^{(j+1)}, \quad i = 0, 1, \dots, n+1. \tag{22.3.7}$$

Here, we define $\gamma_{ni}^{(j)} = 0$ for $i < 0$ and $i > n$. As we did before, let us define

$$P_n^{(j)}(z) = \sum_{i=0}^{n} \gamma_{ni}^{(j)} z^i \quad \text{and} \quad \Gamma_n^{(j)} = \sum_{i=0}^{n} |\gamma_{ni}^{(j)}|. \tag{22.3.8}$$

From the fact that $\lim_{j\to\infty} \sigma_n^{(j)} = \mu^{n+1}$, we reach the following result on stability of column sequences.

Theorem 22.3.3 *Let $\{A_m\}$ and $A_n^{(j)}$ be as in Theorem 22.3.2, and let $P_n^{(j)}(z)$ and $\Gamma_n^{(j)}$ be as in (22.3.8). Then*

$$\lim_{j\to\infty} P_n^{(j)}(z) = \prod_{i=1}^{n} \frac{z - \mu^i}{1 - \mu^i} \equiv \sum_{i=0}^{n} \rho_{ni} z^i \quad \text{and}$$

$$\lim_{j\to\infty} \Gamma_n^{(j)}(z) = \sum_{i=0}^{n} |\rho_{ni}| \leq \prod_{i=1}^{n} \frac{1 + |\mu|^i}{|1 - \mu^i|}. \tag{22.3.9}$$

Therefore, the column sequences are stable. Equality holds in (22.3.9) when μ is real positive.

22.3.2 Analysis of Wimp's Method

It is easy to show that the $A_n^{(j)}$ from the method of Wimp satisfy Theorems 22.3.2 and 22.3.3. In these theorems, we now have $\sigma_n^{(j)} = \Delta A_{j+n+1}/\Delta A_j$ and $\lim_{j\to\infty} \sigma_n^{(j)} = \mu^{n+1}$ again.

These results can be obtained by observing that the method of Wimp is also what we denoted in Chapter 3, a *first generalization of the Richardson extrapolation process*, for which we now have $g_k(l) = (\Delta A_l)^k$, $l = 0, 1, \dots$, and $k = 1, 2, \dots$. We also have $\lim_{j\to\infty} g_k(l+1)/g_k(l) = \mu^k$ for all $k \geq 1$. Thus, Theorems 3.5.3–3.5.6 of Chapter 3 concerning the column sequences $\{A_n^{(j)}\}_{j=0}^{\infty}$ all hold with $c_k = \mu^k$, $k = 1, 2, \ldots$, there. Concerning the convergence of column sequences, we have the result

$$A_n^{(j)} - A \sim \beta_{n+r} \left(\prod_{i=1}^{n} \frac{\mu^{n+r} - \mu^i}{1 - \mu^i} \right) (\Delta A_j)^{n+r} \text{ as } j \to \infty,$$

$$= O(\mu^{(n+r)j}) \text{ as } j \to \infty, \qquad (22.3.10)$$

where r is that integer for which β_{n+r} is the first nonzero β_k with $k \geq n+1$ in the asymptotic expansion

$$A_m \sim A + \sum_{k=1}^{\infty} \beta_k (\Delta A_m)^k \text{ as } m \to \infty, \quad \beta_1 = \alpha_1/(\mu - 1). \qquad (22.3.11)$$

We recall that this asymptotic expansion follows from (22.3.1).

In this case, we can also give a thorough analysis of diagonal sequences $\{A_n^{(j)}\}_{n=0}^{\infty}$ since the $y_l = \Delta A_l$ satisfy $\lim_{l \to \infty}(y_{l+1}/y_l) = \mu$. These results can be obtained from Chapter 8. For example, in case μ is real positive, for any $\varepsilon > 0$ such that $\mu + \varepsilon < 1$, there exists an integer J, such that $y_{l+1}/y_l \leq \mu + \varepsilon < 1$ for all $l \geq J$. Then, with $j \geq J$, $\{A_n^{(j)}\}_{n=0}^{\infty}$ is stable and converges to A, by Theorems 8.6.1 and 8.6.4.

We can also give a different convergence theory for Wimp's method on the diagonal sequences $\{x_n^{(j)}\}_{n=0}^{\infty}$ obtained from the fixed-point iteration sequence $\{x_m\}_{n=0}^{\infty}$ that is based on the error formula

$$x_n^{(j)} - s = (-1)^n h[0, y_j, y_{j+1}, \ldots, y_{j+n}] \prod_{i=0}^{n} y_{j+i}; \quad y_l = \Delta x_l, \ l = 0, 1, \ldots,$$

derived in Chapter 2. In case $\max_{y \in I} |h^{(n)}(y)| = O(e^{\sigma n^\tau})$ as $n \to \infty$ for some $\sigma > 0$ and $\tau < 2$, where I is some interval containing all of the y_l, we have that $x_n^{(j)} - s = O(e^{-\kappa n^2})$ as $n \to \infty$ for some $\kappa > 0$.

In [366, p. 110], the transformations of Wimp and of Overholt are applied to two sequences $\{x_m\}$ from two different fixed-point iteration functions for some polynomial equation $f(x) = 0$. One of these sequences converges to the solution s of $f(x) = 0$, while the other diverges and has two limit points different from s. Numerical results indicate that both methods perform equally well on the convergent sequence, in the sense that both the columns and diagonals of their corresponding tables converge. The method of Overholt diverges (or it is unstable at best) on the divergent sequence along columns and/or diagonals, whereas the method of Wimp appears to converge along diagonals *to the solution s of the equation,* although it too suffers from instability ultimately.

For comparison purposes, we have also applied the Shanks transformation to the same sequences. It appears that the Shanks transformation performs similarly to the methods of Wimp and of Overholt but is inferior to them. In connection with the Shanks transformation on $\{x_m\}$ as in (22.1.1), we recall that column sequences converge. In fact, there holds $\varepsilon_{2n}^{(j)} - s = O(\mu^{(n+1)j})$ as $j \to \infty$, just as is the case for the methods of this chapter.

23

Confluent Transformations

23.1 Confluent Forms of Extrapolation Processes

23.1.1 Derivation of Confluent Forms

In Chapters 1–4, we were concerned with extrapolation methods on functions $A(y)$ that were assumed to be known (or computable) for $y \neq 0$. All these processes made use of $A(y_l)$ for a decreasing sequence $\{y_l\}$. Of course, these y_l are distinct. In this chapter, we are concerned with what happens to the extrapolation processes when the functions $A(y)$ are differentiable as many times as needed and the y_l coalesce. To understand the subject in a simple way, we choose to study it through the first generalization of the Richardson extrapolation process of Chapter 3.

Recall that we assume

$$A(y) \sim A + \sum_{k=1}^{\infty} \alpha_k \phi_k(y) \text{ as } y \to 0+, \tag{23.1.1}$$

where A and the α_k are some scalars independent of y and $\{\phi_k(y)\}$ is an asymptotic sequence as $y \to 0+$; that is, it satisfies

$$\phi_{k+1}(y) = o(\phi_k(y)) \text{ as } y \to 0+, \quad k = 1, 2, \dots . \tag{23.1.2}$$

Here we are interested in A, the limit or antilimit of $A(y)$ as $y \to 0+$. Recall also that $A_n^{(j)}$, the approximation to A, is defined via the linear system

$$A(y_l) = A_n^{(j)} + \sum_{k=1}^{n} \bar{\alpha}_k \phi_k(y_l), \quad j \leq l \leq j + n, \tag{23.1.3}$$

$\bar{\alpha}_1, \dots, \bar{\alpha}_n$ being the additional (auxiliary) unknowns, with $y_0 > y_1 > \cdots$, and $\lim_{l\to\infty} y_l = 0$. Using Cramer's rule, we can express $A_n^{(j)}$ in the determinantal form given in (3.2.1). We reproduce (3.2.1) here explicitly in terms of the functions

$\phi_k(y)$:

$$
A_n^{(j)} = \frac{\begin{vmatrix} \phi_1(y_j) & \cdots & \phi_n(y_j) & A(y_j) \\ \phi_1(y_{j+1}) & \cdots & \phi_n(y_{j+1}) & A(y_{j+1}) \\ \vdots & & \vdots & \vdots \\ \phi_1(y_{j+n}) & \cdots & \phi_n(y_{j+n}) & A(y_{j+n}) \end{vmatrix}}{\begin{vmatrix} \phi_1(y_j) & \cdots & \phi_n(y_j) & 1 \\ \phi_1(y_{j+1}) & \cdots & \phi_n(y_{j+1}) & 1 \\ \vdots & & \vdots & \vdots \\ \phi_1(y_{j+n}) & \cdots & \phi_n(y_{j+n}) & 1 \end{vmatrix}}.
\tag{23.1.4}
$$

By performing elementary row transformations on both determinants in (23.1.4), we obtain

$$
A_n^{(j)} = \frac{\begin{vmatrix} \phi_1(y_j) & \cdots & \phi_n(y_j) & A(y_j) \\ D_1^{(j)}\{\phi_1(y)\} & \cdots & D_1^{(j)}\{\phi_n(y)\} & D_1^{(j)}\{A(y)\} \\ \vdots & & \vdots & \vdots \\ D_n^{(j)}\{\phi_1(y)\} & \cdots & D_n^{(j)}\{\phi_n(y)\} & D_n^{(j)}\{A(y)\} \end{vmatrix}}{\begin{vmatrix} \phi_1(y_j) & \cdots & \phi_n(y_j) & 1 \\ D_1^{(j)}\{\phi_1(y)\} & \cdots & D_1^{(j)}\{\phi_n(y)\} & 0 \\ \vdots & & \vdots & \vdots \\ D_n^{(j)}\{\phi_1(y)\} & \cdots & D_n^{(j)}\{\phi_n(y)\} & 0 \end{vmatrix}},
\tag{23.1.5}
$$

where $D_k^{(s)}\{g(y)\}$ stands for $g[y_s, y_{s+1}, \ldots, y_{s+k}]$, the divided difference of $g(y)$ over the set of points $\{y_s, y_{s+1}, \ldots, y_{s+k}\}$. Let us now assume that $A(y)$ and the $\phi_k(y)$ are all differentiable functions of y, and let all y_l tend to the same value, say $\bar{y}$, simultaneously. Then $D_k^{(s)}\{A(y)\} \to A^{(k)}(\bar{y})/k!$, provided $A(y) \in C^k(I)$, where I is an interval containing $\bar{y}$; similarly, for the $\phi_k(y)$. Consequently, the limits of the determinants in the numerator and the denominator of (23.1.5), and hence the limit of $A_n^{(j)}$, all exist. We summarize this discussion in the following theorem, where we also replace $\bar{y}$ by y for simplicity.

Theorem 23.1.1 *Let $A_n^{(j)}$ be defined via (23.1.3). If $A(y)$ and the $\phi_k(y)$ are n times differentiable in y for y in some right neighborhood of 0, then the limit of $A_n^{(j)}$ as $y_l \to y$, $j \le l \le j + n$, exists. Denoting this limit by $Q_n(y)$, we have $Q_n(y) = \Lambda_n[A(y)]$, where*

$$
\Lambda_n[f(y)] \equiv \frac{\begin{vmatrix} f(y) & \phi_1(y) & \cdots & \phi_n(y) \\ f'(y) & \phi_1'(y) & \cdots & \phi_n'(y) \\ \vdots & \vdots & & \vdots \\ f^{(n)}(y) & \phi_1^{(n)}(y) & \cdots & \phi_n^{(n)}(y) \end{vmatrix}}{\begin{vmatrix} \phi_1'(y) & \cdots & \phi_n'(y) \\ \vdots & & \vdots \\ \phi_1^{(n)}(y) & \cdots & \phi_n^{(n)}(y) \end{vmatrix}}.
\tag{23.1.6}
$$

We call this method that generates the $Q_n(y)$ the *first confluent form* of the first generalization of the Richardson extrapolation process. It is easy to verify that $Q_n(y)$ is also the solution of the linear system

$$A(y) = Q_n(y) + \sum_{k=1}^{n} \bar{\alpha}_k \phi_k(y),$$

$$A^{(i)}(y) = \sum_{k=1}^{n} \bar{\alpha}_k \phi_k^{(i)}(y), \quad i = 1, \ldots, n, \qquad (23.1.7)$$

where the $\bar{\alpha}_k$ are the additional (auxiliary) unknowns. Note that this linear system is obtained by differentiating the asymptotic expansion in (23.1.1) formally term by term i times, truncating the summation at the term $\alpha_n \phi_n(y)$, replacing $\sim$ by $=$, A by $Q_n(y)$, and the α_k by $\bar{\alpha}_k$, and setting $i = 0, 1, \ldots, n$.

The recursive algorithms of Chapter 3 can be used to obtain $Q_n(y)$ of (23.1.6), once we realize that the equations in (23.1.7) can be rewritten in the form $a(l) = A_n^{(0)} + \sum_{k=1}^{n} \bar{\alpha}_k g_k(l)$, $l = 0, 1, \ldots, n$, with $a(l) = \sum_{s=0}^{l} \binom{l}{s} A^{(s)}(y)$ and $g_k(l) = \sum_{s=0}^{l} \binom{l}{s} \phi_k^{(s)}(y)$ and $A_n^{(0)} = Q_n(y)$. See Brezinski and Redivo Zaglia [41, p. 267]. Of course, the most direct way to compute $Q_n(y)$ is by solving the last n of the equations in (23.1.7) numerically for the $\bar{\alpha}_k$ and substituting these in the first equation.

Note that, for computing $Q_n(y) = \Lambda_n[A(y)]$, we need $A(y)$ and its derivatives of order $1, \ldots, n$. In addition, the quality of $Q_n(y)$ improves with increasing n. Therefore, to obtain high accuracy by the first confluent form, we need to compute a large number of derivatives of $A(y)$. Consequently, the first confluent form can be of practical value provided the high-order derivatives of $A(y)$ can be obtained relatively easily.

We now propose another method that requires $A(y)$ and the $\phi_k(y)$ and their first order derivatives only. We derive this new method in a way that is similar to the derivation of the first confluent form via the linear system in (23.1.7).

Let us differentiate (23.1.1) once term by term as before, truncate the summation at the term $\alpha_n \phi_n(y)$, replace $\sim$ by $=$, A by $Q_n^{(j)}(y)$, and the α_k by $\bar{\alpha}_k$, and collocate at the points y_l, $j \leq l \leq j + n - 1$. This results in the linear system

$$A(y) = Q_n^{(j)}(y) + \sum_{k=1}^{n} \bar{\alpha}_k \phi_k(y),$$

$$A'(y_l) = \sum_{k=1}^{n} \bar{\alpha}_k \phi_k'(y_l), \quad j \leq l \leq j + n - 1, \qquad (23.1.8)$$

where y in the first of these equations can take on any value, and the y_l are chosen to satisfy $y_0 > y_1 > \cdots$ and $\lim_{l \to \infty} y_l = 0$ as before.

As we can set $y = y_l$ for $l \in \{j, j + 1, \ldots, j + n - 1\}$ in (23.1.8), we call this method that generates the $Q_n^{(j)}(y)$ the *second confluent form* of the first generalization of the Richardson extrapolation process.

Using Cramer's rule to solve (23.1.8), we have $Q_n^{(j)}(y) = \Lambda_n^{(j)}[A(y)]$, where

$$
\Lambda_n^{(j)}[f(y)] = \frac{\begin{vmatrix} f(y) & \phi_1(y) & \cdots & \phi_n(y) \\ f'(y_j) & \phi_1'(y_j) & \cdots & \phi_n'(y_j) \\ \vdots & \vdots & & \vdots \\ f'(y_{j+n-1}) & \phi_1'(y_{j+n-1}) & \cdots & \phi_n'(y_{j+n-1}) \end{vmatrix}}{\begin{vmatrix} \phi_1'(y_j) & \cdots & \phi_n'(y_j) \\ \vdots & & \vdots \\ \phi_1'(y_{j+n-1}) & \cdots & \phi_n'(y_{j+n-1}) \end{vmatrix}}. \tag{23.1.9}
$$

The following result can be obtained by analyzing the determinantal representation of $Q_n^{(j)}(y)$. We leave its proof to the reader.

Theorem 23.1.2 *Let $Q_n^{(j)}(y)$ be defined via (23.1.8). If $A(y)$ and the $\phi_k(y)$ are n times differentiable in y for y in some right neighborhood of 0, then the limit of $Q_n^{(j)}(y)$ as $y_l \to y$, $j \le l \le j + n - 1$, exists, and satisfies*

$$
\lim_{\substack{y_l \to y \\ j \le l \le j+n-1}} Q_n^{(j)}(y) = Q_n(y), \tag{23.1.10}
$$

where $Q_n(y)$ is the first confluent form defined above.

To determine $Q_n^{(j)}(y)$ numerically we may proceed exactly as proposed in the case of the first confluent form. Thus, we first solve the last n of the equations in (23.1.8) for the $\bar{\alpha}_k$, and substitute these in the first equation to compute $Q_n^{(j)}(y)$ for some y, possibly in the set $\{y_j, y_{j+1}, \ldots, y_{j+n-1}\}$.

As the y_l and y are an important part of the input to the second confluent form, a question that arises naturally is whether one can find a "best" set of these that will give the "highest" accuracy in $Q_n^{(j)}(y)$. This seems to be an interesting research problem.

23.1.2 Convergence Analysis of a Special Case

We now present a convergence theory of the two preceding confluent methods for the special case in which $\phi_k(y) = y^{\sigma_k}$, $k = 1, 2, \ldots$, that was treated in Chapters 1 and 2. This theory is analogous to those we saw in the previous chapters and that pertain to column sequences, and it shows that both of the confluent methods accelerate convergence in this mode.

We first note that

$$
Q_n(y) - A = \Lambda_n[A(y) - A], \quad Q_n^{(j)}(y) - A = \Lambda_n^{(j)}[A(y) - A]. \tag{23.1.11}
$$

Next, in our analysis we assume for each $i = 1, 2, \ldots$, that (i) $\{\phi_k^{(i)}(y)\}_{k=1}^\infty$ is an asymptotic sequence, and (ii) $A^{(i)}(y)$ exists and has an asymptotic expansion that is obtained by differentiating that of $A(y)$ given in (23.1.1) term by term i times. Thus,

$$
A^{(i)}(y) \sim \sum_{k=1}^\infty \alpha_k \phi_k^{(i)}(y) \quad \text{as } y \to 0+, \quad i = 1, 2, \ldots. \tag{23.1.12}
$$

Substituting (23.1.1) and (23.1.12) in (23.1.11), recalling that $\lim_{j \to \infty} y_j = 0$ in $Q_n^{(j)}(y_j)$, we obtain, formally, the asymptotic expansions

$$Q_n(y) - A \sim \sum_{k=n+1}^{\infty} \alpha_k \Lambda_n[\phi_k(y)] \text{ as } y \to 0+,$$

$$Q_n^{(j)}(y_j) - A \sim \sum_{k=n+1}^{\infty} \alpha_k \Lambda_n^{(j)}[\phi_k(y_j)] \text{ as } j \to \infty. \tag{23.1.13}$$

[Note that both summations in (23.1.13) start with $k = n + 1$ since $\Lambda_n[\phi_k(y)] = 0$ and $\Lambda_n^{(j)}[\phi_k(y_j)] = 0$ for $k = 1, \ldots, n$, as is clear from (23.1.6) and (23.1.9).] As usual, it is necessary to prove that these are valid asymptotic expansions. For instance, for $Q_n(y) - A$, it must be shown that (i) $\{\Lambda_n[\phi_k(y)]\}_{k=n+1}^{\infty}$ is an asymptotic scale as $y \to 0+$, and (ii) for each positive integer $N \geq n + 1$, there holds $Q_n(y) - A - \sum_{k=n+1}^{N-1} \alpha_k \Lambda_n[\phi_k(y)] = O(\Lambda_n[\phi_N(y)])$ as $y \to 0+$. Similarly, for $Q_n^{(j)}(y_j) - A$.

Going back to $\phi_k(y) = y^{\sigma_k}$, $k = 1, 2, \ldots$, we first have

$$A(y) \sim A + \sum_{k=1}^{\infty} \alpha_k y^{\sigma_k} \text{ as } y \to 0+, \tag{23.1.14}$$

where

$$\sigma_k \neq 0, \quad k = 1, 2, \ldots; \quad \Re\sigma_1 < \Re\sigma_2 < \cdots; \quad \lim_{k \to \infty} \Re\sigma_k = \infty. \tag{23.1.15}$$

By the fact that $\phi_k^{(i)}(y) = [\sigma_k]_i y^{\sigma_k - i}$, it follows from (23.1.15) that $\{\phi_k^{(i)}(y)\}_{k=1}^{\infty}$, $i = 1, 2, \ldots$, are asymptotic sequences as $y \to 0+$. Here $[x]_0 = 1$ and $[x]_i = x(x-1)\cdots(x-i+1)$ for $i = 1, 2, \ldots$, as usual. We also assume that, for each $i \geq 1$, $A^{(i)}(y)$ has an asymptotic expansion as $y \to 0+$, which can be obtained by differentiating that of $A(y)$ term by term i times. Thus,

$$A^{(i)}(y) \sim \sum_{k=1}^{\infty} [\sigma_k]_i \alpha_k y^{\sigma_k - i} \text{ as } y \to 0+, \quad i = 1, 2, \ldots. \tag{23.1.16}$$

The following lemma concerns $\Lambda_n[\phi_k(y)]$ and $\Lambda_n^{(j)}[\phi_k(y_j)]$. We leave its proof to the reader. (Of course, the condition imposed on the y_l in this lemma is relevant only for the second confluent form.)

Lemma 23.1.3 *With $\phi_k(y) = y^{\sigma_k}$, and the σ_k as in (23.1.15), and with $y_l = y_0\omega^l$, $l = 1, 2, \ldots$, for some $\omega \in (0, 1)$ and y_0, we have*

$$\Lambda_n[\phi_k(y)] = \epsilon_{n,k} y^{\sigma_k}; \quad \epsilon_{n,k} = \prod_{i=1}^{n} \frac{\sigma_i - \sigma_k}{\sigma_i},$$

$$\Lambda_n^{(j)}[\phi_k(y_j)] = \tau_{n,k} y_j^{\sigma_k}; \quad \tau_{n,k} = \frac{\sigma_k}{V(c_1, \ldots, c_n)} \begin{vmatrix} 1/\sigma_k & 1/\sigma_1 & \cdots & 1/\sigma_n \\ c_k^0 & c_1^0 & \cdots & c_n^0 \\ c_k^1 & c_1^1 & \cdots & c_n^1 \\ \vdots & \vdots & & \vdots \\ c_k^{n-1} & c_1^{n-1} & \cdots & c_n^{n-1} \end{vmatrix}. \tag{23.1.17}$$

Thus, $\{\Lambda_n[\phi_k(y)]\}_{k=n+1}^{\infty}$ is an asymptotic scale as $y \to 0+$ and $\{\Lambda_n^{(j)}[\phi_k(y_j)]\}_{k=n+1}^{\infty}$ is an asymptotic scale as $j \to \infty$. Here $c_k = \omega^{\sigma_k}$, $k = 1, 2, \ldots$, and $V(\xi_1, \ldots, \xi_n)$ is the Vandermonde determinant as usual.

We next give the main convergence results that concern the first and second confluent forms.

Theorem 23.1.4 *With $A(y)$ as in (23.1.14)–(23.1.16) and with y_l as in Lemma 23.1.3, $Q_n(y) - A$ and $Q_n^{(j)}(y_j) - A$ have the complete asymptotic expansions given in (23.1.13). That is,*

$$Q_n(y) - A \sim \sum_{k=n+1}^{\infty} \epsilon_{n,k} \alpha_k y^{\sigma_k} \quad as \ y \to 0+,$$

$$Q_n^{(j)}(y_j) - A \sim \sum_{k=n+1}^{\infty} \tau_{n,k} \alpha_k y_j^{\sigma_k} \quad as \ j \to \infty, \quad (23.1.18)$$

with $\epsilon_{n,k}$ and $\tau_{n,k}$ as in (23.1.17). Thus, if $\alpha_{n+\mu}$ is the first nonzero α_k with $k \geq n + 1$, then $Q_n(y) - A$ and $Q_n^{(j)}(y_j) - A$ satisfy the asymptotic equalities

$$Q_n(y) - A \sim \epsilon_{n,n+\mu} \alpha_{n+\mu} y^{\sigma_{n+\mu}} \quad as \ y \to 0+,$$

$$Q_n^{(j)}(y_j) - A \sim \tau_{n,n+\mu} \alpha_{n+\mu} y_j^{\sigma_{n+\mu}} \quad as \ j \to \infty. \quad (23.1.19)$$

We leave the proof of the results in (23.1.18) and (23.1.19) to the reader.

23.2 Confluent Forms of Sequence Transformations

23.2.1 Confluent ε-Algorithm

The idea of confluent methods first appeared in a paper by Wynn [370], and it was based on algorithms for sequence transformations. To illustrate the idea, let us consider the Shanks transformation and the ε-algorithm, as was done by Wynn [370].

We start by replacing the sequence A_m by some function $F(x)$, making the analogy $\lim_{m \to \infty} A_m \leftrightarrow \lim_{x \to \infty} F(x)$, when both limits exist. Let us next replace $\varepsilon_{2n}^{(j)}$ and $\varepsilon_{2n+1}^{(j)}$ in Algorithm 16.2.1 by $\varepsilon_{2n}(x + jh)$ and $h^{-1}\varepsilon_{2n+1}(x + jh)$, respectively. Finally, let us allow $h \to 0$. As a result, we obtain the *confluent ε-algorithm* of Wynn that reads as follows:

Algorithm 23.2.1 (Confluent ε-algorithm)

1. Set

$$\varepsilon_{-1}(x) = 0 \quad and \quad \varepsilon_0(x) = F(x).$$

2. Compute the $\varepsilon_{k+1}(x)$ by the recursion

$$\varepsilon_{k+1}(x) = \varepsilon_{k-1}(x) + \frac{1}{\varepsilon_k'(x)}, \quad k = 0, 1, \ldots .$$

The relevant approximations to $\lim_{x \to \infty} F(x)$ are the $\varepsilon_{2n}(x)$.
The following result is due to Wynn [370].

Theorem 23.2.2 *For $\varepsilon_{2n}(x)$ and $\varepsilon_{2n+1}(x)$ we have*

$$\varepsilon_{2n}(x) = \frac{H_{n+1}^{(0)}(x)}{H_n^{(2)}(x)} \quad and \quad \varepsilon_{2n+1}(x) = \frac{H_n^{(3)}(x)}{H_{n+1}^{(1)}(x)}, \tag{23.2.1}$$

where, for all $j \geq 0$, we define $H_0^{(j)}(x) = 1$ and

$$H_n^{(j)}(x) = \begin{vmatrix} F^{(j)}(x) & F^{(j+1)}(x) & \cdots & F^{(j+n-1)}(x) \\ F^{(j+1)}(x) & F^{(j+2)}(x) & \cdots & F^{(j+n)}(x) \\ \vdots & \vdots & & \vdots \\ F^{(j+n-1)}(x) & F^{(j+n)}(x) & \cdots & F^{(j+2n-2)}(x) \end{vmatrix}, \quad n \geq 1. \tag{23.2.2}$$

Consequently, $\varepsilon_{2n}(x)$ is the solution of the linear system

$$F(x) = \varepsilon_{2n}(x) + \sum_{k=1}^{n} \bar{\alpha}_k F^{(k)}(x),$$

$$F^{(i)}(x) = \sum_{k=1}^{n} \bar{\alpha}_k F^{(k+i)}(x), \quad i = 1, \dots, n. \tag{23.2.3}$$

Note that the determinants $H_n^{(j)}(x)$ are analogous to the Hankel determinants introduced in (16.1.13) in connection with the Shanks transformation. Note also that the linear system in Theorem (23.2.3) is completely analogous to that in (23.1.7), and $\varepsilon_{2n}(x)$ can be computed as the solution of this system. Another way of computing the $\varepsilon_{2n}(x)$ is via (23.2.1), with the $H_n^{(j)}(x)$ being determined from the recursion

$$H_0^{(j)}(x) = 1 \quad and \quad H_1^{(j)}(x) = F^{(j)}(x), \quad j = 0, 1, \dots,$$

$$H_{n+1}^{(j)}(x) H_{n-1}^{(j+2)}(x) = H_n^{(j)}(x) H_n^{(j+2)}(x) - [H_n^{(j+1)}(x)]^2, \quad j \geq 0, \quad n \geq 1. \tag{23.2.4}$$

This can be proved by applying Sylvester's determinant identity to the determinant $H_{n+1}^{(j)}(x)$. For yet another algorithm, see Wynn [376].

Some of the algebraic properties of the confluent ε-algorithm and its application to functions $F(x)$ that are completely monotonic are discussed by Brezinski [33]. See also Brezinski [36] and Brezinski and Redivo Zaglia [41]. Unfortunately, the confluent ε-algorithm cannot be very practical, as it requires knowledge of high-order derivatives of $F(x)$.

23.2.2 Confluent Form of the Higher-Order G-Transformation

The confluent form of the higher-order G-transformation described in Chapter 21 is obtained by letting $h \to 0$ in (21.2.1). By suitable row and column transformations on the determinants in (21.2.1), we first obtain

$$
G_n(x;h) = \frac{
\begin{vmatrix}
F(x) & \Delta F(x)/h & \cdots & \Delta^n F(x)/h^n \\
f(x) & \Delta f(x)/h & \cdots & \Delta^n f(x)/h^n \\
\vdots & \vdots & & \vdots \\
\Delta^{n-1} f(x)/h^{n-1} & \Delta^n f(x)/h^n & \cdots & \Delta^{2n-1} f(x)/h^{2n-1}
\end{vmatrix}
}{
\begin{vmatrix}
\Delta f(x)/h & \cdots & \Delta^n f(x)/h^n \\
\vdots & & \vdots \\
\Delta^n f(x)/h^n & \cdots & \Delta^{2n-1} f(x)/h^{2n-1}
\end{vmatrix}
}, \quad (23.2.5)
$$

where $\Delta g(x) = g(x+h) - g(x)$, $\Delta^2 g(x) = \Delta(\Delta g(x))$, and so on. Next, by letting $h \to 0$ in (23.2.5) and using the fact that $f(x) = F'(x)$, which follows from $F(x) = \int_0^x f(t)\,dt$, and also the fact that $\lim_{h\to 0} \Delta^k g(x)/h^k = g^{(k)}(x)$, we obtain

$$
B_n(x) = \lim_{h\to 0} G_n(x;h) = \frac{H_{n+1}^{(0)}(x)}{H_n^{(2)}(x)}. \quad (23.2.6)
$$

Here $H_n^{(j)}(x)$ is exactly as defined in (23.2.2). Thus, $B_n(x)$ is nothing but $\varepsilon_{2n}(x)$ and can be computed by the confluent ε-algorithm.

All these developments are due to Gray, Atchison, and McWilliams [112], where the convergence acceleration properties of $B_n(x)$ as $x \to \infty$ are also discussed.

23.2.3 Confluent ρ-Algorithm

Wynn [370] applies the preceding approach to the ρ-algorithm. In other words, he replaces $\rho_{2n}^{(j)}$ and $\rho_{2n+1}^{(j)}$ in Algorithm 20.1.1 by $\rho_{2n}(x+jh)$ and $h^{-1}\rho_{2n+1}(x+jh)$ respectively, and lets $h \to 0$. This results in the *confluent ρ-algorithm* of Wynn, which reads as follows:

Algorithm 23.2.3 (Confluent ρ-algorithm)

1. Set

$$
\rho_{-1}(x) = 0 \quad \text{and} \quad \rho_0(x) = F(x).
$$

2. Compute the $\rho_{k+1}(x)$ by the recursion

$$
\rho_{k+1}(x) = \rho_{k-1}(x) + \frac{k+1}{\rho_k'(x)}, \quad k = 0, 1, \ldots .
$$

The relevant approximations to $\lim_{x\to\infty} F(x)$ are the $\rho_{2n}(x)$.
The following result is due to Wynn [370].

Theorem 23.2.4 *For $\rho_{2n}(x)$ and $\rho_{2n+1}(x)$ we have*

$$\rho_{2n}(x) = \frac{\bar{H}_{n+1}^{(0)}(x)}{\bar{H}_{n}^{(2)}(x)} \quad and \quad \rho_{2n+1}(x) = \frac{\bar{H}_{n}^{(3)}(x)}{\bar{H}_{n+1}^{(1)}(x)}, \tag{23.2.7}$$

where, for each $j \geq 0$ and $n \geq 0$, $\bar{H}_{n}^{(j)}(x)$ is obtained from $H_{n}^{(j)}(x)$ by replacing $F^{(i)}(x)$ in the latter by $F^{(i)}(x)/i!$, $i = 0, 1, \ldots$.

It is clear that $\rho_{2n}(x)$ can be computed from (23.2.7), where the $H_{n}^{(j)}(x)$ can be determined by using a recursion relation similar to that in (23.2.4). For further results, see Wynn [376]. See also [41]. It is obvious that the confluent ρ-algorithm, just like the confluent ε-algorithm, is not very practical as it requires knowledge of high-order derivatives of $F(x)$.

23.2.4 Confluent Overholt Method

Applying the technique above to the method of Overholt, Brezinski and Redivo Zaglia [41] derive the following confluent form:

Algorithm 23.2.5 (Confluent Overholt method)

1. Set $V_0(x) = F(x)$.
2. Compute $V_{n+1}(x)$ by the recursion

$$V_{n+1}(x) = V_n(x) - \frac{F'(x)}{F''(x)} \cdot \frac{V_n'(x)}{n+1}, \quad n = 0, 1, \ldots .$$

The relevant approximations to $\lim_{x \to \infty} F(x)$ are the $V_n(x)$.

The following theorem, whose proof can be achieved by induction on n, is given in [41, p. 255, Theorem 5.1].

Theorem 23.2.6 *Assume that $\{[F'(x)]^k\}_{k=1}^{\infty}$ is an asymptotic sequence as $x \to \infty$, and that*

$$F(x) \sim A + \sum_{k=1}^{\infty} \alpha_k [F'(x)]^k \quad as \ x \to \infty. \tag{23.2.8}$$

Assume, in addition, that both sides of (23.2.8) can be differentiated term by term. Then

$$V_n(x) - A \sim \sum_{k=n+1}^{\infty} \alpha_k \left[\prod_{s=1}^{n} (1 - k/s) \right] [F'(x)]^k \quad as \ x \to \infty. \tag{23.2.9}$$

We do not know of another more convenient way of defining the $V_n(x)$ except through the recursion relation of Algorithm 23.2.5. This recursion relation, however, requires us to first obtain $V_n(x)$ in closed form and then differentiate it. Of course, this task can be achieved only in some cases, and by using symbolic computation. Thus, the confluent form of the method of Overholt is not very useful.

The confluent form of the θ-algorithm is obtained similarly in [41]. As mentioned there, this confluent form suffers from the same deficiency as the confluent form of the method of Overholt; hence, it is not very useful.

23.3 Confluent $D^{(m)}$-Transformation

With the approach of the preceding sections, it is possible to derive confluent forms of additional transformations. For example, Levin and Sidi [165, Section 3] give the confluent form of the $D^{(m)}$-transformation for infinite-range integrals $\int_0^\infty f(t)\,dt$ (called the C-transformation in [165]) that is obtained in the same way as the first confluent form of the Richardson extrapolation process. This derivation reads as follows:

$$F(x) = C_n(x) + \sum_{k=1}^m x^k f^{(k-1)}(x) \sum_{i=0}^{n_k-1} \bar{\beta}_{ki} x^{-i},$$

$$F^{(s)}(x) = \frac{d^s}{dx^s}\left\{\sum_{k=1}^m x^k f^{(k-1)}(x) \sum_{i=0}^{n_k-1} \bar{\beta}_{ki} x^{-i}\right\}, \quad s = 1, 2, \ldots, N, \quad (23.3.1)$$

where $F(x) = \int_0^x f(t)\,dt$ hence $F^{(s)}(x) = f^{(s-1)}(x)$ for all $s \geq 1, n = (n_1, \ldots, n_m)$, and $N = \sum_{k=1}^m n_k$. Gray and Wang [114] provide convergence results concerning $C_n(x)$ as $x \to \infty$. See also Gray and Wang [113].

Now, this first confluent form requires the computation of the derivatives of $f(x)$ of order as high as $N + m - 1$, which may be inconvenient as N is a large integer generally. Here, we propose the second confluent form, which circumvents this inconvenience entirely. From the formalism of Section 23.1, this reads as follows:

$$F(x) = C_n^{(j)}(x) + \sum_{k=1}^m x^k f^{(k-1)}(x) \sum_{i=0}^{n_k-1} \bar{\beta}_{ki} x^{-i},$$

$$F'(x_l) = \frac{d}{dx}\left\{\sum_{k=1}^m x^k f^{(k-1)}(x) \sum_{i=0}^{n_k-1} \bar{\beta}_{ki} x^{-i}\right\}\bigg|_{x=x_l}, \quad j \leq l \leq j + N - 1, \quad (23.3.2)$$

where $F(x)$, n, and N are as before and the x_l are chosen to satisfy $0 < x_0 < x_1 < \cdots$, and $\lim_{l\to\infty} x_l = \infty$. Once the $\bar{\beta}_{ki}$ are determined from the last N of the equations in (23.3.2), they can be substituted in the first equation to compute $C_n^{(j)}(x)$, with $x = x_j$, for example.

Obviously, this transformation requires the computation of only one (finite-range) integral $F(x)$, and knowledge of the derivatives $f^{(k)}(x), k = 1, 2, \ldots, m$, independently of the size of N. As before, the question about the "best" choice of x and the x_l is of interest.

23.3.1 Application to the $D^{(1)}$-Transformation and Fourier Integrals

As a special case, let us consider application of the second confluent form of the $D^{(1)}$-transformation to the Fourier integral $\int_0^\infty e^{i\omega t} g(t)\,dt$, where $g \in \mathbf{A}^{(\gamma)}$. Replacing the $k = 1$ term in the summations of the equations in (23.3.2) by $f(x) = e^{i\omega x} g(x)$ (recall that $\rho_0 = 0$ for such integrals), letting $m = 1$, these equations become

$$F(x) = C_n^{(j)}(x) + f(x) \sum_{i=0}^{n-1} \bar{\beta}_i x^{-i}$$

$$f(x_l) = -f(x_l) \sum_{i=1}^{n-1} i \bar{\beta}_i x_l^{-i-1} + f'(x_l) \sum_{i=0}^{n-1} \bar{\beta}_i x^{-i}, \quad j \leq l \leq j+n-1. \quad (23.3.3)$$

Substituting $f'(x) = [i\omega + g'(x)/g(x)] f(x)$ in the last n equations, we see that the $\bar{\beta}_s$ can be obtained by solving the linear system

$$\sum_{s=0}^{n-1} \left[i\omega + \frac{g'(x_l)}{g(x_l)} - \frac{s}{x_l} \right] \frac{1}{x_l^s} \bar{\beta}_s = 1, \quad j \leq l \leq j+n-1. \quad (23.3.4)$$

Obviously, this is a rather inexpensive way of approximating the integral in question, since it requires the computation of only one finite-range integral, namely, the integral $F(x)$ in (23.3.3).

24

Formal Theory of Sequence Transformations

24.1 Introduction

The purpose of this chapter is to present a formal theory of sequence transformations that was begun recently by Germain-Bonne [96], [97]. The theory of Germain-Bonne covers very few cases. It was later extended by Smith and Ford [317] to cover more cases. Unfortunately, even after being extended, so far the formal theory includes a very small number of cases of interest and excludes the most important ones. In addition, for the cases it includes, it has produced results relevant to column sequences only, and these results are quite weak in the sense that they do not give any information about rates of acceleration. Nevertheless, we have chosen to present its present achievements briefly here for the sake of completeness. Our treatment of the subject here follows those of Smith and Ford [317] and Wimp [366, Chapter 5, pp. 101–105].

Let us denote the approximations that result by applying an extrapolation method ExtM to the sequence $\{A_m\}_{m=0}^{\infty}$ by $S_{n,j}$, where

$$S_{n,j} = G_{n,j}(A_j, A_{j+1}, \dots, A_{j+n}) \tag{24.1.1}$$

for some function $G_{n,j}(x_0, x_1, \dots, x_n)$. We assume that this function satisfies

$$G_{n,j}(\alpha x_0 + \beta, \alpha x_1 + \beta, \dots, \alpha x_n + \beta) = \alpha G_{n,j}(x_0, x_1, \dots, x_n) + \beta \tag{24.1.2}$$

for all scalars α and β. This, of course, means that ExtM is a quasi-linear sequence transformation, in the sense defined in the Introduction. It is clear that

$$G_{n,j}(0, 0, \dots, 0) = 0, \quad G_{n,j}(\beta, \beta, \dots, \beta) = \beta.$$

Using (24.1.2), we can write

$$G_{n,j}(x_0, x_1, \dots, x_n) = x_0 + (x_1 - x_0)G_{n,j}(0, Y_1, Y_2, \dots, Y_n), \tag{24.1.3}$$

where we have defined

$$Y_k = \frac{x_k - x_0}{x_1 - x_0}, \quad k = 1, 2, \dots. \tag{24.1.4}$$

Letting

$$\Delta x_k = x_{k+1} - x_k \quad \text{and} \quad X_k = \frac{\Delta x_k}{\Delta x_{k-1}}, \tag{24.1.5}$$

407

we see that

$$Y_1 = 1, \quad Y_k = 1 + \sum_{i=1}^{k-1} X_1 X_2 \cdots X_i, \quad k = 2, 3, \ldots . \qquad (24.1.6)$$

Using (24.1.4)–(24.1.6), and defining the function $g_{n,j}$ through

$$g_{n,j}(X_1, \ldots, X_{n-1}) = G_{n,j}(0, 1, Y_2, \ldots, Y_n), \qquad (24.1.7)$$

we now write (24.1.3) in the form

$$G_{n,j}(x_0, x_1, \ldots, x_n) = x_0 + (\Delta x_0) g_{n,j}(X_1, X_2, \ldots, X_{n-1}). \qquad (24.1.8)$$

Let S stand for the limit or antilimit of $\{A_m\}$, and define

$$R_m = \frac{A_{m+1} - S}{A_m - S}, \quad r_m = \frac{\Delta A_{m+1}}{\Delta A_m}, \quad m = 0, 1, \ldots . \qquad (24.1.9)$$

Going back to $S_{n,j}$ in (24.1.1), and invoking (24.1.8), we thus have that

$$S_{n,j} = A_j + (\Delta A_j) g_{n,j}(r_j, \ldots, r_{j+n-2}), \qquad (24.1.10)$$

and hence

$$\frac{S_{n,j} - S}{A_j - S} = 1 + \frac{\Delta A_j}{A_j - S} g_{n,j}(r_j, \ldots, r_{j+n-2}). \qquad (24.1.11)$$

Before going on, we mention that, in the theory of Germain-Bonne, the functions $G_{n,j}$ and hence $g_{n,j}$ do not depend on j explicitly, that is, they are the same for all j. It is this aspect of the original theory that makes it relevant for only a limited number of cases. The explicit dependence on j that was introduced by Smith and Ford allows more cases to be covered. We present examples of both types of $g_{n,j}$ in the next section.

24.2 Regularity and Acceleration

We now investigate the regularity and acceleration properties of $S_{n,j}$ as $j \to \infty$, while n is being held fixed. The following results are direct consequences of (24.1.10) and (24.1.11).

Proposition 24.2.1

(i) *Provided the sequence $\{g_{n,j}(r_j, \ldots, r_{j+n-2})\}_{j=0}^{\infty}$ is bounded, ExtM is regular on $\{A_m\}$ as $j \to \infty$, that is, $\lim_{j\to\infty} S_{n,j} = \lim_{j\to\infty} A_j = S$.*

(ii) *$\{S_{n,j}\}_{j=0}^{\infty}$ converges faster than $\{A_m\}$, that is, $\lim_{j\to\infty}(S_{n,j} - S)/(A_j - S) = 0$, if and only if*

$$\lim_{j\to\infty} (1 - R_j) g_{n,j}(r_j, \ldots, r_{j+n-2}) = 1. \qquad (24.2.1)$$

24.2.1 Linearly Convergent Sequences

In case the functions $g_{n,j}$ for fixed n are all the same function g_n, and the sequence $\{A_m\}$ converges linearly, the previous proposition can be refined considerably. In the present context, $\{A_m\}$ is linearly converging if, with R_m as defined in (24.1.9), there holds

$$\lim_{m\to\infty} R_m = \lambda \text{ for some } \lambda, \ 0 < |\lambda| < 1. \qquad (24.2.2)$$

As mentioned in the proof of of Theorem 15.3.1, (24.2.2) implies that $\lim_{m\to\infty} r_m = \lambda$ as well, where r_m is as defined in (24.1.9). Note also that the family of sequences $\mathbf{b}^{(1)}/\mathrm{LIN}$ is a subset of the set of linearly convergent sequences.

Using these in Proposition 24.2.1, we can now state the following theorem, whose proof we leave to the reader.

Theorem 24.2.2 *Let $\{A_m\}$ be a linearly convergent sequence as in (24.2.2). In the sequence transformation ExtM, let*

$$\lim_{j\to\infty} g_{n,j}(X_1,\dots,X_{n-1}) = g_n(X_1,\dots,X_{n-1}), \qquad (24.2.3)$$

and assume that g_n is continuous in a neighborhood of $(\lambda,\lambda,\dots,\lambda)$, $\lambda \neq 1$, and satisfies

$$g_n(\lambda,\lambda,\dots,\lambda) = \frac{1}{1-\lambda}. \qquad (24.2.4)$$

Then Proposition 24.2.1 applies, that is, ExtM is regular and accelerates the convergence of $\{A_m\}$ in the sense described there.

Of course, (24.2.3) is automatically satisfied in case $g_{n,j} = g_n$ for every j. When this is the case, under the rest of the conditions of Theorem 24.2.2, the sequence of the partial sums of the geometric series $\sum_{i=0}^{\infty} \lambda^i$, $\lambda \neq 1$, is in the kernel of ExtM, in the sense that $S_{n,j} = S = 1/(1-\lambda)$ for every j.

We now illustrate Theorem 24.2.2 with a few examples.

- The conditions of Theorem 24.2.2 hold when ExtM is the Δ^2-process, because in this case $g_{2,j}(x) = g_2(x) = 1/(1-x)$, as can be shown with the help of the formula $S_{2,j} = A_j - (\Delta A_j)^2/(\Delta^2 A_j)$. (We already proved in Theorem 15.3.1 that the Δ^2-process is regular for and accelerates the convergence of linearly convergent sequences.)
- The conditions of Theorem 24.2.2 hold also when ExtM is the W-transformation of Lubkin. In this case, $g_{3,j}(x,y) = g_3(x,y) = (1-2y+x)/(1-2y+xy)$, and the singularities of this function occur only along $y = 1/(2-x)$, which meets $y = x$ only at $(1,1)$. Therefore, $g_3(x,y)$ will be continuous in any neighborhood of (λ,λ) with $\lambda \neq 1$. This has been stated by Smith and Ford [317]. (We already stated prior to Theorem 15.4.1 that he Lubkin transformation is regular for and accelerates the convergence of linearly convergent sequences.)
- Smith and Ford [317] show that Proposition 24.2.1 applies to the functions $g_{n,j}$ associated with the Levin t-, u-, and v-transformations. Therefore, these transformations are regular on and accelerate the convergence of linearly convergent sequences as $j \to \infty$. Here we treat the u-transformation, the treatment of the rest being similar.

Because $\mathcal{L}_n^{(j)}$ in (19.2.4), with $\omega_m = ma_m$ there, depends on $A_j, A_{j+1}, \ldots, A_{j+n+1}$, we have that

$$g_{n+1,j}(X_1, \ldots, X_n) = \frac{\displaystyle\sum_{i=1}^{n}(-1)^i \binom{n}{i}(j+i+1)^{n-2}\frac{1+\sum_{s=1}^{i-1}X_1\cdots X_s}{X_1 X_2 \cdots X_i}}{\displaystyle\sum_{i=0}^{n}(-1)^i \binom{n}{i}(j+i+1)^{n-2}\frac{1}{X_1 X_2 \cdots X_i}}. \quad (24.2.5)$$

Thus,

$$\lim_{j\to\infty} g_{n+1,j}(X_1, \ldots, X_n) = g_n(X_1, \ldots, X_n) = \frac{\displaystyle\sum_{i=1}^{n}(-1)^i \binom{n}{i}\frac{1+\sum_{s=1}^{i-1}X_1\cdots X_s}{X_1 X_2 \cdots X_i}}{\displaystyle\sum_{i=0}^{n}(-1)^i \binom{n}{i}\frac{1}{X_1 X_2 \cdots X_i}}.$$
$$(24.2.6)$$

Substituting $X_1 = \cdots = X_n = \lambda$ in (24.2.6), it can be shown that (24.2.4) is satisfied, and hence Theorem 24.2.2 applies.

- With the technique we used for the Levin transformations, it can be shown that Proposition 24.2.1 also applies to the functions $g_{n,j}$ associated with the Sidi $\mathcal{S}$-transformation. For this transformation too, the functions $g_{n,j}$ depend on j explicitly. Actually, with $\mathcal{S}_n^{(j)}$ as in (19.3.3), and $\omega_m = ma_m$ there, the function $g_{n+1,j}(X_1, \ldots, X_n)$ is obtained from (24.2.5) by replacing the expression $(j+i+1)^{n-2}$ by $(j+i+1)_{n-2}$ for each i. Consequently, (24.2.6) holds without any change, and hence Theorem 24.2.2 applies.

For additional examples, see Weniger [353].

24.2.2 Logarithmically Convergent Sequences

So far our treatment has been concerned with linearly convergent sequences as defined in (24.2.2). We have seen that convergence acceleration of all such sequences can be achieved by more than one method. We now turn to logarithmically convergent sequences. In this context, a sequence $\{A_m\}$ converges logarithmically if it satisfies

$$\lim_{m\to\infty} R_m = 1 \quad \text{and} \quad \lim_{m\to\infty} r_m = 1. \quad (24.2.7)$$

Here R_m and r_m are as defined in (24.1.9). Note that the family of sequences $\mathbf{b}^{(1)}/\text{LOG}$ is a subset of the set of logarithmically convergent sequences.

Such sequences are very difficult to treat numerically and to analyze analytically. This fact is also reflected in the problems one faces in developing a formal theory for them. The main results pertaining to linearly convergent sequences were given by Delahaye and Germain-Bonne [66], [67]. These authors first define a property they call *generalized remanence*:

Definition 24.2.3 A set M of real convergent sequences is said to possess the property of generalized remanence if the following conditions are satisfied:

1. There exists a convergent sequence $\{\hat{S}_m\}$ with limit $\hat{S}$ such that $\hat{S}_m \neq \hat{S}$ for all m, and such that
   (i) there exists $\{S_m^0\} \in M$ such that $\lim_{m \to \infty} S_m^0 = \hat{S}_0$,
   (ii) for any $m_0 \geq 0$, there exists $p_0 \geq m_0$ and $\{S_m^1\} \in M$ such that $\lim_{m \to \infty} S_m^1 = \hat{S}_1$ and $S_m^1 = S_m^0$ for $m \leq p_0$,
   (iii) for any $m_1 > p_0$, there exists $p_1 \geq m_1$ and $\{S_m^2\} \in M$ such that $\lim_{m \to \infty} S_m^2 = \hat{S}_2$ and $S_m^2 = S_m^1$ for $m \leq p_1$,
   (iv) ....
2. The sequence $\{S_0^0, S_1^0, \dots, S_{p_0}^0, S_{p_0+1}^1, S_{p_0+2}^1, \dots, S_{p_1}^1, \dots\}$ is in M.

(Note that the notion of generalized remanence was given in [67] and it was preceded in [66] by the notion of remanence.)

Delahaye and Germain-Bonne next prove that a sequence set M that has the property of generalized remanence cannot be accelerated, in the sense that there does not exist a sequence transformation that accelerates the convergence of all sequences in M. Following this, they prove that the set of logarithmically convergent sequences possesses the property of generalized remanence and therefore cannot be accelerated.

Techniques other than that involving (generalized) remanence but similar to it have been used to determine further sets of sequences that cannot be accelerated. See Kowalewski [155], [156], and Delahaye [65]. For a list of such sets, see Brezinski and Redivo Zaglia [41, pp. 40–41].

The fact that there does not exist a sequence transformation that can accelerate the convergence of all sequences in a certain set means that the set is too large. This suggests that one should probably investigate the possibility of finding (proper) subsets of this set that can be accelerated. It would be interesting to know what the largest such subsets are. Similarly, it would be interesting to know the smallest subsets that cannot be accelerated. For some progress in this direction, see Delahaye [65] and Osada [224].

24.3 Concluding Remarks

As mentioned in the beginning of this chapter, the formal theory of sequence transformations, despite its elegance, is of limited scope. Indeed, its positive results concern mostly sequences that converge linearly [in the sense of (24.2.2)], whereas its results concerning sequences that converge logarithmically [in the sense of (24.2.7)] are mostly negative.

We have seen that this theory, as it applies to linearly convergent sequences, shows only regularity and acceleration for $\{S_{n,j}\}_{j=0}^\infty$, but gives no information on rates of convergence. Thus, it does not tell anything about the *relative* efficiencies of $S_{n,j}$ and $S_{n+1,j}$, that is, whether $\{S_{n+1,j}\}_{j=0}^\infty$ converges more quickly than $\{S_{n,j}\}_{j=0}^\infty$. In addition, so far no results about the diagonal sequences $\{S_{n,j}\}_{n=0}^\infty$ have been obtained within the framework of the formal theory. (This should be contrasted with the rather complete results we obtained for sequences in $\mathbf{b}^{(1)}/\text{LIN}$ and $\mathbf{b}^{(1)}/\text{LOG}$.) It seems that these subjects may serve as new research problems for the formal theory.

One way of enlarging the scope of the formal theory can be by investigating sets other than those mentioned so far. An interesting set can be, for example, that containing linear combinations of linearly convergent sequences, such as real Fourier series. The reason for this is that such series are not necessarily linearly convergent in the sense of (24.2.2), as can be verified with simple examples. Recall that we considered such sequences in Chapter 6. It would be interesting to see, for example, whether it is possible to obtain positive results analogous to Theorem 24.2.2.

Part III

Further Applications

25

Further Applications of Extrapolation Methods and Sequence Transformations

In Parts I and II of this book, we studied in some detail the Richardson extrapolation and its generalizations and various important sequence transformations. We also mentioned several applications of them. Actually, we discussed in detail the Romberg integration of finite-range integrals of regular integrands, numerical differentiation, and the computation of infinite-range integrals by the D-transformation. We discussed the application of the various generalizations of the D-transformation to the computation of oscillatory infinite-range integrals, including some important integral transforms. We also treated in detail the acceleration of convergence of infinite series, including power series and Fourier series and their generalizations, by the d-transformation and other methods, such as the Shanks transformation, the θ-algorithm, the Baker–Gammel approximants and their extensions, and so on. In connection with acceleration of convergence of power series, we also discussed in some detail the subject of prediction via the d-transformation and mentioned that the approach presented can be used with any sequence transformation. In this chapter, we add further applications of special interest.

We would like to note that extensive surveys and bibliographies covering the application of extrapolation methods to numerical integration can be found in Joyce [145], Davis and Rabinowitz [63], and Rabinowitz [234].

25.1 Extrapolation Methods in Multidimensional Numerical Quadrature

25.1.1 By GREP and d-Transformation

Multidimensional Euler–Maclaurin Expansions

One of the important uses of the Richardson extrapolation process is in computation of finite-range integrals over the s-dimensional hypercube C or hypersimplex S, namely, of the integrals

$$I_C[f] = \int_C f(\mathbf{x})\,d\mathbf{x}, \quad I_S[f] = \int_C f(\mathbf{x})\,d\mathbf{x}; \quad \mathbf{x} = (x_1,\dots,x_s), \quad d\mathbf{x} = \prod_{i=1}^s dx_i.$$

Here C and S are given, respectively, by

$$C = \big\{(x_1,\dots,x_s) : 0 \le x_i \le 1, \ i = 1,\dots,s\big\}$$

415

and

$$S = \{(x_1, \ldots, x_s) : 0 \leq x_i \leq 1, \ i = 1, \ldots, s; \ x_1 + \cdots + x_s \leq 1\}.$$

Let $T_C(h)$ and $T_S(h)$ be approximations to $I_C[f]$ and $I_S[f]$, respectively, that are obtained by applying the trapezoidal rule in each of the variables $x_1, \ldots, x_s$, with stepsize $h = 1/n$, where n is a positive integer. (We restrict our attention to this simple rule for simplicity. Other more sophisticated rules can also be defined.) Thus, $T_C(h)$ and $T_S(h)$ are given by

$$T_C(h) = h^s \sum_{i_1=0}^{n}{}'' \sum_{i_2=0}^{n}{}'' \cdots \sum_{i_s=0}^{n}{}'' f(i_1 h, i_2 h, \ldots, i_s h)$$

and

$$T_S(h) = h^s \sum_{i_1=0}^{n}{}'' \sum_{i_2=0}^{n-i_1}{}'' \cdots \sum_{i_s=0}^{n-i_1-\cdots-i_{s-1}}{}'' f(i_1 h, i_2 h, \ldots, i_s h),$$

where $\sum_{i=0}^{k}{}'' a_i = \frac{1}{2} a_0 + \sum_{i=1}^{k-1} a_i + \frac{1}{2} a_k$. Let us denote C or S by Ω, $I_C[f]$ or $I_S[f]$ by $I[f]$, and $T_C(h)$ or $T_S(h)$ by Q_n.

In case $f(\mathbf{x}) \in C^{\infty}(\Omega)$, the generalized Euler–Maclaurin expansions for Q_n read

$$Q_n \sim I[f] + \sum_{k=1}^{\infty} \alpha_k n^{-2k} \quad \text{as } n \to \infty,$$

for some constants α_k independent of h. Thus, the polynomial Richardson extrapolation can be applied to $\{Q_n\}$ effectively. See Lyness and McHugh [200] and Lyness and Puri [203] for these results. In a recent work, Lyness and Rüde [204] consider double integrals whose integrands involve derivatives of known functions; they derive quadrature formulas Q_n based on function values only and show that Q_n have asymptotic (Euler–Maclaurin) expansions in powers of n^{-2} as well. Thus, the polynomial Richardson extrapolation can be applied to $\{Q_n\}$ to obtain approximations of high accuracy to such integrals too.

In case $f(\mathbf{x})$ is a C^{∞} function in the interior of the set Ω but has algebraic and/or logarithmic singularities at corners and/or along edges and/or on surfaces of Ω, the Euler–Maclaurin expansions assume the following more general form:

$$Q_n \sim I[f] + \sum_{i=1}^{\mu} \sum_{k=0}^{\infty} \frac{\sum_{p=0}^{q_i} \alpha_{ikp} (\log n)^p}{n^{\delta_i + k}} \quad \text{as } n \to \infty, \quad q_i \geq 0 \text{ integers.}$$

Here, Q_n are approximations to $I[f]$ constructed by modifying the rules $T_C(h)$ and $T_S(h)$ suitably by avoiding the singularities of $f(\mathbf{x})$ on the boundary. The δ_i and q_i depend on the nature of the singularities, and the α_{ikp} are constants independent of n. Generally speaking, μ is the number of the different types of singularities of $f(\mathbf{x})$.

We do not go into the details of each expansion here. For specific results about corner singularities, we refer the reader to the original works by Lyness [196] and Lyness and Monegato [201]; for results about line and edge singularities, see Sidi [283], and for those concerning full (line and corner) singularities, see Lyness and de Doncker [199].

Direct Application of GREP to $\{Q_n\}$

In case δ_i and q_i are known, we can apply GREP$^{(m)}$ of Definition 4.2.1 to the sequence $\{Q_n\}$ – rather, to the function $A(y)$, where $A(y) \leftrightarrow Q_n$ – with $m = \sum_{i=1}^{\mu} (q_i + 1)$, $y = n^{-1}$, and with $r_k = 1$ and the $\phi_k(y)$ chosen as $(\log n)^p n^{-\delta_i}, 0 \le p \le q_i, 1 \le i \le \mu$. The method can then be efficiently implemented by the W$^{(m)}$-algorithm. Now, the computation of Q_n for given n requires $O(n^s)$ integrand evaluations. Therefore, to keep the cost of computation in check, it is appropriate to choose $y_l = 1/(l+1), l = 0, 1, \dots$. The numerical instability that results from this choice of the y_l can be controlled by using high-precision floating-point arithmetic, as was discussed in Section 2.1. This approach was originally suggested by Lyness [195] for integrals over hypercubes of functions with a corner singularity. Its use for functions with line singularities was considered by Sidi in [271], an earlier version of [283]. See also Davis and Rabinowitz [63] and Sidi [287] for brief reviews.

Application of the d-Transformation for Infinite Sequences to $\{Q_n\}$

We next propose using the $d^{(m)}$-transformation for infinite sequences (as in Definition 6.2.2) to accelerate the convergence of the sequence $\{Q_n\}$ in case the δ_i and q_i are not all known. We give a theoretical justification of the proposed approach next.

Let us first consider the case in which $q_i = 0$ for each i. In this case, $Q_n = I[f] + \sum_{i=1}^{\mu} H_i(n)$, where $H_i(n) \in \mathbf{A}_0^{(-\delta_i)}, i = 1, \dots, \mu$. Thus, Theorem 6.8.5 applies to $\{Q_n\}$, that is, $Q_n = I[f] + \sum_{k=1}^{\mu} n^k (\Delta^k Q_n) g_k(n)$ for some $g_k \in \mathbf{A}_0^{(0)}, k = 1, \dots, \mu$, and we conclude that the $d^{(\mu)}$-transformation for infinite sequences can be used on this sequence.

In general, when not all q_i are zero, we first observe that $Q_n = I[f] + \sum_{i=1}^{\mu} G_n^{(i)}$, where $G_n^{(i)} = \sum_{p=0}^{q_i} u_{ip}(n)(\log n)^p$ with $u_{ip}(n) \in \mathbf{A}_0^{(-\delta_i)}, 0 \le p \le q_i$. Thus, we have from Example 6.4.9 that, for each i, $\{G_n^{(i)}\}$ is in $\mathbf{b}^{(q_i+1)}$, and so is $\{\Delta G_n^{(i)}\}$ by Proposition 6.1.6. Consequently, by the fact that $\Delta Q_n = \sum_{i=1}^{\mu} \Delta G_n^{(i)}$ and by part (ii) of Heuristic 6.4.1, $\{\Delta Q_n\} \in \mathbf{b}^{(m)}$ with $m = \sum_{i=1}^{\mu} (q_i + 1)$. We can thus approximate $I[f]$ effectively by applying the $d^{(m)}$-transformation for infinite sequences to the sequence $\{Q_n\}$.

To keep the cost of computation under control, again it is appropriate to choose $R_l = l + 1, l = 0, 1, \dots$, in Definition 6.2.2, and use high-precision floating-point arithmetic. Note that the only input needed for this application is the integrand $f(\mathbf{x})$ and the integer m. In case m is not known, we can start with $m = 1$ and increase it if necessary until acceleration takes place. We already know no extra cost is involved in this strategy. Numerical experiments show that this approach produces approximations of high accuracy to $I[f]$.

One may wonder whether other sequence transformations can be used to accelerate the convergence of the sequence $\{Q_n\}$ when the δ_i and q_i are not all known. Judging from the form of the Euler–Maclaurin expansion of Q_n, and invoking Theorems 16.4.6 and 16.4.9, it becomes clear that the only other transformation that is relevant is that of Shanks, provided it is applied to the sequence $\{Q_{r \cdot \sigma^n}\}$, where $r \ge 1$ and $\sigma \ge 2$ are integers. Clearly, this approach is very costly even for moderate s. In view of all this, the $d^{(m)}$-transformation, when applied as proposed here, seems to be the most appropriate sequence transformation for computing singular multidimensional integrals over hypercubes and hypersimplices when the δ_i and q_i in the expansion of Q_n are not available.

Another Use of GREP

Finally, we mention an approach introduced in two papers by Espelid [78], [79] that uses extrapolation in a different way. This approach can be explained in a simple way via the double integral $I[f] = \int_0^1 \int_0^1 f(x, y)\,dx\,dy$, where $f(x, y) = x^\mu g(x, y)$ and $g \in C^\infty([0, 1] \times [0, 1])$. (When μ is not an integer, this integral has an edge singularity along the y-axis.) Here we compute the sequence of nonsingular integrals $F(h) = \int_0^1 [\int_0^{1-h} f(x, y)\,dx]\,dy$ for different values of h. Then $I[f] - F(h) = \int_0^h x^\mu G(x)\,dx$, where $G(x) = \int_0^1 g(x, y)\,dy \in C^\infty[0, 1]$. Expanding $G(x)$ at $x = 0$, we obtain the asymptotic expansion $F(h) \sim I[f] + \sum_{k=1}^\infty \alpha_k h^{\mu+k}$ as $h \to 0$. Now apply the Richardson extrapolation process (GREP$^{(1)}$ in this particular case) to $\{F(h_l)\}$, where $h_0 > h_1 > \cdots$, and $\lim_{l \to \infty} h_l = 0$. For good numerical results, we should compute the integrals $F(h_l)$ with sufficient accuracy. The generalization to arbitrary dimension and edge or corner singularities is now clear. For details and numerical examples, see Espelid [78], [79]. We note that the approach of Espelid seems to be an extension of that of Evans, Hyslop, and Morgan [80] for one-dimensional integrals $\int_0^1 f(x)\,dx$, where $f(x)$ has a singularity at $x = 0$. The approach of [80] is heuristic, and the extrapolation method employed is the ε-algorithm.

25.1.2 Use of Variable Transformations

The performance of GREP and the d-transformation (and the method of Espelid as well) in numerical integration over the hypercube C can be enhanced by first transforming the integral via so-called *variable transformations*, namely, $x_i = \xi_i(t)$, $i = 1, \ldots, s$, where the functions $\psi(t) = \xi_i(t)$, for each i, possess the following properties:

(a) $\psi(t) \in C^\infty[0, 1]$ maps $[0, 1]$ to $[0, 1]$, such that $\psi(0) = 0$ and $\psi(1) = 1$ and $\psi'(t) > 0$ on $(0, 1)$.
(b) $\psi'(1 - t) = \psi'(t)$ so that $\psi(1 - t) = 1 - \psi(t)$.
(c) $\psi^{(p)}(0) = \psi^{(p)}(1) = 0$, $p = 1, \ldots, \bar{p}$, for some $\bar{p}$, $1 \leq \bar{p} \leq \infty$. In other words, when $\bar{p}$ is finite,

$$\psi(t) \sim \sum_{i=0}^\infty c_i t^{\bar{p}+i+1} \text{ as } t \to 0+, \quad \psi(t) \sim 1 - \sum_{i=0}^\infty c_i (1 - t)^{\bar{p}+i+1} \text{ as } t \to 1-,$$

while for $\bar{p} = \infty$ these asymptotic expansions are empty.

Following these variable transformations, we obtain

$$I[f] = \int_C g(\mathbf{t})\,d\mathbf{t}, \quad g(\mathbf{t}) = f(\xi_1(t_1), \ldots, \xi_s(t_s)) \prod_{i=1}^s \xi_i'(t_i).$$

with $\mathbf{t} = (t_1, \ldots, t_s)$ and $d\mathbf{t} = \prod_{i=1}^s dt_i$, as usual. Now let $h = 1/n$ and approximate the transformed integral $\int_C g(\mathbf{t})\,d\mathbf{t}$ (using the trapezoidal rule in each of the variables t_i) by

$$\hat{Q}_n = \hat{T}_C(h) = h^s \sum_{i_1=1}^{n-1} \sum_{i_2=1}^{n-1} \cdots \sum_{i_s=1}^{n-1} g(i_1 h, i_2 h, \ldots, i_s h).$$

[Here we have used the fact that $g(\mathbf{t}) = 0$ when, for each i, $t_i = 0$ or $t_i = 1$, because $\xi_i'(0) = \xi_i'(1) = 0$.] Then, $\hat{Q}_n$ has the asymptotic expansion

$$\hat{Q}_n \sim I[f] + \sum_{i=1}^{\mu} \sum_{k=0}^{\infty} \frac{\sum_{p=0}^{q_i} \hat{\alpha}_{ikp}(\log n)^p}{n^{\hat{\delta}_i + k}} \quad \text{as } n \to \infty, \quad q_i \geq 0 \text{ integers.}$$

By property (c) of the $\xi_i(t)$, the new integrand $g(\mathbf{t})$ will be such that $\Re\hat{\delta}_i > \Re\delta_i$, so that $\hat{Q}_n - I[f]$ tends to 0 faster than $Q_n - I[f]$. This also means that, when GREP or the d-transformation is used to accelerate the convergence of $\{\hat{Q}_n\}$, the computational effort spent to obtain a required level of accuracy will be smaller than that spent on $\{Q_n\}$ for the same purpose. In case $\xi_i^{(p)}(0) = \xi_i^{(p)}(1) = 0$ for all $p \geq 1$, the asymptotic expansion of $\hat{Q}_n - I[f]$ is even empty. In such a case, the convergence of $\{\hat{Q}_n\}$ to $I[f]$ is very quick and extrapolation does not improve things.

The Euler–Maclaurin expansions for the $\hat{Q}_n$ and the application of GREP to the sequences $\{\hat{Q}_n\}$ have been considered recently by Verlinden, Potts, and Lyness [347].

We close this section by giving a list of useful variable transformations $x = \psi(t)$.

1. The Korobov [154] transformation:

$$\psi(t) = (2m + 1)\binom{2m}{m} \int_0^t [u(1 - u)]^m \, du.$$

Thus, $\psi^{(p)}(0) = \psi^{(p)}(1) = 0$ for $p = 1, \ldots, m$. Analysis of the trapezoidal and midpoint rules following the Korobov transformation was given by Sidi [293] for regular integrands and by Verlinden, Potts, and Lyness [347] for integrands with endpoint singularities.

2. The Sag and Szekeres [254] tanh-transformation:

$$\psi(t) = \frac{1}{2}\tanh\left[-\frac{c}{2}\left(\frac{1}{t} - \frac{1}{1-t}\right)\right] + \frac{1}{2}, \quad c > 0.$$

In this case, we have $\psi^{(p)}(0) = \psi^{(p)}(1) = 0$ for all $p \geq 1$.

3. The IMT-transformation of Iri, Moriguti, and Takasawa [138]:

$$\psi(t) = \frac{\int_0^t \phi(u)\, du}{\int_0^1 \phi(u)\, du}, \quad \phi(t) = \exp\left[-\frac{c}{t(1-t)}\right], \quad c > 0.$$

In this case too, we have $\psi^{(p)}(0) = \psi^{(p)}(1) = 0$ for all $p \geq 1$. For a most elegant theory of this transformation, we refer the reader to Iri, Moriguti, and Takasawa [138].

4. The double exponential transformation of Mori [214]:

$$\psi(t) = \frac{1}{2}\tanh\left\{a\sinh\left[b\left(\frac{1}{1-t} - \frac{1}{t}\right)\right]\right\} + \frac{1}{2}, \quad a, b > 0.$$

In this case too, $\psi^{(p)}(0) = \psi^{(p)}(1) = 0$ for all $p \geq 1$.

5. The Sidi [293] class $\mathcal{S}_m$ transformations: These transformations satisfy properties (a) and (b). They also satisfy property (c) with $\bar{p} = m$ and with the refinement

$$\psi'(t) \sim \sum_{i=0}^{\infty} c_i t^{m+2i} \quad \text{as } t \to 0+, \quad \psi'(t) \sim \sum_{i=0}^{\infty} c_i (1-t)^{m+2i} \quad \text{as } t \to 1-.$$

It is this refinement that makes class $\mathcal{S}_m$ transformations very useful, especially when m is chosen to be an even integer. For the rigorous analysis of the trapezoidal and midpoint rules following these transformations see [293].

A special representative of class $\mathcal{S}_m$ transformations, given already by Sidi [293], is the $\sin^m$-transformation:

$$\psi(t) = \frac{\int_0^t \phi(u)\, du}{\int_0^1 \phi(u)\, du}, \quad \phi(t) = \left(\sin \pi t\right)^m.$$

Denoting $\psi(t)$ by $\psi_m(t)$, it is easy to show by integration parts that the $\psi_m(t)$ can be computed recursively via

$$\psi_0(t) = t, \quad \psi_1(t) = \left(\sin \frac{\pi t}{2}\right)^2,$$

$$\psi_m(t) = \psi_{m-2}(t) - \frac{\Gamma(\frac{m}{2})}{2\sqrt{\pi}\,\Gamma(\frac{m+1}{2})}(\sin \pi t)^{m-1} \cos \pi t, \quad m = 2, 3, \ldots .$$

In particular, $\psi_2(t) = t - (\sin 2\pi t)/(2\pi)$ is quite effective.

Additional transformations in $\mathcal{S}_m$ for some values of m can be obtained by composition of several transformations in $\mathcal{S}_{m_i}$. In particular, if $\Psi = \psi_1 \circ \psi_2 \circ \cdots \circ \psi_r$ with $\psi_i \in \mathcal{S}_{m_i}$, then $\Psi \in \mathcal{S}_M$ with $M = \prod_{i=1}^r (m_i + 1) - 1$. Also, M is even if and only if m_i are all even. See [293].

Another transformation that is in the class $\mathcal{S}_m$ and similar to the $\sin^m$-transformation (with even m only) was recently given by Elliott [76], and the analysis given by Sidi [293] covers it completely. (Note that the lowest-order transformation of [76] is nothing but the $\sin^2$-transformation of [293].) Finally, Laurie [160] has given a polynomial transformation that satisfies the refined property (c) partially. We refer the interested reader to the original works for details.

It follows from the results of Sidi [293, p. 369, Remarks] that, when applied to regular integrands, class $\mathcal{S}_m$ variable transformations with even m are much more effective than the Korobov transformation with the same m, even though they both behave in the same (polynomial) fashion asymptotically as $t \to 0+$ and as $t \to 1-$. In addition, numerical work with class $\mathcal{S}_m$ variable transformations is less prone to overflows and underflows than that with the tanh-, IMT-, and the double exponential transformations that behave exponentially as $t \to 0+$ and as $t \to 1-$. See [293] for details.

Finally, we would like to mention that variable transformations were originally considered within the context of multidimensional integration over the hypercube by the so-called *lattice methods*. These methods can be viewed as generalizations of the one-dimensional trapezoidal rule to higher dimensions. It turns out that the $\sin^m$-transformations with even m are quite effective in these applications too. See, for example, Sloan and Joe [316] and Hill and Robinson [133] and Robinson and Hill [239]. Of these, [316] is an excellent source of information for lattice methods.

25.2 Extrapolation of Numerical Solutions of Ordinary Differential Equations

Another useful application of the Richardson extrapolation process is to numerical so-lution of ordinary differential equations by finite-difference methods. We explain the general idea via the solution of initial value problems by linear multistep methods.

Consider the initial value problem $Y' = f(x, Y)$ with $Y(a) = y_0$, whose solution we denote $Y(x)$. Let $x_n = a + nh$, $n = 0, 1, \ldots$, for some $h > 0$, and let $\{y_n\}$ be the nu-merical solution to this problem obtained by a linear multistep method with the fixed stepsize h. That is, $y_n \approx Y(x_n)$, $n = 1, 2, \ldots$.

Let us recall that a *linear multistep method (of stepnumber k)* is of the form

$$\sum_{j=0}^{k} \alpha_j y_{n+j} = h \sum_{j=0}^{k} \beta_j f_{n+j}; \quad f_m \equiv f(x_m, y_m),$$

for which the initial values $y_1, \ldots, y_{k-1}$ should be provided by the user with sufficient accuracy. Here α_j and β_j are fixed constants and $\alpha_k \neq 0$ and $|\alpha_0| + |\beta_0| \neq 0$. When $\beta_k = 0$, the method is said to be *explicit*; otherwise, it is *implicit*.

Let us fix x such that $x = a + mh$. Let us also assume, for simplicity, that $f(x, y)$ is infinitely differentiable with respect to x and y. Provided the linear multistep method satisfies certain suitable conditions, there holds

$$y_m \sim Y(x) + \sum_{i=0}^{\infty} c_i(x)h^{p+i} \quad \text{as } m \to \infty \text{ and } h \to 0, \quad \lim_{\substack{m \to \infty \\ h \to 0}} mh = x - a. \quad (25.2.1)$$

Here p is some positive integer that depends on the local error of the linear multistep method and the $c_i(x)$ are independent of h. Provided the method satisfies some "sym-metry" condition, (25.2.1) assumes the refined form

$$y_m \sim Y(x) + \sum_{i=0}^{\infty} d_i(x)h^{p+2i} \quad \text{as } m \to \infty \text{ and } h \to 0, \quad \lim_{\substack{m \to \infty \\ h \to 0}} mh = x - a. \quad (25.2.2)$$

When $\{y_n\}$ is generated by the Euler method, that is, $y_{n+1} = y_n + hf(x_n, y_n)$, (25.2.1) holds with $p = 1$. When $\{y_n\}$ is generated by the "trapezoidal rule", that is, $y_{n+1} = y_n + \frac{h}{2}[f(x_n, y_n) + f(x_{n+1}, y_{n+1})]$, then (25.2.2) holds with $p = 2$. The same is true when the "implicit midpoint rule" is used, that is, when $y_{n+1} = y_{n-1} + hf(\frac{x_n + x_{n+1}}{2}, \frac{y_n + y_{n+1}}{2})$. (Note that this last method is not a linear multistep method.)

The existence of the asymptotic expansions in (25.2.1) and (25.2.2) immediately suggests that the Richardson extrapolation process can be used to improve the accuracy of the numerical solutions. Let us first denote the y_n by $y_n(h)$. Now we start with a stepsize h_0 and $x_n = a + nh_0$, $n = 0, 1, \ldots$, and apply the linear multistep method to compute the approximations $y_n(h_0)$. We next apply the same method with stepsizes $h_i = h_0/2^i$, $i = 1, 2, \ldots$, to obtain the approximations $y_n(h_i)$. [Obviously, $y_{2^i n}(h_i) \approx Y(x_n)$ and $\lim_{i \to \infty} y_{2^i n}(h_i) = Y(x_n)$ for each fixed n.] Finally, for each $n = 1, 2, \ldots$, we apply the Richardson extrapolation process to the sequence $\{y_{2^i n}(h_i)\}_{i=0}^{\infty}$, to obtain better and better approximations to $Y(x_n)$, $n = 1, 2, \ldots$.

Clearly, the most important research topics of this subject are (i) classification of those difference methods that give rise to expansions of the forms described in (25.2.1) and (25.2.2), and (ii) explicit construction of these asymptotic expansions. The rest is immediate.

This interesting line of research was initiated by Gragg [104], [105]. [Before the work of Gragg, the existence of the asymptotic expansions in (25.2.1) and (25.2.2) was tacitly assumed.] Important contributions to this topic have been made by several authors. See, for example, the works Stetter [323], [324], [325], and Hairer and Lubich [121]. See also Stoer and Bulirsch [326], Marchuk and Shaidurov [206], and Hairer, Nørsett, and Wanner [122]. For introductions and summaries, see also Lambert [157, Chapter 6] and Walz [349, Chapter 3].

Asymptotic expansions of the forms given in (25.2.1) and (25.2.2) have also been derived for difference solutions of two-point boundary value problems in ordinary differential equations and linear and nonlinear integral and integro-differential equations of Volterra and Fredholm types. Again, the resulting numerical solutions can be improved by applying the Richardson extrapolation process precisely as described here. We do not consider these problems here but refer the reader to the relevant literature.

25.3 Romberg-Type Quadrature Formulas for Periodic Singular and Weakly Singular Fredholm Integral Equations

25.3.1 Description of Periodic Integral Equations

The numerical solution of Fredholm integral equations

$$\omega f(t) + \int_a^b K(t, x) f(x)\, dx = g(t), \quad a \le t \le b, \tag{25.3.1}$$

is of practical interest in different disciplines. (Such equations are of the first or the second kind depending on whether $\omega = 0$ or $\omega = 1$, respectively.) In this section, we consider a special class of such equations that arise from so-called *boundary integral formulation* of two-dimensional elliptic boundary value problems in a bounded domain Ω. The integral term in (25.3.1) in such a case is actually a line integral along the boundary curve $\partial\Omega$ of the domain Ω. These equations have the following important features: (i) Their kernel functions $K(t, x)$ are singular along the line $x = t$. (ii) The input functions $K(t, x)$ and $g(t)$ and the solution $f(t)$ are all periodic with period $T = b - a$. (iii) When the curve $\partial\Omega$ is infinitely smooth and the function g is infinitely smooth along $\partial\Omega$, that is, when $g(t) \in C^\infty(-\infty, \infty)$, so is the solution f.

In case $K(t, x)$ has an integrable singularity across $x = t$, (25.3.1) is said to be *weakly singular*. In case $K(t, x) \sim c/(x - t)$ as $x \to t$ for some constant $c \ne 0$, and the integral $\int_a^b K(t, x) f(x)\, dx$ is defined only in the Cauchy principal value sense, it is said to be *singular*.

Here we consider those integral equations with the following properties:

(i) The kernel $K(t, x)$ is periodic both in t and in x with period $T = b - a$ and is infinitely differentiable in $(-\infty, \infty) \setminus \{t + kT\}_{k=-\infty}^{\infty}$. It either has a polar singularity

(PS) and can be expressed as in

$$K(t, x) = \frac{H_1(t, x)}{x - t} + H_2(t, x), \quad (\text{PS}), \tag{25.3.2}$$

or it has an algebraic singularity (AS) and can be expressed as in

$$K(t, x) = H_1(t, x)|t - x|^s + H_2(t, x), \quad \Re s > -1, \quad (\text{AS}), \tag{25.3.3}$$

or it has a logarithmic singularity (LS) and can be expressed as in

$$K(t, x) = H_1(t, x) \log |t - x| + H_2(t, x), \quad (\text{LS}), \tag{25.3.4}$$

or has an algebraic-logarithmic singularity (ALS) and can be expressed as in

$$K(t, x) = H_1(t, x)|t - x|^s \log |t - x| + H_2(t, x), \quad \Re s > -1, \quad (\text{ALS}), \tag{25.3.5}$$

and $H_1(t, x)$ and $H_2(t, x)$ in all cases are infinitely differentiable for all $t, x \in [a, b]$ (including $t = x$), but are not necessarily periodic. In all cases, we assume that $H_1(t, t) \not\equiv 0$.

(ii) The function $g(t)$ is periodic in t with period T and infinitely differentiable in $(-\infty, \infty)$.

(iii) Finally, we assume that the solution $f(t)$ exists uniquely and is periodic in t with period T and infinitely differentiable in $(-\infty, \infty)$. (This assumption seems to hold always. For a heuristic justification, see Sidi and Israeli [310, Section 1].)

A common approach to the numerical solution of integral equations is via so-called *quadrature methods*, where one replaces the integral in (25.3.1) by a numerical quadrature formula and then collocates the resulting equation at the abscissas x_i of this formula, thus obtaining a set of equations for the approximations to the $f(x_i)$. When treating singular and weakly singular integral equations, such an approach produces low accuracy because of the singularity in the kernel. Here we would like to describe an approach due to Sidi and Israeli [310] that is suitable precisely for such (periodic) integral equations and that makes use of appropriate Romberg-type numerical quadrature formulas of high accuracy. (See also Section D.6 of Appendix D, where the relevant Euler–Maclaurin expansions are summarized.)

25.3.2 *"Corrected" Quadrature Formulas*

Let n be a positive integer and set $h = T/n$ and $x_i = a + ih, i = 1, \ldots, n$. In addition, let $t \in \{x_1, \ldots, x_n\}$. Following Section D.6 of Appendix D, let us now define the "corrected" trapezoidal rule approximations $I[h; t]$ to the integral $I[t] = \int_a^b K(t, x) f(x) \, dx$ as in

$$I[h; t] = \sum_{i=1}^n w_n(t, x_i) f(x_i), \tag{25.3.6}$$

where

$$w_n(t, x) = h K(t, x) \text{ for } x \neq t, \quad h = T/n, \tag{25.3.7}$$

and

$$w_n(t,t) = \begin{cases} 0 & \text{(PS)} \\ h[H_2(t,t) - 2\zeta(-s)H_1(t,t)h^s] & \text{(AS)} \\ h[H_2(t,t) + H_1(t,t)\log(\frac{h}{2\pi})] & \text{(LS)} \\ h[H_2(t,t) + 2\{\zeta'(-s) - \zeta(-s)\log h\}H_1(t,t)h^s] & \text{(ALS)} \end{cases} \quad (25.3.8)$$

[Here, $\zeta(-s)$ and $\zeta'(-s)$ can be computed by accelerating the convergence of the series $\sum_{k=1}^{\infty} k^{-z}$ and $\sum_{k=1}^{\infty} k^{-z}\log k$ for suitable z and by using Riemann's reflection formula when necessary. See Appendix E.] These $I[h;t]$ have the asymptotic expansions

$$I[h;t] = I[t] + \gamma(t)h + O(h^\mu) \text{ as } h \to 0, \text{ for every } \mu > 0, \quad \text{(PS)} \quad (25.3.9)$$

$$I[h;t] \sim I[t] + \sum_{k=1}^{\infty} \alpha_k(t;s)h^{s+2k+1} \text{ as } h \to 0, \quad \text{(AS)} \quad (25.3.10)$$

$$I[h;t] \sim I[t] + \sum_{k=1}^{\infty} \beta_k(t;0)h^{2k+1} \text{ as } h \to 0, \quad \text{(LS)} \quad (25.3.11)$$

$$I[h;t] \sim I[t] + \sum_{k=1}^{\infty} [\alpha_k(t;s)\log h + \beta_k(t;s)]h^{s+2k+1} \text{ as } h \to 0, \text{ (ALS)} \quad (25.3.12)$$

where $\gamma(t)$, the $\alpha_k(t;s)$, and the $\beta_k(t;s)$ depend only on t but are independent of h:

$$\gamma(t) = -H_2(t,t)f(t) - \frac{\partial}{\partial x}[H_1(t,x)f(x)]\Big|_{x=t},$$

$$\alpha_k(t;s) = 2\frac{\zeta(-s-2k)}{(2k)!}\frac{\partial^{2k}}{\partial x^{2k}}[H_1(t,x)f(x)]\Big|_{x=t}, \quad k=1,2,\dots, \quad (25.3.13)$$

$$\beta_k(t;s) = -2\frac{\zeta'(-s-2k)}{(2k)!}\frac{\partial^{2k}}{\partial x^{2k}}[H_1(t,x)f(x)]\Big|_{x=t}, \quad k=1,2,\dots.$$

From these expansions, it follows that $I[h;t] - I[t]$ is $O(h)$ for PS, $O(h^{s+3})$ for AS, $O(h^3)$ for LS, and $O(h^{s+3}\log h)$ for ALS.

When s is a positive even integer in (25.3.5) (ALS), say $s = 2p$ with $p = 1,2,\dots,$ we have $\alpha_k(t;2p) = 0$ for all k. Therefore, in this case,

$$w_n(t,t) = h[H_2(t,t) + 2\zeta'(-2p)H_1(t,t)h^{2p}], \quad (25.3.14)$$

$$I[h;t] \sim I[t] + \sum_{k=1}^{\infty} \beta_k(t;2p)h^{2p+2k+1} \text{ as } h \to 0, \quad (25.3.15)$$

from which we also have $I[h;t] - I[t] = O(h^{2p+3})$ as $h \to 0$. [The case that normally arises in applications is that with $s = 2$, and the (ALS) formula $I[h;t]$ in (25.3.6) with (25.3.7) and (25.3.14), and with $p = 1$, has been used in a recent paper by Christiansen [52]. In this case, we have $I[h;t] - I[t] = O(h^5)$ as $h \to 0$.]

Note that, in constructing $I[h;t]$ for the weakly singular cases, we do not need to know $H_1(t,x)$ and $H_2(t,x)$ for all t and x, but only for $x = t$. We can obtain $H_1(t,t)$ and $H_2(t,t)$ simply by expanding $K(t,x)$ for $x \to t$. For the singular case, neither $H_1(t,t)$ nor $H_2(t,t)$ needs to be known.

The quadrature method based on any of the rules $I[h; t]$ is now defined by the equations

$$\omega \tilde{f}_k + I[h; x_k] = g(x_k), \quad k = 1, 2, \ldots, n. \tag{25.3.16}$$

More explicitly, these equations are

$$\omega \tilde{f}_k + \sum_{i=1}^{n} w_n(x_k, x_i) \tilde{f}_i = g(x_k), \quad k = 1, 2, \ldots, n, \tag{25.3.17}$$

where $\tilde{f}_i$ is the approximation to $f(x_i)$. In general, the accuracy of the $\tilde{f}_i$ is the same as that of the underlying numerical quadrature formula, which is $I[h; t]$ in this case. We can increase the accuracy of the quadrature method by increasing that of $I[h; t]$, which we propose to achieve by using extrapolation. What makes this possible is the periodicity of the integrand $K(t, x) f(x)$ as a function of x. We turn to this subject next.

25.3.3 Extrapolated Quadrature Formulas

Treatment of the Singular Case

We start by illustrating this point for the case of PS, where one extrapolation suffices to remove the single term $\gamma(t)h$ from the expansion of $I[h; t]$. Let us choose $h = T/n$ for some even integer n and let $x_i = a + ih$, $i = 0, 1, \ldots, n$. Performing this single extrapolation, we obtain the Romberg-type quadrature rule

$$J[h; t] = 2I[h; t] - I[2h; t] \tag{25.3.18}$$

as the new approximation to $I[t]$. Consequently, we also have

$$J[h; t] = I[t] + O(h^\mu) \text{ as } h \to 0, \quad \text{for every } \mu > 0. \tag{25.3.19}$$

That is, as $n \to \infty$, the error in $J[h; t]$ tends to zero faster than any negative power of n. The quadrature method for (25.3.1) based on $J[h; t]$ is thus

$$\omega \tilde{f}_k + J[h; x_k] = g(x_k), \quad k = 1, 2, \ldots, n. \tag{25.3.20}$$

More explicitly,

$$\omega \tilde{f}_k + 2h \sum_{i=1}^{n} \epsilon_{k,i} K(x_k, x_i) \tilde{f}_i = g(x_k), \quad k = 1, 2, \ldots, n, \tag{25.3.21}$$

where we have defined

$$\epsilon_{k,i} = \begin{cases} 1 \text{ if } k - i \text{ odd,} \\ 0 \text{ if } k - i \text{ even.} \end{cases} \tag{25.3.22}$$

Surprisingly, $J[h; t]$ is the midpoint rule approximation (with $n/2$ abscissas) for the integral

$$I[t] = \int_{t-T/2}^{t+T/2} G(x) \, dx = \int_{-T/2}^{T/2} G(t + \xi) \, d\xi; \quad G(x) \equiv K(t, x) f(x).$$

From symmetry, this can also be written as

$$I[t] = \int_{-T/2}^{T/2} G_e(\xi)\,d\xi; \quad G_e(\xi) \equiv \frac{1}{2}[G(t+\xi) + G(t-\xi)].$$

Note that $G_e(\xi)$ is periodic with period T and has no singularities in $(-\infty, \infty)$. In view of this, the error term given in (25.3.19) can be improved substantially in case the integrand $G(x)$ possesses certain analyticity properties. This improvement was given by Sidi and Israeli [310, Theorem 9], and we include it here for completeness.

Theorem 25.3.1 *Let* $K(t,z)f(z) \equiv G(z)$, *for fixed* t *and variable complex* z, *be analytic in the strip* $|\Im z| < \sigma$ *for some* $\sigma > 0$, *except at the simple poles* $t + kT$, $k = 0, \pm1, \pm2, \dots$. *Define*

$$M(\tau) = \max\left\{ \max_{-\infty < x < \infty} |G_e(x + i\tau)|, \max_{-\infty < x < \infty} |G_e(x - i\tau)| \right\},$$

where $G_e(\xi)$ *is as defined above. Then*

$$|J[h;t] - I[t]| \le 2TM(\sigma')\frac{\exp(-n\pi\sigma'/T)}{1 - \exp(-n\pi\sigma'/T)}, \quad \text{for every } \sigma' \in (0,\sigma).$$

Simply put, this theorem says that the error in $J[h;t]$ tends to zero as $n \to \infty$ exponentially in n, like $e^{-n\pi\sigma/T}$, for all practical purposes.

Treatment of the Weakly Singular Case

For the case of LS, we start by using only one extrapolation to eliminate the term $\beta_1(t;0)h^3$ from the asymptotic expansion of $I[h;t]$. Let us choose $h = T/n$ for some even integer n and let $x_i = a + ih, i = 0, 1, \dots, n$. Performing this single extrapolation, we obtain the Romberg-type quadrature rule

$$J_1[h;t] = \frac{8}{7}I[h;t] - \frac{1}{7}I[2h;t] \tag{25.3.23}$$

as the new approximation to $I[t]$. We also have

$$J_1[h;t] \sim I[t] + \sum_{k=2}^{\infty} \frac{2^3 - 2^{2k+1}}{7}\beta_k(t;0)h^{2k+1} \quad \text{as } h \to 0, \tag{25.3.24}$$

hence $J_1[h;t] - I[t] = O(h^5)$ as $h \to 0$. The quadrature method for (25.3.1) based on $J_1[h;t]$ is thus

$$\omega\tilde{f}_k + J_1[h;x_k] = g(x_k), \quad k = 1, 2, \dots, n. \tag{25.3.25}$$

More explicitly,

$$\omega\tilde{f}_k + \sum_{i=1}^{n}\left[\frac{8}{7}w_n(x_k,x_i) - \frac{1}{7}\epsilon_{k,i}^{(1)}w_{n/2}(x_k,x_i)\right]\tilde{f}_i = g(x_k), \quad k = 1, 2, \dots, n,$$

$$\tag{25.3.26}$$

where

$$\epsilon_{k,i}^{(1)} = \begin{cases} 1 \text{ if } k-i \text{ even,} \\ 0 \text{ if } k-i \text{ odd.} \end{cases} \tag{25.3.27}$$

By applying two extrapolations, we can remove the terms $\beta_k(t;0)h^{2k+1}, k = 1, 2$, from the asymptotic expansion of $I[h;t]$. This time we choose $h = T/n$ for an integer n that is divisible by 4, and let $x_i = a + ih, i = 0, 1, \ldots, n$. Performing the two extrapolations, we obtain the Romberg-type quadrature rule

$$J_2[h;t] = \frac{32}{31} J_1[h;t] - \frac{1}{31} J_1[2h;t]$$

$$= \frac{256}{217} I[h;t] - \frac{40}{217} I[2h;t] + \frac{1}{217} I[4h;t] \tag{25.3.28}$$

as the new approximation to $I[t]$. We also have

$$J_2[h;t] \sim I[t] + \sum_{k=3}^{\infty} \frac{2^3 - 2^{2k+1}}{7} \cdot \frac{2^5 - 2^{2k+1}}{31} \beta_k(t;0)h^{2k+1} \text{ as } h \to 0, \tag{25.3.29}$$

hence $J_2[h;t] - I[t] = O(h^7)$ as $h \to 0$. The quadrature method for (25.3.1) based on $J_2[h;t]$ is thus

$$\omega \tilde{f}_k + J_2[h;x_k] = g(x_k), \quad k = 1, 2, \ldots, n. \tag{25.3.30}$$

More explicitly,

$$\omega \tilde{f}_k + \sum_{i=1}^{n} \left[\frac{256}{217} w_n(x_k, x_i) - \frac{40}{217} \epsilon_{k,i}^{(1)} w_{n/2}(x_k, x_i) + \frac{1}{217} \epsilon_{k,i}^{(2)} w_{n/4}(x_k, x_i) \right] \tilde{f}_i = g(x_k),$$

$$k = 1, 2, \ldots, n, \tag{25.3.31}$$

where $\epsilon_{k,i}^{(1)}$ are as before and

$$\epsilon_{k,i}^{(2)} = \begin{cases} 1 \text{ if } k-i \text{ divisible by 4,} \\ 0 \text{ otherwise.} \end{cases} \tag{25.3.32}$$

For the development of Romberg-type formulas of all orders for all types of weak singularities, we refer the reader to Sidi and Israeli [310].

25.3.4 Further Developments

Once the $\tilde{f}_i$ have been obtained, we can construct a trigonometric polynomial $P_n(t)$ in $\cos(2\pi kt/T)$, $\sin(2\pi kt/T)$, $k = 0, 1, \ldots$, that satisfies the interpolation conditions $P_n(x_i) = \tilde{f}_i$, $i = 1, 2, \ldots, n$. As the $\tilde{f}_i$ are good approximations to the $f(x_i)$ for all $i = 1, \ldots, n$, we expect $P_n(t)$ to be a good approximation to $f(t)$ throughout $[a, b]$.

It is easy to see that the methodology presented here can be applied to systems of periodic integral equations in several unknown functions, where the integral terms may contain both singular and weakly singular kernels of the forms discussed. All these kernels have their singularities only along $x = t$. Such systems occur very frequently in applications.

In case $g(t) \equiv 0$ in (25.3.1), we have an eigenvalue problem. It is clear that our methodology can be applied without any changes to such problems too.

The approach of Sidi and Israeli [310] to the solution of periodic singular and weakly singular Fredholm integral equations outlined partially here has been used successfully in different applications involving boundary integral equation formulations. See, for example, Almgren, Dai, and Hakim [5], Coifman et al. [55], Fainstein et al. [82], Haroldsen and Meiron [124], Hou, Lowengrub, and Shelley [136], McLaughlin, Muraki, and Shelley [209], Nie and Tian [219], Nitsche [221], Shelley, Tian, and Wlodarski [266], and Tyvand and Landrini [342].

For the case of LS [with a special kernel $K(t, x)$ only], Christiansen [51] derived a numerical quadrature rule that has the same appearance as the rule $I[h; t]$ given through (25.3.6) and (25.3.7). It differs from our $I[h; t]$ in its $w_n(t, t)$, which is more complicated than our $w_n(t, t)$ in (25.3.8) (LS). The asymptotic expansion of the error in Christiansen's rule was given in Sidi [289], where it was shown that this rule too has an error that is $O(h^3)$ and its asymptotic expansion contains all the powers h^{3+k}, $k = 0, 1, \ldots$. This should be compared with the error of our $I[h; t]$, whose asymptotic expansion has only the odd powers h^{3+2k}, $k = 0, 1, \ldots$. We are thus led to conclude that the Romberg-type quadrature formulas based on the $I[h; t]$ presented here will be more effective than those based on the rule of Christiansen [51]. This conclusion has also been verified numerically in Sidi [289].

25.4 Derivation of Numerical Schemes for Time-Dependent Problems from Rational Approximations

Rational approximations have been used in deriving numerical schemes (finite differences, finite elements, etc.) for time-dependent problems. Let us consider the solution of the problem

$$\frac{\partial U}{\partial t} = AU, \tag{25.4.1}$$

where t denotes time and A is an operator independent of t. At least formally, we can write

$$U(t + \Delta t) = \sum_{k=0}^{\infty} \left(\frac{\partial^k}{\partial t^k} U(t) \right) \frac{(\Delta t)^k}{k!} = \exp\left(\Delta t \cdot \frac{\partial}{\partial t} \right) U(t) = \exp(\Delta t \cdot A)U(t). \tag{25.4.2}$$

Let us now approximate $\exp(z)$ by a rational function $r(z) = p(z)/q(z)$, and use this to replace (25.4.2) by the approximate equation

$$U(t + \Delta t) \approx r(\Delta t \cdot A)U(t) \quad \Rightarrow \quad q(\Delta t \cdot A)U(t + \Delta t) \approx p(\Delta t \cdot A)U(t). \tag{25.4.3}$$

If we write $p(z) = \sum_{i=0}^{K} a_i z^i$ and $q(z) = \sum_{i=0}^{L} b_i z^i$, and denote $U^{(k)} \equiv U(k\Delta t)$, then (25.4.3) can be written in the form

$$\sum_{i=0}^{L} b_i (\Delta t)^i A^i U^{(k+1)} \approx \sum_{i=0}^{K} a_i (\Delta t)^i A^i U^{(k)}. \tag{25.4.4}$$

Finally, replacing the operator A by a suitable approximation $\hat{A}$ if necessary, we obtain from the approximate equation in (25.4.4) the numerical scheme

$$\sum_{i=0}^{L} b_i (\Delta t)^i \hat{A}^i u^{(k+1)} = \sum_{i=0}^{K} a_i (\Delta t)^i \hat{A}^i u^{(k)}, \tag{25.4.5}$$

where $u^{(k)}$ is an approximation for $U^{(k)}$. We expect the numerical solution $u^{(k)}$ to have high accuracy, provided $r(z)$ and $\hat{A}$ are high-accuracy approximations for $\exp(z)$ and A, respectively.

As an example, let us consider the one-dimensional heat equation

$$\frac{\partial U}{\partial t} = \kappa \frac{\partial^2 U}{\partial x^2}, \quad 0 < x < 1, \; t > 0, \tag{25.4.6}$$

subject to the boundary and initial conditions

$$U(0, t) = f_0(t), \quad U(1, t) = f_1(t); \quad U(x, 0) = g(x). \tag{25.4.7}$$

First, choosing $r(z) = (1 + \frac{1}{2}z)/(1 - \frac{1}{2}z)$, the $[1/1]$ Padé approximant to $\exp(z)$ for which $\exp(z) - r(z) = O(z^3)$, we obtain [cf. (25.4.4)]

$$\left(I - \frac{1}{2}(\Delta t)A \right) U^{(k+1)} \approx \left(I + \frac{1}{2}(\Delta t)A \right) U^{(k)}; \quad A \equiv \kappa \frac{\partial^2}{\partial x^2}. \tag{25.4.8}$$

Letting $\Delta x = 1/N$ and $x_i = i\Delta x$, $i = 0, 1, \ldots, N$, and denoting $U(x_i, t) = U_i(t)$ for short, we next approximate A via the central difference of order 2

$$\frac{\partial^2}{\partial x^2} U_i(t) \approx \frac{U_{i+1}(t) - 2U_i(t) + U_{i-1}(t)}{(\Delta x)^2}, \quad i = 1, \ldots, N - 1. \tag{25.4.9}$$

Consequently, the equations in (25.4.5) become

$$u_i^{(k+1)} - \frac{1}{2}\omega\left(u_{i+1}^{(k+1)} - 2u_i^{(k+1)} + u_{i-1}^{(k+1)}\right) = u_i^{(k)} + \frac{1}{2}\omega\left(u_{i+1}^{(k)} - 2u_i^{(k)} + u_{i-1}^{(k)}\right),$$

$$i = 1, \ldots, N - 1, \; k = 0, 1, \ldots; \; \omega \equiv \kappa \frac{\Delta t}{(\Delta x)^2}, \tag{25.4.10}$$

with the boundary and initial conditions

$$u_0^{(k)} = f_0(k\Delta t), \quad u_N^{(k)} = f_1(k\Delta t); \quad u_i^{(0)} = g(x_i), \; i = 0, 1, \ldots, N. \tag{25.4.11}$$

Here $u_i^{(k)}$ is an approximation to $U_i^{(k)} = U_i(k\Delta t) = U(i\Delta x, k\Delta t)$. The resulting (implicit) finite-difference method is known as the *Crank–Nicolson method* and can be found in standard books on numerical solution of partial differential equations. See, for example, Ames [10] or Iserles [141]. It is unconditionally stable, that is, Δx and Δt can be reduced independently. Let us express the difference equations in (25.4.10) in the form

$$F_{i,k}(u) = 0,$$

where, with $v_i^{(k)} = v(i \, \Delta x, k \, \Delta t)$ for arbitrary $v(x, t)$,

$$F_{i,k}(v) \equiv \frac{v_i^{(k+1)} - v_i^{(k)}}{\Delta t} + \frac{\kappa}{2} \left[\frac{v_{i+1}^{(k)} - 2v_i^{(k)} + v_{i-1}^{(k)}}{(\Delta x)^2} + \frac{v_{i+1}^{(k+1)} - 2v_i^{(k+1)} + v_{i-1}^{(k+1)}}{(\Delta x)^2} \right].$$

Then, the local truncation error of the Crank–Nicolson method is given as

$$F_{i,k}(U) = O((\Delta x)^2) + O((\Delta t)^2).$$

[Recall that $U(x, t)$ is the exact solution of (25.4.6).] Thus, the method is second-order-accurate in both Δx and Δt.

Rational approximations, Padé approximants and others, to $\exp(z)$ have been used to derive numerical methods for the solution of ordinary differential equations as well. For such problems, the relevant rational approximations $r(z)$ are required to be *A-acceptable*, that is, they are required to satisfy $|r(z)| < 1$ for $\Re z < 0$. (This is the same as requiring that the corresponding numerical method be *A-stable*.) An important technique that has been used in the study of A-acceptability is that of *order stars*, developed first in Wanner, Hairer, and Nørsett [350]. These authors applied the technique to Padé approximants and showed that the $[m/n]$ Padé approximants are A-acceptable if and only if $m \leq n \leq m + 2$. These results were generalized to approximations other than Padé by Iserles [140]. See also Iserles and Nørsett [142].

25.5 Derivation of Numerical Quadrature Formulas from Sequence Transformations

Let $\mu(t)$ be a nondecreasing function on an interval $[a, b]$ with infinitely many points of increase there, and let the integral

$$I[f] = \int_a^b f(x) \, d\mu(x) \tag{25.5.1}$$

be defined in the sense of Stieltjes. Let us approximate this integral by the n-point numerical quadrature formula

$$I_n[f] = \sum_{k=1}^n w_{n,k} f(x_{n,k}), \tag{25.5.2}$$

where the $x_{n,k}$ and $w_{n,k}$ are, respectively, the abscissas and weights of this formula. Furthermore, $x_{n,k} \in [a, b]$. One way of viewing the numerical quadrature formula $I_n[f]$ is as follows: Let $f(z)$ be analytic in a domain Ω of the z-plane that contains the interval $[a, b]$ in its interior. Then, we can write

$$I[f] = \frac{1}{2\pi i} \int_C H(z) f(z) \, dz, \quad I_n[f] = \frac{1}{2\pi i} \int_C H_n(z) f(z) \, dz, \tag{25.5.3}$$

where C is a closed contour that is in the interior of Ω and that contains $[a, b]$ in its interior, and the functions $H(z)$ and $H_n(z)$ are defined via

$$H(z) = \int_a^b \frac{d\mu(x)}{z - x}, \quad H_n(z) = \sum_{k=1}^n \frac{w_{n,k}}{z - x_{n,k}}. \quad (25.5.4)$$

We see that $H(z)$ is analytic in the z-plane cut along the interval $[a, b]$ and that $H_n(z)$ is a rational function of z, its numerator and denominator polynomials being of degree $n - 1$ and n, respectively. From the error expression

$$I[f] - I_n[f] = \frac{1}{2\pi i} \int_C [H(z) - H_n(z)] f(z) \, dz, \quad (25.5.5)$$

it is clear that, for $I_n[f]$ to converge to $I[f]$, $H_n(z)$ must converge to $H(z)$ in the z-plane cut along the interval $[a, b]$.

Now, $H(z)$ has the asymptotic expansion

$$H(z) \sim \sum_{i=0}^\infty \mu_i z^{-i-1} \text{ as } z \to \infty; \quad \mu_i = \int_a^b x^i \, d\mu(x), \quad i = 0, 1, \dots . \quad (25.5.6)$$

In view of (25.5.6), one effective way of approximating $H(z)$ is via Padé approximants, that is, by choosing $H_n(z) = z^{-1} \hat{H}_{n-1,n}(z^{-1})$, where $\hat{H}_{m,n}(\zeta)$ is the $[m/n]$ Padé approximant from the power series $\sum_{i=0}^\infty \mu_i \zeta^i$. As we saw in Section 17.9, the approximation $I_n[f]$ that results from this is nothing but the n-point Gaussian quadrature formula for $I[f]$, for which $I_n[f] = I[f]$ for all $f(x)$ that are polynomials of degree at most $2n - 1$.

Recall that Padé approximants from $\sum_{i=0}^\infty \mu_i \zeta^i$ are also obtained by applying the Shanks transformation to the sequence (of partial sums) $\{\sum_{i=0}^n \mu_i \zeta^i\}_{n=0}^\infty$. This means that numerical quadrature formulas other than Gaussian can be derived by applying to this sequence suitable sequence transformations other than that of Shanks. For example, we may use the Levin transformations or appropriate modifications of them for this purpose, provided $\{\mu_n\} \in \mathbf{b}^{(1)}$.

Let us first observe that

$$H(z) - A_{n-1}(z) = \frac{1}{z^{n-1}} \int_a^b \frac{x^{n-1}}{z - x} \, d\mu(x); \quad A_k(z) = \sum_{i=0}^{k-1} \mu_i z^{-i-1}, \quad k = 1, 2, \dots .$$
$$(25.5.7)$$

If the integral on the right-hand side of (25.5.7) has an asymptotic expansion of the form $c_n \sum_{i=0}^\infty \delta_i n^{-i}$ as $n \to \infty$, then we can approximate $H(z)$ by applying GREP$^{(1)}$ to the sequence $\{A_n(z)\}$ via the equations

$$A_{r-1}(z) = A_n^{(j)}(z) + \frac{c_r}{z^{r-1}} \sum_{i=0}^{n-1} \frac{\bar{\delta}_i}{r^i}, \quad r = j+1, \dots, j+n+1. \quad (25.5.8)$$

[Here we have defined $A_0(z) = 0$.] The approximation $A_n^{(0)}(z)$ to $H(z)$ is given by

$$A_n^{(0)}(z) = \frac{\sum_{i=0}^n \lambda_i z^i A_i(z)}{\sum_{i=0}^n \lambda_i z^i}; \quad \lambda_i = (-1)^i \binom{n}{i} \frac{(i+1)^{n-1}}{c_{i+1}}, \quad i = 0, 1, \dots, n. \quad (25.5.9)$$

Obviously, when the c_i are independent of z, so are the λ_i, and hence $A_n^{(0)}(z)$ is a rational function whose numerator and denominator polynomials are of degree $n-1$ and n, respectively. Therefore, we can choose $H_n(z)$ to be $A_n^{(0)}(z)$. Then, the abscissas $x_{n,k}$ of $I_n[f]$ are the poles of $H_n(z)$ [equivalently, the zeros of the polynomial $\sum_{i=0}^{n} \lambda_i z^i$], while the weights $w_{n,k}$ of $I_n[f]$ are the residues of $A_n^{(0)}(z)$ at the $x_{n,k}$, namely,

$$w_{n,k} = \left.\frac{\sum_{i=1}^{n} \lambda_i z^i \left(\sum_{r=0}^{i-1} \mu_r z^{-r-1}\right)}{\sum_{i=1}^{n} i\lambda_i z^{i-1}}\right|_{z=x_{n,k}}, \quad k = 1, \dots, n, \qquad (25.5.10)$$

provided the poles $x_{n,k}$ are simple. This approach to the derivation of numerical quadrature formulas was proposed by Sidi [275] for finite-range integrals, and it was extended to infinite-range integrals in Sidi [279], [282]. See also Davis and Rabinowitz [63, p. 307]. We show in the following how this approach is applied to finite-range integrals with algebraic-logarithmic endpoint singularities.

When $[a, b] = [0, 1]$ and $d\mu(x) = w(x)dx$, where $w(x) = (1-x)^{\alpha} x^{\beta}(-\log x)^{\nu}$, $\alpha + \nu > -1$, $\beta > -1$, we have that $\{\mu_n\} \in \mathbf{b}^{(1)}$ indeed. By making the transformation of variable $x = e^{-t}$ in (25.5.7), we obtain

$$H(z) - A_{n-1}(z) = \frac{1}{z^{n-1}} \int_0^{\infty} e^{-nt} \frac{t^{\nu}(1-e^{-t})^{\alpha} e^{-\beta t}}{z - e^{-t}} \, dt. \qquad (25.5.11)$$

which, upon applying Watson's lemma, yields the asymptotic expansion

$$H(z) - A_{n-1}(z) \sim \frac{1}{z^{n-1} n^{\alpha+\nu+1}} \sum_{i=0}^{\infty} u_i(z) n^{-i} \quad \text{as } n \to \infty, \qquad (25.5.12)$$

where $u_0(z) = \frac{\Gamma(\alpha+\nu+1)}{z-1}$. Note that this asymptotic expansion is valid for all complex $z \notin [0, 1]$. Thus, we can construct $H_n(z) = A_n^{(0)}(z)$ as in (25.5.9) with $c_n = n^{-(\alpha+\nu+1)}$. As is shown by Sidi [275] for integer $\alpha + \nu$ and by Sidi and Lubinsky [314] for arbitrary $\alpha + \nu$, the polynomial $\sum_{i=0}^{n} \lambda_i z^i$ of (25.5.9) has n simple real zeros $x_{n,k}$ in the open interval $(0, 1)$. Furthermore, these zeros are clustered near 0. (There are similar formulas for integrals of the form $\int_{-1}^{1} w(x) f(x) \, dx$, with $w(x) = (1-x^2)^{\alpha}$, for example, that have their abscissas in $(-1, 1)$ and located symmetrically with respect to 0. See Sidi [275] for details.)

The resulting formulas $I_n[f]$ are remarkable in that their abscissas are independent of β and depend only on $\alpha + \nu$. Thus, by setting $c_n = n^{-1}$ in (25.5.9) and (25.5.10), we are able to obtain numerical quadrature formulas for $w(x) = x^{\beta}$ for any β. Furthermore, $c_n = n^{-1}$ can also be used to obtain formulas for $w(x) = x^{\beta}(-\log x)^{\nu}$, where ν is a small integer such as $1, 2, \dots$. All these formulas have the same set of abscissas. More generally, by setting $c_n = n^{-\gamma}$ in (25.5.9) and (25.5.10), we are able to obtain numerical quadrature formulas for $w(x) = (1-x)^{\alpha} x^{\beta}(-\log x)^{\nu}$, where β is arbitrary and $\alpha + \nu + 1 = \gamma + s$, where s is a small integer like $1, 2, \dots$. All these formulas too have the same set of abscissas. Numerical experiments show that these rules are very effective. Their accuracy appears to be comparable to that of the corresponding Gaussian formulas, at least for moderate n.

When we apply the formulas corresponding to $w(x) = 1$ [that is, with $c_n = n^{-1}$ and $\mu_n = (n+1)^{-1}$] to integrals $I[f] = \int_0^1 f(x)\,dx$, where $f(x)$ has an integrable singularity at $x = 0$ and is infinitely differentiable everywhere in $(0, 1]$, the new rules seem to be more effective than the corresponding Gaussian rules, as suggested by the numerical examples in [275]. In view of this observation, let us transform the variable x via $x = 2\psi(u/2)$, where $\psi(t)$ is precisely as in Subsection 25.1.2. This results in $I[f] = \int_0^1 g(u)\,du$, where $g(u) \equiv f(2\psi(u/2))\psi'(u/2)$. In case $\psi(t)$ is infinitely differentiable on $[0, 1]$, the transformed integrand $g(u)$ is infinitely differentiable on $(0, 1]$ and may have a singularity at $u = 0$ that is weaker than the singularity of $f(x)$ at $x = 0$. Thus, the new rules can be much more effective on $\int_0^1 g(u)\,du$ than on $\int_0^1 f(x)\,dx$. (This idea was proposed by Johnston [143], who used Gaussian quadrature. Johnston [143] also demonstrated via numerical experiments that class $\mathcal{S}_m$ variable transformations are very appropriate for this purpose.)

25.6 Computation of Integral Transforms with Oscillatory Kernels

25.6.1 Via Numerical Quadrature Followed by Extrapolation

In Chapter 11, we considered in detail the application of various powerful extrapolation methods such as the $\tilde{D}$-, $\bar{D}$-, sD-, and mW-transformations and their variants to integrals of the form $I[f] = \int_0^\infty f(t)\,dt$, where $f(x) = u(x)K(x)$, $u(x)$ either does not oscillate at infinity or oscillates very slowly there, and the kernel function $K(x)$ oscillates about zero infinitely many times as $x \to \infty$ with its phase of oscillation being polynomial ultimately. Here we recall briefly the main points of Chapter 11 on this topic.

To set the background, we begin with the use of (the variant of) the mW-transformation on such integrals. In this method, we first choose a sequence of points x_l, $0 < x_0 < x_1 < \cdots$, as the consecutive zeros of $K(x)$ or of $K'(x)$. We next compute the integrals $\psi(x_l) = \int_{x_l}^{x_{l+1}} f(t)\,dt$ and $F(x_l) = \int_0^{x_l} f(t)\,dt$, $l = 0, 1, \ldots$. Finally, we define the approximations $W_n^{(j)}$ to $I[f]$ via the linear systems

$$F(x_l) = W_n^{(j)} + \psi(x_l) \sum_{i=0}^{n-1} \frac{\bar{\beta}_i}{x_l^i}, \quad j \le l \le j+n.$$

These systems can be solved for the $W_n^{(j)}$ very efficiently via the W-algorithm, as described in Subsection 11.1.1.

Since (in case of convergence) $I[f] = \sum_{k=0}^\infty c_k$, where $c_0 = \int_0^{x_0} f(t)\,dt$ and $c_k = \psi(x_{k-1})$, $k = 1, 2, \ldots$, we can approximate $I[f]$ by accelerating the convergence of the series $\sum_{k=0}^\infty c_k$. In view of the fact that this series is alternating, several sequence transformations are effective for this purpose. For example, we can apply the iterated Δ^2-process, the iterated Lubkin transformation, the Shanks transformation (the ε-algorithm), and the Levin t- and u-transformations successfully. This line of research was begun by Longman [170], [171], who used the Euler transformation. Later the use of the iterated Shanks transformations was demonstrated by Alaylioglu, Evans, and Hyslop [3], and the use of the Shanks and Levin transformations was demonstrated in the survey paper by Blakemore, Evans, and Hyslop [27].

In their surveys on the numerical computation of Bessel function integrals [the case $K(x) = J_\nu(\rho x), \nu > 0, \rho > 0$], Lucas and Stone [189] and Michalski [211] test a large battery of methods and conclude that the mW-transformation and its variants are among the most effective. In particular, the mW-transformation with equidistant x_l, namely, $x_l = x_0 + l\pi/\rho, l = 0, 1, \ldots$, produces very good results for moderate values of ν. For large values of ν, the mW-transformation, with the x_l chosen as the zeros of $K(x)$ or $K'(x)$, produces very high accuracy. The work of Lucas and Stone [189] is concerned with the case of large values of ν. The numerical experiments of Sidi [299] show the $\bar{D}$-transformation and its variants to be equally effective for small and for large ν. In connection with Bessel function integrals, we would like to also mention that the use of equidistant x_l as above was first proposed by Sidi [274], [281] in connection with application of the $\tilde{D}$- and W-transformations. The same x_l were later used by Lyness [197] in connection with application of the Euler transformation.

Finally, we recall the variant of the mW-transformation given in Subsection 11.8.4, which is defined via the linear systems

$$F(x_l) = W_n^{(j)} + \psi(x_l) \sum_{i=0}^{n-1} \frac{\bar{\beta}_i}{(l+1)^{i/m}}, \quad j \le l \le j + n,$$

when $x_l \sim \sum_{i=0}^{\infty} a_i l^{(1-i)/m}$ as $l \to \infty$, $a_0 > 0$. Here x_l and $\psi(x_l)$ are as before. As mentioned in Subsection 11.8.4, this method is comparable to other variants of the mW-transformation as far as its performance is concerned, and it can also be implemented via the W-algorithm.

25.6.2 Via Extrapolation Followed by Numerical Quadrature

We continue with other uses of sequence transformations that are relevant to Fourier and inverse Laplace transforms. Let us consider the case in which $K(x)$ is periodic with period 2τ and assume that $K(x + \tau) = -K(x)$ for all $x \ge 0$. [This is the case when $K(x) = \sin(\pi x/\tau)$, for example.] Let us choose $x_l = (l+1)\tau, l = 0, 1, \ldots$. We can write

$$I[f] = \sum_{k=0}^{\infty} (-1)^k \int_0^\tau u(k\tau + \xi) K(\xi) \, d\xi.$$

Upon interchanging the order of integration and summation, we obtain

$$I[f] = \int_0^\tau S(\xi) K(\xi) \, d\xi, \quad S(\xi) = \sum_{k=0}^{\infty} (-1)^k u(k\tau + \xi).$$

Being an infinite series that is ultimately alternating, $S(\xi)$ can be summed by an appropriate sequence transformation. Following that, the integral $\int_0^\tau S(\xi) K(\xi) \, d\xi$ can be approximated by a low-order Gaussian quadrature formula, for example.

The x_l can be chosen to be consecutive zeros of $K(x)$ or of $K'(x)$ too. Obviously, $x_{l+1} - x_l = \tau$ for all l. In this case, we can write

$$I[f] = \int_0^{x_0} u(\xi) K(\xi) \, d\xi + \sum_{k=0}^{\infty} (-1)^k \int_0^\tau u(x_0 + k\tau + \xi) K(x_0 + \xi) \, d\xi.$$

Upon interchanging the order of integration and summation, we obtain

$$I[f] = \int_0^{x_0} u(\xi) K(\xi) \, d\xi + \int_0^{\tau} S(\xi) K(x_0 + \xi) \, d\xi,$$

$$S(\xi) = \sum_{k=0}^{\infty} (-1)^k u(x_0 + k\tau + \xi).$$

Now we proceed as before.

25.6.3 Via Extrapolation in a Parameter

An extrapolation method for (convergent) oscillatory integrals $I[f] = \int_0^{\infty} f(t) \, dt$, completely different from the ones discussed earlier, was given by Toda and Ono [335]. In this method, one first computes the integral $A(\sigma) = \int_0^{\infty} e^{-\sigma t} f(t) \, dt$, $\sigma > 0$, for a decreasing sequence of values σ_k, $k = 0, 1, \ldots$, tending to zero. (Toda and Ono choose $\sigma_k = 2^{-k}$.) Next, one applies the polynomial Richardson extrapolation to the sequence $\{A(\sigma_k)\}$ assuming that $A(\sigma)$ has the asymptotic expansion

$$A(\sigma) \sim I[f] + \sum_{k=1}^{\infty} \alpha_k \sigma^k \quad \text{as } \sigma \to 0+; \quad \alpha_k \text{ constants.} \qquad (25.6.1)$$

Intuitively, this method is likely to be useful because, for $\sigma > 0$, the integral $\int_0^x e^{-\sigma t} f(t) \, dt$ converges as $x \to \infty$ more quickly than $\int_0^x f(t) \, dt$, thus allowing the integral $A(\sigma)$ to be computed more easily than $I[f]$ itself.

Toda and Ono showed by example that their method works well for integrals such as $\int_0^{\infty} \sin t / t \, dt$, $\int_0^{\infty} \cos t / t \, dt$, and $\int_0^{\infty} (\cos t - \cos 2t) / t \, dt$. As part of their method, Toda and Ono also suggest that $A(\sigma)$ be approximated as follows: Letting $\sigma t = u$, first rewrite $A(\sigma)$ in the form $A(\sigma) = \int_0^{\infty} e^{-u} f(u/\sigma) \, du / \sigma$. Next, transform this integral by letting $u = \psi(v)$, where $\psi(v) = \exp(v - e^{-v})$, one of the double exponential formulas of Takahasi and Mori [333]. This results in $A(\sigma) = \int_{-\infty}^{\infty} Q(v) \, dv$, where $Q(v) = \exp(-\psi(v)) f(\psi(v)/\sigma) \psi'(v) / \sigma$. Finally, this integral is approximated by the trapezoidal rule $T(h) = h \sum_{m=-\infty}^{\infty} Q(mh)$. It is clear that the doubly infinite summation defining $T(h)$ converges quickly for all small h.

An approach similar to that of Toda and Ono [335] was later used by Lund [194] in computing Hankel transforms.

The theoretical justification of the method of Toda and Ono was given in a paper by Sugihara [331]. This paper shows that $\lim_{h \to 0} T(h) = A(\sigma)$ and that the asymptotic expansion in (25.6.1) is valid for a quite general class of kernel functions that includes, for example, $K(x) = e^{\pm ix}$ and $K(x) = J_{\nu}(x)$ with integer ν. Here we provide a different proof of the validity of (25.6.1) in the context of Abel summability, whether the integral $\int_0^{\infty} f(t) \, dt$ converges or not. This proof is simpler than the one given by Sugihara [331]. We recall only the definition of Abel summability: If $\lim_{\epsilon \to 0+} \int_a^{\infty} e^{-\epsilon t} g(t) \, dt = \gamma$ for some finite γ, then we say that the integral $\int_a^{\infty} g(t) \, dt$ exists in the sense of Abel summability and γ is its Abel sum. Of course, in case $\int_a^{\infty} g(t) \, dt$ exists in the ordinary sense, its value and its Abel sum are the same. We begin with the following simple lemma.

Lemma 25.6.1 *Let $f \in C[a, b]$ and $f' \in C(a, b]$ such that $\lim_{x \to a+} f'(x) = \delta$. Then $f \in C^1[a, b]$ as well, and $f'(a) = \delta$.*

Proof. By the mean value theorem, we have

$$\frac{f(x) - f(a)}{x - a} = f'(\hat{x}), \quad \text{for some } \hat{x} \in (a, x).$$

Letting $x \to a+$ on both sides of this equality, and noting that $\lim_{x \to a+} \hat{x} = a$, we have

$$\lim_{x \to a+} \frac{f(x) - f(a)}{x - a} = \delta,$$

from which it follows that $f'(x)$ is continuous at $x = a$ from the right and that $f'(a) = \delta$. ∎

Theorem 25.6.2 *Let the function $f(t)$ be such that $|f(t)|$ is integrable on any finite interval (a, x), $a \geq 0$, and $f(t) = O(t^c)$ as $t \to \infty$, for some real constant c. In addition, for each $k = 0, 1, \ldots$, assume that the integrals $\int_a^\infty t^k f(t) \, dt$ exist in the sense of Abel summability, and denote their corresponding Abel sums by $(-1)^k \beta_k$. Define also $A(\sigma) = \int_a^\infty e^{-\sigma t} f(t) \, dt$, $\sigma > 0$. Then*

$$A(\sigma) \sim I[f] + \sum_{k=1}^\infty \frac{\beta_k}{k!} \sigma^k \quad \text{as } \sigma \to 0+; \quad I[f] = \beta_0. \tag{25.6.2}$$

Proof. First, we realize that $A(\sigma)$ is the Laplace transform of $H(t - a)f(t)$, where $H(x)$ is the Heaviside unit step function, and that $A(\sigma)$ is an analytic function of σ for $\Re \sigma > 0$. Next, we recall that $A^{(k)}(\sigma) = (-1)^k \int_a^\infty e^{-\sigma t} t^k f(t) \, dt$ for $\Re \sigma > 0$ and all k. It thus follows from the Abel summability of the integrals $\int_a^\infty t^k f(t) \, dt$ that $\lim_{\sigma \to 0+} A^{(k)}(\sigma) = \beta_k$, $k = 0, 1, \ldots$. Let us now define $A(0) = \beta_0$. This makes $A(\sigma)$ continuous at $\sigma = 0$ from the right. Applying now Lemma 25.6.1 to $A(\sigma)$, we conclude that $A'(\sigma)$ is also continuous at $\sigma = 0$ from the right with $A'(0) = \beta_1$. By repeated application of this lemma, we conclude that $A(\sigma)$ is infinitely differentiable at $\sigma = 0$ from the right and that $A^{(k)}(0) = \beta_k$ for each k. Therefore, $A(\sigma)$ has the Maclaurin series

$$A(\sigma) = \sum_{k=0}^{N-1} \frac{\beta_k}{k!} \sigma^k + \frac{A^{(k)}(\sigma')}{N!} \sigma^N, \quad \text{for some } \sigma' \in (0, \sigma). \tag{25.6.3}$$

By the fact that N is arbitrary and that $\lim_{\sigma \to 0+} \sigma' = 0$, we have that

$$\lim_{\sigma \to 0+} \left[A(\sigma) - \sum_{k=0}^{N-1} \frac{\beta_k}{k!} \sigma^k \right] \sigma^{-N} = \frac{A^{(N)}(0)}{N!} = \frac{\beta_N}{N!}, \tag{25.6.4}$$

from which the result in (25.6.2) follows. ∎

It is easy to see that Theorem 25.6.2 is valid for all functions $f(x)$ in the class $\hat{\mathbf{B}}$ defined in Section 11.8.

A similar treatment for the case in which $K(x) = e^{iqx}$, with real q, was given by Lugannani and Rice [190]. In this method, one first computes the integral $B(\sigma) = \int_a^\infty e^{-\sigma^2 t^2/2} f(t) \, dt$ for a decreasing sequence of values σ_k, $k = 0, 1, \ldots$, tending to zero. Again, because, for $\sigma > 0$, the integral $\int_0^x e^{-\sigma^2 t^2/2} f(t) \, dt$ converges as $x \to \infty$ more quickly than $\int_0^x f(t) \, dt$, the integral $B(\sigma)$ can be computed more easily than $I[f]$ itself. Again, it can be shown that

$$B(\sigma) \sim I[f] + \sum_{k=1}^{\infty} \alpha_k \sigma^{2k} \quad \text{as } \sigma \to 0+; \quad \alpha_k \text{ constants.} \tag{25.6.5}$$

As a result, we can apply the polynomial Richardson extrapolation process to the sequence $\{B(\sigma_k)\}$ to approximate $I[f]$. See Lugannani and Rice [190] for details.

25.7 Computation of Inverse Laplace Transforms

In this section, we discuss various methods, based on extrapolation processes and sequence transformations, that have been used in the numerical inversion of Laplace transforms.

25.7.1 Inversion by Extrapolation of the Bromwich Integral

First, we concentrate on the extrapolation methods used in computing the Bromwich integral. In Section 11.8, we showed that if $\hat{u}(z)$ is the Laplace transform of $u(t)$, that is, if $\hat{u}(z) = \int_0^{\infty} e^{-zt} u(t) \, dt$, then the Bromwich integral for the inverse transform $u(t)$ can be expressed as in

$$\frac{u(t+) + u(t-)}{2} = \frac{e^{ct}}{2\pi} \left[\int_0^{\infty} e^{i\xi t} \hat{u}(c + i\xi) \, d\xi + \int_0^{\infty} e^{-i\xi t} \hat{u}(c - i\xi) \, d\xi \right].$$

Here c is real and we have assumed that $\hat{u}(z)$ is analytic for $\Re z \geq c$. We mentioned in Section 11.8 that, in case $\hat{u}(z) = e^{-zt_0} \hat{u}_0(z)$ with $\hat{u}_0(z) \in \mathbf{A}^{(\gamma)}$ for some γ [and even in more general cases where $\hat{u}_0(z)$ oscillates very slowly as $\Im z \to \pm\infty$], the mW-transformation with $\xi_l = (l+1)\pi/(t-t_0)$, $l = 0, 1, \ldots$, is very effective.

The P-transformation developed by Levin [162] is very similar to the mW-transformation but uses the points $\xi_l = (l+1)$, $l = 0, 1, \ldots$. Because these points do not take into consideration the fact that the phase of oscillation of the integrands $e^{\pm i\xi t} \hat{u}(c \pm i\xi)$ is proportional to $t - t_0$, this transformation will be costly when $|t - t_0|$ is too small or too large.

25.7.2 Gaussian-Type Quadrature Formulas for the Bromwich Integral

Expressing the Bromwich integral in the form

$$u(t) = \frac{u(t+) + u(t-)}{2} = \frac{1}{2\pi \mathrm{t} \mathrm{i}} \int_{ct-i\infty}^{ct+i\infty} e^{\zeta} \hat{u}(\zeta/t) \, d\zeta, \tag{25.7.1}$$

we can develop numerical quadrature formulas of Gaussian type for this integral. In case $\hat{u}(z) \in \mathbf{A}^{(-s)}$, let us write $\hat{u}(\zeta/t) = t\zeta^{-s} V(\zeta)$. Of course, $V(\zeta) \in \mathbf{A}^{(0)}$, and we can rewrite (25.7.1) in the form

$$u(t) = \frac{1}{2\pi i} \int_{ct-i\infty}^{ct+i\infty} w(\zeta) V(\zeta) \, d\zeta, \quad w(\zeta) = e^{\zeta} \zeta^{-s}. \tag{25.7.2}$$

Here $w(\zeta)$ is viewed as a (complex-valued) weight function. Salzer [255], [256], [257] showed that it is possible to obtain numerical quadrature formulas of Gaussian type for this integral that are of the form

$$u_n(t) = \sum_{k=1}^{n} w_{n,k} V(\zeta_{n,k}) = t^{-1} \sum_{k=1}^{n} w_{n,k} \zeta_{n,k}^s \hat{u}(\zeta_{n,k}/t) \tag{25.7.3}$$

with complex abscissas $\zeta_{n,k}$ and weights $w_{n,k}$. These formulas are obtained by demanding that

$$\sum_{k=1}^{n} w_{n,k} p(\zeta_{n,k}) = \frac{1}{2\pi i} \int_{ct-i\infty}^{ct+i\infty} w(\zeta) p(\zeta) \, d\zeta, \tag{25.7.4}$$

for all $p(\zeta)$ that are polynomials of degree at most $2n - 1$ in ζ^{-1}. As shown by Zakian [377], [378], their abscissas and weights are related to the partial fraction expansions of the $[n-1/n]$ Padé approximants $W_{n-1,n}(\tau)$ from the power series $W(\tau) := \sum_{k=0}^{\infty} \mu_k \tau^k$, where

$$\mu_k = \frac{1}{2\pi i} \int_{ct-i\infty}^{ct+i\infty} w(\zeta) \zeta^{-k} \, d\zeta = \frac{1}{\Gamma(s+k)}, \quad k = 0, 1, \dots . \tag{25.7.5}$$

[Note that $W(\tau)$ converges for all τ and hence represents an entire function, which we denote $W(\tau)$ as well. For $s = 1$, we have $W(\tau) = e^{\tau}$.] Actually, there holds

$$W_{n-1,n}(\tau) = \sum_{k=1}^{n} \frac{w_{n,k}}{1 - \tau/\zeta_{n,k}}. \tag{25.7.6}$$

It turns out that the abscissas $\zeta_{n,k}$ are in the right half plane $\Re z > 0$. For even n, they are all complex and come in conjugate pairs. For odd n, only one of the abscissas is real and the rest appear in conjugate pairs. See Piessens [230].

Now the assumption that $\hat{u}(z) \in \mathbf{A}^{(-s)}$ means that $u(t) \sim \sum_{i=0}^{\infty} c_i t^{s-1+i}$ as $t \to 0+$. With this $u(t)$, we actually have that $\hat{u}(z) \sim \sum_{i=0}^{\infty} c_i \Gamma(s+i) z^{-s-i}$ as $\Re z \to \infty$, as can be verified with the help of Watson's lemma. Substituting this in (25.7.3), and invoking (25.7.5), it is easy to show that

$$u_n(t) - u(t) = O(t^{s+2n-1}) \quad \text{as } t \to 0+. \tag{25.7.7}$$

In other words, the Gaussian formulas produce good approximations to $u(t)$ for small to moderate values of t.

For an extensive list of references on this subject, see Davis and Rabinowitz [63, pp. 266–270].

Note that rational approximations other than Padé approximants can be used to derive numerical quadrature formulas of the form (25.7.3). Because the sequence $\{\mu_k\}$ with μ_k as in (25.7.5) is in $\mathbf{b}^{(1)}$, we can use the Levin t- or the Sidi S-transformation to replace the Padé approximants from the series $W(\tau)$. It turns out that the abscissas and weights of the formulas that are obtained from the t-transformation grow in modulus very quickly, which ultimately causes numerical instabilities in finite-precision arithmetic. The ones obtained from the $\mathcal{S}$-transformation grow much more slowly; actually, their growth rates are similar to those of the Gaussian-type formulas. (Interestingly, for $s = 1$ these formulas are the same as the corresponding Gaussian-type formulas.) The abscissas and weights that result from the $\mathcal{S}$-transformation on the series $W(\tau)$ satisfy

$$\mathcal{S}_n^{(0)}(\tau) = \frac{\sum_{i=0}^n \lambda_i \tau^{n-i} \sum_{r=0}^{i-1} \mu_r \tau^r}{\sum_{i=0}^n \lambda_i \tau^{n-i}} = \sum_{k=1}^n \frac{w_{n,k}}{1 - \tau/\zeta_{n,k}}; \quad \lambda_i = (-1)^i \binom{n}{i} \frac{(i+1)_{n-1}}{\mu_i}.$$
(25.7.8)

In other words, the $\zeta_{n,k}$ are the zeros of the polynomial $\sum_{i=0}^n \lambda_i \tau^{n-i}$ [equivalently, the poles of $\mathcal{S}_n^{(0)}(\tau)$], and the $w_{n,k}$ are given by

$$w_{n,k} = -\frac{1}{\tau} \cdot \left. \frac{\sum_{i=0}^n \lambda_i \tau^{n-i} \sum_{r=0}^{i-1} \mu_r \tau^r}{\sum_{i=0}^n (n-i)\lambda_i \tau^{n-i-1}} \right|_{\tau = \zeta_{n,k}}.$$
(25.7.9)

This approach to Laplace transform inversion was used in the M.Sc. thesis of Shelef [265], which was done under the supervision of the author.

25.7.3 Inversion via Rational Approximations

We continue by describing a method for computing inverse Laplace transforms that is based on rational approximations in the complex plane. Let $\hat{u}(z)$ be the Laplace transform of $u(t)$. One way of approximating the inverse transform $u(t)$ is by generating a sequence of rational approximations $\{\hat{\phi}_n(z)\}$ to $\hat{u}(z)$ and inverting each $\hat{\phi}_n(z)$ exactly to obtain the sequence $\{\phi_n(t)\}$. Of course, we hope that $\{\phi_n(t)\}$ will converge to $u(t)$ quickly. For this to happen, $\{\hat{\phi}_n(z)\}$ should tend to $\hat{u}(z)$ quickly in the z-plane. Also, as $\lim_{z\to\infty} \hat{u}(z) = 0$, we must require that $\lim_{z\to\infty} \hat{\phi}_n(z) = 0$ as well. This means that the degree of the numerator polynomial $P_n(z)$ of $\hat{\phi}_n(z)$ should be strictly smaller than that of its denominator polynomial $Q_n(z)$ for each n. Let the degree of $Q_n(z)$ be exactly n and assume, for simplicity, that the zeros $\alpha_1, \ldots, \alpha_n$ of $Q_n(z)$ are simple. Then, $\hat{\phi}_n(z)$ has the partial fraction expansion

$$\hat{\phi}_n(z) = \sum_{k=1}^n \frac{A_k}{z - \alpha_k}; \quad A_k = \frac{P_n(\alpha_k)}{Q_n'(\alpha_k)}, \quad k = 1, \ldots, n,$$
(25.7.10)

and $\phi_n(t)$ is given by

$$\phi_n(t) = \sum_{k=1}^n A_k e^{\alpha_k t}.$$
(25.7.11)

The rational functions $\hat{\phi}_n(z)$ can be obtained by applying a sequence transformation to the Taylor series of $\hat{u}(z)$ at a point z_0 in the right half plane where $\hat{u}(z)$ is analytic. Thus, we can use the $[n - 1/n]$ Padé approximants for this purpose. We can also use the Sidi–Levin rational d-approximants when appropriate. In particular, we can use the rational approximations obtained by applying the Levin $\mathcal{L}$-transformation or the Sidi $\mathcal{S}$-transformation whenever this is possible. If $z_0 = 0$ and $\sum_{i=0}^\infty c_i z^i$ is the Taylor series of $\hat{u}(z)$, then the appropriate $\mathcal{L}$-approximants are given by

$$\mathcal{L}_n^{(0)}(z) = \frac{\sum_{i=0}^n \lambda_i z^{n-i} \sum_{r=0}^{i-1} c_r z^r}{\sum_{i=0}^n \lambda_i z^{n-i}}; \quad \lambda_i = (-1)^i \binom{n}{i} \frac{(i+1)^{n-1}}{c_i}. \quad (25.7.12)$$

Because the $\hat{\phi}_n(z)$ are obtained from information on $\hat{u}(z)$ at $z = 0$, the $\phi_n(t)$ will produce good approximations for large t.

Sometimes $\hat{u}(z)$ assumes the form $\hat{u}(z) = z^{-1} v(z)$, where $v(z)$ is analytic at $z = 0$. In this case, we choose $\hat{\phi}_n(z) = z^{-1} v_{n,n}(z)$, where $v_{n,n}(z)$ are the $[n/n]$ Padé approximants from the expansion of $v(z)$ at $z = 0$. Again, other rational approximations can be used instead of Padé approximants.

If $\hat{u}(z) \sim \sum_{i=0}^\infty d_i z^{-i-1}$ as $z \to \infty$, then we can choose $\hat{\phi}_n(z) = z^{-1} U_{n-1,n}(z^{-1})$, where $U_{n-1,n}(\tau)$ is the $[n-1/n]$ Padé approximant from the power series $U(\tau) := \sum_{i=0}^\infty d_i \tau^i$. Because in this case $u(t) \sim \sum_{i=0}^\infty d_i t^i / i!$ as $t \to 0+$, it follows that

$$\phi_n(t) - u(t) = O(t^{2n}) \text{ as } t \to 0+. \quad (25.7.13)$$

We can also choose $\hat{\phi}_n(z)$ as a two-point Padé approximant from the expansions $\hat{u}(z) = \sum_{i=0}^\infty c_i z^i$ and $\hat{u}(z) \sim \sum_{i=0}^\infty d_i z^{-i-1}$ as $z \to \infty$ and invert these exactly to obtain the $\phi_n(t)$. In this case, we expect the $\phi_n(t)$ to approximate $u(t)$ well both for small and for large values of t.

The preceding methods should be very practical in case $\hat{u}(z)$ is difficult to compute accurately, but its expansions as $z \to 0$ and/or $z \to \infty$ can be obtained relatively easily. The general ideas were suggested by Luke and were developed further and used extensively (through continued fractions, Padé approximants, and rational approximations from the Levin transformation) by Longman in the solution of problems in theoretical seismology and electrical circuit theory. See, for example, Luke [191], [192], and Longman and [173], [175], [176], [177].

Two-point Padé approximants were used also by Grundy [117] but in a different manner. In Grundy's approach, the approximations $\phi_n(t)$ are not sums of exponential functions but two-point Padé approximants obtained from two asymptotic expansions of the inverse transform $u(t)$ in some fixed power of t. One of these expansions is at $t = 0$ and is obtained from that of $\hat{u}(z)$ at $z = \infty$, which we have already discussed. The other expansion is at some finite point t_0 that is determined by the singularity structure of $\hat{u}(z)$. For details and examples, see Grundy [117].

25.7.4 Inversion via the Discrete Fourier Transform

Finally, we mention a method that was originally introduced by Dubner and Abate [70] and improved by Durbin [71] that uses the discrete Fourier transform. Taking $u(t)$ to be real for simplicity, Durbin gives the following approximation to $u(t)$:

$$u_T(t) = \frac{e^{at}}{T} \left[\frac{1}{2} \hat{u}(a) + \sum_{k=1}^\infty \left\{ \Re \hat{u}\left(a + i\frac{k\pi}{T}\right) \cos\left(\frac{k\pi t}{T}\right) - \Im \hat{u}\left(a + i\frac{k\pi}{T}\right) \sin\left(\frac{k\pi t}{T}\right) \right\} \right]. \quad (25.7.14)$$

Here a is a real constant chosen such that $\hat{u}(z)$ is analytic for $\Re z \geq a$. The error in this approximation is given by

$$u_T(t) - u(t) = \sum_{k=1}^{\infty} e^{-2kaT} u(2kT + t), \qquad (25.7.15)$$

and, provided $|u(t)| \leq C$ for all t, it satisfies

$$|u_T(t) - u(t)| \leq Ce^{-2aT} \quad \text{for all } t \in [0, 2T]. \qquad (25.7.16)$$

We see from (25.7.16) that, by choosing aT large, we make the error in $u_T(t)$ small. However, in general we can only approximate the sum of the Fourier series inside the square brackets in (25.7.14), and the error we commit in this approximation is magnified by the factor e^{at}, which may be large when t is of order T. This suggests that we should obtain the sum of this series as accurately as possible. This can be achieved by using appropriate sequence transformations. Crump [58] uses the Shanks transformation. For the same problem Kiefer and Weiss [151] also use summation by parts. Of course, the d-transformation with the complex series approach and APS can be used to give very high accuracy, as was done by Sidi [294]. [The latter is discussed in Chapter 12.] In this case, it is easy to see that (25.7.14) can be rewritten as in

$$u_T(t) = \frac{e^{at}}{T} \Re \left[\frac{1}{2} \hat{u}(a) + \sum_{k=1}^{\infty} \hat{u}\left(a + i\frac{k\pi}{T}\right) \zeta^k \right], \quad \zeta \equiv \exp\left(i\frac{\pi t}{T}\right), \quad (25.7.17)$$

and this is very convenient for the complex series approach. When $\hat{u}(z) \in \mathbf{A}^{(-s)}$ for some s, we can use the $d^{(1)}$-transformation and other sequence transformations such as the θ-algorithm. For more complicated $\hat{u}(z)$, however, we can use only the $d^{(m)}$- and Shanks transformations.

25.8 Simple Problems Associated with $\{a_n\} \in \mathbf{b}^{(1)}$

In this section, we treat two simple problems associated with sequences $\{a_n\}$ that are in $\mathbf{b}^{(1)}$.

1. Given that $a_n = h(n) \in \mathbf{A}_0^{(\gamma)}$ strictly for some $\gamma \neq 0$, we would like to find γ. We can easily show that

$$\frac{\Delta a_n}{a_n} = \frac{\Delta h(n)}{h(n)} \sim \sum_{i=0}^{\infty} \frac{c_i}{n^{i+1}} \quad \text{as } n \to \infty; \quad c_0 = \gamma.$$

Thus, $U_n = n(\Delta a_n / a_n)$ satisfies

$$U_n \sim \gamma + \sum_{i=1}^{\infty} \frac{c_i}{n^i} \quad \text{as } n \to \infty.$$

[Compare this with the asymptotic expansion given in (15.3.18).]

2. Given that $a_n = \zeta^n h(n)$, with $\zeta \neq 1$ and $h(n) \in \mathbf{A}_0^{(\gamma)}$ strictly for some γ, we would like to find ζ and γ. This time, we have that

$$V_n = \frac{a_{n+1}}{a_n} = \zeta \frac{h(n+1)}{h(n)} = \zeta \left(1 + \frac{\Delta h(n)}{h(n)}\right) \sim \zeta + \sum_{i=0}^{\infty} \frac{\zeta c_i}{n^{i+1}} \quad \text{as } n \to \infty.$$

Similarly, we can show that

$$-\frac{\Delta V_n}{V_n} = -\frac{\Delta(h(n+1)/h(n))}{h(n+1)/h(n)} \sim \sum_{i=0}^{\infty} \frac{e_i}{n^{i+2}} \quad \text{as } n \to \infty; \quad e_0 = \gamma.$$

Thus, $W_n = -n^2(\Delta V_n/V_n)$ satisfies

$$W_n \sim \gamma + \sum_{i=1}^{\infty} \frac{e_i}{n^i} \quad \text{as } n \to \infty.$$

It is thus clear that γ in the first problem and ζ and γ in the second problem can be determined by applying the polynomial Richardson extrapolation to the sequences $\{U_n\}$, $\{V_n\}$, and $\{W_n\}$. They can also be determined via the Levin u-transformation, the iterated Lubkin transformation, the ρ-algorithm, and the θ-algorithm.

25.9 Acceleration of Convergence of Infinite Series with Special Sign Patterns

Let us first consider infinite series of the form $\sum_{k=0}^{\infty}(-1)^{\lfloor k/q \rfloor} v_k$, where q is a positive integer and $\{v_n\} \in \mathbf{b}^{(r)}$ for some r. Note that the sequence of the factors $(-1)^{\lfloor k/q \rfloor}$ comprises q $(+1)$s, followed by q (-1)s, followed by q $(+1)$s, and so on. We have already seen one such example (with $q = 2$ and $r = 1$), namely, Lubkin's series $\sum_{k=0}^{\infty}(-1)^{\lfloor k/2 \rfloor}/(k+1)$, in our discussion of the Δ^2-process in Chapter 15.

In case $q \geq 2$ or $r \geq 2$ or both, such series cannot be handled by the iterated Δ^2-process, the iterated Lubkin transformation, the Levin transformations, and the θ-algorithm. When the term $(-1)^{\lfloor n/q \rfloor} v_n$ has no nonoscillatory part in its asymptotic expansion as $n \to \infty$, the transformations of Euler and of Shanks can be applied effectively. Otherwise, they are not useful. The $d^{(m)}$-transformation with $m = qr$ is always effective, however. We now discuss the reason for this.

Denote $c_k = (-1)^{\lfloor k/q \rfloor}$. Then the c_k satisfy the linear $(q+1)$-term recursion relation $c_{n+q} + c_n = 0$, hence the linear difference equation $c_n = -\sum_{k=1}^{q} \frac{1}{2}\binom{q}{k}\Delta^k c_n$. Therefore, $\{c_n\} \in \mathbf{b}^{(q)}$. In fact, c_n is of the form $c_n = \sum_{s=1}^{q} \alpha_s \omega_s^n$, where $\omega_s = \exp[i(2s-1)\pi/q]$, $s = 1, \ldots, q$, and the constants α_s are determined from the initial conditions $c_0 = \cdots = c_{q-1} = 1$. As a result of this, we have that $\{(-1)^{\lfloor n/q \rfloor} v_n\} \in \mathbf{b}^{(m)}$ for some $m \leq qr$, by Heuristic 6.4.1. Hence, the $d^{(qr)}$-transformation is effective.

To illustrate this, let us consider Lubkin's series again. For this example, $c_n = (-1)^{\lfloor n/2 \rfloor} = \sqrt{2} \cos(n\pi/2 - \pi/4)$. Therefore, the term $(-1)^{\lfloor n/2 \rfloor}/(n+1)$ has no nonoscillatory part as $n \to \infty$. Consequently, the Euler, Shanks, and $d^{(2)}$-transformations are effective on this example. Others are not, as mentioned in Wimp [366, p. 171].

Incidentally, after some complicated manipulations involving the functions $\beta(x) = \sum_{k=0}^{\infty} \frac{(-1)^k}{x+k}$ and $\psi(x) = \frac{d}{dx} \log \Gamma(x)$, it can be shown for the generalization of Lubkin's series, where $(-1)^{\lfloor k/2 \rfloor}$ is replaced by $(-1)^{\lfloor k/q \rfloor}$, that

$$\sum_{k=0}^{\infty} \frac{(-1)^{\lfloor k/q \rfloor}}{k+1} = \frac{1}{q} \log 2 + \frac{\pi}{2q} \sum_{k=1}^{q-1} \tan \frac{k\pi}{2q}, \quad q = 1, 2, \ldots, \tag{25.9.1}$$

which seems to be new. This series can be summed very efficiently by the $d^{(q)}$-transformation.

25.9.1 Extensions

Let us now consider the more general case of the series $\sum_{k=0}^{\infty} c_k v_k$, where $\{v_n\} \in \mathbf{b}^{(r)}$ for some r, as before, and the c_n are either $+1$ or -1 [but not necessarily $(-1)^{\lfloor n/q \rfloor}$], such that $c_{n+q} = -c_n$. This means that $c_0, c_1, \ldots, c_{q-1}$ take on the values $+1$ or -1 in any order, and the remaining c_k satisfy $c_{n+q} = -c_n$. (For example, with $q = 4$, the c_k may be such that $|c_k| = 1$ and have the sign pattern $+ - + + - + - - + - + + - + - - \cdots$.) Since these c_n satisfy the recursion $c_{n+q} + c_n = 0$ too, we see that $\{c_n v_n\} \in \mathbf{b}^{(m)}$ for some $m \le qr$, as before. Therefore, the $d^{(m)}$-transformation can be applied with no changes to these series.

Finally, the argument of the preceding paragraph is valid for those series $\sum_{k=0}^{\infty} c_k v_k$, where $\{v_n\} \in \mathbf{b}^{(r)}$ for some r, as before, and the c_n are either $+1$ or -1, such that $c_{n+q} = c_n$. This means that $c_0, c_1, \ldots, c_{q-1}$ take on the values $+1$ or -1 in any order, and the remaining c_k satisfy $c_{n+q} = c_n, n = 0, 1, \ldots$. (For example, with $q = 4$, the c_k may be such that $|c_k| = 1$ and have the sign pattern $+ - + + + - + + \cdots$.) Since these c_n satisfy the recursion $c_{n+q} - c_n = 0$ too, we see that $\{c_n v_n\} \in \mathbf{b}^{(m)}$ for some $m \le qr$, as before. Therefore, the $d^{(m)}$-transformation can be applied with no changes to these series as well.

25.10 Acceleration of Convergence of Rearrangements of Infinite Series

An interesting series that was considered by Lubkin [187, Example 2] is $S := \sum_{k=0}^{\infty} a_k = -1/2 + 1 - 1/4 + 1/3 - 1/6 + 1/5 - \cdots$, which is a rearrangement of the series $T := \sum_{k=0}^{\infty} u_k = \sum_{k=0}^{\infty} (-1)^k/(k+1) = 1 - 1/2 + 1/3 - 1/4 + \cdots$, the sum of T being $\log 2$. Despite the fact that T converges conditionally, it is easy to see that the rearrangement S has $\log 2$ as its sum too.

Since $\{(-1)^n/(n+1)\} \in \mathbf{b}^{(1)}$, the series T can be summed very accurately by most sequence transformations we discussed earlier, such as the iterated Δ^2-process, the iterated Lubkin transformation, the transformations of Euler, Shanks, and Levin, and the θ-algorithm. None of these methods accelerates the convergence of the rearrangement series S, as numerical computations show, however. The only method that is effective on S appears to be the $d^{(2)}$-transformation. The reason for this is that the sequence $\{a_n\}$ is in $\mathbf{b}^{(2)}$, as we show for a more general case next.

Let $\{u_n\} \in \mathbf{b}^{(1)}$, and consider the rearrangement series $S := \sum_{k=0}^{\infty} a_k = u_1 + u_0 + u_3 + u_2 + u_5 + u_4 + \cdots$. Thus,

$$a_{2k} = u_{2k+1}, \quad \text{and} \quad a_{2k+1} = u_{2k}, \quad k = 0, 1, \ldots.$$

We would like to show that $\{a_n\} \in \mathbf{b}^{(2)}$ in the relaxed sense of Section 6.4, that is, that there exists a 3-term recursion relation of the form

$$a_{n+2} + \mu(n)a_{n+1} + \nu(n)a_n = 0, \quad n = 0, 1, \dots, \tag{25.10.1}$$

with $\mu(n) \in \mathbf{A}_0^{(r)}$ and $\nu(n) \in \mathbf{A}_0^{(s)}$ for some integers r and s. Letting $n = 2k$ and $n = 2k + 1$ in (25.10.1), the latter can be written in terms of the u_n as follows:

$$u_{2k+3} + \mu(2k)u_{2k} + \nu(2k)u_{2k+1} = 0, \quad n = 2k,$$

$$u_{2k+2} + \mu(2k+1)u_{2k+3} + \nu(2k+1)u_{2k} = 0, \quad n = 2k + 1. \tag{25.10.2}$$

Now, we know that $u_{n+1} = c(n)u_n$ for some $c(n) \in \mathbf{A}_0^{(q)}$, q an integer. Therefore, $u_{n+2} = c(n+1)c(n)u_n$, $u_{n+3} = c(n+2)c(n+1)c(n)u_n$, etc. Substituting these in (25.10.2), we obtain the equations

$$c(2k+2)c(2k+1)c(2k) + \mu(2k) + \nu(2k)c(2k) = 0,$$

$$c(2k+1)c(2k) + \mu(2k+1)c(2k+2)c(2k+1)c(2k) + \nu(2k+1) = 0. \tag{25.10.3}$$

Letting $2k = n$ in the first of these equations, and $2k + 1 = n$ in the second, we finally obtain the following linear system for $\mu(n)$ and $\nu(n)$:

$$\mu(n) + c(n)\nu(n) = -c(n+2)c(n+1)c(n)$$
$$c(n+1)c(n)c(n-1)\mu(n) + \nu(n) = -c(n)c(n-1) \tag{25.10.4}$$

Since $c(n) \in \mathbf{A}_0^{(q)}$, q an integer, the elements of the matrix of these equations and the right-hand-side vector are also in $\mathbf{A}_0^{(\sigma)}$ for some integers σ. By Cramer's rule, so are $\mu(n)$ and $\nu(n)$, which implies that $\{a_n\} \in \mathbf{b}^{(2)}$.

25.10.1 Extensions

We can use this approach to treat other rearrangement series, $S := \sum_{k=0}^{\infty} a_k$, such as that obtained alternately from p terms of $T := \sum_{k=0}^{\infty} u_k$ with even index followed by q terms with odd index, namely, $S := u_0 + u_2 + \cdots + u_{2p-2} + u_1 + u_3 + \cdots + u_{2q-1} + u_{2p} + \cdots$. Using the same technique, we can now show that $\{a_n\} \in \mathbf{b}^{(p+q)}$ when $\{u_n\} \in \mathbf{b}^{(1)}$. As a result, the only acceleration method that is effective on such series is the $d^{(p+q)}$-transformation. Thus, the $d^{(3)}$-transformation sums the series $1 + 1/3 - 1/2 + 1/5 + 1/7 - 1/4 + \cdots$ very efficiently to $\frac{3}{2} \log 2$. The $d^{(5)}$-transformation sums the series $1 + 1/3 + 1/5 - 1/2 - 1/4 + 1/7 + 1/9 + 1/11 - 1/6 - 1/8 + \cdots$ very efficiently to $\log 2 + \frac{1}{2} \log(3/2)$.

All this can be generalized in a straightforward manner to the case in which $\{u_n\} \in \tilde{\mathbf{b}}^{(m)}$. We leave the details to the reader.

25.11 Acceleration of Convergence of Infinite Products

We now consider the efficient computation of some classes of infinite products by extrapolation. Let $\{A_n\}$ be the sequence of the partial products of the convergent infinite product $A = \prod_{k=1}^{\infty}(1 + \nu_k)$. That is to say, $A_n = \prod_{k=1}^{n}(1 + \nu_k)$, $n = 1, 2, \dots$. We propose to accelerate the convergence of the sequence $\{A_n\}$. As we show next, this can be achieved provided the v_k have suitable properties.

First, the infinite product converges if and only if $\sum_{k=1}^{\infty} v_k$ converges. Next, let us assume that $v_n = w(n) \in \tilde{\mathbf{A}}_0^{(-s/m,m)}$ strictly for some integer s, that is, $v_n \sim \sum_{i=0}^{\infty} \epsilon_i n^{-(s+i)/m}$ as $n \to \infty$, $\epsilon_0 \neq 0$, and convergence implies that $s \geq m + 1$.

Now, $A_n = A_{n-1}(1 + v_n)$, from which we obtain $\Delta A_{n-1} = v_n A_{n-1}$. If we let $a_1 = A_1$ and $a_n = A_n - A_{n-1}$, $n \geq 2$, so that $A_n = \sum_{k=1}^{n} a_k$, $n \geq 1$, then the latter equality becomes $a_n/v_n = A_{n-1}$. Applying Δ to both sides of this last equality, we have $\Delta(a_n/v_n) = a_n$, which can be written in the form $a_n = p(n)\Delta a_n$, where $p(n) = (v_{n+1} + \Delta v_n/v_n)^{-1}$. A careful analysis shows that $p(n) \in \tilde{\mathbf{A}}_0^{(1,m)}$ strictly, with $p(n) \sim -(m/s)n$ as $n \to \infty$. By Theorem 6.6.5, this implies that $\{a_n\} \in \tilde{\mathbf{b}}^{(m)}$; hence, both the $\tilde{d}^{(m)}$- and the $d^{(m)}$-transformations can be applied successfully. In addition, for best numerical results we can use GPS. The iterated Lubkin transformation and the θ-algorithm can also be applied to $\{A_n\}$.

25.11.1 Extensions

We now extend this approach to infinite products of the form $A = \prod_{k=1}^{\infty}(1 + c_k v_k)$, where $v_n = w(n) \in \tilde{\mathbf{A}}_0^{(-s/m,m)}$ strictly for some integer s, as before, and the c_n are such that $c_{n+q} = c_n$ for all n. This means that $c_1, \ldots, c_q$ take on arbitrary nonzero values, and the remaining c_k are determined by $c_{n+q} = c_n$.

As an illustration, let us consider the case in which the c_n are either $+1$ or -1, such that $c_{n+q} = c_n$. Two simple examples of this are $c_n = (-1)^n$ for $q = 2$, and $c_n = (-1)^{\lfloor n/2 \rfloor}$ for $q = 4$. Another more complex example with $q = 5$ is one in which the c_k are such that $|c_k| = 1$ and have the sign pattern $+ + - + - + + - + - \cdots$.

In this case, we have $A_{n+q} = A_n \prod_{k=n+1}^{n+q}(1 + c_k v_k)$. Replacing n by qn, denoting $A_{qn} = A'_n$, and using the fact that $c_{qn+k} = c_k$ for all k, this becomes

$$A'_{n+1} = A'_n \prod_{k=qn+1}^{qn+q}(1 + c_k v_k) = A'_n \prod_{k=1}^{q}(1 + c_k v_{qn+k}).$$

Now, by the fact that $v_n = w(n) \in \tilde{\mathbf{A}}_0^{(-s/m,m)}$, there holds

$$\prod_{k=1}^{q}(1 + c_k v_{qn+k}) = 1 + v'_n, \quad v'_n = w'(n) \in \tilde{\mathbf{A}}_0^{(-s'/m,m)} \quad \text{strictly,}$$

for some integer $s' \geq s$. (Note that $s' = s$ when $\sum_{k=1}^{q} c_k \neq 0$, but $s' > s$ for $\sum_{k=1}^{q} c_k = 0$.) Thus, we have shown that $\{A'_n\}$ is the sequence of partial products of the infinite product $A = \prod_{k=1}^{\infty}(1 + v'_k)$ and that this infinite product is precisely of the form treated in the beginning of this section. Therefore, we can accelerate the convergence of $\{A'_n\}$ in exactly the same form described there.

New nonlinear methods for accelerating the convergence of infinite products of the forms $\prod_{k=1}^{\infty}(1 + v_k)$ and $\prod_{k=1}^{\infty}[1 + (-1)^k v_k]$, with $v_n = w(n) \in \mathbf{A}_0^{(-s)}$ for some positive integer s, have recently been devised in a paper by Cohen and Levin [54]. These methods are derived by using an approach analogous to that used in deriving the

$\mathcal{L}$- and the $d^{(2)}$-transformations. Unlike the d-transformations, they require knowledge of s, however.

One interesting problem treated by Cohen and Levin [54] is that of approximating the limit of the product $A(z) = \prod_{k=1}^{\infty}(1 + v_k z)$, with $v_n = w(n) \in \tilde{\mathbf{A}}_0^{(-s/m,m)}$ strictly for some integer $s \geq m + 1$, as before. Here $A(z)$, as a function of z, vanishes at the points $z_k = -1/v_k$, $k = 1, 2, \ldots$, and we would like to find approximations to $A(z)$ that will vanish at the first μ zeros $z_1, \ldots, z_\mu$. Such approximations can be obtained as follows: First, accelerate the convergence of the sequence of partial products of $\prod_{k=\mu+1}^{\infty}(1 + v_k z)$ via the $d^{(m)}$- or $\tilde{d}^{(m)}$-transformations. Call the resulting approximations $A_n^{(j)}(z; \mu)$. Next, approximate $A(z)$ by $[\prod_{k=1}^{\mu}(1 + v_k z)]A_n^{(j)}(z; \mu)$. This procedure is analogous to the one proposed in [54].

25.12 Computation of Infinite Multiple Integrals and Series

The problem of accelerating the convergence of infinite multiple series and integrals by extrapolation methods has been of some interest recently.

The first work in this field seems to be due to Streit [327] who considered the generalization of the Aitken Δ^2-process to double series.

The first paper in multiple power series acceleration was published by Chisholm [48]. In this paper, Chisholm defines the diagonal Padé approximants to double power series $f(x, y) := \sum_{i=0}^{\infty} \sum_{j=0}^{\infty} c_{ij} x^i y^j$. These approximants are of the form $f_{n,n}(x, y) = \sum_{i=0}^{n} \sum_{j=0}^{n} u_{ij} x^i y^j / \sum_{i=0}^{n} \sum_{j=0}^{n} v_{ij} x^i y^j$.

The nondiagonal approximants $f_{m,n}(x, y)$ were later defined by Graves-Morris, Hughes Jones, and Makinson [109]. The diagonal approximants of [48] and the nondiagonal ones of [109] were generalized to power series in an arbitrary number of variables by Chisholm and McEwan [50] and by Hughes Jones [137], respectively. General order Padé approximants for multiple power series were defined by Levin [163] and further developed by Cuyt [59], [60], [61].

A general discussion on accelerating the convergence of infinite double series and integrals was presented by Levin [164], who also gave a generalization of the Shanks transformation. Generalizations of the Levin and the Shanks transformations to multiple infinite series were also considered in Albertsen, Jacobsen, and Sørensen [4].

A recent paper by Greif and Levin [115] combines the general idea of Levin [164] with an approach based on the D-transformation for one-dimensional infinite-range integrals and the d-transformation for one-dimensional infinite series. Earlier, Sidi [268, Chapter 4, Section 6] proposed an approach in which one uses of the d-transformation sequentially in summation of multiple series and provided some theoretical justification for it at the same time. The same approach can be used to compute multiple infinite-range integrals. In a more recent work by Levin and Sidi [166], multidimensional generalizations of GREP and of the D- and d-transformations are reviewed and some new methods are derived. We do not go into these methods here, as this would require new definitions and notation. However, we do describe the approach of Sidi [268], because it is simple to explain and use; it is very effective as well. Our description here is precisely that given by Levin and Sidi [166].

25.12.1 Sequential D-Transformation for s-D Integrals

Let us consider the s-dimensional (s-D) integral $I[f] = \int_{I\!\!R_0^s} f(\mathbf{t}) \, d\mathbf{t}$, where we have denoted $I\!\!R_0^s = \{(t_1, \ldots, t_s) : t_i \geq 0, \ i = 1, \ldots, s\}, \mathbf{t} = (t_1, \ldots, t_s)$, and $d\mathbf{t} = \prod_{j=1}^s dt_j$, and let us define

$$H_1(t_1, \ldots, t_s) = f(\mathbf{t}) = f(t_1, \ldots, t_s),$$

$$H_{k+1}(t_{k+1}, \ldots, t_s) = \int_0^\infty H_k(t_k, \ldots, t_s) \, dt_k, \quad k = 1, \ldots, s-1.$$

Then $I[f] = \int_0^\infty H_s(t_s) \, dt_s$. Let us now assume that, for each k and for fixed $t_{k+1}, \ldots, t_s$, and as a function of t_k, $H_k(t_k, \ldots, t_s) \in \mathbf{B}^{(m_k)}$ for some integer m_k. [This assumption seems to hold when $f(\mathbf{t})$, as a function of the variable t_k – the rest of the variables being held fixed – is in $\mathbf{B}^{(m_k)}$.] This means we can compute $H_{k+1}(t_{k+1}, \ldots, t_s)$ by applying the $D^{(m_k)}$-transformation to the integral $\int_0^\infty H_k(t_k, \ldots, t_s) \, dt_k$. The computation of $I[f]$ can thus be completed by applying the $D^{(m_s)}$-transformation to the integral $\int_0^\infty H_s(t_s) \, dt_s$.

It is very easy to see that the preceding assumption is automatically satisfied when $f(\mathbf{x}) = \prod_{j=1}^s f_j(x_j)$, with $f_j \in \mathbf{B}^{(m_j)}$ for some integers m_j. This then serves as the motivation for sequential use of the D-transformation.

As an example, consider the function $f(x, y) = e^{-ax} u(y)/(x + g(y))$, where a is a constant with $\Re a > 0$, $u(y) \in \mathbf{B}^{(q)}$, and $g(y) \in \mathbf{A}^{(r)}$ for some positive integer r and $g(y) > 0$ for all large y. [We have $q = 2$ when $u(y) = \cos by$ or $u(y) = J_\nu(by)$, for example.] First, $f(x, y)$ is in $\mathbf{B}^{(1)}$ as a function of x (with fixed y) and $f(x, y)$ is in $\mathbf{B}^{(q)}$ as a function of y (with fixed x). Next, invoking the relation $1/c = \int_0^\infty e^{-c\xi} \, d\xi, \Re c > 0$, we can show that

$$H_2(y) = \int_0^\infty f(x, y) \, dx = u(y) \int_0^\infty e^{-\xi g(y)}/(a + \xi) \, d\xi.$$

Applying Watson's lemma to this integral, we see that $H_2(y)$ has an asymptotic expansion of the form

$$H_2(y) \sim u(y) \sum_{i=0}^\infty \alpha_i [g(y)]^{-i-1} \sim u(y) \sum_{i=0}^\infty \delta_i y^{-i-r} \quad \text{as } y \to \infty.$$

This implies that $H_2(y) \in \mathbf{B}^{(q)}$.

25.12.2 Sequential d-Transformation for s-D Series

Sequential use of the d-transformation for computing s-D infinite series is analogous to the use of the D-transformation for s-D integrals. Let us now consider the s-D infinite series $S(\{a_\mathbf{i}\}) = \sum_{\mathbf{i} \in \mathbb{Z}_+^s} a_\mathbf{i}$, where $\mathbb{Z}_+^s = \{(i_1, \ldots, i_s) : i_j \text{ integer} \geq 1, \ j = 1, \ldots, s\}$, and define

$$L_1(i_1, \ldots, i_s) = a_\mathbf{i} = a_{i_1, \ldots, i_s},$$

$$L_{k+1}(i_{k+1}, \ldots, i_s) = \sum_{i_k=1}^\infty L_k(i_k, \ldots, i_s), \quad k = 1, \ldots, s-1.$$

Table 25.12.1: *Numerical results by the*
sequential d-transformation on the double
power series of Example 25.12.1

x	y	approximation
-0.5	-0.5	0.3843515211843
-1.0	-1.0	0.3149104237
-1.5	-1.5	0.26744390
-2.0	-2.0	0.2337732
-2.5	-2.5	0.207640

Hence, $S(\{a_{\mathbf{i}}\}) = \sum_{i_s=1}^{\infty} L_s(i_s)$. Let us assume that, for each k and for fixed $i_{k+1}, \ldots, i_s$, the sequence $\{L_k(i_k, \ldots, i_s)\}_{i_k=1}^{\infty}$ is in $\mathbf{b}^{(m_k)}$ for some integer m_k. [This assumption seems to hold when $\{a_{\mathbf{i}}\}_{i_k=1}^{\infty} \in \mathbf{b}^{(m_k)}$, for each k and for $i_{k+1}, \ldots, i_s$ fixed.] Therefore, we can compute $L_{k+1}(i_{k+1}, \ldots, i_s)$ by applying the $d^{(m_k)}$-transformation to the series $\sum_{i_k=1}^{\infty} L_k(i_k, \ldots, i_s)$, and the computation of $S(\{a_{\mathbf{i}}\})$ can be completed by applying the $d^{(m_s)}$-transformation to the series $\sum_{i_s=1}^{\infty} L_s(i_s)$.

What motivates this approach to the summation of s-D series is the fact that the preceding assumption is automatically satisfied when $a_{\mathbf{i}} = \prod_{j=1}^{s} a_{i_j}^{(j)}$, with $\{a_i^{(j)}\}_{i=1}^{\infty} \in \mathbf{b}^{(m_j)}$ for some integers m_j.

As an example, consider the double series $\sum_{j=1}^{\infty} \sum_{k=1}^{\infty} a_{j,k}$, where $a_{j,k} = x^j u_k / (j + g(k))$, where $|x| < 1$, $\{u_k\} \in \mathbf{b}^{(q)}$, and $g(k) \in \mathbf{A}_0^{(r)}$ for some positive integer r and $g(k) > 0$ for all large k. [We have $q = 2$ when $u_k = \cos k\theta$ or $u_k = P_k(y)$, the kth Legendre polynomial, for example.] First, $\{a_{j,k}\}_{j=1}^{\infty} \in \mathbf{b}^{(1)}$ with fixed k, while $\{a_{j,k}\}_{k=1}^{\infty} \in \mathbf{b}^{(q)}$ with fixed j. Next, invoking the relation $1/c = \int_0^{\infty} e^{-c\xi} \, d\xi$, $\Re c > 0$, we can show that

$$L_2(k) = \sum_{j=1}^{\infty} a_{j,k} = x u_k \int_0^{\infty} e^{-\xi g(k)}/(e^{\xi} - x) \, d\xi.$$

Applying Watson's lemma, we can show that $L_2(k)$ has the asymptotic expansion

$$L_2(k) \sim u_k \sum_{i=0}^{\infty} \alpha_i [g(k)]^{-i-1} \sim u_k \sum_{i=0}^{\infty} \delta_i k^{-i-r} \quad \text{as } k \to \infty.$$

Therefore, $\{L_2(k)\} \in \mathbf{b}^{(q)}$.

The following examples have been taken from Sidi [268]:

Example 25.12.1 Consider the double power series

$$\sum_{j=1}^{\infty} \sum_{k=1}^{\infty} c_{j,k} x^{j-1} y^{k-1}, \quad c_{j,k} = \frac{1}{j^2 + k^3}.$$

Because $\{c_{j,k} x^{j-1} y^{k-1}\}_{j=1}^{\infty} \in \mathbf{b}^{(1)}$ and $\{c_{j,k} x^{j-1} y^{k-1}\}_{k=1}^{\infty} \in \mathbf{b}^{(1)}$, we apply the sequential d-transformation with $p = 1$ and $q = 1$. Using about 100 terms of the series, in double-precision arithmetic, this method produces the results shown in Table 25.12.1.

Note that the series diverges when $|x| > 1$ or $|y| > 1$, but the method produces its sum very efficiently. (The accuracy decreases as the rate of divergence increases, since the absolute errors in the partial sums of the series increase in finite-precision arithmetic in this case.) The series converges very slowly when $|x| = 1$ or $|y| = 1$, and the method produces very accurate results for such x and y.

Example 25.12.2 Consider the double Fourier sine series

$$U(x, y) = \sum_{j=1}^{\infty} \sum_{k=1}^{\infty} c_{j,k} \sin\left(\frac{j\pi x}{a}\right) \sin\left(\frac{k\pi y}{b}\right), \quad c_{j,k} = \frac{32}{\pi^4} \cdot \frac{1}{jk(j^2/a^2 + k^2/b^2)}.$$

The function $U(x, y)$ is the solution of the 2-D Poisson equation $\Delta U = -2$ for $(x, y) \in R$, where $R = \{(x, y) : 0 < x < a, \ 0 < y < b\}$, with homogeneous boundary conditions on ∂R. Obviously, this double series converges very slowly. It is easy to see that $\{c_{j,k} \sin(j\pi x/a) \sin(k\pi y/b)\}_{j=1}^{\infty} \in \mathbf{b}^{(2)}$ and $\{c_{j,k} \sin(j\pi x/a) \sin(k\pi y/b)\}_{k=1}^{\infty} \in \mathbf{b}^{(2)}$. Therefore, we apply the sequential d-transformation with $p = 2$ and $q = 2$. Using about 400 terms of this series, we can obtain its sum to 13-digit accuracy in double-precision arithmetic. The exact value of $U(x, y)$ can easily be obtained from the simple series

$$U(x, y) = x(a - x) - \frac{8a^2}{\pi^3} \sum_{\substack{n=1 \\ n \text{ odd}}}^{\infty} \frac{\cosh[n\pi(2y - b)/(2a)]}{n^3 \cosh[n\pi b/(2a)]} \sin\left(\frac{n\pi x}{a}\right),$$

which converges very quickly for $0 < y < b$.

25.13 A Hybrid Method: The Richardson–Shanks Transformation

In Section 16.4, we saw via Theorems 16.4.3, 16.4.6, and 16.4.9 that the Shanks transformation is a very effective extrapolation method for sequences $\{A_m\}$ that satisfy (16.4.1) and (16.4.2), namely, for

$$A_m \sim A + \sum_{k=1}^{\infty} \alpha_k \lambda_k^m \quad \text{as } m \to \infty, \tag{25.13.1}$$

$$\lambda_k \text{ distinct}, \quad \lambda_k \neq 1 \text{ for all } k; \quad |\lambda_1| \geq |\lambda_2| \geq \cdots; \quad \lim_{k \to \infty} \lambda_k = 0. \tag{25.13.2}$$

We now describe a hybrid approach that reduces the cost of this transformation when some of the largest λ_k are available. We assume that $\lambda_1, \ldots, \lambda_s$ are available.

The first step of this approach consists of eliminating the known λ_k explicitly from the asymptotic expansion of A_m in (25.13.1), to obtain a new sequence $\{\tilde{A}_m\}$. This is achieved by applying to $\{A_m\}$ the Richardson extrapolation process as follows:

$$x_0^{(j)} = A_j, \quad j = 0, 1, \ldots,$$

$$x_p^{(j)} = \frac{x_{p-1}^{(j+1)} - \lambda_p x_{p-1}^{(j)}}{1 - \lambda_p}, \quad j = 0, 1, \ldots, \quad p = 1, 2, \ldots, s.$$

$$\tilde{A}_j = x_s^{(j)}, \quad j = 0, 1, \ldots. \tag{25.13.3}$$

From Theorem 1.5.1, it follows that the new sequence satisfies

$$\tilde{A}_m \sim A + \sum_{k=s+1}^{\infty} \tilde{\alpha}_k \lambda_k^m \text{ as } m \to \infty; \quad \tilde{\alpha}_k = \alpha_k \prod_{i=1}^{s} \frac{\lambda_k - \lambda_i}{1 - \lambda_i}. \quad (25.13.4)$$

Thus, the Richardson extrapolation process functions as a genuine linear filter in this step.

The second step consists of applying the Shanks transformation to $\{\tilde{A}_m\}$, which will be effective judging from (25.13.4). Let us denote the resulting approximations $\varepsilon_{2n}^{(j)}$ by $\tilde{\varepsilon}_{2n}^{(j)}(\{A_k\})$. Then, Theorem 16.4.3 says that

$$\tilde{\varepsilon}_{2n}^{(j)}(\{A_k\}) - A = O(\lambda_{s+n+1}^j) \text{ as } j \to \infty, \quad (25.13.5)$$

provided $|\lambda_{s+n}| > |\lambda_{s+n+1}|$.

It is clear that, formally speaking, the "elimination" of the terms $\alpha_k \lambda_k^m$, $k = 1, \dots, s + n$, from (25.13.1) by this procedure requires using of A_m, $j \leq m \leq j + s + 2n$, whereas application of the Shanks transformation directly to $\{A_m\}$ for "eliminating" the same terms requires using A_m, $j \leq m \leq j + 2s + 2n$.

We denote the procedure developed here the *Richardson–Shanks transformation* (R-EPS). This approach and its application as in the next subsection are due to Sidi and Israeli [311]. Obviously, it can be applied using APS when necessary.

25.13.1 An Application

An interesting application of R-EPS is to problems in which a sequence $\{B_m\}$ satisfies

$$B_m \sim B + \sum_{k=1}^{s} \beta_k \mu_k^m + \sum_{k=1}^{\infty} \delta_k \nu_k^m \text{ as } m \to \infty; \quad |\nu_1| \geq |\nu_2| \geq \cdots, \quad (25.13.6)$$

such that $\mu_1, \dots, \mu_s$ are known, and we are required to determine

$$\tilde{B}(m) \equiv B + \sum_{r=1}^{s} \beta_r \mu_r^m. \quad (25.13.7)$$

This means we need to find $\beta_1, \dots, \beta_s$, in addition to B.

First, B can be approximated by applying R-EPS with $A_m = B_m$ and $\lambda_k = \mu_k$ for $k = 1, \dots, s$, in (25.13.1) and (25.13.2). Thus, the approximations to B produced by R-EPS are $\tilde{\varepsilon}_{2n}^{(j)}(\{B_k\})$.

To approximate β_r, $r \in \{1, \dots, s\}$, we propose the following course of action: First, multiply B_m by μ_r^{-m}, $r = 1, 2, \dots$, and denote $B_m \mu_r^{-m} = A_m$ and $\beta_r = A$. Therefore,

$$A_m \sim A + B\left(\frac{1}{\mu_r}\right)^m + \sum_{\substack{k=1 \\ k \neq r}}^{s} \beta_k \left(\frac{\mu_k}{\mu_r}\right)^m + \sum_{k=1}^{\infty} \delta_k \left(\frac{\nu_k}{\mu_r}\right)^m \text{ as } m \to \infty. \quad (25.13.8)$$

Next, apply R-EPS to the sequence $\{A_m\}$ with $\lambda_k = \mu_k/\mu_r$ for $1 \leq k \leq s$, $k \neq r$, and $\lambda_r = 1/\mu_r$, to obtain the approximations $\tilde{\varepsilon}_{2n}^{(j)}(\{B_k \mu_r^{-k}\})$. Then,

$$\tilde{\varepsilon}_{2n}^{(j)}(\{B_k \mu_r^{-k}\}) - \beta_r = O(|\nu_{n+1}/\mu_r|^j) \text{ as } j \to \infty, \quad (25.13.9)$$

provided $|\nu_n| > |\nu_{n+1}|$.

Once B and the β_r are approximated, we form

$$\tilde{B}_n^{(j)}(m) \equiv \tilde{\varepsilon}_{2n}^{(j)}(\{B_k\}) + \sum_{r=1}^{s} \tilde{\varepsilon}_{2n}^{(j)}(\{B_k \mu_r^{-k}\}) \mu_r^m \qquad (25.13.10)$$

as an approximation to $\tilde{B}(m)$. When $|v_n| > |v_{n+1}|$, we also have

$$\tilde{B}_n^{(j)}(m) - \tilde{B}(m) = O(|v_{n+1}/\rho|^j) \text{ as } j \to \infty; \quad \rho = \min\{1, |\mu_1|, \dots, |\mu_s|\}. \qquad (25.13.11)$$

Sequences of the type described here arise, for example, when solving numerically time-dependent problems with steady-state solutions that are periodic in time. This may be the result of the periodicity being built directly into the associated equations and/or of the presence of boundary conditions that are periodic in time. (The numerical solutions involve marching in time with a fixed time increment.) For such problems, it turns out that $|\mu_k| = 1$ for $k = 1, \dots, s$, and $|v_k| < 1$ for all $k = 1, 2, \dots$, so that $B_m = B_m^{\text{steady}} + o(1)$ as $m \to \infty$, where $B_m^{\text{steady}} = B + \sum_{k=1}^{s} \beta_k \mu_k^m$ is the numerical steady-state solution, and we are interested in determining B_m^{steady}.

Important Remark. In some problems, we may be given not the whole sequence $\{B_m\}$ but only $\{B_m\}_{m=j}^{\infty}$ for some possibly unknown j, and we may be asked to approximate $\tilde{B}(m)$. This can be achieved by applying R-EPS to the sequence $\{B_k'\}_{k=0}^{\infty}$, where $B_k' = B_{j+k}, k = 0, 1, \dots$, exactly as before. Then, with $\tilde{B}_n^{(j)}(m)$ as defined in (25.13.10), there holds

$$\tilde{B}_n^{(j)}(j + p) = \tilde{\varepsilon}_{2n}^{(0)}(\{B_k'\}) + \sum_{r=1}^{s} \tilde{\varepsilon}_{2n}^{(0)}(\{B_k' \mu_r^{-k}\}) \mu_r^p. \qquad (25.13.12)$$

The key to this is the fact that R-EPS, being a composition of two quasi-linear processes, the Richardson extrapolation and the Shanks transformation, is itself quasi-linear too. That is, for any two constants $a \neq 0$ and b, there holds $\tilde{\varepsilon}_{2n}^{(j)}(\{a A_k + b\}) = a\tilde{\varepsilon}_{2n}^{(j)}(\{A_k\}) + b$. Therefore,

$$\tilde{\varepsilon}_{2n}^{(j)}(\{B_k \mu_r^{-k}\}) = \mu_r^{-j} \tilde{\varepsilon}_{2n}^{(j)}(\{B_k \mu_r^{j-k}\}) = \mu_r^{-j} \tilde{\varepsilon}_{2n}^{(0)}(\{B_k' \mu_r^{-k}\}), \quad r = 1, \dots, s. \quad (25.13.13)$$

In addition,

$$\tilde{\varepsilon}_{2n}^{(j)}(\{B_k\}) = \tilde{\varepsilon}_{2n}^{(0)}(\{B_k'\}). \qquad (25.13.14)$$

The validity of (25.13.12) is now a consequence of (25.13.10), (25.13.13), and (25.13.14).

We would like emphasize that the procedure described here does not require knowledge of j because $\{B_k' \mu_r^{-k}\}_{k=0}^{\infty} = \{B_j, \mu_r^{-1} B_{j+1}, \mu_r^{-2} B_{j+2}, \dots\}$ in (25.13.12).

Example 25.13.1 As a model, consider the linear system of ordinary differential equations

$$\mathbf{y}'(t) = C\mathbf{y}(t) + \mathbf{f}(t), \quad t > 0; \quad \mathbf{y}(0) = \mathbf{y}_0. \qquad (25.13.15)$$

Here $\mathbf{y}(t) = [y_1(t), \dots, y_r(t)]^T$, and C is a constant $r \times r$ matrix whose eigenvalues have negative real parts, and the "forcing function" $\mathbf{f}(t) = [f_1(t), \dots, f_r(t)]^T$ is periodic with period τ. Using the fact that

$$\mathbf{y}(t) = e^{Ct}\mathbf{y}_0 + \int_0^t e^{C(t-s)}\mathbf{f}(s)\, ds, \tag{25.13.16}$$

it can be shown that $\mathbf{y}(t) = \mathbf{y}^{\text{trans}}(t) + \mathbf{y}^{\text{steady}}(t)$, where $\mathbf{y}^{\text{trans}}(t)$ is the transient, that is, it satisfies $\lim_{t\to\infty} \mathbf{y}^{\text{trans}}(t) = 0$, and $\mathbf{y}^{\text{steady}}(t)$ is periodic with period τ. Let us now solve this system numerically by a linear multistep method with a fixed time step $\Delta t = h$. For simplicity of presentation, we use the Euler method, namely,

$$\mathbf{y}_{m+1} = Q\mathbf{y}_m + h\mathbf{f}_m, \quad m = 0, 1, \ldots; \quad Q = I + hC, \quad \mathbf{f}_m = \mathbf{f}(mh), \quad \mathbf{y}_m \approx \mathbf{y}(mh). \tag{25.13.17}$$

Here we choose $h = \tau/M$, $M = 2K + 1$ for some integer K. Then, $\mathbf{f}_{m+pM} = \mathbf{f}_m$ for integers p. We also choose h small enough so that all eigenvalues of Q are smaller than unity in modulus; hence, the matrix $cI - Q$, where $|c| = 1$, is nonsingular. By induction, it follows that

$$\mathbf{y}_m = Q^m\mathbf{y}_0 + h\sum_{j=0}^{m-1} Q^{m-1-j}\mathbf{f}_j, \quad m = 0, 1, \ldots. \tag{25.13.18}$$

Letting $\{\mathbf{g}_k\}_{k=-K}^K$ be the discrete Fourier transform of $\{\mathbf{f}_j\}_{j=0}^{2K}$, we can write

$$\mathbf{f}_j = \sum_{k=-K}^K \mathbf{g}_k \omega^{jk}, \quad j = 0, 1, \ldots; \quad \omega = \exp\left(i\frac{2\pi}{2K+1}\right). \tag{25.13.19}$$

Of course, just like $\mathbf{f}_j$, $\mathbf{g}_k$ are r-dimensional vectors. Substituting (25.13.19) in (25.13.18), we can show after some manipulation that

$$\mathbf{y}_m = \mathbf{y}_m^{\text{trans}} + \mathbf{y}_m^{\text{steady}};$$

$$\mathbf{y}_m^{\text{trans}} = Q^m\left[\mathbf{y}_0 - h\sum_{k=-K}^K (\omega^k I - Q)^{-1}\mathbf{g}_k\right], \quad \mathbf{y}_m^{\text{steady}} = h\sum_{k=-K}^K (\omega^k I - Q)^{-1}\mathbf{g}_k\omega^{km}. \tag{25.13.20}$$

Obviously, $\mathbf{y}_m^{\text{trans}}$ is the transient since $\lim_{m\to\infty}\mathbf{y}_m^{\text{trans}} = 0$ by the fact that the eigenvalues of Q are all less than unity in modulus, and $\mathbf{y}_m^{\text{steady}}$ is the periodic steady-state solution because $\mathbf{y}_{m+pM}^{\text{steady}} = \mathbf{y}_m^{\text{steady}}$ for every integer p. If we assume, for simplicity, that C is diagonalizable, then Q is diagonalizable as well. Let us denote its distinct eigenvalues by v_k, $k = 1, \ldots, q$. Then we have $\mathbf{y}_m^{\text{trans}} = \sum_{k=1}^q \mathbf{d}_k v_k^m$, where $\mathbf{d}_k$ is an eigenvector corresponding to v_k. We also have $\mathbf{y}_m^{\text{steady}} = \sum_{k=-K}^K \mathbf{b}_k(\omega^k)^m$, where $\mathbf{b}_k = h(\omega^k I - Q)^{-1}\mathbf{g}_k$ for each k. The vectors $\mathbf{b}_k$ and $\mathbf{d}_k$ are independent of m. Combining all this in (25.13.20), we finally have that

$$\mathbf{y}_m = \sum_{k=-K}^K \mathbf{b}_k(\omega^k)^m + \sum_{k=1}^q \mathbf{d}_k v_k^m. \tag{25.13.21}$$

We now apply R-EPS to the sequence $\{\mathbf{y}_m\}$ componentwise to determine the vectors $\mathbf{b}_k$, $-K \le k \le K$, hence $\mathbf{y}_m^{\text{steady}}$. It is clear that, for errors that behave like $|v_{n+1}|^j$, we need to store and use the vectors $\mathbf{y}_m$, $m = j, j+1, \ldots, j+2K+2n$.

Let us consider the special case in which

$$\mathbf{f}(t) = \sum_{k=-L}^{L} \mathbf{c}_k \exp(\mathrm{i}\, 2k\pi t/\tau)$$

$$= \mathbf{c}_0 + \sum_{k=1}^{L} (\mathbf{c}_k + \mathbf{c}_{-k}) \cos(2k\pi t/\tau) + \sum_{k=1}^{L} \mathrm{i}\,(\mathbf{c}_k - \mathbf{c}_{-k}) \sin(2k\pi t/\tau),$$

for some integer L. Choosing $K \geq L$, we have in this case $\mathbf{f}_j = \sum_{k=-L}^{L} \mathbf{c}_k \omega^{jk}$, $j = 0, 1, \ldots$, with ω as in (25.13.19). Consequently,

$$\mathbf{y}_m = \sum_{k=-L}^{L} \mathbf{b}_k (\omega^k)^m + \sum_{k=1}^{q} \mathbf{d}_k v_k^m. \qquad (25.13.22)$$

If L is considerably small compared with K, then the cost of applying R-EPS to the sequence $\{\mathbf{y}_m\}$ in (25.13.22) is much smaller than that incurred by applying R-EPS to the sequence in (25.13.21). For errors that behave like $|v_{n+1}|^j$, this time we need to store and use the vectors $\mathbf{y}_m$, $m = j, j+1, \ldots, j+2L+2n$.

Such a situation can also arise (approximately) in case $\mathbf{f}(t)$ is infinitely differentiable on $(-\infty, \infty)$. Now, $\mathbf{f}(t)$ can be approximated with very high accuracy via a small number of terms of its Fourier series, namely, $\mathbf{f}(t) \approx \sum_{k=-L}^{L} \mathbf{c}_k \exp(\mathrm{i}\, 2k\pi t/\tau)$ with a small L, so that

$$\mathbf{y}_m \approx \sum_{k=-L}^{L} \mathbf{b}_k (\omega^k)^m + \sum_{k=1}^{q} \mathbf{d}_k v_k^m.$$

A different approach to the problem described here was suggested by Skelboe [315]. In this approach, $\mathbf{y}_m^{\text{steady}}$ is obtained by applying the Shanks transformation to the sequences $\{\mathbf{y}_{Mm+i}\}_{m=0}^{\infty}$, $i = 0, 1, \ldots, M - 1$. The limits of these sequences are, of course, $\mathbf{y}_i^{\text{steady}}$, $i = 0, 1, \ldots, M - 1$, and these determine $\mathbf{y}_m^{\text{steady}}$ completely. With large M, this approach seems to have a larger cost than the R-EPS approach we have proposed.

25.14 Application of Extrapolation Methods to Ill-Posed Problems

An interesting application of extrapolation methods to ill-posed problems appears in Brezinski and Redivo Zaglia [41, Section 6.3]. Consider the linear system

$$Ax = b, \qquad (25.14.1)$$

where the matrix A is nonsingular. As is well known, the numerical solution of this system will have poor accuracy if the condition number of A is large. To avoid this problem partially, we solve instead of (25.14.1) the perturbed problem

$$(A + \epsilon B)x(\epsilon) = b \qquad (25.14.2)$$

with some small ϵ, such that the matrix $A + \epsilon B$ is better conditioned. This is the so-called *Tikhonov regularization technique*. Of course, $x(\epsilon)$ may be quite different from $x = x(0)$. The problem now is how to apply this technique and still obtain a reasonable numerical approximation to x.

From (25.14.2), we have

$$x(\epsilon) = (A + \epsilon B)^{-1}b = (I + \epsilon M)^{-1}x, \quad M = A^{-1}B. \tag{25.14.3}$$

Provided $\epsilon < 1/\rho(M)$, where $\rho(M)$ is the spectral radius of M, we obtain from (25.14.3) the convergent expansion

$$x(\epsilon) = x + \sum_{k=1}^{\infty}[(-1)^k M^k x]\epsilon^k. \tag{25.14.4}$$

It is now easy to see from (25.14.4) that we can apply the polynomial Richardson extrapolation or the rational extrapolation to $x(\epsilon)$ componentwise to obtain approximations to x. Practically, the methods are applied to a sequence $\{x(\epsilon_m)\}$, where $\epsilon_0 > \epsilon_1 > \cdots$, and $\lim_{m\to\infty} \epsilon_m = 0$.

Ill-posed problems arise frequently from the discrete solution of Fredholm integral equations of the first kind. They may also arise in the solution of the least squares problems

$$\text{minimize } \|Ax - b\|,$$

where, this time, A may be a rectangular matrix of full column rank. If this problem is replaced by a regularized one of the form

$$\text{minimize } (\|Ax - b\|^2 + \epsilon\|Bx\|^2),$$

where B is some suitable matrix (for example, $B = I$), and $\|y\| = \sqrt{y^*y}$ is the standard Euclidean norm, then the solution $x(\epsilon)$ to this problem can be shown to satisfy the linear system

$$(A^*A + \epsilon B^*B)x(\epsilon) = A^*b$$

that is precisely of the form discussed above.

For numerical examples and for additional applications of similar nature, including the relevant bibliography, we refer the reader to Brezinski and Redivo Zaglia [41].

25.15 Logarithmically Convergent Fixed-Point Iteration Sequences

We close this chapter by mentioning a rare and yet interesting type of sequences that arise from fixed-point iterative methods applied to a nonlinear equation $x = g(x)$, for which $g'(s) = 1$, where s denotes the solution. (Recall that we discussed the case in which $|g'(s)| < 1$ when we studied the Δ^2-process.) These sequences have been studied in detail in de Bruijn [42, pp. 153–175].

Let us pick x_0 and compute $x_1, x_2, \ldots$, according to $x_{m+1} = g(x_m)$. Let us denote $e_m = x_m - s, m = 0, 1, \ldots$. Assume that $g(x)$ is infinitely differentiable in a neighborhood of s, so that

$$g(x) - s = u + \sum_{i=1}^{\infty}\alpha_{p+i}u^{p+i}, \quad p \geq 1 \text{ integer}, \quad \alpha_{p+1} < 0, \quad u \equiv x - s > 0.$$

$$\tag{25.15.1}$$

Then, provided $x_0 > s$ is sufficiently close to s, $\{x_m\}$ converges to s. Actually, $\{x_m\}$ converges logarithmically, and there holds $x_m - s \sim (-p\alpha_{p+1}m)^{-1/p}$ as $m \to \infty$.

As follows from [42, p. 175, Exercise 8.11], the equation

$$\psi(\bar{g}(u)) = 1 + \psi(u), \quad \bar{g}(u) \equiv g(s+u) - s, \qquad (25.15.2)$$

has a unique solution $\psi(u)$ that has an asymptotic expansion of the form

$$\psi(u) \sim \sum_{i=1}^{p} c_{-i}u^{-i} + c_0 \log u + \sum_{i=1}^{\infty} c_i u^i \quad \text{as } u \to 0+, \quad c_{-p} \neq 0. \quad (25.15.3)$$

The coefficients c_i are determined by the expansion of $g(x)$ in (25.15.1) and from (25.15.2).

The asymptotic behavior of x_m can be obtained by analyzing the relation

$$\psi(e_m) = m + \psi(e_0), \quad m = 1, 2, \ldots . \qquad (25.15.4)$$

Taking $s = 0$ for simplicity, Sablonnière [246] gives the following results for the cases $p = 1, 2$:

When $p = 1$,

$$x_m \sim s + \sum_{k=1}^{\infty} \beta_k m^{-k} \quad \text{as } m \to \infty; \quad (c_0 = 0),$$

$$x_m \sim s + \sum_{k=1}^{\infty} \left[\sum_{i=0}^{k-1} \beta_{k,i}(\log m)^i\right] m^{-k} \quad \text{as } m \to \infty; \quad (c_0 \neq 0),$$

and when $p = 2$ and $g(x)$ is odd,

$$x_m \sim s + \sum_{k=1}^{\infty} \beta_k m^{-k+1/2} \quad \text{as } m \to \infty; \quad (c_0 = 0),$$

$$x_m \sim s + \sum_{k=1}^{\infty} \left[\sum_{i=0}^{k-1} \beta_{k,i}(\log m)^i\right] m^{-k+1/2} \quad \text{as } m \to \infty; \quad (c_0 \neq 0).$$

When $p = 2$ and $\alpha_4 \neq 0$, hence $g(x)$ is not necessarily odd,

$$x_m \sim s + \sum_{k=1}^{\infty} \left\{\left[\sum_{i=0}^{k-1} \beta'_{k,i}(\log m)^i\right] m^{-k+1/2}\right.$$

$$\left. + \left[\sum_{i=0}^{k-1} \beta''_{k,i}(\log m)^i\right] m^{-k}\right\} \quad \text{as } m \to \infty; \quad (c_0 \neq 0).$$

The β_k, $\beta_{k,i}$, etc. depend on the α_k. Explicit expressions for the first few of them are given by Sablonnière [246].

We see that whether c_0 vanishes or not makes a big difference in the nature of the asymptotic expansion of x_m. When $c_0 \neq 0$, x_m has powers of $\log m$ in its asymptotic expansion, and this will surely cause problems when convergence acceleration methods are applied to $\{x_m\}$. Surprisingly, numerical and theoretical results given by Sablonnière [246], [247] and Sedogbo [262] show that, when $c_0 \neq 0$, the iterated Lubkin transformation and the θ-algorithm are very effective when applied to $\{x_m\}$. For example, when

$p = 1$, the kth column in these methods will "eliminate" all the terms in the summation $[\sum_{i=0}^{k-1} \beta_{k,i} (\log m)^i] m^{-k}$ simultaneously. The iterated Δ^2-process and the ε-algorithm, when modified suitably as described in [246], [247], and [262], perform similarly. We refer the reader to these papers for more results and details.

Before we end, we recall that $g'(s) = 1$ implies that s is a multiple root of the equation $f(x) = x - g(x) = 0$. It is known that, when applied to $f(x) = 0$, the Newton-Raphson method converges linearly. It can even be modified slightly to converge quadratically. See, for example, Ralston and Rabinowitz [235]. [The sequence $\{x_m\}$ generated via the fixed-point iterations $x_{m+1} = g(x_m)$ treated here converges logarithmically, hence it is inferior to that generated by the Newton-Raphson method.]

Part IV

Appendices

A

Review of Basic Asymptotics

As the subject of extrapolation methods depends heavily on asymptotics, it is important and useful to review some of the fundamentals of the latter. It is correct to say that, without a good understanding of asymptotics, it is impossible to appreciate the beauty, relevance, and intricacies, both mathematical and practical, of the research that has been done in the area of extrapolation methods.

For thorough treatments of asymptotic methods, we refer the reader to the books by Olver [223], Murray [215], Erdélyi [77], de Bruijn [42], and Bender and Orszag [21]. See also Henrici [132, Chapter 11].

A.1 The O, o, and $\sim$ Symbols

Definition A.1.1 Let $x_0 \geq 0$ be finite or infinite. Let $f(x)$ and $g(x)$ be two functions defined on an interval I that is (i) of the form $(x_0, x_0 + \delta)$, $\delta > 0$, when x_0 is finite and (ii) of the form (X, ∞), $X > 0$, when x_0 is infinite. Finally, let $\lim_{x \to x_0}$ stand for $\lim_{x \to x_0+}$ when x_0 is finite and for $\lim_{x \to \infty}$ when x_0 is infinite.

(i) We write

$$f(x) = O(g(x)) \ \text{ as } x \to x_0 \ \text{ if } |f(x)/g(x)| \text{ is bounded on } I.$$

In words, *f is of order not exceeding g*.

(ii) We write

$$f(x) = o(g(x)) \ \text{ as } x \to x_0 \ \text{ if } \lim_{x \to x_0} f(x)/g(x) = 0.$$

In words, *f is of order less than g*.

(iii) We write

$$f(x) \sim g(x) \ \text{ as } x \to x_0 \ \text{ if } \lim_{x \to x_0} f(x)/g(x) = 1.$$

In words, *f is asymptotically equal to g*.

Special cases of these definitions are $f(x) = O(1)$ as $x \to x_0$, which means that $f(x)$ is bounded as $x \to x_0$, and $f(x) = o(1)$ as $x \to x_0$, which means that $f(x) \to 0$ as $x \to x_0$.

The notations $O(g)$ and $o(g)$ can also be used to denote the *classes* of functions f that satisfy $f = O(g)$ and $f = o(g)$, respectively, or to denote generically *unspecified* functions f with these properties.

Thus, in terms of the o-notation,

$$f(x) \sim g(x) \ \text{ as } x \to x_0 \ \text{ if and only if } \ f(x) = g(x)[1 + o(1)] \ \text{ as } x \to x_0.$$

Here are some important consequences of Definition A.1.1:

1. Obviously, $f = o(g)$ implies $f = O(g)$, but the converse of this is not true.
2. We must also realize that $f = O(g)$ does *not* imply $g = O(f)$. As a consequence, $f = O(g)$ does *not* imply $1/f = O(1/g)$ either. The latter holds provided $g = O(f)$.

459

3. On the other hand, if $f \sim g$, then $g \sim f$ too. Of course, we have $1/f \sim 1/g$ as well. Clearly, $f \sim g$ implies both $f = O(g)$ and $g = O(f)$, and hence both $1/f = O(1/g)$ and $1/g = O(1/f)$.
4. From what we have seen so far, we understand that the amount of information contained in $f \sim g$ is largest; this is followed by $f = O(g)$ *and* $g = O(f)$ simultaneously, which is followed by $f = O(g)$ only. Also, $f = o(g)$ contains more information than $f = O(g)$. In case $g = o(h)$, the relation $f = O(g)$ contains more information than $f = o(h)$. Thus, as $x \to 0+$, $f(x) \sim x^3$ is better than $cx^3 \le |f(x)| \le dx^3$ for some $c > 0$ and $d > 0$, which is better than $f(x) = O(x^3)$, which is better than $f(x) = o(x^2)$.

If the functions $f(x)$ and $g(x)$ in Definition A.1.1 are replaced by the sequences $\{a_n\}$ and $\{b_n\}$, we can analogously define $a_n = O(b_n)$, $a_n = o(b_n)$, and $a_n \sim b_n$ as $n \to \infty$. Only this time, we should keep in mind that $n \to \infty$ through integer values only.

A.2 Asymptotic Expansions

Definition A.2.1 The sequence of functions $\{\phi_k(x)\}_{k=0}^\infty$ is called an *asymptotic sequence* or an *asymptotic scale* as $x \to x_0$ if

$$\frac{\phi_{k+1}(x)}{\phi_k(x)} = o(1) \text{ as } x \to x_0, \quad k = 0, 1, \dots .$$

Two most important examples of asymptotic sequences are (i) $\{(x - x_0)^k\}$ when x_0 is finite, and (ii) $\{x^{-k}\}$ when $x_0 = \infty$.

Definition A.2.2 We say that the (convergent or divergent) series $\sum_{k=0}^\infty a_k \phi_k(x)$ *represents the function $f(x)$ asymptotically as $x \to x_0$* [or that it is an *asymptotic expansion of $f(x)$ as $x \to x_0$*], and write

$$f(x) \sim \sum_{k=0}^\infty a_k \phi_k(x) \text{ as } x \to x_0,$$

if $\{\phi_k(x)\}_{k=0}^\infty$ is an asymptotic sequence and if for every integer $n \ge 0$ there holds

$$f(x) - \sum_{k=0}^{n-1} a_k \phi_k(x) = O(\phi_n(x)) \text{ as } x \to x_0.$$

Theorem A.2.3 *Let $\{\phi_k(x)\}_{k=0}^\infty$ be an asymptotic sequence. Then the following three statements are equivalent:*

(i) *The infinite expansion $\sum_{k=0}^\infty a_k \phi_k(x)$ represents the function $f(x)$ asymptotically as $x \to x_0$; i.e.,*

$$f(x) \sim \sum_{k=0}^\infty a_k \phi_k(x) \text{ as } x \to x_0.$$

(ii) *For every integer $n \ge 0$, there holds*

$$f(x) - \sum_{k=0}^n a_k \phi_k(x) = o(\phi_n(x)) \text{ as } x \to x_0.$$

(iii) *For every integer $n \ge 0$,*

$$\lim_{x \to x_0} \frac{f(x) - \sum_{k=0}^{n-1} a_k \phi_k(x)}{\phi_n(x)} \text{ exists and equals } a_n.$$

When x_0 is finite, the sequence $\{(x - x_0)^k\}_{k=0}^\infty$ is asymptotic as $x \to x_0$, as already mentioned. If $f \in C^\infty[x_0, x_0 + \delta]$, $\delta > 0$, then its Taylor series about x_0, namely, $\sum_{k=0}^\infty \frac{f^{(k)}(x_0)}{k!}(x - x_0)^k$, *whether convergent or not*, represents $f(x)$ asymptotically as $x \to x_0+$. This can be proved by using the Taylor series with the remainder. As is known, if $f(x)$ is analytic at x_0, then its Taylor series at x_0 converges when $|x - x_0| < \delta$ for some $\delta > 0$ and $f(x)$ is equal to the sum of this series when $|x - x_0| < \delta$. In case $f(x)$ is not analytic at x_0, its Taylor series at x_0 diverges for all $x \neq x_0$.

All this remains valid when $x_0 = \infty$, provided we replace the sequence $\{(x - x_0)^k\}_{k=0}^{\infty}$ by $\{x^{-k}\}_{k=0}^{\infty}$, and make other suitable modifications.

An immediate consequence of the preceding theorem is that, if the series $\sum_{k=0}^{\infty} a_k \phi_k(x)$ is an asymptotic expansion of the function $f(x)$ as $x \to x_0$, then it is unique. (The converse is not necessarily true; that is, more than one function may be represented by the same asymptotic expansion. The question of the uniqueness of $f(x)$ when $f(x) \sim \sum_{k=0}^{\infty} a_k x^{-k}$ as $x \to \infty$ is discussed at length in Hardy [123, Section 8.11], where it is assumed that $f(z)$ is analytic in some sector of the z-plane.)

A.3 Operations with Asymptotic Expansions

The following results on the algebra of asymptotic expansions can be proved with Theorem A.2.3.

Theorem A.3.1 *Let* $f(x) \sim \sum_{k=0}^{\infty} a_k \phi_k(x)$ *and* $g(x) \sim \sum_{k=0}^{\infty} b_k \phi_k(x)$ *as* $x \to x_0$. *Then, for arbitrary constants* α *and* β, *there holds*

$$\alpha f(x) + \beta g(x) \sim \sum_{k=0}^{\infty} (\alpha a_k + \beta b_k)\phi_k(x) \ \ as \ x \to x_0.$$

Theorem A.3.2 *Let* $\phi_k(x) = (x - x_0)^k$ *when* x_0 *is finite and* $\phi_k(x) = x^{-k}$ *when* $x_0 = \infty$. *Let* $f(x) \sim \sum_{k=0}^{\infty} a_k \phi_k(x)$ *and* $g(x) \sim \sum_{k=0}^{\infty} b_k \phi_k(x)$ *as* $x \to x_0$. *Then the following are true:*

(i) *The product* $f(x)g(x)$ *is represented asymptotically as* $x \to x_0$ *by the Cauchy product of the asymptotic expansions of* $f(x)$ *and* $g(x)$. *That is,*

$$f(x)g(x) \sim \sum_{k=0}^{\infty} \left(\sum_{i=0}^{k} a_{k-i} b_i \right) \phi_k(x) \ \ as \ x \to x_0.$$

(ii) *If* $a_0 \neq 0$, *then the function* $g(x)/f(x)$ *has an asymptotic representation given by*

$$g(x)/f(x) \sim \sum_{k=0}^{\infty} d_k \phi_k(x) \ \ as \ x \to x_0,$$

where the d_k *are determined recursively from the equations*

$$\sum_{i=0}^{k} a_{k-i} d_i = b_k, \ \ k = 0, 1, \dots .$$

In particular, the function $1/f(x)$ *has an asymptotic representation given by*

$$1/f(x) \sim \sum_{k=0}^{\infty} c_k \phi_k(x) \ \ as \ x \to x_0,$$

where the c_k *are determined recursively from the equations*

$$c_0 a_0 = 1 \ \ and \ \sum_{i=0}^{k} a_{k-i} c_i = 0, \ \ k = 1, 2, \dots .$$

In the next theorem, we discuss the asymptotic expansion of the composition of two functions. For the sake of clarity, we restrict ourselves only to the case in which $x_0 = \infty$ and $\phi_k(x) = x^{-k}$, $k = 0, 1, \dots .$

Theorem A.3.3 *Assume that $f(x) \sim a_1 x^{-1} + a_2 x^{-2} + \cdots$ as $x \to \infty$. Assume also that $g(w) \sim b_0 + b_1 w + b_2 w^2 + \cdots$ as $w \to 0$. Then the function $g(f(x))$ admits an asymptotic expansion of the form*

$$g(f(x)) \sim \sum_{k=0}^{\infty} d_k x^{-k} \quad \text{as } x \to \infty.$$

for some constants d_k. In particular, we have $d_0 = b_0$, $d_1 = b_1 a_1$, $d_2 = b_1 a_2 + b_2 a_1^2$, etc.

We now make a few important remarks about other operations with asymptotic expansions such as integration and differentiation.

Theorem A.3.4 *Let $f(x) \sim \sum_{k=2}^{\infty} a_k x^{-k}$ as $x \to \infty$. Then*

$$\int_x^{\infty} f(t)\, dt \sim \sum_{k=2}^{\infty} \frac{a_k}{k-1} x^{-k+1} \quad \text{as } x \to \infty.$$

If $f(x) \sim \sum_{k=0}^{\infty} a_k x^{-k}$ as $x \to \infty$, then

$$\int_c^x f(t)\, dt \sim K + a_0 x + a_1 \log x + \sum_{k=2}^{\infty} \frac{a_k}{-k+1} x^{-k+1} \quad \text{as } x \to \infty,$$

where $c > 0$ and $K = \int_c^{\infty} [f(t) - a_0 - a_1 t^{-1}]\, dt - a_0 c - a_1 \log c$. Thus, term-by-term integration of the asymptotic expansion of $f(x)$ is permitted.

Term-by-term differentiation of asymptotic expansions is not always permitted. Concerning this problem, we do have the following results, however.

Theorem A.3.5 *Let $f(x) \sim \sum_{k=0}^{\infty} a_k x^{-k}$ as $x \to \infty$. Assume that $f'(x)$ is continuous for all large x and that $f'(x) \sim \sum_{k=1}^{\infty} b_k x^{-k-1}$ as $x \to \infty$ holds. Then $b_k = -k a_k$, $k = 1, 2, \ldots$. Thus, term-by-term differentiation of the asymptotic expansion of $f(x)$ is permitted in this case.*

Theorem A.3.6 *Let $f(z)$ be analytic in the closed annular sector $S = \{z : |z| \geq \rho > 0, \alpha \leq \arg z \leq \beta\}$, and let $f(z) \sim \sum_{k=0}^{\infty} a_k z^{-k}$ as $z \to \infty$ in S. This asymptotic expansion can be differentiated term-by-term any number of times in the interior of S.*

Other operations on asymptotic expansions likewise are not straightforward and need to be carried out with care. Similarly, if L is an operator and $f = O(g)$, for example, then $L(f) = O(L(g))$ does not necessarily hold. This remark applies equally to $f = o(g)$ and to $f \sim g$.

To illustrate this point, consider the behavior as $x \to \infty$ of $L(f; x)$ defined by

$$L(f; x) = \begin{vmatrix} f(x) & \Delta f(x) \\ \Delta f(x) & \Delta^2 f(x) \end{vmatrix}; \quad \Delta f(x) = f(x+1) - f(x), \quad \Delta^2 f(x) = \Delta(\Delta f(x)).$$

If $g(x) = x^{\gamma}$, γ real and not an integer for simplicity, then we have $L(g; x) \sim -\gamma x^{2\gamma-2}$. If $f(x) \sim x^{\gamma}$, we cannot be sure that $L(f; x) \sim -\gamma x^{2\gamma-2}$, however. (i) When $f(x) = x^{\gamma} + O(x^{\gamma-1})$, the best we can say is that $L(f; x) = O(x^{2\gamma-1})$. (ii) When $f(x) = x^{\gamma} + a x^{\gamma-1} + O(x^{\gamma-2})$, the best we can say is that $L(f; x) = O(x^{2\gamma-2})$. (iii) Only when $f(x) = x^{\gamma} + a x^{\gamma-1} + b x^{\gamma-2} + o(x^{\gamma-2})$ do we have that $L(f; x) \sim -\gamma x^{2\gamma-2}$. Observe that, in all three cases, $f(x) \sim x^{\gamma}$.

B

The Laplace Transform and Watson's Lemma

B.1 The Laplace Transform

Definition B.1.1 The *Laplace transform* $\mathcal{L}[f(t); z] \equiv \hat{f}(z)$ of the function $f(t)$ is defined by the integral

$$\hat{f}(z) = \int_0^\infty e^{-zt} f(t) \, dt \qquad (B.1.1)$$

whenever this integral exists.

For simplicity of treatment, in this book we assume that

(i) $f(t)$ is integrable in the sense of Riemann on any interval $(0, T)$, $T > 0$, and that
(ii) there exist positive constants M, a, and t_0 for which

$$|f(t)| \leq M e^{at} \quad \text{for every } t \geq t_0.$$

Functions with these properties are said to be *of exponential order*.

With these restrictions on $f(t)$, the integral in (B.1.1) converges absolutely for each z that satisfies $\Re z > a$. In addition, $\hat{f}(z)$ is analytic in the right half plane $\Re z > a$, and $\lim_{\Re z \to \infty} \hat{f}(z) = 0$. The derivatives of $\hat{f}(z)$ are given by

$$\hat{f}^{(n)}(z) = (-1)^n \int_0^\infty e^{-zt} t^n f(t) \, dt, \qquad (B.1.2)$$

and these integral representations converge for $\Re z > a$ too.

If $f^{(k)}(t)$, $k = 1, \ldots, n - 1$, are continuous for all $t \geq 0$, and $f^{(n)}(t)$ is of exponential order, then the Laplace transform of $f^{(n)}(t)$ is given in terms of $\hat{f}(z)$ by

$$\mathcal{L}[f^{(n)}(t); z] = z^n \hat{f}(z) - \sum_{i=0}^{n-1} z^{n-i-1} f^{(i)}(0+).$$

The following *inversion formula* is known as the *Bromwich integral*:

$$\frac{f(t+) + f(t-)}{2} = \frac{1}{2\pi i} \int_{c-i\infty}^{c+i\infty} e^{zt} \hat{f}(z) \, dz,$$

where the path of integration is the straight line $\Re z = c$, which is parallel to the $\Im z$-axis, and $\hat{f}(z)$ has all its singularities to the left of $\Re z = c$.

For an extensive treatment of Laplace transforms, see Sneddon [319].

B.2 Watson's Lemma

One of the most important tools of asymptotic analysis that we make use of in this book is *Watson's lemma*. This lemma forms the basis of the method of Laplace and the method of

463

steepest descent for the asymptotic expansion of integrals. For the proof, see Olver [223] or Murray [215].

Theorem B.2.1 *Let the function $f(t)$ be of exponential order and satisfy*

$$f(t) \sim \sum_{k=0}^{\infty} c_k t^{\alpha_k} \quad as \ t \to 0+,$$

where $-1 < \Re\alpha_0 < \alpha_1 < \cdots$, *and* $\lim_{k\to\infty} \Re\alpha_k = +\infty$. *Then, for any* $\delta > 0$,

$$\hat{f}(z) = \int_0^{\infty} e^{-zt} f(t)\, dt \sim \sum_{k=0}^{\infty} \frac{c_k \Gamma(\alpha_k + 1)}{z^{\alpha_k + 1}} \quad as \ z \to \infty, \ \ |\arg z| < \pi/2 - \delta.$$

C

The Gamma Function

The *Gamma function* $\Gamma(z)$ is defined as in

$$\Gamma(z) = \int_0^\infty e^{-t} t^{z-1} \, dt, \quad \Re z > 0,$$

and hence is analytic in the right half plane $\Re z > 0$.

For positive integer values of z, it assumes the special values

$$\Gamma(n) = (n-1)!, \quad n = 1, 2, \ldots .$$

By integrating by parts m times, we obtain from the preceding integral representation the equality

$$\Gamma(z) = \frac{\Gamma(z+m)}{z(z+1)\cdots(z+m-1)}.$$

But the right-hand side of this equality is defined and analytic as well in the right half plane $\Re z > -m$ except at $z = 0, -1, -2, \ldots, -m+1$. Thus, through this equality, $\Gamma(z)$ can be continued analytically to the whole complex plane except the points $z = -n$, $n = 0, 1, \ldots$, where it has simple poles, and the residue at $z = -n$ is $(-1)^n / n!$ for each such n.

Another representation of $\Gamma(z)$ that is also valid for *all* z is *Euler's limit formula*

$$\Gamma(z) = \lim_{n \to \infty} \frac{n! \, n^z}{z(z+1)(z+2)\cdots(z+n)}.$$

Two important identities that result from Euler's limit formula are the *reflection formula*

$$\Gamma(z)\Gamma(1-z) = \frac{\pi}{\sin \pi z}, \quad z \neq 0, \pm 1, \pm 2, \ldots,$$

and the *duplication formula*

$$\Gamma(2z) = \frac{2^{2z-1}}{\pi^{1/2}} \Gamma(z)\Gamma(z + \tfrac{1}{2}), \quad 2z \neq 0, -1, -2, \ldots .$$

The following infinite product representation is valid for *all* z:

$$\frac{1}{\Gamma(z)} = z e^{Cz} \prod_{k=1}^{\infty} \left\{ \left(1 + \frac{z}{k}\right) e^{-z/k} \right\}.$$

Here, $C = 0.5772156\cdots$ is *Euler's constant* defined by

$$C = \lim_{n \to \infty} \left(\sum_{k=1}^{n} \frac{1}{k} - \log n \right).$$

From this product representation, it is obvious that the function $1/\Gamma(z)$ is entire; i.e., it is analytic everywhere in the complex plane. Therefore, $\Gamma(z)$ has no zeros.

The logarithmic derivative of $\Gamma(z)$, namely, $\Gamma'(z)/\Gamma(z)$, is called the *Psi function* and is denoted $\psi(z)$. From the product representation, it follows that

$$\psi(z) = -C - \frac{1}{z} + \sum_{k=1}^{\infty}\left(\frac{1}{k} - \frac{1}{k+z}\right), \quad z \neq 0, -1, -2, \dots,$$

from which it also follows that

$$\Gamma'(1) = \psi(1) = -C.$$

The following asymptotic expansion is known as *Stirling's formula*:

$$\Gamma(z) \sim \sqrt{2\pi}\, z^{z-1/2} e^{-z}\left(1 + \frac{1}{12z} + \frac{1}{288z^2} - \cdots\right) \quad \text{as } z \to \infty, \ |\arg z| < \pi.$$

In particular, we have

$$\Gamma(x+1) = \sqrt{2\pi}\, x^{x+1/2} \exp\left(-x + \frac{\theta(x)}{12x}\right) \quad \text{for } x > 0; \quad (0 < \theta(x) < 1).$$

One of the useful consequences of Stirling's formula is the asymptotic expansion

$$\frac{\Gamma(z+a)}{\Gamma(z+b)} \sim z^{a-b}\left(1 + \sum_{k=1}^{\infty} c_k z^{-k}\right) \quad \text{as } z \to \infty,$$

along any curve joining $z = 0$ and $z = \infty$ providing $z + a$ and $z + b$ are different from $0, -1, -2, \dots$. Here c_k are some constants that depend only a and b.

For more details, see Olver [223] or Abramowitz and Stegun [1].

D

Bernoulli Numbers and Polynomials and the Euler–Maclaurin Formula

D.1 Bernoulli Numbers and Polynomials

The *Bernoulli numbers* B_n, $n = 0, 1, \ldots$, are defined via

$$\frac{t}{e^t - 1} = \sum_{n=0}^{\infty} B_n \frac{t^n}{n!},$$

i.e., $B_n = \frac{d^n}{dt^n} \left(\frac{t}{e^t - 1} \right) |_{t=0}$, $n = 0, 1, \ldots$. They can be computed from the recursion relation

$$B_0 = 1 \quad \text{and} \quad \sum_{k=0}^{n} \binom{n+1}{k} B_k = 0, \quad n = 1, 2, \ldots .$$

The first few of the Bernoulli numbers are thus $B_0 = 1$, $B_1 = -1/2$, $B_2 = 1/6$, $B_4 = -1/30, \ldots$, and $B_3 = B_5 = B_7 = \cdots = 0$. The Bernoulli numbers B_{2n} are related to the Riemann Zeta function $\zeta(z)$ as in

$$\zeta(2n) = \sum_{k=1}^{\infty} \frac{1}{k^{2n}} = (-1)^{n-1} \frac{(2\pi)^{2n} B_{2n}}{2(2n)!}, \quad n = 1, 2, \ldots ,$$

which also shows that $(-1)^{n-1} B_{2n} > 0$, $n = 1, 2, \ldots$, and also that $B_{2n}/(2n)! = O((2\pi)^{-2n})$ as $n \to \infty$.

The *Bernoulli polynomials* $B_n(x)$, $n = 0, 1, \ldots$, are defined via

$$\frac{t e^{xt}}{e^t - 1} = \sum_{n=0}^{\infty} B_n(x) \frac{t^n}{n!},$$

and they can be computed from

$$B_n(x) = \sum_{k=0}^{n} \binom{n}{k} B_k x^{n-k}, \quad n = 0, 1, \ldots .$$

Thus, $B_0(x) = 1$, $B_1(x) = x - \frac{1}{2}$, $B_2(x) = x^2 - x + \frac{1}{6}$, etc., and $B_n(0) = B_n$, $n = 0, 1, \ldots$. They satisfy

$$B_n(x + 1) - B_n(x) = n x^{n-1} \quad \text{and} \quad B_n(1 - x) = (-1)^n B_n(x), \quad n = 1, 2, \ldots ,$$

from which

$$B_n(1) = (-1)^n B_n, \quad n = 1, 2, \ldots ,$$

and $B_1(\frac{1}{2}) = B_3(\frac{1}{2}) = B_5(\frac{1}{2}) = \cdots = 0$. They also satisfy

$$B_{2n}(\tfrac{1}{2}) = -(1 - 2^{1-2n}) B_{2n}, \quad n = 1, 2, \ldots ,$$

and

$$B_n'(x) = n B_{n-1}(x), \quad \int_0^1 B_n(x)\,dx = 0, \quad n = 1, 2, \ldots .$$

Finally,

$$|B_{2n}(x)| < |B_{2n}| \quad \text{when} \ x \in (0, 1), \quad n = 1, 2, \ldots ,$$

so that, for $n = 1, 2, \ldots ,$ the sign of $B_{2n} - B_{2n}(x)$ on $(0, 1)$ is the same as that of B_{2n}; thus, $(-1)^{n-1}[B_{2n} - B_{2n}(x)] \geq 0$ on $[0, 1]$. Actually, $B_{2n} - B_{2n}(x) = 0$ only for $x = 0$ and $x = 1$. In addition, $(-1)^{n-1}[B_{2n} - B_{2n}(x)]$ achieves its maximum on $[0, 1]$ at $x = 1/2$, as a result of which, $(-1)^n[B_{2n}(\frac{1}{2}) - B_{2n}(x)] \geq 0$ on $[0, 1]$ too.

For each n, the *periodic Bernoullian function* $\bar{B}_n(x)$ is defined to be the 1-periodic extension of the Bernoulli polynomial $B_n(x)$. That is,

$$\bar{B}_n(x) = B_n(x - i), \quad \text{when} \ x \in [i, i + 1), \quad i = 0, \pm 1, \pm 2, \ldots .$$

Thus, $\bar{B}_1(x)$ is a piecewise linear sawtooth function, and $\bar{B}_n(x) \in C^{n-2}(-\infty, \infty), n = 2, 3, \ldots .$ They also satisfy

$$\bar{B}'_n(x) = n\bar{B}_{n-1}(x) \ \text{and} \ \int_a^{a+1} \bar{B}_n(x) \, dx = 0 \ \text{for every} \ a, \quad n = 1, 2, \ldots .$$

For more details, see Abramowitz and Stegun [1] and Steffensen [322].

D.2 The Euler–Maclaurin Formula

D.2.1 The Euler–Maclaurin Formula for Sums

An important tool of asymptotic analysis is the famous *Euler–Maclaurin formula*. In the following theorem, we give the most general form of this formula with its remainder.

Theorem D.2.1 *Let $F(t) \in C^m[r, \infty)$, where r is an integer, and let $\theta \in [0, 1]$ be fixed. Then, for any integer $n > r$,*

$$\sum_{i=r}^{n-1} F(i + \theta) = \int_r^n F(t) \, dt + \sum_{k=1}^m \frac{B_k(\theta)}{k!} \left[F^{(k-1)}(n) - F^{(k-1)}(r)\right] + R_m(n; \theta),$$

where the remainder term $R_m(n; \theta)$ is given by

$$R_m(n; \theta) = -\int_r^n F^{(m)}(t) \frac{\bar{B}_m(\theta - t)}{m!} \, dt .$$

For the important special cases in which $\theta = 0$ and $\theta = 1/2$, Theorem D.2.1 assumes the following forms:

Theorem D.2.2 *Let $F(t) \in C^{2m}[r, \infty)$, where r is an integer. Then, for any integer $n > r$,*

$$\sum_{i=r}^n {}'' F(i) = \int_r^n F(t) \, dt + \sum_{k=1}^{m-1} \frac{B_{2k}}{(2k)!} \left[F^{(2k-1)}(n) - F^{(2k-1)}(r)\right] + R_m(n),$$

where the double prime on the summation on the left-hand side means that the first and last terms in the summation are halved, and the remainder term $R_m(n)$ is given by

$$R_m(n) = \int_r^n F^{(2m)}(t) \frac{B_{2m} - \bar{B}_{2m}(t)}{(2m)!} \, dt$$

$$= (n - r) \frac{B_{2m}}{(2m)!} F^{(2m)}(\eta_{m,n}) \ \text{for some} \ \eta_{m,n} \in (r, n).$$

Theorem D.2.3 *Let $F(t) \in C^{2m}[r, \infty)$, where r is an integer. Then, for any integer $n > r$,*

$$\sum_{i=r}^{n-1} F(i + \tfrac{1}{2}) = \int_r^n F(t) \, dt + \sum_{k=1}^{m-1} \frac{B_{2k}(\frac{1}{2})}{(2k)!} \left[F^{(2k-1)}(n) - F^{(2k-1)}(r)\right] + R_m(n),$$

where the remainder term $R_m(n)$ is given by

$$R_m(n) = \int_r^n F^{(2m)}(t) \frac{B_{2m}(\frac{1}{2}) - \bar{B}_{2m}(t + \frac{1}{2})}{(2m)!} dt$$

$$= (n - r) \frac{B_{2m}(\frac{1}{2})}{(2m)!} F^{(2m)}(\eta_{m,n}) \text{ for some } \eta_{m,n} \in (r, n).$$

These theorems can be used to find the asymptotic expansion of the sum $\sum_{i=r}^n F(i + \theta)$ as $n \to \infty$ or to approximate its limit as $n \to \infty$ when this limit exists. We give one such example in the next section.

D.2.2 The Euler–Maclaurin Formula for Integrals

Taking $r = 0$ and making the transformation of variable $t = n(x - a)/(b - a)$ in Theorem D.2.1, and defining $f(x) \equiv F(n(x - a)/(b - a))$, we obtain the following Euler–Maclaurin formula for the *offset trapezoidal rule* approximation of integrals:

Theorem D.2.4 *Let $f(x) \in C^m[a, b]$ and let $\theta \in [0, 1]$ be fixed. Set $h = (b - a)/n$ for n integer and define*

$$\tilde{T}_n(\theta) = h \sum_{i=0}^{n-1} f(a + ih + \theta h) \text{ and } I = \int_a^b f(x)\, dx.$$

Then,

$$\tilde{T}_n(\theta) = I + \sum_{k=1}^m \frac{B_k(\theta)}{k!} \left[f^{(k-1)}(b) - f^{(k-1)}(a) \right] h^k + U_m(h; \theta),$$

where the remainder term $U_m(h; \theta)$ is given by

$$U_m(h; \theta) = -h^m \int_a^b f^{(m)}(x) \frac{\bar{B}_m \left(\theta - n\frac{x-a}{b-a} \right)}{m!} dx.$$

Obviously, $U_m(h; \theta) = O(h^m)$ as $h \to 0$ [equivalently, $U_m(h; \theta) = O(n^{-m})$ as $n \to \infty$].

If we let $\theta = 0$ and $\theta = 1/2$ in Theorem D.2.4, then half the terms in the summation on k there disappear, and we obtain the following special forms of the Euler–Maclaurin expansion for the trapezoidal and midpoint rule approximations of $\int_a^b f(x)\, dx$:

Theorem D.2.5 *Let $f(x) \in C^{2m}[a, b]$ and define*

$$T_n = \tilde{T}_n(0) - B_1[f(b) - f(a)]h = h \sum_{i=0}^n{}'' f(a + ih) \text{ and } I = \int_a^b f(x)\, dx,$$

where the double prime on the summation on the right-hand side means that the first and last terms in this summation are halved. Then,

$$T_n = I + \sum_{k=1}^{m-1} \frac{B_{2k}}{(2k)!} \left[f^{(2k-1)}(b) - f^{(2k-1)}(a) \right] h^{2k} + U_m(h),$$

with the remainder term $U_m(h)$ being given by

$$U_m(h) = h^{2m} \int_a^b f^{(2m)}(x) \frac{B_{2m} - \bar{B}_{2m}(n\frac{x-a}{b-a})}{(2m)!} dx$$

$$= (b - a) \frac{B_{2m}}{(2m)!} f^{(2m)}(\xi_{m,n}) h^{2m} \text{ for some } \xi_{m,n} \in (a, b).$$

Theorem D.2.6 *Let $f(x) \in C^{2m}[a, b]$ and define*

$$M_n = \tilde{T}_n(\tfrac{1}{2}) = h \sum_{i=0}^{n-1} f(a + ih + \tfrac{1}{2}h) \text{ and } I = \int_a^b f(x)\,dx.$$

Then,

$$M_n = I + \sum_{k=1}^{m-1} \frac{B_{2k}(\tfrac{1}{2})}{(2k)!} \left[f^{(2k-1)}(b) - f^{(2k-1)}(a) \right] h^{2k} + U_m(h),$$

where the remainder term $U_m(h)$ is given by

$$U_m(h) = h^{2m} \int_a^b f^{(2m)}(x) \frac{B_{2m}(\tfrac{1}{2}) - \bar{B}_{2m}(\tfrac{1}{2} + n\frac{x-a}{b-a})}{(2m)!}\,dx$$

$$= (b - a) \frac{B_{2m}(\tfrac{1}{2})}{(2m)!} f^{(2m)}(\xi_{m,n}) h^{2m} \text{ for some } \xi_{m,n} \in (a, b).$$

When $f(x) \in C^{\infty}[a, b]$ in Theorems D.2.4–D.2.6, the finite summations on k can be replaced by the infinite summations $\sum_{k=1}^{\infty}$ and the remainder can be deleted, and this gives asymptotic expansions of the approximations to $\int_a^b f(x)\,dx$ as $h \to 0$ (equivalently, as $n \to \infty$). For example, the result of Theorem D.2.4 becomes

$$\tilde{T}_n(\theta) \sim I + \sum_{k=1}^{\infty} \frac{B_k(\theta)}{k!} \left[f^{(k-1)}(b) - f^{(k-1)}(a) \right] h^k \text{ as } h \to 0.$$

For more details, see Steffensen [322].

D.3 Applications of Euler–Maclaurin Expansions

D.3.1 *Application to Harmonic Numbers*

We continue by applying Theorem D.2.2 to the function $F(t) = 1/t$ to derive an asymptotic expansion for the harmonic number $H_n = \sum_{i=1}^n 1/i$ as $n \to \infty$. Taking $r = 1$ in Theorem D.2.2, we first obtain

$$H_n = \log n + \frac{1}{2} + \frac{1}{2n} + \sum_{k=1}^{m-1} \frac{B_{2k}}{2k} - \sum_{k=1}^{m-1} \frac{B_{2k}}{2k} n^{-2k} + R_m(n),$$

where

$$R_m(n) = \int_1^n \frac{B_{2m} - \bar{B}_{2m}(t)}{t^{2m+1}}\,dt.$$

Because $B_{2m} - \bar{B}_{2m}(t)$ is bounded for all t, $R_m(\infty)$ exists, and hence

$$R_m(n) = R_m(\infty) - \int_n^{\infty} \frac{B_{2m} - \bar{B}_{2m}(t)}{t^{2m+1}}\,dt = R_m(\infty) + O(n^{-2m}) \text{ as } n \to \infty.$$

By the well-known fact that $H_n - \log n \to C$ as $n \to \infty$, where $C = 0.5772156 \cdots$ is *Euler's constant*, we realize that $R_m(\infty) + \sum_{k=1}^{m-1} B_{2k}/(2k) + 1/2 = C$, independently of m. Thus, we have obtained the asymptotic expansion

$$H_n \sim \log n + C + \frac{1}{2n} - \sum_{k=1}^{\infty} \frac{B_{2k}}{2k} n^{-2k} \text{ as } n \to \infty.$$

D.3.2 Application to Cauchy Principal Value Integrals

Let $f(x) = g(x)/(x - t)$ for some $t \in (a, b)$, and assume that $g(t) \neq 0$. Then the integral $\int_a^b f(x)\,dx$ does not exist in the ordinary sense. Provided $g(x)$ is differentiable at $x = t$, it is defined in the *Cauchy principal value* sense; that is,

$$I = \lim_{\epsilon \to 0} \left[\int_a^{t-\epsilon} + \int_{t+\epsilon}^b \right] f(x)\,dx$$

exists. For such integrals, the following result has been given by Sidi and Israeli [310, Theorem 4 and Corollary]:

Theorem D.3.1 *Let $h = (b - a)/n$ for some integer $n \geq 1$ and $x_i = a + ih$, $i = 0, 1, \ldots, n$. Let t be fixed and $t \in \{x_1, \ldots, x_{n-1}\}$. Define the trapezoidal rule approximation to I by*

$$\hat{T}_n = h \sum_{\substack{i=0 \\ x_i \neq t}}^{n} {}'' f(x_i) + g'(t)h.$$

Provided $g(x) \in C^{2m}[a, b]$, there holds

$$\hat{T}_n = I + \sum_{k=1}^{m-1} \frac{B_{2k}}{(2k)!} \left[f^{(2k-1)}(b) - f^{(2k-1)}(a) \right] h^{2k} + \hat{U}_m(h),$$

where $\hat{U}_m(h) = O(h^{2m})$ as $h \to 0$ (such that $t \in \{x_1, \ldots, x_{n-1}\}$ always), and $\hat{U}_m(h)$ is given by the Cauchy principal value integral

$$\hat{U}_m(h) = h^{2m} \int_a^b f^{(2m)}(x) \frac{B_{2m} - \bar{B}_{2m}(n\frac{x-a}{b-a})}{(2m)!}\,dx.$$

This result follows in a nontrivial fashion from Theorem D.2.5.

D.4 A Further Development

We now use Theorems D.2.4–D.2.6 to prove the following interesting result that appears to be new. We take $[a, b] = [0, 1]$ for simplicity.

Theorem D.4.1 *Set $t = n^{-2}$ in Theorem D.2.5. If $f \in C^{2m}[0, 1]$, then the trapezoidal rule approximation T_n can be continued to a function $w(t)$ defined for all real $t \in [0, 1]$ [i.e., $w(n^{-2}) = T_n$, $n = 1, 2, \ldots$] such that $w(t) \in C^q[0, 1]$, where $q = \lceil 2m/3 \rceil - 1$; i.e., q is the greatest integer less than $2m/3$. Thus, if $f \in C^\infty[0, 1]$, then $w(t) \in C^\infty[0, 1]$ as well.*

The same holds when T_n is replaced by the midpoint rule approximation M_n of Theorem D.2.6.

A similar result holds for the general offset trapezoidal rule approximation $\tilde{T}_n(\theta)$ of Theorem D.2.4 (with $\theta \neq 0, 1/2, 1$). Only this time, $t = n^{-1}$, and $q = \lceil m/2 \rceil - 1$ when $f(x) \in C^m[0, 1]$.

Proof. We give the proof for T_n only. That for M_n is identical and that for $\tilde{T}_n(\theta)$ is similar.

By Theorem D.2.5, it seems natural to try

$$w(t) = \sum_{k=0}^m c_k t^k + W_m(t),$$

where $c_0 = \int_0^1 f(x)\,dx$ and $c_k = B_{2k}[f^{(2k-1)}(1) - f^{(2k-1)}(0)]/(2k)!$, $k = 1, 2, \ldots$, and

$$W_m(t) = -t^m \int_0^1 f^{(2m)}(x) \frac{\bar{B}_{2m}(xt^{-1/2})}{(2m)!}\,dx.$$

Obviously, $\sum_{k=0}^m c_k t^k$, the polynomial part of $w(t)$, is in $C^\infty[0, 1]$. By the fact that $\bar{B}_{2m}(x) \in C^{2m-2}(-\infty, \infty)$, $W_m(t)$ is in $C^{2m-2}(0, 1]$; hence, we have to analyze its behavior at $t = 0$. First, provided $m \geq 1$, $W_m(t)$ is continuous at $t = 0$ because $\lim_{t\to 0} W_m(t) = 0$. Next, provided $m \geq 2$, we can differentiate the integral expression with respect to t for $t > 0$. Using also the fact that $\bar{B}'_k(x) = k\bar{B}_{k-1}(x)$, we obtain

$$W'_m(t) = -mt^{m-1} \int_0^1 f^{(2m)}(x) \frac{\bar{B}_{2m}(xt^{-1/2})}{(2m)!} dx + t^{m-3/2} \int_0^1 xf^{(2m)}(x) \frac{\bar{B}_{2m-1}(xt^{-1/2})}{2(2m-1)!} dx,$$

from which $\lim_{t\to 0} W'_m(t) = 0$, implying that $W'_m(t)$ is continuous at $t = 0$. Continuing this way, and provided $m > 3p/2$, we obtain $W^{(p)}_m(t) = \hat{W}(t) + \tilde{W}(t)$, where

$$\hat{W}(t) = (-1)^{p+1} t^{m-3p/2} \int_0^1 x^p f^{(2m)}(x) \frac{\bar{B}_{2m-p}(xt^{-1/2})}{2^p(2m-p)!} dx,$$

hence $\lim_{t\to 0} \hat{W}(t) = 0$, and $\tilde{W}(t)$ is smoother than $\hat{W}(t)$ at $t = 0$. This completes the proof. ∎ ■

D.5 The Euler–Maclaurin Formula for Integrals with Endpoint Singularities

Theorems D.2.4–D.2.6 concerned integrands $f(x)$ that were sufficiently smooth on $[a, b]$. We now want to treat functions $f(x)$ that are smooth on (a, b) but have singularities at one or both endpoints $x = a$ and $x = b$. Here we give Navot's extensions of the Euler–Maclaurin formula for integrals with algebraic and/or logarithmic end point singularities. Navot's results were rederived by Lyness and Ninham [202] by a different technique that uses generalized functions. For a generalization of Navot's results to integrands with other singularities, see Waterman, Yos, and Abodeely [351]. For an extension to Hadamard finite part integrals, see Monegato and Lyness [213].

These results have been extended to multidimensional integrals over hypercubes and hypersimplices in the works of Lyness [196], Lyness and Monegato [201], Lyness and de Doncker [199], and Sidi [283]. Recently, numerical integration over curved surfaces and accompanying Euler–Maclaurin expansions have been considered. See the works by Georg [95], Verlinden and Cools [346], and Lyness [198], for example.

D.5.1 Algebraic Singularity at One Endpoint

We start by stating a result of fundamental importance that is due to Navot [216] and that generalizes Theorem D.2.4 to the case in which $f(x)$ has an algebraic singularity at $x = a$. As before, $h = (b - a)/n$ throughout this section too.

Theorem D.5.1 *Let $g(x) \in C^m[a, b]$ and let α be a constant such that $\alpha \neq 0, 1, 2, \dots$, and $\Re\alpha > -1$. Let also $\theta \in (0, 1]$ be fixed. Set $f(x) = (x - a)^\alpha g(x)$. Then, with $\tilde{T}_n(\theta)$ as in Theorem D.2.4, and with $p = \lceil \Re\alpha + m \rceil$, $s_m(x) = \sum_{j=0}^{m-1} \frac{g^{(j)}(a)}{j!}(x - a)^{j+\alpha}$ and $\Delta_m(x) = f(x) - s_m(x)$, we have*

$$\tilde{T}_n(\theta) = I + \sum_{k=1}^{m} \frac{B_k(\theta)}{k!} f^{(k-1)}(b) h^k + \sum_{j=0}^{m-1} \zeta(-\alpha - j, \theta) \frac{g^{(j)}(a)}{j!} h^{\alpha+j+1} + U_{m,p}(h; \theta),$$

where the remainder term $U_{m,p}(h; \theta)$ is given by

$$U_{m,p}(h; \theta) = -h^m \int_a^b \Delta_m^{(m)}(x) \frac{\bar{B}_m(\theta - n\frac{x-a}{b-a})}{m!} dx$$

$$+ \sum_{k=m+1}^{p} \frac{B_k(\theta)}{k!} s_m^{(k-1)}(b) h^k + h^p \int_b^\infty s_m^{(p)}(x) \frac{\bar{B}_p(\theta - n\frac{x-a}{b-a})}{p!} dx,$$

where $\zeta(z, \theta)$ is the generalized Zeta function defined by $\zeta(z, \theta) = \sum_{n=0}^{\infty}(n + \theta)^{-z}$ for $\Re z > 1$ and continued analytically to the complex z-plane.

The generalized Zeta function has properties very similar to those of the "standard" Riemann Zeta function, and both are considered in some detail in Appendix E.

Obviously, $U_{m,p}(h; \theta) = O(h^m)$ as $h \to 0$ [equivalently, $U_{m,p}(h; \theta) = O(n^{-m})$ as $n \to \infty$]. Also, when $-1 < \Re\alpha < 0$, we can take $p = m$.

Using the fact that

$$\zeta(-j, \theta) = -\frac{B_{j+1}(\theta)}{j+1}, \quad j = 0, 1, \ldots,$$

it can be shown that the result of Theorem D.5.1 reduces precisely to that of Theorem D.2.4 as $\alpha \to 0$.

When $g(x) \in C^\infty[a, b]$, we have the asymptotic expansion

$$\tilde{T}_n(\theta) \sim I + \sum_{k=1}^\infty \frac{B_k(\theta)}{k!} f^{(k-1)}(b) h^k + \sum_{j=0}^\infty \zeta(-\alpha - j, \theta) \frac{g^{(j)}(a)}{j!} h^{\alpha+j+1} \quad \text{as } h \to 0.$$

If we let $\theta = 1$ and $\theta = 1/2$ in Theorem D.5.1, half the terms in the summation on k disappear. Let us now define

$$T'_n = \tilde{T}_n(1) - B_1(1) f(b) h = h \sum_{i=1}^n {}' f(a + ih),$$

where the prime on the summation on the right-hand side means that the last term in the summation is halved.

Thus, when $g(x) \in C^\infty[a, b]$, the asymptotic expansions of T'_n and M_n become, respectively,

$$T'_n \sim I + \sum_{k=1}^\infty \frac{B_{2k}}{(2k)!} f^{(2k-1)}(b) h^{2k} + \sum_{j=0}^\infty \zeta(-\alpha - j) \frac{g^{(j)}(a)}{j!} h^{\alpha+j+1} \quad \text{as } h \to 0,$$

because $\zeta(z, 1)$ is nothing but $\zeta(z)$, and

$$M_n \sim I + \sum_{k=1}^\infty \frac{B_{2k}(\frac{1}{2})}{(2k)!} f^{(2k-1)}(b) h^{2k} + \sum_{j=0}^\infty \zeta(-\alpha - j, \tfrac{1}{2}) \frac{g^{(j)}(a)}{j!} h^{\alpha+j+1} \quad \text{as } h \to 0.$$

D.5.2 Algebraic-Logarithmic Singularity at One Endpoint

Theorem D.5.1 enables us to treat integrals of functions $u(x)$ with an algebraic-logarithmic singularity at $x = a$ of the form $u(x) = (\log(x - a))(x - a)^\alpha g(x)$ with $g(x) \in C^m[a, b]$.

We first observe that, if $J = \int_a^b u(x)\,dx$ and $I = \int_a^b (x - a)^\alpha g(x)\,dx$, then $J = \frac{d}{d\alpha} I$. This suggests that we differentiate with respect to α the Euler–Maclaurin expansion associated with I given in Theorem D.5.1. A careful analysis reveals that $\frac{d}{d\alpha} U_{m,p}(h; \theta) = O(h^m)$ as $h \to 0$. Thus, if we define

$$\tilde{S}_n(\theta) = h \sum_{i=0}^{n-1} u(a + ih + \theta h),$$

then

$$\tilde{S}_n(\theta) = J + \sum_{k=1}^m \frac{B_k(\theta)}{k!} u^{(k-1)}(b) h^k$$

$$+ \sum_{j=0}^{m-1} \frac{g^{(j)}(a)}{j!} \left[\zeta(-\alpha - j, \theta) \log h - \zeta'(-\alpha - j, \theta) \right] h^{\alpha+j+1} + O(h^m) \quad \text{as } h \to 0,$$

where $\zeta'(z, \theta) = \frac{d}{dz}\zeta(z, \theta)$. When $g(x) \in C^{\infty}[a, b]$, we have the asymptotic expansion

$$\tilde{S}_n(\theta) \sim J + \sum_{k=1}^{\infty} \frac{B_k(\theta)}{k!} u^{(k-1)}(b) h^k$$
$$+ \sum_{j=0}^{\infty} \frac{g^{(j)}(a)}{j!} \left[\zeta(-\alpha - j, \theta) \log h - \zeta'(-\alpha - j, \theta) \right] h^{\alpha+j+1} \quad \text{as } h \to 0.$$

Simplifications take place in this expansion when we take $\theta = 1$ and $\theta = 1/2$ as before. Further simplifications take place when $\alpha = 0$. All these developments too are due to Navot [217].

Needless to say, this approach can be used to treat those integrals $J = \int_a^b u(x)\,dx$ with $u(x) = (\log(x - a))^r (x - a)^{\alpha} g(x)$, where r is a positive integer. In case $g(x) \in C^{\infty}[a, b]$, this results in an asymptotic expansion of the form

$$\tilde{S}_n(\theta) \sim J + \sum_{k=1}^{\infty} \frac{B_k(\theta)}{k!} u^{(k-1)}(b) h^k$$
$$+ \sum_{j=0}^{\infty} \frac{g^{(j)}(a)}{j!} \left(\sum_{i=0}^{r} c_{ji} (\log h)^i \right) h^{\alpha+j+1} \quad \text{as } h \to 0,$$

where c_{ji} are constants independent of $g(x)$ and n; they depend on $\zeta^{(s)}(-\alpha - j, \theta), s = 0, 1, \ldots, r$. Here, $\zeta^{(s)}(z, \theta) = \frac{d^s}{dz^s}\zeta(z, \theta)$.

D.5.3 Algebraic-Logarithmic Singularities at Both Endpoints

We end by discussing briefly the case in which $f(x) = (x - a)^{\alpha}(b - x)^{\beta} g(x)$, where $\alpha, \beta \neq 0, 1, \ldots$, and $\Re\alpha, \Re\beta > -1$ and $g(x)$ is sufficiently smooth on $[a, b]$. For simplicity, we assume that $g(x) \in C^{\infty}[a, b]$. With $\theta \in (0, 1)$ and $\tilde{T}_n(\theta)$ as in Theorem D.2.4, we then have

$$\tilde{T}_n(\theta) \sim I + \sum_{j=0}^{\infty} \zeta(-\alpha - j, \theta) \frac{g_a^{(j)}(a)}{j!} h^{\alpha+j+1}$$
$$+ \sum_{j=0}^{\infty} (-1)^j \zeta(-\beta - j, 1 - \theta) \frac{g_b^{(j)}(b)}{j!} h^{\beta+j+1} \quad \text{as } h \to 0,$$

where we have defined $g_a(x) = (b - x)^{\beta} g(x)$ and $g_b(x) = (x - a)^{\alpha} g(x)$.

This result was first stated by Lyness and Ninham [202]. It can be obtained by applying Theorem D.5.1 to the integrals $\int_a^c u(x)\,dx$ and $\int_c^b u(x)\,dx$, where $c = a + kh$ for some integer $k \approx n/2$, and by adding the resulting Euler–Maclaurin expansions, and recalling that $B_k(1 - \theta) = (-1)^k B_k(\theta)$ for all k. We leave the details to the interested reader.

When $\theta = 1$, this result should be modified as follows: Define

$$\check{T}_n = h \sum_{i=1}^{n-1} f(a + ih).$$

Then,

$$\check{T}_n \sim I + \sum_{j=0}^{\infty} \zeta(-\alpha - j) \frac{g_a^{(j)}(a)}{j!} h^{\alpha+j+1}$$
$$+ \sum_{j=0}^{\infty} (-1)^j \zeta(-\beta - j) \frac{g_b^{(j)}(b)}{j!} h^{\beta+j+1} \quad \text{as } h \to 0.$$

By repeated differentiation with respect to α and β, we can extend these results to the case in which the integrand is of the form

$$u(x) = (\log(x - a))^r (x - a)^{\alpha} (\log(b - x))^s (b - x)^{\beta} g(x),$$

where r and s are some nonnegative integers.

D.6 Application to Singular Periodic Integrands

An interesting application of the results of the previous sections is to integrals $I = \int_a^b f(x)\,dx$, where $f(x)$ is *periodic* with period $T = b - a$, and infinitely differentiable on $(-\infty, \infty)$ except at the points $x = t + kT$, $k = 0, \pm 1, \pm 2, \dots$, where it may have polar or algebraic and/or logarithmic singularities.

(i) When $f(x)$ has a polar singularity at $t \in (a, b)$, it is of the form $f(x) = g(x)/(x - t) + \tilde{g}(x)$ for $x \in [a, b]$, and $g(t) \neq 0$.

(ii) When $f(x)$ has an algebraic singularity at $t \in (a, b)$, it is of the form $f(x) = g(x)|x - t|^s + \tilde{g}(x)$, $\Re s > -1$, for $x \in [a, b]$.

(iii) When $f(x)$ has a logarithmic singularity in $t \in (a, b)$, it is of the form $f(x) = g(x)\log|x - t| + \tilde{g}(x)$ for $x \in [a, b]$,

(iv) When $f(x)$ has an algebraic-logarithmic singularity at $t \in (a, b)$, it is of the form $f(x) = g(x)|x - t|^s \log|x - t| + \tilde{g}(x)$, $\Re s > -1$, for $x \in [a, b]$.

In the case of the polar singularity, $\int_a^b f(x)\,dx$ does not exist as an ordinary integral but is defined in the Cauchy principal value sense. In the remaining cases, $\int_a^b f(x)\,dx$ exists because $f(x)$ is integrable across $x = t$, and we say that these singularities are weak.

Of course, in all four cases, $g, \tilde{g} \in C^\infty[a, b]$, but $g, \tilde{g} \notin C^\infty(-\infty, \infty)$, and they are not necessarily periodic. Such integrals arise as part of periodic singular and/or weakly singular Fredholm integral equations. For these integrals, Sidi and Israeli [310, Theorem 7] derived "corrected" trapezoidal rule approximations $T(h; t)$, and obtained asymptotic expansions for $T(h; t)$ as $h \to 0$, where $h = (b - a)/n = T/n$ as usual. (See also Sidi [289] for further considerations and comparisons with other works.)

(i) In case of a polar singularity,

$$T(h; t) = h \sum_{i=1}^{n-1} f(t + ih),$$

$$T(h; t) = I - [\tilde{g}(t) + g'(t)]h + O(h^\mu) \text{ as } h \to 0, \text{ for every } \mu > 0.$$

(ii) In case of an algebraic singularity,

$$T(h; t) = h \sum_{i=1}^{n-1} f(t + ih) + \tilde{g}(t)h - 2\zeta(-s)g(t)h^{s+1},$$

$$T(h; t) \sim I + 2 \sum_{k=1}^{\infty} \frac{\zeta(-s - 2k)}{(2k)!} g^{(2k)}(t)h^{s+2k+1} \text{ as } h \to 0.$$

(iii) In case of a logarithmic singularity,

$$T(h; t) = h \sum_{i=1}^{n-1} f(t + ih) + \tilde{g}(t)h + g(t)h \log\left(\frac{h}{2\pi}\right),$$

$$T(h; t) \sim I - 2 \sum_{k=1}^{\infty} \frac{\zeta'(-2k)}{(2k)!} g^{(2k)}(t)h^{2k+1} \text{ as } h \to 0.$$

(iv) In case of an algebraic-logarithmic singularity,

$$T(h; t) = h \sum_{i=1}^{n-1} f(t + ih) + \tilde{g}(t)h + 2[\zeta'(-s) - \zeta(-s)\log h]g(t)h^{s+1},$$

$$T(h; t) \sim I + 2 \sum_{k=1}^{\infty} \frac{\zeta(-s - 2k)\log h - \zeta'(-s - 2k)}{(2k)!} g^{(2k)}(t)h^{s+2k+1} \text{ as } h \to 0.$$

Here we used the fact that $\int_a^b f(x)\,dx = \int_t^{t+T} f(x)\,dx$. This is so because, by the periodicity of $f(x)$, $\int_a^b f(x)\,dx = \int_{a'}^{b'} f(x)\,dx$ for any a' and b' such that $b' - a' = T$. We refer the reader to Sidi and Israeli [310] for details and for further results.

Three types of the integrals treated here occur commonly in applications:

1. Those with a polar singularity, for which we have $T(h;t) - I = -[\tilde{g}(t) + g'(t)]h + O(h^\mu)$ as $h \to 0$, for every $\mu > 0$.
2. Those with a logarithmic singularity, for which we have $T(h;t) - I = O(h^3)$ as $h \to 0$.
3. Those with an algebraic-logarithmic singularity with $s = 2$. [In this case too the singularity is only logarithmic in nature, because $|x - t|^s = (x - t)^s$ is analytic for all x including $x = t$ when s is a positive even integer.] Now in Appendix E it is stated that $\zeta(-2j) = 0$ for $j = 1, 2, \ldots$. Therefore, for $s = 2p$, $p > 0$ an integer, $T(h;t)$ and its asymptotic expansion become

$$T(h;t) = h \sum_{i=1}^{n-1} f(t + ih) + \tilde{g}(t)h + 2\zeta'(-2p)g(t)h^{2p+1},$$

$$T(h;t) \sim I - 2 \sum_{k=1}^{\infty} \frac{\zeta'(-2p - 2k)}{(2k)!} g^{(2k)}(t)h^{2p+2k+1} \quad \text{as } h \to 0.$$

Consequently, $T(h;t) - I = O(h^{2p+3})$ as $h \to 0$. In particular, when $s = 2$, i.e., when $p = 1$, we have $T(h;t) - I = O(h^5)$ as $h \to 0$.

When $s = 1, 2, \ldots$, in $f(x) = g(x)|x - t|^s + \tilde{g}(x)$, we take $T(h;t) = h \sum_{i=0}^{n-1} f(t + ih)$; that is, $T(h;t)$ is simply the trapezoidal rule approximation to $\int_a^b f(x)\,dx$. When s is an even integer, $f(x)$ has no singularity across $x = t$. In this case, $T(h;t) - I$ has an empty asymptotic expansion, which implies that $T(h;t) = I + O(h^\mu)$ as $h \to 0$, for every $\mu > 0$. When s is an odd integer, the derivatives of $f(x)$ have finite jump discontinuities across $x = t$, and $T(h;t)$ has an asymptotic expansion of the form $T(h;t) \sim I + \sum_{k=1}^{\infty} \alpha_k(t)h^{2k}$ as $h \to 0$, as has been shown by Averbuch et al. [14].

E

The Riemann Zeta Function and the Generalized Zeta Function

E.1 Some Properties of $\zeta(z)$

The *Riemann Zeta function* $\zeta(z)$ is defined to be the sum of the (convergent) Dirichlet series $\sum_{k=1}^{\infty} k^{-z}$ for $\Re z > 1$. From the theory of Dirichlet series, it is known that $\zeta(z)$ is an analytic function of z for $\Re z > 1$. It turns out that $\zeta(z)$ can be continued analytically to the whole z-plane except $z = 1$, where it has a simple pole with residue 1, and we denote its analytic continuation again by $\zeta(z)$. The *reflection formula* due to Riemann, namely,

$$\zeta(1 - z) = 2^{1-z}\pi^{-z}\cos(\tfrac{1}{2}\pi z)\Gamma(z)\zeta(z),$$

can be used to continue $\zeta(z)$ analytically. It is interesting to note the following special values that this analytic continuation assumes:

$$\zeta(0) = -\frac{1}{2}; \quad \zeta(-2n) = 0 \ \text{ and } \ \zeta(1 - 2n) = -\frac{B_{2n}}{2n}, \quad n = 1, 2, \ldots .$$

We also have

$$\zeta'(0) = -\frac{1}{2}\log(2\pi),$$

$$\zeta'(-2n) = (-1)^n 2^{-2n-1}\pi^{-2n}\Gamma(2n + 1)\zeta(2n + 1), \quad n = 1, 2, \ldots .$$

Let us, in addition, recall the connection between $\zeta(2n)$ and B_{2n}, namely,

$$\zeta(2n) = \sum_{k=1}^{\infty} \frac{1}{k^{2n}} = (-1)^{n-1}\frac{(2\pi)^{2n}B_{2n}}{2(2n)!}, \quad n = 1, 2, \ldots .$$

The Zeta function has no zeros in the right half plane $\Re z > 1$, and its only zeros in the left half plane $\Re z < 0$ are $-2, -4, -6, \ldots$. As for the zeros in the remaining strip $0 \le \Re z \le 1$, a famous and still unproved conjecture by Riemann claims that they all lie on the midline $\Re z = \frac{1}{2}$.

E.2 Asymptotic Expansion of $\sum_{k=0}^{n-1}(k + \theta)^{-z}$

We now derive the asymptotic expansion as $n \to \infty$ of the partial sum $S_n(z) = \sum_{k=0}^{n-1}(k + \theta)^{-z}$ with $\theta > 0$ whether $\sum_{k=0}^{\infty}(k + \theta)^{-z}$ converges or not. Simultaneously, we also obtain the analytic structure of the *generalized Zeta function* $\zeta(z, \theta)$, which is defined via $\zeta(z, \theta) = \sum_{k=0}^{\infty}(k + \theta)^{-z}$ for $\Re z > 1$ and then continued analytically to the complex z-plane, and hence of the Riemann Zeta function $\zeta(z)$ as well. [Observe that $\zeta(z, 1) = \zeta(z)$.] This can be achieved by using the Euler–Maclaurin expansion, as was done in Section D.3 of Appendix D. Our approach here is quite different in that it involves the Laplace transform and thus enables us to conclude also that the partial sum $S_n(z)$ is associated with a function $A(y)$ in $\mathbf{F}_\infty^{(1)}$.

Using the fact that $a^{-z} = \left(\int_0^\infty e^{-at} t^{z-1}\, dt \right) / \Gamma(z)$, $a > 0$ and $\Re z > 0$, we obtain

$$S_n(z) = \sum_{k=0}^{n-1} (k+\theta)^{-z} = \frac{1}{\Gamma(z)} \int_0^\infty t^{z-1} \left(\sum_{k=0}^{n-1} e^{-(k+\theta)t} \right) dt,$$

which, by the additional fact that $\sum_{k=0}^{n-1} \tau^k = (1 - \tau^n)/(1 - \tau)$, gives

$$S_n(z) = \frac{1}{\Gamma(z)} \int_0^\infty \frac{t^{z-1} e^{(1-\theta)t}}{e^t - 1}\, dt - \frac{1}{\Gamma(z)} \int_0^\infty e^{-nt} \left(\frac{t^{z-1} e^{(1-\theta)t}}{e^t - 1} \right) dt, \quad \Re z > 1. \quad \text{(E.2.1)}$$

Because $t^{z-1} e^{(1-\theta)t}/(e^t - 1) = t^{z-2} + O(t^{z-1})$ as $t \to 0+$, the second integral in (E.2.1) is $O(n^{-z+1})$, hence tends to 0 as $n \to \infty$, as can be shown by using Watson's lemma. Therefore, we conclude that

$$\zeta(z, \theta) = \frac{1}{\Gamma(z)} \int_0^\infty \frac{t^{z-1} e^{(1-\theta)t}}{e^t - 1}\, dt, \quad \Re z > 1,$$

and hence that

$$\zeta(z, \theta) = S_n(z) + \frac{1}{\Gamma(z)} \int_0^\infty e^{-nt} \left(\frac{t e^{(1-\theta)t}}{e^t - 1} \right) t^{z-2}\, dt, \quad \Re z > 1. \quad \text{(E.2.2)}$$

Now, from Appendix D, we have for any positive integer m,

$$\frac{t e^{(1-\theta)t}}{e^t - 1} = \sum_{s=0}^{m-1} \frac{B_s(1-\theta)}{s!} t^s + W_m(t), \quad W_m(t) = O(t^m) \text{ as } t \to 0.$$

Substituting this in (E.2.2) and making use of the facts that $\Gamma(z+k)/\Gamma(z) = (z)_k = z(z+1)\cdots (z+k-1)$ and $B_k(1-\theta) = (-1)^k B_k(\theta)$ for all k, we obtain for $\Re z > 1$

$$\zeta(z, \theta) = S_n(z) + \frac{n^{-z+1}}{z-1} + \sum_{s=1}^{m-1} (-1)^s \frac{(z)_{s-1} B_s(\theta)}{s!} n^{-s-z+1} + \frac{1}{\Gamma(z)} \int_0^\infty e^{-nt} W_m(t) t^{z-2}\, dt.$$

$$\text{(E.2.3)}$$

The first term on the right-hand side of (E.2.3), namely, the term $S_n(z)$, is an entire function of z. The second term, $n^{-z+1}/(z-1)$, is analytic everywhere except at $z = 1$, where it has a simple pole with residue 1. The third term, namely the summation $\sum_{s=1}^{m-1}$, being a polynomial in z multiplied by n^{-z}, is entire just as the first one. Finally, because $W_m(t) t^{z-2}$ is (i) $O(t^{m+z-3})$ as $t \to \infty$, and (ii) $O(t^{m+z-2})$ as $t \to 0$, and because $1/\Gamma(z)$ is entire, the last term, $\left[\int_0^\infty e^{-nt} W_m(t) t^{z-2}\, dt \right] / \Gamma(z)$, exists and is analytic for $\Re z > -m + 1$. In addition, it is $O(n^{-m-z+1})$ as $n \to \infty$, by Watson's lemma. Putting everything together, we conclude that the right-hand side of (E.2.3) is analytic in the half plane $\Re z > -m + 1$ except at $z = 1$, where it has a simple pole with residue 1, and it is thus the analytic continuation of the left-hand side. We have thus shown that $\zeta(z, \theta)$ can be continued analytically to the whole complex plane except $z = 1$, where it has a simple pole with residue 1.

When $z = 0, -1, -2, \ldots, -m + 2$, the integral term on the right-hand side of (E.2.3) vanishes because $1/\Gamma(z) = 0$ for these values of z. Thus, for $z = -p$, $p = 0, 1, \ldots$, (E.2.3) reduces to

$$\sum_{k=0}^{n-1} (k+\theta)^p = \frac{n^{p+1}}{p+1} - \sum_{s=1}^{p+1} (-1)^s \frac{(-p)_{s-1} B_s(\theta)}{s!} n^{p-s+1} + \zeta(-p, \theta).$$

Comparing this with the Euler–Maclaurin formula for the sum $\sum_{k=0}^{n-1} (k+\theta)^p$, which is obtained by applying Theorem D.2.1 to the function $F(t) = t^p$, we obtain the following well-known result:

$$\zeta(-p, \theta) = -\frac{B_{p+1}(\theta)}{p+1}, \quad p = 0, 1, \ldots . \quad \text{(E.2.4)}$$

[Note that the relations $\zeta(-2n) = 0$ and $\zeta(1 - 2n) = -B_{2n}/(2n), n = 1, 2, \ldots$, mentioned in the preceding section can be obtained by letting $\theta = 1$ in (E.2.4).]

We have also shown that, for all other values of z, $S_n(z)$ has the asymptotic expansion

$$S_n(z) \sim \zeta(z, \theta) - \frac{n^{-z+1}}{z-1} - \sum_{s=1}^{\infty}(-1)^s \frac{(z)_{s-1}B_s(\theta)}{s!}n^{-z-s+1} \quad \text{as } n \to \infty, \quad z \neq 1, \quad (\text{E.2.5})$$

whether $\sum_{k=0}^{\infty}(k+\theta)^{-z}$ converges or not.

Note that, when $\sum_{k=0}^{\infty}(k+\theta)^{-z}$ converges, that is, when $\Re z > 1$, this asymptotic expansion can be obtained by applying Watson's lemma to the second integral in (E.2.1).

Let $z \neq 1$ and $\Re z > -m + 1$ for some positive integer m. We can rewrite (E.2.3) in the form $S_n(z) = \zeta(z, \theta) + n^{-z+1}g(n)$, where $g(x)$ is given by

$$g(x) = -(z-1)^{-1} - \sum_{s=1}^{m-1}(-1)^s \frac{(z)_{s-1}B_s(\theta)}{s!}x^{-s} - \frac{x^{z-1}}{\Gamma(z)}\int_0^{\infty} e^{-xt} W_m(t)t^{z-2}\, dt,$$

and hence $g(x) \in \mathbf{A}_0^{(0)}$. It is clear from one of the examples of functions in the class $\mathbf{A}^{(\gamma)}$ that are given in Section 5.1 that $g(x) \in \mathbf{A}^{(0)}$. This shows that $A(y) \equiv \zeta(z, \theta) + y^{z-1}g(y^{-1})$ is not only in $\mathbf{F}^{(1)}$ but also in $\mathbf{F}_\infty^{(1)}$.

We continue our approach to show that the asymptotic expansion of $\frac{d}{dz}S_n(z) = -\sum_{k=0}^{n-1}(k+\theta)^{-z}\log(k+\theta)$ as $n \to \infty$ can be obtained by differentiating that of $S_n(z)$ given in (E.2.5) term by term. This can be done by differentiating both sides of (E.2.3), and realizing that

$$\frac{d}{dz}\int_0^{\infty} e^{-nt} W_m(t)t^{z-2}\, dt = \int_0^{\infty} e^{-nt} \log t\, W_m(t)t^{z-2}\, dt = O(n^{-m-z+1}\log n) \quad \text{as } n \to \infty.$$

This last assertion follows from the fact that, when $f(t)$ is of exponential order and also satisfies $f(t) = O(t^p)$ as $t \to 0+$, $p > -1$, then $\int_0^{\infty} e^{-\lambda t}\log t f(t)\, dt = O(\lambda^{-p+1}\log \lambda)$ as $\lambda \to \infty$, $|\arg \lambda| < \pi/2 - \delta$, for any $\delta > 0$. Thus,

$$S_n'(z) \sim \zeta'(z, \theta) + \sum_{s=0}^{\infty}\left[c_s(z)\log n - c_s'(z)\right]n^{-z-s+1} \quad \text{as } n \to \infty, \quad \text{for any } z \neq 1, \quad (\text{E.2.6})$$

where $c_0(z) = 1/(z-1)$ and $c_s(z) = (-1)^s(z)_{s-1}B_s(\theta)/s!$, $s = 1, 2, \ldots$, again whether $\sum_{k=0}^{\infty}(k+\theta)^{-z}\log(k+\theta)$ converges or not. Here, $g'(z)$ stands for $\frac{d}{dz}g(z)$ in general.

For more details on the Zeta function, see Olver [223], Abramowitz and Stegun [1], or Titchmarsh [334].

F

Some Highlights of Polynomial Approximation Theory

F.1 Best Polynomial Approximations

Let $f(x)$ be defined on the finite interval $[a, b]$, and let Π_n denote the set of all polynomials of degree at most n. The *best polynomial approximation of degree at most n* in the L_∞-norm on $[a, b]$ to $f(x)$ is that polynomial $p^*(x) \in \Pi_n$ that satisfies

$$\max_{x \in [a,b]} |f(x) - p^*(x)| \le \max_{x \in [a,b]} |f(x) - p(x)| \text{ for all } p \in \Pi_n.$$

It is known that, when $f \in C[a, b]$, $p^*(x)$ exists and is unique.

Let us denote

$$E_n(f) = \max_{x \in [a,b]} |f(x) - p^*(x)|.$$

What interests us here are the convergence properties of the sequence of best polynomial approximations when $f(x)$ is infinitely smooth. We recall one of Jackson's theorems:

Theorem F.1.1 *Let $\{p_n^*(x)\}$ be the sequence of best polynomial approximations to $f(x)$ on $[a, b]$, where $p_n^*(x) \in \Pi_n$. If $f \in C^k[a, b]$, then*

$$E_n(f) = O(n^{-k}) \text{ as } n \to \infty.$$

The following result is an immediate consequence of Jackson's theorem.

Theorem F.1.2 *Let $\{p_n^*(x)\}$ be the sequence of best polynomial approximations to $f(x)$ on $[a, b]$, where $p_n^*(x) \in \Pi_n$. If $f \in C^\infty[a, b]$, then*

$$E_n(f) = O(n^{-\mu}) \text{ as } n \to \infty, \text{ for every } \mu > 0.$$

A theorem of Bernstein says that, if $f(z)$ is analytic in an open domain that contains the interval $[a, b]$ in its interior, then the result of Theorem F.1.2 can be improved to read

$$E_n(f) = O(e^{-\alpha n}) \text{ as } n \to \infty, \text{ for some } \alpha > 0.$$

α depends on the location of the singularity of $f(z)$ that is closest to $[a, b]$ in some well-defined sense, which we omit.

If $f(z)$ is an entire function, that is, it is analytic in the entire z-plane, an additional improvement takes place, and we have

$$E_n(f) = O(e^{-\alpha n}) \text{ as } n \to \infty, \text{ for every } \alpha > 0.$$

F.2 Chebyshev Polynomials and Expansions

Let $x = \cos \theta$ for $\theta \in [0, \pi]$. The Chebyshev polynomials are defined by

$$T_n(x) = \cos(n\theta), \quad n = 0, 1, \dots .$$

Thus, $T_0(x) = 1$, $T_1(x) = x$, $T_2(x) = 2x^2 - 1$, ... From this definition, it follows that the $T_n(x)$ satisfy the three-term recursion relation

$$T_{n+1}(x) = 2x T_n(x) - T_{n-1}(x), \quad n = 1, 2, \ldots,$$

from which it can be shown that $T_n(x)$ is a polynomial in x of degree exactly n and that $T_n(x)$ is even or odd depending on whether n is an even or an odd integer.

Because the transformation $x = \cos\theta$ is one-to-one between $-1 \leq x \leq 1$ and $0 \leq \theta \leq \pi$, we see from the definition of the $T_n(x)$ that

$$|T_n(x)| \leq 1 \text{ for } x \in [-1, 1].$$

In other words, for $x \in [-1, 1]$, $T_n(x)$ assumes values between -1 and 1 only.

When x is real and $|x| > 1$, the $T_n(x)$ can be shown to satisfy

$$T_n(x) = (\text{sign } x)^n \cosh(n\phi), \quad \text{where } \phi > 0 \text{ and } e^\phi = |x| + \sqrt{x^2 - 1} > 1.$$

As a result, we have that

$$T_n(x) \sim \frac{1}{2}(\text{sign } x)^n e^{n\phi} = \frac{1}{2}(\text{sign } x)^n (|x| + \sqrt{x^2 - 1})^n \text{ as } n \to \infty.$$

In other words, when $|x| > 1$, the sequence $\{|T_n(x)|\}$ increases to infinity like $e^{n\phi}$ for some $\phi > 0$ that depends on x.

The $T_n(x)$ also satisfy the orthogonality property

$$\int_{-1}^{1} \frac{T_m(x) T_n(x)}{\sqrt{1 - x^2}} \, dx = \begin{cases} 0 & \text{if } m \neq n \\ \frac{1}{2}\pi & \text{if } m = n \neq 0. \\ \pi & \text{if } m = n = 0 \end{cases}$$

If $f(x)$ satisfies $\int_{-1}^{1}(1 - x^2)^{-1/2}|f(x)|^2 \, dx < \infty$, then it can be expanded in a series of Chebyshev polynomials in the form

$$f(x) = \sum_{n=0}^{\infty} {}' a_n T_n(x); \quad a_n = \frac{2}{\pi} \int_{-1}^{1} \frac{f(x) T_n(x)}{\sqrt{1 - x^2}} \, dx, \quad n = 0, 1, \ldots,$$

where the prime on the summation indicates that the $n = 0$ term is to be multiplied by $1/2$. It is known that the sequence of the partial sums converges to $f(x)$ in the sense that

$$\lim_{n \to \infty} \int_{-1}^{1} \frac{1}{\sqrt{1 - x^2}} \left[f(x) - \sum_{i=0}^{n} {}' a_i T_i(x) \right]^2 dx = 0.$$

As we are interested in approximation of functions on an arbitrary finite interval $[a, b]$, we consider the Chebyshev expansions of functions on such an interval in the sequel. First, it is clear that the expansion of $f(x)$ in a Chebyshev series on $[a, b]$ can be achieved by transforming $[a, b]$ to $[-1, 1]$ by the linear transformation $x = \xi(t) = (b + a)/2 + (b - a)t/2$, so that $t \in [-1, 1]$ when $x \in [a, b]$. Then, with $\tau(x) = (2x - a - b)/(b - a)$, we have

$$f(x) = \sum_{n=0}^{\infty} {}' a_n T_n(\tau(x)); \quad a_n = \frac{2}{\pi} \int_{-1}^{1} \frac{f(\xi(t)) T_n(t)}{\sqrt{1 - t^2}} \, dt, \quad n = 0, 1, \ldots.$$

Let us denote

$$W_n(f) = \max_{x \in [a,b]} \left| f(x) - \sum_{i=0}^{n} {}' a_i T_i(\tau(x)) \right|.$$

Theorem F.2.1 *Let $f \in C^k[a, b]$. Then*

$$a_n = o(n^{-k}) \text{ and } W_n(f) = o(n^{-k+1}) \text{ as } n \to \infty.$$

Thus, for $k \geq 2$ $f(x)$ has a Chebyshev expansion that converges uniformly on $[a, b]$.

The following result is an immediate consequence of Theorem F.2.1.

Theorem F.2.2 *Let* $f \in C^{\infty}[a, b]$. *Then*

$$a_n = O(n^{-\mu}) \text{ and } W_n(f) = O(n^{-\mu}) \text{ as } n \to \infty, \text{ for every } \mu > 0.$$

If $f(z)$ is analytic in an open domain that contains the interval $[a, b]$ in its interior, then the result of Theorem F.2.2 can be improved to read

$$a_n = O(e^{-\alpha n}) \text{ and } W_n(f) = O(e^{-\alpha n}) \text{ as } n \to \infty, \text{ for some } \alpha > 0.$$

α depends on the location of the singularity of $f(z)$ that is closest to $[a, b]$ in some well-defined sense, which we omit.

If $f(z)$ is an entire function, an additional improvement takes place, and we have

$$a_n = O(e^{-\alpha n}) \text{ and } W_n(f) = O(e^{-\alpha n}) \text{ as } n \to \infty, \text{ for every } \alpha > 0.$$

For the preceding and more, see, for example, the books by Cheney [47] and Davis [62].

G

A Compendium of Sequence Transformations

For the sake of convenience, we collect here the definitions of the various sequence transformations and the algorithms used in their implementation. $\{A_m\}$ is the sequence we wish to extrapolate.

1. Richardson Extrapolation Process (REP)

Given the distinct scalars $c_1, c_2, \ldots$, such that $c_k \neq 1$ for all k, define $A_n^{(j)}$ through

$$A_r = A_n^{(j)} + \sum_{k=1}^{n} \bar{\alpha}_k c_k^r, \quad j \le r \le j+n.$$

The $A_n^{(j)}$ can be computed recursively via Algorithm 1.3.1; that is,

$$A_0^{(j)} = A_j, \quad j \ge 0; \quad A_n^{(j)} = \frac{A_{n-1}^{(j+1)} - c_n A_{n-1}^{(j)}}{1 - c_n}, \quad j \ge 0, \; n \ge 1.$$

2. Polynomial Richardson Extrapolation Process [REP-POL(γ)]

Given the scalar $\gamma \neq 0$, choose integers R_l such that $R_0 < R_1 < R_2 < \cdots$, and define $A_n^{(j)}$ through

$$A_{R_l} = A_n^{(j)} + \sum_{k=1}^{n} \frac{\bar{\alpha}_k}{R_l^{k\gamma}}, \quad j \le r \le j+n.$$

Letting $\lambda_n^{(j)} = (R_j/R_{j+n})^\gamma$, the $A_n^{(j)}$ can be computed recursively via Algorithm 2.2.1; that is,

$$A_0^{(j)} = A_j, \quad j \ge 0; \quad A_n^{(j)} = \frac{A_{n-1}^{(j+1)} - \lambda_n^{(j)} A_{n-1}^{(j)}}{1 - \lambda_n^{(j)}}, \quad j \ge 0, \; n \ge 1.$$

3. Richardson Extrapolation Process with Confluence (REP-CONF)

Given the distinct scalars $c_1, c_2, \ldots$, such that $c_k \neq 1$ for all k, and the nonnegative integers $q_1, q_2, \ldots$, first let

$$\lambda_i = c_1, \qquad g_i(m) = c_1^m m^{i-1}, \quad 1 \le i \le \nu_1 \equiv q_1 + 1,$$

$$\lambda_{\nu_1+i} = c_2, \qquad g_{\nu_1+i}(m) = c_2^m m^{i-1}, \quad 1 \le i \le \nu_2 \equiv q_2 + 1,$$

$$\lambda_{\nu_1+\nu_2+i} = c_3, \qquad g_{\nu_1+\nu_2+i}(m) = c_2^m m^{i-1}, \quad 1 \le i \le \nu_3 \equiv q_3 + 1,$$

and so on. Then, define $A_n^{(j)}$ through

$$A_r = A_n^{(j)} + \sum_{k=1}^{n} \bar{\alpha}_k g_k(r), \quad j \le r \le j+n.$$

483

The $A_n^{(j)}$ can be computed recursively via the SGRom-algorithm (Algorithm 14.1.1); that is,

$$A_0^{(j)} = A_j, \quad j \geq 0; \quad A_n^{(j)} = \frac{A_{n-1}^{(j+1)} - \lambda_n A_{n-1}^{(j)}}{1 - \lambda_n}, \quad j \geq 0, \ n \geq 1.$$

4. Iterated Δ^2-Process (IDELTA)

Define first

$$\phi_m(\{A_s\}) = A_{m+1} - \frac{(\Delta A_m)(\Delta A_{m+1})}{\Delta^2 A_m}.$$

Next, define $B_n^{(j)}$ recursively from

$$B_0^{(j)} = A_j, \quad j \geq 0; \quad B_{n+1}^{(j)} = \phi_j(\{B_n^{(s)}\}), \quad j, n \geq 0.$$

5. Iterated $\Delta^2(\gamma)$- and $\Delta^2(\gamma, p)$-Processes [IMDELTA(γ, p)]

Given the scalar γ and the integer $p \geq 1$, define first

$$\psi_m(\{A_s\}; \gamma) = A_{m+1} - \frac{\gamma - 1}{\gamma} \frac{(\Delta A_m)(\Delta A_{m+1})}{\Delta^2 A_m}.$$

Next, define $B_n^{(j)}$ recursively from

$$B_0^{(j)} = A_j, \quad j \geq 0; \quad B_{n+1}^{(j)} = \psi_j(\{B_n^{(s)}\}; \gamma - 2n), \quad j, n \geq 0, \quad \text{for } p = 1,$$

$$B_0^{(j)} = A_j, \quad j \geq 0; \quad B_{n+1}^{(j)} = \psi_j(\{B_n^{(s)}\}; \gamma - n/p), \quad j, n \geq 0, \quad \text{for } p \geq 2.$$

For $p \geq 2$, IMDELTA(γ, p) is the iterated $\Delta^2(\gamma, p)$-process [see (15.7.2)], whereas for $p = 1$, it is the iterated $\Delta^2(\gamma)$-process [see (15.3.15)].

6. Iterated Lubkin Transformation (ILUBKIN)

Define first

$$W_m(\{A_s\}) = \frac{\Delta^2(A_m/\Delta A_m)}{\Delta^2(1/\Delta A_m)}.$$

Next, define $B_n^{(j)}$ recursively from

$$B_0^{(j)} = A_j, \quad j \geq 0; \quad B_{n+1}^{(j)} = W_j(\{B_n^{(s)}\}), \quad j, n \geq 0.$$

7. Shanks Transformation (EPSILON)

Define $e_n(A_j)$ through

$$A_r = e_n(A_j) + \sum_{k=1}^{n} \bar{\alpha}_k \Delta A_{r+k-1}, \quad j \leq r \leq j + n.$$

Next, define the $\varepsilon_k^{(j)}$ recursively via the ε-algorithm (Algorithm 16.2.1); that is,

$$\varepsilon_{-1}^{(j)} = 0, \quad \varepsilon_0^{(j)} = A_j, \quad j \geq 0; \quad \varepsilon_{k+1}^{(j)} = \varepsilon_{k-1}^{(j+1)} + \frac{1}{\varepsilon_k^{(j+1)} - \varepsilon_k^{(j)}}, \quad j, k \geq 0.$$

Then, $e_n(A_j) = \varepsilon_{2n}^{(j)}$ for all j and n.

8. Higher-Order G-Transformation (G-TRAN)

Given the scalars $u_0, u_1, \dots$, define $A_n^{(j)}$ through

$$A_r = A_n^{(j)} + \sum_{k=1}^{n} \bar{\alpha}_k u_{r+k-1}, \quad j \leq r \leq j + n.$$

The $A_n^{(j)}$ can be computed with the help of the rs-algorithm (Algorithm 21.3.1) or the FS/qd-algorithm (Algorithm 21.3.2). Here are the steps of the FS/qd-algorithm:

$$e_0^{(j)} = 0, \quad q_1^{(j)} = \frac{u_{j+1}}{u_j}, \quad M_0^{(j)} = \frac{A_j}{u_j}, \quad N_0^{(j)} = \frac{1}{u_j}, \quad j \geq 0.$$

$$e_n^{(j)} = q_n^{(j+1)} - q_n^{(j)} + e_{n-1}^{(j+1)}, \quad q_{n+1}^{(j)} = \frac{e_n^{(j+1)}}{e_n^{(j)}} q_n^{(j+1)},$$

$$M_n^{(j)} = \frac{M_{n-1}^{(j+1)} - M_{n-1}^{(j)}}{e_n^{(j)}}, \quad N_n^{(j)} = \frac{N_{n-1}^{(j+1)} - N_{n-1}^{(j)}}{e_n^{(j)}}, \quad A_n^{(j)} = \frac{M_n^{(j)}}{N_n^{(j)}}, \quad j \geq 0, \ n \geq 1.$$

9. The $\mathcal{L}$-Transformation (L-TRAN)

Define $\mathcal{L}_n^{(j)}$ through

$$A_r = \mathcal{L}_n^{(j)} + \omega_r \sum_{i=0}^{n-1} \frac{\bar{\beta}_i}{r^i}, \quad J \leq r \leq J + n; \quad J = j + 1.$$

Then, $\mathcal{L}_n^{(j)}$ is given by the closed-form expression

$$\mathcal{L}_n^{(j)} = \frac{\Delta^n \left(J^{n-1} A_J / \omega_J \right)}{\Delta^n \left(J^{n-1} / \omega_J \right)} = \frac{\sum_{i=0}^{n}(-1)^i \binom{n}{i}(J+i)^{n-1} A_{J+i}/\omega_{J+i}}{\sum_{i=0}^{n}(-1)^i \binom{n}{i}(J+i)^{n-1}/\omega_{J+i}}; \quad J = j+1.$$

Here the choices $\omega_m = \Delta A_{m-1}$ and $\omega_m = m \Delta A_{m-1}$ give rise to the t- and u-transformations, respectively. The choice $\omega_m = m \Delta A_{m-1}$ (u-transformation) is the more useful.

10. The $\mathcal{S}$-Transformation (S-TRAN)

Define $\mathcal{S}_n^{(j)}$ through

$$A_r = \mathcal{S}_n^{(j)} + \omega_r \sum_{i=0}^{n-1} \frac{\bar{\beta}_i}{(r)_i}, \quad J \leq r \leq J + n; \quad J = j + 1.$$

Then, $\mathcal{S}_n^{(j)}$ is given by the closed-form expression

$$\mathcal{S}_n^{(j)} = \frac{\Delta^n \left((J)_{n-1} A_J / \omega_J \right)}{\Delta^n \left((J)_{n-1} / \omega_J \right)} = \frac{\sum_{i=0}^{n}(-1)^i \binom{n}{i}(J+i)_{n-1} A_{J+i}/\omega_{J+i}}{\sum_{i=0}^{n}(-1)^i \binom{n}{i}(J+i)_{n-1}/\omega_{J+i}}; \quad J = j+1.$$

Here too we can make the choices $\omega_m = \Delta A_{m-1}$ or $\omega_m = m \Delta A_{m-1}$ to obtain transformations analogous to, respectively, the t- or the u-transformation.

11. The $d^{(1)}$-Transformation ($d^{(1)}$-TRAN)

Pick integers R_l such that $1 \leq R_0 < R_1 < R_2 < \cdots$, and define $d_n^{(1,j)} = A_n^{(j)}$ through

$$A_{R_l} = A_n^{(j)} + R_l (\Delta A_{R_l - 1}) \sum_{i=0}^{n-1} \frac{\bar{\beta}_i}{R_l^i}, \quad j \leq l \leq j + n.$$

The $A_n^{(j)}$ and the corresponding $\Gamma_n^{(j)}$ and $\Lambda_n^{(j)}$ can be computed recursively via the W-algorithm (Algorithm 7.2.4). We give here the steps of the resulting algorithm:
1. For $j \geq 0$, set

$$M_0^{(j)} = \frac{A_{R_j}}{R_j(\Delta A_{R_j - 1})}, \quad N_0^{(j)} = \frac{1}{R_j(\Delta A_{R_j - 1})}, \quad H_0^{(j)} = (-1)^j |N_0^{(j)}|, \quad K_0^{(j)} = (-1)^j |M_0^{(j)}|.$$

2. For $j \geq 0$ and $n \geq 1$, compute $M_n^{(j)}$, $N_n^{(j)}$, $H_n^{(j)}$, and $K_n^{(j)}$ recursively from

$$Q_n^{(j)} = \frac{Q_{n-1}^{(j+1)} - Q_{n-1}^{(j)}}{R_{j+n}^{-1} - R_j^{-1}},$$

where $Q_n^{(j)}$ stand for $M_n^{(j)}$ or $N_n^{(j)}$ or $H_n^{(j)}$ or $K_n^{(j)}$.

3. For $j, n \geq 0$, set

$$A_n^{(j)} = \frac{M_n^{(j)}}{N_n^{(j)}}, \quad \Gamma_n^{(j)} = \left| \frac{H_n^{(j)}}{N_n^{(j)}} \right|, \quad \Lambda_n^{(j)} = \left| \frac{K_n^{(j)}}{N_n^{(j)}} \right|.$$

When $R_l = l + 1$, $l = 0, 1, \ldots$, we have $d_n^{(1,j)} = \mathcal{L}_n^{(j)}$ with $\omega_m = m\Delta A_{m-1}$, that is, the $d^{(1)}$-transformation reduces to the u-transformation.

12. The $d^{(m)}$-Transformation ($d^{(m)}$-TRAN)

Pick integers R_l such that $1 \leq R_0 < R_1 < R_2 < \cdots$, let $n = (n_1, n_2, \ldots, n_m)$, and define $d_n^{(m,j)}$ through

$$A_{R_l} = d_n^{(m,j)} + \sum_{k=1}^{m} R_l^k (\Delta^k A_{R_l-1}) \sum_{i=0}^{n_k-1} \frac{\bar{\beta}_{ki}}{R_l^i}, \quad j \leq l \leq j + N; \quad N = \sum_{k=1}^{m} n_k.$$

The $d_n^{(m,j)}$ can be computed recursively via the $W^{(m)}$-algorithm (Algorithm 7.3.5) when $m > 1$.

13. The $\tilde{d}^{(m)}$-Transformation ($\tilde{d}^{(m)}$-TRAN)

Pick integers R_l such that $1 \leq R_0 < R_1 < R_2 < \cdots$, and define $\tilde{d}_n^{(m,j)} = A_n^{(j)}$ through

$$A_{R_l} = A_n^{(j)} + R_l (\Delta A_{R_l-1}) \sum_{i=0}^{n-1} \frac{\bar{\beta}_{ki}}{R_l^{i/m}}, \quad j \leq l \leq j + n.$$

The $A_n^{(j)}$ and the corresponding $\Gamma_n^{(j)}$ and $\Lambda_n^{(j)}$ can be computed recursively via the W-algorithm (Algorithm 7.2.4). We give here the steps of the resulting algorithm:
1. For $j \geq 0$, set

$$M_0^{(j)} = \frac{A_{R_j}}{R_j (\Delta A_{R_j-1})}, \quad N_0^{(j)} = \frac{1}{R_j (\Delta A_{R_j-1})}, \quad H_0^{(j)} = (-1)^j |N_0^{(j)}|, \quad K_0^{(j)} = (-1)^j |M_0^{(j)}|.$$

2. For $j \geq 0$ and $n \geq 1$, compute $M_n^{(j)}$, $N_n^{(j)}$, $H_n^{(j)}$, and $K_n^{(j)}$ recursively from

$$Q_n^{(j)} = \frac{Q_{n-1}^{(j+1)} - Q_{n-1}^{(j)}}{R_{j+n}^{-1/m} - R_j^{-1/m}},$$

where $Q_n^{(j)}$ stand for $M_n^{(j)}$ or $N_n^{(j)}$ or $H_n^{(j)}$ or $K_n^{(j)}$.
3. For $j, n \geq 0$, set

$$A_n^{(j)} = \frac{M_n^{(j)}}{N_n^{(j)}}, \quad \Gamma_n^{(j)} = \left| \frac{H_n^{(j)}}{N_n^{(j)}} \right|, \quad \Lambda_n^{(j)} = \left| \frac{K_n^{(j)}}{N_n^{(j)}} \right|.$$

14. The ρ-, $\rho(\gamma)$-, and $\rho(\gamma, p)$-Algorithms [RHO(γ, p)]

Given the scalar γ and the positive integer p, let

$$C_k^{(\gamma,p)} = -\gamma + k \text{ if } p = 1,$$

$$C_{2n}^{(\gamma,p)} = -\gamma + \frac{n}{p} \text{ and } C_{2n+1}^{(\gamma,p)} = -\gamma + \frac{n}{p} + 1 \text{ if } p \geq 2,$$

and define $\hat{\rho}_k^{(j)}$ recursively via the recursion relation

$$\hat{\rho}_{-1}^{(j)} = 0, \quad \hat{\rho}_0^{(j)} = A_j, \quad j \geq 0; \quad \hat{\rho}_{k+1}^{(j)} = \hat{\rho}_{k-1}^{(j+1)} + \frac{C_k^{(\gamma,p)}}{\hat{\rho}_k^{(j+1)} - \hat{\rho}_k^{(j)}}, \quad j, k \geq 0.$$

Here, the relevant quantities are the $\hat{\rho}_{2n}^{(j)}$. For $p \geq 2$, RHO(γ, p) is the $\rho(\gamma, p)$-algorithm (Algorithm 20.1.6), and for $p = 1$ it reduces to the $\rho(\gamma)$-algorithm (Algorithm 20.1.3). For $p = 1$ and $\gamma = -1$, RHO(γ, p) reduces to the ρ-algorithm (Algorithm 20.1.1).

15. The θ-Algorithm (THETA)

Define $\theta_k^{(j)}$ recursively via Algorithm 20.2.1; that is,

$$\theta_{-1}^{(j)} = 0, \quad \theta_0^{(j)} = A_j, \quad j \geq 0;$$

$$\theta_{2n+1}^{(j)} = \theta_{2n-1}^{(j+1)} + D_{2n}^{(j)}; \quad D_k^{(j)} = 1/\Delta\theta_k^{(j)} \quad \text{for all } j, k \geq 0,$$

$$\theta_{2n+2}^{(j)} = \theta_{2n}^{(j+1)} - \frac{\Delta\theta_{2n}^{(j+1)}}{\Delta D_{2n+1}^{(j)}} D_{2n+1}^{(j)}, \quad j, n \geq 0.$$

Here, the relevant quantities are the $\theta_{2n}^{(j)}$.

16. The Overholt Transformation (OVERH)

Let $\lambda_n^{(j)} = (\Delta A_{j+n}/\Delta A_{j+n-1})^n$, and define $A_n^{(j)}$ recursively as in (22.1.4); that is,

$$A_0^{(j)} = A_j, \quad j \geq 0; \quad A_n^{(j)} = \frac{A_{n-1}^{(j+1)} - \lambda_n^{(j)} A_{n-1}^{(j)}}{1 - \lambda_n^{(j)}}, \quad j \geq 0, \ n \geq 1.$$

17. The Wimp Transformation (WIMP)

Let $\lambda_n^{(j)} = \Delta A_{j+n}/\Delta A_j$, and define $A_n^{(j)}$ recursively as in (22.2.2); that is,

$$A_0^{(j)} = A_j, \quad j \geq 0; \quad A_n^{(j)} = \frac{A_{n-1}^{(j+1)} - \lambda_n^{(j)} A_{n-1}^{(j)}}{1 - \lambda_n^{(j)}}, \quad j \geq 0, \ n \geq 1.$$

H

Efficient Application of Sequence Transformations: Summary

In this appendix, we summarize the types of sequences $\{A_m\}$ and series $\sum_{k=1}^{\infty} a_k$ treated in this book and point to the sequence transformations appropriate for each type. Our conclusions are based on numerical comparisons carried out in double- and quadruple-precision arithmetic. It is worth emphasizing that the differences between the various methods become more pronounced in quadruple-precision arithmetic. Therefore, we urge the reader to use quadruple-precision arithmetic in comparing the methods.

For convenience, we adopt the shorthand names introduced in Appendix G. In addition, by L-TRAN and S-TRAN, we mean the $\mathcal{L}$- and $\mathcal{S}$-transformations with $\omega_r = r a_r$.

1. Exponential Sequences

Let $\{A_m\}$ be such that

$$A_m \sim A + \sum_{k=1}^{\infty} \alpha_k c_k^m \text{ as } m \to \infty,$$

where α_k and c_k are scalars independent of m and

$$c_k \text{ distinct}, \quad c_k \neq 1 \text{ for all } k; \quad |c_1| \geq |c_2| \geq \cdots; \quad \lim_{k \to \infty} c_k = 0.$$

In case the c_k are not known, the only methods that can be used to accelerate the convergence of $\{A_m\}$ are EPSILON and IDELTA. We must remember, however, that application of the latter may be problematic in some cases, as exemplified in Theorem 15.3.5.

If the c_k are known, the more appropriate method is REP, because it is less expensive than EPSILON and IDELTA.

2. Exponential Sequences with Confluence

Let $\{A_m\}$ be such that

$$A_m \sim A + \sum_{k=1}^{\infty} P_k(m) c_k^m \text{ as } m \to \infty,$$

where $P_k(m)$ are polynomials in m, and c_k are constants independent of m satisfying

$$c_k \text{ distinct}, \quad c_k \neq 1 \text{ for all } k; \quad |c_1| \geq |c_2| \geq \cdots; \quad \lim_{k \to \infty} c_k = 0.$$

In case the c_k are not known, the only method that can be used to accelerate the convergence of $\{A_m\}$ is EPSILON.

If the c_k and the degrees of the polynomials $P_k(m)$ (or some upper bounds for them), say q_k, are known, the more appropriate method is REP-CONF, because it is less costly than EPSILON.

488

3. Exponential Sequences (Cont'd)

$$A_m \sim A + \sum_{k=1}^{\infty} \alpha_k \lambda^{km} \ \text{ as } m \to \infty,$$

where α_k are constants independent of m and $|\lambda| < 1$.

If λ is not known, only OVERH, WIMP, EPSILON, and IDELTA can be used effectively. Of these, OVERH and WIMP appear to be the most effective.

If λ is known, we can also apply REP with $c_k = \lambda^k$.

4. Asymptotically Polynomial Sequences

Let $\{A_m\}$ be such that

$$A_m \sim A + \sum_{k=1}^{\infty} \alpha_k m^{-k\gamma} \ \text{ as } m \to \infty, \ \ \Re\gamma > 0,$$

and γ is known. The appropriate method for this case is REP-POL(γ).

5. Sequences in $\mathbf{b}^{(1)}/$LOG

Let $\{A_m\}$ be such that

$$A_m \sim A + \sum_{i=0}^{\infty} \alpha_i m^{\gamma-i} \ \text{ as } m \to \infty, \ \ \gamma \neq 0, 1, \ldots, \ \alpha_0 \neq 0.$$

Such sequences also arise as partial sums of infinite series $\sum_{k=1}^{\infty} a_k$, where

$$a_m \sim \sum_{i=0}^{\infty} e_i m^{\gamma-i-1} \ \text{ as } m \to \infty.$$

If γ is not known, the methods appropriate for such sequences are ILUBKIN, THETA, and L-TRAN. Here L-TRAN appears to be the most effective. When γ is known, we can also apply IMDELTA$(\gamma, 1)$ and RHO$(\gamma, 1)$.

When approximations with highest possible accuracy in finite-precision arithmetic are desired, $d^{(1)}$-TRAN in the GPS mode, that is, with R_l chosen as in

$$R_0 \geq 1, \ \ R_l = \begin{cases} R_{l-1} + 1 \text{ if } \lfloor \sigma R_{l-1} \rfloor = R_{l-1}, \\ \lfloor \sigma R_{l-1} \rfloor \text{ otherwise,} \end{cases} \ \ l = 1, 2, \ldots; \ \ \sigma > 1,$$

produces the best results.

6. Sequences in $\mathbf{b}^{(1)}/$LIN

Let $\{A_m\}$ be such that

$$A_m \sim A + \zeta^m \sum_{i=0}^{\infty} \alpha_i m^{\gamma-i} \ \text{ as } m \to \infty, \ \ \zeta \neq 1, \ \alpha_0 \neq 0.$$

Such sequences also arise as partial sums of infinite series $\sum_{k=1}^{\infty} a_k$, where

$$a_m \sim \zeta^m \sum_{i=0}^{\infty} e_i m^{\gamma-i} \ \text{ as } m \to \infty.$$

The methods appropriate for such sequences are IDELTA, ILUBKIN, EPSILON, THETA, L-TRAN, and S-TRAN. Of these, the last two appear to be the most effective.

When approximations with highest possible accuracy in finite-precision arithmetic are desired, all these methods can be applied in the APS mode; that is, they can be applied to the subsequences $\{A_{\kappa m+\eta}\}$, where $\kappa \geq 1$ and $\eta \geq 0$ are appropriate integers. For the same purpose, $d^{(1)}$-TRAN in the APS mode, that is, with $R_l = \kappa(l+1)$, $\kappa \geq 1$ integer, produces the best results. For $d^{(1)}$-TRAN, it is not necessary to choose κ as an integer, however. Excellent results are produced in the APS mode also when $\kappa \geq 1$ is not necessarily an integer, and the R_l are chosen as follows:

$$R_l = \lfloor \kappa(l+1) \rfloor, \quad l = 0, 1, \ldots .$$

7. Sequences in $\mathbf{b}^{(1)}$/FAC

Let $\{A_m\}$ be such that

$$A_m \sim A + (m!)^{-r} \zeta^m \sum_{i=0}^{\infty} \alpha_i m^{\gamma-i} \quad \text{as } m \to \infty, \quad r > 0 \text{ integer}, \quad \alpha_0 \neq 0.$$

Such sequences also arise as partial sums of infinite series $\sum_{k=1}^{\infty} a_k$, where

$$a_m \sim (m!)^{-r} \zeta^m \sum_{i=0}^{\infty} e_i m^{\gamma-i+r} \quad \text{as } m \to \infty.$$

The methods appropriate for such sequences are IDELTA, ILUBKIN, EPSILON, THETA, L-TRAN, and S-TRAN. Of these, the last two appear to be the most effective.

8. Factorially Divergent Sequences in $\mathbf{b}^{(1)}$/FACD

Let $\{A_m\}$ be the sequence of the partial sums of the infinite series $\sum_{k=1}^{\infty} a_k$, where

$$a_m = (m!)^r \zeta^m h(m), \quad h(m) \sim \sum_{i=0}^{\infty} e_i m^{\gamma-i} \quad \text{as } m \to \infty, \quad r > 0 \text{ integer}, \quad e_0 \neq 0.$$

Then, $\sum_{k=1}^{\infty} a_k$ is divergent, and

$$A_m \sim (m!)^r \zeta^m \sum_{i=0}^{\infty} \alpha_i m^{\gamma-i} \quad \text{as } m \to \infty.$$

When $h(m)$ is independent of ζ and $h(m) = m^\omega \int_0^\infty e^{-mt} \varphi(t)\, dt$ for some integer $\omega \geq 0$ and some $\varphi(t)$ of exponential order, the divergent series $\sum_{k=1}^{\infty} a_k$ has a (generalized) Borel sum, which, as a function of ζ, is analytic in the ζ-plane cut along the real interval $[0, +\infty)$. In such a case, IDELTA, ILUBKIN, EPSILON, THETA, L-TRAN, and S-TRAN can be applied to produce approximations to the (generalized) Borel sum. Of these transformations, S-TRAN appears to produce the best results and is followed by L-TRAN.

9. Sums of Sequences in $\mathbf{b}^{(1)}$/LOG/LIN/FAC

When $\{A_m\}$ is the sum of different sequences in $\mathbf{b}^{(1)}$/LOG, $\mathbf{b}^{(1)}$/LIN, and $\mathbf{b}^{(1)}$/FAC, the only appropriate methods that produce good results appear to be $d^{(p)}$-TRAN with some $p > 1$ always and EPSILON in certain cases. (IDELTA, ILUBKIN, THETA, L-TRAN, and S-TRAN do not produce any acceleration in general.)

Important special cases of such sequences are

(i) $A_m = A + \displaystyle\sum_{k=1}^{p} h_k(m), \quad h_k(m) \in \mathbf{A}_0^{(\gamma_k)}, \quad \gamma_k \text{ distinct}, \quad \gamma_k \neq 0, 1, \ldots ,$

(ii) $A_m = A + \displaystyle\sum_{k=1}^{p} \zeta_k^m h_k(m), \quad h_k(m) \in \mathbf{A}_0^{(\gamma_k)}, \quad \zeta_k \text{ distinct}, \quad \zeta_k \neq 1.$

EPSILON is effective in case (ii) but does not produce any acceleration in case (i). It is also not effective when $\{A_m\}$ is a sum of sequences from both $\mathbf{b}^{(1)}$/LOG and $\mathbf{b}^{(1)}$/LIN.

The APS and GPS strategies can be used effectively wherever necessary. Highest possible accuracy for sequences described in case (i), for example, can be achieved only by $d^{(p)}$-TRAN with GPS.

Now, sequences $\{A_m\}$ for which

$$A_m \sim A + \sum_{i=0}^{\infty} \alpha_i m^{\gamma - i/p} \quad \text{as } m \to \infty, \quad \gamma \neq \frac{i}{p}, \quad i = 0, 1, \ldots, \quad p > 0 \text{ integer}, \quad \alpha_0 \neq 0,$$

are special instances of case (i) with $\gamma_k = \gamma - (k-1)/p$, $k = 1, \ldots, p$. Such sequences also arise as partial sums of infinite series $\sum_{k=1}^{\infty} a_k$, where

$$a_m \sim \sum_{i=0}^{\infty} e_i m^{\gamma - 1 - i/p} \quad \text{as } m \to \infty.$$

The sequence $\{a_m\}$ is also in $\tilde{\mathbf{b}}^{(p)}$, which is discussed in the following. For these sequences, ILUBKIN and THETA are also effective. When p is known, $\tilde{d}^{(p)}$-TRAN is effective as well. When γ as well as p are known, IMDELTA(γ, p) and RHO(γ, p) are successful too. Highest possible accuracy for such sequences too can be achieved only by $d^{(p)}$-TRAN and $\tilde{d}^{(p)}$-TRAN with GPS.

10. Series $\sum_{k=1}^{\infty} a_k$ with $\{a_k\} \in \tilde{\mathbf{b}}^{(p)}$

The most appropriate methods for accelerating the convergence of the series $\sum_{k=1}^{\infty} a_k$ with $\{a_k\} \in \tilde{\mathbf{b}}^{(p)}$, $p > 1$, that is, with

$$a_m \sim (m!)^{r/p} e^{Q(m)} \sum_{i=0}^{\infty} e_i m^{\gamma - i/p} \quad \text{as } m \to \infty, \quad r \text{ integer}, \quad Q(m) = \sum_{i=0}^{p-1} \theta_i m^{1-i/p},$$

seem to be $\tilde{d}^{(p)}$-TRAN and $d^{(p)}$-TRAN. When (i) $r < 0$, or (ii) $r = 0$ and $\lim_{m\to\infty} \Re Q(m) < \infty$, or (iii) $r = 0$ and $Q(m) \equiv 0$ and $\gamma \neq -1 + i/p$, $i = 0, 1, \ldots$, these a_m give rise to partial sums A_m that satisfy

$$A_m \sim A + (m!)^{r/p} e^{Q(m)} \sum_{i=0}^{\infty} \alpha_i m^{\gamma + 1 - i/p} \quad \text{as } m \to \infty.$$

11. Series $\sum_{k=1}^{\infty} a_k$ with $\{a_k\} \in \mathbf{b}^{(p)}$

The only methods that are useful in accelerating the convergence of the series $\sum_{k=1}^{\infty} a_k$ with $\{a_k\} \in \mathbf{b}^{(p)}$, $p > 1$, appear to be $d^{(p)}$-TRAN always and EPSILON in certain cases.

In many instances, a_m can be shown to be the sum of products of terms that form sequences in $\mathbf{b}^{(p_i)}$ for various values of p_i. Heuristics 6.4.1–6.4.3 can be used to determine the p_i and p conveniently.

12. Power Series $\sum_{k=1}^{\infty} c_k z^{k-1}$ with $\{c_k\} \in \mathbf{b}^{(p)}$

The only methods that are useful in accelerating the convergence of the series $\sum_{k=1}^{\infty} c_k z^{k-1}$ with $\{c_k\} \in \mathbf{b}^{(p)}$, $p > 1$, appear to be $d^{(p)}$-TRAN (i.e., rational d-approximants) and EPSILON (i.e., Padé approximants).

IDELTA, ILUBKIN, THETA, L-TRAN, and S-TRAN do not produce any acceleration when $p > 1$, in general. They can be used when $p = 1$.

APS can be used effectively wherever necessary.

13. (Generalized) Fourier Series and Series of Special Functions

Series of the form $\sum_{k=1}^{\infty} c_k \phi_k(x)$, where $\{c_k\} \in \mathbf{b}^{(p)}$ for some p, and $\{\phi_k(x)\}$ is a sequence of trigonometric functions [such as $\phi_k(x) = \sin kx$] or of special functions [such as $\phi_k(x) = P_k(x)$, Legendre polynomials], which is in $\mathbf{b}^{(2)}$ in most cases of interest, can be treated directly by $d^{(2p)}$-TRAN and EPSILON. Using the complex series approach of Chapter 13, they can be treated at half the cost by $d^{(p)}$-TRAN and EPSILON. (This cost can be reduced further by using the extended complex series approach when possible.)

IDELTA, ILUBKIN, THETA, L-TRAN, and S-TRAN do not produce any acceleration when $p > 1$, in general. When $p = 1$, they can be used only with the complex series approach.

APS can be used effectively wherever necessary.

When $\phi_k(x)$ are orthogonal polynomials, the special-purpose methods of Chapter 18, such as Padé approximants and Baker–Gammel approximants from orthogonal polynomial expansions, and their generalizations, can also be used. Their use may be computationally more involved, however.

Concluding Remarks on Sequence Transformations

We end by collecting some of the information about those sequence transformations that require no information except the elements of the sequences to be transformed. These are $d^{(m)}$-TRAN, $\tilde{d}^{(m)}$-TRAN, with $m = 1, 2, \ldots$, L-TRAN, S-TRAN, EPSILON, RHO($-1, 1$), THETA, IDELTA, ILUBKIN, OVERH, and WIMP.

1. For $\{A_m\} \in \mathbf{b}^{(1)}$/LOG: L-TRAN is the best, followed by ILUBKIN and THETA, while S-TRAN is inferior. RHO($-1, 1$) is effective only for sequences that satisfy $\{\Delta A_m\} \in \mathbf{A}_0^{(-2)}$. The rest of the methods are useless. For such sequences, all relevant methods are numerically unstable. For numerically stable results that have very high accuracy, $d^{(1)}$-TRAN with GPS should be used.

2. For $\{A_m\} \in \mathbf{b}^{(1)}$/LIN/FAC: L-TRAN and S-TRAN are the best, followed by THETA and ILUBKIN, which are followed by EPSILON and IDELTA. In case $\{A_m\} \in \mathbf{b}^{(1)}$/LIN, all these methods are numerically unstable near points of singularity of the limit or antilimit. In such situations, $d^{(1)}$-TRAN with APS should be used or L-TRAN and S-TRAN should be applied to some subsequence $\{A_{\kappa m + \eta}\}$, where κ and η are integers.

3. For $\{A_m\} \in \mathbf{b}^{(1)}$/FACD: S-TRAN is the best, followed by L-TRAN. THETA, ILUBKIN, EPSILON, and IDELTA are inferior.

4. For $\{\Delta A_m\} \in \tilde{\mathbf{b}}^{(p)}$, $\lim_{m \to \infty} \Delta A_{m+1}/\Delta A_m = 1$, $p \geq 2$: $d^{(p)}$-TRAN and $\tilde{d}^{(p)}$-TRAN are the best, followed by THETA and ILUBKIN. The other methods are useless. Again, numerically stable and high-accuracy approximations are obtained from $d^{(p)}$-TRAN and $\tilde{d}^{(p)}$-TRAN with GPS.

5. For $\{A_m\}$, with $A_m = \sum_{k=1}^{p} H_m^{(k)}$, $\{H_m^{(k)}\} \in \mathbf{b}^{(1)}$/LOG, $p \geq 2$: $d^{(p)}$-TRAN is the only useful method.

6. For $\{A_m\}$, with $A_m = \sum_{k=1}^{p} H_m^{(k)}$, $\{H_m^{(k)}\} \in \mathbf{b}^{(1)}$/LIN, $p \geq 2$: $d^{(p)}$-TRAN is the best, followed by EPSILON. The other methods are useless.

7. For general $\{\Delta A_m\} \in \mathbf{b}^{(p)}$, $p \geq 2$: $d^{(p)}$-TRAN is the only useful method.

8. For general exponential sequences (with or without confluence): EPSILON is the only useful method. For exponential sequences for which $A_m \sim A + \sum_{k=1}^{\infty} \alpha_k \lambda^{km}$, $|\lambda| < 1$, such as those arising from the fixed-point iterative solution of nonlinear equations, WIMP and OVERH may be preferable. For all types of exponential sequences, the remaining methods are useless.

It is clear that, to accelerate the convergence of the different classes of sequences mentioned here, it is sufficient to have available the methods $d^{(m)}$-TRAN, $\tilde{d}^{(m)}$-TRAN, with $m = 1, 2, \ldots$, EPSILON, L-TRAN, S-TRAN, and OVERH or WIMP. Recall also that L-TRAN (as the u-transformation) is already $d^{(1)}$-TRAN (and also $\tilde{d}^{(1)}$-TRAN), with $R_l = l + 1$.

When additional quantitative information on the nature of $\{A_m\}$ is available, provided they are applicable, REP, REP-CONF, GREP$^{(m)}$, and REP-POL(γ) are more economical than others.

I

FORTRAN 77 Program for the $d^{(m)}$-Transformation

I.1 General Description

In this appendix, we provide a FORTRAN 77 code that implements the $d^{(m)}$-transformation on *real* infinite series of the form $\sum_{k=1}^{\infty} a_k$ via the $W^{(m)}$-algorithm. After modifying the type declarations in a suitable fashion, this code can be adapted to *complex* series as well. Here we use the notation and terminology of Section 7.3.

The most important part of the code we give here is SUBROUTINE WMALG, which implements the $W^{(m)}$-algorithm with the normal ordering of the $\varphi_k(t)t^i$. This subroutine is exactly the same as that given originally in Ford and Sidi [87, Appendix B] and can be included in any implementation of GREP$^{(m)}$, such as the $D^{(m)}$-, $\bar{D}^{(m)}$-, and $\tilde{D}^{(m)}$-transformations. (Recall that, when $m = 1$, the $W^{(m)}$-algorithm becomes the W-algorithm, as mentioned in Chapter 7. Recall also that, when $m = 1$ and $R_l = l + 1$, $l = 0, 1, \ldots$, the $d^{(m)}$-transformation reduces to the Levin u-transformation.) SUBROUTINE WMALG produces the approximations APPROX(J,P), $0 \leq$ J+P $\leq$ LMAX, where APPROX(J,P)=$A_{\mathrm{P}}^{(\mathrm{J})}$.

Computation of the R_l and $g_k(l)$ is carried out in SUBROUTINE MLTAG, which is somewhat different from that of Ford and Sidi [87, Appendix B] with regard to how the $\varphi_k(t)$ are ordered and how the R_l are prescribed:

- To avoid (as much as possible) division by vanishing $g_1(l)$, which may arise when some of the a_k may be zero (this may happen with Fourier series, for example), the $\varphi_k(t)$ are ordered as follows:

$$\varphi_k(t) = n^{m-k+1}\Delta^{m-k}a_n, \quad k = 1, \ldots, m.$$

Thus, $\varphi_1(t) = n^m \Delta^{m-1}a_n$, with $t = 1/n$.
- The R_l in the $d^{(m)}$-transformation are assigned in one of the following two ways:
  1. **Arithmetic Progression Sampling (APS).** For some $\kappa \geq 1$, set

$$R_l = \lfloor \kappa(l + 1) \rfloor, \quad l = 0, 1, \ldots . \tag{I.1.1}$$

  2. **Geometric Progression Sampling (GPS).** For some $\sigma > 1$, set

$$R_0 = 1, \quad R_l = \begin{cases} R_{l-1} + 1 & \text{if } \lfloor \sigma R_{l-1} \rfloor = R_{l-1}, \\ \lfloor \sigma R_{l-1} \rfloor & \text{otherwise,} \end{cases} \quad l = 1, 2, \ldots . \tag{I.1.2}$$

Finally, the elements a_k, $k = 1, 2, \ldots$, of the infinite series are computed by FUNCTION CF via $a_k = $ CF(K).

The user is to supply the parameters m, κ, and σ. In the computer program, we have $m = $ M, $\kappa = $ KAPPA, and $\sigma = $ SIGMA. We note that, to invoke APS via (I.1.1), SIGMA $= 1$ must be prescribed. When invoking GPS via (I.1.2), KAPPA can be assigned any value.

The computer code we give here has in it the following convergent series as test cases, with the appropriate m, κ, and σ.

(1) $\sum_{n=1}^{\infty} \frac{1}{n^2} = \frac{\pi^2}{6}$ $m = 1$

(2) $\sum_{n=1}^{\infty} \left(\frac{1}{n^{3/2}} + \frac{1}{n^2}\right) = 4.2573094155337147798209827345 7009 \cdots$ $m = 2$

(3) $\sum_{n=1}^{\infty} \frac{\cos n\theta}{n} = -\log\left|2 \sin \frac{\theta}{2}\right|$ $m = 2$

(4) $\sum_{n=1}^{\infty} \frac{(-1)^{n-1}}{n} = \log 2$ $m = 1$

(5) $\sum_{n=1}^{\infty} \frac{(-1)^{\lfloor (n-1)/2 \rfloor}}{n} = S_2$ $m = 2$

(6) $\sum_{n=1}^{\infty} \frac{(-1)^{\lfloor (n-1)/3 \rfloor}}{n} = S_3$ $m = 3$

(7) $\sum_{n=1}^{\infty} \frac{(-1)^{\lfloor (n-1)/4 \rfloor}}{n} = S_4$ $m = 4$

(8) $\sum_{n=1}^{\infty} (c_n - c_{n+1}) = c_1 = \frac{e^{-0.1}}{2}; \quad c_n = \frac{\exp(-0.1n^{1/2})}{1+n^{1/2}}$ $m = 2$

(9) $\sum_{n=1}^{\infty} (c_n - c_{n+1}) = c_1 = \frac{e^{0.1}}{2}; \quad c_n = \frac{\exp(-0.1n^{2/3}+0.2n^{1/3})}{1+n^{1/3}}$ $m = 3$

(10) $-\frac{1}{2} + 1 - \frac{1}{4} + \frac{1}{3} - \frac{1}{6} + \frac{1}{5} - \cdots = \log 2$ $m = 2$

(11) $1 + \frac{1}{3} - \frac{1}{2} + \frac{1}{5} + \frac{1}{7} - \frac{1}{4} + \cdots = \frac{3}{2}\log 2$ $m = 3$

(12) $1 + \frac{1}{3} + \frac{1}{5} - \frac{1}{2} - \frac{1}{4} + \frac{1}{7} + \frac{1}{9} + \frac{1}{11} - \frac{1}{6} - \frac{1}{8} + \cdots = \frac{1}{2}\log 6.$ $m = 5$

In series (3), the value of θ should be provided by the user through the variable THETA in the main program.

In the series (5)–(7), S_2, S_3, and S_4 are special cases of (25.9.1), namely,

$$S_q = \sum_{k=0}^{\infty} \frac{(-1)^{\lfloor k/q \rfloor}}{k+1} = \frac{1}{q}\log 2 + \frac{\pi}{2q}\sum_{k=1}^{q-1} \tan \frac{k\pi}{2q}, \quad q = 1, 2, \ldots.$$

The series (10)–(12) are rearrangements of series (4), which converges conditionally.

For gaining experience with the use of the d-transformation, we urge the reader to run this code in both double- and quadruple-precision arithmetic, with the values of KAPPA and SIGMA as recommended in the main program and with other values as well. (As the code is already in double precision, it can be run in quadruple precision without any changes.)

Following these series, we suggest that the reader apply the code to other series that diverge, such as

(13) $\sum_{n=1}^{\infty} n^{1/2}$ antilimit: $\zeta(-\frac{1}{2})$ $m = 1$

(14) $\sum_{n=1}^{\infty} (-1)^{n-1} n^{1/2}$ antilimit: $(1 - 2^{3/2})\zeta(-\frac{1}{2})$ $m = 1$

(15) $\sum_{n=0}^{\infty} \frac{(2n+1)(s)_n}{2^s (1-s)_{n+1}} P_n(x), \quad s \geq \frac{3}{4}$ antilimit: $(1 - x)^{-s}$ $m = 2$

In series (15), $P_n(x)$ are the Legendre polynomials and, as before,

$$(u)_0 = 1, \quad (u)_n = u(u + 1) \cdots (u + n - 1), \quad n = 1, 2, \ldots.$$

For $s < 3/4$, this series converges to $(1 - x)^{-s}$ provided $x \in [-1, 1)$. For all other x in the complex plane cut along $[1, +\infty)$, it diverges for every s with antilimit $(1 - x)^{-s}$. See Davis [62, p. 327, Exercise 16].

I.2 The Code

```
CCCCCCCCCCCCCCCCCCCCCCCCCCCCCCCCCCCCCCCCCCCCCCCCCCCCCCCCCCCCCCCCCCCCCCCCCCC
C     THIS PROGRAM APPLIES THE LEVIN-SIDI d^(m)-TRANSFORMATION TO INFINITE
C     SERIES (SUM OF CF(I) FOR I=1 UP TO I=INFINITY) OF VARIOUS FORMS.
C
C     THE d^(m)-TRANSFORMATION IS IMPLEMENTED BY THE W^(m)-ALGORITHM OF
C     FORD AND SIDI.
C
C     THE FOLLOWING ARE NEEDED AS INPUT AND SHOULD BE PROVIDED BY THE
C     USER VIA THE PARAMETER STATEMENTS: NP, M, LMAX, POSITIVE INTEGERS,
C     KAPPA, SIGMA, DOUBLE PRECISION CONSTANTS .GE. 1D0.
```

```
C     WHEN NP=3 THETA SHOULD ALSO BE PROVIDED.
C
C     THE USER IS TO PROVIDE CF(I) AS A FUNCTION SUBPROGRAM.
C
C     MDIM.GE.M AND LDIM.GE.LMAX SHOULD ALWAYS BE SATISFIED. MDIM AND
C     LDIM CAN BE INCREASED BY THE USER TO ACCOMMODATE LARGE VALUES OF
C     M AND LMAX.
CCCCCCCCCCCCCCCCCCCCCCCCCCCCCCCCCCCCCCCCCCCCCCCCCCCCCCCCCCCCCCCCCCCCCCCC
C     THE FOLLOWING CASES ARE INCLUDED IN THE PRESENT CF(I):
C     IF NP=1,  THEN M=1, KAPPA=1D0, SIGMA=1.3D0
C     IF NP=2,  THEN M=2, KAPPA=1D0, SIGMA=1.3D0
C     IF NP=3,  THEN M=2, KAPPA=?, SIGMA=1D0  (THIS IS A FOURIER SERIES.)
C               (KAPPA DEPENDS ON X=ABS(THETA), WHERE THETA IS THE ANGLE.)
C               (FOR X LARGE KAPPA=1D0 IS ENOUGH.)
C               (FOR X SMALL TRY KAPPA>1D0)
C               (E.G., FOR X=PI/3, KAPPA=1D0)
C               (E.G., FOR X=PI/6, KAPPA=2D0)
C               (E.G., FOR X=PI/50, KAPPA=20D0)
C     IF NP=4,  THEN M=1, KAPPA=1D0, SIGMA=1D0
C     IF NP=5,  THEN M=2, KAPPA=1D0, SIGMA=1D0
C     IF NP=6,  THEN M=3, KAPPA=1D0, SIGMA=1D0
C     IF NP=7,  THEN M=4, KAPPA=1D0, SIGMA=1D0
C     IF NP=8,  THEN M=2, KAPPA=1D0, SIGMA=1.15D0
C     IF NP=9,  THEN M=3, KAPPA=1D0, SIGMA=1.15D0
C     IF NP=10, THEN M=2, KAPPA=1D0, SIGMA=1D0
C     IF NP=11, THEN M=3, KAPPA=2D0, SIGMA=1D0
C     IF NP=12, THEN M=5, KAPPA=4D0, SIGMA=1D0
CCCCCCCCCCCCCCCCCCCCCCCCCCCCCCCCCCCCCCCCCCCCCCCCCCCCCCCCCCCCCCCCCCCCCCCC
C     NOTE THAT IRL(L)=R_L , L=0,1,...,  OF THE d-TRANSFORMATION, AND
C     APPROX(J,P) IS THE APPROXIMATION RETURNED BY THE d^(M)-TRANSFORMATION
C     THAT IS OBTAINED FROM THE PARTIAL SUMS A_{R_0}, A_{R_1},...,
C     A_{R_{J+P}} OF THE GIVEN SERIES.
C
C     IN THE OUTPUT ERROR(L,0) IS THE ABSOLUTE ERROR IN THE PARTIAL SUM
C     A_{R_L}, WHILE ERROR(0,L) IS THE ABSOLUTE ERROR IN APPROX(0,L).
C     THE EXACT SUM OF THE SERIES IS CONTAINED IN THE VARIABLE RESULT.
CCCCCCCCCCCCCCCCCCCCCCCCCCCCCCCCCCCCCCCCCCCCCCCCCCCCCCCCCCCCCCCCCCCCCCCC
      IMPLICIT DOUBLE PRECISION (A-H,O-Z)
      DOUBLE PRECISION KAPPA,KAPPAP
      PARAMETER (MDIM=6,LDIM=120,EPSDIV=1D-77)
      PARAMETER (NP=12,M=5,LMAX=100,KAPPA=4D0,SIGMA=1.0D0)
      DIMENSION G(MDIM),PSIAI(0:LDIM,2,2),BIGPSI(0:LDIM,MDIM,2)
      DIMENSION PSIG(0:MDIM,2:MDIM+1,2),APPROX(0:LDIM,0:LDIM)
      EXTERNAL MLTAG
      COMMON /SIGKAP/SIGMAP,KAPPAP
      COMMON /NP/NPP/THETA/THETA
      COMMON /RL/IRL(0:LDIM)
      WRITE(6,111)
 111  FORMAT('SUMMATION OF CF(I) FROM I=1 TO I=INFINITY',//)
      NPP=NP
      PI=DACOS(-1D0)
      THETA=PI/50D0
      SIGMAP=SIGMA
      KAPPAP=KAPPA
      CALL EXACT(RESULT)
      CALL WMALGM(MDIM,LDIM,M,LMAX,MLTAG,G,PSIAI,BIGPSI,PSIG,
     *            APPROX,EPSDIV)
      WRITE(6,121)
```

```
121   FORMAT(/,3X,'L',5X,'R_L',7X,'ERROR(L,0)',5X,'ERROR(0,L)')
      DO 20 L=0,LMAX
      ER1=DABS(APPROX(L,0)-RESULT)
      ER2=DABS(APPROX(0,L)-RESULT)
      WRITE(6,131) L,IRL(L),ER1,ER2
 20   CONTINUE
131   FORMAT(I4,2X,I6,2X,1P,2D15.3)
      WRITE (6,*) 'EXACT SUM =',RESULT
      STOP
      END

      SUBROUTINE WMALGM(MDIM,LDIM,M,LMAX,MLTAG,G,PSIAI,BIGPSI,PSIG,
     *                  APPROX,EPSDIV)
CCCCCCCCCCCCCCCCCCCCCCCCCCCCCCCCCCCCCCCCCCCCCCCCCCCCCCCCCCCCCCCCCCCCCCCC
C     THIS SUBROUTINE GIVES THE IMPLEMENTATION OF SIDI'S GENERALIZED
C     RICHARDSON EXTRAPOLATION PROCESS GREP^(m) VIA THE W^(m)-ALGORITHM
C     OF FORD AND SIDI.
C
C     THE APPROXIMATIONS TO THE LIMIT OR ANTILIMIT OF THE SEQUENCE IN
C     QUESTION ARE CONTAINED IN THE ARRAY APPROX. IN PARTICULAR,
C     APPROX(0,L), L=0,1,..., SEEM TO BE THE BEST IN GENERAL.
CCCCCCCCCCCCCCCCCCCCCCCCCCCCCCCCCCCCCCCCCCCCCCCCCCCCCCCCCCCCCCCCCCCCCCCC
      IMPLICIT DOUBLE PRECISION(A-H,O-Z)
      INTEGER CUR,TEMP,P,PM,Q,QP
      DIMENSION G(MDIM),PSIAI(0:LDIM,2,2)
      DIMENSION BIGPSI(0:LDIM,MDIM,2),PSIG(0:MDIM,2:MDIM+1,2)
      DIMENSION APPROX(0:LDIM,0:LDIM)
      CUR=1
      TEMP=2
      CALL MLTAG(M,0,T,A,G)
      APPROX(0,0)=A
      PSIAI(0,1,CUR)=A/G(1)
      PSIAI(0,2,CUR)=1D0/G(1)
      BIGPSI(0,1,CUR)=1D0/T
      DO 10 K=2,M
      PSIG(0,K,CUR)=G(K)/G(1)
 10   CONTINUE
      PSIG(0,M+1,CUR)=T
      DO 80 L=1,LMAX
      CALL MLTAG(M,L,T,A,G)
      APPROX(L,0)=A
      PSIAI(0,1,TEMP)=A/G(1)
      PSIAI(0,2,TEMP)=1D0/G(1)
      BIGPSI(0,1,TEMP)=1D0/T
      DO 20 K=2,M
      PSIG(0,K,TEMP)=G(K)/G(1)
 20   CONTINUE
      PSIG(0,M+1,TEMP)=T
      SIGN=-1D0
      DO 60 P=1,L
      IF (P.LE.M) THEN
         D=PSIG(P-1,P+1,TEMP)-PSIG(P-1,P+1,CUR)
         DO 30 I=P+2,M+1
         PSIG(P,I,TEMP)=(PSIG(P-1,I,TEMP)-PSIG(P-1,I,CUR))/D
 30      CONTINUE
      END IF
      IF (P.LT.M) THEN
         BIGPSI(P,P+1,TEMP)=SIGN/PSIG(P,M+1,TEMP)
```

```
      SIGN=-SIGN
      END IF
      PM=MINO(P-1,M-1)
      DO 40 Q=1,PM
      PS=BIGPSI(P-2,Q,CUR)
      DQ=PS/BIGPSI(P-1,Q,CUR)-PS/BIGPSI(P-1,Q,TEMP)
      QP=Q+1
      BIGPSI(P,QP,TEMP)=(BIGPSI(P-1,QP,TEMP)-BIGPSI(P-1,QP,CUR))/DQ
40    CONTINUE
      IF (P.GT.M) THEN
         PS=BIGPSI(P-2,M,CUR)
         D=PS/BIGPSI(P-1,M,CUR)-PS/BIGPSI(P-1,M,TEMP)
      END IF
      BIGPSI(P,1,TEMP)=(BIGPSI(P-1,1,TEMP)-BIGPSI(P-1,1,CUR))/D
      DO 50 I=1,2
      PSIAI(P,I,TEMP)=(PSIAI(P-1,I,TEMP)-PSIAI(P-1,I,CUR))/D
50    CONTINUE
60    CONTINUE
      DO 70 P=1,L
      J=L-P
      IF (DABS(PSIAI(P,2,TEMP)).GE.EPSDIV) THEN
         APPROX(J,P)=PSIAI(P,1,TEMP)/PSIAI(P,2,TEMP)
      ELSE
         APPROX(J,P)=1D75
         WRITE(6,101)J,P
101      FORMAT(1X,'APPROX(',I3,',',I3,') IS NOT DEFINED')
      END IF
70    CONTINUE
      JJ=CUR
      CUR=TEMP
      TEMP=JJ
80    CONTINUE
      RETURN
      END


      SUBROUTINE MLTAG(M,L,T,A,G)
CCCCCCCCCCCCCCCCCCCCCCCCCCCCCCCCCCCCCCCCCCCCCCCCCCCCCCCCCCCCCCCCCCCCCCC
C     THIS SUBROUTINE IS CALLED BY SUBROUTINE WMALGM AND PROVIDES THE
C     LATTER WITH THE NECESSARY INPUT FOR THE d-TRANSFORMATION.
C     THE CONSTANT LDIM IN THE PARAMETER STATEMENT BELOW MUST BE THE SAME
C     AS THAT IN THE MAIN PROGRAM.
CCCCCCCCCCCCCCCCCCCCCCCCCCCCCCCCCCCCCCCCCCCCCCCCCCCCCCCCCCCCCCCCCCCCCCC
      IMPLICIT DOUBLE PRECISION(A-H,O-Z)
      DOUBLE PRECISION KAPPA
      PARAMETER (LDIM=120)
      DIMENSION G(M)
      COMMON /SIGKAP/SIGMA,KAPPA
      COMMON /RL/IRL(0:LDIM)
      IF (SIGMA.EQ.1D0) THEN
         LSUMP=KAPPA*L+1D-10
         LSUM=KAPPA*(L+1)+1D-10
      END IF
      IF (SIGMA.GT.1D0) THEN
         IF (L.EQ.0) THEN
            LSUMP=0
            LSUM=1
         ELSE
            LSUM=1
```

```
            LSUMP=1
            DO 10 I=1,L
            IR=SIGMA*LSUM+1D-10
            IF (IR.LE.LSUM) THEN
                LSUM=LSUM+1
            ELSE
                LSUM=IR
            END IF
            IF (I.EQ.L-1) LSUMP=LSUM
10          CONTINUE
         END IF
      END IF
      IRL(L)=LSUM
      IF (L.EQ.0) A=0
      DO 20 I=LSUMP+1,LSUM
      A=A+CF(I)
20    CONTINUE
      P=LSUM
      T=1D0/P
      DO 30 K=1,M
      G(K)=CF(LSUM+K-1)
30    CONTINUE
      DO 50 I=2,M
      DO 40 J=M,I,-1
      G(J)=G(J)-G(J-1)
40    CONTINUE
50    CONTINUE
      DO 60 K=1,M
      G(K)=G(K)*P
      P=P*LSUM
60    CONTINUE
      DO 70 K=1,M/2
      ST=G(K)
      G(K)=G(M-K+1)
      G(M-K+1)=ST
70    CONTINUE
      RETURN
      END


      FUNCTION CF(I)
CCCCCCCCCCCCCCCCCCCCCCCCCCCCCCCCCCCCCCCCCCCCCCCCCCCCCCCCCCCCCCCCCCCCCCCC
C     CF(I) IS THE I-TH TERM OF THE INFINITE SERIES, I=1,2,...
CCCCCCCCCCCCCCCCCCCCCCCCCCCCCCCCCCCCCCCCCCCCCCCCCCCCCCCCCCCCCCCCCCCCCCCC
      IMPLICIT DOUBLE PRECISION (A-H,O-Z)
      COMMON /NP/NP/THETA/THETA
      UU(X)=DEXP(-0.1D0*DSQRT(X))/(1+DSQRT(X))
      VV(X)=DEXP(-0.1D0*X**(2D0/3D0)+0.2D0*X**(1D0/3D0))
     *      /(1+X**(1D0/3D0))
      FI=DFLOAT(I)
      II=I-1
      IF (NP.EQ.1) CF=1D0/(FI)**2
      IF (NP.EQ.2) CF=1D0/(FI)**1.5D0+1D0/(FI)**2
      IF (NP.EQ.3) CF=DCOS((FI)*THETA)/(FI)
      IF (NP.EQ.4) CF=(-1)**(II)/(FI)
      IF (NP.EQ.5) CF=(-1)**(II/2)/(FI)
      IF (NP.EQ.6) CF=(-1)**(II/3)/(FI)
      IF (NP.EQ.7) CF=(-1)**(II/4)/(FI)
      IF (NP.EQ.8) CF=UU(FI)-UU(FI+1)
```

```
      IF (NP.EQ.9) CF=VV(FI)-VV(FI+1)
      IF (NP.EQ.10) THEN
         IF (I-2*(I/2).EQ.1) CF=-1D0/(I+1)
         IF (I-2*(I/2).EQ.0) CF=1D0/(I-1)
      END IF
      IF (NP.EQ.11) THEN
         IF (I-3*(I/3).EQ.1) CF=3D0/(4*I-1)
         IF (I-3*(I/3).EQ.2) CF=3D0/(4*I+1)
         IF (I-3*(I/3).EQ.0) CF=-3D0/(2*I)
      END IF
      IF (NP.EQ.12) THEN
         IF (I-5*(I/5).EQ.1) CF=5D0/(6*I-1)
         IF (I-5*(I/5).EQ.2) CF=5D0/(6*I+3)
         IF (I-5*(I/5).EQ.3) CF=5D0/(6*I+7)
         IF (I-5*(I/5).EQ.4) CF=-5D0/(4*I-6)
         IF (I-5*(I/5).EQ.0) CF=-5D0/(4*I)
      END IF
      RETURN
      END


      SUBROUTINE EXACT(RESULT)
CCCCCCCCCCCCCCCCCCCCCCCCCCCCCCCCCCCCCCCCCCCCCCCCCCCCCCCCCCCCCCCCCCCCCC
C     RESULT IS THE EXACT SUM OF THE INFINITE SERIES.
CCCCCCCCCCCCCCCCCCCCCCCCCCCCCCCCCCCCCCCCCCCCCCCCCCCCCCCCCCCCCCCCCCCCCC
      IMPLICIT DOUBLE PRECISION (A-H,O-Z)
      COMMON /NP/NP/THETA/THETA
      PI=DACOS(-1D0)
      IF (NP.EQ.1) RESULT=PI**2/6D0
      IF (NP.EQ.2) RESULT=4.2573094155337147798209827345700 9D0
      IF (NP.EQ.3) RESULT=-DLOG(DABS(2D0*DSIN(THETA/2)))
      IF (NP.EQ.4) RESULT=DLOG(2D0)
      IF (NP.EQ.5) RESULT=DLOG(2D0)/2+PI/4
      IF (NP.EQ.6) RESULT=DLOG(2D0)/3+2*PI/(3D0**1.5D0)
      IF (NP.EQ.7) RESULT=DLOG(2D0)/4
     *   +PI/8*(DTAN(PI/8)+DTAN(PI/4)+DTAN(3*PI/8))
      IF (NP.EQ.8) RESULT=DEXP(-0.1D0)/2
      IF (NP.EQ.9) RESULT=DEXP(0.1D0)/2
      IF (NP.EQ.10) RESULT=DLOG(2D0)
      IF (NP.EQ.11) RESULT=1.5D0*DLOG(2D0)
      IF (NP.EQ.12) RESULT=DLOG(6D0)/2
      RETURN
      END
```

Bibliography

[1] M. Abramowitz and I.A. Stegun. *Handbook of Mathematical Functions with Formulas, Graphs, and Mathematical Tables*. Number 55 in Nat. Bur. Standards Appl. Math. Series. US Government Printing Office, Washington, D.C., 1964.

[2] A.C. Aitken. On Bernoulli's numerical solution of algebraic equations. *Proc. Roy. Soc. Edinburgh*, 46:289–305, 1926.

[3] A. Alaylioglu, G.A. Evans, and J. Hyslop. The evaluation of oscillatory integrals with infinite limits. *J. Comp. Phys.*, 13:433–438, 1973.

[4] N.C. Albertsen, G. Jacobsen, and S.B. Sørensen. Nonlinear transformations for accelerating the convergence of M-dimensional series. *Math. Comp.*, 41:623–634, 1983.

[5] R. Almgren, W.-S. Dai, and V. Hakim. Scaling behavior in anisotropic Hele-Shaw flow. *Phys. Rev. Lett.*, 71:3461–3464, 1993.

[6] A. Ambroladze and H. Wallin. Convergence rates of Padé and Padé-type approximants. *J. Approx. Theory*, 86:310–319, 1996.

[7] A. Ambroladze and H. Wallin. Convergence of rational interpolants with preassigned poles. *J. Approx. Theory*, 89:238–256, 1997.

[8] A. Ambroladze and H. Wallin. Padé-type approximants of Markov and meromorphic functions. *J. Approx. Theory*, 88:354–369, 1997.

[9] A. Ambroladze and H. Wallin. Extremal polynomials with preassigned zeros and rational approximants. *Constr. Approx.*, 14:209–229, 1998.

[10] W.F. Ames. *Numerical Methods for Partial Differential Equations*. Academic Press, New York, second edition, 1977.

[11] R.J. Arms and A. Edrei. The Padé tables and continued fractions generated by totally positive sequences. In H. Shankar, editor, *Mathematical Essays*, pages 1–21, Athens, Ohio, 1970. Ohio University Press.

[12] T.A. Atchison and H.L. Gray. Nonlinear transformations related to the evaluation of improper integrals II. *SIAM J. Numer. Anal.*, 5:451–459, 1968.

[13] K.E. Atkinson. *An Introduction to Numerical Analysis*. Wiley, New York, second edition, 1989.

[14] A. Averbuch, E. Braverman, R. Coifman, M. Israeli, and A. Sidi. Efficient computation of oscillatory integrals via adaptive multiscale local Fourier bases. *Appl. Comput. Harmonic Anal.*, 9:19–53, 2000.

[15] G.A. Baker, Jr. *Essentials of Padé Approximants*. Academic Press, New York, 1975.

[16] G.A. Baker, Jr. and P.R. Graves-Morris. *Padé Approximants*. Cambridge University Press, Cambridge, second edition, 1996.

[17] M.N. Barber and C.J. Hamer. Extrapolation of sequences using a generalized epsilon-algorithm. *J. Austral. Math. Soc.*, Series B, 23:229–240, 1982.

[18] F.L. Bauer. The quotient-difference and epsilon algorithms. In R.E. Langer, editor, *On Numerical Approximation*, pages 361–370. The University of Wisconsin Press, 1959.

[19] F.L. Bauer. Nonlinear sequence transformations. In H.L. Garabedian, editor, *Approximation of Functions*, pages 134–151. Elsevier, 1965.

501

[20] F.L. Bauer, H. Rutishauser, and E. Stiefel. New aspects in numerical quadrature. In *Experimental Arithmetic, High Speed Computing, and Mathematics*, pages 199–218, Providence, Rhode Island, 1963. American Mathematical Society.

[21] C.M. Bender and S.A. Orszag. *Advanced Mathematical Methods for Scientists and Engineers*. McGraw-Hill, New York, 1978.

[22] R. Bhattacharya, D. Roy, and S. Bhowmick. On the regularity of the Levin u-transform. *Comput. Phys. Comm.*, 55:297–301, 1989.

[23] S. Bhowmick, R. Bhattacharya, and D. Roy. Iterations of convergence accelerating nonlinear transforms. *Comput. Phys. Comm.*, 54:31–46, 1989.

[24] W.G. Bickley and J.C.P. Miller. The numerical summation of slowly convergent series of positive terms. *Phil. Mag.*, 7th Ser., 22:754–767, 1936.

[25] G.D. Birkhoff and W.J. Trjitzinsky. Analytic theory of singular difference equations. *Acta Math.*, 60:1–89, 1932.

[26] P.E. Bjørstad, G. Dahlquist, and E.H. Grosse. Extrapolation of asymptotic expansions by a modified Aitken δ^2-formula. *BIT*, 21:56–65, 1981.

[27] M. Blakemore, G.A. Evans, and J. Hyslop. Comparison of some methods for evaluating infinite range oscillatory integrals. *J. Comp. Phys.*, 22:352–376, 1976.

[28] P.B. Borwein. Padé approximants for the q-elementary functions. *Constr. Approx.*, 4:391–402, 1988.

[29] J.P. Boyd. The rate of convergence of Chebyshev polynomials for functions which have asymptotic power series about one endpoint. *Math. Comp.*, 37:189–195, 1981.

[30] J.P. Boyd. A Chebyshev polynomial rate-of-convergence theorem for Stieltjes functions. *Math. Comp.*, 39:201–206, 1982.

[31] C. Brezinski. Application du ρ-algorithme á la quadrature numérique. *C. R. Acad. Sci. Paris*, 270 A:1252–1253, 1970.

[32] C. Brezinski. Accélération de suites à convergence logarithmique. *C. R. Acad. Sci. Paris*, 273 A:727–730, 1971.

[33] C. Brezinski. Convergence d'une forme confluente de l'ε-algorithme. *C. R. Acad. Sci. Paris*, 273 A:582–585, 1971.

[34] C. Brezinski. Études sur les ϵ- et ρ-algorithmes. *Numer. Math.*, 17:153–162, 1971.

[35] C. Brezinski. L'ε-algorithme et les suites totalement monotones et oscillantes. *C. R. Acad. Sci. Paris*, 276 A:305–308, 1973.

[36] C. Brezinski. *Accélération de la Convergence en Analyse Numérique*. Springer-Verlag, Berlin, 1977.

[37] C. Brezinski. A general extrapolation algorithm. *Numer. Math.*, 35:175–187, 1980.

[38] C. Brezinski. *History of Continued Fractions and Padé Approximants*. Springer-Verlag, Berlin, 1990.

[39] C. Brezinski. *A Bibliography on Continued Fractions, Padé Approximation, Extrapolation and Related Subjects*. Prensas Universitarias de Zaragoza, Zaragoza, 1991.

[40] C. Brezinski and M. Crouzeix. Remarques sur le procédé δ^2 d'Aitken. *C. R. Acad. Sci. Paris*, 270 A:896–898, 1970.

[41] C. Brezinski and M. Redivo Zaglia. *Extrapolation Methods: Theory and Practice*. North-Holland, Amsterdam, 1991.

[42] N.G. de Bruijn. *Asymptotic Methods in Analysis*. North-Holland, Amsterdam, 1970.

[43] R. Bulirsch and J. Stoer. Fehlerabschätzungen und Extrapolation mit rationalen Funktionen bei Verfahren vom Richardson-Typus. *Numer. Math.*, 6:413–427, 1964.

[44] R. Bulirsch and J. Stoer. Asymptotic upper and lower bounds for results of extrapolation methods. *Numer. Math.*, 8:93–104, 1966.

[45] R. Bulirsch and J. Stoer. Numerical treatment of ordinary differential equations by extrapolation methods. *Numer. Math.*, 8:1–13, 1966.

[46] R. Bulirsch and J. Stoer. Numerical quadrature by extrapolation. *Numer. Math.*, 9:271–278, 1967.

[47] E.W. Cheney. *Introduction to Approximation Theory*. McGraw-Hill, New York, 1966.

[48] J.S.R. Chisholm. Rational approximants defined from double power series. *Math. Comp.*, 27:841–848, 1973.

[49] J.S.R. Chisholm, A. Genz, and G.E. Rowlands. Accelerated convergence of sequences of quadrature approximations. *J. Comp. Phys.*, 10:284–307, 1972.

[50] J.S.R. Chisholm and J. McEwan. Rational approximants defined from power series in n variables. *Proc. Roy. Soc. London*, A 336:421–452, 1974.

[51] S. Christiansen. Numerical solution of an integral equation with a logarithmic kernel. *BIT*, 11:276–287, 1971.

[52] S. Christiansen. Derivation and investigation of fifth order quadrature formulas for biharmonic boundary integral operators. *Appl. Numer. Math.*, 37:145–159, 2001.

[53] C.W. Clenshaw and K. Lord. Rational approximations from Chebyshev series. In B.K.P. Scaife, editor, *Studies in Numerical Analysis*, pages 95–113, London, 1974. Academic Press.

[54] A.M. Cohen and D. Levin. Accelerating infinite products. *Numer. Algorithms*, 22:157–165, 1999.

[55] R. Coifman, M. Goldberg, T. Hrycak, M. Israeli, and V. Rokhlin. An improved operator expansion algorithm for direct and inverse scattering computations. *Waves Random Media*, 9:441–457, 1999.

[56] F. Cordellier. Caractérisation des suites que la première étape du θ-algorithme. *C. R. Acad. Sci. Paris*, 284 A:389–392, 1977.

[57] F. Cordellier. Démonstration algébrique de l'extension de l'identité de Wynn aux tables de Padé non normales. In L. Wuytack, editor, *Padé Approximation and Its Applications*, number 765 in Lecture Notes in Mathematics, pages 36–60, Berlin, 1977. Springer-Verlag.

[58] K.S. Crump. Numerical inversion of Laplace transforms using a Fourier series approximation. *Journal of the ACM*, 23:89–96, 1976.

[59] A. Cuyt. *Padé Approximations for Operators: Theory and Applications*. Springer-Verlag, Berlin, 1984.

[60] A. Cuyt. A multivariate convergence theorem of the de Montessus de Ballore type. *J. Comp. Appl. Math.*, 32:47–57, 1990.

[61] A. Cuyt. Multivariate partial Newton–Padé and Newton–Padé type approximations. *J. Approx. Theory*, 72:301–316, 1993.

[62] P.J. Davis. *Interpolation and Approximation*. Dover, New York, 1975.

[63] P.J. Davis and P. Rabinowitz. *Methods of Numerical Integration*. Academic Press, New York, second edition, 1984.

[64] R. de Montessus de Ballore. Sur les fractions continue algébriques. *Bull. Soc. Math. France*, 30:28–36, 1902.

[65] J.P. Delahaye. *Sequence Transformations*. Springer-Verlag, Berlin, 1988.

[66] J.P. Delahaye and B. Germain-Bonne. Résultats négatifs en accélération de la convergence. *Numer. Math.*, 35:443–457, 1980.

[67] J.P. Delahaye and B. Germain-Bonne. The set of logarithmically convergent sequences cannot be accelerated. *SIAM J. Numer. Anal.*, 19:840–844, 1982.

[68] J.E. Drummond. A formula for accelerating the convergence of a general series. *Bull. Austral. Math. Soc.*, 6:69–74, 1972.

[69] J.E. Drummond. Summing a common type of slowly convergent series of positive terms. *J. Austral. Math. Soc.*, Series B, 19:416–421, 1976.

[70] R. Dubner and J. Abate. Numerical inversion of Laplace transforms by relating them to the finite Fourier cosine transform. *Journal of the ACM*, 15:115–123, 1968.

[71] F. Durbin. Numerical inversion of Laplace transforms: an efficient improvement of Dubner and Abate's method. *Computer J.*, 17:371–376, 1968.

[72] A. Edrei. Proof of a conjecture of Schönberg on the generating function of a totally positive sequence. *Canad. J. Math.*, 5:86–94, 1953.

[73] U.T. Ehrenmark. A note on an extension of extrapolative techniques for a class of infinite oscillatory integrals. *BIT*, 30:152–155, 1990.

[74] U.T. Ehrenmark. On computing uniformly valid approximations for viscous waves on a plane beach. *J. Comp. Appl. Math.*, 50:263–281, 1994.

[75] U.T. Ehrenmark. The numerical inversion of two classes of Kontorovich–Lebedev transforms by direct quadrature. *J. Comp. Appl. Math.*, 61:43–72, 1995.

[76] D. Elliott. Sigmoidal transformations and the trapezoidal rule. *J. Austral. Math. Soc.*, Series B (E), 40:E77–E137, 1998.

[77] A. Erdélyi. *Asymptotic Expansions*. Dover, New York, 1956.

[78] T.O. Espelid. Integrating singularities using non-uniform subdivision and extrapolation. In H. Brass and G. Hämmerlin, editors, *Numerical Integration IV*, number 112 in ISNM, pages 77–89, Basel, 1993. Birkhäuser.

[79] T.O. Espelid. On integrating vertex singularities using extrapolation. *BIT*, 34:62–79, 1994.

[80] G.A. Evans and J. Hyslop A.P.G. Morgan. An extrapolation procedure for the evaluation of singular integrals. *Intern. J. Computer Math.*, 12:251–265, 1983.

[81] E. Fabry. Sur les points singuliers d'une fonction donnée par son développement en série et sur l'impossibilité du prolongement analytique dans les cas très génèraux. *Ann. Ecole Normale (3)*, 13:107–114, 1896.

[82] G. Fainstein, A. Sidi, M. Israeli, and Y. Tsur-Lavie. Application of boundary integral equations to the solution of stresses around a shallow circular hole. In I.W. Farmer, J.J.K. Daemen, C.S. Desai, C.E. Glass, and S.P. Neumann, editors, *28th U.S. Symposium in Rock Mechanics, Tucson, Arizona*, pages 745–755, 1987.

[83] T. Fessler, W.F. Ford, and D.A. Smith. HURRY: An acceleration algorithm for scalar sequences and series. *ACM Trans. Math. Software*, 9:346–354, 1983.

[84] J. Fleischer. Analytic continuation of scattering amplitudes and Padé approximants. *Nuclear Phys. B*, 37:59–76, 1972.

[85] J. Fleischer. Generalizations of Padé approximants. In P.R. Graves-Morris, editor, *Padé Approximants: Lectures delivered at a summer school held at the University of Kent, July 1972*, pages 126–131, Bath, 1973. Pitman Press.

[86] J. Fleischer. Nonlinear Padé approximants for Legendre series. *J. Math. Phys.*, 14:246–248, 1973.

[87] W.F. Ford and A. Sidi. An algorithm for a generalization of the Richardson extrapolation process. *SIAM J. Numer. Anal.*, 24:1212–1232, 1987.

[88] G. Freud. *Orthogonal Polynomials*. Pergamon Press, New York, 1971.

[89] P.A. Frost and E.Y. Harper. An extended Padé procedure for constructing global approximations from asymptotic expansions: An application with examples. *SIAM Rev.*, 18:62–91, 1976.

[90] B. Gabutti. An algorithm for computing generalized Euler's transformations of series. *Computing*, 34:107–116, 1985.

[91] J.L. Gammel. Review of two recent generalizations of the Padé approximant. In P.R. Graves-Morris, editor, *Padé Approximants and Their Applications*, pages 3–9, New York, 1973. Academic Press.

[92] C.R. Garibotti and F.F. Grinstein. Recent results relevant to the evaluation of infinite series. *J. Comp. Appl. Math.*, 9:193–200, 1983.

[93] K.O. Geddes. Block structure in the Chebyshev–Padé table. *SIAM J. Numer. Anal.*, 18:844–861, 1981.

[94] A. Genz. Application of the ε-algorithm to quadrature problems. In P.R. Graves-Morris, editor, *Padé Approximants and Their Applications*, pages 105–116, New York, 1973. Academic Press.

[95] K. Georg. Approximation of integrals for boundary element methods. *SIAM J. Sci. Statist. Comput.*, 12:443–453, 1991.

[96] B. Germain-Bonne. Transformations de suites. *RAIRO*, R1:84–90, 1973.

[97] B. Germain-Bonne. *Estimation de la limite de suites et formalisation de procédés d'accélération de la convergence*. PhD thesis, Université de Lille I, 1978.

[98] J. Gilewicz. Numerical detection of the best Padé approximant and determination of the Fourier coefficients of the insufficiently sampled functions. In P.R. Graves-Morris, editor, *Padé Approximants and Their Applications*, pages 99–103, New York, 1973. Academic Press.

[99] J. Gilewicz. *Approximants de Padé*. Number 667 in Lecture Notes in Mathematics. Springer-Verlag, New York, 1978.

[100] M. Golomb. Zeros and poles of functions defined by power series. *Bull. Amer. Math. Soc.*, 49:581–592, 1943.

[101] A.A. Gončar. Poles of rows of the Padé table and meromorphic continuation of functions. *Math. USSR-Sbornik*, 43:527–546, 1982.

[102] A.A. Gončar, E.A. Rakhmanov, and S.P. Suetin. On the rate of convergence of Padé approximants of orthogonal expansions. In A.A. Gončar and E.B. Saff, editors, *Progress in Approximation Theory*, pages 169–190, New York, 1992. Springer-Verlag.

[103] S. Graffi, V. Grecchi, and B. Simon. Borel summability: Application to the anharmonic oscillator. *Phys. Lett.*, 32B:631–634, 1970.

[104] W.B. Gragg. *Repeated extrapolation to the limit in the numerical solution of ordinary differential equations*. PhD thesis, University of California at Los Angeles, 1964. Supervised by P. Henrici.

[105] W.B. Gragg. On extrapolation algorithms for ordinary initial value problems. *Journal of the Society for Industrial and Applied Mathematics: Series B, Numerical Analysis*, 2:384–403, 1965.

[106] W.B. Gragg. The Padé table and its relation to certain algorithms of numerical analysis. *SIAM Rev.*, 14:1–62, 1972.

[107] W.B. Gragg. Laurent, Fourier and Chebyshev–Padé tables. In E.B. Saff and R.S. Varga, editors, *Padé and Rational Approximation*, pages 61–72, New York, 1977. Academic Press.

[108] W.B. Gragg and A.S. Householder. On a theorem of König. *Numer. Math.*, 8:465–468, 1966.

[109] P.R. Graves-Morris, R. Hughes Jones, and G.J. Makinson. The calculation of rational approximants in two variables. *J. Inst. Maths. Applics.*, 13:311–320, 1974.

[110] H.L. Gray and T.A. Atchison. Nonlinear transformations related to the evaluation of improper integrals I. *SIAM J. Numer. Anal.*, 4:363–371, 1967.

[111] H.L. Gray and T.A. Atchison. The generalized G-transform. *Math. Comp.*, 22:595–605, 1968.

[112] H.L. Gray, T.A. Atchison, and G.V. McWilliams. Higher order G-transformations. *SIAM J. Numer. Anal.*, 8:365–381, 1971.

[113] H.L. Gray and S. Wang. An extension of the Levin–Sidi class of nonlinear transformations for accelerating convergence of infinite integrals and series. *Appl. Math. Comp.*, 33:75–87, 1989.

[114] H.L. Gray and S. Wang. A new method for approximating improper integrals. *SIAM J. Numer. Anal.*, 29:271–283, 1992.

[115] C. Greif and D. Levin. The d_2-transformation for infinite double series and the D_2-transformation for infinite double integrals. *Math. Comp.*, 67:695–714, 1998.

[116] J. Grotendorst. A Maple package for transforming sequences and functions. *Comput. Phys. Comm.*, 67:325–342, 1991.

[117] R.E. Grundy. Laplace transform inversion using two-point rational approximants. *J. Inst. Maths. Applics.*, 20:299–306, 1977.

[118] A.J. Guttmann. Asymptotic analysis of power-series expansions. In C. Domb and J.L. Lebowitz, editors, *Phase Transitions and Critical Phenomena*, volume 13, pages 1–234, New York, 1989. Academic Press.

[119] A.J. Guttmann and G.S. Joyce. On a new method of series analysis in lattice statistics. *J. Phys. A*, 5:L81–L84, 1972.

[120] J. Hadamard. Essai sur l'étude des fonctions données par leur développement de Taylor. *J. Math. Pures Appl. (4)*, 8:101–186, 1892.

[121] E. Hairer and Ch. Lubich. Asymptotic expansions of the global error of fixed-stepsize methods. *Numer. Math.*, 45:345–360, 1984.

[122] E. Hairer, S.P. Nørsett, and G. Wanner. *Solving Ordinary Differential Equations I: Nonstiff Problems*. Springer-Verlag, New York, second edition, 1993.

[123] G.H. Hardy. *Divergent Series*. Clarendon Press, Oxford, 1949.

[124] D.J. Haroldsen and D.I. Meiron. Numerical calculation of three-dimensional interfacial potential flows using the point vortex method. *SIAM J. Sci. Comput.*, 20:648–683, 1988.

[125] J.F. Hart, E.W. Cheney, C.L. Lawson, H.J. Maehly, C.K. Mesztenyi, J.R. Rice, H.J. Thacher, Jr., and C. Witzgall, editors. *Computer Approximations*. SIAM Series in Applied Mathematics. Wiley, New York, 1968.

[126] T. Hasegawa. Numerical integration of functions with poles near the interval of integration. *J. Comp. Appl. Math.*, 87:339–357, 1997.

[127] T. Hasegawa and A. Sidi. An automatic integration procedure for infinite range integrals involving oscillatory kernels. *Numer. Algorithms*, 13:1–19, 1996.

[128] T. Hasegawa and T. Torii. Indefinite integration of oscillatory functions by the Chebyshev series expansion. *J. Comp. Appl. Math.*, 17:21–29, 1987.

[129] T. Håvie. Generalized Neville type extrapolation schemes. *BIT*, 19:204–213, 1979.

[130] P. Henrici. *Elements of Numerical Analysis*. Wiley, New York, 1964.

[131] P. Henrici. *Applied and Computational Complex Analysis*, volume 1. Wiley, New York, 1974.

[132] P. Henrici. *Applied and Computational Complex Analysis*, volume 2. Wiley, New York, 1977.

[133] M. Hill and I. Robinson. d2lri: A nonadaptive algorithm for two-dimensional cubature. *J. Comp. Appl. Math.*, 112:121–145, 1999.

[134] J.T. Holdeman, Jr. A method for the approximation of functions defined by formal series expansions in orthogonal polynomials. *Math. Comp.*, 23:275–287, 1969.

[135] H.H.H. Homeier. A Levin-type algorithm for accelerating the convergence of Fourier series. *Numer. Algorithms*, 3:245–254, 1992.

[136] T.Y. Hou, J.S. Lowengrub, and M.J. Shelley. Boundary integral methods for multicomponent fluids and multiphase materials. *J. Comp. Phys.*, 169:302–362, 2001.

[137] R. Hughes Jones. General rational approximants in N variables. *J. Approx. Theory*, 16:201–233, 1976.

[138] M. Iri, S. Moriguti, and Y. Takasawa. On a certain quadrature formula. *Kokyuroku of Res. Inst. for Math. Sci. Kyoto Univ.*, 91:82–118, 1970. In Japanese. English translation in *J. Comp. Appl. Math.*, 17:3–20, (1987).

[139] A. Iserles. A note on Padé approximations and generalized hypergeometric functions. *BIT*, 19:543–545, 1979.

[140] A. Iserles. Generalized order star theory. In M.G. de Bruin and H. van Rossum, editors, *Padé Approximation and Its Applications Amsterdam 1980*, pages 228–238, New York, 1981. Springer-Verlag.

[141] A. Iserles. *A First Course in the Numerical Analysis of Differential Equations*. Cambridge Texts in Applied Mathematics. Cambridge University Press, Cambridge, 1996.

[142] A. Iserles and S.P. Nørsett. *Order Stars*. Chapman & Hall, London, 1991.

[143] P.R. Johnston. Semi-sigmoidal transformations for evaluating weakly singular boundary element integrals. *Intern. J. Numer. Methods Engrg.*, 47:1709–1730, 2000.

[144] W.B. Jones and W.J. Thron. *Continued Fractions: Analytic Theory and Applications*. Addison-Wesley, London, 1980.

[145] D.C. Joyce. Survey of extrapolation processes in numerical analysis. *SIAM Rev.*, 13:435–490, 1971.

[146] G.S. Joyce and A.J. Guttmann. A new method of series analysis. In P.R. Graves-Morris, editor, *Padé Approximants and Their Applications*, pages 163–167, New York, 1973. Academic Press.

[147] D.K. Kahaner. Numerical quadrature by the ϵ-algorithm. *Math. Comp.*, 26:689–693, 1972.

[148] M. Kaminski and A. Sidi. Solution of an integer programming problem related to convergence of rows of Padé approximants. *Appl. Numer. Math.*, 8:217–223, 1991.

[149] J. Karlsson and E.B. Saff. Singularities of functions determined by the poles of Padé approximants. In M.G. de Bruin and H. van Rossum, editors, *Padé Approximation and Its Applications Amsterdam 1980*, pages 239–254, New York, 1981. Springer-Verlag.

[150] J. Karlsson and H. Wallin. Rational approximation by an interpolation procedure in several variables. In E.B. Saff and R.S. Varga, editors, *Padé and Rational Approximation*, pages 83–100, New York, 1977. Academic Press.

[151] J.E. Kiefer and G.H. Weiss. A comparison of two methods for accelerating the convergence of Fourier series. *Comp. & Maths. with Applics.*, 7:527–535, 1981.

[152] K. Knopp. *Theory and Application of Infinite Series*. Hafner, New York, 1947.

[153] J. Koenig. Über eine Eigenschaft der Potenzreichen. *Math. Ann.*, 23:447–449, 1884.

[154] N.M. Korobov. *Number-Theoretic Methods of Approximate Analysis*. GIFL, Moscow, 1963. Russian.

[155] C. Kowalewski. Accélération de la convergence pour certain suites à convergence logarithmique. In M.G. de Bruin and H. van Rossum, editors, *Padé Approximation and Its Applications Amsterdam 1980*, pages 263–272, New York, 1981. Springer-Verlag.

[156] C. Kowalewski. *Possibilités d'accélération de la convergence logarithmique*. PhD thesis, Université de Lille I, 1981.

[157] J.D. Lambert. *Computational Methods in Ordinary Differential Equations*. Wiley, New York, 1973.

[158] P.-J. Laurent. Un théorème de convergence pour le procédé d'extrapolation de Richardson. *C. R. Acad. Sci. Paris*, 256:1435–1437, 1963.

[159] D.P. Laurie. Propagation of initial rounding error in Romberg-like quadrature. *BIT*, 15:277–282, 1975.

[160] D.P. Laurie. Periodizing transformations for numerical integration. *J. Comp. Appl. Math.*, 66:337–344, 1996.

[161] D. Levin. Development of non-linear transformations for improving convergence of sequences. *Intern. J. Computer Math.*, B3:371–388, 1973.

[162] D. Levin. Numerical inversion of the Laplace transform by accelerating the convergence of Bromwich's integral. *J. Comp. Appl. Math.*, 1:247–250, 1975.

[163] D. Levin. General order Padé-type rational approximants defined from double power series. *J. Inst. Maths. Applics.*, 18:1–8, 1976.

[164] D. Levin. On accelerating the convergence of infinite double series and integrals. *Math. Comp.*, 35:1331–1345, 1980.

[165] D. Levin and A. Sidi. Two new classes of nonlinear transformations for accelerating the convergence of infinite integrals and series. *Appl. Math. Comp.*, 9:175–215, 1981. Originally appeared as a Tel Aviv University preprint in 1975.

[166] D. Levin and A. Sidi. Extrapolation methods for infinite multiple series and integrals. *J. Comp. Meth. Sci. & Engrg.*, 1:167–184, 2001.

[167] M.J. Lighthill. *Introduction to Fourier Analysis and Generalized Functions*. Cambridge University Press, Cambridge, 1978.

[168] Jing Lin. Extension of Wilson's theorem. *Computations of Mathematics*, 1:43–48, 1987. In Chinese.

[169] Xiaoyan Liu and E.B. Saff. Intermediate rows of the Walsh array of best rational approximations to meromorphic functions. *Methods Appl. Anal.*, 2:269–284, 1995.

[170] I.M. Longman. Note on a method for computing infinite integrals of oscillatory functions. *Proc. Cambridge Phil. Soc.*, 52:764–768, 1956.

[171] I.M. Longman. Tables for the rapid and accurate numerical evaluation of certain infinite integrals involving Bessel functions. *Mathematical Tables and Other Aids to Computation*, 11:166–180, 1957.

[172] I.M. Longman. A short table of $\int_x^\infty J_0(t)t^{-n}dt$ and $\int_x^\infty J_1(t)t^{-n}dt$ (in Technical Notes and Short Papers). *Mathematical Tables and Other Aids to Computation*, 13:306–311, 1959.

[173] I.M. Longman. The application of rational approximations to the solution of problems in theoretical seismology. *Bull. Seism. Soc. Amer.*, 56:1045–1065, 1966.

[174] I.M. Longman. Computation of the Padé table. *Intern. J. Computer Math.*, Section B, 3:53–64, 1971.

[175] I.M. Longman. Approximate Laplace transform inversion applied to a problem in electrical network theory. *SIAM J. Appl. Math.*, 23:439–445, 1972.

[176] I.M. Longman. Numerical Laplace transform inversion of a function arising in viscoelasticity. *J. Comp. Phys.*, 10:224–231, 1972.

[177] I.M. Longman. On the generation of rational approximations for Laplace transform inversion with an application to viscoelasticity. *SIAM J. Appl. Math.*, 24:429–440, 1973.

[178] I.M. Longman. Difficulties in convergence acceleration. In M.G. de Bruin and H. van Rossum, editors, *Padé Approximation and Its Applications Amsterdam 1980*, pages 273–289, New York, 1981. Springer-Verlag.

[179] L. Lorentzen and H. Waadeland. *Continued Fractions with Applications*. North-Holland, Amsterdam, 1992.

[180] D.S. Lubinsky. Padé tables of a class of entire functions. *Proc. Amer. Math. Soc.*, 94:399–406, 1985.

[181] D.S. Lubinsky. Padé tables of entire functions of very slow and smooth growth. *Constr. Approx.*, 1:349–358, 1985.

[182] D.S. Lubinsky. Uniform convergence of rows of the Padé table for functions with smooth Maclaurin series coefficients. *Constr. Approx.*, 3:307–330, 1987.

[183] D.S. Lubinsky. Padé tables of entire functions of very slow and smooth growth II. *Constr. Approx.*, 4:321–339, 1988.

[184] D.S. Lubinsky. On the diagonal Padé approximants of meromorphic functions. *Indagationes Mathematicae*, 7:97–110, 1996.

[185] D.S. Lubinsky and A. Sidi. Convergence of linear and nonlinear Padé approximants from series of orthogonal polynomials. *Trans. Amer. Math. Soc.*, 278:333–345, 1983.

[186] D.S. Lubinsky and A. Sidi. Strong asymptotics for polynomials biorthogonal to powers of $\log x$. *Analysis*, 14:341–379, 1994.

[187] S. Lubkin. A method of summing infinite series. *J. Res. Nat. Bur. Standards*, 48:228–254, 1952.

[188] S.K. Lucas. Evaluating infinite integrals involving products of Bessel functions of arbitrary order. *J. Comp. Appl. Math.*, 64:269–282, 1995.

[189] S.K. Lucas and H.A. Stone. Evaluating infinite integrals involving Bessel functions of arbitrary order. *J. Comp. Appl. Math.*, 64:217–231, 1995.

[190] R. Lugannani and S. Rice. Use of Gaussian convergence factors in numerical evaluation of slowly convergent integrals. *J. Comp. Phys.*, 37:264–267, 1980.

[191] Y.L. Luke. On the approximate inversion of some Laplace transforms. In *Proceedings of 4th U.S. National Congress on Applied Mechanics*, pages 269–276, 1962.

[192] Y.L. Luke. Approximate inversion of a class of Laplace transforms applicable to supersonic flow problems. *Quart. J. Mech. Appl. Math.*, 17:91–103, 1964.

[193] Y.L. Luke. *Algorithms for the Computation of Special Functions*. Academic Press, New York, 1977.

[194] J. Lund. Bessel transforms and rational extrapolation. *Numer. Math.*, 47:1–14, 1985.

[195] J.N. Lyness. Applications of extrapolation techniques to multidimensional quadrature of some integrand functions with a singularity. *J. Comp. Phys.*, 20:346–364, 1976.

[196] J.N. Lyness. An error functional expansion for N-dimensional quadrature with an integrand function singular at a point. *Math. Comp.*, 30:1–23, 1976.

[197] J.N. Lyness. Integrating some infinite oscillating tails. *J. Comp. Appl. Math.*, 12/13:109–117, 1985.

[198] J.N. Lyness. Quadrature over curved surfaces by extrapolation. *Math. Comp.*, 63:727–740, 1994.

[199] J.N. Lyness and E. de Doncker. Quadrature error expansions II. The full corner singularity. *Numer. Math.*, 64:355–370, 1993.

[200] J.N. Lyness and B.J. McHugh. Integration over multidimensional hypercubes. *Computer J.*, 6:264–270, 1963.

[201] J.N. Lyness and G. Monegato. Quadrature error functional expansions for the simplex when the integrand function has singularities at vertices. *Math. Comp.*, 34:213–225, 1980.

[202] J.N. Lyness and B.W. Ninham. Numerical quadrature and asymptotic expansions. *Math. Comp.*, 21:162–178, 1967.

[203] J.N. Lyness and K.K. Puri. The Euler–Maclaurin expansion for the simplex. *Math. Comp.*, 27:273–293, 1973.

[204] J.N. Lyness and U. Rüde. Cubature of integrands containing derivatives. *Numer. Math.*, 78:439–461, 1998.

[205] H.J. Maehly. Rational approximations for transcendental functions. In *Proceedings of the International Conference on Information Processing*, pages 57–62, Butterworth, London, 1960.

[206] G.I. Marchuk and V.V. Shaidurov. *Difference methods and their extrapolations*. Springer-Verlag, New York, 1983.

[207] A.C. Matos and M. Prévost. Acceleration property for the columns of the E-algorithm. *Numer. Algorithms*, 2:393–408, 1992.

[208] J.H. McCabe and J.A. Murphy. Continued fractions which correspond to power series at two points. *J. Inst. Maths. Applics.*, 17:233–247, 1976.

[209] D.W. McLaughlin, D.J. Muraki, and M.J. Shelley. Self-focussed optical structures in a nematic liquid crystal. *Physica D*, 97:471–497, 1996.

[210] G. Meinardus. Über das asymptotische Verhalten von Iterationsfolgen. *Z. Angew. Math. Mech.*, 63:70–72, 1983.

[211] K.A. Michalski. Extrapolation methods for Sommerfeld integral tails. *IEEE Trans. Antennas Propagat.*, 46:1405–1418, 1998.

[212] G.F. Miller. On the convergence of the Chebyshev series for functions possessing a singularity in the range of representation. *SIAM J. Numer. Anal.*, 3:390–409, 1966.

[213] G. Monegato and J.N. Lyness. The Euler–Maclaurin expansion and finite-part integrals. *Numer. Math.*, 81:273–291, 1998.

[214] M. Mori. An IMT-type double exponential formula for numerical integration. *Publ. Res. Inst. Math. Sci. Kyoto Univ.*, 14:713–729, 1978.

[215] J.D. Murray. *Asymptotic Analysis*. Clarendon Press, Oxford, 1974.

[216] I. Navot. An extension of the Euler–Maclaurin summation formula to functions with a branch singularity. *J. Math. and Phys.*, 40:271–276, 1961.

[217] I. Navot. A further extension of the Euler–Maclaurin summation formula. *J. Math. and Phys.*, 41:155–163, 1962.

[218] G. Németh. Chebyshev expansions for Fresnel integrals. *Numer. Math.*, 7:310–312, 1965.

[219] Q. Nie and F.R. Tian. Singularities in Hele–Shaw flows driven by a multipole. *SIAM J. Appl. Math.*, 62:385–406, 2001.

[220] W. Niethammer. Numerical application of Euler's series transformation and its generalizations. *Numer. Math.*, 34:271–283, 1980.

[221] M. Nitsche. Singularity formation in a cylindrical and a spherical vortex sheet. *J. Comp. Phys.*, 173:208–230, 2001.

[222] N.E. Nörlund. *Vorlesungen über Differenzenrechnung*. Springer-Verlag, Berlin, 1924.

[223] F.W.J. Olver. *Asymptotics and Special Functions*. Academic Press, New York, 1974.

[224] N. Osada. Accelerable subsets of logarithmic sequences. *J. Comp. Appl. Math.*, 32:217–227, 1990.

[225] N. Osada. A convergence acceleration method for some logarithmically convergent sequences. *SIAM J. Numer. Anal.*, 27:178–189, 1990.

[226] N. Osada. The E-algorithm and the Ford-Sidi algorithm. *J. Comp. Appl. Math.*, 122:223–230, 2000.

[227] K.J. Overholt. Extended Aitken acceleration. *BIT*, 5:122–132, 1965.

[228] B.N. Parlett. Global convergence of the basic QR algorithm on Hessenberg matrices. *Math. Comp.*, 22:803–817, 1968.

[229] O. Perron. *Die Lehre von den Kettenbrüchen*. Teubner, Stuttgart, third edition, 1957. In two volumes.

[230] R. Piessens. Gaussian quadrature formulas for the numerical integration of Bromwich's integral and the inversion of the Laplace transforms. *J. Engrg. Math.*, 5:1–9, 1971.

[231] R.E. Powell and S.M. Shah. *Summability Theory and Its Applications*. Van Nostrand Rheinhold, London, 1972.

[232] R. de Prony. Essai expérimental et analytique: sur les lois de la dilatabilité de fluides élastiques et sur celles de la force expansive de la vapeur de l'eau et de la vapeur de l'alkool, à différentes températures. *Journal de l'École Polytechnique Paris*, 1:24–76, 1795.

[233] W.C. Pye and T.A. Atchison. An algorithm for the computation of the higher order G-transformation. *SIAM J. Numer. Anal.*, 10:1–7, 1973.

[234] P. Rabinowitz. Extrapolation methods in numerical integration. *Numer. Algorithms*, 3:17–28, 1992.

[235] A. Ralston and P. Rabinowitz. *A First Course in Numerical Analysis*. McGraw-Hill, New York, second edition, 1978.

[236] L.F. Richardson. The approximate arithmetical solution by finite differences of physical problems involving differential equations, with an application to the stress in a masonry dam. *Phil. Trans. Roy. Soc. London*, Series A, 210:307–357, 1910.

[237] L.F. Richardson. Theory of the measurement of wind by shooting spheres upward. *Phil. Trans. Roy. Soc. London*, Series A, 223:345–382, 1923.

[238] L.F. Richardson. The deferred approach to the limit, I: Single lattice. *Phil. Trans. Roy. Soc. London*, Series A, 226:299–349, 1927.

[239] I. Robinson and M. Hill. Algorithm 816: r2d2lri: An algorithm for automatic two-dimensional cubature. *ACM Trans. Math. Software*, 28:75–100, 2002.

[240] W. Romberg. Vereinfachte numerische Integration. *Det. Kong. Norske Videnskabers Selskab Forhandlinger (Trondheim)*, 28:30–36, 1955.

[241] J.B. Rosser. Transformations to speed the convergence of series. *J. Res. Nat. Bur. Standards*, 46:56–64, 1951.

[242] H. Rutishauser. Anwendungen des Quotienten-Differenzen-Algorithmus. *Z. Angew. Math. Phys.*, 5:496–508, 1954.

[243] H. Rutishauser. Der Quotienten-Differenzen-Algorithmus. *Z. Angew. Math. Phys.*, 5:233–251, 1954.

[244] H. Rutishauser. *Der Quotienten-Differenzen-Algorithmus*. Birkhäuser, Basel, 1957.

[245] H. Rutishauser. Ausdehnung des Rombergschen Prinzips. *Numer. Math.*, 5:48–54, 1963.

[246] P. Sablonnière. Convergence acceleration of logarithmic fixed point sequences. *J. Comp. Appl. Math.*, 19:55–60, 1987.

[247] P. Sablonnière. Comparison of four algorithms accelerating the convergence of a subset of logarithmic fixed point sequences. *Numer. Algorithms*, 1:177–197, 1991.

[248] P. Sablonnière. Asymptotic behaviour of iterated modified Δ^2 and θ_2 transforms on some slowly convergent sequences. *Numer. Algorithms*, 3:401–409, 1992.

[249] E.B. Saff. An extension of Montessus de Ballore theorem on the convergence of interpolating rational functions. *J. Approx. Theory*, 6:63–67, 1972.

[250] H. Safouhi and P.E. Hoggan. Efficient evaluation of Coulomb integrals: the nonlinear D- and $\bar{D}$-transformations. *J. Phys. A: Math. Gen.*, 31:8941–8951, 1998.

[251] H. Safouhi and P.E. Hoggan. Non-linear transformations for rapid and efficient evaluation of multicenter bielectronic integrals over B functions. *J. Math. Chem.*, 25:259–280, 1999.

[252] H. Safouhi and P.E. Hoggan. Three-center two-electron Coulomb and hybrid integrals evaluated using non-linear D- and $\bar{D}$-transformations. *J. Phys. A: Math. Gen.*, 32:6203–6217, 1999.

[253] H. Safouhi, D. Pinchon, and P.E. Hoggan. Efficient evaluation of integrals for density functional theory: nonlinear D transformations to evaluate three-center nuclear attraction integrals over B functions. *Intern. J. Quantum Chem.*, 70:181–188, 1998.

[254] T.W. Sag and G. Szekeres. Numerical evaluation of high-dimensional integrals. *Math. Comp.*, 18:245–253, 1964.

[255] H.E. Salzer. Orthogonal polynomials arising in the numerical evaluation of inverse Laplace transforms. *Mathematical Tables and Other Aids to Computation*, 9:164–177, 1955.

[256] H.E. Salzer. Tables for numerical calculation of inverse Laplace transforms. *J. Math. and Phys.*, 37:89–109, 1958.

[257] H.E. Salzer. Additional formulas and tables for orthogonal polynomials originating from inversion integrals. *J. Math. and Phys.*, 40:72–86, 1961.

[258] J.R. Schmidt. On the numerical solution of linear simultaneous equations by an iterative method. *Phil. Mag.*, 7:369–383, 1941.

[259] C. Schneider. Vereinfachte Rekursionen zur Richardson-Extrapolation in Spezialfällen. *Numer. Math.*, 24:177–184, 1975.

[260] I.J. Schönberg. On the Pólya frequency functions I: The totally positive functions and their Laplace transforms. *J. d'Anal. Math.*, 1:331–374, 1951.

[261] R.E. Scraton. The practical use of the Euler transformation. *BIT*, 29:356–360, 1989.

[262] G.A. Sedogbo. Convergence acceleration of some logarithmic sequences. *J. Comp. Appl. Math.*, 32:253–260, 1990.

[263] R.E. Shafer. On quadratic approximation. *SIAM J. Numer. Anal.*, 11:447–460, 1974.

[264] D. Shanks. Nonlinear transformations of divergent and slowly convergent sequences. *J. Math. and Phys.*, 34:1–42, 1955.

[265] R. Shelef. New numerical quadrature formulas for Laplace transform inversion by Bromwich's integral. Master's thesis, Technion–Israel Institute of Technology, 1987. In Hebrew. Supervised by A. Sidi.

[266] M.J. Shelley, F.-R. Tian, and K. Wlodarski. Hele-Shaw flow and pattern formation in a time-dependent gap. *Nonlinearity*, 10:1471–1495, 1997.

[267] A. Sidi. Computation of the Chebyshev–Padé table. *J. Comp. Appl. Math.*, 1:69–71, 1975.

[268] A. Sidi. *Exponential function approximation to Laplace transform inversion and development of non-linear methods for accelerating the convergence of infinite integrals and series*. PhD thesis, Tel Aviv University, 1977. In Hebrew. Supervised by I.M. Longman.

[269] A. Sidi. Uniqueness of Padé approximants from series of orthogonal polynomials. *Math. Comp.*, 31:738–739, 1977.

[270] A. Sidi. Convergence properties of some nonlinear sequence transformations. *Math. Comp.*, 33:315–326, 1979.

[271] A. Sidi. Euler–Maclaurin expansions for a double integral with a line of singularities in the domain of integration. Technical Report 161, Computer Science Dept., Technion–Israel Institute of Technology, 1979.

[272] A. Sidi. Some properties of a generalization of the Richardson extrapolation process. *J. Inst. Maths. Applics.*, 24:327–346, 1979.

[273] A. Sidi. Analysis of convergence of the T-transformation for power series. *Math. Comp.*, 35:833–850, 1980.

[274] A. Sidi. Extrapolation methods for oscillatory infinite integrals. *J. Inst. Maths. Applics.*, 26:1–20, 1980.

[275] A. Sidi. Numerical quadrature and nonlinear sequence transformations; unified rules for efficient computation of integrals with algebraic and logarithmic endpoint singularities. *Math. Comp.*, 35:851–874, 1980.

[276] A. Sidi. Some aspects of two-point Padé approximants. *J. Comp. Appl. Math.*, 6:9–17, 1980.

[277] A. Sidi. A new method for deriving Padé approximants for some hypergeometric functions. *J. Comp. Appl. Math.*, 7:37–40, 1981.

[278] A. Sidi. An algorithm for a special case of a generalization of the Richardson extrapolation process. *Numer. Math.*, 38:299–307, 1982.

[279] A. Sidi. Converging factors for some asymptotic moment series that arise in numerical quadrature. *J. Austral. Math. Soc.*, Series B, 24:223–233, 1982.

[280] A. Sidi. Interpolation at equidistant points by a sum of exponential functions. *J. Approx. Theory*, 34:194–210, 1982.

[281] A. Sidi. The numerical evaluation of very oscillatory infinite integrals by extrapolation. *Math. Comp.*, 38:517–529, 1982.

[282] A. Sidi. Numerical quadrature rules for some infinite range integrals. *Math. Comp.*, 38:127–142, 1982.

[283] A. Sidi. Euler–Maclaurin expansions for integrals over triangles and squares of functions having algebraic/logarithmic singularities along an edge. *J. Approx. Theory*, 39:39–53, 1983.

[284] A. Sidi. Interpolation by a sum of exponential functions when some exponents are preassigned. *J. Math. Anal. Appl.*, 112:151–164, 1985.

[285] A. Sidi. Borel summability and converging factors for some everywhere divergent series. *SIAM J. Math. Anal.*, 17:1222–1231, 1986.

[286] A. Sidi. Extrapolation methods for divergent oscillatory infinite integrals that are defined in the sense of summability. *J. Comp. Appl. Math.*, 17:105–114, 1987.

[287] A. Sidi. Generalizations of Richardson extrapolation with applications to numerical integration. In H. Brass and G. Hämmerlin, editors, *Numerical Integration III*, number 85 in ISNM, pages 237–250, Basel, 1988. Birkhäuser.

[288] A. Sidi. A user-friendly extrapolation method for oscillatory infinite integrals. *Math. Comp.*, 51:249–266, 1988.

[289] A. Sidi. Comparison of some numerical quadrature formulas for weakly singular periodic Fredholm integral equations. *Computing*, 43:159–170, 1989.

[290] A. Sidi. On a generalization of the Richardson extrapolation process. *Numer. Math.*, 57:365–377, 1990.

[291] A. Sidi. On rates of acceleration of extrapolation methods for oscillatory infinite integrals. *BIT*, 30:347–357, 1990.

[292] A. Sidi. Quantitative and constructive aspects of the generalized Koenig's and de Montessus's theorems for Padé approximants. *J. Comp. Appl. Math.*, 29:257–291, 1990.

[293] A. Sidi. A new variable transformation for numerical integration. In H. Brass and G. Hämmerlin, editors, *Numerical Integration IV*, number 112 in ISNM, pages 359–373, Basel, 1993. Birkhäuser.

[294] A. Sidi. Acceleration of convergence of (generalized) Fourier series by the d-transformation. *Annals Numer. Math.*, 2:381–406, 1995.

[295] A. Sidi. Convergence analysis for a generalized Richardson extrapolation process with an application to the $d^{(1)}$-transformation on convergent and divergent logarithmic sequences. *Math. Comp.*, 64:1627–1657, 1995.

[296] A. Sidi. Extension and completion of Wynn's theory on convergence of columns of the epsilon table. *J. Approx. Theory*, 86:21–40, 1996.

[297] A. Sidi. Further results on convergence and stability of a generalization of the Richardson extrapolation process. *BIT Numerical Mathematics*, 36:143–157, 1996.

[298] A. Sidi. A complete convergence and stability theory for a generalized Richardson extrapolation process. *SIAM J. Numer. Anal.*, 34:1761–1778, 1997.

[299] A. Sidi. Computation of infinite integrals involving Bessel functions of arbitrary order by the $\bar{D}$-transformation. *J. Comp. Appl. Math.*, 78:125–130, 1997.

[300] A. Sidi. Further convergence and stability results for the generalized Richardson extrapolation process GREP[(1)] with an application to the $D^{(1)}$-transformation for infinite integrals. *J. Comp. Appl. Math.*, 112:269–290, 1999.

[301] A. Sidi. Extrapolation methods and derivatives of limits of sequences. *Math. Comp.*, 69:305–323, 2000.

[302] A. Sidi. The generalized Richardson extrapolation process GREP[(1)] and computation of derivatives of limits of sequences with applications to the $d^{(1)}$-transformation. *J. Comp. Appl. Math.*, 122:251–273, 2000.

[303] A. Sidi. A new algorithm for the higher order G-transformation. Preprint, Computer Science Dept., Technion–Israel Institute of Technology, 2000.

[304] A. Sidi. Numerical differentiation, polynomial interpolation, and Richardson extrapolation. Preprint, Computer Science Dept., Technion–Israel Institute of Technology, 2000.

[305] A. Sidi. The Richardson extrapolation process with a harmonic sequence of collocation points. *SIAM J. Numer. Anal.*, 37:1729–1746, 2000.

[306] A. Sidi. New convergence results on the generalized Richardson extrapolation process GREP[(1)] for logarithmic sequences. *Math. Comp.*, 71:1569–1596, 2002. Published electronically on November 28, 2001.

[307] A. Sidi. A convergence and stability study of the iterated Lubkin transformation and the θ-algorithm. *Math. Comp.*, 72:419–433, 2003. Published electronically on May 1, 2002.

[308] A. Sidi and J. Bridger. Convergence and stability analyses for some vector extrapolation methods in the presence of defective iteration matrices. *J. Comp. Appl. Math.*, 22:35–61, 1988.

[309] A. Sidi, W.F. Ford, and D.A. Smith. Acceleration of convergence of vector sequences. *SIAM J. Numer. Anal.*, 23:178–196, 1986. Originally appeared as NASA TP-2193, (1983).

[310] A. Sidi and M. Israeli. Quadrature methods for periodic singular and weakly singular Fredholm integral equations. *J. Sci. Comput.*, 3:201–231, 1988. Originally appeared as Technical Report No. 384, Computer Science Dept., Technion–Israel Institute of Technology, (1985), and also as ICASE Report No. 86-50 (1986).

[311] A. Sidi and M. Israeli. A hybrid extrapolation method: The Richardson-Shanks transformation, 1989. Unpublished research.

[312] A. Sidi and D. Levin. Rational approximations from the d-transformation. *IMA J. Numer. Anal.*, 2:153–167, 1982.

[313] A. Sidi and D. Levin. Prediction properties of the t-transformation. *SIAM J. Numer. Anal.*, 20:589–598, 1983.

[314] A. Sidi and D.S. Lubinsky. On the zeros of some polynomials that arise in numerical quadrature and convergence acceleration. *SIAM J. Numer. Anal.*, 20:400–405, 1983.

[315] S. Skelboe. Computation of the periodic steady-state response of nonlinear networks by extrapolation methods. *IEEE Trans. Circuits and Systems*, 27:161–175, 1980.

[316] I.H. Sloan and S. Joe. *Lattice Methods in Multiple Integration*. Clarendon Press, Oxford, 1994.

[317] D.A. Smith and W.F. Ford. Acceleration of linear and logarithmic convergence. *SIAM J. Numer. Anal.*, 16:223–240, 1979.

[318] D.A. Smith and W.F. Ford. Numerical comparisons of nonlinear convergence accelerators. *Math. Comp.*, 38:481–499, 1982.

[319] I.H. Sneddon. *The Use of Integral Transforms*. McCraw-Hill, New York, 1972.

[320] H. Stahl. On the divergence of certain Padé approximants and the behavior of the associated orthogonal polynomials. In C. Brezinski, A. Ronveaux, A. Draux, A.P. Magnus, and P. Maroni, editors, *Polynomes Orthogonaux et Applications, Proceedings, Bar-le-Duc, 1984*, number 1171 in Springer Lecture Notes in Mathematics, pages 321–330, New York, 1985. Springer-Verlag.

[321] H. Stahl and V. Totik. *General Orthogonal Polynomials*. Cambridge University Press, Cambridge, 1992.

[322] J.F. Steffensen. *Interpolation*. Chelsea, New York, 1950.

[323] H.J. Stetter. Asymptotic expansions for the error of discretization algorithms for non-linear functional equations. *Numer. Math.*, 7:18–31, 1965.

[324] H.J. Stetter. Symmetric two-step algorithms for ordinary differential equations. *Computing*, 5:267–280, 1970.

[325] H.J. Stetter. *Analysis of Discretization Methods for Ordinary Differential Equations*. Springer-Verlag, Berlin, 1973.

[326] J. Stoer and R. Bulirsch. *Introduction to Numerical Analysis*. Springer-Verlag, New York, 1980.

[327] R.F. Streit. The evaluation of double series. *BIT*, 12:400–408, 1972.

[328] S.P. Suetin. On the convergence of rational approximations to polynomial expansions in domains of meromorphy. *Math. USSR-Sbornik*, 34:367–381, 1978.

[329] S.P. Suetin. On the poles of the mth row of a Padé table. *Math. USSR-Sbornik*, 48:493–497, 1984.

[330] S.P. Suetin. On the inverse problem for the mth row of the Padé table. *Math. USSR-Sbornik*, 52:231–244, 1985.

[331] M. Sugihara. Methods of numerical integration of oscillatory functions by the DE-formula with the Richardson extrapolation. *J. Comp. Appl. Math.*, 17:47–68, 1987.

[332] G. Szegő. *Orthogonal Polynomials*. American Mathematical Society, Providence, Rhode Island, 1939.

[333] H. Takahasi and M. Mori. Double exponential formulas for numerical integration. *Publ. Res. Inst. of Math. Sci. Kyoto Univ.*, 9:721–741, 1974.

[334] E.C. Titchmarsh. *Theory of the Riemann Zeta Function*. Cambridge University Press, Cambridge, 1930.

[335] H. Toda and H. Ono. Some remarks for efficient usage of the double exponential formulas. *Kokyuroku of Res. Inst. for Math. Sci. Kyoto Univ.*, 339:74–109, 1978. In Japanese.

[336] J. Todd, editor. *Survey of Numerical Analysis*. McGraw-Hill, New York, 1962.

[337] L.N. Trefethen and M.H. Gutknecht. On convergence and degeneracy in rational Padé and Chebyshev approximation. *SIAM J. Math. Anal.*, 16:198–210, 1985.

[338] L.N. Trefethen and M.H. Gutknecht. Padé, stable Padé, and Chebyshev–Padé approximation. In J.C. Mason and M.G. Cox, editors, *Algorithms for Approximation*, pages 227–264, Oxford, 1987. Oxford University Press.

[339] R.R. Tucker. The δ^2-process and related topics. *Pacific J. Math.*, 22:349–359, 1967.

[340] R.R. Tucker. The δ^2-process and related topics II. *Pacific J. Math.*, 28:455–463, 1969.

[341] R.R. Tucker. A geometric derivation of Daniel Shanks e_k transform. *Faculty Rev. Bull. N.C. A and T State Univ.*, 65:60–63, 1973.

[342] P.A. Tyvand and M. Landrini. Free-surface flow of a fluid body with an inner circular cylinder in impulsive motion. *J. Engrg. Math.*, 40:109–140, 2001.

[343] A.H. Van Tuyl. Acceleration of convergence of a family of logarithmically convergent sequences. *Math. Comp.*, 63:229–246, 1994.

[344] J.-M. Vanden Broeck and L.W. Schwartz. A one-parameter family of sequence transformations. *SIAM J. Math. Anal.*, 10:658–666, 1979.

[345] D. Vekemans. Algorithm for the E-prediction. *J. Comp. Appl. Math.*, 85:181–202, 1997.

[346] P. Verlinden and R. Cools. Proof of a conjectured asymptotic expansion for the approximation of surface integrals. *Math. Comp.*, 63:717–725, 1994.

[347] P. Verlinden, D.M. Potts, and J.N. Lyness. Error expansions for multidimensional trapezoidal rules with Sidi transformations. *Numer. Algorithms*, 16:321–347, 1997.

[348] H.S. Wall. *Analytic Theory of Continued Fractions*. Van Nostrand, New York, 1948.

[349] G. Walz. *Asymptotics and Extrapolation*. Akademie Verlag, Berlin, 1996.

[350] G. Wanner, E. Hairer, and S.P. Nørsett. Order stars and stability theorems. *BIT*, 18:475–489, 1978.

[351] P.C. Waterman, J.M. Yos, and R.J. Abodeely. Numerical integration of non-analytic functions. *J. Math. and Phys.*, 43:45–50, 1964.

[352] L. Weiss and R. McDonough. Prony's method, Z-transforms, and Padé approximation. *SIAM Rev.*, 5:145–149, 1963.

[353] E.J. Weniger. Nonlinear sequence transformations for the acceleration of convergence and the summation of series. *Comput. Phys. Rep.*, 10:189–371, 1989.

[354] E.J. Weniger. Interpolation between sequence transformations. *Numer. Algorithms*, 3:477–486, 1992.

[355] E.J. Weniger. Prediction properties of Aitken iterated Δ^2 process, of Wynn's epsilon algorithm, and of Brezinski's iterated theta algorithm. *J. Comp. Appl. Math.*, 122:329–356, 2000.

[356] E.J. Weniger, J. Čížek, and F. Vinette. The summation of the ordinary and renormalized perturbation series for the ground state energy of the quartic, sextic, and octic anharmonic oscillators using nonlinear sequence transformations. *J. Math. Phys.*, 34:571–609, 1993.

[357] D.V. Widder. *The Laplace Transform*. Princeton University Press, Princeton, 1946.

[358] R. Wilson. Divergent continued fractions and polar singularities. *Proc. London Math. Soc.*, 26:159–168, 1927.

[359] R. Wilson. Divergent continued fractions and polar singularities II. Boundary pole multiple. *Proc. London Math. Soc.*, 27:497–512, 1928.

[360] R. Wilson. Divergent continued fractions and polar singularities III. Several boundary poles. *Proc. London Math. Soc.*, 28:128–144, 1928.

[361] J. Wimp. Derivative-free iteration processes. *SIAM J. Numer. Anal.*, 7:329–334, 1970.

[362] J. Wimp. The summation of series whose terms have asymptotic representations. *J. Approx. Theory*, 10:185–198, 1974.

[363] J. Wimp. Toeplitz arrays, linear sequence transformations and orthogonal polynomials. *Numer. Math.*, 23:1–17, 1974.

[364] J. Wimp. Acceleration methods. In *Encyclopedia of Computer Science and Technology*, volume I, pages 181–210, New York, 1975. Dekker.

[365] J. Wimp. New methods for accelerating the convergence of sequences arising in Laplace transform theory. *SIAM J. Numer. Anal.*, 14:194–204, 1977.

[366] J. Wimp. *Sequence Transformations and Their Applications*. Academic Press, New York, 1981.

[367] L. Wuytack. A new technique for rational extrapolation to the limit. *Numer. Math.*, 17:215–221, 1971.

[368] P. Wynn. On a device for computing the $e_m(S_n)$ transformation. *Mathematical Tables and Other Aids to Computation*, 10:91–96, 1956.

[369] P. Wynn. On a procrustean technique for the numerical transformation of slowly convergent sequences and series. *Proc. Cambridge Phil. Soc.*, 52:663–671, 1956.

[370] P. Wynn. Confluent forms of certain nonlinear algorithms. *Arch. Math.*, 11:223–234, 1960.

[371] P. Wynn. On the convergence and stability of the epsilon algorithm. *SIAM J. Numer. Anal.*, 3:91–122, 1966.

[372] P. Wynn. Upon systems of recursions which obtain among the quotients of the Padé table. *Numer. Math.*, 8:264–269, 1966.

[373] P. Wynn. A general system of orthogonal polynomials. *Quart. J. Math. Oxford*, 18:81–96, 1967.

[374] P. Wynn. Transformations to accelerate the convergence of Fourier series. In *Gertrude Blanche Anniversary Volume*, pages 339–379, Wright Patterson Air Force Base, 1967.

[375] P. Wynn. A note on the generalized Euler transformation. *Computer J.*, 14:437–441, 1971.

[376] P. Wynn. Upon some continuous prediction algorithms. I. *Calcolo*, 9:197–234, 1972.

[377] V. Zakian. Properties of I_{MN} approximants. In P.R. Graves-Morris, editor, *Padé Approximants and Their Applications*, pages 141–144, New York, 1973. Academic Press.

[378] V. Zakian. Properties of I_{MN} and J_{MN} approximants and applications to numerical inversion of Laplace transforms. *J. Math. Anal. Appl.*, 50:191–222, 1975.

Index

515